KINEMATIC CHAINS AND MACHINE COMPONENTS DESIGN

KINEMATIC CHAINS AND MACHINE COMPONENTS DESIGN

Dan B. Marghitu

Department of Mechanical Engineering, Auburn University, Auburn, AL

ELSEVIER
ACADEMIC
PRESS

AMSTERDAM • BOSTON • HEIDELBERG • LONDON
NEW YORK • OXFORD • PARIS • SAN DIEGO
SAN FRANCISCO • SINGAPORE • SYDNEY • TOKYO

Elsevier Academic Press
30 Corporate Drive, Suite 400, Burlington, MA 01803, USA
525 B Street, Suite 1900, San Diego, California 92101-4495, USA
84 Theobald's Road, London WC1X 8RR, UK

This book is printed on acid-free paper. ∞

Library of Congress Cataloging-in-Publication Data

Marghitu, Dan B.
    Kinematic chains and machine components design / Dan Marghitu.
        p. cm.
    Includes bibliographical references and index.
    ISBN 0-12-471352-1 (alk. paper)
    1. Machinery, Kinematics of. 2. Machinery–Design and construction.   I. Title.

 TJ175.M243 2005
 621.8′11–dc22

                                                                    2004061907

British Library Cataloguing in Publication Data
A catalogue record for this book is available from the British Library.

ISBN: 0-12-471352-1

For information on all Elsevier Academic Press publications
visit our Web site at www.books.elsevier.com

Printed in the United States of America
05   06   07   08   09   10   9   8   7   6   5   4   3   2   1

to my beloved family

Table of Contents

Preface

A number of bodies linked by joints form a kinematic chain. On the basis of the presence of loops in a mechanical structure it can be distinguished closed kinematic chains, if there are one or more loops so that each link and each joint is contained in at least one of them. A closed kinematic chain have no open attachment point. An open kinematic chain contains no loop. Kinematic chains design is a vital component of modern machine design practice. Kinematic chains are used to transmit forces and moments and to manipulate objects. A knowledge of the kinematic and dynamic properties of these machines is crucial for their design and control. A feature of this book and its main distinction from other books is that it presents a different method for kinematic and dynamic force analysis of kinematic chains. The other important feature of the approach used here is the attention given to the solution of the problems using the symbolical software Mathematica. Methods, algorithms and software packages for the solution of classical mechanical problems are presented. The book presents texts that are teachable and computer-oriented.

The book will assist all those interested in the design of mechanisms, manipulators, building machines, textile machines, vehicles, aircraft, satellites, ships, biomechanical systems (vehicle simulators, barrier tests, human motion studies, etc.), controlled mechanical systems, mechatronical devices and many others.

This book is appropriate for use as a text for undergraduate or graduate courses in mechanical engineering dealing with the subjects of the analysis and design of mechanisms, vehicle dynamics, mechatronics and multibody systems and machine components design. A basic knowledge of mechanics and calculus is assumed. The book may also be useful for practicing engineers and researchers in the fields of machine design and dynamics, and also biomechanics and mechatronics.

About the Author

Dan Marghitu is currently a professor at AUBURN UNIVERSITY, Mechanical Engineering Department, involved in teaching and research activities.

He received a D.E.A. from Paul Sabatier University and a Ph.D. from Southern Methodist University.

Part I   Kinematic Chains

I.1   Introduction

I.1.1   Vector Algebra

Vector Terminology

Scalars are mathematics quantities that can be fully defined by specifying their magnitude in suitable units of measure. The mass is a scalar and can be expressed in kilograms, the time is a scalar and can be expressed seconds, and the temperature can be expressed in degrees.

Vectors are quantities that require the specification of magnitude, orientation, and sense. The characteristics of a vector are the magnitude, the orientation, and the sense.

The *magnitude* of a vector is specified by a positive number and a unit having appropriate dimensions. No unit is stated if the dimensions are those of a pure number. The *orientation* of a vector is specified by the relationship between the vector and given reference lines and/or planes. The *sense* of a vector is specified by the order of two points on a line parallel to the vector.

Orientation and sense together determine the *direction* of a vector. The *line of action* of a vector is a hypothetical infinite straight line collinear with the vector. Displacement, velocity, and force are examples of vectors.

To distinguish vectors from scalars it is customary to denote vectors by boldface letters. Thus, the vector shown in Figure I.1.1(a) is denoted by $\mathbf{r}$ or $\mathbf{r}_{AB}$. The symbol $|\mathbf{r}| = r$ represents the magnitude (or module, or absolute value) of the vector $\mathbf{r}$. In handwritten work a distinguishing mark is used for vectors, such as an arrow over the symbol, $\vec{r}$ or $\overrightarrow{AB}$, a line over the symbol, $\bar{r}$, or an underline, $\underline{r}$.

The vectors are depicted by either straight or curved arrows. A vector represented by a straight arrow has the direction indicated by the arrow. The direction of a vector represented by a curved arrow is the same as the direction in which a right-handed screw moves when the axis of the screw is normal to the plane in which the arrow is drawn and the screw is rotated as indicated by the arrow.

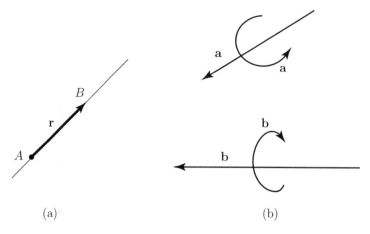

FIGURE I.1.1   *Vector representations: (a) straight arrow and (b) straight and curved arrows.*

Figure I.1.1(b) shows representations of vectors. Sometimes vectors are represented by means of a straight or curved arrow together with a measure number. In this case the vector is regarded as having the direction indicated by the arrow if the measure number is positive, and the opposite direction if it is negative.

A *bound* (or *fixed*) vector is a vector associated with a particular point P in space (Fig. I.1.2). The point P is the *point of application* of the vector, and the line passing through P and parallel to the vector is the line of action of the vector. The point of application can be represented as the tail [Fig. I.1.2(a)] or the head of the vector arrow [Fig. I.1.2(b)].

A *free* vector is not associated with a particular point or line in space. A *transmissible* (or *sliding*) vector is a vector that can be moved along its line of action without change of meaning.

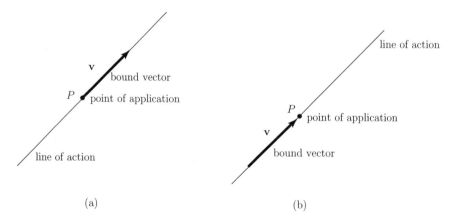

FIGURE I.1.2   *Bound or fixed vector: (a) point of application represented as the tail of the vector arrow and (b) point of application represented as the head of the vector arrow.*

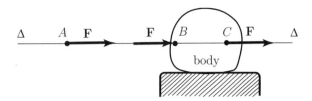

FIGURE I.1.3 *Transmissible vector: the force vector* **F** *can be applied anywhere along the line* Δ.

To move the body in Figure I.1.3 the force vector **F** can be applied anywhere along the line Δ or may be applied at specific points A, B, C. The force vector **F** is a transmissible vector because the resulting motion is the same in all cases.

The force **F** applied at B will cause a different deformation of the body than the same force **F** applied at a different point C. The points B and C are on the body. If one is interested in the deformation of the body, the force **F** positioned at C is a bound vector.

The operations of vector analysis deal only with the characteristics of vectors and apply, therefore, to both bound and free vectors. Vector analysis is a branch of mathematics that deals with quantities that have both magnitude and direction.

Vector Equality

Two vectors **a** and **b** are said to be equal to each other when they have the same characteristics

$$\mathbf{a} = \mathbf{b}.$$

Equality does not imply physical equivalence. For instance, two forces represented by equal vectors do not necessarily cause identical motions of a body on which they act.

Product of a Vector and a Scalar
Definition
The product of a vector **v** and a scalar s, $s\mathbf{v}$ or $\mathbf{v}s$, is a vector having the following characteristics:

1. Magnitude.

$$|s\mathbf{v}| \equiv |\mathbf{v}s| = |s||\mathbf{v}|,$$

   where $|s|$ denotes the absolute value (or magnitude, or module) of the scalar s.
2. Orientation. $s\mathbf{v}$ is parallel to **v**. If $s = 0$, no definite orientation is attributed to $s\mathbf{v}$.
3. Sense. If $s > 0$, the sense of $s\mathbf{v}$ is the same as that of **v**. If $s < 0$, the sense of $s\mathbf{v}$ is opposite to that of **v**. If $s = 0$, no definite sense is attributed to $s\mathbf{v}$.

Zero Vectors
Definition
A *zero vector* is a vector that does not have a definite direction and whose magnitude is equal to zero. The symbol used to denote a zero vector is **0**.

Unit Vectors
Definition
A *unit vector* (versor) is a vector with the magnitude equal to 1. Given a vector $\mathbf{v}$, a unit vector $\mathbf{u}$ having the same direction as $\mathbf{v}$ is obtained by forming the quotient of $\mathbf{v}$ and $|\mathbf{v}|$:

$$\mathbf{u} = \frac{\mathbf{v}}{|\mathbf{v}|}.$$

Vector Addition
The sum of a vector $\mathbf{v}_1$ and a vector $\mathbf{v}_2$: $\mathbf{v}_1 + \mathbf{v}_2$ or $\mathbf{v}_2 + \mathbf{v}_1$ is a vector whose characteristics are found by either graphical or analytical processes. The vectors $\mathbf{v}_1$ and $\mathbf{v}_2$ add according to the parallelogram law: $\mathbf{v}_1 + \mathbf{v}_2$ is equal to the diagonal of a parallelogram formed by the graphical representation of the vectors [(Fig. I.1.4(a))]. The vector $\mathbf{v}_1 + \mathbf{v}_2$ is called the *resultant* of $\mathbf{v}_1$ and $\mathbf{v}_2$. The vectors can be added by moving them successively to parallel positions so that the head of one vector connects to the tail of the next vector. The resultant is the vector whose tail connects to the tail of the first vector, and whose head connects to the head of the last vector [(Fig. I.1.4(b))].

The sum $\mathbf{v}_1 + (-\mathbf{v}_2)$ is called the *difference* of $\mathbf{v}_1$ and $\mathbf{v}_2$ and is denoted by $\mathbf{v}_1 - \mathbf{v}_2$ [(Figs. I.1.4(c) and I.1.4 (d))].

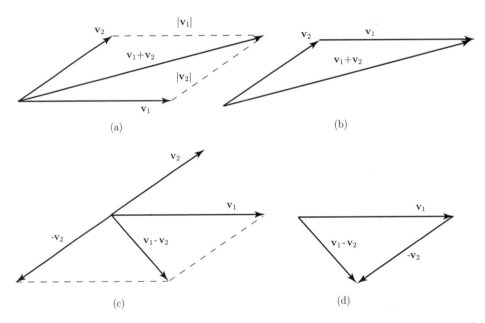

FIGURE I.1.4 *Vector addition: (a) parallelogram law, (b) moving the vectors successively to parallel positions. Vector difference: (c) parallelogram law, (d) moving the vectors successively to parallel positions.*

The sum of n vectors $\mathbf{v}_i$, $i = 1, \ldots, n$,

$$\sum_{i=1}^{n} \mathbf{v}_i \quad \text{or} \quad \mathbf{v}_1 + \mathbf{v}_2 + \cdots + \mathbf{v}_n$$

is called the *resultant* of the vectors $\mathbf{v}_i$, $i = 1, \ldots, n$.

The vector addition is:

1. Commutative. The characteristics of the resultant are independent of the order in which the vectors are added (commutativity):

$$\mathbf{v}_1 + \mathbf{v}_2 = \mathbf{v}_2 + \mathbf{v}_1.$$

2. Associative. The characteristics of the resultant are not affected by the manner in which the vectors are grouped (associativity):

$$\mathbf{v}_1 + (\mathbf{v}_2 + \mathbf{v}_3) = (\mathbf{v}_1 + \mathbf{v}_2) + \mathbf{v}_3.$$

3. Distributive. The vector addition obeys the following laws of distributivity:

$$\mathbf{v} \sum_{i=1}^{p} s_i = \sum_{i=1}^{p} (\mathbf{v} s_i), \quad \text{for } s_i \neq 0, \ s_i \in \mathcal{R},$$

$$s \sum_{i=1}^{n} \mathbf{v}_i = \sum_{i=1}^{n} (s \mathbf{v}_i), \quad \text{for } s \neq 0, \ s \in \mathcal{R},$$

where $\mathcal{R}$ is the set of real numbers.

Every vector can be regarded as the sum of n vectors ($n = 2, 3, \ldots$) of which all but one can be selected arbitrarily.

Resolution of Vectors and Components

Let $\mathbf{1}_1, \mathbf{1}_2, \mathbf{1}_3$ be any three unit vectors not parallel to the same plane (noncollinear vectors):

$$|\mathbf{1}_1| = |\mathbf{1}_2| = |\mathbf{1}_3| = 1$$

For a given vector $\mathbf{v}$ (Fig. I.1.5), there are three unique scalars, v_1, v_2, v_3, such that $\mathbf{v}$ can be expressed as:

$$\mathbf{v} = v_1 \mathbf{1}_1 + v_2 \mathbf{1}_2 + v_3 \mathbf{1}_3$$

The opposite action of addition of vectors is the *resolution* of vectors. Thus, for the given vector $\mathbf{v}$ the vectors $v_1 \mathbf{1}_1$, $v_2 \mathbf{1}_2$, and $v_3 \mathbf{1}_3$ sum to the original vector. The vector $v_k \mathbf{1}_k$ is called the $\mathbf{1}_k$ *component* of $\mathbf{v}$ and v_k is called the $\mathbf{1}_k$ *scalar component* of $\mathbf{v}$, where $k = 1, 2, 3$. A vector is often replaced by its components since the components are equivalent to the original vector.

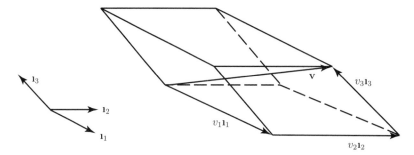

FIGURE I.1.5 *Resolution of a vector* **v** *and components.*

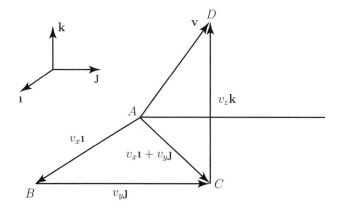

FIGURE I.1.6 *Cartesian reference frame and the orthogonal scalar components* v_x, v_y, v_z.

Every vector equation $\mathbf{v} = \mathbf{0}$, where $\mathbf{v} = v_1\mathbf{I}_1 + v_2\mathbf{I}_2 + v_3\mathbf{I}_3$, is equivalent to three scalar equations: $v_1 = 0, \quad v_2 = 0, \quad v_3 = 0$.

If the unit vectors $\mathbf{I}_1, \mathbf{I}_2, \mathbf{I}_3$ are mutually perpendicular they form a *Cartesian reference frame*. For a Cartesian reference frame the following notation is used (Fig. I.1.6):

$$\mathbf{I}_1 \equiv \mathbf{I}, \quad \mathbf{I}_2 \equiv \mathbf{J}, \quad \mathbf{I}_3 \equiv \mathbf{k},$$

and

$$\mathbf{I} \perp \mathbf{J}, \quad \mathbf{I} \perp \mathbf{k}, \quad \mathbf{J} \perp \mathbf{k}.$$

The symbol $\perp$ denotes perpendicular.

When a vector $\mathbf{v}$ is expressed in the form $\mathbf{v} = v_x\mathbf{I} + v_y\mathbf{J} + v_z\mathbf{k}$, where $\mathbf{I}, \mathbf{J}, \mathbf{k}$ are mutually perpendicular unit vectors (Cartesian reference frame or orthogonal reference frame), the magnitude of $\mathbf{v}$ is given by

$$|\mathbf{v}| = \sqrt{v_x^2 + v_y^2 + v_z^2}$$

The vectors $\mathbf{v}_x = v_x\mathbf{ı}$, $\mathbf{v}_y = v_y\mathbf{J}$, and $\mathbf{v}_z = v_z\mathbf{k}$ are the *orthogonal* or *rectangular component vectors* of the vector $\mathbf{v}$. The measures v_x, v_y, v_z are the *orthogonal* or *rectangular scalar components* of the vector $\mathbf{v}$.

If $\mathbf{v}_1 = v_{1x}\mathbf{ı} + v_{1y}\mathbf{J} + v_{1z}\mathbf{k}$ and $\mathbf{v}_2 = v_{2x}\mathbf{ı} + v_{2y}\mathbf{J} + v_{2z}\mathbf{k}$, then the sum of the vectors is

$$\mathbf{v}_1 + \mathbf{v}_2 = (v_{1x} + v_{2x})\mathbf{ı} + \left(v_{1y} + v_{2y}\right)\mathbf{J} + (v_{1z} + v_{2z})\,v_{1z}\mathbf{k}.$$

Angle Between Two Vectors

Two vectors $\mathbf{a}$ and $\mathbf{b}$ are considered. One can move either vector parallel to itself (leaving its sense unaltered) until their initial points (tails) coincide. The *angle* between $\mathbf{a}$ and $\mathbf{b}$ is the angle θ in Figures I.1.7(a) and I.1.7(b). The angle between $\mathbf{a}$ and $\mathbf{b}$ is denoted by the symbols $(\mathbf{a}, \mathbf{b})$ or $(\mathbf{b}, \mathbf{a})$. Figure I.1.7(c) represents the case $(\mathbf{a}, \mathbf{b}) = 0$, and Figure I.1.7(d) represents the case $(\mathbf{a}, \mathbf{b}) = 180°$.

The direction of a vector $\mathbf{v} = v_x\mathbf{ı} + v_y\mathbf{J} + v_z\mathbf{k}$ relative to a Cartesian reference, $\mathbf{ı}$, $\mathbf{J}$, $\mathbf{k}$, is given by the cosines of the angles formed by the vector and the respective unit vectors.

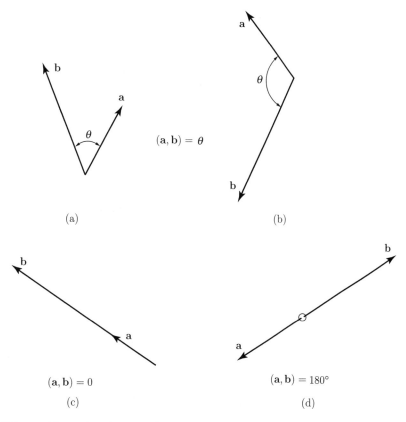

FIGURE I.1.7 *The angle θ between the vectors $\mathbf{a}$ and $\mathbf{b}$: (a) $0 < \theta < 90°$, (b) $90° < \theta < 180°$, (c) $\theta = 0°$, and (d) $\theta = 180°$.*

These are called *direction cosines* and are denoted as (Fig. I.1.8):

$$\cos(\mathbf{v}, \mathbf{\imath}) = \cos\alpha = l; \;\; \cos(\mathbf{v}, \mathbf{\jmath}) = \cos\beta = m; \;\; \cos(\mathbf{v}, \mathbf{k}) = \cos\gamma = n.$$

The following relations exist:

$$v_x = |\mathbf{v}|\cos\alpha; \;\; v_y = |\mathbf{v}|\cos\beta; \;\; v_z = |\mathbf{v}|\cos\gamma,$$

$$l^2 + m^2 + n^2 = 1, \;\; (v_x^2 + v_y^2 + v_z^2 = v^2).$$

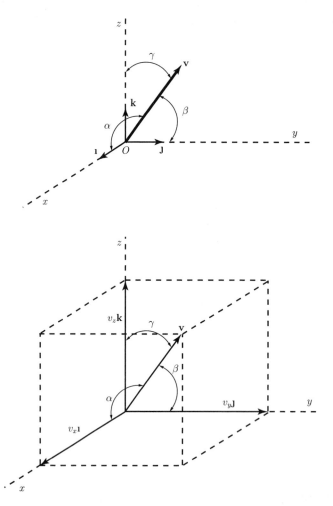

FIGURE I.1.8 *Direction cosines.*

Scalar (Dot) Product of Vectors

Definition

The scalar (dot) product of a vector **a** and a vector **b** is

$$\mathbf{a} \cdot \mathbf{b} = \mathbf{b} \cdot \mathbf{a} = |\mathbf{a}|\,|\mathbf{b}|\cos(\mathbf{a}, \mathbf{b}).$$

For any two vectors **a** and **b** and any scalar s

$$(s\mathbf{a})\cdot\mathbf{b} = s(\mathbf{a} \cdot \mathbf{b}) = \mathbf{a}\cdot(s\mathbf{b}) = s\mathbf{a} \cdot \mathbf{b}.$$

If

$$\mathbf{a} = a_x\mathbf{\iota} + a_y\mathbf{J} + a_z\mathbf{k},$$

and

$$\mathbf{b} = b_x\mathbf{\iota} + b_y\mathbf{J} + b_z\mathbf{k},$$

where $\mathbf{\iota}$, $\mathbf{J}$, $\mathbf{k}$ are mutually perpendicular unit vectors, then

$$\mathbf{a} \cdot \mathbf{b} = a_xb_x + a_yb_y + a_zb_z\,.$$

The following relationships exist:

$$\mathbf{\iota} \cdot \mathbf{\iota} = \mathbf{J} \cdot \mathbf{J} = \mathbf{k} \cdot \mathbf{k} = 1,$$
$$\mathbf{\iota} \cdot \mathbf{J} = \mathbf{J} \cdot \mathbf{k} = \mathbf{k} \cdot \mathbf{\iota} = 0.$$

Every vector **v** can be expressed in the form

$$\mathbf{v} = \mathbf{\iota} \cdot \mathbf{v}\,\mathbf{\iota} + \mathbf{J} \cdot \mathbf{v}\,\mathbf{J} + \mathbf{k} \cdot \mathbf{v}\,\mathbf{k}.$$

The vector **v** can always be expressed as

$$\mathbf{v} = v_x\mathbf{\iota} + v_y\mathbf{J} + v_z\mathbf{k}.$$

Dot multiply both sides by $\mathbf{\iota}$

$$\mathbf{\iota} \cdot \mathbf{v} = v_x\mathbf{\iota} \cdot \mathbf{\iota} + v_y\mathbf{\iota} \cdot \mathbf{J} + v_z\mathbf{\iota} \cdot \mathbf{k}.$$

But,

$$\mathbf{\iota} \cdot \mathbf{\iota} = 1, \quad \text{and} \quad \mathbf{\iota} \cdot \mathbf{J} = \mathbf{\iota} \cdot \mathbf{k} = 0.$$

Hence,

$$\mathbf{\iota} \cdot \mathbf{v} = v_x.$$

Similarly,

$$\mathbf{J} \cdot \mathbf{v} = v_y \quad \text{and} \quad \mathbf{k} \cdot \mathbf{v} = v_z.$$

The associative, commutative, and distributive laws of elementary algebra are valid for the dot multiplication (product) of vectors.

Vector (Cross) Product of Vectors

Definition

The vector (cross) product of a vector $\mathbf{a}$ and a vector $\mathbf{b}$ is the vector (Fig. I.1.9):

$$\mathbf{a} \times \mathbf{b} = |\mathbf{a}| \, |\mathbf{b}| \sin(\mathbf{a}, \mathbf{b}) n$$

where $\mathbf{n}$ is a unit vector whose direction is the same as the direction of advance of a right-handed screw rotated from $\mathbf{a}$ toward $\mathbf{b}$, through the angle $(\mathbf{a}, \mathbf{b})$, when the axis of the screw is perpendicular to both $\mathbf{a}$ and $\mathbf{b}$.

The magnitude of $\mathbf{a} \times \mathbf{b}$ is given by

$$|\mathbf{a} \times \mathbf{b}| = |\mathbf{a}| \, |\mathbf{b}| \sin(\mathbf{a}, \mathbf{b}).$$

If $\mathbf{a}$ is parallel to $\mathbf{b}$, $\mathbf{a} || \mathbf{b}$, then $\mathbf{a} \times \mathbf{b} = 0$. The symbol $||$ denotes parallel. The relation $\mathbf{a} \times \mathbf{b} = \mathbf{0}$ implies only that the product $|\mathbf{a}| \, |\mathbf{b}| \sin(\mathbf{a}, \mathbf{b})$ is equal to zero, and this is the case whenever $|\mathbf{a}| = 0$, or $|\mathbf{b}| = 0$, or $\sin(\mathbf{a}, \mathbf{b}) = 0$.

For any two vectors $\mathbf{a}$ and $\mathbf{b}$ and any real scalar s,

$$(s\mathbf{a}) \times \mathbf{b} = s(\mathbf{a} \times \mathbf{b}) = \mathbf{a} \times (s\mathbf{b}) = s\mathbf{a} \times \mathbf{b}.$$

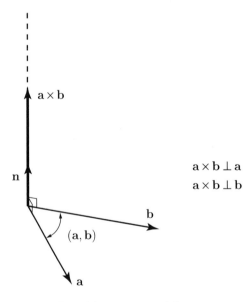

FIGURE I.1.9 *Vector (cross) product of the vector $\mathbf{a}$ and the vector $\mathbf{b}$.*

The sense of the unit vector $\mathbf{n}$ which appears in the definition of $\mathbf{a} \times \mathbf{b}$ depends on the order of the factors $\mathbf{a}$ and $\mathbf{b}$ in such a way that

$$\mathbf{b} \times \mathbf{a} = -\mathbf{a} \times \mathbf{b}.$$

Vector multiplication obeys the following law of distributivity (Varignon theorem):

$$\mathbf{a} \times \sum_{i=1}^{n} \mathbf{v}_i = \sum_{i=1}^{n} (\mathbf{a} \times \mathbf{v}_i).$$

The cross product is not commutative, but the associative law and the distributive law are valid for cross products.

A set of mutually perpendicular unit vectors $\mathbf{i}$, $\mathbf{J}$, $\mathbf{k}$ is called *right-handed* if $\mathbf{i} \times \mathbf{J} = \mathbf{k}$. A set of mutually perpendicular unit vectors $\mathbf{i}$, $\mathbf{J}$, $\mathbf{k}$ is called *left-handed* if $\mathbf{i} \times \mathbf{J} = -\mathbf{k}$.

If

$$\mathbf{a} = a_x \mathbf{i} + a_y \mathbf{J} + a_z \mathbf{k},$$

and

$$\mathbf{b} = b_x \mathbf{i} + b_y \mathbf{J} + b_z \mathbf{k},$$

where $\mathbf{i}$, $\mathbf{J}$, $\mathbf{k}$ are mutually perpendicular unit vectors, then $\mathbf{a} \times \mathbf{b}$ can be expressed in the following determinant form:

$$\mathbf{a} \times \mathbf{b} = \begin{vmatrix} \mathbf{i} & \mathbf{J} & \mathbf{k} \\ a_x & a_y & a_z \\ b_x & b_y & b_z \end{vmatrix}.$$

The determinant can be expanded by minors of the elements of the first row:

$$\begin{vmatrix} \mathbf{i} & \mathbf{J} & \mathbf{k} \\ a_x & a_y & a_z \\ b_x & b_y & b_z \end{vmatrix} = \mathbf{i}\begin{vmatrix} a_y & a_z \\ b_y & b_z \end{vmatrix} - \mathbf{J}\begin{vmatrix} a_x & a_z \\ b_x & b_z \end{vmatrix} + \mathbf{k}\begin{vmatrix} a_x & a_y \\ b_x & b_y \end{vmatrix}$$

$$= \mathbf{i}(a_y b_z - a_z b_y) - \mathbf{J}(a_x b_z - a_z b_x) + \mathbf{k}(a_x b_y - a_y b_x)$$

$$= (a_y b_z - a_z b_y)\mathbf{i} + (a_z b_x - a_x b_z)\mathbf{J} + (a_x b_y - a_y b_x)\mathbf{k}.$$

Scalar Triple Product of Three Vectors

Definition
The scalar triple product of three vectors $\mathbf{a}$, $\mathbf{b}$, $\mathbf{c}$ is

$$[\mathbf{a}, \mathbf{b}, \mathbf{c}] \equiv \mathbf{a} \cdot (\mathbf{b} \times \mathbf{c}) = \mathbf{a} \cdot \mathbf{b} \times \mathbf{c}.$$

It does not matter whether the dot is placed between **a** and **b**, and the cross between **b** and **c**, or vice versa, that is,

$$[\mathbf{a}, \mathbf{b}, \mathbf{c}] = \mathbf{a} \cdot \mathbf{b} \times \mathbf{c} = \mathbf{a} \times \mathbf{b} \cdot \mathbf{c}.$$

A change in the order of the factors appearing in a scalar triple product at most changes the sign of the product, that is,

$$[\mathbf{b}, \mathbf{a}, \mathbf{c}] = -[\mathbf{a}, \mathbf{b}, \mathbf{c}],$$

and

$$[\mathbf{b}, \mathbf{c}, \mathbf{a}] = [\mathbf{a}, \mathbf{b}, \mathbf{c}].$$

If **a**, **b**, **c** are parallel to the same plane, or if any two of the vectors **a**, **b**, **c** are parallel to each other, then $[\mathbf{a}, \mathbf{b}, \mathbf{c}] = 0$.

The scalar triple product $[\mathbf{a}, \mathbf{b}, \mathbf{c}]$ can be expressed in the following determinant form

$$[\mathbf{a}, \mathbf{b}, \mathbf{c}] = \begin{vmatrix} a_x & a_y & a_z \\ b_x & b_y & b_z \\ c_x & c_y & c_z \end{vmatrix}.$$

Vector Triple Product of Three Vectors

Definition

The vector triple product of three vectors **a**, **b**, **c** is the vector $\mathbf{a} \times (\mathbf{b} \times \mathbf{c})$. The parentheses are essential because $\mathbf{a} \times (\mathbf{b} \times \mathbf{c})$ is not, in general, equal to $(\mathbf{a} \times \mathbf{b}) \times \mathbf{c}$.

For any three vectors **a**, **b**, and **c**,

$$\mathbf{a} \times (\mathbf{b} \times \mathbf{c}) = \mathbf{a} \cdot \mathbf{cb} - \mathbf{a} \cdot \mathbf{bc}.$$

Derivative of a Vector

The derivative of a vector is defined in exactly the same way as is the derivative of a scalar function. The derivative of a vector has some of the properties of the derivative of a scalar function.

The derivative of the sum of two vector functions **a** and **b** is

$$\frac{d}{dt}(\mathbf{a} + \mathbf{b}) = \frac{d\mathbf{a}}{dt} + \frac{d\mathbf{b}}{dt}.$$

The time derivative of the product of a scalar function f and a vector function **a** is

$$\frac{d(f\mathbf{a})}{dt} = \frac{df}{dt}\mathbf{a} + f\frac{d\mathbf{a}}{dt}.$$

I.1.2 Centroids

Position Vector

The position vector of a point P relative to a point M is a vector $\mathbf{r}_{MP}$ having the following characteristics (Fig. I.1.10):

- magnitude ($|\mathbf{r}_{MP}| = r_{MP}$) the length of line MP;
- orientation parallel to line MP;
- sense MP (from point M to point P).

The vector $\mathbf{r}_{MP}$ is shown as an arrow connecting M to P. The position of a point P relative to P is a zero vector.

Let $\mathbf{\imath}$, $\mathbf{\jmath}$, $\mathbf{k}$ be mutually perpendicular unit vectors (Cartesian reference frame) with the origin at O (Fig. I.1.10). The axes of the Cartesian reference frame are x, y, z. The unit vectors $\mathbf{\imath}$, $\mathbf{\jmath}$, $\mathbf{k}$ are parallel to x, y, z, and they have the senses of the positive x, y, z axes. The coordinates of the origin O are $x = y = z = 0$, i.e., $O(0, 0, 0)$. The coordinates of a point P are $x = x_P$, $y = y_P$, $z = z_P$, i.e., $P(x_P, y_P, z_P)$. The position vector of P relative to the origin O is

$$\mathbf{r}_{OP} = \mathbf{r}_P = x_P\,\mathbf{\imath} + y_P\,\mathbf{\jmath} + z_P\,\mathbf{k}.$$

The position vector of the point P relative to a point M, $M \neq O$ of coordinates (x_M, y_M, z_M) is

$$\mathbf{r}_{MP} = (x_P - x_M)\,\mathbf{\imath} + (y_P - y_M)\,\mathbf{\jmath} + (z_P - z_M)\,\mathbf{k}.$$

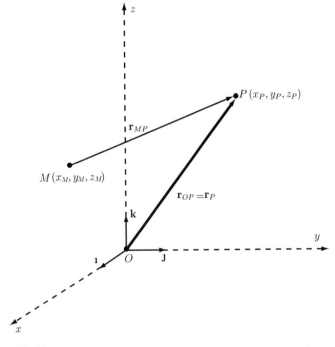

FIGURE I.1.10 *Position vector.*

The distance d between P and M is given by

$$d = |\mathbf{r}_P - \mathbf{r}_M| = |\mathbf{r}_{MP}| = \sqrt{(x_P - x_M)^2 + (y_P - y_M)^2 + (z_P - z_M)^2}.$$

First Moment

The position vector of a point P relative to a point O is $\mathbf{r}_P$ and a scalar associated with P is s, e.g., the mass m of a particle situated at P. The *first moment* of a point P with respect to a point O is the vector $\mathbf{M} = s\,\mathbf{r}_P$. The scalar s is called the *strength* of P.

Centroid of a Set of Points

The set of n points P_i, $i = 1, 2, \ldots, n$, is $\{S\}$ (Fig. I.1.11):

$$\{S\} = \{P_1, P_2, \ldots, P_n\} = \{P_i\}_{i=1,2,\ldots,n}.$$

The strengths of the points P_i are s_i, $i = 1, 2, \ldots, n$, i.e., n scalars, all having the same dimensions, and each associated with one of the points of $\{S\}$.

The *centroid* of the set $\{S\}$ is the point C with respect to which the sum of the first moments of the points of $\{S\}$ is equal to zero.

The position vector of C relative to an arbitrarily selected reference point O is $\mathbf{r}_C$ (Fig. I.1.11). The position vector of P_i relative to O is $\mathbf{r}_i$. The position vector of P_i relative to C is $\mathbf{r}_i - \mathbf{r}_C$. The sum of the first moments of the points P_i with respect to C is $\sum_{i=1}^{n} s_i(\mathbf{r}_i - \mathbf{r}_C)$. If C is to be centroid of $\{S\}$, this sum is equal to zero:

$$\sum_{i=1}^{n} s_i(\mathbf{r}_i - \mathbf{r}_C) = \sum_{i=1}^{n} s_i\mathbf{r}_i - \mathbf{r}_C \sum_{i=1}^{n} s_i = 0.$$

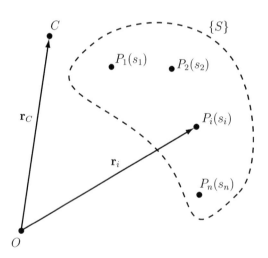

FIGURE I.1.11   *Centroid of a set of points.*

The position vector $\mathbf{r}_C$ of the centroid C, relative to an arbitrarily selected reference point O, is given by

$$\mathbf{r}_C = \frac{\sum_{i=1}^{n} s_i \mathbf{r}_i}{\sum_{i=1}^{n} s_i}.$$

If $\sum_{i=1}^{n} s_i = 0$, the centroid is not defined.

The centroid C of a set of points of given strength is a unique point, its location being independent of the choice of reference point O.

The Cartesian coordinates of the centroid $C(x_C, y_C, z_C)$ of a set of points P_i, $i = 1, \ldots, n$, of strengths s_i, $i = 1, \ldots, n$, are given by the expressions

$$x_C = \frac{\sum_{i=1}^{n} s_i x_i}{\sum_{i=1}^{n} s_i}, \quad y_C = \frac{\sum_{i=1}^{n} s_i y_i}{\sum_{i=1}^{n} s_i}, \quad z_C = \frac{\sum_{i=1}^{n} s_i z_i}{\sum_{i=1}^{n} s_i}.$$

The *plane of symmetry* of a set is the plane where the centroid of the set lies, the points of the set being arranged in such a way that corresponding to every point on one side of the plane of symmetry there exists a point of equal strength on the other side, the two points being equidistant from the plane.

A set $\{S'\}$ of points is called a *subset* of a set $\{S\}$ if every point of $\{S'\}$ is a point of $\{S\}$. The centroid of a set $\{S\}$ may be located using the *method of decomposition*:

- divide the system $\{S\}$ into subsets;
- find the centroid of each subset;
- assign to each centroid of a subset a strength proportional to the sum of the strengths of the points of the corresponding subset;
- determine the centroid of this set of centroids.

Centroid of a Curve, Surface, or Solid

The position vector of the centroid C of a curve, surface, or solid relative to a point O is

$$\mathbf{r}_C = \frac{\int_D \mathbf{r} \, d\tau}{\int_D d\tau},$$

where D is a curve, surface, or solid, $\mathbf{r}$ denotes the position vector of a typical point of D, relative to O, and $d\tau$ is the length, area, or volume of a differential element of D. Each of the two limits in this expression is called an "integral over the domain D (curve, surface, or solid)."

The integral $\int_D d\tau$ gives the total length, area, or volume of D, that is

$$\int_D d\tau = \tau.$$

The position vector of the centroid is

$$\mathbf{r}_C = \frac{1}{\tau} \int_D \mathbf{r} \, d\tau.$$

Let $\mathbf{i}, \mathbf{j}, \mathbf{k}$ be mutually perpendicular unit vectors (Cartesian reference frame) with the origin at O. The coordinates of C are x_C, y_C, z_C and

$$\mathbf{r}_C = x_C \mathbf{i} + y_C \mathbf{j} + z_C \mathbf{k}.$$

It results that

$$x_C = \frac{1}{\tau} \int_D x \, d\tau, \quad y_C = \frac{1}{\tau} \int_D y \, d\tau, \quad z_C = \frac{1}{\tau} \int_D z \, d\tau.$$

Mass Center of a Set of Particles

The *mass center* of a set of particles $\{S\} = \{P_1, P_2, \ldots, P_n\} = \{P_i\}_{i=1,2,\ldots,n}$ is the centroid of the set of points at which the particles are situated with the strength of each point being taken equal to the mass of the corresponding particle, $s_i = m_i$, $i = 1, 2, \ldots, n$. For the system of n particles the following relation can be written

$$\left(\sum_{i=1}^{n} m_i \right) \mathbf{r}_C = \sum_{i=1}^{n} m_i \mathbf{r}_i,$$

and the position vector of the mass center C is

$$\mathbf{r}_C = \frac{\displaystyle\sum_{i=1}^{n} m_i \mathbf{r}_i}{m}, \tag{I.1.1}$$

where m is the total mass of the system.

Mass Center of a Curve, Surface, or Solid

The position vector of the mass center C of a continuous body D, curve, surface, or solid, relative to a point O is

$$\mathbf{r}_C = \frac{1}{m} \int_D \mathbf{r}\rho \, d\tau,$$

or using the orthogonal Cartesian coordinates

$$x_C = \frac{1}{m} \int_D x\rho \, d\tau, \quad y_C = \frac{1}{m} \int_D y\rho \, d\tau, \quad z_C = \frac{1}{m} \int_D z\rho \, d\tau,$$

where ρ is the mass density of the body: mass per unit of length if D is a curve, mass per unit area if D is a surface, and mass per unit of volume if D is a solid; $\mathbf{r}$ is the position vector of a typical point of D, relative to O, $d\tau$ is the length, area, or volume of a differential element of D, $m = \int_D \rho \, d\tau$ is the total mass of the body, and x_C, y_C, z_C are the coordinates of C.

If the mass density ρ of a body is the same at all points of the body, $\rho = $ constant, the density, as well as the body, are said to be *uniform*. The mass center of a uniform body coincides with the centroid of the figure occupied by the body.

The *method of decomposition* may be used to locate the mass center of a continuous body B:

- divide the body B into a number of bodies, which may be particles, curves, surfaces, or solids;
- locate the mass center of each body;
- assign to each mass center a strength proportional to the mass of the corresponding body (e.g., the weight of the body);
- locate the centroid of this set of mass centers.

First Moment of an Area

A planar surface of area A and a reference frame xOy in the plane of the surface are shown in Figure I.1.12. The first moment of area A about the x axis is

$$M_x = \int_A y \, dA, \tag{I.1.2}$$

and the first moment about the y axis is

$$M_y = \int_A x \, dA. \tag{I.1.3}$$

The first moment of area gives information of the shape, size, and orientation of the area.

The entire area A can be concentrated at a position $C(x_C, y_C)$, the centroid. The coordinates x_C and y_C are the centroidal coordinates. To compute the centroidal coordinates the

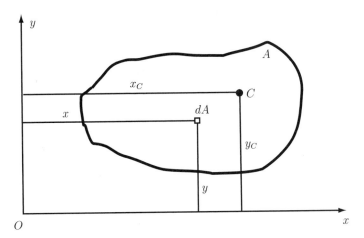

FIGURE I.1.12 *Centroid of a planar surface of area.*

moments of the distributed area are equated with that of the concentrated area about both axes:

$$A\,y_C = \int_A y\,dA, \quad \Longrightarrow \quad y_C = \frac{\int_A y\,dA}{A} = \frac{M_x}{A}, \qquad (\text{I.1.4})$$

$$A\,x_C = \int_A x\,dA, \quad \Longrightarrow \quad x_C = \frac{\int_A x\,dA}{A} = \frac{M_y}{A}. \qquad (\text{I.1.5})$$

The location of the centroid of an area is independent of the reference axes employed, i.e., the centroid is a property only of the area itself.

If the axes xy have their origin at the centroid, $O \equiv C$, then these axes are called *centroidal axes*. The first moments about centroidal axes are zero. All axes going through the centroid of an area are called centroidal axes for that area, and the first moments of an area about any of its centroidal axes are zero. The perpendicular distance from the centroid to the centroidal axis must be zero.

Figure I.1.13 shows a plane area with the axis of symmetry collinear with the axis y. The area A can be considered as composed of area elements in symmetric pairs as shown in the figure. The first moment of such a pair about the axis of symmetry y is zero. The entire

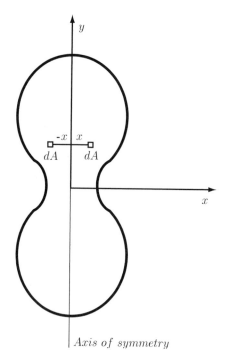

FIGURE I.1.13 *Plane area with axis of symmetry.*

area can be considered as composed of such symmetric pairs and the coordinate x_C is zero:

$$x_C = \frac{1}{A} \int_A x \, dA = 0.$$

Thus, *the centroid of an area with one axis of symmetry must lie along the axis of symmetry*. The axis of symmetry then is a centroidal axis, which is another indication that the first moment of area must be zero about the axis of symmetry. With two orthogonal axes of symmetry, the centroid must lie at the intersection of these axes. For such areas as circles and rectangles, the centroid is easily determined by inspection.

In many problems, the area of interest can be considered formed by the addition or subtraction of simple areas. For simple areas the centroids are known by inspection. The areas made up of such simple areas are *composite* areas. For composite areas

$$x_C = \frac{\sum_i A_i x_{Ci}}{A},$$

$$y_C = \frac{\sum_i A_i y_{Ci}}{A},$$

where x_{Ci} and y_{Ci} (with proper signs) are the centroidal coordinates to simple area A_i, and where A is the total area.

I.1.3 Moments and Couples

Moment of a Bound Vector About a Point
Definition
The moment of a bound vector $\mathbf{v}$ about a point A is the vector

$$\mathbf{M}_A^{\mathbf{v}} = \mathbf{r}_{AB} \times \mathbf{v}, \tag{I.1.6}$$

where $\mathbf{r}_{AB}$ is the position vector of B relative to A, and B is any point of line of action, Δ, of the vector $\mathbf{v}$ (Fig. I.1.14).

The vector $\mathbf{M}_A^{\mathbf{v}} = \mathbf{0}$ if and only the line of action of $\mathbf{v}$ passes through A or $\mathbf{v} = \mathbf{0}$. The magnitude of $\mathbf{M}_A^{\mathbf{v}}$ is

$$|\mathbf{M}_A^{\mathbf{v}}| = M_A^{\mathbf{v}} = |\mathbf{r}_{AB}| \, |\mathbf{v}| \, \sin \theta,$$

where θ is the angle between $\mathbf{r}_{AB}$ and $\mathbf{v}$ when they are placed tail to tail. The perpendicular distance from A to the line of action of $\mathbf{v}$ is

$$d = |\mathbf{r}_{AB}| \, \sin \theta,$$

and the magnitude of $\mathbf{M}_A^{\mathbf{v}}$ is

$$|\mathbf{M}_A^{\mathbf{v}}| = M_A^{\mathbf{v}} = d \, |\mathbf{v}|.$$

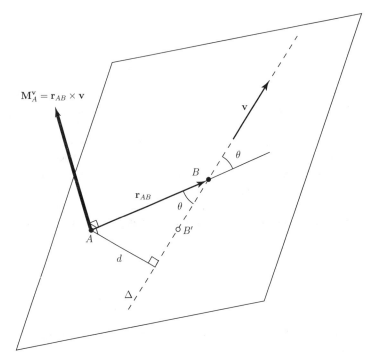

FIGURE I.1.14   *Moment of a bound vector about a point.*

The vector $\mathbf{M}_A^{\mathbf{v}}$ is perpendicular to both $\mathbf{r}_{AB}$ and $\mathbf{v}$:

$$\mathbf{M}_A^{\mathbf{v}} \perp \mathbf{r}_{AB} \quad \text{and} \quad \mathbf{M}_A^{\mathbf{v}} \perp \mathbf{v}.$$

The vector $\mathbf{M}_A^{\mathbf{v}}$ being perpendicular to $\mathbf{r}_{AB}$ and $\mathbf{v}$ is perpendicular to the plane containing $\mathbf{r}_{AB}$ and $\mathbf{v}$.

The moment given by Eq. (I.1.6) does not depend on the point B of the line of action of $\mathbf{v}$, Δ, where $\mathbf{r}_{AB}$ intersects Δ. Instead of using the point B the point B' (Fig. I.1.14) can be used. The position vector of B relative to A is $\mathbf{r}_{AB} = \mathbf{r}_{AB'} + \mathbf{r}_{B'B}$ where the vector $\mathbf{r}_{B'B}$ is parallel to $\mathbf{v}$, $\mathbf{r}_{B'B} \| \mathbf{v}$. Therefore,

$$\mathbf{M}_A^{\mathbf{v}} = \mathbf{r}_{AB} \times \mathbf{v} = (\mathbf{r}_{AB'} + \mathbf{r}_{B'B}) \times \mathbf{v} = \mathbf{r}_{AB'} \times \mathbf{v} + \mathbf{r}_{B'B} \times \mathbf{v} = \mathbf{r}_{AB'} \times \mathbf{v},$$

because $\mathbf{r}_{B'B} \times \mathbf{v} = \mathbf{0}$.

Moment of a Bound Vector About a Line
Definition
The moment $\mathbf{M}_\Omega^{\mathbf{v}}$ of a bound vector $\mathbf{v}$ about a line Ω is the Ω resolute (Ω component) of the moment $\mathbf{v}$ about any point on Ω (Fig. I.1.15).

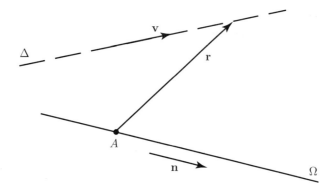

FIGURE I.1.15 *Moment of a bound vector about a line.*

The $\mathbf{M}_{\Omega}^{\mathbf{v}}$ is the Ω resolute of $\mathbf{M}_A^{\mathbf{v}}$

$$\mathbf{M}_{\Omega}^{\mathbf{v}} = \mathbf{n}\cdot\mathbf{M}_A^{\mathbf{v}}\,\mathbf{n}$$
$$= \mathbf{n}\cdot(\mathbf{r}\times\mathbf{v})\,\mathbf{n}$$
$$= [\mathbf{n},\mathbf{r},\mathbf{v}]\,\mathbf{n},$$

where $\mathbf{n}$ is a unit vector parallel to Ω, and $\mathbf{r}$ is the position vector of a point on the line of action of $\mathbf{v}$ relative to a point on Ω.

The magnitude of $\mathbf{M}_{\Omega}^{\mathbf{v}}$ is given by

$$\left|\mathbf{M}_{\Omega}^{\mathbf{v}}\right| = |[\mathbf{n},\mathbf{r},\mathbf{v}]|.$$

The moment of a vector about a line is a free vector.

If a line Ω is parallel to the line of action Δ of a vector $\mathbf{v}$, then $[\mathbf{n},\mathbf{r},\mathbf{v}]\mathbf{n} = \mathbf{0}$ and $\mathbf{M}_{\Omega}^{\mathbf{v}} = \mathbf{0}$.

If a line Ω intersects the line of action Δ of $\mathbf{v}$, then $\mathbf{r}$ can be chosen in such a way that $\mathbf{r} = \mathbf{0}$ and $\mathbf{M}_{\Omega}^{\mathbf{v}} = \mathbf{0}$.

If a line Ω is perpendicular to the line of action Δ of a vector $\mathbf{v}$, and d is the shortest distance between these two lines, then

$$\left|\mathbf{M}_{\Omega}^{\mathbf{v}}\right| = d|\mathbf{v}|.$$

Moments of a System of Bound Vectors

Definition

The moment of a system $\{S\}$ of bound vectors $\mathbf{v}_i$, $\{S\} = \{\mathbf{v}_1,\mathbf{v}_2,\ldots,\mathbf{v}_n\} = \{\mathbf{v}_i\}_{i=1,2,\ldots,n}$ about a point A is

$$\mathbf{M}_A^{\{S\}} = \sum_{i=1}^{n}\mathbf{M}_A^{\mathbf{v}_i}.$$

Definition

The moment of a system $\{S\}$ of bound vectors $\mathbf{v}_i$, $\{S\} = \{\mathbf{v}_1, \mathbf{v}_2, \ldots, \mathbf{v}_n\} = \{\mathbf{v}_i\}_{i=1,2,\ldots,n}$ about a line Ω is

$$\mathbf{M}_\Omega^{\{S\}} = \sum_{i=1}^{n} \mathbf{M}_\Omega^{\mathbf{v}_i}.$$

The moments $\mathbf{M}_A^{\{S\}}$ and $\mathbf{M}_P^{\{S\}}$ of a system $\{S\}$, $\{S\} = \{\mathbf{v}_i\}_{i=1,2,\ldots,n}$, of bound vectors, $\mathbf{v}_i$, about two points A and P, are related to each other as follows:

$$\mathbf{M}_A^{\{S\}} = \mathbf{M}_P^{\{S\}} + \mathbf{r}_{AP} \times \mathbf{R}, \qquad (I.1.7)$$

where $\mathbf{r}_{AP}$ is the position vector of P relative to A, and $\mathbf{R}$ is the resultant of $\{S\}$.

Proof

Let B_i a point on the line of action of the vector $\mathbf{v}_i$, $\mathbf{r}_{ABi}$ and $\mathbf{r}_{PBi}$ the position vectors of B_i relative to A and P (Fig. I.1.16). Thus,

$$\mathbf{M}_A^{\{S\}} = \sum_{i=1}^{n} \mathbf{M}_A^{\mathbf{v}_i} = \sum_{i=1}^{n} \mathbf{r}_{ABi} \times \mathbf{v}_i$$

$$= \sum_{i=1}^{n} (\mathbf{r}_{AP} + \mathbf{r}_{PBi}) \times \mathbf{v}_i = \sum_{i=1}^{n} (\mathbf{r}_{AP} \times \mathbf{v}_i + \mathbf{r}_{PBi} \times \mathbf{v}_i)$$

$$= \sum_{i=1}^{n} \mathbf{r}_{AP} \times \mathbf{v}_i + \sum_{i=1}^{n} \mathbf{r}_{PBi} \times \mathbf{v}_i$$

$$= \mathbf{r}_{AP} \times \sum_{i=1}^{n} \mathbf{v}_i + \sum_{i=1}^{n} \mathbf{r}_{PBi} \times \mathbf{v}_i$$

$$= \mathbf{r}_{AP} \times \mathbf{R} + \sum_{i=1}^{n} \mathbf{M}_P^{\mathbf{v}_i}$$

$$= \mathbf{r}_{AP} \times \mathbf{R} + \mathbf{M}_P^{\{S\}}.$$

If the resultant $\mathbf{R}$ of a system $\{S\}$ of bound vectors is not equal to zero, $\mathbf{R} \neq \mathbf{0}$, the points about which $\{S\}$ has a minimum moment $\mathbf{M}_{min}$ lie on a line called *central axis*, (CA), of $\{S\}$, which is parallel to $\mathbf{R}$ and passes through a point P whose position vector $\mathbf{r}$ relative to an arbitrarily selected reference point O is given by

$$\mathbf{r} = \frac{\mathbf{R} \times \mathbf{M}_O^{\{S\}}}{\mathbf{R}^2}.$$

The minimum moment $\mathbf{M}_{min}$ is given by

$$\mathbf{M}_{min} = \frac{\mathbf{R} \cdot \mathbf{M}_O^{\{S\}}}{\mathbf{R}^2} \mathbf{R}.$$

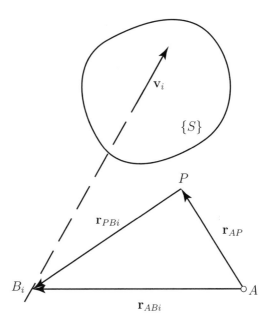

FIGURE I.1.16 *Moments of a system of bound vectors.*

Couples

Definition

A *couple* is a system of bound vectors whose resultant is equal to zero and whose moment about some point is not equal to zero.

A system of vectors is not a vector, therefore couples are not vectors.

A couple consisting of only two vectors is called a *simple couple*. The vectors of a simple couple have equal magnitudes, parallel lines of action, and opposite senses. Writers use the word "couple" to denote the simple couple.

The moment of a couple about a point is called the *torque* of the couple, $\mathbf{M}$ or $\mathbf{T}$. The moment of a couple about one point is equal to the moment of the couple about any other point, i.e., it is unnecessary to refer to a specific point. The moment of a couple is a free vector.

The torques are vectors and the magnitude of a torque of a simple couple is given by

$$|\mathbf{M}| = d|\mathbf{v}|,$$

where d is the distance between the lines of action of the two vectors comprising the couple, and $\mathbf{v}$ is one of these vectors.

Proof

In Figure I.1.17, the torque $\mathbf{M}$ is the sum of the moments of $\mathbf{v}$ and $-\mathbf{v}$ about any point. The moments about point A are

$$\mathbf{M} = \mathbf{M}_A^{\mathbf{v}} + \mathbf{M}_A^{-\mathbf{v}} = \mathbf{r} \times \mathbf{v} + \mathbf{0}.$$

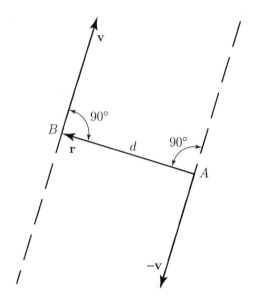

FIGURE I.1.17 *Couple.*

Hence,

$$|\mathbf{M}| = |\mathbf{r} \times \mathbf{v}| = |\mathbf{r}||\mathbf{v}| \sin(\mathbf{r}, \mathbf{v}) = d|\mathbf{v}|.$$

The direction of the torque of a simple couple can be determined by inspection: $\mathbf{M}$ is perpendicular to the plane determined by the lines of action of the two vectors comprising the couple, and the sense of $\mathbf{M}$ is the same as that of $\mathbf{r} \times \mathbf{v}$.

The moment of a couple about a line Ω is equal to the Ω resolute of the torque of the couple. The moments of a couple about two parallel lines are equal to each other.

Equivalence of Systems

Definition

Two systems $\{S\}$ and $\{S'\}$ of bound vectors are said to be *equivalent* when:

1. The resultant of $\{S\}$, $\mathbf{R}$, is equal to the resultant of $\{S'\}$, $\mathbf{R}'$

$$\mathbf{R} = \mathbf{R}'$$

2. There exists at least one point about which $\{S\}$ and $\{S'\}$ have equal moments

$$\text{exists} \quad P: \quad \mathbf{M}_P^{\{S\}} = \mathbf{M}_P^{\{S'\}}.$$

Figures I.1.18(a) and I.1.18(b) each show a rod subjected to the action of a pair of forces. The two pairs of forces are equivalent, but their effects on the rod are different from each other. The word "equivalence" is not to be regarded as implying physical equivalence.

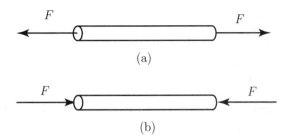

FIGURE I.1.18 *Equivalent systems (not physical equivalence): (a) tension and (b) compression.*

For given a line Ω and two equivalent systems $\{S\}$ and $\{S'\}$ of bound vectors, the sum of the Ω resolutes of the vectors in $\{S\}$ is equal to the sum of the Ω resolutes of the vectors in $\{S'\}$.

The moments of two equivalent systems of bound vectors, about point, are equal to each other.

The moments of two equivalent systems $\{S\}$ and $\{S'\}$ of bound vectors, about *any* line Ω, are equal to each other.

Transitivity of the equivalence relation. If $\{S\}$ is equivalent to $\{S'\}$, and $\{S'\}$ is equivalent to $\{S''\}$, then $\{S\}$ is equivalent to $\{S''\}$.

Every system $\{S\}$ of bound vectors with the resultant $\mathbf{R}$ can be replaced with a system consisting of a couple C and a single bound vector $\mathbf{v}$ whose line of action passes through an arbitrarily selected *base point* O. The torque $\mathbf{M}$ of C depends on the choice of base point $\mathbf{M} = \mathbf{M}_O^{\{S\}}$. The vector $\mathbf{v}$ is independent of the choice of base point, $\mathbf{v} = \mathbf{R}$.

A couple C can be replaced with any system of couples, the sum of whose torque is equal to the torque of C.

When a system of bound vectors consists of a couple of torque $\mathbf{M}$ and a single vector parallel to $\mathbf{M}$, it is called a *wrench*.

Force Vector and Moment of a Force

Force is a vector quantity, having both magnitude and direction. Force is commonly explained in terms of Newton's three laws of motion set forth in his *Principia Mathematica* (1687). Newton's first principle: a body that is at rest or moving at a uniform rate in a straight line will remain in that state until some force is applied to it. Newton's second law of motion states that a particle acted on by forces whose resultant is not zero will move in such a way that the time rate of change of its momentum will at any instant be proportional to the resultant force. Newton's third law states that when one body exerts a force on another body, the second body exerts an equal force on the first body. This is the principle of action and reaction.

Because force is a vector quantity it can be represented graphically as a directed line segment. The representation of forces by vectors implies that they are concentrated either at a single point or along a single line. The force of gravity is invariably distributed throughout the volume of a body. Nonetheless, when the equilibrium of a body is the primary consideration, it is generally valid as well as convenient to assume that the forces are concentrated at a single point. In the case of gravitational force, the total weight of a body may be assumed to be concentrated at its center of gravity.

Force is measured in newtons (N); a force of 1 N will accelerate a mass of one kilogram at a rate of one meter per second. The newton is a unit of the International System (SI) used for measuring force.

Using the English system, the force is measured in pounds. One pound of force imparts to a one-pound object an acceleration of 32.17 feet per second squared.

The force vector $\mathbf{F}$ can be expressed in terms of a Cartesian reference frame, with the unit vectors $\mathbf{i}$, $\mathbf{j}$, and $\mathbf{k}$ [Fig. I.1.19(a)]:

$$\mathbf{F} = F_x \mathbf{i} + F_y \mathbf{j} + F_z \mathbf{k}. \tag{I.1.8}$$

The components of the force in the x, y, and z directions are F_x, F_y, and F_z. The resultant of two forces $\mathbf{F}_1 = F_{1x}\mathbf{i} + F_{1y}\mathbf{j} + F_{1z}\mathbf{k}$ and $\mathbf{F}_2 = F_{2x}\mathbf{i} + F_{2y}\mathbf{j} + F_{2z}\mathbf{k}$ is the vector sum of those forces:

$$\mathbf{R} = \mathbf{F}_1 + \mathbf{F}_2 = (F_{1x} + F_{2x})\mathbf{i} + (F_{1y} + F_{2y})\mathbf{j} + (F_{1z} + F_{2z})\mathbf{k}. \tag{I.1.9}$$

A moment is defined as the moment of a force about (with respect to) a point. The moment of the force $\mathbf{F}$ about the point O is the cross product vector

$$\mathbf{M}_O^{\mathbf{F}} = \mathbf{r} \times \mathbf{F}$$

$$= \begin{vmatrix} \mathbf{i} & \mathbf{j} & \mathbf{k} \\ r_x & r_y & r_z \\ F_x & F_y & F_z \end{vmatrix}$$

$$= (r_y F_z - r_z F_y)\mathbf{i} + (r_z F_x - r_x F_z)\mathbf{j} + (r_x F_y - r_y F_x)\mathbf{k}. \tag{I.1.10}$$

where $\mathbf{r} = r_x \mathbf{i} + r_y \mathbf{j} + r_z \mathbf{k}$ is a position vector directed from the point about which the moment is taken (O in this case) to any point A on the line of action of the force [Fig. I.1.19(a)]. If the coordinates of O are x_O, y_O, z_O and the coordinates of A are x_A, y_A, z_A, then

$\mathbf{r} = \mathbf{r}_{OA} = (x_A - x_O)\mathbf{i} + (y_A - y_O)\mathbf{j} + (z_A - z_O)\mathbf{k}$ and the moment of the force $\mathbf{F}$ about the point O is

$$\mathbf{M}_O^{\mathbf{F}} = \mathbf{r}_{OA} \times \mathbf{F} = \begin{vmatrix} \mathbf{i} & \mathbf{j} & \mathbf{k} \\ x_A - x_O & y_A - y_O & z_A - z_O \\ F_x & F_y & F_z \end{vmatrix}.$$

The magnitude of $\mathbf{M}_O^{\mathbf{F}}$ is

$$|\mathbf{M}_O^{\mathbf{F}}| = M_O^{\mathbf{F}} = r F |\sin \theta|,$$

where $\theta = \angle(\mathbf{r}, \mathbf{F})$ is the angle between vectors $\mathbf{r}$ and $\mathbf{F}$, and $r = |\mathbf{r}|$ and $F = |\mathbf{F}|$ are the magnitudes of the vectors.

The line of action of $\mathbf{M}_O^{\mathbf{F}}$ is perpendicular to the plane containing $\mathbf{r}$ and $\mathbf{F}$ ($\mathbf{M}_O^{\mathbf{F}} \perp \mathbf{r}$ & $\mathbf{M}_O^{\mathbf{F}} \perp \mathbf{F}$) and the sense is given by the right-hand rule.

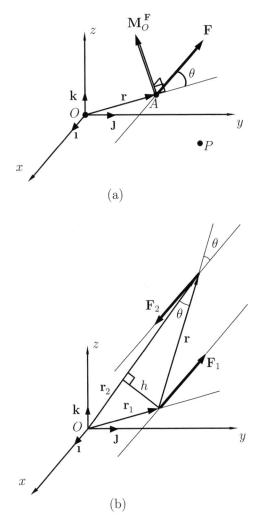

(a)

(b)

FIGURE I.1.19 *Moment of a force: (a) moment of a force about a point and (b) torque of the couple.*

The moment of the force **F** about another point P is

$$\mathbf{M}_P^{\mathbf{F}} = \mathbf{r}_{PA} \times \mathbf{F} = \begin{vmatrix} \mathbf{1} & \mathbf{J} & \mathbf{k} \\ x_A - x_P & y_A - y_P & z_A - z_P \\ F_x & F_y & F_z \end{vmatrix},$$

where x_P, y_P, z_P are the coordinates of the point P.

The system of two forces, $\mathbf{F}_1$ and $\mathbf{F}_2$, which have equal magnitudes $|\mathbf{F}_1| = |\mathbf{F}_2|$, opposite senses $\mathbf{F}_1 = -\mathbf{F}_2$, and parallel directions $(\mathbf{F}_1 \| \mathbf{F}_2)$ is a couple. The resultant force of a couple is zero $\mathbf{R} = \mathbf{F}_1 + \mathbf{F}_2 = \mathbf{0}$. The resultant moment $\mathbf{M} \neq \mathbf{0}$ about an arbitrary point is

$$\mathbf{M} = \mathbf{r}_1 \times \mathbf{F}_1 + \mathbf{r}_2 \times \mathbf{F}_2,$$

or

$$\mathbf{M} = \mathbf{r}_1 \times (-\mathbf{F}_2) + \mathbf{r}_2 \times \mathbf{F}_2 = (\mathbf{r}_2 - \mathbf{r}_1) \times \mathbf{F}_2 = \mathbf{r} \times \mathbf{F}_2, \qquad (\text{I.1.11})$$

where $\mathbf{r} = \mathbf{r}_2 - \mathbf{r}_1$ is a vector from any point on the line of action of $\mathbf{F}_1$ to any point of the line of action of $\mathbf{F}_2$. The direction of the torque of the couple is perpendicular to the plane of the couple and the magnitude is given by [Fig. I.1.19(b)]:

$$|\mathbf{M}| = M = r F_2 |\sin\theta| = h F_2, \qquad (\text{I.1.12})$$

where $h = r|\sin\theta|$ is the perpendicular distance between the lines of action. The resultant moment of a couple is independent of the point with respect to which moments are taken.

Representing Systems by Equivalent Systems

To simplify the analysis of the forces and moments acting on a given system one can represent the system by an equivalent, less complicated one. The actual forces and moments can be replaced with a total force and a total moment.

Figure I.1.20 shows an arbitrary system of forces and moments, {system 1}, and a point P. This system can be represented by a system, {system 2}, consisting of a single force $\mathbf{F}$ acting at P and a single couple of torque $\mathbf{M}$. The conditions for equivalence are

$$\sum \mathbf{F}^{\{\text{system 2}\}} = \sum \mathbf{F}^{\{\text{system 1}\}} \implies \mathbf{F} = \sum \mathbf{F}^{\{\text{system 1}\}},$$

and

$$\sum \mathbf{M}_P^{\{\text{system 2}\}} = \sum \mathbf{M}_P^{\{\text{system 1}\}} \implies \mathbf{M} = \sum \mathbf{M}_P^{\{\text{system 1}\}}.$$

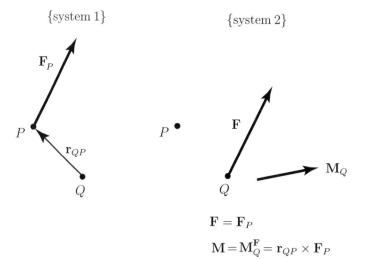

$$\mathbf{F} = \mathbf{F}_P$$
$$\mathbf{M} = \mathbf{M}_Q^{\mathbf{F}} = \mathbf{r}_{QP} \times \mathbf{F}_P$$

FIGURE I.1.20 *Equivalent systems.*

These conditions are satisfied if $\mathbf{F}$ equals the sum of the forces in {system 1}, and $\mathbf{M}$ equals the sum of the moments about P in {system 1}. Thus, no matter how complicated a system of forces and moments may be, it can be represented by a single force acting at a given point and a single couple. Three particular cases occur frequently in practice.

1. *Force represented by a force and a couple.* A force $\mathbf{F}_P$ acting at a point P {system 1} in Figure I.1.20 can be represented by a force $\mathbf{F}$ acting at a different point Q and a couple of torque $\mathbf{M}$, {system 2}. The moment of {system 1} about point Q is $\mathbf{r}_{QP} \times \mathbf{F}_P$, where $\mathbf{r}_{QP}$ is the vector from Q to P. The conditions for equivalence are

$$\sum \mathbf{M}_P^{\{system\ 2\}} = \sum \mathbf{M}_P^{\{system\ 1\}} \implies \mathbf{F} = \mathbf{F}_P,$$

and

$$\sum \mathbf{M}_Q^{\{system\ 2\}} = \sum \mathbf{M}_Q^{\{system\ 1\}} \implies \mathbf{M} = \mathbf{M}_Q^{\mathbf{F}_P} = \mathbf{r}_{QP} \times \mathbf{F}_P.$$

The systems are equivalent if the force $\mathbf{F}$ equals the force $\mathbf{F}_P$ and the couple of torque $\mathbf{M}_Q^{\mathbf{F}_P}$ equals the moment of $\mathbf{F}_P$ about Q.

2. *Concurrent forces represented by a force.* A system of concurrent forces whose lines of action intersect at a point P {system 1}, in Figure I.1.21(a), can be represented by a single force whose line of action intersects P, {system 2}. The sums of the forces in the two systems are equal if

$$\mathbf{F} = \mathbf{F}_1 + \mathbf{F}_2 + \cdots + \mathbf{F}_n.$$

The sum of the moments about P equals zero for each system, so the systems are equivalent if the force $\mathbf{F}$ equals the sum of the forces in {system 1}.

3. *Parallel forces represented by a force.* A system of parallel forces whose sum is not zero can be represented by a single force $\mathbf{F}$ shown in Figure I.1.21(b).

4. *System represented by a wrench.* In general any system of forces and moments can be represented by a single force acting at a given point and a single couple. Figure I.1.22 shows an arbitrary force $\mathbf{F}$ acting at a point P and an arbitrary couple of torque $\mathbf{M}$, {system 1}. This system can be represented by a simpler one, i.e., one may represent the force $\mathbf{F}$ acting at a different point Q and the component of $\mathbf{M}$ that is parallel to $\mathbf{F}$. A coordinate system is chosen so that $\mathbf{F}$ is along the y axis

$$\mathbf{F} = F\mathbf{J},$$

and $\mathbf{M}$ is contained in the xy plane

$$\mathbf{M} = M_x\mathbf{I} + M_y\mathbf{J}.$$

The equivalent system, {system 2}, consists of the force $\mathbf{F}$ acting at a point Q on the z axis

$$\mathbf{F} = F\mathbf{J},$$

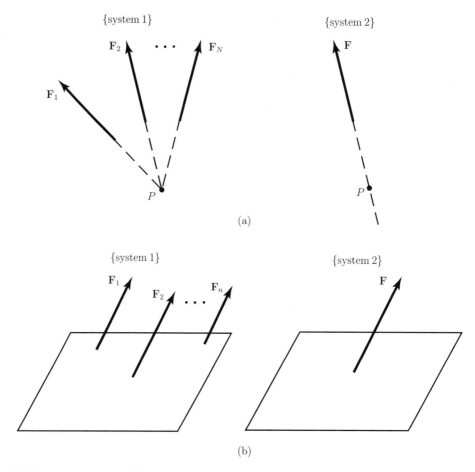

(a)

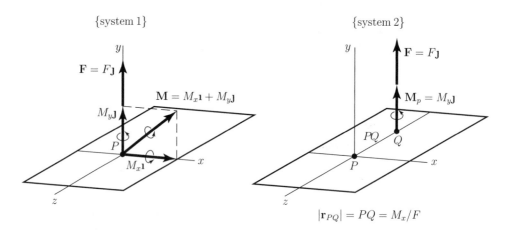

(b)

FIGURE I.1.21 *System of forces: (a) concurrent forces, and (b) parallel forces.*

{system 1}

$\mathbf{F} = F\mathbf{J}$

$\mathbf{M} = M_x\mathbf{1} + M_y\mathbf{J}$

$M_y\mathbf{J}$

$M_x\mathbf{1}$

{system 2}

$\mathbf{F} = F\mathbf{J}$

$\mathbf{M}_p = M_y\mathbf{J}$

$|\mathbf{r}_{PQ}| = PQ = M_x/F$

FIGURE I.1.22 *System represented by a wrench.*

and the component of $\mathbf{M}$ parallel to $\mathbf{F}$

$$\mathbf{M}_p = M_y\mathbf{J}.$$

The distance PQ is chosen so that $|\mathbf{r}_{PQ}| = PQ = M_x/F$. The {system 1} is equivalent to {system 2}. The sum of the forces in each system is the same $\mathbf{F}$. The sum of the moments about P in {system 1} is $\mathbf{M}$, and the sum of the moments about P in {system 2} is

$$\sum \mathbf{M}_P^{\{\text{system 2}\}} = \mathbf{r}_{PQ} \times \mathbf{F} + M_y\mathbf{J} = (-PQ\mathbf{k}) \times (F\mathbf{J}) + M_y\mathbf{J} = M_x\mathbf{1} + M_y\mathbf{J} = \mathbf{M}.$$

The system of the force $\mathbf{F} = F\mathbf{J}$ and the couple $\mathbf{M}_p = M_y\mathbf{J}$ that is parallel to $\mathbf{F}$ is a wrench. A wrench is the simplest system that can be equivalent to an arbitrary system of forces and moments.

The representation of a given system of forces and moments by a wrench requires the following steps:

1. Choose a convenient point P and represent the system by a force $\mathbf{F}$ acting at P and a couple $\mathbf{M}$ [Fig. I.1.23(a)].

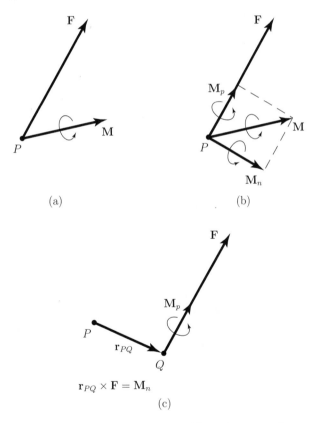

(a)

(b)

(c)

FIGURE I.1.23 *Steps required to represent a system of forces by a wrench.*

2. Determine the components of **M** parallel and normal to **F** [Fig. I.1.23(b)]

$$\mathbf{M} = \mathbf{M}_p + \mathbf{M}_n, \quad \text{where} \quad \mathbf{M}_p \| \mathbf{F}.$$

3. The wrench consists of the force **F** acting at a point Q and the parallel component **M**$_p$ [Fig. I.1.23(c)]. For equivalence, the following condition must be satisfied:

$$\mathbf{r}_{PQ} \times \mathbf{F} = \mathbf{M}_n,$$

where **M**$_n$ is the normal component of **M**.

In general, the {system 1} cannot be represented by a force **F** alone.

I.1.4   Equilibrium

Equilibrium Equations

A body is in *equilibrium* when it is stationary or in steady translation relative to an inertial reference frame. The following conditions are satisfied when a body, acted upon by a system of forces and moments, is in equilibrium:

1. The sum of the forces is zero:

$$\sum \mathbf{F} = 0. \tag{I.1.13}$$

2. The sum of the moments about any point is zero:

$$\sum \mathbf{M}_P = 0, \quad \forall P. \tag{I.1.14}$$

If the sum of the forces acting on a body is zero and the sum of the moments about one point is zero, then the sum of the moments about every point is zero.

Proof

The body shown in Figure I.1.24 is subjected to forces $\mathbf{F}_{Ai}$, $i = 1, \ldots, n$, and moments $\mathbf{M}_j$, $j = 1, \ldots, m$. The sum of the forces is zero,

$$\sum \mathbf{F} = \sum_{i=1}^{n} \mathbf{F}_{Ai} = \mathbf{0},$$

and the sum of the moments about a point P is zero,

$$\sum \mathbf{M}_P = \sum_{i=1}^{n} \mathbf{r}_{PAi} \times \mathbf{F}_{Ai} + \sum_{j=1}^{m} \mathbf{M}_j = \mathbf{0},$$

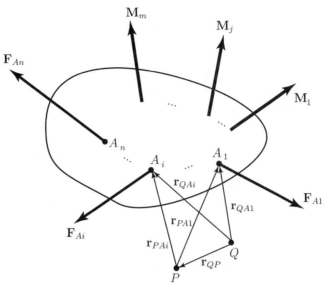

$$\mathbf{r}_{QAi} = \mathbf{r}_{QP} + \mathbf{r}_{PAi}$$

FIGURE I.1.24 *Forces and moments acting on a body.*

where $\mathbf{r}_{PAi} = \vec{P}A_i$, $i = 1, \ldots, n$. The sum of the moments about any other point Q is

$$\sum \mathbf{M}_Q = \sum_{i=1}^{n} \mathbf{r}_{QAi} \times \mathbf{F}_{Ai} + \sum_{j=1}^{m} \mathbf{M}_j$$

$$= \sum_{i=1}^{n} \left(\mathbf{r}_{QP} + \mathbf{r}_{PAi} \right) \times \mathbf{F}_{Ai} + \sum_{j=1}^{m} \mathbf{M}_j$$

$$= \mathbf{r}_{QP} \times \sum_{i=1}^{n} \mathbf{F}_{Ai} + \sum_{i=1}^{n} \mathbf{r}_{PAi} \times \mathbf{F}_{Ai} + \sum_{j=1}^{m} \mathbf{M}_j$$

$$= \mathbf{r}_{QP} \times \mathbf{0} + \sum_{i=1}^{n} \mathbf{r}_{PAi} \times \mathbf{F}_{Ai} + \sum_{j=1}^{m} \mathbf{M}_j$$

$$= \sum_{i=1}^{n} \mathbf{r}_{PAi} \times \mathbf{F}_{Ai} + \sum_{j=1}^{m} \mathbf{M}_j = \sum \mathbf{M}_P = \mathbf{0}.$$

A body is subjected to concurrent forces $\mathbf{F}_1, \mathbf{F}_2, \ldots, \mathbf{F}_n$ and no couples. If the sum of the concurrent forces is zero,

$$\mathbf{F}_1 + \mathbf{F}_2 + \cdots + \mathbf{F}_n = \mathbf{0},$$

the sum of the moments of the forces about the concurrent point is zero, so the sum of the moments about every point is zero. The only condition imposed by equilibrium on a set of concurrent forces is that their sum is zero.

Free-Body Diagrams

Free-body diagrams are used to determine forces and moments acting on simple bodies in equilibrium.

The beam in Figure I.1.25(a) has a pin support at the left end A and a roller support at the right end B. The beam is loaded by a force F and a moment M at C. To obtain the free-body diagram first the beam is isolated from its supports. Next, the reactions exerted on the beam by the supports are shown on the the free-body diagram [Fig. I.1.25(b)]. Once the free-body diagram is obtained one can apply the equilibrium equations.

The steps required to determine the reactions on bodies are:

1. Draw the free-body diagram, isolating the body from its supports and showing the forces and the reactions.
2. Apply the equilibrium equations to determine the reactions.

For two-dimensional systems, the forces and moments are related by three scalar equilibrium equations:

$$\sum F_x = 0, \tag{I.1.15}$$

$$\sum F_y = 0, \tag{I.1.16}$$

$$\sum M_P = 0, \quad \forall P. \tag{I.1.17}$$

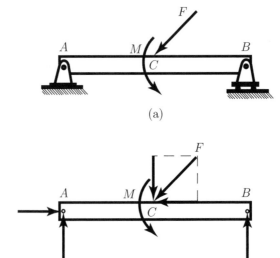

(a)

(b)

FIGURE I.1.25 *Free-body diagram: (a) beam with supports and (b) free-body diagram of the beam.*

One can obtain more than one equation from Eq. (I.1.17) by evaluating the sum of the moments about more than one point. The additional equations will not be independent of Eqs. (I.1.15)–(I.1.17). One cannot obtain more than three independent equilibrium equations from a two-dimensional free-body diagram, which means one can solve for at most three unknown forces or moments.

For three-dimensional systems, the forces and moments are related by six scalar equilibrium equations:

$$\sum F_x = 0 \tag{I.1.18}$$

$$\sum F_y = 0, \tag{I.1.19}$$

$$\sum F_z = 0, \tag{I.1.20}$$

$$\sum M_x = 0, \tag{I.1.21}$$

$$\sum M_y = 0, \tag{I.1.22}$$

$$\sum M_z = 0. \tag{I.1.23}$$

The sums of the moments about any point can be evaluated. Although one can obtain other equations by summing the moments about additional points, they will not be independent of these equations. For a three-dimensional free-body diagram, six independent equilibrium equations are obtained and one can solve for at most six unknown forces or moments.

A body has *redundant supports* when the body has more supports than the minimum number necessary to maintain it in equilibrium. Redundant supports are used whenever possible for strength and safety. Each support added to a body results in additional reactions. The difference between the number of reactions and the number of independent equilibrium equations is called the *degree of redundancy.*

A body has *improper supports* if it will not remain in equilibrium under the action of the loads exerted on it. The body with improper supports will move when the loads are applied.

Two-force and three-force members

A body is a *two-force member* if the system of forces and moments acting on the body is equivalent to two forces acting at different points.

For example, a body is subjected to two forces, $\mathbf{F}_A$ and $\mathbf{F}_B$, at A and B. If the body is in equilibrium, the sum of the forces equals zero only if $\mathbf{F}_A = -\mathbf{F}_B$. Furthermore, the forces $\mathbf{F}_A$ and $-\mathbf{F}_B$ form a couple, so the sum of the moments is not zero unless the lines of action of the forces lie along the line through the points A and B. Thus for equilibrium the two forces are equal in magnitude, are opposite in direction, and have the same line of action.

A body is a *three-force member* if the system of forces and moments acting on the body is equivalent to three forces acting at different points. If a three-force member is in equilibrium, the three forces are coplanar and the three forces are either parallel or concurrent.

Proof

Let the forces $\mathbf{F}_1$, $\mathbf{F}_2$, and $\mathbf{F}_3$ acting on the body at A_1, A_2, and A_3. Let π be the plane containing the three points of application A_1, A_2, and A_3. Let $\Delta = A_1 A_2$ be the line through the points of application of $\mathbf{F}_1$ and $\mathbf{F}_2$. Since the moments due to $\mathbf{F}_1$ and $\mathbf{F}_2$ about Δ are zero, the moment due to $\mathbf{F}_3$ about Δ must equal zero,

$$[\mathbf{n} \cdot (\mathbf{r} \times \mathbf{F}_3)] \, \mathbf{n} = [\mathbf{F}_3 \cdot (\mathbf{n} \times \mathbf{r})] \, \mathbf{n} = \mathbf{0},$$

where $\mathbf{n}$ is the unit vector of Δ. This equation requires that $\mathbf{F}_3$ be perpendicular to $\mathbf{n} \times \mathbf{r}$, which means that $\mathbf{F}_3$ is contained in π. The same procedure can be used to show that $\mathbf{F}_1$ and $\mathbf{F}_2$ are contained in π, so the forces $\mathbf{F}_1$, $\mathbf{F}_2$, and $\mathbf{F}_3$ are coplanar.

If the three coplanar forces are not parallel, there will be points where their lines of action intersect. Suppose that the lines of action of two forces $\mathbf{F}_1$ and $\mathbf{F}_2$ intersect at a point P. Then the moments of $\mathbf{F}_1$ and $\mathbf{F}_2$ about P are zero. The sum of the moments about P is zero only if the line of action of the third force, $\mathbf{F}_3$, also passes through P. Therefore either the forces are concurrent or they are parallel.

The analysis of a body in equilibrium can often be simplified by recognizing the two-force or three-force member.

I.1.5 Dry Friction

If a body rests on an incline plane, the friction force exerted on it by the surface prevents it from sliding down the incline. The question is, what is the steepest incline on which the body can rest?

A body is placed on a horizontal surface. The body is pushed with a small horizontal force F. If the force F is sufficiently small, the body does not move. Figure I.1.26 shows the free-body diagram of the body, where the force W is the weight of the body and N is the normal force exerted by the surface. The force F is the horizontal force, and F_f is the friction force exerted by the surface. Friction force arises in part from the interactions of the roughness, or asperities, of the contacting surfaces. The body is in equilibrium and $F_f = F$.

The force F is slowly increased. As long as the body remains in equilibrium, the friction force F_f must increase correspondingly, since it equals the force F. The body slips on the surface. The friction force, after reaching the maximum value, cannot maintain the body in equilibrium. The force applied to keep the body moving on the surface is smaller than the force required to cause it to slip. Why more force is required to start the body sliding on a

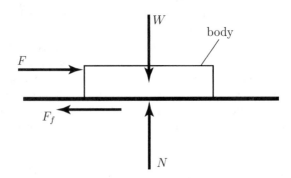

FIGURE I.1.26   *Friction force F_f exerted by a surface on a body.*

TABLE I.1.1 Typical Values of the Static Coefficient of Friction

Materials	μ_s
metal on metal	0.15 – 0.20
metal on wood	0.20 – 0.60
metal on masonry	0.30 – 0.70
wood on wood	0.25 – 0.50
masonry on masonry	0.60 – 0.70
rubber on concrete	0.50 – 0.90

surface than to keep it sliding is explained in part by the necessity to break the asperities of the contacting surfaces before sliding can begin.

The theory of dry friction, or *Coulomb friction*, predicts:

- the maximum friction forces that can be exerted by dry, contacting surfaces that are stationary relative to each other;
- the friction forces exerted by the surfaces when they are in relative motion, or sliding.

Static Coefficient of Friction

The magnitude of the maximum friction force, F_f, that can be exerted between two plane dry surfaces in contact is

$$F_f = \mu_s N, \tag{I.1.24}$$

where μ_s is a constant, the *static coefficient of friction*, and N is the normal component of the contact force between the surfaces. The value of the static coefficient of friction, μ_s, depends on:

- the materials of the contacting surfaces;
- the conditions of the contacting surfaces, namely smoothness and degree of contamination.

Typical values of μ_s for various materials are shown in Table I.1.1.

Equation (I.1.24) gives the maximum friction force that the two surfaces can exert without causing it to slip. If the static coefficient of friction μ_s between the body and the surface is known, the largest value of F one can apply to the body without causing it to slip is $F = F_f = \mu_s N$. Equation (I.1.24) determines the magnitude of the maximum friction force but not its direction. The friction force resists the impending motion.

Kinetic Coefficient of Friction

The magnitude of the friction force between two plane dry contacting surfaces that are in motion relative to each other is

$$F_f = \mu_k N, \tag{I.1.25}$$

where μ_k is the *kinetic coefficient of friction* and N is the normal force between the surfaces. The value of the kinetic coefficient of friction is generally smaller than the value of the static coefficient of friction, μ_s.

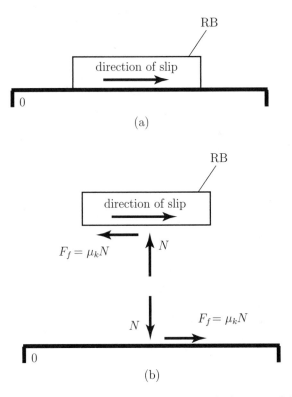

FIGURE I.1.27 *(a) Body moving on a surface and (b) free-body diagrams of the body and of the surface.*

To keep the body in Figure I.1.26 in uniform motion (sliding on the surface), the force exerted must be $F = F_f = \mu_k N$. The friction force resists the relative motion when two surfaces are sliding relative to each other.

The body RB shown in Figure I.1.27(a) is moving on the fixed surface 0. The direction of motion of RB is the positive axis x. The friction force on the body RB acts in the direction opposite to its motion, and the friction force on the fixed surface is in the opposite direction as shown in Figure I.1.27(b).

Angles of Friction

The *angle of friction*, θ, is the angle between the friction force, $F_f = |\mathbf{F}_f|$, and the normal force, $N = |\mathbf{N}|$, to the surface Fig. (I.1.28). The magnitudes of the normal force and friction force, and θ are related by

$$F_f = R \sin \theta,$$
$$N = R \cos \theta,$$

where $R = |\mathbf{R}| = |\mathbf{N} + \mathbf{F}_f|$.

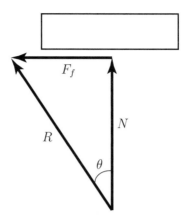

FIGURE I.1.28 *The angle of friction.*

The value of the angle of friction when slip is impending is called the *static angle of friction, θ_s,*

$$\tan \theta_s = \mu_s.$$

The value of the angle of friction when the surfaces are sliding relative to each other is called the *kinetic angle of friction, θ_k,*

$$\tan \theta_k = \mu_k.$$

I.1.6   Problems

I.1.1 (a) Find the angle made by the vector $\mathbf{v} = -10\mathbf{\imath} + 5\mathbf{\imath}$ with the positive x-axis and determine the unit vector in the direction of $\mathbf{v}$. (b) Determine the magnitude of the resultant $\mathbf{v} = \mathbf{v}_1 + \mathbf{v}_2$ and the angle which $\mathbf{v}$ makes with the positive x-axis, where the vectors $\mathbf{v}_1$ and $\mathbf{v}_2$ are shown in Figure I.1.29. The magnitudes of the vectors are $|\mathbf{v}_1| = v_1 = 5$, $|\mathbf{v}_2| = v_2 = 10$, and the angles of the vectors with the positive x-axis are $\theta_1 = 30°$, $\theta_2 = 60°$.

I.1.2 The planar vectors $\mathbf{a}$, $\mathbf{b}$, and $\mathbf{c}$ are given in xOy plane as shown in Figure I.1.30. The magnitude of the vectors are $a = P$, $b = 2P$, and $c = P\sqrt{2}$. The angles in the figure are $\alpha = 45°$, $\beta = 120°$, and $\gamma = 30°$. Determine the magnitude of the resultant $\mathbf{v} = \mathbf{a} + \mathbf{b} + \mathbf{c}$ and the angle that $\mathbf{v}$ makes with the positive x-axis.

I.1.3 The cube in Figure I.1.31 has the sides equal to l. Find the direction cosines of the resultant $\mathbf{v} = \mathbf{v}_1 + \mathbf{v}_2 + \mathbf{v}_3 + \mathbf{v}_4$.

I.1.4 The following spatial vectors are given: $\mathbf{v}_1 = -3\mathbf{\imath} + 4\mathbf{\jmath} - 3\mathbf{k}$, $\mathbf{v}_2 = 3\mathbf{\imath} + 3\mathbf{k}$, and $\mathbf{v}_3 = 1\mathbf{\imath} + 2\mathbf{\jmath} + 3\mathbf{k}$. Find the expressions $\mathbf{E}_1 = \mathbf{v}_1 + \mathbf{v}_2 + \mathbf{v}_3$, $\mathbf{E}_2 = \mathbf{v}_1 + \mathbf{v}_2 - \mathbf{v}_3$, $\mathbf{E}_3 = (\mathbf{v}_1 \times \mathbf{v}_2) \times \mathbf{v}_3$, and $E_4 = (\mathbf{v}_1 \times \mathbf{v}_2) \cdot \mathbf{v}_3$.

I.1.5 Find the angle between the vectors $\mathbf{v}_1 = 2\mathbf{\imath} - 4\mathbf{\jmath} + 4\mathbf{k}$ and $\mathbf{v}_2 = 4\mathbf{\imath} + 2\mathbf{\jmath} + 4\mathbf{k}$. Find the expressions $\mathbf{v}_1 \times \mathbf{v}_2$ and $\mathbf{v}_1 \cdot \mathbf{v}_2$.

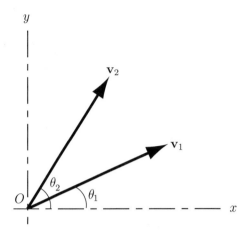

FIGURE I.1.29 *Vectors for Problem I.1.1.*

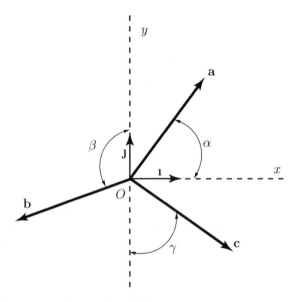

FIGURE I.1.30 *Planar vectors for Problem I.1.2.*

I.1.6 The following vectors are given $\mathbf{v}_1 = 2\,\mathbf{\imath} + 4\,\mathbf{j} + 6\,\mathbf{k}$, $\mathbf{v}_2 = 1\,\mathbf{\imath} + 3\,\mathbf{j} + 5\,\mathbf{k}$, and $\mathbf{v}_3 = -2\,\mathbf{\imath} + 2\,\mathbf{k}$. Find the vector triple product of $\mathbf{v}_1$, $\mathbf{v}_2$, and $\mathbf{v}_3$, and explain the result.

I.1.7 Solve the vectorial equation $\mathbf{x} \times \mathbf{a} = \mathbf{x} \times \mathbf{b}$, where $\mathbf{a}$ and $\mathbf{b}$ are two known given vectors.

I.1.8 Solve the vectorial equation $\mathbf{v} = \mathbf{a} \times \mathbf{x}$, where $\mathbf{v}$ and $\mathbf{a}$ are two known given vectors.

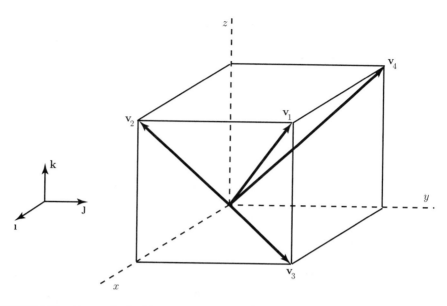

FIGURE I.1.31 *Vectors for Problem I.1.3.*

I.1.9 Solve the vectorial equation $\mathbf{a} \cdot \mathbf{x} = m$, where $\mathbf{a}$ is a known given vector and m is a known given scalar.

I.1.10 The forces $\mathbf{F}_1$, $\mathbf{F}_2$, $\mathbf{F}_3$, and $\mathbf{F}_4$, shown in Figure I.1.32, act on the sides of a cube (the side of the cube is l). The magnitude of the forces are $\mathbf{F}_1 = \mathbf{F}_2 = F$, and $\mathbf{F}_3 = \mathbf{F}_4 = F\sqrt{2}$. Represent the given system of forces by an equivalent system at O.

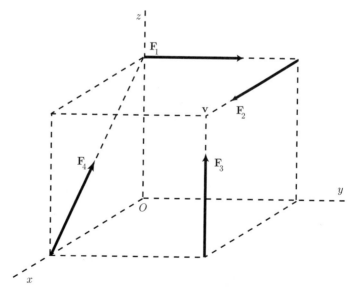

FIGURE I.1.32 *Forces for Problem I.1.10.*

I.1.11 Figure I.1.33 represents the vectors $\mathbf{v}_1$, $\mathbf{v}_2$, $\mathbf{v}_3$, and $\mathbf{v}_4$ acting on a cube with the side l. The magnitude of the forces are $\mathbf{v}_1 = V$ and $\mathbf{v}_2 = \mathbf{v}_3 = \mathbf{v}_4 = 2\,V$. Find the equivalent system at O.

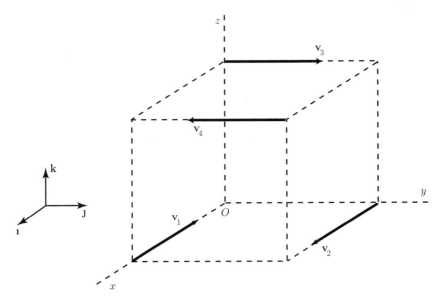

FIGURE I.1.33 *Vectors for Problem I.1.11.*

I.1.12 Repeat the previous problem for Figure I.1.34.

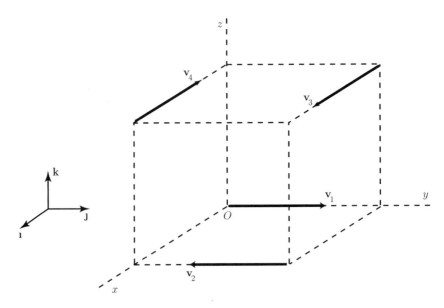

FIGURE I.1.34 *Vectors for Problem I.1.12.*

I.1.13 The parallelepiped shown in Figure I.1.35 has the sides $l = 1$ m, $w = 2$ m, and $h = 3$ m. The magnitude of the forces are $\mathbf{F}_1 = \mathbf{F}_2 = 10$ N, and $\mathbf{F}_3 = \mathbf{F}_4 = 20$ N. Find the equivalent wrench of the system.

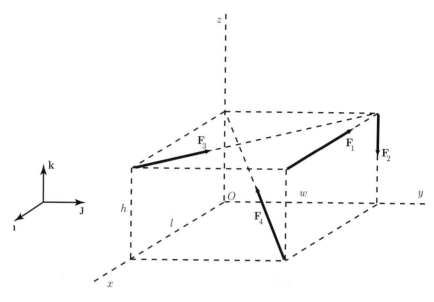

FIGURE I.1.35 *Forces for Problem I.1.13.*

I.1.14 A uniform rectangular plate of length l and width w is held open by a cable (Fig. I.1.36). The plate is hinged about an axis parallel to the plate edge of length l. Points A and B are at the extreme ends of this hinged edge. Points D and C are at the ends of the other edge of length l and are respectively adjacent to points A and B. Points D and C move as the plate opens. In the closed position, the plate is in a horizontal plane. When held open by a cable, the plate has rotated through an angle θ relative to the closed position. The supporting cable runs from point D to point E where point E is located a height h directly above the point B on the hinged edge of the plate. The cable tension required to hold the plate open is T. Find the projection of the tension force onto the diagonal axis AC of the plate. Numerical application: $l = 1.0$ m, $w = 0.5$ m, $\theta = 45°$, $h = 1.0$ m, and $T = 100$ N.

I.1.15 A smooth sphere of mass m is resting against a vertical surface and an inclined surface that makes an angle θ with the horizontal, as shown in Figure I.1.37. Find the forces exerted on the sphere by the two contacting surfaces. Numerical application: (a) $m = 10$ kg, $\theta = 30°$, and $g = 9.8$ m/s^2; (b) $m = 2$ slugs, $\theta = 60°$, and $g = 32.2$ ft/sec^2.

I.1.16 The links 1 and 2 shown in Figure I.1.38 are each connected to the ground at A and C, and to each other at B using frictionless pins. The length of link 1 is $AB = l$. The angle between the links is $\angle ABC = \theta$. A force of magnitude P is applied at the point D ($AD = 2l/3$) of the link 1. The force makes an angle θ with the horizontal. Find the force exerted by the lower link 2 on the upper link 1. Numerical application: (a) $l = 1$ m, $\theta = 30°$, and $P = 1000$ N; (b) $l = 2$ ft, $\theta = 45°$, and $P = 500$ lb.

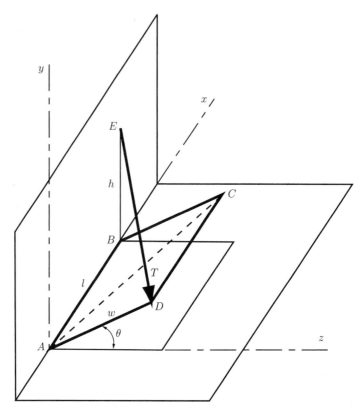

FIGURE I.1.36 *Rectangular plate for Problem I.1.14.*

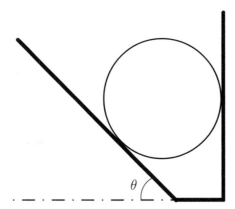

FIGURE I.1.37 *Smooth sphere for Problem I.1.15.*

I.1.17 The block of mass *m* rests on a rough horizontal surface and is acted upon by a force, *F*, that makes an angle θ with the horizontal, as shown in Figure I.1.39.

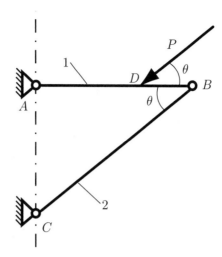

FIGURE I.1.38 *Two links connected for Problem I.1.16.*

The coefficient of static friction between the surface and the block is μ_s. Find the magnitude of the force F required to cause the block to begin to slide. Numerical application: (a) $m = 2$ kg, $\theta = 60°$, $\mu_s = 0.4$, and $g = 9.8$ m/s²; (b) $m = 10$ slugs, $\theta = 30°$, $\mu_s = 0.3$, and $g = 32.2$ ft/sec².

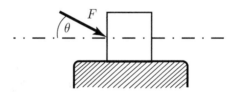

FIGURE I.1.39 *Block on a rough surface for Problem I.1.17.*

I.1.18 Find the x-coordinate of the centroid of the plane region bounded by the curves $y = x^2$ and $y = \sqrt{x}$, $(x > 0)$.

I.1.19 The shaft shown in Figure I.1.40 turns in the bearings A and B. The dimensions of the shaft are $a = 6$ in. and $b = 3$ in. The forces on the gear attached to the shaft are $F_t = 900$ lb and $F_r = 500$ lb. The gear forces act at a radius $R = 4$ in. from the axis of the shaft. Find the loads applied to the bearings.

I.1.20 The shaft shown in Figure I.1.41 turns in the bearings A and B. The dimensions of the shaft are $a = 120$ mm and $b = 30$ mm. The forces on the gear attached to the shaft are $F_t = 4500$ N, $F_r = 2500$ N, and $F_a = 1000$ N. The gear forces act at a radius $R = 100$ mm from the shaft axis. Determine the bearings loads.

I.1.21 The dimensions of the shaft shown Figure I.1.42 are $a = 2$ in. and $l = 5$ in. The force on the disk with the radius $r_1 = 5$ in. is $F_1 = 600$ lb and the force on the disk with the radius $r_2 = 2.5$ in. is $F_2 = 1200$ lb. Determine the forces on the bearings at A and B.

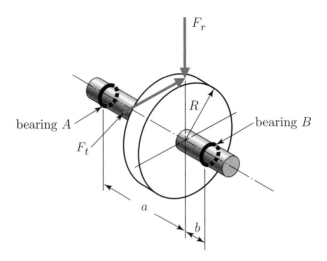

FIGURE I.1.40   *Shaft with gear for Problem I.1.19.*

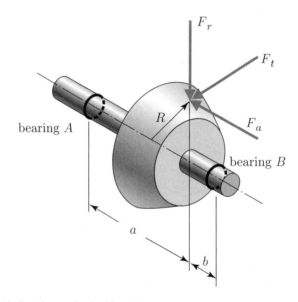

FIGURE I.1.41   *Shaft with gear for Problem I.1.20.*

I.1.22 The dimensions of the shaft shown Figure I.1.43 are $a = 50$ mm and $l = 120$ mm. The force on the disk with the radius $r_1 = 50$ mm is $F_1 = 2000$ N and the force on the disk with the radius $r_2 = 100$ mm is $F_2 = 4000$ N. Determine the bearing loads at A and B.

I.1.23 The force on the gear in Figure I.1.44 is $F = 1.5$ kN and the radius of the gear is $R = 60$ mm. The dimensions of the shaft are $l = 300$ mm and $a = 60$ mm. Determine the bearing loads at A and B.

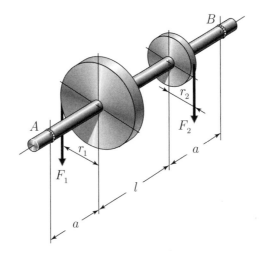

FIGURE I.1.42   *Shaft with two disks for Problem I.1.21.*

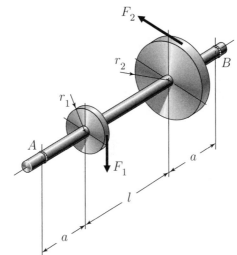

FIGURE I.1.43   *Shaft with two disks for Problem I.1.22.*

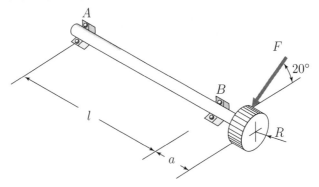

FIGURE I.1.44   *Shaft with gear for Problem I.1.23.*

I.2   Fundamentals

I.2.1   Degrees of Freedom and Motion

The *number of degrees of freedom* (DOF) of a system is equal to the number of independent parameters (measurements) that are needed to uniquely define its position in space at any instant of time. The number of DOF is defined with respect to a reference frame.

Figure I.2.1 shows a rigid body (RB) lying in a plane. The rigid body is assumed to be incapable of deformation and the distance between two particles on the rigid body is constant at any time. If this rigid body always remains in the plane, three parameters (three DOF) are required to completely define its position: two linear coordinates (x, y) to define the position of any one point on the rigid body, and one angular coordinate θ to define the angle of the body with respect to the axes. The minimum number of measurements needed to define its position are shown in the figure as x, y, and θ. A rigid body in a plane then has three degrees of freedom. Note that the particular parameters chosen to define its position are not unique. Any alternative set of three parameters could be used. There is an infinity of sets of parameters possible, but in this case there must always be three parameters per set, such as two lengths and an angle, to define the position because a rigid body in plane motion has three DOF.

Six parameters are needed to define the position of a free rigid body in a three-dimensional (3-D) space. One possible set of parameters which could be used are three lengths, (x, y, z), plus three angles $(\theta_x, \theta_y, \theta_z)$. Any free rigid body in 3-D space has six degrees of freedom.

A rigid body free to move in a reference frame will, in the general case, have complex motion, which is simultaneously a combination of rotation and translation. For simplicity, only the two-dimensional (2-D) or planar case will be presented. For planar motion the following terms will be defined, Figure I.2.2:

- Pure rotation in which the body possesses one point (center of rotation) which has no motion with respect to a "fixed" reference frame [Fig. I.2.2(a)]. All other points on the body describe arcs about that center.
- Pure translation in which all points on the body describe parallel paths [Fig. I.2.2(b)].

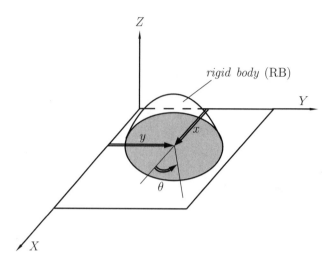

FIGURE I.2.1 *Rigid body in planar motion with three DOF: translation along the x axis, translation along the y axis, and rotation, θ, about the z.*

- Complex motion that exhibits a simultaneous combination of rotation and translation [Fig. I.2.2(c)]. With general plane motion, points on the body will travel nonparallel paths, and there will be, at every instant, a center of rotation, which will continuously change location.

Translation and rotation represent independent motions of the body. Each can exist without the other. For a 2-D coordinate system, as shown in Figure I.2.1, the x and y terms represent the translation components of motion, and the θ term represents the rotation component.

I.2.2   Links and Joints

Linkages are basic elements of all mechanisms. Linkages are made up of links and joints. A *link*, sometimes known as an *element* or a *member*, is an (assumed) rigid body which possesses nodes. *Nodes* are defined as points at which links can be attached. A link connected to its neighboring elements by s nodes is an element of *degree s*. A link of degree 1 is also called unary [Fig. I.2.3(a)], of degree 2, binary [Fig. I.2.3(b)], and of degree 3, ternary [Fig. I.2.3(c)], etc.

A *joint* is a connection between two or more links (at their nodes). A joint allows some relative motion between the connected links. Joints are also called *kinematic pairs*.

The number of independent coordinates that uniquely determine the relative position of two constrained links is termed *degree of freedom* of a given joint. Alternatively the term *joint class* is introduced. A kinematic pair is of the jth class if it diminishes the relative motion of linked bodies by j degrees of freedom; i.e., j scalar constraint conditions correspond to the given kinematic pair. It follows that such a joint has $(6-j)$ independent coordinates. The number of DOF is the fundamental characteristic quantity of joints. One of the links of a system is usually considered to be the reference link, and the position of other RBs is

rotation

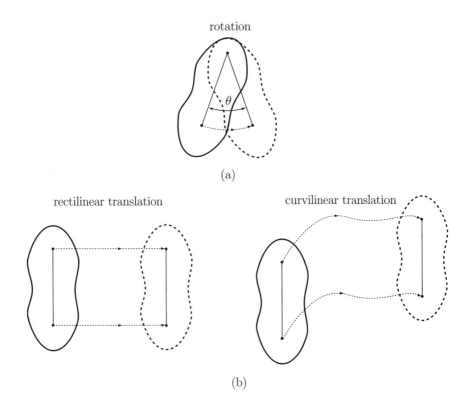

(a)

rectilinear translation   curvilinear translation

(b)

general plane motion

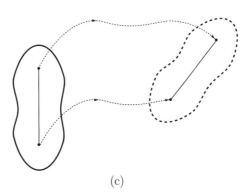

(c)

FIGURE I.2.2 *Rigid body in motion: (a) pure rotation, (b) pure translation, and (c) general motion.*

determined in relation to this reference body. If the reference link is stationary, the term *frame* or *ground* is used.

The coordinates in the definition of DOF can be linear or angular. Also the coordinates used can be absolute (measured with regard to the frame) or relative. Figures I.2.4–I.2.9

Link   Schematic representation

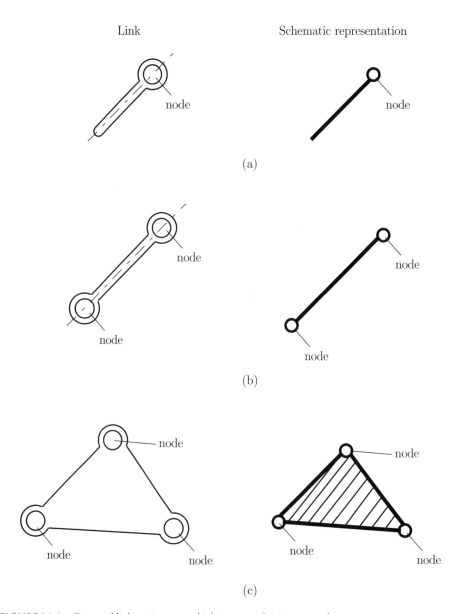

(a)

(b)

(c)

FIGURE I.2.3 *Types of links: (a) unary, (b) binary, and (c) ternary elements.*

show examples of joints commonly found in mechanisms. Figures I.2.4(a) and I.2.4(b) show two forms of a planar, one DOF joint, namely a rotating pin joint and a translating slider joint. These are both typically referred to as *full joints* and are of the 5th class. The pin joint allows one rotational (R) DOF, and the slider joint allows one translational (T) DOF between the joined links. These are both special cases of another common, one DOF joint, the screw and nut [Fig. I.2.5(a)]. Motion of either the nut or the screw relative to the other results in helical motion. If the helix angle is made zero [Fig. I.2.5(b)], the nut rotates

Type of full joint

Schematic representation

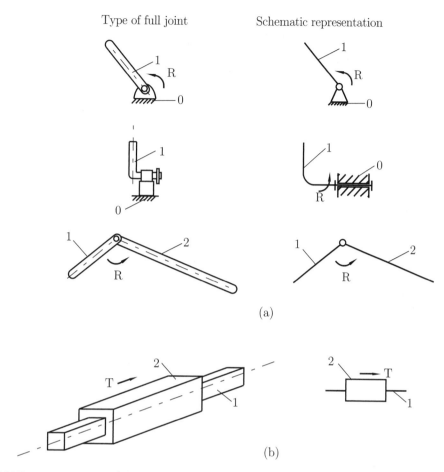

(a)

(b)

FIGURE I.2.4 *One DOF joint, full joint (5th class): (a) pin joint, and (b) slider joint.*

without advancing and it becomes a pin joint. If the helix angle is made 90°, the nut will translate along the axis of the screw, and it becomes a slider joint.

Figure I.2.6 shows examples of two DOF joints, which simultaneously allow two independent, relative motions, namely translation (T) and rotation (R), between the joined links. A two DOF joint is usually referred to as a *half joint* and is of the 4th class. A half joint is sometimes also called a roll-slide joint because it allows both rotation (rolling) and translation (sliding).

A joystick, ball-and-socket joint, or sphere joint [Fig. I.2.7(a)], is an example of a three DOF joint (3rd class), which allows three independent angular motions between the two links that are joined. This ball joint would typically be used in a 3-D mechanism, one example being the ball joints used in automotive suspension systems. A plane joint [Fig. I.2.7(b)] is also an example of a three DOF joint, which allows two translations and one rotation.

Note that to visualize the DOF of a joint in a mechanism, it is helpful to "mentally disconnect" the two links that create the joint from the rest of the mechanism. It is easier to see how many DOF the two joined links have with respect to one another. Figure I.2.8

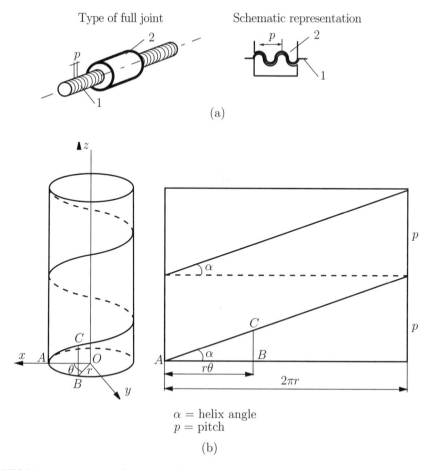

Type of full joint   Schematic representation

(a)

α = helix angle
p = pitch

(b)

FIGURE I.2.5 *(a) Screw and nut joint; (b) helical motion.*

shows an example of a 2nd class joint (cylinder on plane) and Figure I.2.9 represents a 1st class joint (sphere on plane).

The type of contact between the elements can be point (P), curve (C), or surface (S). The term *lower joint* was coined by Reuleaux to describe joints with surface contact. He used the term *higher joint* to describe joints with point or curve contact. The main practical advantage of lower joints over higher joints is their ability to better trap lubricant between their enveloping surfaces. This is especially true for the rotating pin joint.

A *closed joint* is a joint that is kept together or closed by its geometry. A pin in a hole or a slider in a two-sided slot are forms of closed joints. A *force closed joint*, such as a pin in a half-bearing or a slider on a surface, requires some external force to keep it together or closed. This force could be supplied by gravity, by a spring, or by some external means. In linkages, closed joints are usually preferred, and are easy to accomplish. For cam-follower systems force closure is often preferred.

The *order of a joint* is defined as the number of links joined minus one. The simplest joint combination of two links has order one and it is a single joint [Fig. I.2.10(a)]. As additional

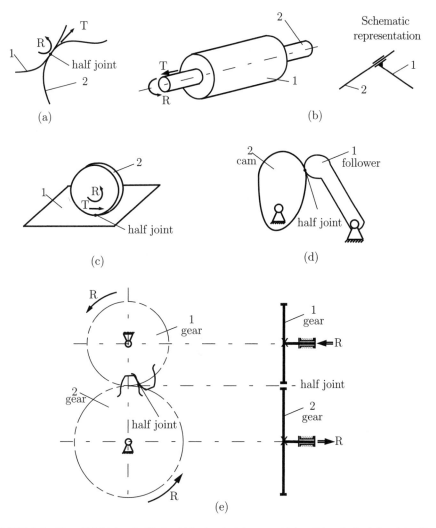

FIGURE I.2.6 *Two DOF joint, half joint (4th class): (a) general joint, (b) cylinder joint, (c) roll and slide disk, (d) cam-follower joint, and (e) gear joint.*

links are placed on the same joint, the order is increased on a one for one basis [Fig. I.2.10(b)]. Joint order has significance in the proper determination of overall DOF for an assembly.

Bodies linked by joints form a *kinematic chain*. Simple kinematic chains are shown in Figure I.2.11. A *contour* or *loop* is a configuration described by a polygon consisting of links connected by joints [Fig. I.2.11(a)]. The presence of loops in a mechanical structure can be used to define the following types of chains:

- *Closed kinematic chains* have one or more loops so that each link and each joint is contained in at least one of the loops [Fig. I.2.11(a)]. A closed kinematic chain has no open attachment point.

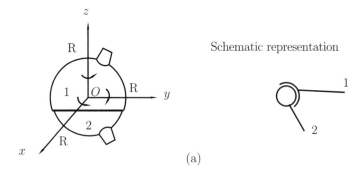

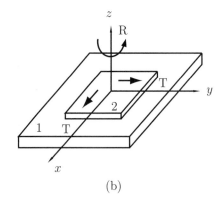

FIGURE I.2.7 *Three DOF joint (3rd class): (a) ball and socket joint, and (b) plane joint.*

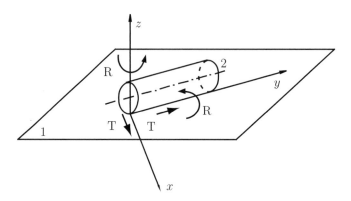

FIGURE I.2.8 *Four DOF joint (2nd class) cylinder on a plane.*

- *Open kinematic chains* contain no closed loops [Fig. I.2.11(b)]. A common example of an open kinematic chain is an industrial robot.
- *Mixed kinematic chains* are a combination of closed and open kinematic chains.

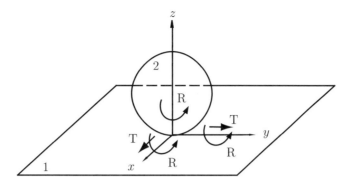

FIGURE I.2.9 *Five DOF joint (1st class) sphere on a plane.*

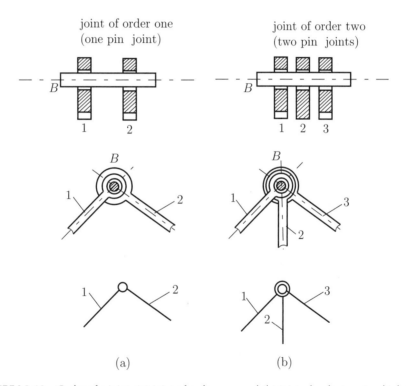

FIGURE I.2.10 *Order of a joint: (a) joint of order one, and (b) joint of order two (multiple joints).*

Another classification is also useful:

- *Simple chains* contain only binary elements.
- *Complex chains* contain at least one element of degree 3 or higher.

A *mechanism* is defined as a kinematic chain in which at least one link has been "grounded" or attached to the frame [Figs. I.2.11(a) and I.2.12]. Using Reuleaux's definition,

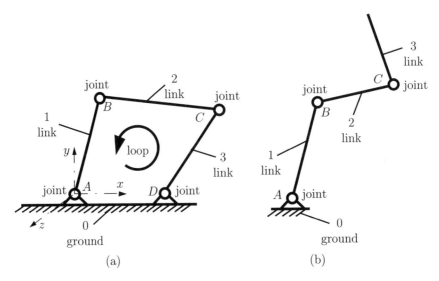

FIGURE I.2.11 *Kinematic chains: (a) closed kinematic chain, and (b) open kinematic chain.*

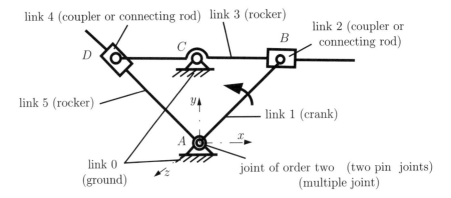

FIGURE I.2.12 *Complex mechanism with five moving links.*

a *machine* is a collection of mechanisms arranged to transmit forces and do work. He viewed all energy, or force-transmitting devices as machines that utilize mechanisms as their building blocks to provide the necessary motion constraints.

The following terms can be defined (Fig. I.2.12):

- A *crank* is a link that makes a complete revolution about a fixed grounded pivot.
- A *rocker* is a link that has oscillatory (back and forth) rotation and is fixed to a grounded pivot.
- A *coupler* or connecting rod is a link that has complex motion and is not fixed to ground.

Ground is defined as any link or links that are fixed (nonmoving) with respect to the reference frame. Note that the reference frame may in fact itself be in motion.

I.2.3 Family and Degrees of Freedom

The concept of *number of degrees of freedom* is fundamental to the analysis of mechanisms. It is usually necessary to be able to determine quickly the number of DOF of any collection of links and joints that may be used to solve a problem.

The number of DOF or the *mobility* of a system can be defined as:

- the number of inputs that need to be provided in order to create a predictable system output, or
- the number of independent coordinates required to define the position of the system.

The *family f* of a mechanism is the number of DOF that are eliminated from all the links of the system. Every free body in space has six degrees of freedom. A system of family f consisting of n movable links has $(6 - f)n$ degrees of freedom. Each joint of class j diminishes the freedom of motion of the system by $j - f$ degrees of freedom. Designating the number of joints of class k as c_k, it follows that the number of degrees of freedom of the particular system is

$$M = (6 - f)n - \sum_{j=f+1}^{5} (j - f) c_j. \tag{I.2.1}$$

This is referred to in the literature on mechanisms as the Dobrovolski formula.

A *driver* link is that part of a mechanism that causes motion. An example is a crank. The number of driver links is equal to the number of DOF of the mechanism. A *driven* link or *follower* is that part of a mechanism whose motion is affected by the motion of the driver.

Mechanisms of family $f = 1$

The family of a mechanism can be computed with the help of a mobility table (Table I.2.1). Consider the mechanism, shown in Figure I.2.13, that can be used to measure the weight of postal envelopes. The translation along the i-axis is denoted by T_i, and the rotation about the i-axis is denoted by R_i, where $i = x, y, z$. Every link in the mechanism is analyzed in terms of its translation and rotation about the reference frame xyz. For example the link 0 (ground) has no translations, $T_i = $ No, and no rotations, $R_i = $ No. The link 1 has a rotation

TABLE I.2.1 Mobility Table for the Mechanism Shown in Figure I.2.13

Link	T_x	T_y	T_z	R_x	R_y	R_z
0	No	No	No	No	No	No
1	No	No	No	No	No	Yes
2	Yes	Yes	No	No	No	Yes
3	No	Yes	No	No	No	No
4	No	Yes	Yes	Yes	No	No
5	No	No	No	Yes	No	No
					No	

For all links $R_y = $ No $\Longrightarrow f = 1$.

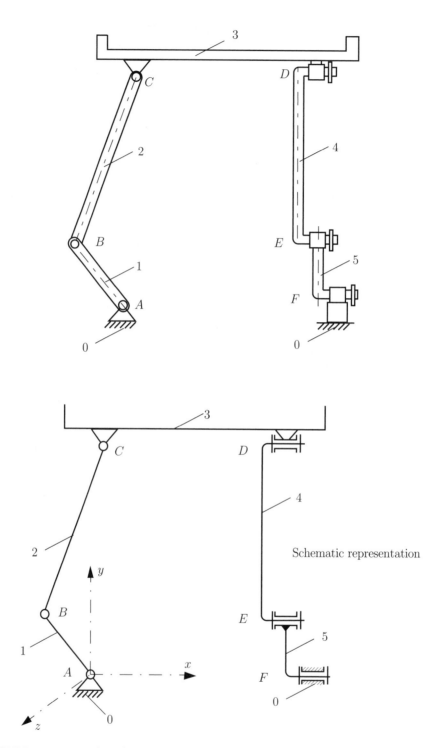

Schematic representation

FIGURE I.2.13 *Spatial mechanism of family f = 1.*

motion about the z-axis, $R_z =$ Yes. The link 2 has a planar motion (xy is the plane of motion) with a translation along the x-axis, $T_x =$ Yes, a translation along the y-axis, $T_y =$ Yes, and a rotation about the z-axis, $R_z =$ Yes. The link 3 has a translation along y, $T_y =$ Yes. The link 4 has a planar motion (yz, the plane of motion) with a translation along y, $T_y =$ Yes, a translation along z, $T_z =$ Yes, and a rotation about x, $R_x =$ Yes. The link 5 has a rotation about the x-axis, $R_x =$ Yes. The results of this analysis are presented with the help of a mobility table (Table I.2.1).

From Table I.2.1 it can be seen that link i, $i = 0, 1, 2, 3, 4, 5$, has no rotation about the y-axis, i.e., there is no rotation about the y-axis for any of the links of the mechanism ($R_y =$ No). The family of the mechanism is $f = 1$ because there is one DOF, rotation about y, which is eliminated from all the links.

There are six joints of class 5 (rotational joints) in the system at A, B, C, D, E, and F. The number of DOF for the mechanism in Figure I.2.13, which is of $f = 1$ family is given by

$$M = 5n - \sum_{j=2}^{5} (j-1)\, c_j = 5n - 4\, c_5 - 3\, c_4 - 2\, c_3 - c_2 = 5\,(5) - 4\,(6) = 1.$$

The mechanism has one DOF (one driver link).

Mechanisms of family $f = 2$

A mobility table for a mechanism of family $f = 2$ (Fig. I.2.14) is given in Table I.2.2.

The number of DOF for the $f = 2$ family mechanism is given by

$$M = 4n - \sum_{j=3}^{5} (j-2)\, c_j = 4n - 3\, c_5 - 2\, c_4 - c_3.$$

The mechanism in Figure I.2.14 has four moving links ($n = 4$), four rotational joints (A, B, D, E) and one screw and nut joint (C); i.e., there are five joints of class 5 ($c_5 = 5$). The number of DOF for this mechanism is

$$M = 4n - 3\, c_5 - 2\, c_4 - c_3 = 4\,(4) - 3\,(5) = 1.$$

TABLE I.2.2 Mobility Table for the Mechanism Shown in Figure I.2.14

Link	T_x	T_y	T_z	R_x	R_y	R_z
0	No	No	No	No	No	No
1	No	No	No	No	No	Yes
2	Yes	Yes	No	Yes	No	Yes
3	Yes	Yes	No	No	No	Yes
4	No	No	No	No	No	Yes
			No		No	

For all links $T_z =$ No & $R_y =$ No $\Longrightarrow f = 2$.

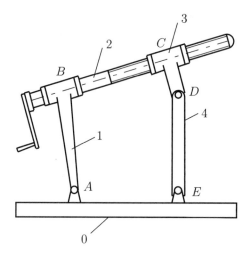

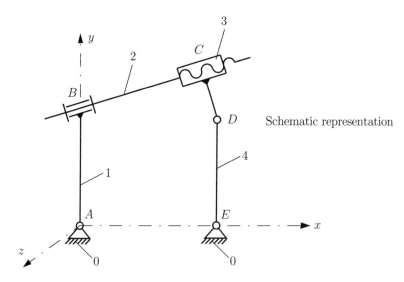

Schematic representation

FIGURE I.2.14 *Spatial mechanism of family f = 2.*

Mechanisms of family $f = 3$

The number of DOF for mechanisms of family $f = 3$ is given by

$$M = 3n - \sum_{j=4}^{5} (j - 3) c_j = 3n - 2c_5 - c_4.$$

For the mechanism in Figure I.2.11(a) the mobility table is given in Table I.2.3.

TABLE I.2.3 Mobility Table for the Mechanism Shown in Figure I.2.11(a)

Link	T_x	T_y	T_z	R_x	R_y	R_z
0	No	No	No	No	No	No
1	No	No	No	No	No	Yes
2	Yes	Yes	No	No	No	Yes
3	No	No	No	No	No	Yes
			No	No	No	

For all links T_z = No & R_x = No & R_y = No $\implies f = 3$.

TABLE I.2.4 Mobility Table for the Mechanism Shown in Figure I.2.12

Link	T_x	T_y	T_z	R_x	R_y	R_z
0	No	No	No	No	No	No
1	No	No	No	No	No	Yes
2	Yes	Yes	No	No	No	Yes
3	No	No	No	No	No	Yes
4	Yes	Yes	No	No	No	Yes
5	No	No	No	No	No	Yes
			No	No	No	

For all links T_z = No & R_x = No & R_y = No $\implies f = 3$.

The mechanism in Figure I.2.11(a) has three moving links ($n = 3$) and four rotational joints at A, B, C, and D, ($c_5 = 4$). The number of DOF for this mechanism is given by

$$M = 3\,n - 2\,c_5 - c_4 = 3\,(3) - 2\,(4) = 1.$$

The mobility table for the mechanism shown in Figure I.2.12 is given in Table I.2.4.

There are seven joints of class 5 ($c_5 = 7$) in the system:

- at A there is one rotational joint between link 0 and link 1;
- at B there is one rotational joint between link 1 and link 2;
- at B there is one translational joint between link 2 and link 3;
- at C there is one rotational joint between link 0 and link 3;
- at D there is one rotational joint between link 3 and link 4;
- at D there is one translational joint between link 4 and link 5;
- at A there is one rotational joint between link 5 and link 0.

The number of moving links is five ($n = 5$). The number of DOF for this mechanism is given by

$$M = 3\,n - 2\,c_5 - c_4 = 3\,(5) - 2\,(7) = 1,$$

and this mechanism has one driver link.

Mechanisms of family $f = 4$

The number of DOF for mechanisms of family $f = 4$ is given by

$$M = 2n - \sum_{j=5}^{5} (j - 4) c_j = 2n - c_5.$$

For the mechanism shown in Figure I.2.15 the mobility table is given in Table I.2.5. There are three translational joints of class 5 ($c_5 = 3$) in the system:

- at B there is one translational joint between link 0 and link 1;
- at C there is one translational joint between link 1 and link 2;
- at D there is one translational joint between link 2 and link 0.

The number of DOF for this mechanism with two moving links ($n = 2$) is given by

$$M = 2n - c_5 = 2(2) - (3) = 1.$$

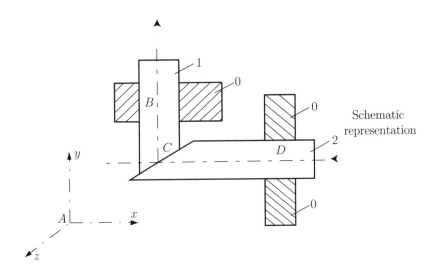

FIGURE I.2.15 *Spatial mechanism of family $f = 4$.*

TABLE I.2.5 Mobility Table for the Mechanism Shown in Figure I.2.15

Link	T_x	T_y	T_z	R_x	R_y	R_z
0	No	No	No	No	No	No
1	No	Yes	No	No	No	No
2	Yes	No	No	No	No	No
			No	No	No	No

For all links $T_z =$ No & $R_x =$ No & $R_y =$ No & $R_z =$ No $\Longrightarrow f = 4$.

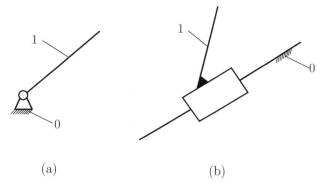

(a)   (b)

FIGURE I.2.16 *Spatial mechanism of family f = 5: (a) driver link with rotational motion, and (b) driver link with translational motion.*

Mechanisms of family $f = 5$

The number of DOF for mechanisms of family $f = 5$ is equal with the number of moving links:

$$M = n.$$

The driver link with rotational motion [Fig. I.2.16(a)] and the driver link with translational motion [Fig. I.2.16(b)] are in the $f = 5$ category.

I.2.4 Planar Mechanisms

For the special case of planar mechanisms $(f = 3)$ the Eq. (I.2.1) has the form,

$$M = 3n - 2c_5 - c_4, \tag{I.2.2}$$

where n is the number of moving links, c_5 is the number of full joints (one DOF), and c_4 is the number of half joints (two DOF).

There is a special significance to kinematic chains which do not change their DOF after being connected to an arbitrary system. Kinematic chains defined in this way are called *system groups* or *fundamental kinematic chains*. Connecting them to or disconnecting them from a given system enables given systems to be modified or structurally new systems to be created while maintaining the original DOF. The term *system group* has been introduced for the classification of planar mechanisms used by Assur and further investigated by Artobolevski. Limited to planar systems from Eq. (I.2.2), it can be obtained

$$3n - 2c_5 = 0, \tag{I.2.3}$$

according to which the number of system group links n is always even. In Eq. (I.2.3) there are no two DOF joints because a half joint, c_4, can be substituted with two full joints and an extra link (see Section I.2.15).

I.2.5 Dyads

The simplest fundamental kinematic chain is the binary group with two links ($n = 2$) and three full joints ($c_5 = 3$). The binary group is also called a *dyad*. The sets of links shown in Figure I.2.17 are dyads and one can distinguish the following classical types:

- rotation rotation rotation (dyad RRR) or dyad of type one $D10$ [Fig. I.2.17(a)];
- rotation rotation translation (dyad RRT) or dyad of type two $D20$ [Fig. I.2.17(b)];
- rotation translation rotation (dyad RTR) or dyad of type three $D30$ Fig. I.2.17(c)];
- translation rotation translation (dyad TRT) or dyad of type four $D40$ [Fig. I.2.17(d)];
- translation translation rotation (dyad TTR) or dyad of type five $D50$ [Fig. I.2.17(e)].

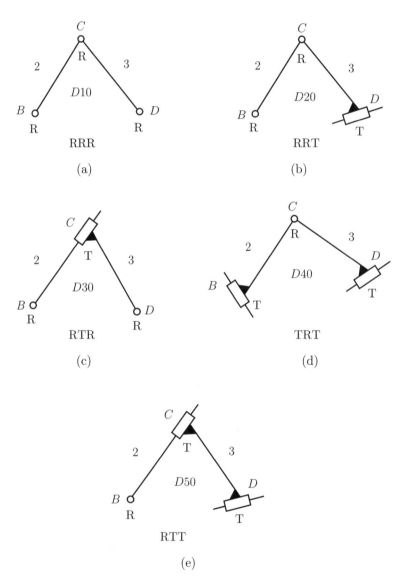

FIGURE I.2.17 *Types of dyads: (a) RRR, (b) RRT, (c) RTR, (d) TRT, and (e) TTR.*

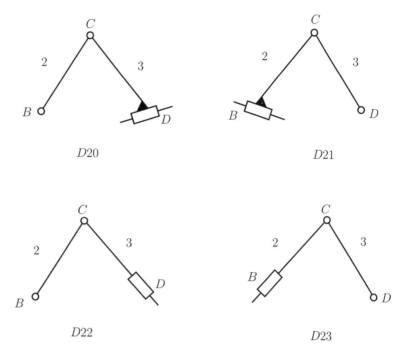

FIGURE I.2.18   *RRT dyads.*

The advantage of the group classification of a system lies in its simplicity. The solution of the whole system can then be obtained by composing partial solutions. Different versions of dyads exist for each classical dyad [40, 41, 42].

For the classical dyad RRT or $D20$ there are three more different versions, $D21$, $D22$, $D23$, as shown in Figure I.2.18. For the classical dyad RTR or $D30$ there is one different version, $D31$, as shown in Figure I.2.19. Figure I.2.20 shows three different versions, $D41$, $D42$, $D43$, of the dyad TRT or $D40$. Figure I.2.21 shows seven different versions, $D51, D52, \ldots, D57$, of the dyad TTR or $D50$. In this way 19 dyads, Dij, can be obtained where i represents the type and j represents the version.

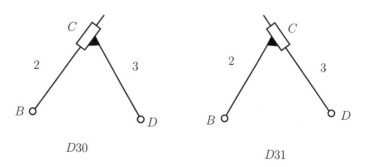

FIGURE I.2.19   *RTR dyads.*

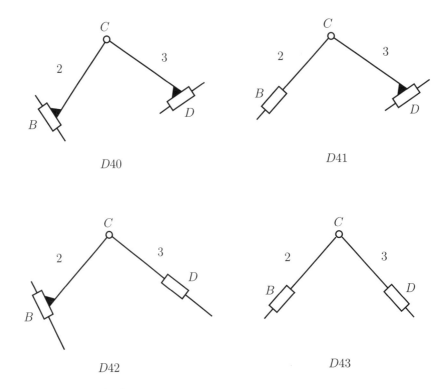

FIGURE I.2.20 *TRT dyads.*

I.2.6 Mechanisms with One Dyad

One can connect a dyad to a driver link to create a mechanism with one DOF. The driver link 1 (link AB) can have rotational (R) or translational motion (T). The driver link is connected to a first dyad comprised of the links 2 and 3, and with three joints at B, C, and D. The driver link 1 and the last link 3 are connected to the ground 0.

The closed chain R-$D42$ represents a mechanism with a driver link 1, with rotational motion (R) and one dyad $D42$ [Fig. I.2.22(a)]. Figure I.2.22(b) shows a mechanism R-$D20$ where the dyad $D20$ has the length $l_3 = 0$.

Figure I.2.23 shows a mechanism T-$D54$. The mechanism has one contour with one rotational joint at A and three translational joints at A, B, and C. The angles α and β are constant angles. From the relations

$$\alpha = \phi + \beta = \text{constant} \quad \text{and} \quad \beta = \text{constant},$$

it results in the angle $\phi = \text{constant}$. With $\phi = \text{constant}$ the link 2 has a translational motion in plane. The mechanism has the family $f = 4$ and it is a *degenerate mechanism*. In general, the planar mechanisms (Fig. I.2.23) have the family $f = 3$ with two translations and one rotation. The rotational joint at A is superfluous. For a closed chain to function as a family $f = 3$ mechanism there must be at least two rotational joints for each contour.

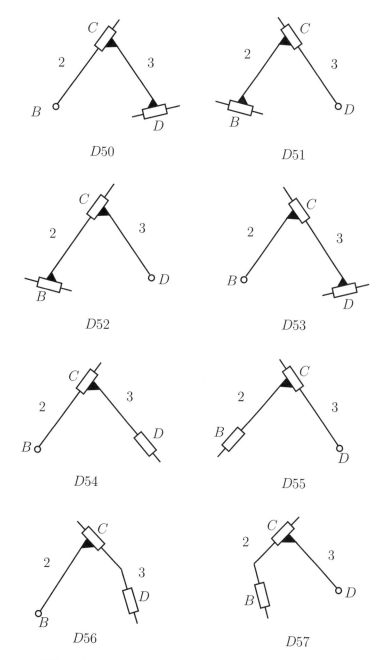

FIGURE I.2.21 *TTR dyads.*

I.2.7 Mechanisms with Two Dyads

There are also mechanisms with one driver link and two dyads. The second dyad is comprised of the links 4 and 5 and three joints at B', C', and D'.

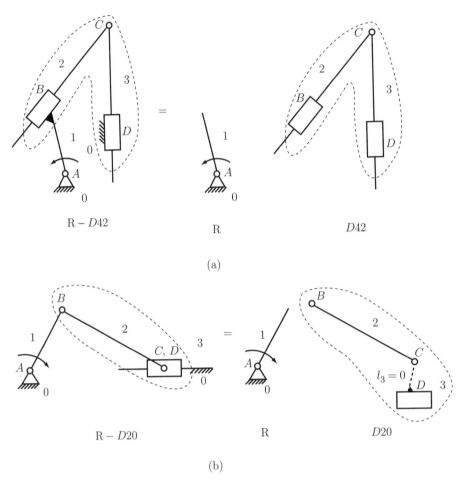

FIGURE I.2.22 *Planar mechanisms: (a) R–D42 and (b) R–D20.*

Figure I.2.24 represents the ways the second dyad can be connected to the initial mechanism with one driver and one dyad. For simplification only rotational joints are considered for the following mechanism examples, R–D10–D10.

Figure I.2.24(a) shows the first link of the second dyad, link 4, connected to the driver link 1, and the second link of the second dyad, link 5, connected to ground 0. The symbolization of the dyad connection is $1 + 0$.

Figure I.2.24(b) shows the first link of the second dyad, link 4, connected to the driver link 1, and the second link of the second dyad, link 5, connected to link 2 of the first dyad. The symbolization of the dyad connection is $1 + 2$.

Figure I.2.24(c) shows the first link of the second dyad, link 4, connected to link 2 of the first dyad, and the second link of the second dyad, link 5, connected to ground 0. The symbolization of the dyad connection is $2 + 0$.

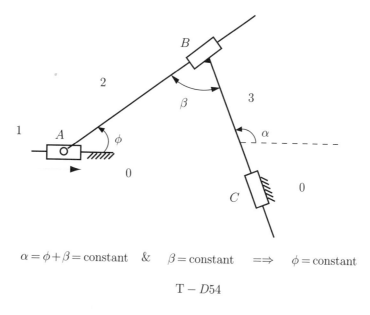

$$\alpha = \phi + \beta = \text{constant} \quad \& \quad \beta = \text{constant} \quad \Longrightarrow \quad \phi = \text{constant}$$

$$\text{T} - D54$$

FIGURE I.2.23 *Planar T-D54 mechanism with f = 4.*

Figure I.2.24(d) shows the first link of the second dyad, link 4, connected to link 2 of the first dyad, and the second link of the second dyad, link 5, connected to link 3 of the first dyad. The symbolization of the dyad connection is $2 + 3$.

Figure I.2.24(e) shows the first link of the second dyad, link 4, connected to link 3 of the first dyad, and the second link of the second dyad, link 5, connected to ground 0. The symbolization of the dyad connection is $3 + 0$.

Figure I.2.24(f) shows the first link of the second dyad, link 4, connected to the driver link 1, and the second link of the second dyad, link 5, connected to link 3 of the first dyad. The symbolization of the dyad connection is $1 + 3$.

Figure I.2.25 represents mechanisms with two dyads with rotational and translational joints and their symbolization. Figure I.2.25(a) shows a rotational driver link, R, connected to a first dyad, $D21$. The first link 4 of the second dyad $D30$ is connected to the driver link 1 at B', and the second link 5 of the second dyad $D30$ is connected to link 3 at D'. The symbolization of the mechanism is R–$D21$–$D30$–1+3.

Figure I.2.25(b) shows a rotational driver link, R, connected to a first dyad, $D43$. The first link 4 of the second dyad $D50$ is connected to link 2 at B', and the second link 5 of the second dyad $D30$ is connected to ground 0 at D'. The symbolization of the mechanism is R–$D43$–$D50$–2+0.

Figure I.2.25(c) shows a mechanism R–$D31$–$D20$–3+0. The driver link 1, with rotational motion is connected to the first dyad $D31$. The first link 4 of the second dyad $D20$ is connected to the link 3, and the second link 5 is connected to the ground 0.

Figure I.2.26 shows a mechanism T–$D21$–$D50$–2+0. There are two contours: 0–1–2–3–0 and 0–1–2–4–5–0. The first contour, 0–1–2–3–0, has translational joints at A and B

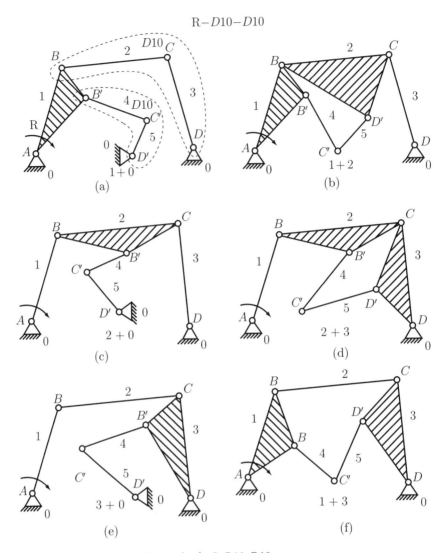

FIGURE I.2.24 *Mechanism with two dyads: R–D10–D10.*

and rotational joints at C and D. The family of this contour is $f_I = 3$. The second contour, 0–1–2–4–5–0, has translational joints at A, B, C', and D' and one rotational joint at B'.

- The angle $\phi = $ constant and the angle $\lambda_1 = $ constant. Then the angle $\alpha = \phi - \lambda_1 = $ constant.
- The angle $\lambda_2 = $ constant. Then the angle $\gamma = \alpha + \lambda_2 = $ constant.
- The angle $\theta = $ constant and the angle $\lambda_3 = $ constant. Then the angle $\delta = \theta + \lambda_3 = $ constant.

With $\gamma = $ constant and $\delta = $ constant, the links 2 and 4 have a translational motion in plane. The second contour has the family $f_{II} = 4$ and the mechanism is a degenerate

---

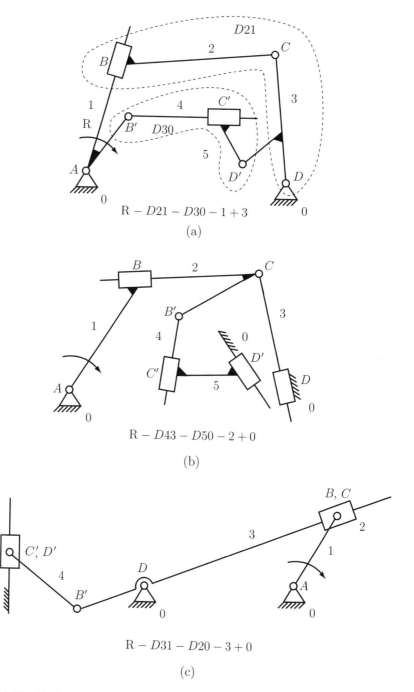

FIGURE I.2.25 *Mechanisms with two dyads: (a) R–D21–D30–1+3, (b) R–D43–D50–2+0, and (c) R–D31–D20–3+0.*

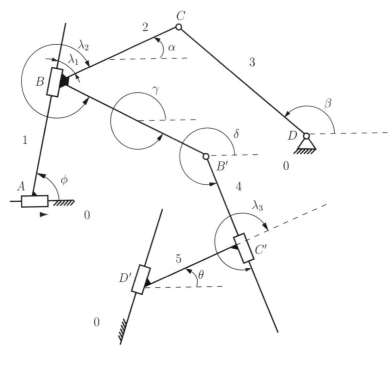

$$T - D21 - D50 - 2 + 0$$

FIGURE I.2.26 *T–D21–D50–2+0 mechanism.*

mechanism. The closed contours that do not have a minimum of two rotational joints are contours of family $f = 4$.

To calculate the number of degrees of freedom for kinematic chains with different families the following formula is introduced [1]:

$$M = (6 - f_a)n - \sum_{j=f+1}^{5} (j - f_a)c_j, \qquad (I.2.4)$$

where f_a is the *apparent family*.

For the mechanism in Figure I.2.26 with two contours, the apparent family is

$$f_a = \frac{f_I + f_{II}}{2} = \frac{3 + 4}{2} = \frac{7}{2},$$

and the number of degrees of freedom of the degenerate mechanism is

$$M = (6 - f_a)n - (5 - f_a)c_5 = \left(6 - \frac{7}{2}\right)4 - \left(5 - \frac{7}{2}\right)6 = 1. \qquad (I.2.5)$$

There are $n = 4$ moving links (link 2 and link 4 form one moving link) and $c_5 = 6$.

I.2.8 Mechanisms with Three Dyads

There are also mechanisms with one driver link and three dyads. The second dyad is comprised of the links 4 and 5 and three joints at B', C', and D'. The third dyad has the links 6 and 7 and three joints at B'', C'', and D''.

Figure I.2.27 represents some ways of connection for the the third dyad. For simplification, only rotational joints are considered, i.e., the mechanism $R - D10 - D10 - D10$ is presented.

Figure I.2.27(a) shows the first link of the second dyad, link 4, connected to the driver link 1, and the second link of the second dyad, link 5, connected to link 2. The connection symbolization for the second dyad is $1+2$. The first link of the third dyad, link 6, is connected to link 4 and the second link of the third dyad, link 7, is connected to link 3. The connection symbolization for the third dyad is $4+3$. The connection symbolization for the mechanism is $1+2-4+3$.

Figure I.2.27(b) represents the first link of the second dyad, link 4, connected to the driver link 1, and the second link of the second dyad, link 5, connected to link 2. The connection symbolization for the second dyad is $1+2$. The first link of the third dyad, link

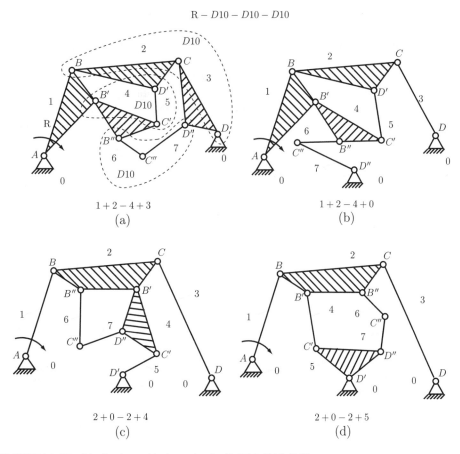

FIGURE I.2.27 *Mechanism with three dyads: R–D10–D10–D10.*

6, is connected to link 4 and the second link of the third dyad, link 7, is connected to ground 0. The connection symbolization for the third dyad is 4+0. The connection symbolization for the mechanism is 1+2−4+0.

Figure I.2.27(c) shows the first link of the second dyad, link 4, connected to the driver link 2, and the second link of the second dyad, link 5, connected to ground 0. The connection symbolization for the second dyad is 2+0. The first link of the third dyad, link 6, is connected to link 2 and the second link of the third dyad, link 7, is connected to link 4. The connection symbolization for the third dyad is 2+4. The connection symbolization for the mechanism is 2+0−2+4.

Figure I.2.27(d) shows the first link of the second dyad, link 4, connected to the driver link 2, and the second link of the second dyad, link 5, connected to ground 0. The connection symbolization for the second dyad is 2+0. The first link of the third dyad, link 6, is connected to link 2 and the second link of the third dyad, link 7, is connected to link 5. The connection symbolization for the third dyad is 2+5. The connection symbolization for the mechanism is 2+0−2+5.

Mechanisms with three dyads with rotational and translational joints and their symbolization are shown in Figure I.2.28. Figure I.2.28(a) shows a rotational driver link, R, connected to a first dyad, $D42$. The first link 4 of the second dyad $D30$ is connected to the driver link 1, and the second link 5 of the second dyad $D30$ is connected to link 2. The first link 6 of the third dyad $D21$ is connected to link 4, and the second link 7 of the second dyad $D21$ is connected to ground 0. The symbolization of the mechanism is R–$D42$–$D30$–$D21$–1+2–4+0.

Figure I.2.28(b) presents a mechanism T–$D22$–$D30$–$D20$–1+2–4+5. The slider driver link, T, is connected to a first dyad, $D22$. The first link 4 of the second dyad $D30$ is connected to the driver link 1, and the second link 5 is connected to link 2. The first link 6 of the third dyad $D20$ is connected to link 4, and the second link 7 is connected to link 5.

I.2.9 Independent Contours

A contour is a configuration described by a polygon consisting of links connected by joints. A contour with at least one link that is not included in any other contour of the chain is called an *independent contour*. The number of independent contours, N, of a kinematic chain can be computed as

$$N = c - n, \qquad (I.2.6)$$

where c is the number of joints, and n is the number of moving links.

Planar kinematic chains are presented in Figure I.2.29. The kinematic chain shown in Figure I.2.29(a) has two moving links, 1 and 2 ($n = 2$), three joints ($c = 3$), and one independent contour ($N = c - n = 3 - 2 = 1$). This kinematic chain is a dyad. In Figure I.2.29(b), a new kinematic chain is obtained by connecting the free joint of link 1 to the ground (link 0). In this case, the number of independent contours is also $N = c - n = 3 - 2 = 1$. The kinematic chain shown in Figure I.2.29(c) has three moving links, 1, 2, and 3 ($n = 3$), four joints ($c = 4$), and one independent contour ($N = c - n = 4 - 3 = 1$). A closed chain with three moving links, 1, 2, and 3 ($n = 3$), and one fixed link 0, connected by four joints ($c = 4$) is shown in Figure I.2.29(d). This is a

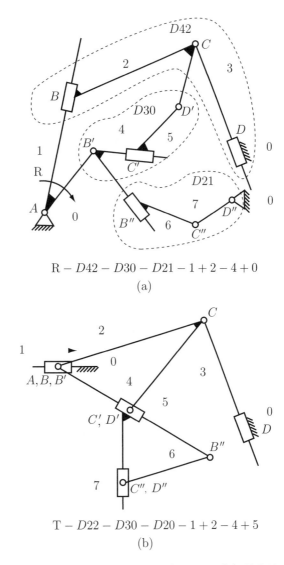

$$R - D42 - D30 - D21 - 1 + 2 - 4 + 0$$

(a)

$$T - D22 - D30 - D20 - 1 + 2 - 4 + 5$$

(b)

FIGURE I.2.28 *(a) R–D42–D30–D21–1+2–4+0 mechanism and (b) T–D22–D30–D20–1+2–4+5 mechanism.*

four-bar mechanism. In order to find the number of independent contours, only the moving links are considered. Thus, there is one independent contour ($N = c - n = 4 - 3 = 1$). The kinematic chain presented in Figure I.2.29(e) has four moving links, 1, 2, 3, and 4 ($n = 4$), and six joints ($c = 6$). There are three contours: 1-2-3, 1-2-4, and 3-2-4. Only two contours are independent contours ($N = 6 - 4 = 2$).

Spatial kinematic chains are depicted in Figure I.2.30. The kinematic chain shown in Figure I.2.30(a) has five links, 1, 2, 3, 4, and 5 ($n = 5$), six joints ($c = 6$), and one independent contour ($N = c - n = 6 - 5 = 1$). For the spatial chain shown in Figure I.2.30(b), there are six links, 1, 2, 3, 4, 5, and 6 ($n = 6$), eight joints ($c = 8$), and three contours,

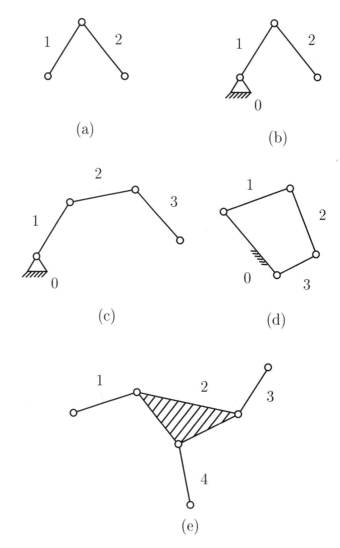

FIGURE I.2.29 *Planar kinematic chains.*

1-2-3-4-5, 1-2-3-6, and 5-4-3-6. In this case, two of the contours are independent contours $(N = c - n = 8 - 6 = 2)$.

I.2.10 Spatial System Groups

The system groups for spatial mechanisms can be determined by analogy to the system groups for the planar mechanisms. The system groups have the degree of freedom $M = 0$. All possible system groups can be determined for each family of chains [40, 41, 43].

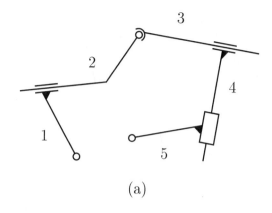

(a)

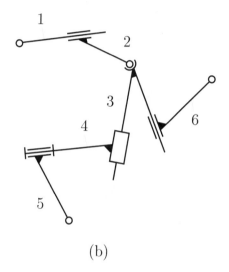

(b)

FIGURE I.2.30 *Spatial kinematic chains.*

For the family $f = 0$, for system groups, from Eqs. (I.2.1) and (I.2.6) the mobility is

$$M = 6n - 5c_5 - 4c_4 - 3c_3 - 2c_2 - c_1 = 0, \qquad (\text{I.2.7})$$

and the number of moving links is

$$n = c - N. \qquad (\text{I.2.8})$$

From Eqs. (I.2.7) and (I.2.8) the number of joints of class 5 is

$$c_5 = 6N - 5c_1 - 4c_2 - 3c_3 - 2c_4, \qquad (\text{I.2.9})$$

TABLE I.2.6 The Number of Configurations of System Groups with One, Two and Three Independent Contours ($N = 1, 2$, and 3)

w	0	1	2	3	4	5	6	7	8	9	10	20	30	40
$N = 1$	5	18	22	23	23	23	23	23	23	23	23	23	23	23
$N = 2$	5	30	62	76	82	84	85	85	85	85	85	85	85	85
$N = 3$	5	31	100	158	190	205	214	218	218	220	220	220	220	220

and the number of moving links is

$$n = -N + c_1 + c_2 + c_3 + c_4 + c_5. \qquad (I.2.10)$$

For the family $f = 1$, $c_1 = 0$ it results:

$$c_5 = 5N - 4c_2 - 3c_3 - 2c_4, \quad n = -N + c_2 + c_3 + c_4 + c_5. \qquad (I.2.11)$$

For the family $f = 2$, $c_1 = 0$, $c_2 = 0$ it results:

$$c_5 = 4N - 3c_3 - 2c_4, \quad n = -N + c_3 + c_4 + c_5. \qquad (I.2.12)$$

For the family $f = 3$, $c_1 = 0$, $c_2 = 0$, $c_3 = 0$ it results:

$$c_5 = 3N - 2c_4, \quad n = -N + c_4 + c_5. \qquad (I.2.13)$$

For the family $f = 4$, $c_1 = 0$, $c_2 = 0$, $c_3 = 0$, $c_4 = 0$ it results:

$$c_5 = 2N, \quad n = -N + c_5. \qquad (I.2.14)$$

Using the above conditions, all the possible solutions for spatial system groups can be determined. The number of joints, c_1, c_2, c_3, and c_4, are cycled from 0 to w, where w is a positive integer for system groups with one or more independent contours ($N \geq 1$). The number of joints, c_5, and the number of moving links, n, are computed for each system group. An acceptable solution has to verify the conditions $n > 0$ and $c_5 > 0$. In Table I.2.6, the number of possible solutions is presented for some values of w between 0 and 40 and for kinematic chains with one contour ($N = 1$), two contours ($N = 2$), and three contours ($N = 3$). For $N = 1$ and $w \geq 3$, there are 23 possible solutions. For $N = 2$, there are 85 solutions for $w \geq 6$, and for $N = 3$ there are 220 solutions for $w \geq 9$.

I.2.11 Spatial System Groups with One Independent Contour

The combinations of spatial system groups with one independent contour ($N = 1$) are presented in Table I.2.7. The number of joints, c_1, c_2, c_3, and c_4, are cycled from 0 to 3, and the number of joints, c_5, and the number of moving links, n, are computed. System groups from Table I.2.7 are exemplified next for each of the families $f = 0, 1, 2, 3$, and 4.

TABLE I.2.7 The Configurations of System Groups with One Independent Contour ($N = 1$)

Index	f	c_1	c_2	c_3	c_4	c_5	n
1	0	0	0	0	0	6	5
2	0	0	0	0	1	4	4
3	0	0	0	0	2	2	3
4	0	0	0	0	3	0	2
5	0	0	0	1	0	3	3
6	0	0	0	1	1	1	2
7	0	0	0	2	0	0	1
8	0	0	1	0	0	2	2
9	0	0	1	0	1	0	1
10	0	1	0	0	0	1	1
11	1	0	0	0	0	5	4
12	1	0	0	0	1	3	3
13	1	0	0	0	2	1	2
14	1	0	0	1	0	2	2
15	1	0	0	1	1	0	1
16	1	0	1	0	0	1	1
17	2	0	0	0	0	4	3
18	2	0	0	0	1	2	2
19	2	0	0	0	2	0	1
20	2	0	0	1	0	1	1
21	3	0	0	0	0	3	2
22	3	0	0	0	1	1	1
23	4	0	0	0	0	2	1

For the family $f = 0$, four system groups are illustrated in Figure I.2.31. The values c_5 and n are computed from Eqs. (I.2.9) and (I.2.10), respectively. A spatial system group with no joints of class 1, 2, 3, and 4 ($c_1 = c_2 = c_3 = c_4 = 0$) is shown in Figure I.2.31(a). The system group has six joints of class 5 ($c_5 = 6(1) = 6$), and five moving links ($n = -1 + 6 = 5$). A system group with one joint of class 4 ($c_4 = 1$) and no joints of class 1, 2, and 3 ($c_1 = c_2 = c_3 = 0$) is shown in Figure I.2.31(b). The system group has four joints of class 5 ($c_5 = 6(1) - 2(1) = 4$), and four moving links ($n = -1 + 1 + 4 = 4$). A system group with two joints of class 4 ($c_4 = 2$) and no joints of class 1, 2, and 3 ($c_1 = c_2 = c_3 = 0$) is shown in Figure I.2.31(c). The system group has two joints of class 5 ($c_5 = 6(1) - 2(2) = 2$), and three moving links ($n = -1 + 2 + 2 = 3$). A system group with one joint of class 3 ($c_3 = 1$) and no joints of class 1, 2, and 4 ($c_1 = c_2 = c_4 = 0$) is shown in Figure I.2.31(d). The system group has three joints of class 5 ($c_5 = 6(1) - 3(1) = 3$), and three moving links ($n = -1 + 1 + 3 = 3$).

The spatial mechanism presented in Figure I.2.32 is built from the system group shown in Figure I.2.31(b). The mechanism has one DOF ($M = 6n - 5c_5 - 4c_4 - 3c_3 - 2c_2 - c_1 = 6(5) - 5(5) - 4(1) = 1$). The driver link is link 5.

For the family $f = 1$, three systems groups are depicted in Figure I.2.33. The values c_5 and n are computed from Eq. (I.2.11). The missing translations and rotations with respect

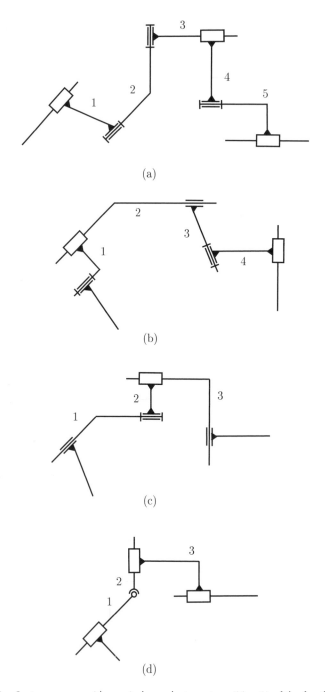

(a)

(b)

(c)

(d)

FIGURE I.2.31 *System groups with one independent contour (N = 1) of the family f = 0.*

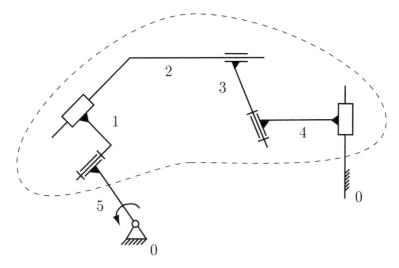

FIGURE I.2.32 *Spatial mechanism with one independent contour and a system group of the family* $f = 0$.

to the axis of the reference frame $xOyz$ are specified further on for each system group. For a Cartesian reference frame $xOyz$ the rotations about the axis are represented by R and the translations along the axis are represented by T.

A spatial system group with no joints of class 1, 2, 3, and 4 ($c_1 = c_2 = c_3 = c_4 = 0$) is shown in Figure I.2.33(a). The system group has five joints of class 5 ($c_5 = 5(1) = 5$), and four moving links ($n = -1 + 5 = 4$). There are no rotations R_x (rotation about x-axis) for the links of the system group. A system group with no joints of class 1, 2, and 3 ($c_1 = c_2 = c_3 = 0$) and one joint of class 4 ($c_4 = 1$) is shown in Figure I.2.33(b). The system group has three joints of class 5 ($c_5 = 5(1) - 2(1) = 3$), and three moving links ($n = -1 + 1 + 3 = 3$). There are no translations T_z (translation along z-axis) for the links. A system group with one joint of class 3 ($c_3 = 1$) and no joints of class 1, 2, and 4 ($c_1 = c_2 = c_4 = 0$) is shown in Figure I.2.33(c). The system group has two joints of class 5 ($c_5 = 5(1) - 3(1) = 2$), and two moving links ($n = -1 + 3 = 2$). There are no translations T_y for the links.

For the family $f = 2$, four system groups are presented in Figure I.2.34. The values c_5 and n are computed from Eq. (I.2.12). Two spatial system groups with no joints of class 1, 2, 3, and 4 ($c_1 = c_2 = c_3 = c_4 = 0$) are shown in Figures I.2.34(a) and I.2.34(b). The system groups have four joints of class 5 ($c_5 = 4(1) = 4$), and three moving links ($n = -1 + 4 = 3$). For the system group in Figure I.2.34(a), there are no translations T_x and no rotations R_y for the links. For the system group in Figure I.2.34(b), there are no translations T_y and no rotations R_x for the links. A system group with no joints of class 1, 2, and 3 ($c_1 = c_2 = c_3 = 0$) and one joint of class 4 ($c_4 = 1$) is shown in Figure I.2.34(c). The system group has two joints of class 5 ($c_5 = 4(1) - 2(1) = 2$), and two moving links ($n = -1 + 1 + 2 = 2$). There are no translations T_z and no rotations R_y for the links. A system group with one joint of class 3 ($c_3 = 1$) and no joints of class 1, 2, and 4 ($c_1 = c_2 = c_4 = 0$) is shown in Figure I.2.34(d). The system group has one joint of class 5 ($c_5 = 4(1) - 3(1) = 1$), and one moving link ($n = -1 + 1 + 1 = 1$). There are no translations T_x and T_z for the links.

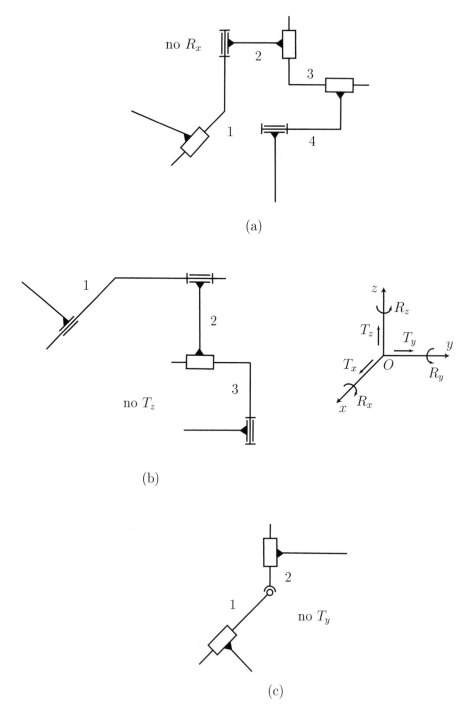

FIGURE I.2.33 *System groups with one independent contour (N = 1) of the family f = 1.*

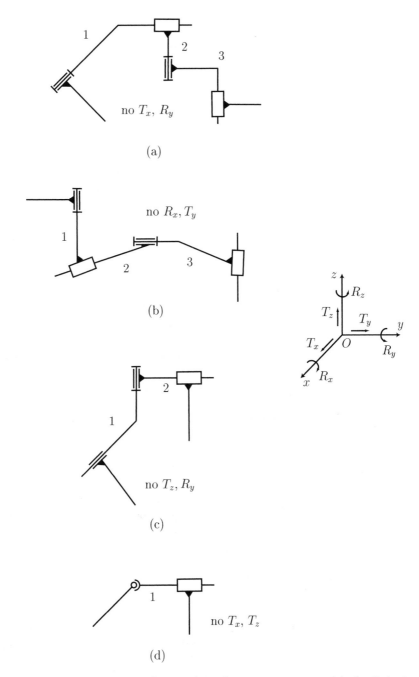

FIGURE I.2.34 *System groups with one independent contour (N = 1) of the family f = 2.*

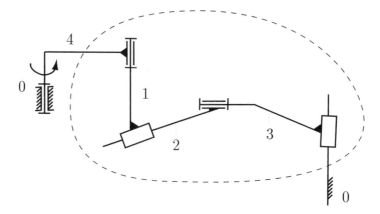

FIGURE I.2.35 *Spatial mechanism with one independent contour and a system group of the family* $f = 2$.

The spatial mechanism presented in Figure I.2.35 is derived from the system group shown in Figure I.2.33(b). The mechanism has one degree of freedom ($M = 4n - 3c_5 - 2c_4 - c_3 = 4(4) - 3(5) = 1$). The link 4 is the driver link.

For the family $f = 3$, three system groups are presented in Figure I.2.36. The values c_5 and n are computed from Eq. (I.2.13). Three system groups with no joints of class 1, 2, 3, and 4 ($c_1 = c_2 = c_3 = c_4 = 0$) are shown in Figure I.2.36. The system groups have three joints of class 5 ($c_5 = 3(1) = 3$), and two moving links ($n = -1 + 3 = 2$). There are no translation T_x and no rotations R_y and R_z for the system group in Figure I.2.36(a). For the system group in Figure I.2.36(b) there are no translation T_x and no rotations R_x and R_z. There are no translation T_z and no rotations R_x and R_y for the system group shown in Figure I.2.36(c).

For the family $f = 4$, two planar system groups with no joints of class 1, 2, 3, and 4 ($c_1 = c_2 = c_3 = c_4 = 0$) are shown in Figure I.2.37. The values c_5 and n are computed from Eq. (I.2.14) for each system group; there are two joints of class 5 ($c_5 = 2(1) = 2$), and one moving link ($n = -1 + 2 = 1$). Also, there are two planar translations for the links and thus the family of the system is $f = 6 - 2 = 4$.

I.2.12 Spatial System Groups with Two Independent Contours

There are also spatial system groups with two independent contours ($N = 2$). The number of joints, c_1, c_2, c_3, and c_4, are cycled, and the number of joints, c_5, and the number of moving links, n, are computed. Examples of system groups with $N = 2$ are described next for each of the families $f = 1, 2, 3$, and 4.

For the family $f = 1$, a system group is depicted in Figure I.2.38. The system group has no joints of class 1, 2, 3, and 4 ($c_1 = c_2 = c_3 = c_4 = 0$). There are ten joints of class 5 ($c_5 = 5(2) = 10$), and eight moving links ($n = -2 + 10 = 8$). There is no translation T_x for the links.

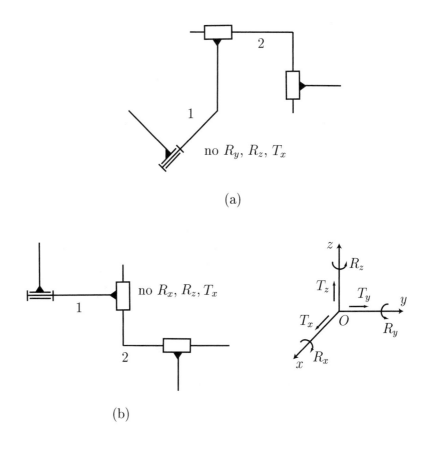

(a)

no R_y, R_z, T_x

no R_x, R_z, T_x

(b)

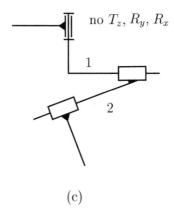

no T_z, R_y, R_x

(c)

FIGURE I.2.36 *System groups with one independent contour (N = 1) of the family f = 3.*

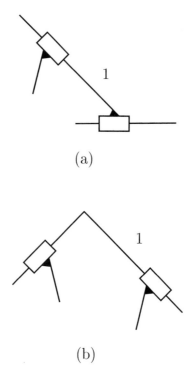

(a)

(b)

FIGURE I.2.37 *System groups with one independent contour (N = 1) of the family f = 4.*

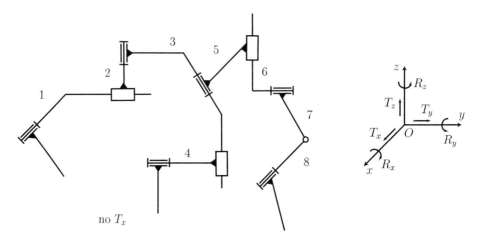

FIGURE I.2.38 *System group with two independent contours (N = 2) of the family f = 1.*

For the family $f = 2$, two system groups are illustrated in Figure I.2.39. A system group with no joints of class 1, 2 and 3 ($c_1 = c_2 = c_3 = 0$) and one joint of class 4 ($c_4 = 1$) is shown in Figure I.2.39(a). The system group has six joints of class 5 ($c_5 = 4(2) - 2(1) = 6$), and five moving links ($n = -2 + 1 + 6 = 5$). There are no translation T_x and no rotation

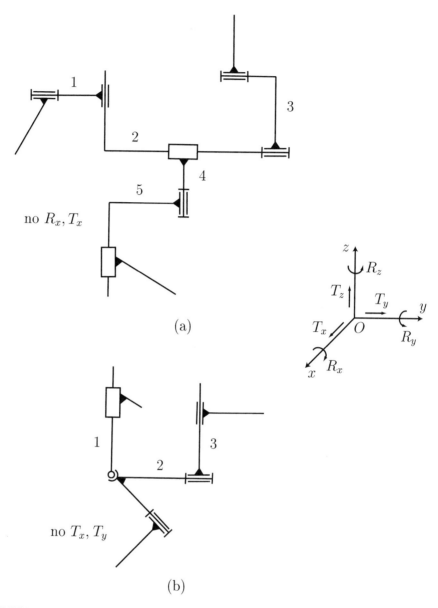

no R_x, T_x

(a)

no T_x, T_y

(b)

FIGURE I.2.39 *System groups with two independent contours (N = 2) of the family f = 2.*

R_x for the links. A system group with no joints of class 1 and 2 ($c_1 = c_2 = 0$), one joint of class 3 ($c_3 = 1$), and one joint of class 4 ($c_4 = 1$) is shown in Figure I.2.39(b). The system group has three joints of class 5 ($c_5 = 4(2) - 3(1) - 2(1) = 3$), and three moving links ($n = -2 + 1 + 1 + 3 = 3$). There are no translations T_x and T_y for the links.

For the family $f = 3$, three system groups are presented in Figure I.2.40. A system group with no joints of class 1, 2, 3, and 4 ($c_1 = c_2 = c_3 = c_4 = 0$) is shown in

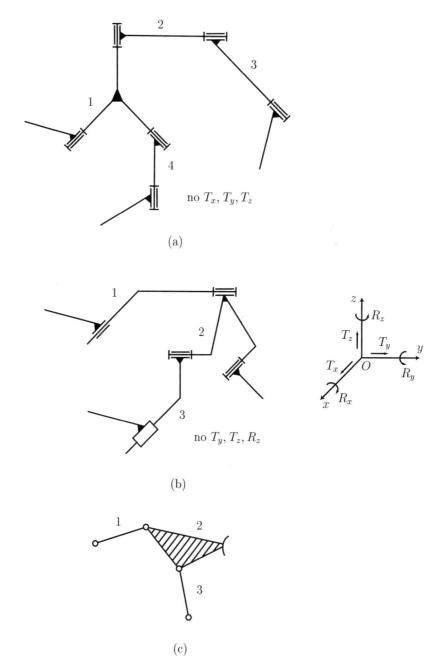

FIGURE I.2.40 *System groups with two independent contours (N = 2) of the family f = 3.*

Figure I.2.40(a). The system group has six joints of class 5 ($c_5 = 3(2) = 6$), and four moving links ($n = -2 + 6 = 4$). There are no translations T_x, T_y, and T_z for the links. A spatial system group and a planar system group with no joints of class 1, 2, and 3 ($c_1 = c_2 = c_3 = 0$) and one joint of class 4 ($c_4 = 1$) are shown in Figures I.2.40(b) and

---

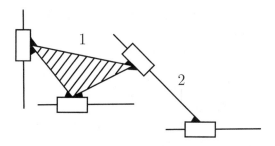

FIGURE I.2.41 *System group with two independent contours (N = 2) of the family f = 4.*

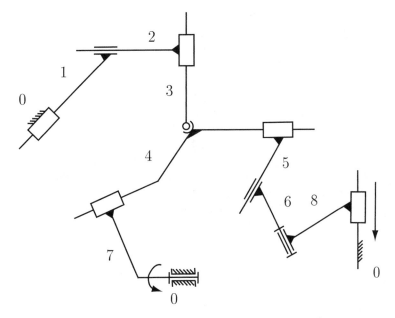

FIGURE I.2.42 *Spatial mechanism with two independent contours and a system group of the family f = 0.*

I.2.40(c), respectively. The system groups have four joints of class 5 ($c_5 = 3(2) - 2(1) = 4$), and three moving links ($n = -2 + 1 + 4 = 3$). There are no translations T_y, T_z and no rotations R_z for the spatial system in Figure I.2.40(b).

For the family $f = 4$, a planar system group with no joints of class 1, 2, 3, and 4 ($c_1 = c_2 = c_3 = c_4 = 0$) is shown in Figure I.2.41. The system group has four joints of class 5 ($c_5 = 2(2) = 4$), and two moving links ($n = -2 + 4 = 2$).

The spatial mechanism shown in Figure I.2.42 contains a system group of the family $f = 0$ that has $c_1 = c_2 = 0$, $c_3 = 1$, $c_4 = 2$, $c_5 = 6(2) - 3(1) - 2(2) = 5$, and $n = -2 + 1 + 2 + 5 = 6$. The mechanism has two degrees of freedom $M = 6n - 5c_5 - 4c_4 - 3c_3 - 2c_2 - c_1 = 6(8) - 5(7) - 4(2) - 3(1) = 2$. The links 7 and 8 are driver links.

I.2.13 Spatial System Groups with Three Independent Contours

There are also spatial system groups with three independent contours ($N = 3$). The number of joints, c_1, c_2, c_3, and c_4, are cycled and the number of joints, c_5, and the number of moving links, n, are computed. System groups with $N = 3$ are exemplified for each of the families $f = 2$, 3, and 4.

For the family $f = 2$, a spatial system group with no joints of class 1 and 2 ($c_1 = c_2 = 0$), one joint of class 3 ($c_3 = 1$), and one joint of class 4 ($c_4 = 1$) is shown in Figure I.2.43. The system group has seven joints of class 5 ($c_5 = 4(3) - 3(1) - 2(1) = 7$), and six moving links ($n = -3 + 1 + 1 + 7 = 6$). There are no translations T_x and T_z for the links.

For the family $f = 3$, a planar system group with no joints of class 1, 2, and 3 ($c_1 = c_2 = c_3 = 0$) and one joint of class 4 ($c_4 = 1$) is depicted in Figure I.2.44. The system group has seven joints of class 5 ($c_5 = 3(3) - 2(1) = 7$), and five moving links ($n = -3 + 1 + 7 = 5$).

For the family $f = 4$, a planar system group with no joints of class 1, 2, 3, and 4 ($c_1 = c_2 = c_3 = c_4 = 0$) is shown in Figure I.2.45. The system group has six joints of class 5 ($c_5 = 2(3) = 6$), and three moving links ($n = -3 + 6 = 3$).

The spatial mechanism presented in Figure I.2.46 contains a system group of the family $f = 0$ that has $c_1 = c_2 = 0$, $c_3 = 3$, $c_4 = 4$, $c_5 = 6(3) - 3(3) - 2(4) = 1$, and $n = -3 + 3 + 4 + 1 = 5$. The mechanism has three degrees of freedom $M = 6n - 5c_5 - 4c_4 - 3c_3 - 2c_2 - c_1 = 6(8) - 5(4) - 4(4) - 3(3) = 3$. The links 6, 7, and 8 are driver links.

The method [40, 41] presented is based essentially on system group formation using the number of independent contours and joints as inputs. The number of joints of different classes are cycled for different families and several structures of spatial system groups with one, two, or more independent contours are obtained. For a given family, different configurations of system groups with the same number of independent contours can be obtained. Spatial mechanisms can be structured based on spatial system groups.

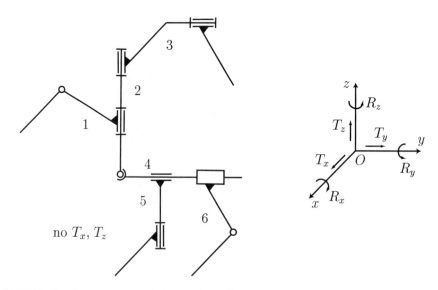

FIGURE I.2.43 *System group with three independent contours (N = 3) of the family f = 2.*

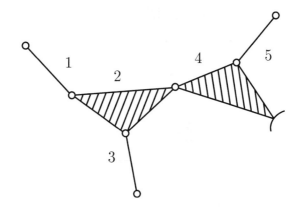

FIGURE I.2.44 *System group with three independent contours (N = 3) of the family f = 3.*

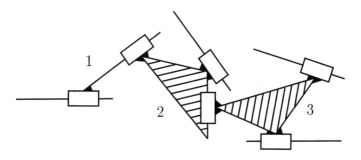

FIGURE I.2.45 *System group with three independent contours (N = 3) of the family f = 4.*

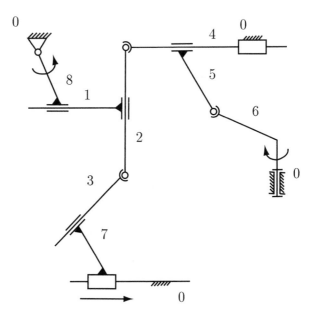

FIGURE I.2.46 *Spatial mechanism with three independent contours and a system group of the family f = 0.*

I.2.14 Decomposition of Kinematic Chains

A planar mechanism is shown in Figure I.2.47(a). This kinematic chain can be decomposed into system groups and driver links. The mobility of the mechanism will be determined first. The number of DOF for this mechanism is given by $M = 3n - 2c_5 - c_4 = 3n - 2c_5$.

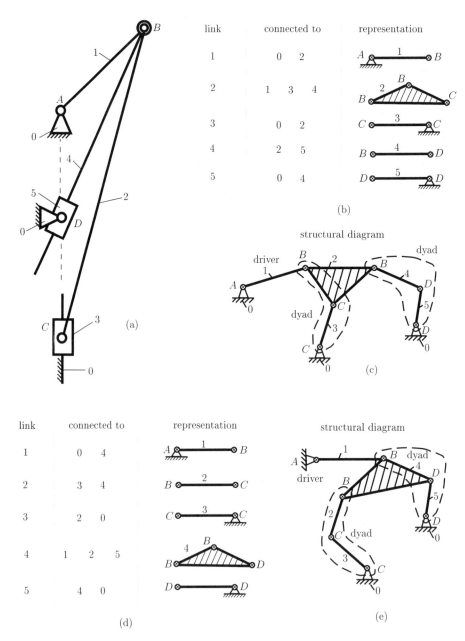

FIGURE I.2.47   *Planar R-RRT-RTR mechanism.*

The mechanism has five moving links ($n = 3$). To find the number of c_5 a *connectivity table* will be used, Figure I.2.47(b). The links are represented with bars (binary links) or triangles (ternary links). The one DOF joints (rotational joint or translation joint) are represented with a cross circle. The first column has the number of the current link, the second column shows the links connected to the current link, and the last column contains the graphical representation. The link 1 is connected to ground 0 at A and to link 2 at B [Fig. I.2.47(b)]. Next, link 2 is connected to link 1 at B, link 3 at C, and link 4 at B. Link 2 is a ternary link because it is connected to three links. At B there is a multiple joint, two rotational joints, one joint between link 1 and link 2, and one joint between link 2 and link 4. Link 3 is connected to ground 0 at C and to link 2 at C. At C there is a joint between link 3 and link 0 and a joint between link 3 and link 2. Link 4 is connected to link 2 at B and to link 5 at D. The last link, 5, is connected to link 4 at D and to ground 0 at D. In this way the table in Figure I.2.47(b) is obtained. The *structural diagram* is obtained using the graphical representation of the table connecting all the links Figure I.2.47(c). The c_5 joints (with cross circles), all the links, and the way the links are connected are all represented on the structural diagram. The number of one DOF joints is given by the number of cross circles. From Figure I.2.47(c) it results $c_5 = 7$. The number of DOF for the mechanism is $M = 3\,(5) - 2\,(7) = 1$. If $M = 1$, there is just one driver link. One can choose link 1 as the driver link of the mechanism. Once the driver link is taken away from the mechanism the remaining kinematic chain (links 2, 3, 4, 5) has the mobility equal to zero. The dyad is the simplest system group and has two links and three joints. On the structural diagram one can notice that links 2 and 3 represent a dyad and links 4 and 5 represent another dyad. The mechanism has been decomposed into a driver link (link 1) and two dyads (links 2 and 3, and links 4 and 5).

The connectivity table and the structural diagram are not unique for this mechanism. The new connectivity table can be obtained in Figure I.2.47(d). Link 1 is connected to ground 0 at A and to link 4 at B. Link 2 is connected to link 3 at C and to link 4 at B. Link 3 is connected to link 2 at C and to ground 0 at C. The link 4 is connected to link 1 at B, to link 2 at B, and to link 5 at D. This time link 4 is the ternary link. Link 5 is connected to link 4 at D and to ground 0 at D. The structural diagram is shown in Figure I.2.47(e). Using this structural diagram the mechanism can be decomposed into a driver link (link 1) and two dyads (links 2 and 3, and links 4 and 5).

If the driver link is link 1, the mechanism has the same structure no matter what structural diagram [Fig. I.2.47(c) or Fig. I.2.47(e)] is used.

Next, the driver link with rotational motion (R) and the dyads are represented as shown in Figure I.2.48(a). The first dyad (BCC) has the length between 2 and 3 equal to zero, $l_{CC} = 0$. The second dyad (BDD) has the length between 5 and 0 equal to zero, $l_{DD} = 0$. Figure I.2.48(b) shows the dyads with the lengths l_{CC} and l_{DD} different than zero. Using Figure I.2.48(b), the first dyad (BCC) has a rotational joint at B(R), a rotational joint at C(R), and a translational joint at C(T). The first dyad (BCC) is a rotation rotation translation dyad (dyad RRT). Using Figure I.2.48(b), the second dyad (BDD) has a rotational joint at B(R), a translational joint at D(T), and a rotational joint at D(R). The second dyad (BDD) is a rotation translation rotation dyad (dyad RTR). The mechanism is an R-RRT-RTR mechanism.

The mechanism in Figure I.2.49(a) is formed by a driver 1 with rotational motion R [Fig. I.2.49(b)], a dyad RTR [Fig. I.2.49(c)], and a dyad RTR [Fig. I.2.49(d)]. The mechanism in Figure I.2.49(a) is an R-RTR-RTR mechanism. The connectivity table is shown in Figure I.2.50(a) and the structural diagram is represented in Figure I.2.50(b).

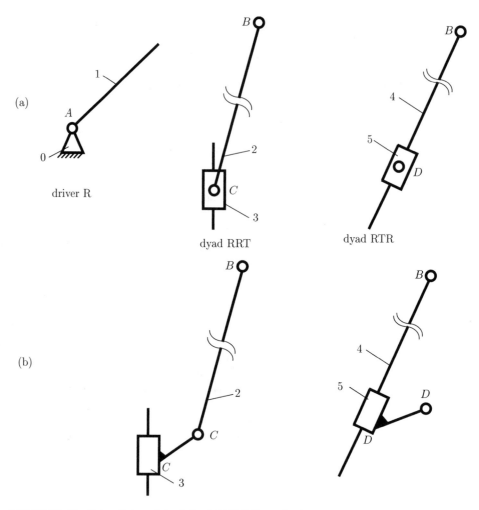

(a)

1

A

0

driver R

B

2

C

3

dyad RRT

B

4

5

D

dyad RTR

(b)

B

2

C

C

3

B

4

5

D

D

FIGURE I.2.48   *Driver link and dyads for R-RRT-RTR mechanism.*

I.2.15   Linkage Transformation

For planar mechanisms the half joints can be substituted, and in this way mechanisms with just full joints are obtained. The transformed mechanism has to be equivalent with the initial mechanism from a kinematical point of view. The number of DOF of the transformed mechanism has to be equal to the number of DOF of the initial mechanism. The relative motion of the links of the transformed mechanism has to be the same as the relative motion of the links of the initial mechanism.

A half joint constrains the possibility of motion of the connected links in motion. A constraint equation can be written and the number of degrees of freedom of a half joint is $M = -1$. To have the same number of degrees of freedom for a kinematic chain with n moving links and c_5 full jointS, the following equation is obtained

$$M = 3n - 2c_5 = -1. \qquad (I.2.15)$$

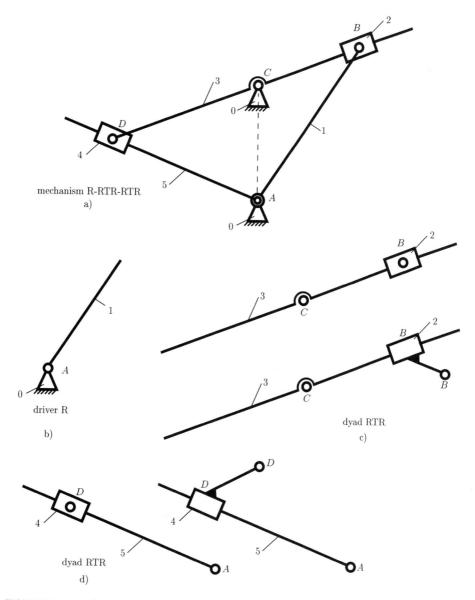

mechanism R-RTR-RTR
a)

driver R
b)

dyad RTR
c)

dyad RTR
d)

FIGURE I.2.49 *Planar R-RTR-RTR mechanism.*

The relation between the number of full joints and the number of moving links is obtained from Eq. (I.2.15).

$$c_5 = \frac{3n + 1}{2}. \tag{I.2.16}$$

A half joint can be substituted with one link ($n = 1$) and two full joints ($c_5 = 2$).

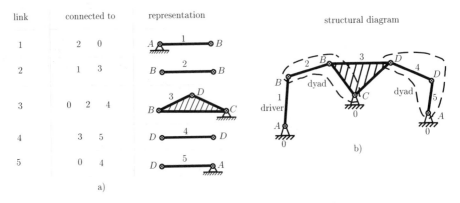

link	connected to		representation
1	2	0	A —1— B
2	1	3	B —2— B
3	0 2 4		B —3— D C
4	3	5	D —4— D
5	0	4	D —5— A

a)

structural diagram

b)

FIGURE I.2.50 *Connectivity table and structural diagram for R-RTR-RTR mechanism.*

Figure I.2.51(a) shows a cam and follower mechanism. There is a half joint at the contact point C between the links 1 and 2. One can substitute the half joint at C with one link, link 3, and two full at C and D as shown in Figure I.2.51(b). To have the same relative motion, the length of link 3 has to be equal to the radius of curvature ρ of the cam at the contact point C.

In this way the half joint at the contact point can be substituted for two full joints, C and D, and an extra link 3, between links 1 and 2. The mechanism still has one DOF, and the cam and follower system (0, 1, and 2) is in fact a four-bar mechanism (0, 1, 2, and 3) in another disguise.

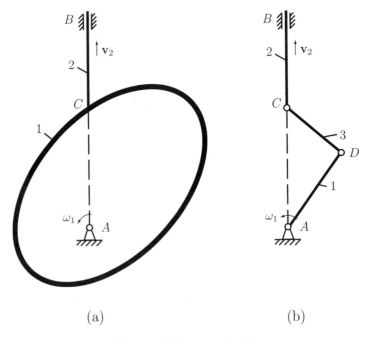

(a)   (b)

FIGURE I.2.51 *Transformation of cam and follower mechanism.*

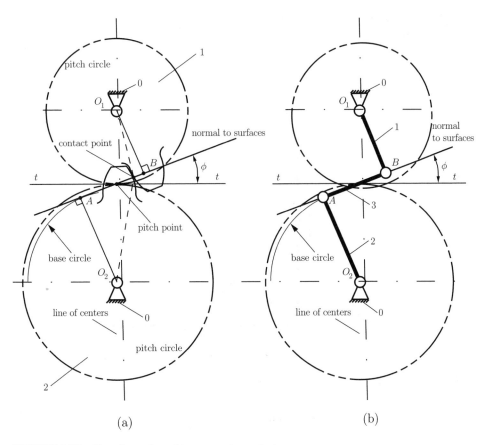

FIGURE I.2.52 *Transformation of two gears in contact.*

The half joint at the contact point of two gears in motion can be substituted for two full joints, *A* and *B*, and an extra link 3, between gears 1 and 2 (Fig. I.2.52). The mechanism still has one DOF, and the two-gear system (0, 1, and 2) [Fig. I.2.52(a)] is in fact a four-bar mechanism (0, 1, 2, and 3) in another disguise [Fig. I.2.52(b)]. The following relations can be written

$$O_1O_2 = \frac{m}{2}\,(N_1 + N_2),$$

$$O_1A = r_1 \cos \phi,$$

$$O_2B = r_2 \cos \phi,$$

$$AB = AP + PB = \frac{mN_1}{2}\sin \phi + \frac{mN_2}{2}\sin \phi = \frac{m}{2}\,(N_1 + N_2)\sin \phi,$$

where *m* is the module, *N* is the number of teeth, *r* is the pitch radius, and ϕ is the pressure angle. Because *m*, N_1, N_2, and ϕ are constants, the links of the four-bar mechanism [Fig. I.2.52(b)] are constant as well.

I.2.16 Problems

I.2.1 Determine the number of degrees of freedom (DOF) of the planar elipsograph mechanism in Figure I.2.53.

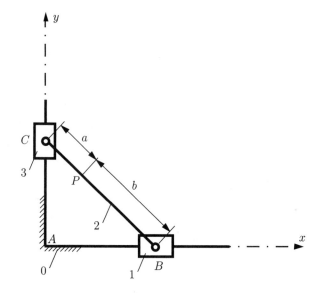

FIGURE I.2.53 *Elipsograph mechanism for Problem I.2.1.*

I.2.2 Find the mobility of the planar mechanism represented in Figure I.2.54.

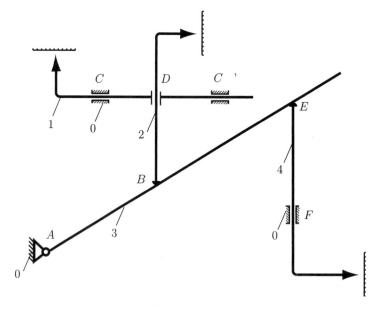

FIGURE I.2.54 *Planar mechanism for Problem I.2.2.*

I.2.3 Determine the family and the number of DOF for the mechanism depicted in Figure I.2.55.

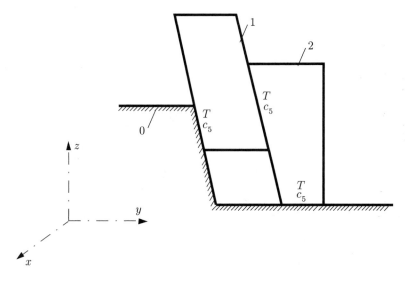

FIGURE I.2.55 *Mechanism for Problem I.2.3.*

I.2.4 Roller 2 of the mechanism in Figure I.2.56 undergoes an independent rotation about its axis which does not influence the motion of link 3. The purpose of element 2 is to substitute the sliding friction with a rolling friction. From a kinematical point of view, roller 2 is a passive element. Find the number of DOF.

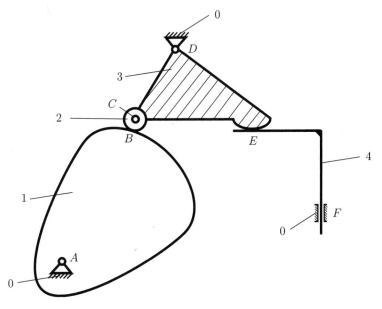

FIGURE I.2.56 *Mechanism with cam for Problem I.2.4.*

I.2.5 Find the family and the number of DOF of the mechanism in Figure I.2.57.

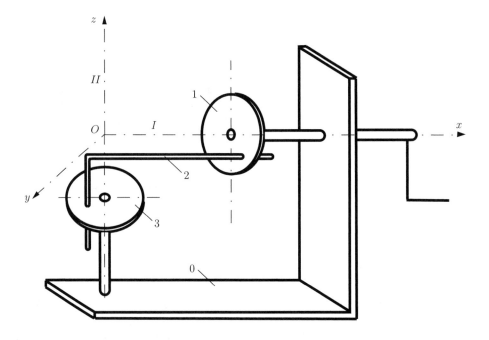

Schematic representation

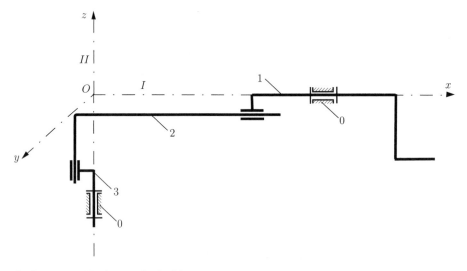

FIGURE I.2.57 *Mechanism for Problem I.2.5.*

I.2.6 Determine the number of DOF for the mechanism in Figure I.2.58.

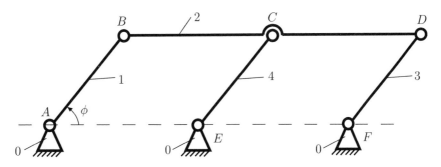

FIGURE I.2.58 *Mechanism for Problem I.2.6.*

I.2.7 Find the family, the number of DOF, and draw the structural diagram, and find the dyads for the mechanism shown in Figure I.2.59.

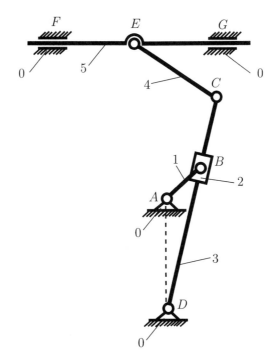

FIGURE I.2.59 *Mechanism for Problem I.2.7.*

I.2.8 Determine the family and the number of DOF for the mechanism in Figure I.2.60.

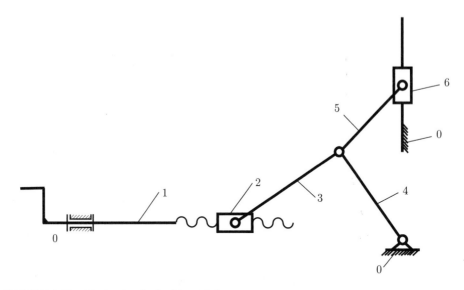

FIGURE I.2.60 *Mechanism for Problem I.2.8.*

I.2.9 Find the family and the number of DOF for the mechanism shown in Figure I.2.61.

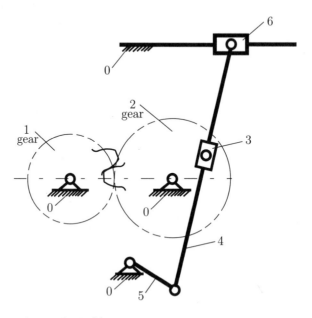

FIGURE I.2.61 *Mechanism for Problem I.2.9.*

I.2.10 Determine the number of DOF for the cam mechanism in Figure I.2.62.

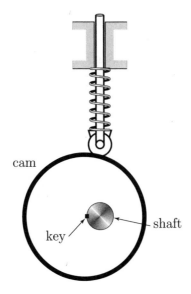

cam

key

shaft

FIGURE I.2.62 *Cam mechanism for Problem I.2.10.*

I.2.11 Find the number of DOF for the planetary gear train in Figure I.2.63.

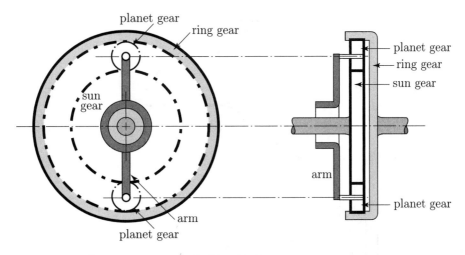

planet gear

ring gear

sun
gear

arm

planet gear

planet gear

ring gear

sun gear

arm

planet gear

FIGURE I.2.63 *Planetary gear train for Problem I.2.11.*

I.2.12 Determine the number of DOF for the Geneva mechanism in Figure I.2.64.

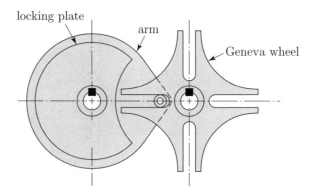

FIGURE I.2.64 *Geneva mechanism for Problem I.2.12.*

I.2.13 Find the number of DOF for the planetary gear train in Figure I.2.65.

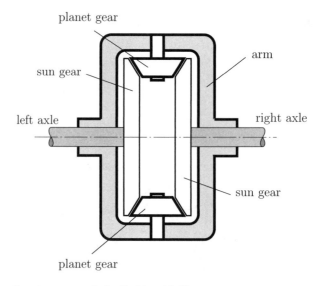

FIGURE I.2.65 *Planetary gear train for Problem I.2.13.*

I.3   Position Analysis

I.3.1   Absolute Cartesian Method

The position analysis of a kinematic chain requires the determination of the joint positions and/or the position of the center of gravity (CG) of the link. A planar link with the end nodes A and B is considered in Figure I.3.1. Let (x_A, y_A) be the coordinates of joint A with respect to the reference frame xOy, and (x_B, y_B) be the coordinates of joint B with the same reference frame. Using Pythagoras the following relation can be written:

$$(x_B - x_A)^2 + (y_B - y_A)^2 = AB^2 = L_{AB}^2, \tag{I.3.1}$$

where L_{AB} is the length of the link AB.

Let ϕ be the angle of the link AB with the horizontal axis Ox. Then, the slope m of the link AB is defined as

$$m = \tan \phi = \frac{y_B - y_A}{x_B - x_A}. \tag{I.3.2}$$

Let b be the intercept of AB with the vertical axis Oy. Using the slope m and the y intercept b, the equation of the straight link (line), in the plane, is

$$y = mx + b, \tag{I.3.3}$$

where x and y are the coordinates of any point on this link.

Two lines are perpendicular to each other if and only if the slope of one is the negative reciprocal of the slope of the other. Thus, if m and n are the slopes of two perpendicular lines

$$m = -\frac{1}{n} \quad \text{and} \quad mn = -1. \tag{I.3.4}$$

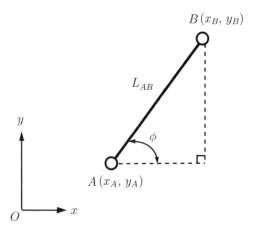

FIGURE I.3.1 *Planar link with two end nodes A and B.*

If two distinct points $A(x_A, y_A)$ and $B(x_B, y_B)$ are on a straight line then the equation of the straight line can be written in the forms

$$\frac{x - x_A}{x_B - x_A} = \frac{y - y_A}{y_B - y_A} \quad \text{and} \quad \begin{vmatrix} x & y & 1 \\ x_A & y_A & 1 \\ x_B & y_B & 1 \end{vmatrix} = 0. \tag{I.3.5}$$

Given two points $P(x_P, y_P)$ and $Q(x_Q, y_Q)$ and a real number k, $k \in \mathcal{R} - \{-\infty\}$, the coordinates of a point $R(x_R, y_R)$ on the line segment PQ, whose distance from P bears to the distance from R to Q the ratio k $(PR = k\,RQ)$, are

$$x_R = \frac{x_P + k\,x_Q}{1 + k} \quad \text{and} \quad y_R = \frac{y_P + k\,y_Q}{1 + k}. \tag{I.3.6}$$

The symbol $\in$ means "belongs to".

For $k = 1$ the above formulas become

$$x_R = \frac{x_P + x_Q}{2} \quad \text{and} \quad y_R = \frac{y_P + y_Q}{2}. \tag{I.3.7}$$

These give the coordinates of the midpoint of the interval from P to Q.

For $k > 0$ the point R is interior to the segment PQ and for $k < 0$ the point R is exterior to the segment PQ.

For a link with a translational joint (Fig. I.3.2) the sliding direction (Δ) is given by the equation

$$x \cos \alpha + y \sin \alpha - p = 0, \tag{I.3.8}$$

where p is the distance from the origin O to the sliding line (Δ). The position function for the joint $A(x_A, y_A)$ is

$$x_A \cos \alpha + y_A \sin \alpha - p = \pm d, \tag{I.3.9}$$

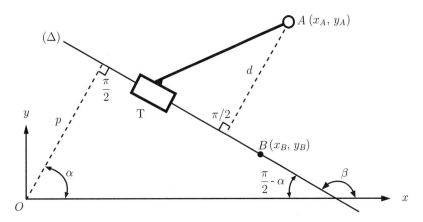

FIGURE I.3.2 *Link with a translational joint.*

where d is the distance from A to the sliding line. The relation between the joint A and a point B on the sliding direction, $B \in (\Delta)$, is

$$(x_A - x_B) \sin \beta + (y_A - y_B) \cos \beta = \pm d, \qquad (I.3.10)$$

where $\beta = \alpha + \dfrac{\pi}{2}$.

If $A x + B y + C = 0$ is the linear equation of the line (Δ) then the distance d is (Fig. I.3.2):

$$d = \frac{|A x_A + B y_A + C|}{\sqrt{A^2 + B^2}}. \qquad (I.3.11)$$

For a driver link in rotational motion [Fig. I.3.3(a)] the following relations can be written:

$$x_B = x_A + L_{AB} \cos \phi \quad \text{and} \quad y_B = y_A + L_{AB} \sin \phi. \qquad (I.3.12)$$

From Figure I.3.3(b), for a driver link in translational motion, one can have

$$x_B = x_A + s \cos \phi + L_1 \cos(\phi + \alpha),$$
$$y_B = y_A + s \sin \phi + L_1 \sin(\phi + \alpha). \qquad (I.3.13)$$

For the RRR dyad (Fig. I.3.4) there are two quadratic equations of the form

$$(x_A - x_C)^2 + (y_A - y_C)^2 = AC^2 = L_{AC}^2 = L_2^2,$$
$$(x_B - x_C)^2 + (y_B - y_C)^2 = BC^2 = L_{BC}^2 = L_3^2, \qquad (I.3.14)$$

where the coordinates of the joint C, x_C and y_C, are the unknowns. With x_C and y_C determined, the angles ϕ_2 and ϕ_3 are computed from the relations

$$\tan \phi_2 = \frac{y_C - y_A}{x_C - x_A} \quad \text{and} \quad \tan \phi_3 = \frac{y_C - y_B}{x_C - x_B}. \qquad (I.3.15)$$

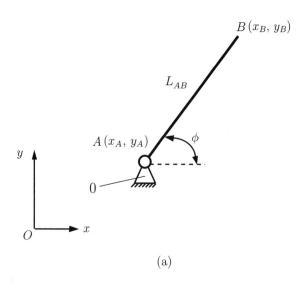

(a)

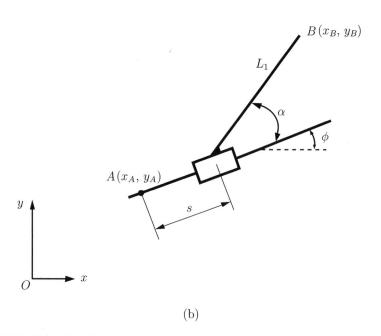

(b)

FIGURE I.3.3 *Driver link: (a) in rotational motion, and (b) in translational motion.*

The following relations can be written for the RRT dyad [Fig. I.3.5(a)]:

$$(x_A - x_C)^2 + (y_A - y_C)^2 = AC^2 = L_{AC}^2 = L_2^2,$$
$$(x_C - x_B)\sin\alpha - (y_C - y_B)\cos\alpha = \pm h. \qquad (I.3.16)$$

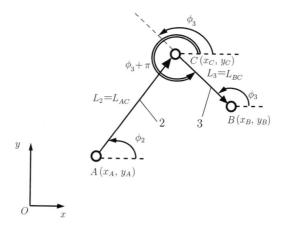

FIGURE I.3.4 *RRR dyad.*

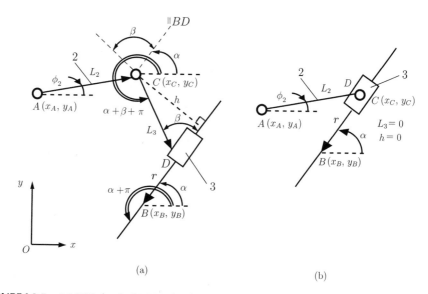

(a)   (b)

FIGURE I.3.5 *(a) RRT dyad; (b) RRT dyad, particular case, $L_3 = h = 0$.*

From the two above equations the two unknowns x_C and y_C are computed. Figure I.3.5(b) depicts the particular case for the RRT dyad when $L_3 = h = 0$ and the position equations are

$$(x_A - x_C)^2 + (y_A - y_C)^2 = L_2^2 \quad \text{and} \quad \tan\alpha = \frac{y_C - y_B}{x_C - x_B}. \tag{I.3.17}$$

For the RTR dyad [Fig. I.3.6(a)] the known data are: the positions of joints A and B, x_A, y_A, x_B, y_B, the angle α, and the length L_2 ($h = L_2 \sin\alpha$). There are four unknowns in the position of $C(x_C, y_C)$ and in the equation for the sliding line (Δ) : $y = mx + b$.

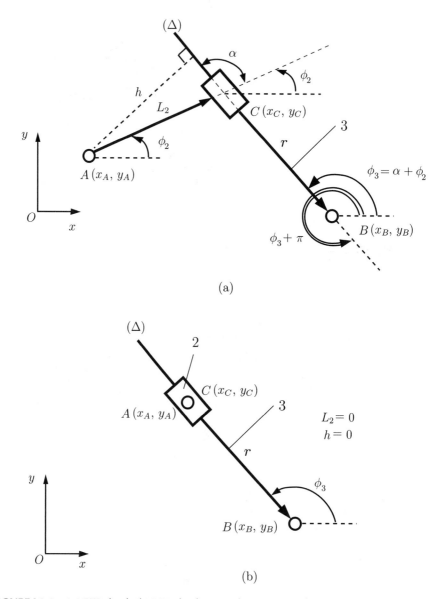

FIGURE I.3.6 *(a) RTR dyad; (b) RTR dyad, particular case, $L_2 = h = 0$.*

The unknowns in the sliding line m and b are computed from the relations

$$L_2 \sin \alpha = \frac{|m\,x_A - y_A + b|}{\sqrt{m^2 + 1}} \quad \text{and} \quad y_B = m\,x_B + b. \qquad \text{(I.3.18)}$$

The coordinates of joint C can be obtained using the equations

$$(x_A - x_C)^2 + (y_A - y_C)^2 = L_2^2 \quad \text{and} \quad y_C = m\,x_C + b. \qquad \text{(I.3.19)}$$

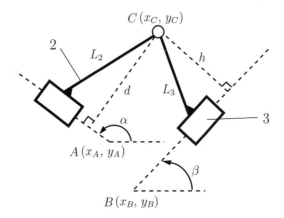

FIGURE I.3.7 *TRT dyad.*

In Figure I.3.6(b), the particular case when $L_1 = h = 0$ is shown, the position equation is

$$\tan \phi_2 = \tan \phi_3 = \frac{y_A - y_B}{x_A - x_B}. \tag{I.3.20}$$

To compute coordinates of joint C for the TRT dyad (Fig. I.3.7) two equations can be written:

$$(x_C - x_A)\sin \alpha - (y_C - y_A)\cos \alpha = \pm d,$$
$$(x_C - x_B)\sin \beta - (y_C - y_B)\cos \beta = \pm h. \tag{I.3.21}$$

The input data are x_A, y_A, x_B, y_B, α, β, d, h and the output data are x_C, y_C.

Consider the mechanism shown in Figure I.3.8. The angle of link 1 with the horizontal axis Ax is ϕ, $\phi = \angle(AB, Ax)$, and it is known. The following dimensions are given: $AB = l_1$, $CD = l_3$, $CE = l_4$, $AD = d$, and h is the distance from the slider 5 to the horizontal axis Ax.

Next the positions of the joints and the angles of the links will be calculated.

The origin of the system is at A, $A \equiv O$, $x_A = y_A = 0$. The coordinates of the rotational joint at B are

$$x_B = l_1 \cos \phi \quad \text{and} \quad y_B = l_1 \sin \phi.$$

The coordinates of the rotational joint at D are

$$x_D = d \quad \text{and} \quad y_D = 0.$$

For the dyad DBC (RTR) the angle $\phi_2 = \phi_3$ of link 2 or link 3 with the horizontal axis is calculated from the equation

$$\tan \phi_2 = \tan \phi_3 = \frac{y_B - y_D}{x_B - x_D} = \frac{l_1 \sin \phi}{l_1 \cos \phi - d}. \tag{I.3.22}$$

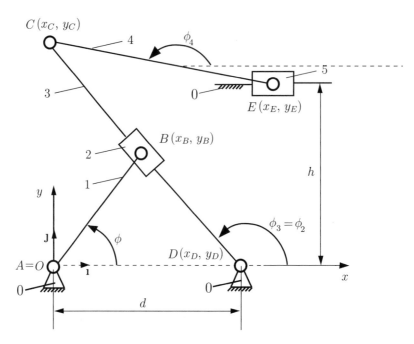

FIGURE I.3.8 *Planar mechanism.*

The joints $C(x_C, y_C)$ and D are on link 3 (straight line DBC) and

$$\tan \phi_3 = \frac{y_C - y_D}{x_C - x_D} = \frac{y_C}{x_C - d}. \tag{I.3.23}$$

Equations (I.3.22) and (I.3.23) give

$$\frac{y_B - y_D}{x_B - x_D} = \frac{y_C - y_D}{x_C - x_D} \quad \text{or} \quad \frac{l_1 \sin \phi}{l_1 \cos \phi - d} = \frac{y_C}{x_C - d}. \tag{I.3.24}$$

The length of link 3 is $CD = l_3$ (constant) and the distance from C to D is

$$(x_C - x_D)^2 + (y_C - y_D)^2 = l_3^2 \quad \text{or} \quad (x_C - d)^2 + y_C^2 = l_3^2. \tag{I.3.25}$$

The coordinates x_C and y_C of joint C result from Eq. (I.3.24) and Eq. (I.3.25).

Because of the quadratic equation, two solutions are obtained for x_C and y_C. For continuous motion of the mechanism there are constraint relations for choosing the correct solution: $x_C < x_B < x_D$ and $y_C > 0$.

For the last dyad CEE (RRT) a position function can be written for joint E ($CE = l_4 =$ constant) as:

$$(x_C - x_E)^2 + (y_C - h)^2 = l_4^2.$$

It results in values x_{E1} and x_{E2}, and the solution $x_E > x_C$ will be selected for continuous motion of the mechanism.

The angle ϕ_4 of link 4 with the horizontal axis is obtained from

$$\tan \phi_4 = \frac{y_C - y_E}{x_C - x_E} = \frac{y_C - h}{x_C - x_E}.$$ (I.3.26)

I.3.2 Vector Loop Method

First the independent closed loops are identified. A vector equation corresponding to each independent loop is established. The vector equation gives rise to two scalar equations, one for the horizontal axis x, and one for the vertical axis y.

For an open kinematic chain (Fig. I.3.9) with general joints (pin joints, slider joints, etc.), a vector loop equation can be considered:

$$\mathbf{r}_A + \mathbf{r}_1 + \cdots + \mathbf{r}_n = \mathbf{r}_B,$$ (I.3.27)

or

$$\sum_{k=1}^{n} \mathbf{r}_k = \mathbf{r}_B - \mathbf{r}_A.$$ (I.3.28)

The vectorial Eq. (I.3.28) can be projected on the reference frame xOy:

$$\sum_{k=1}^{n} r_k \cos \phi_k = x_B - x_A \quad \text{and} \quad \sum_{k=1}^{n} r_k \sin \phi_k = y_B - y_A.$$ (I.3.29)

RRR Dyad

The input data are: the position of A is (x_A, y_A), the position of B is (x_B, y_B), the length of AC is $L_{AC} = L_2$, and the length of BC is $L_{BC} = L_3$ (Fig. I.3.4).

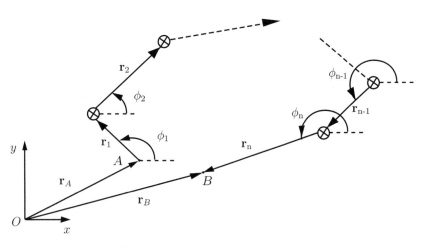

FIGURE I.3.9 *Kinematic chain.*

The unknown data are: the position of $C(x_C, y_C)$ and the angles ϕ_2 and ϕ_3.

The position equation for the RRR dyad is $\mathbf{r}_{AC} + \mathbf{r}_{CB} = \mathbf{r}_B - \mathbf{r}_A$, or

$$L_2 \cos \phi_2 + L_3 \cos(\phi_3 + \pi) = x_B - x_A,$$
$$L_2 \sin \phi_2 + L_3 \sin(\phi_3 + \pi) = y_B - y_A. \tag{I.3.30}$$

The angles ϕ_2 and ϕ_3 can be computed from Eq. (I.3.30). The position of C can be computed using the known angle ϕ_2:

$$x_C = x_A + L_2 \cos \phi_2 \quad \text{and} \quad y_C = y_A + L_2 \sin \phi_2. \tag{I.3.31}$$

RRT Dyad

The input data are: the position of A is (x_A, y_A), the position of B is (x_B, y_B), the length of AC is L_2, the length of CD is L_3, and the angles α and β are constants [Fig. I.3.5(a)]. The unknown data are: the position of $C(x_C, y_C)$, the angle ϕ_2, and the distance $r = DB$.

The vectorial equation for this kinematic chain is $\mathbf{r}_{AC} + \mathbf{r}_{CD} + \mathbf{r}_{DB} = \mathbf{r}_B - \mathbf{r}_A$, or

$$L_2 \cos \phi_2 + L_3 \cos(\alpha + \beta + \pi) + r \cos(\alpha + \pi) = x_B - x_A,$$
$$L_2 \sin \phi_2 + L_3 \sin(\alpha + \beta + \pi) + r \sin(\alpha + \pi) = y_B - y_A. \tag{I.3.32}$$

One can compute r and ϕ_2 from Eq. (I.3.32). The position of C can be found with Eq. (I.3.31).

Particular case $L_3 = 0$ [Fig. I.3.5(b)]
In this case Eq. (I.3.32) can be written as

$$L_2 \cos \phi_2 + r \cos(\alpha + \pi) = x_B - x_A,$$
$$L_2 \sin \phi_2 + r \sin(\alpha + \pi) = y_B - y_A. \tag{I.3.33}$$

RTR Dyad

The input data are: the position of A is (x_A, y_A), the position of B is (x_B, y_B), the length of AC is L_2, and the angle α is constant [Fig. I.3.6(a)]. The unknown data are: the distance $r = CB$ and the angles ϕ_2 and ϕ_3. The vectorial loop equation can be written as $\mathbf{r}_{AC} + \mathbf{r}_{CB} = \mathbf{r}_B - \mathbf{r}_A$, or

$$L_2 \cos \phi_2 + r \cos(\alpha + \phi_2 + \pi) = x_B - x_A,$$
$$L_2 \sin \phi_2 + r \sin(\alpha + \phi_2 + \pi) = y_B - y_A. \tag{I.3.34}$$

One can compute r and ϕ_2 from Eq. (I.3.34). The angle ϕ_3 can be written as:

$$\phi_3 = \phi_2 + \alpha. \tag{I.3.35}$$

Particular case $L_2 = 0$ [Fig. I.3.6(b)]

In this case from Eqs. (I.3.34) and (I.3.35) one can obtain

$$r \cos \phi_3 = x_B - x_A \quad \text{and} \quad r \sin \phi_3 = y_B - y_A. \qquad \text{(I.3.36)}$$

The method is illustrated through the following example. Figure I.3.10(a) shows a four-bar mechanism (R-RRR mechanism) with link lengths r_0, r_1, r_2, and r_3. Find the angles ϕ_2 and ϕ_3 as functions of the driver link angle $\phi = \phi_1$.

The links are denoted as vectors $\mathbf{r}_0$, $\mathbf{r}_1$, $\mathbf{r}_2$, and $\mathbf{r}_3$, ($|\mathbf{r}_i| = r_i$, $i = 0, 1, 2, 3$), and the angles are measured counterclockwise from the x-axis [Fig. I.3.10(b)]. For the closed loop $ABCD$, a vectorial equation can be written as:

$$\mathbf{r}_0 + \mathbf{r}_1 + \mathbf{r}_2 + \mathbf{r}_3 = \mathbf{0}. \qquad \text{(I.3.37)}$$

By projecting the above vectorial equation onto x and y, two scalar equations are obtained:

$$r_0 + r_1 \cos \phi_1 + r_2 \cos \phi_2 - r_3 \cos \phi_3 = 0, \qquad \text{(I.3.38)}$$

and

$$r_1 \sin \phi_1 + r_2 \sin \phi_2 - r_3 \sin \phi_3 = 0. \qquad \text{(I.3.39)}$$

Equations (I.3.38) and (I.3.39) represent a set of nonlinear equations in two unknowns, ϕ_2 and ϕ_3. The solution of these two equations solves the position analysis.

Rearranging Eqs. (I.3.38) and (I.3.39):

$$r_2 \cos \phi_2 = (r_3 \cos \phi_3 - r_0) - r_1 \cos \phi_1, \qquad \text{(I.3.40)}$$

and

$$r_2 \sin \phi_2 = r_3 \sin \phi_3 - r_1 \sin \phi_1. \qquad \text{(I.3.41)}$$

Squaring both sides of the above equations and adding

$$r_2^2 = r_0^2 + r_1^2 + r_3^2 - 2r_3 \cos \phi_3 (r_0 + r_1 \cos \phi_1)$$
$$- 2r_1 r_3 \sin \phi_1 \sin \phi_3 + 2r_0 r_1 \cos \phi_1,$$

or

$$a \sin \phi_3 + b \cos \phi_3 = c, \qquad \text{(I.3.42)}$$

where

$$a = \sin \phi_1, \quad b = \cos \phi_1 + (r_0/r_1), \quad \text{and}$$

$$c = (r_0/r_3) \cos \phi_1 + [(r_0^2 + r_1^2 + r_3^2 - r_2^2)/(2r_1 r_3)]. \qquad \text{(I.3.43)}$$

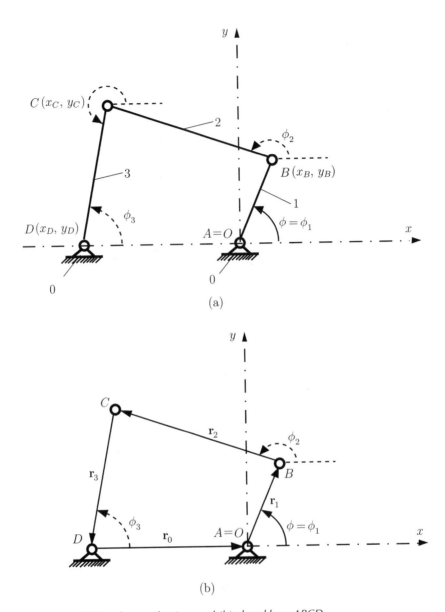

FIGURE I.3.10 *(a) Four-bar mechanism, and (b) closed loop ABCD.*

Using the relations

$$\sin \phi_3 = 2 \tan(\phi_3/2)[1 + \tan^2(\phi_3/2)],$$

and

$$\cos \phi_3 = [1 - \tan^2(\phi_3/2)]/[1 + \tan^2(\phi_3/2)], \qquad (I.3.44)$$

in Eq. (I.3.42), the following relation is obtained:

$$(b + c) \tan^2(\phi_3/2) - 2a \tan(\phi_3/2) + (c - b) = 0,$$

which gives

$$\tan(\phi_3/2) = (a \pm \sqrt{a^2 + b^2 - c^2})/(b + c). \qquad \text{(I.3.45)}$$

Thus, for each given value of ϕ_1 and the length of the links, two distinct values of the angle ϕ_3 are obtained:

$$\phi_{3(1)} = 2 \tan^{-1}[(a + \sqrt{a^2 + b^2 - c^2})/(b + c)],$$

$$\phi_{3(2)} = 2 \tan^{-1}[(a - \sqrt{a^2 + b^2 - c^2})/(b + c)]. \qquad \text{(I.3.46)}$$

The two values of ϕ_3 correspond to the two different positions of the mechanism.

The angle ϕ_2 can be eliminated from Eqs. (I.3.38) and (I.3.39) to give ϕ_1 in a similar way to that just described.

I.3.3 Examples

EXAMPLE I.3.1:

Figure I.3.11(a) shows a quick-return shaper mechanism. Given the lengths $AB = 0.20$ m, $AD = 0.40$ m, $CD = 0.70$ m, $CE = 0.30$ m, and the input angle $\phi = \phi_1 = 45°$, obtain the positions of all the other joints. The distance from the slider 5 to the horizontal axis Ax is $y_E = 0.35$ m.

Solution The coordinates of the joint B are

$$x_B = AB \cos \phi = 0.20 \sin 45° = 0.141 \text{ m},$$

$$y_B = AB \sin \phi = 0.20 \cos 45° = 0.141 \text{ m}.$$

The vector diagram Figure I.3.11(b) is drawn by representing the RTR (*BBD*) dyad. The vector equation, corresponding to this loop, is written as:

$$\mathbf{r}_B + \mathbf{r} - \mathbf{r}_D = 0 \quad \text{or} \quad \mathbf{r} = \mathbf{r}_D - \mathbf{r}_B,$$

where $\mathbf{r} = \mathbf{r}_{BD}$ and $|\mathbf{r}| = r$. Projecting the above vectorial equation on x- and y-axis, two scalar equations are obtained:

$$r \cos(\pi + \phi_3) = x_D - x_B = -0.141 \text{ m},$$

$$r \sin(\pi + \phi_3) = y_D - y_B = -0.541 \text{ m},$$

Continued

EXAMPLE I.3.1: *Cont'd*

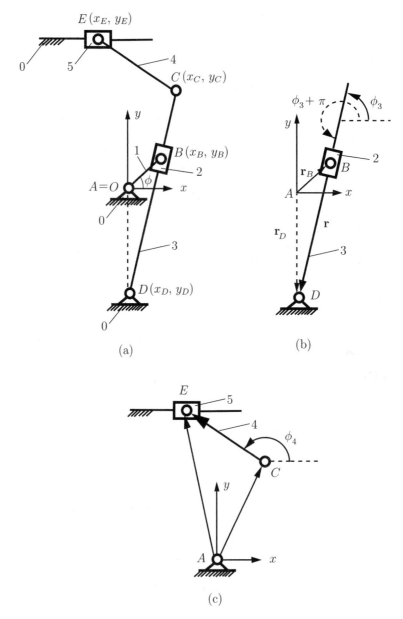

(a)

(b)

(c)

FIGURE I.3.11 *(a) Quick-return shaper mechanism, (b) vector diagram representing the RTR (BBD) dyad, and (c) vector diagram representing the RRT (CEE) dyad.*

The angle ϕ_3 is obtained by solving the system equations

$$\tan \phi_3 = \frac{y_D - y_B}{x_D - x_B} = \frac{0.541}{0.141} \implies \phi_3 = 75.36°.$$

EXAMPLE I.3.1: *Cont'd*

The distance r is

$$r = \frac{x_D - x_B}{\cos(\pi + \phi_3)} = 0.56 \text{ m}.$$

The coordinates of the joint C are

$$x_C = CD \sin \phi_3 = 0.17 \text{ m},$$
$$y_C = AB \cos \phi_3 - AD = 0.26 \text{ m}.$$

For the next dyad RRT (*CEE*) [Fig. I.3.11(c)], one can write

$$CE \cos(\pi - \phi_4) = x_E - x_C,$$
$$CE \sin(\pi - \phi_4) = y_E - y_C.$$

Solving this system, the unknowns ϕ_4 and x_E are obtained:

$$\phi_4 = 165.9° \quad \text{and} \quad x_E = -0.114 \text{ m}.$$

EXAMPLE I.3.2: R-RTR-RRT mechanism

The planar R-RTR-RRT mechanism is considered in Figure I.3.12. The driver is the rigid link 1 (the element *AB*) and makes an angle $\phi = \phi_1 = \pi/6$ with the horizontal. The length of the links are $AB = 0.02$ m, $BC = 0.03$ m, and $CD = 0.06$ m. The following dimensions are given: $AE = 0.05$ m and $L_a = 0.02$ m. Find the positions of the joints and the angles of the links.

Solution   *Position of joint A*: A Cartesian reference frame $xOyz$ with the unit vectors $[\imath, \jmath, \mathbf{k}]$ is selected, as shown in Figure I.3.12. Since joint A is in the origin of the reference system $A \equiv O$, then

$$x_A = y_A = 0.$$

Position of joint E: Then the coordinates of joint E are

$$x_E = -AE = -0.05 \text{ m} \text{ and } y_E = 0.$$

Position of joint B: Because joint A is fixed and the angle ϕ is known, the coordinates of joint B are computed with

$$x_B = AB \cos \phi = 0.02 \cos \pi/6 = 0.017 \text{ m},$$
$$y_B = AB \sin \phi = 0.02 \sin \pi/6 = 0.010 \text{ m}.$$

Continued

EXAMPLE I.3.2: R-RTR-RRT mechanism–*Cont'd*

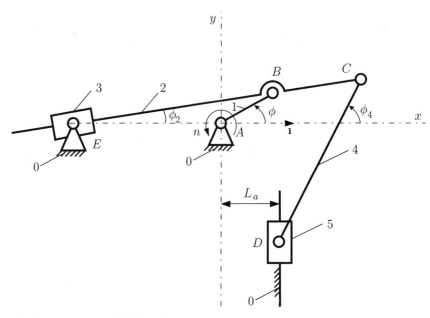

FIGURE I.3.12 *R-RTR-RRT mechanism.*

Position of joint C: Joints E, B, and C are located on the same straight line, EBC. The slope of this straight line is

$$m = \frac{y_B - y_E}{x_B - x_E} = \frac{y_C - y_E}{x_C - x_E} \quad \text{or} \quad \frac{0.010}{0.017 - (-0.05)} = \frac{y_C}{x_C - (-0.05)}. \quad \text{(I.3.47)}$$

The length of the link BC is constant and a quadratic equation can be written as:

$$(x_C - x_B)^2 + (y_C - y_B)^2 = BC^2 \quad \text{or}$$
$$(x_C - 0.017)^2 + (y_C - 0.01)^2 = 0.03^2. \quad \text{(I.3.48)}$$

Solving Eq. (I.3.47) and Eq. (I.3.48) two sets of solutions are found for the position of joint C. These solutions are

$$x_{C_1} = -0.012 \text{ m}, \quad y_{C_1} = 0.005 \text{ m},$$
$$x_{C_2} = 0.046 \text{ m}, \quad y_{C_2} = 0.014 \text{ m}.$$

The points C_1 and C_2 are the intersections of the circle of radius BC (with its center at B), with the straight line EC, as shown in Figure I.3.13. To determine the position of joint C for this position of the mechanism ($\phi = \pi/6$), an additional constraint condition is needed: $x_C > x_B$. With this constraint the coordinates of joint C have the following numerical values:

$$x_C = x_{C_2} = 0.046 \text{ m} \quad \text{and} \quad y_C = y_{C_2} = 0.014 \text{ m}.$$

EXAMPLE I.3.3: R-RTR-RRT mechanism–*Cont'd*

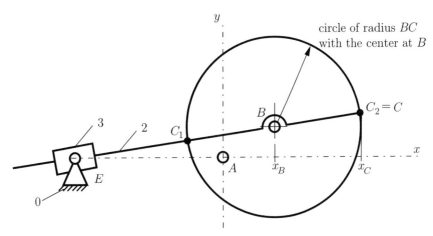

FIGURE I.3.13 *Position of joint C.*

Position of joint D: The *x*-coordinate of *D* is $x_D = L_a = 0.02$ m. The length of the link *CD* is constant and a quadratic equation can be written:

$$(x_D - x_C)^2 + (y_D - y_C)^2 = CD^2 \quad \text{or} \quad (0.02 - 0.046)^2 + (y_D - 0.014)^2 = 0.06^2. \tag{I.3.49}$$

Solving Eq. (I.3.49), two sets of solutions are found for the position of the joint *D*. These solutions are

$$y_{D_1} = -0.039 \text{ m} \quad \text{and} \quad y_{D_2} = 0.067 \text{ m}.$$

The points D_1 and D_2 are the intersections of the circle of radius *CD* (with its center at *C*) with the vertical line $x = L_a$, as shown in Figure I.3.14. To determine the correct position of joint *D* for the angle $\phi = \pi/6$, an additional constraint condition is needed: $y_D < y_C$. With this constraint the coordinates of joint *D* are

$$x_D = 0.02 \text{ m} \quad \text{and} \quad y_D = y_{D_1} = -0.039 \text{ m}.$$

Angle ϕ_2: The angle of link 2 (or link 3) with the horizontal axis is calculated from the slope of the straight line *EB*:

$$\phi_2 = \phi_3 = \arctan \frac{y_B - y_E}{x_B - x_E} = \arctan \frac{0.010}{0.017 - (-0.050)} = 0.147 \text{ rad} = 8.449°.$$

Angle ϕ_4: The angle of link 4 with the horizontal axis is obtained from the slope of the straight line *CD*:

$$\phi_4 = \arctan \frac{y_C - y_D}{x_C - x_D} = \arctan \frac{0.014 + 0.039}{0.046 - 0.020} = 1.104 \text{ rad} = 63.261°.$$

Continued

EXAMPLE I.3.2: R-RTR-RRT mechanism–*Cont'd*

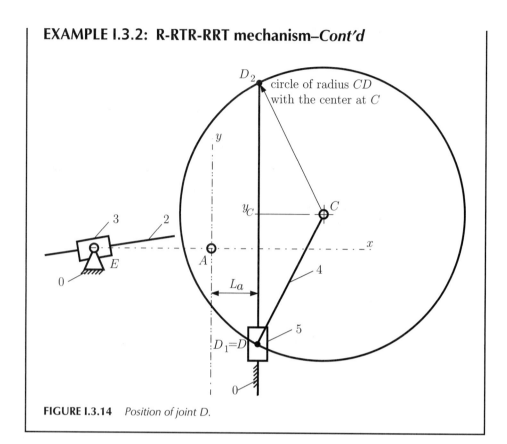

FIGURE I.3.14 *Position of joint D.*

EXAMPLE I.3.3: R-TRR-RRT mechanism

The mechanism is shown in Figure I.3.15. The following data are given: $AC = 0.100$ m, $BC = 0.300$ m, $BD = 0.900$ m, and $L_a = 0.100$ m. If the angle of link 1 with the horizontal axis is $\phi = 45°$, find the positions of joint D.

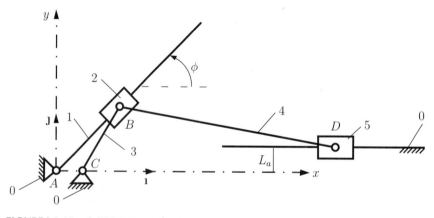

FIGURE I.3.15 *R-TRR-RRT mechanism.*

EXAMPLE I.3.3: R-TRR-RRT mechanism–*Cont'd*

Solution *Position of joint A*: A Cartesian reference frame with the origin at A is selected. The coordinates of the joint A are

$$x_A = y_A = 0.$$

Position of joint C: The coordinates of joint C are

$$x_C = AC = 0.100 \text{ m} \quad \text{and} \quad y_C = 0.$$

Position of joint B: The slope of the line AB is

$$\tan \phi = \frac{y_B}{x_B} \quad \text{or} \quad \tan 45° = \frac{y_B}{x_B}. \tag{I.3.50}$$

The length of the link BC is constant and the following equation can be written:

$$(x_B - x_C)^2 + (y_B - y_C)^2 = BC^2 \quad \text{or} \quad (x_B - 0.1)^2 + y_B^2 = 0.3^2. \tag{I.3.51}$$

Equations (I.3.50) and (I.3.51) form a system of two equations with the unknowns x_B and y_B. The following numerical results are obtained:

$$x_{B_1} = -0.156 \text{ m}, \quad y_{B_1} = -0.156 \text{ m},$$
$$x_{B_2} = 0.256 \text{ m}, \quad y_{B_2} = 0.256 \text{ m}.$$

To determine the correct position of the joint B for the angle $\phi = 45°$, an additional constraint condition is needed: $x_B > x_C$. With this constraint the coordinates of joint B are

$$x_B = x_{B_2} = 0.256 \text{ m} \quad \text{and} \quad y_B = y_{B_2} = 0.256 \text{ m}.$$

Position of joint D: The slider 5 has a translational motion in the horizontal direction and $y_D = L_a$. There is only one unknown, x_D, for joint D. The following expression can be written:

$$(x_B - x_D)^2 + (y_B - y_D)^2 = BD^2 \quad \text{or} \quad (0.256 - x_D)^2 + (0.256 - 0.1)^2 = 0.9^2 \tag{I.3.52}$$

Solving Eq. (I.3.49), two numerical values are obtained:

$$x_{D_1} = -0.630 \text{ m}, \quad x_{D_2} = 1.142 \text{ m}. \tag{I.3.53}$$

For continuous motion of the mechanism, a geometric constraint $x_D > x_B$ has to be selected. Using this relation the coordinates of joint D are

$$x_D = 1.142 \text{ m} \quad \text{and} \quad y_D = 0.100 \text{ m}.$$

I.3.4 Problems

I.3.1 Find the analytical expression of any point P of the elipsograph mechanism in Figure I.2.53.

I.3.2 The following data are given for the four-bar mechanism shown in Figure I.3.16: $AB = CD = 0.04$ m and $AD = BC = 0.09$ m. Find the trajectory of the point M located on the link BC for the case (a) $BM = MC$ and (b) $MC = 2\,BM$.

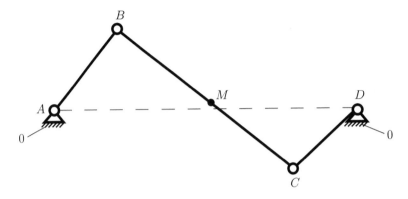

FIGURE I.3.16 *Four-bar mechanism for Problem I.3.2.*

I.3.3 The planar four-bar mechanism depicted in Figure I.3.17 has dimensions $AB = 0.03$ m, $BC = 0.065$ m, $CD = 0.05$ m, $BM = 0.09$ m, and $CM = 0.12$ m. Find the trajectory described by the point M.

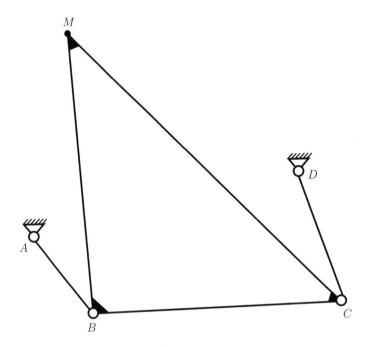

FIGURE I.3.17 *Four-bar mechanism for Problem I.3.3.*

I.3.4 The mechanism shown in Figure I.3.18 has dimensions $AB = 0.03$ m, $BC = 0.12$ m, $CD = 0.12$ m, $DE = 0.07$ m, $CF = 0.17$ m, $R_1 = 0.04$ m, $R_4 = 0.08$ m, $L_a = 0.025$ m, and $L_b = 0.105$ m. Find the trajectory of the joint C.

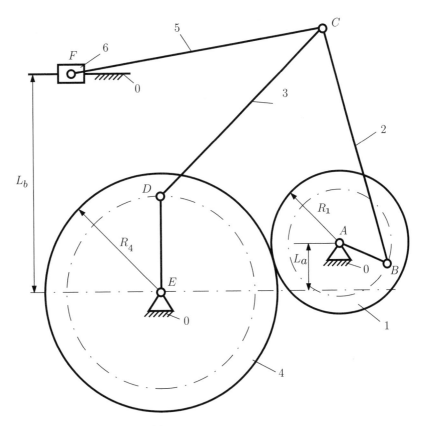

FIGURE I.3.18 *Mechanism for Problem I.3.4.*

I.3.5 The planar R-RRR-RRT mechanism considered is depicted in Figure I.3.19. The driver link is the rigid link 1 (the element AB). The following data are given: $AB = 0.150$ m, $BC = 0.400$ m, $CD = 0.370$ m, $CE = 0.230$ m, $EF = CE$, $L_a = 0.300$ m, $L_b = 0.450$ m, and $L_c = CD$. The angle of driver link 1 with the horizontal axis is $\phi = \phi_1 = 45°$. Find the positions of the joints and the angles of the links.

I.3.6 The R-RRR-RTT mechanism is shown in Figure I.3.20. The following data are given: $AB = 0.080$ m, $BC = 0.350$ m, $CD = 0.150$ m, $CE = 0.200$ m, $L_a = 0.200$ m, $L_b = 0.350$ m, $L_c = 0.040$ m. The angle of the driver element (link AB) with the horizontal axis is $\phi = 135°$. Determine the positions of the joints and the angles of the links.

I.3.7 The mechanism shown in Figure I.3.21 has the following dimensions: $AB = 40$ mm, $AD = 150$ mm, $BC = 100$ mm, $CE = 30$ mm, $EF = 120$ mm, and $a = 90$ mm.

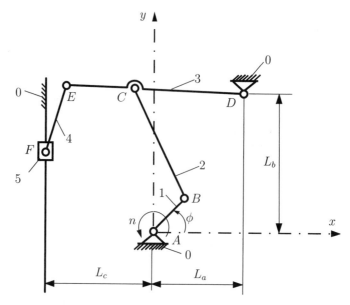

FIGURE I.3.19 *R-RRR-RRT mechanism for Problem I.3.5.*

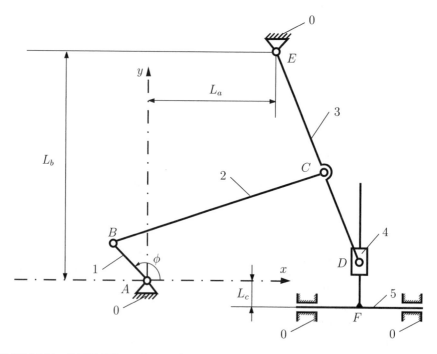

FIGURE I.3.20 *R-RRR-RTT mechanism for Problem I.3.6.*

The angle of driver link 1 with the horizontal axis is $\phi = \phi_1 = 30°$. Find the positions of the joints and the angles of the links.

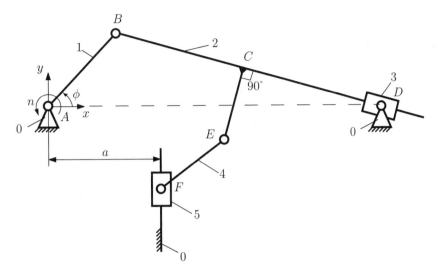

FIGURE I.3.21 *Mechanism for Problem I.3.7.*

I.3.8 The dimensions for the mechanism shown in Figure I.3.22 are: $AB = 250$ mm, $BD = 670$ mm, $DE = 420$ mm, $AE = 640$ mm, $BC = 240$ mm, $CD = 660$ mm, $CF = 850$ mm, and $b = 170$ mm. The angle of driver link 1 with the horizontal axis is $\phi = \phi_1 = 30°$. Find the positions of the joints and the angles of the links.

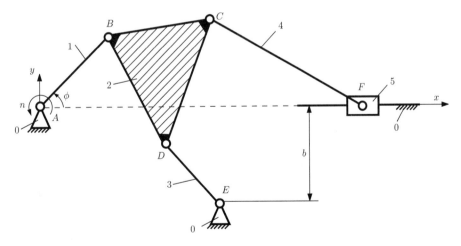

FIGURE I.3.22 *Mechanism for Problem I.3.8.*

I.3.9 The mechanism in Figure I.3.23 has the dimensions: $AB = 120$ mm, $AC = 60$ mm, $BD = 240$ mm, $DE = 330$ mm, $EF = 190$ mm, $L_a = 300$ mm, and $L_b = 70$ mm.

The angle of driver link 1 with the horizontal axis is $\phi = \phi_1 = 150°$. Find the positions of the joints and the angles of the links.

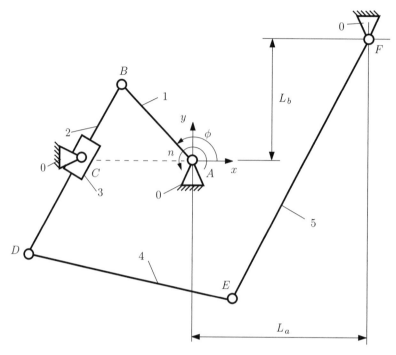

FIGURE I.3.23 *Mechanism for Problem I.3.9.*

I.3.10 The dimensions for the mechanism shown in Figure I.3.24 are: $AB = 100$ mm, $BC = 260$ mm, $AD = 240$ mm, $CD = 140$ mm, $DE = 80$ mm, $EF = 250$ mm, and

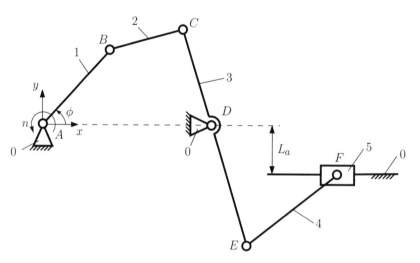

FIGURE I.3.24 *Mechanism for Problem I.3.10.*

$L_a = 20$ mm. The angle of driver link 1 with the horizontal axis is $\phi = \phi_1 = 45°$. Find the positions of the joints and the angles of the links.

I.3.11 The mechanism in Figure I.3.25 has the dimensions: $AB = 150$ mm, $AC = 450$ mm, $BD = 700$ mm, $L_a = 100$ mm, and $L_b = 200$ mm. The angle of driver link 1 with the horizontal axis is $\phi = \phi_1 = 120°$. Find the positions of the joints and the angles of the links.

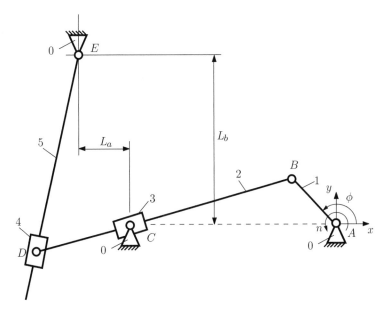

FIGURE I.3.25 *Mechanism for Problem I.3.11.*

I.3.12 Figure I.3.26 shows a mechanism with the following dimensions: $AB = 180$ mm, $BD = 700$ mm, and $L_a = 210$ mm. The angle of the driver link 1 with the horizontal axis is $\phi = \phi_1 = 135°$. Find the positions of the joints and the angles of the links.

I.3.13 The mechanism in Figure I.3.27 has the dimensions: $AB = 100$ mm, $AC = 240$ mm, $BD = 400$ mm, $DE = 250$ mm, $EF = 135$ mm, $L_a = 35$ mm, and $L_b = 170$ mm. The angle of driver link 1 with the horizontal axis is $\phi = \phi_1 = 150°$. Find the positions of the joints and the angles of the links.

I.3.14 Figure I.3.28 shows a mechanism with the following dimensions: $AB = 120$ mm, $BC = 450$ mm, $CD = DE = 180$ mm, $EF = 300$ mm, $L_a = 450$ mm, $L_b = 150$ mm, and $L_c = 140$ mm. The angle of driver link 1 with the horizontal axis is $\phi = \phi_1 = 120°$. Find the positions of the joints and the angles of the links.

I.3.15 Figure I.3.29 shows a mechanism with the following dimensions: $AB = 140$ mm, $BC = 650$ mm, $CE = 250$ mm, $CD = 400$ mm, $EF = 350$ mm, $L_a = 370$ mm, $L_b = 550$ mm, and $L_c = 700$ mm. The angle of driver link 1 with the horizontal axis is $\phi = \phi_1 = 150°$. Find the positions of the joints and the angles of the links.

I.3.16 Figure I.3.30 shows a mechanism with the following dimensions: $AB = 60$ mm, $BC = 160$ mm, $CF = 150$ mm, $CD = 60$ mm, $DE = 180$ mm, $L_a = 210$ mm,

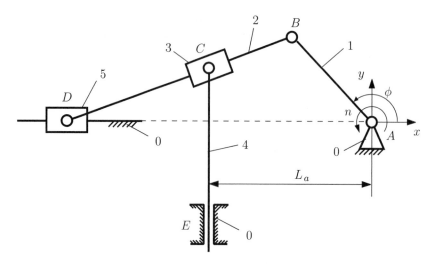

FIGURE I.3.26   *Mechanism for Problem I.3.12.*

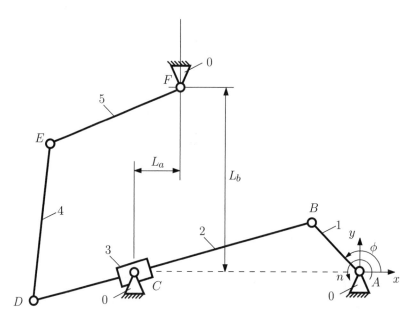

FIGURE I.3.27   *Mechanism for Problem I.3.13.*

$L_b = 120$ mm, and $L_c = 65$ mm. The angle of driver link 1 with the horizontal axis is $\phi = \phi_1 = 30°$. Find the positions of the joints and the angles of the links.

I.3.17 Figure I.3.31 shows a mechanism with the following dimensions: $AB = 20$ mm, $BC = 50$ mm, $AD = 25$ mm, and $BE = 60$ mm. The angle of driver link 1 with the horizontal axis is $\phi = \phi_1 = 60°$. Find the positions of the joints and the angles of the links.

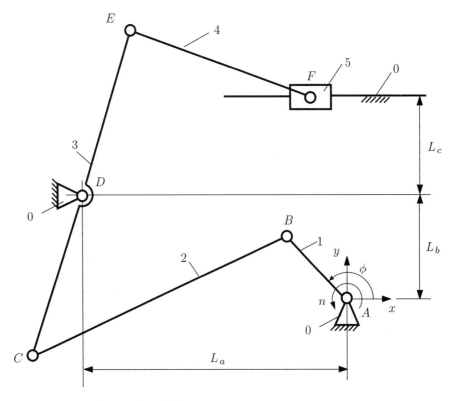

FIGURE I.3.28 *Mechanism for Problem I.3.14.*

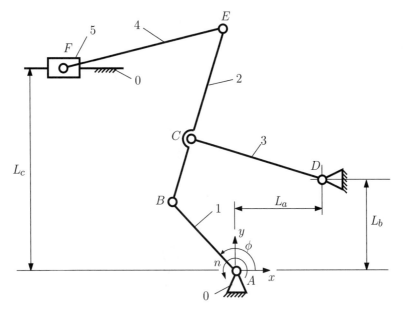

FIGURE I.3.29 *Mechanism for Problem I.3.15.*

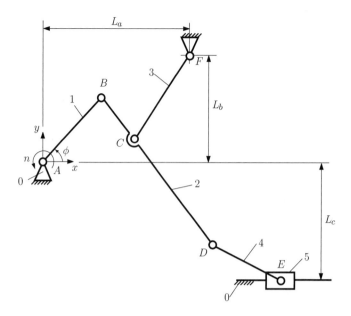

FIGURE I.3.30 *Mechanism for Problem I.3.16.*

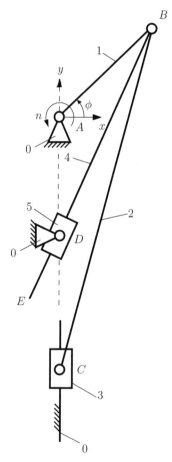

FIGURE I.3.31 *Mechanism for Problem I.3.17.*

I.3.18 The dimensions of the mechanism shown in Figure I.3.32 are: $AB = 150$ mm, $BC = 300$ mm, $BE = 600$ mm, $CE = 850$ mm, $CD = 330$ mm, $EF = 1200$ mm, $L_a = 350$ mm, $L_b = 200$ mm, and $L_c = 100$ mm. The angle of driver link 1 with the horizontal axis is $\phi = \phi_1 = 120°$. Find the positions of the joints and the angles of the links.

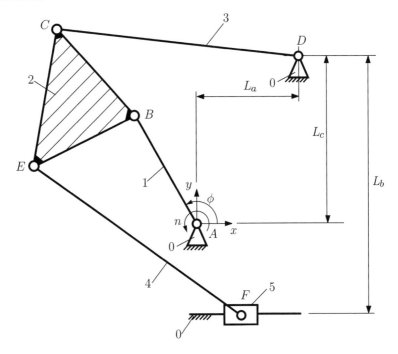

FIGURE I.3.32   *Mechanism for Problem I.3.18.*

I.3.19 The dimensions of the mechanism shown in Figure I.3.33 are: $AB = 150$ mm, $AC = 220$ mm, $CD = 280$ mm, $DE = 200$ mm, and $L_a = 230$ mm. The angle of driver link 1 with the horizontal axis is $\phi = \phi_1 = 60°$. Find the positions of the joints and the angles of the links.

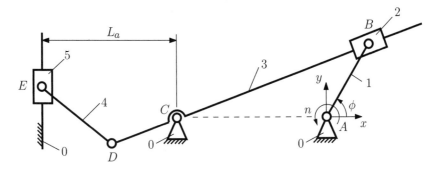

FIGURE I.3.33   *Mechanism for Problem I.3.19.*

I.3.20 The dimensions of the mechanism shown in Figure I.3.34 are: $AB = 200$ mm, $AC = 60$ mm, $CD = 200$ mm, and $DE = 500$ mm. The angle of driver link 1 with the horizontal axis is $\phi = \phi_1 = 45°$. Find the positions of the joints and the angles of the links.

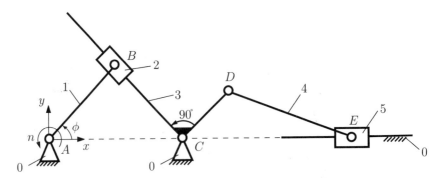

FIGURE I.3.34 *Mechanism for Problem I.3.20.*

I.3.21 The dimensions of the mechanism shown in Figure I.3.35 are: $AB = 120$ mm, $AC = 200$ mm, $CD = 380$ mm, and $b = 450$ mm. The angle of driver link 1 with

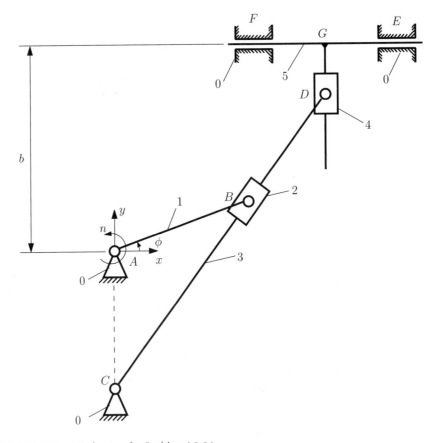

FIGURE I.3.35 *Mechanism for Problem I.3.21.*

the horizontal axis is $\phi = \phi_1 = 30°$. Find the positions of the joints and the angles of the links.

I.3.22 The dimensions of the mechanism shown in Figure I.3.36 are: $AB = 160$ mm, $AC = 90$ mm, and $CD = 160$ mm. The angle of driver link 1 with the horizontal axis is $\phi = \phi_1 = 30°$. Find the positions of the joints and the angles of the links.

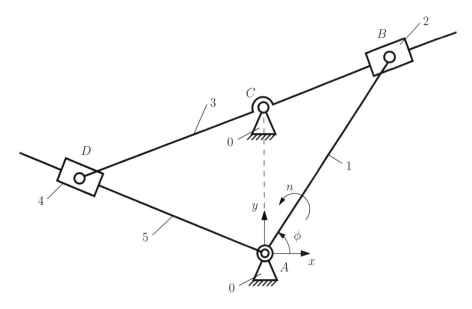

FIGURE I.3.36 *Mechanism for Problem I.3.22.*

I.3.23 The dimensions of the mechanism shown in Figure I.3.37 are: $AB = 100$ mm, $AC = 280$ mm, $BD = L_a = 470$ mm, and $DE = 220$ mm. The angle of driver link 1 with the horizontal axis is $\phi = \phi_1 = 30°$. Find the positions of the joints and the angles of the links.

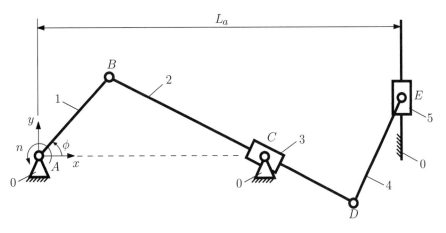

FIGURE I.3.37 *Mechanism for Problem I.3.23.*

I.3.24 The dimensions of the mechanism shown in Figure I.3.38 are: $AB = 250$ mm, $AD = 700$ mm, $BC = 300$ mm, and $a = 650$ mm. The angle of driver link 1 with the horizontal axis is $\phi = \phi_1 = 145°$. Find the positions of the joints and the angles of the links.

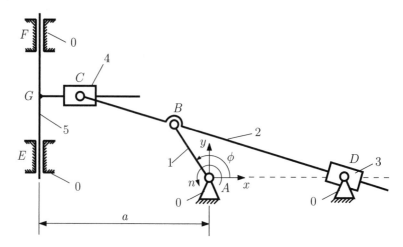

FIGURE I.3.38 *Mechanism for Problem I.3.24.*

I.4   Velocity and Acceleration Analysis

I.4.1   Kinematics of the Rigid Body

The motion of a rigid body (RB) is defined when the position vector, velocity, and acceleration of all points of the rigid body are defined as functions of time with respect to a fixed reference frame with the origin at O_0.

Let $\mathbf{i}_0$, $\mathbf{j}_0$, and $\mathbf{k}_0$ be the constant unit vectors of a fixed orthogonal Cartesian reference frame $O_0 x_0 y_0 z_0$ (primary reference frame). The unit vectors $\mathbf{i}_0$, $\mathbf{j}_0$, and $\mathbf{k}_0$ of the primary reference frame are constant with respect to time. Let $\mathbf{i}$, $\mathbf{j}$, and $\mathbf{k}$ be the unit vectors of a mobile orthogonal Cartesian reference frame $Oxyz$ (Fig. I.4.1). A reference frame that moves with the rigid body is a *body fixed* (or mobile) reference frame. The unit vectors $\mathbf{i}$, $\mathbf{j}$, and $\mathbf{k}$ of the body fixed reference frame are not constant, because they rotate with the body fixed reference frame. The location of the point O is arbitrary.

The position vector of a point M [$M \in$(RB)], with respect to the fixed reference frame $O_0 x_0 y_0 z_0$, is denoted by $\mathbf{r}_1 = \mathbf{r}_{O_0 M}$ and, with respect to the mobile reference frame $Oxyz$, is denoted by $\mathbf{r} = \mathbf{r}_{OM}$. The location of the origin O of the mobile reference frame, with respect to the fixed point O_0, is defined by the position vector $\mathbf{r}_O = \mathbf{r}_{O_0 O}$. Thus the relation between the vectors $\mathbf{r}_1$, $\mathbf{r}$, and $\mathbf{r}_0$ is given by

$$\mathbf{r}_1 = \mathbf{r}_O + \mathbf{r} = \mathbf{r}_O + x\mathbf{i} + y\mathbf{j} + z\mathbf{k}, \tag{I.4.1}$$

where x, y, and z represent the projections of the vector $\mathbf{r}$ on the mobile reference frame. The magnitude of the vector $\mathbf{r} = \mathbf{r}_{OM}$ is a constant, as the distance between the points O and M is constant [$O \in$(RB) and $M \in$ (RB)]. Thus, the x, y, and z components of the vector $\mathbf{r}$, with respect to the mobile reference frame, are constant. The unit vectors $\mathbf{i}$, $\mathbf{j}$, and $\mathbf{k}$ are time-dependent vector functions. The vectors $\mathbf{i}$, $\mathbf{j}$, and $\mathbf{k}$ are the unit vector of an orthogonal Cartesian reference frame. Thus, the following relations can be written as

$$\mathbf{i} \cdot \mathbf{i} = 1, \quad \mathbf{j} \cdot \mathbf{j} = 1, \quad \mathbf{k} \cdot \mathbf{k} = 1, \tag{I.4.2}$$

$$\mathbf{i} \cdot \mathbf{j} = 0, \quad \mathbf{j} \cdot \mathbf{k} = 0, \quad \mathbf{k} \cdot \mathbf{i} = 0. \tag{I.4.3}$$

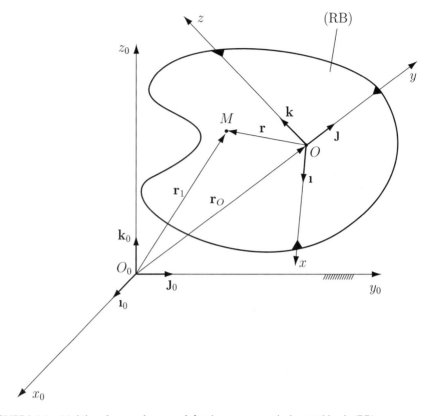

FIGURE I.4.1 *Mobile reference frame* $(\mathbf{\imath}, \mathbf{\jmath}, \mathbf{k})$ *that moves with the rigid body* (RB).

Velocity of a Point on the Rigid Body

The velocity of an arbitrary point M of the rigid body, with respect to the fixed reference frame $Ox_0y_0z_0$, is the derivative with respect to time of the position vector $\mathbf{r}_1$:

$$\mathbf{v} = \frac{d\mathbf{r}_1}{dt} = \dot{\mathbf{r}}_1 = \dot{\mathbf{r}}_O + \dot{\mathbf{r}} = \mathbf{v}_O + x\mathbf{i} + y\mathbf{\jmath} + z\dot{\mathbf{k}} + \dot{x}\mathbf{\imath} + \dot{y}\mathbf{\jmath} + \dot{z}\mathbf{k}, \qquad (\text{I.4.4})$$

where $\mathbf{v}_O = \dot{\mathbf{r}}_O$ represents the velocity of the origin of the mobile reference frame $O_1x_1y_1z_1$ with respect to the fixed reference frame $Oxyz$. Because all the points in the rigid body maintain their relative position, their velocity relative to the mobile reference frame $Oxyz$ is zero: $\dot{x} = \dot{y} = \dot{z} = 0$. The velocity of the point M is

$$\mathbf{v} = \mathbf{v}_O + x\mathbf{i} + y\mathbf{\jmath} + z\dot{\mathbf{k}}.$$

The derivative of the Eqs. (I.4.2) and (I.4.3) with respect to time gives

$$\mathbf{i} \cdot \mathbf{\imath} = 0, \quad \mathbf{\jmath} \cdot \mathbf{\jmath} = 0, \quad \dot{\mathbf{k}} \cdot \mathbf{k} = 0, \qquad (\text{I.4.5})$$

and

$$\mathbf{\dot{i}} \cdot \mathbf{j} + \mathbf{j} \cdot \mathbf{\dot{i}} = 0, \quad \mathbf{\dot{j}} \cdot \mathbf{k} + \mathbf{\dot{k}} \cdot \mathbf{j} = 0, \quad \mathbf{\dot{k}} \cdot \mathbf{i} + \mathbf{\dot{i}} \cdot \mathbf{k} = 0. \tag{I.4.6}$$

For Eq. (I.4.6) the following convention is introduced:

$$\begin{aligned}
\mathbf{\dot{i}} \cdot \mathbf{j} &= -\mathbf{\dot{j}} \cdot \mathbf{i} = \omega_z, \\
\mathbf{\dot{j}} \cdot \mathbf{k} &= -\mathbf{\dot{k}} \cdot \mathbf{j} = \omega_x, \\
\mathbf{\dot{k}} \cdot \mathbf{i} &= -\mathbf{\dot{i}} \cdot \mathbf{k} = \omega_y,
\end{aligned} \tag{I.4.7}$$

where ω_x, ω_y and ω_z can be considered as the projections of a vector $\boldsymbol{\omega}$:

$$\boldsymbol{\omega} = \omega_x \mathbf{i} + \omega_y \mathbf{j} + \omega_z \mathbf{k}.$$

To calculate $\mathbf{\dot{i}}$, $\mathbf{\dot{j}}$, $\mathbf{\dot{k}}$ the following formula is introduced for an arbitrary vector, $\mathbf{d}$,

$$\mathbf{d} = d_x \mathbf{i} + d_y \mathbf{j} + d_z \mathbf{k} = (\mathbf{d} \cdot \mathbf{i}) \mathbf{i} + (\mathbf{d} \cdot \mathbf{j}) \mathbf{j} + (\mathbf{d} \cdot \mathbf{k}) \mathbf{k}. \tag{I.4.8}$$

Using Eq. (I.4.8) and the results from Eqs. (I.4.5) and (I.4.6), it results

$$\begin{aligned}
\mathbf{\dot{i}} &= (\mathbf{\dot{i}} \cdot \mathbf{i}) \mathbf{i} + (\mathbf{\dot{i}} \cdot \mathbf{j}) \mathbf{j} + (\mathbf{\dot{i}} \cdot \mathbf{k}) \mathbf{k} \\
&= (0) \mathbf{i} + (\omega_z) \mathbf{j} - (\omega_y) \mathbf{k} \\
&= \begin{vmatrix} \mathbf{i} & \mathbf{j} & \mathbf{k} \\ \omega_x & \omega_y & \omega_z \\ 1 & 0 & 0 \end{vmatrix} = \boldsymbol{\omega} \times \mathbf{i}, \\[2mm]
\mathbf{\dot{j}} &= (\mathbf{\dot{j}} \cdot \mathbf{i}) \mathbf{i} + (\mathbf{\dot{j}} \cdot \mathbf{j}) \mathbf{j} + (\mathbf{\dot{j}} \cdot \mathbf{k}) \mathbf{k} \\
&= (-\omega_z) \mathbf{i} + (0) \mathbf{j} + (\omega_x) \mathbf{k} \\
&= \begin{vmatrix} \mathbf{i} & \mathbf{j} & \mathbf{k} \\ \omega_x & \omega_y & \omega_z \\ 0 & 1 & 0 \end{vmatrix} = \boldsymbol{\omega} \times \mathbf{j}, \\[2mm]
\mathbf{\dot{k}} &= (\mathbf{\dot{k}} \cdot \mathbf{i}) \mathbf{i} + (\mathbf{\dot{k}} \cdot \mathbf{j}) \mathbf{j} + (\mathbf{\dot{k}} \cdot \mathbf{k}) \mathbf{k} \\
&= (\omega_y) \mathbf{i} - (\omega_x) \mathbf{j} + (0) \mathbf{k} \\
&= \begin{vmatrix} \mathbf{i} & \mathbf{j} & \mathbf{k} \\ \omega_x & \omega_y & \omega_z \\ 0 & 0 & 1 \end{vmatrix} = \boldsymbol{\omega} \times \mathbf{k}.
\end{aligned} \tag{I.4.9}$$

The relations

$$\mathbf{\dot{i}} = \boldsymbol{\omega} \times \mathbf{i}, \quad \mathbf{\dot{j}} = \boldsymbol{\omega} \times \mathbf{j}, \quad \mathbf{\dot{k}} = \boldsymbol{\omega} \times \mathbf{k}. \tag{I.4.10}$$

are known as *Poisson formulas*.

Using Eqs. (I.4.4) and (I.4.10), the velocity of M is

$$\mathbf{v} = \mathbf{v}_O + x\boldsymbol{\omega} \times \mathbf{i} + y\boldsymbol{\omega} \times \mathbf{j} + z\boldsymbol{\omega} \times \mathbf{k} = \mathbf{v}_O + \boldsymbol{\omega} \times (x\mathbf{i} + y\mathbf{j} + z\mathbf{k}),$$

or

$$\mathbf{v} = \mathbf{v}_O + \boldsymbol{\omega} \times \mathbf{r}. \tag{I.4.11}$$

Combining Eqs. (I.4.4) and (I.4.11), it results

$$\dot{\mathbf{r}} = \boldsymbol{\omega} \times \mathbf{r}. \tag{I.4.12}$$

Using Eq. (I.4.11), the components of the velocity are

$$v_x = v_{Ox} + z\omega_y - y\omega_z,$$
$$v_y = v_{Oy} + x\omega_z - z\omega_x,$$
$$v_z = v_{Oz} + y\omega_x - x\omega_y.$$

Acceleration of a Point on the Rigid Body

The acceleration of an arbitrary point $M \in$ (RB), with respect to a fixed reference frame $O_0 x_0 y_0 z_0$, represents the double derivative with respect to time of the position vector $\mathbf{r}_1$:

$$\mathbf{a} = \ddot{\mathbf{r}}_1 = \dot{\mathbf{v}} = \frac{d\mathbf{v}}{dt} = \frac{d}{dt}(\mathbf{v}_O + \boldsymbol{\omega} \times \mathbf{r}) = \frac{d}{dt}\mathbf{v}_O + \frac{d}{dt}\boldsymbol{\omega} \times \mathbf{r} + \boldsymbol{\omega} \times \frac{d}{dt}\mathbf{r}$$
$$= \dot{\mathbf{v}}_O + \dot{\boldsymbol{\omega}} \times \mathbf{r} + \boldsymbol{\omega} \times \dot{\mathbf{r}}. \tag{I.4.13}$$

The acceleration of the point O, with respect to the fixed reference frame $O_0 x_0 y_0 z_0$, is

$$\mathbf{a}_O = \dot{\mathbf{v}}_O = \ddot{\mathbf{r}}_O. \tag{I.4.14}$$

The derivative of the vector $\boldsymbol{\omega}$, with respect to the time, is the vector $\boldsymbol{\alpha}$ given by

$$\boldsymbol{\alpha} = \dot{\boldsymbol{\omega}} = \dot{\omega}_x \mathbf{i} + \dot{\omega}_y \mathbf{j} + \dot{\omega}_z \mathbf{k} + \omega_x \dot{\mathbf{i}} + \omega_y \dot{\mathbf{j}} + \omega_z \dot{\mathbf{k}} \tag{I.4.15}$$
$$= \alpha_x \mathbf{i} + \alpha_y \mathbf{j} + \alpha_z \mathbf{k} + \omega_x \boldsymbol{\omega} \times \mathbf{i} + \omega_y \boldsymbol{\omega} \times \mathbf{j} + \omega_z \boldsymbol{\omega} \times \mathbf{k}$$
$$= \alpha_x \mathbf{i} + \alpha_y \mathbf{j} + \alpha_z \mathbf{k} + \boldsymbol{\omega} \times \boldsymbol{\omega} = \alpha_x \mathbf{i} + \alpha_y \mathbf{j} + \alpha_z \mathbf{k}.$$

where $\alpha_x = \dot{\omega}_x, \alpha_y = \dot{\omega}_y$, and $\alpha_z = \dot{\omega}_z$. In the previous expression the Poisson formulas have been used.

Using Eqs. (I.4.13), (I.4.14), and (I.4.15), the acceleration of the point M is

$$\mathbf{a} = \mathbf{a}_O + \boldsymbol{\alpha} \times \mathbf{r} + \boldsymbol{\omega} \times (\boldsymbol{\omega} \times \mathbf{r}). \tag{I.4.16}$$

The components of the acceleration are

$$a_x = a_{Ox} + \left(z\alpha_y - y\alpha_z\right) + \omega_y\left(y\omega_x - x\omega_y\right) + \omega_z\left(x\omega_x - x\omega_z\right),$$

$$a_y = a_{Oy} + (x\alpha_z - z\alpha_x) + \omega_z\left(z\omega_y - y\omega_z\right) + \omega_x\left(x\omega_y - y\omega_z\right),$$

$$a_z = a_{Oz} + \left(y\alpha_x - x\alpha_y\right) + \omega_x\left(x\omega_z - z\omega_x\right) + \omega_y\left(y\omega_z - z\omega_y\right).$$

The vector $\boldsymbol{\omega}$ characterizes the rotational motion of the rigid body and is called the *angular velocity*. The vector $\boldsymbol{\alpha}$ is called the *angular acceleration*.

The angular velocity can also be introduced in another way. If the orientation of a rigid body RB in a reference frame RF_0 depends on only a single scalar variable ζ, there exists for each value of ζ a vector $\boldsymbol{\omega}$ such that the derivative with respect to ζ in RF_0 of every vector $\mathbf{c}$ fixed in rigid body RB is given by

$$\frac{d\mathbf{c}}{d\zeta} = \boldsymbol{\omega} \times \mathbf{c}, \tag{I.4.17}$$

where the vector $\boldsymbol{\omega}$ is the rate of change of orientation of the rigid body RB in the reference frame RF_0 with respect to ζ. The vector $\boldsymbol{\omega}$ is given by

$$\boldsymbol{\omega} = \frac{\dfrac{d\mathbf{a}}{d\zeta} \times \dfrac{d\mathbf{b}}{d\zeta}}{\dfrac{d\mathbf{a}}{d\zeta} \cdot \mathbf{b}}, \tag{I.4.18}$$

where $\mathbf{a}$ and $\mathbf{b}$ are any two nonparallel vectors fixed in the rigid body RB.

Proof
The vectors $\mathbf{a}$ and $\mathbf{b}$ are fixed in the rigid body. The magnitudes $\mathbf{a} \cdot \mathbf{a}$, $\mathbf{b} \cdot \mathbf{b}$, and the angle between $\mathbf{a}$ and $\mathbf{b}$ are independent of ζ

$$\frac{d\left(\mathbf{a} \cdot \mathbf{a}\right)}{d\zeta} = 0, \quad \frac{d(\mathbf{b} \cdot \mathbf{b})}{d\zeta} = 0, \quad \frac{d\left(\mathbf{a} \cdot \mathbf{b}\right)}{d\zeta} = 0,$$

or

$$\frac{d\mathbf{a}}{d\zeta} \cdot \mathbf{a} = 0, \quad \frac{d\mathbf{b}}{d\zeta} \cdot \mathbf{b} = 0, \quad \frac{d\mathbf{a}}{d\zeta} \cdot \mathbf{b} + \mathbf{a} \cdot \frac{d\mathbf{b}}{d\zeta} = 0.$$

Using the vector triple product of three vectors $\mathbf{p}$, $\mathbf{q}$, $\mathbf{t}$, it results

$$\mathbf{p} \times (\mathbf{q} \times \mathbf{t}) = \mathbf{p} \cdot \mathbf{t}\, \mathbf{q} - \mathbf{p} \cdot \mathbf{q}\, \mathbf{t}, \quad (\mathbf{p} \times \mathbf{q}) \times \mathbf{t} = \mathbf{t} \cdot \mathbf{p}\, \mathbf{q} - \mathbf{t} \cdot \mathbf{q}\, \mathbf{t}.$$

From these expressions it follows that

$$
\frac{\dfrac{d\mathbf{a}}{d\zeta} \times \dfrac{d\mathbf{b}}{d\zeta}}{\dfrac{d\mathbf{a}}{d\zeta} \cdot \mathbf{b}} \times \mathbf{a} = \frac{\left(\dfrac{d\mathbf{a}}{d\zeta} \times \dfrac{d\mathbf{b}}{d\zeta}\right) \times \mathbf{a}}{\dfrac{d\mathbf{a}}{d\zeta} \cdot \mathbf{b}} = \frac{\mathbf{a} \cdot \dfrac{d\mathbf{a}}{d\zeta}\dfrac{d\mathbf{b}}{d\zeta} - \mathbf{a} \cdot \dfrac{d\mathbf{b}}{d\zeta}\dfrac{d\mathbf{a}}{d\zeta}}{\dfrac{d\mathbf{a}}{d\zeta} \cdot \mathbf{b}}
$$

$$
= \frac{-\mathbf{a} \cdot \dfrac{d\mathbf{b}}{d\zeta}\dfrac{d\mathbf{a}}{d\zeta}}{\dfrac{d\mathbf{a}}{d\zeta} \cdot \mathbf{b}} = \frac{\dfrac{d\mathbf{a}}{d\zeta} \cdot \mathbf{b}\dfrac{d\mathbf{a}}{d\zeta}}{\dfrac{d\mathbf{a}}{d\zeta} \cdot \mathbf{b}} = \frac{d\mathbf{a}}{d\zeta}, \tag{I.4.19}
$$

and

$$
\frac{\dfrac{d\mathbf{a}}{d\zeta} \times \dfrac{d\mathbf{b}}{d\zeta}}{\dfrac{d\mathbf{a}}{d\zeta} \cdot \mathbf{b}} \times \mathbf{b} = \frac{\left(\dfrac{d\mathbf{a}}{d\zeta} \times \dfrac{d\mathbf{b}}{d\zeta}\right) \times \mathbf{b}}{\dfrac{d\mathbf{a}}{d\zeta} \cdot \mathbf{b}} = \frac{\mathbf{b} \cdot \dfrac{d\mathbf{a}}{d\zeta}\dfrac{d\mathbf{b}}{d\zeta} - \mathbf{b} \cdot \dfrac{d\mathbf{b}}{d\zeta}\dfrac{d\mathbf{a}}{d\zeta}}{\dfrac{d\mathbf{a}}{d\zeta} \cdot \mathbf{b}}
$$

$$
= \frac{\mathbf{b} \cdot \dfrac{d\mathbf{a}}{d\zeta}\dfrac{d\mathbf{b}}{d\zeta}}{\dfrac{d\mathbf{a}}{d\zeta} \cdot \mathbf{b}} = \frac{d\mathbf{b}}{d\zeta}. \tag{I.4.20}
$$

The following vector is defined as

$$
\boldsymbol{\omega} = \frac{\dfrac{d\mathbf{a}}{d\zeta} \times \dfrac{d\mathbf{b}}{d\zeta}}{\dfrac{d\mathbf{a}}{d\zeta} \cdot \mathbf{b}},
$$

and the Eqs. (I.4.19) and (I.4.20) can be written as

$$
\frac{d\mathbf{a}}{d\zeta} = \boldsymbol{\omega} \times \mathbf{a}, \quad \frac{d\mathbf{b}}{d\zeta} = \boldsymbol{\omega} \times \mathbf{b}.
$$

In general a given vector $\mathbf{d}$ can be expressed as

$$
\mathbf{d} = d_1\mathbf{n}_1 + d_2\mathbf{n}_2 + d_3\mathbf{n}_3,
$$

where $\mathbf{n}_1$, $\mathbf{n}_2$, $\mathbf{n}_3$ are three unit vectors not parallel to the same plane, and d_1, d_2, d_3 are three scalars.

Any vector $\mathbf{c}$ fixed in the rigid body RB can be expressed as

$$
\mathbf{c} = c_1\mathbf{a} + c_2\mathbf{b} + c_3\mathbf{a} \times \mathbf{b}, \tag{I.4.21}
$$

where c_1, c_2, and c_3 are constant and independent of ζ. Differentiating Eq. (I.4.21) with respect to ζ, the following expression is obtained:

$$
\begin{aligned}
\frac{d\,\mathbf{c}}{d\zeta} &= c_1\frac{d\,\mathbf{a}}{d\zeta} + c_2\frac{d\,\mathbf{b}}{d\zeta} + c_3\frac{d\,\mathbf{a}}{d\zeta} \times \mathbf{b} + c_3\mathbf{a} \times \frac{d\,\mathbf{b}}{d\zeta} \\
&= c_1\boldsymbol{\omega} \times \mathbf{a} + c_2\boldsymbol{\omega} \times \mathbf{b} + c_3\left[(\boldsymbol{\omega} \times \mathbf{a}) \times \mathbf{b} + \mathbf{a} \times (\boldsymbol{\omega} \times \mathbf{b})\right] \\
&= c_1\boldsymbol{\omega} \times \mathbf{a} + c_2\boldsymbol{\omega} \times \mathbf{b} + c_3\left[\mathbf{b} \cdot \boldsymbol{\omega}\,\mathbf{a} - \mathbf{b} \cdot \mathbf{a}\,\boldsymbol{\omega} + \mathbf{a} \cdot \mathbf{b}\,\boldsymbol{\omega} - \mathbf{a} \cdot \boldsymbol{\omega}\,\mathbf{b}\right] \\
&= c_1\boldsymbol{\omega} \times \mathbf{a} + c_2\boldsymbol{\omega} \times \mathbf{b} + c_3\left[\boldsymbol{\omega} \cdot \mathbf{b}\,\mathbf{a} - \mathbf{a} \cdot \mathbf{b}\,\boldsymbol{\omega} + \mathbf{a} \cdot \mathbf{b}\,\boldsymbol{\omega} - \mathbf{a} \cdot \boldsymbol{\omega}\,\mathbf{b}\right] \\
&= c_1\boldsymbol{\omega} \times \mathbf{a} + c_2\boldsymbol{\omega} \times \mathbf{b} + c_3\left[\boldsymbol{\omega} \cdot \mathbf{b}\,\mathbf{a} - \boldsymbol{\omega} \cdot \mathbf{a}\,\mathbf{b}\right] \\
&= c_1\boldsymbol{\omega} \times \mathbf{a} + c_2\boldsymbol{\omega} \times \mathbf{b} + c_3\boldsymbol{\omega} \times (\mathbf{a} \times \mathbf{b}) \\
&= \boldsymbol{\omega} \times (c_1\mathbf{a} + c_2\mathbf{b} + c_3\mathbf{a} \times \mathbf{b}) \\
&= \boldsymbol{\omega} \times \mathbf{c}.
\end{aligned}
\tag{I.4.22}
$$

The vector $\boldsymbol{\omega}$ is not associated with any particular point. With the help of $\boldsymbol{\omega}$ the process of differentiation is replaced with that of cross multiplication.

The vector $\boldsymbol{\omega}$ can be expressed in a symmetrical relation in $\mathbf{a}$ and $\mathbf{b}$:

$$
\boldsymbol{\omega} = \frac{1}{2}\left(\frac{\dfrac{d\,\mathbf{a}}{d\zeta} \times \dfrac{d\,\mathbf{b}}{d\zeta}}{\dfrac{d\,\mathbf{a}}{d\zeta} \cdot \mathbf{b}} + \frac{\dfrac{d\,\mathbf{b}}{d\zeta} \times \dfrac{d\,\mathbf{a}}{d\zeta}}{\dfrac{d\,\mathbf{b}}{d\zeta} \cdot \mathbf{a}}\right).
\tag{I.4.23}
$$

The first derivatives of a vector $\mathbf{p}$, with respect to a scalar variable ζ in two reference frames RF_i and RF_j, are related as follows:

$$
\frac{^{(j)}d\,\mathbf{p}}{d\zeta} = \frac{^{(i)}d\,\mathbf{p}}{d\zeta} + \boldsymbol{\omega}_{ij} \times \mathbf{p},
\tag{I.4.24}
$$

where $\boldsymbol{\omega}_{ij}$ is the rate of change of orientation of RF_i in RF_j with respect to ζ and $\dfrac{^{(j)}d\,\mathbf{p}}{d\zeta}$ is the total derivative of $\mathbf{p}$ with respect to ζ in RF_j.

Proof

The vector $\mathbf{p}$ can be expressed as

$$
\mathbf{p} = p_1\mathbf{l}_1 + p_2\mathbf{l}_2 + p_3\mathbf{l}_3,
$$

where $\mathbf{l}_1, \mathbf{l}_2, \mathbf{l}_3$ are three unit vectors not parallel to the same plane fixed in RF_i, and p_x, p_y, p_z are the scalar measure numbers of $\mathbf{p}$. Differentiating in RF_j, the following expression is obtained:

$$\frac{^{(j)}d\,\mathbf{p}}{d\zeta} = \frac{^{(j)}d}{d\zeta}\,(p_1\mathbf{l}_1 + p_2\mathbf{l}_2 + p_3\mathbf{l}_3)$$

$$= \frac{^{(j)}d\,p_2}{d\zeta}\mathbf{l}_1 + \frac{^{(j)}d\,p_2}{d\zeta}\mathbf{l}_2 + \frac{^{(j)}d\,p_3}{d\zeta}\mathbf{l}_3 + p_1\frac{^{(j)}d\,\mathbf{l}_1}{d\zeta} + p_2\frac{^{(j)}d\,\mathbf{l}_2}{d\zeta} + p_3\frac{^{(j)}d\,\mathbf{l}_3}{d\zeta}$$

$$= \frac{d\,p_2}{d\zeta}\mathbf{l}_1 + \frac{d\,p_2}{d\zeta}\mathbf{l}_2 + \frac{d\,p_3}{d\zeta}\mathbf{l}_3 + p_1\boldsymbol{\omega}_{ij}\times\mathbf{l}_1 + p_2\boldsymbol{\omega}_{ij}\times\mathbf{l}_2 + p_3\boldsymbol{\omega}_{ij}\times\mathbf{l}_3$$

$$= \frac{^{(i)}d\,p_2}{d\zeta}\mathbf{l}_1 + \frac{^{(i)}d\,p_2}{d\zeta}\mathbf{l}_2 + \frac{^{(i)}d\,p_3}{d\zeta}\mathbf{l}_3 + \boldsymbol{\omega}_{ij}\times(p_1\mathbf{l}_1 + p_2\mathbf{l}_2 + p_3\mathbf{l}_3)$$

$$= \frac{^{(i)}d\,\mathbf{p}}{d\zeta} + \boldsymbol{\omega}_{ij}\times\mathbf{p}.$$

The *angular velocity* of a rigid body RB in a reference frame RF_0 is the rate of change of orientation with respect to the time t

$$\boldsymbol{\omega} = \frac{1}{2}\left(\frac{\dfrac{d\,\mathbf{a}}{dt}\times\dfrac{d\,\mathbf{b}}{dt}}{\dfrac{d\,\mathbf{a}}{dt}\cdot\mathbf{b}} + \frac{\dfrac{d\,\mathbf{b}}{dt}\times\dfrac{d\,\mathbf{a}}{dt}}{\dfrac{d\,\mathbf{b}}{dt}\cdot\mathbf{a}}\right) = \frac{1}{2}\left(\frac{\dot{\mathbf{a}}\times\dot{\mathbf{b}}}{\dot{\mathbf{a}}\cdot\mathbf{b}} + \frac{\dot{\mathbf{b}}\times\dot{\mathbf{a}}}{\dot{\mathbf{b}}\cdot\mathbf{a}}\right). \qquad (I.4.25)$$

The direction of $\boldsymbol{\omega}$ is related to the direction of the rotation of the rigid body through a right-hand rule.

Let RF_i, $i = 1, 2, \ldots, n$ be n reference frames. The angular velocity of a rigid body r in the reference frame RF_n, can be expressed as

$$\boldsymbol{\omega}_{rn} = \boldsymbol{\omega}_{r1} + \boldsymbol{\omega}_{12} + \boldsymbol{\omega}_{23} + \cdots + \boldsymbol{\omega}_{r,n-1}. \qquad (I.4.26)$$

Proof

Let $\mathbf{p}$ be any vector fixed in the rigid body. Then

$$\frac{^{(i)}d\,\mathbf{p}}{dt} = \boldsymbol{\omega}_{ri}\times\mathbf{p}$$

$$\frac{^{(i-1)}d\,\mathbf{p}}{dt} = \boldsymbol{\omega}_{r,i-1}\times\mathbf{p}.$$

On the other hand:

$$\frac{^{(i)}d\,\mathbf{p}}{dt} = \frac{^{(i-1)}d\,\mathbf{p}}{dt} + \boldsymbol{\omega}_{i,i-1}\times\mathbf{p}.$$

Hence,

$$\boldsymbol{\omega}_{ri}\times\mathbf{p} = \boldsymbol{\omega}_{r,i-1}\times\mathbf{p} + \boldsymbol{\omega}_{i,i-1}\times\mathbf{p},$$

as this equation is satisfied for all **p** fixed in the rigid body:

$$\omega_{ri} = \omega_{r,i-1} + \omega_{i,i-1}. \tag{I.4.27}$$

With $i = n$, Eq. (I.4.27) gives

$$\omega_{rn} = \omega_{r,n-1} + \omega_{n,n-1}. \tag{I.4.28}$$

With $i = n - 1$, Eq. (I.4.27) gives

$$\omega_{r,n-1} = \omega_{r,n-2} + \omega_{n-1,n-2}. \tag{I.4.29}$$

Substitute Eq. (I.4.29) into Eq. (I.4.28) and

$$\omega_{rn} = \omega_{r,n-2} + \omega_{n-1,n-2} + \omega_{n,n-1}.$$

Next use Eq. (I.4.27) with $i = n - 2$, then with $i = n - 3$, and so forth.

Motion of a Point that Moves Relative to a Rigid Body

A reference frame that moves with the rigid body is a body fixed reference frame. Figure I.4.2 shows a rigid body (RB) in motion relative to a primary reference frame with its origin at point O_0, $O_0 x_0 y_0 z_0$. The primary reference frame is a fixed reference frame or an earth fixed reference frame. The unit vectors $\mathbf{1}_0, \mathbf{J}_0$, and $\mathbf{k}_0$ of the primary reference frame are constant.

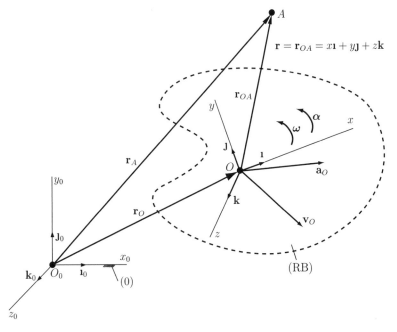

FIGURE I.4.2 *Motion of a point A that moves relative to a rigid body (RB).*

The body fixed reference frame (mobile reference frame), $Oxyz$, has its origin at a point O of the rigid body $[O \in (RB)]$, and is a moving reference frame relative to the primary reference. The unit vectors $\imath, \jmath$, and $\mathbf{k}$ of the body fixed reference frame are not constant, because they rotate with the body fixed reference frame.

The position vector of a point P of the rigid body $[P \in (RB)]$ relative to the origin O of the body fixed reference frame is the vector $\mathbf{r}_{OP}$. The velocity of P relative to O is

$$\frac{d\mathbf{r}_{OP}}{dt} = \boldsymbol{\omega} \times \mathbf{r}_{OP},$$

where $\boldsymbol{\omega}$ is the angular velocity vector of the rigid body.

The position vector of a point A (the point A is not assumed to be a point of the rigid body, as shown in Fig. I.4.2) relative to the origin O_0 of the primary reference frame is

$$\mathbf{r}_A = \mathbf{r}_O + \mathbf{r},$$

where

$$\mathbf{r} = \mathbf{r}_{OA} = x\imath + y\jmath + z\mathbf{k},$$

is the position vector of A relative to the origin O of the body fixed reference frame, and x, y, and z are the coordinates of A in terms of the body fixed reference frame. The velocity of the point A is the time derivative of the position vector $\mathbf{r}_A$:

$$\mathbf{v}_A = \frac{d\mathbf{r}_O}{dt} + \frac{d\mathbf{r}}{dt} = \mathbf{v}_O + \mathbf{v}_{AO}$$

$$= \mathbf{v}_O + \frac{dx}{dt}\imath + x\frac{d\imath}{dt} + \frac{dy}{dt}\jmath + y\frac{d\jmath}{dt} + \frac{dz}{dt}\mathbf{k} + z\frac{d\mathbf{k}}{dt}.$$

Using Poisson formulas, the total derivative of the position vector $\mathbf{r}$ is

$$\frac{d\mathbf{r}}{dt} = \dot{\mathbf{r}} = \dot{x}\imath + \dot{y}\jmath + \dot{z}\mathbf{k} + \boldsymbol{\omega} \times \mathbf{r}.$$

The velocity of A relative to the body fixed reference frame is a derivative in the body fixed reference frame:

$$\mathbf{v}_{Arel} = \frac{^{(RB)}d\mathbf{r}}{dt} = \frac{dx}{dt}\imath + \frac{dy}{dt}\jmath + \frac{dz}{dt}\mathbf{k} = \dot{x}\imath + \dot{y}\jmath + \dot{z}\mathbf{k}, \qquad (I.4.30)$$

A general formula for the total derivative of a moving vector $\mathbf{r}$ can be written as

$$\frac{d\mathbf{r}}{dt} = \frac{^{(RB)}d\mathbf{r}}{dt} + \boldsymbol{\omega} \times \mathbf{r}, \qquad (I.4.31)$$

where $\dfrac{d\mathbf{r}}{dt} = \dfrac{^{(0)}d\mathbf{r}}{dt}$ is the derivative in the fixed reference frame (0) $(O_0 x_0 y_0 z_0)$, and $\dfrac{^{(RB)}d\mathbf{r}}{dt}$ is the derivative in the mobile reference frame (body fixed reference frame).

The velocity of the point A relative to the primary reference frame is

$$\mathbf{v}_A = \mathbf{v}_O + \mathbf{v}_{Arel} + \boldsymbol{\omega} \times \mathbf{r},$$

(I.4.32)

Equation (I.4.32) expresses the velocity of a point A as the sum of three terms:

- the velocity of a point O of the rigid body,
- the velocity $\mathbf{v}_{Arel}$ of A relative to the rigid body, and
- the velocity $\boldsymbol{\omega} \times \mathbf{r}$ of A relative to O due to the rotation of the rigid body.

The acceleration of the point A relative to the primary reference frame is obtained by taking the time derivative of Eq. (I.4.32)

$$\begin{aligned} \mathbf{a}_A &= \mathbf{a}_O + \mathbf{a}_{AO} \\ &= \mathbf{a}_O + \mathbf{a}_{Arel} + 2\boldsymbol{\omega} \times \mathbf{v}_{Arel} + \boldsymbol{\alpha} \times \mathbf{r} + \boldsymbol{\omega} \times (\boldsymbol{\omega} \times \mathbf{r}), \end{aligned}$$

(I.4.33)

where

$$\mathbf{a}_{Arel} = \frac{^{(RB)}d^2\,\mathbf{r}}{dt^2} = \frac{d^2x}{dt^2}\mathbf{i} + \frac{d^2y}{dt^2}\mathbf{j} + \frac{d^2z}{dt^2}\mathbf{k},$$

(I.4.34)

is the acceleration of A relative to the body fixed reference frame or relative to the rigid body. The term

$$\mathbf{a}_{Cor} = 2\boldsymbol{\omega} \times \mathbf{v}_{Arel}.$$

is called the *Coriolis acceleration*.

In the case of planar motion, Eq. (I.4.33) becomes

$$\mathbf{a}_A = \mathbf{a}_O + \mathbf{a}_{Arel} + 2\boldsymbol{\omega} \times \mathbf{v}_{Arel} + \boldsymbol{\alpha} \times \mathbf{r} - \omega^2 \mathbf{r}.$$

(I.4.35)

The velocity $\mathbf{v}_A$ and the acceleration $\mathbf{a}_A$ of a point A are relative to the primary reference frame. The terms $\mathbf{v}_{Arel}$ and $\mathbf{a}_{Arel}$ are the velocity and acceleration of point A relative to the body fixed reference frame, i.e., they are the velocity and acceleration measured by an observer moving with the rigid body (Fig. I.4.3).

If A is a point of the rigid body, $A \in$ RB, $\mathbf{v}_{Arel} = \mathbf{0}$, and $\mathbf{a}_{Arel} = \mathbf{0}$.

Inertial Reference Frames

A reference frame is inertial if Newton's second law is applied in the form $\sum \mathbf{F} = m\mathbf{a}$. Figure I.4.4 shows a nonaccelerating, nonrotating reference frame with the origin at O_0, and a secondary nonrotating, earth-centered reference frame with the origin at O. The nonaccelerating, nonrotating reference frame with the origin at O_0 is assumed to be an inertial reference. The acceleration of the earth, due to the gravitational attractions of the sun, moon, etc., is $\mathbf{g}_O$. The earth-centered reference frame has the acceleration $\mathbf{g}_O$, too.

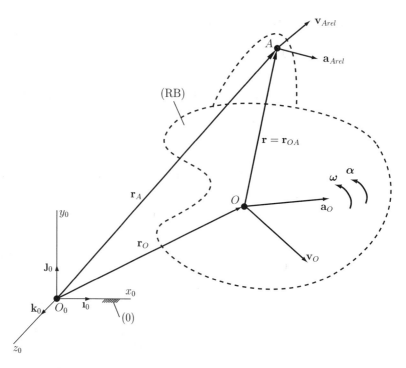

FIGURE I.4.3 *Velocity ($\mathbf{v}_{Arel}$) and acceleration ($\mathbf{a}_{Arel}$) of A relative to the rigid body.*

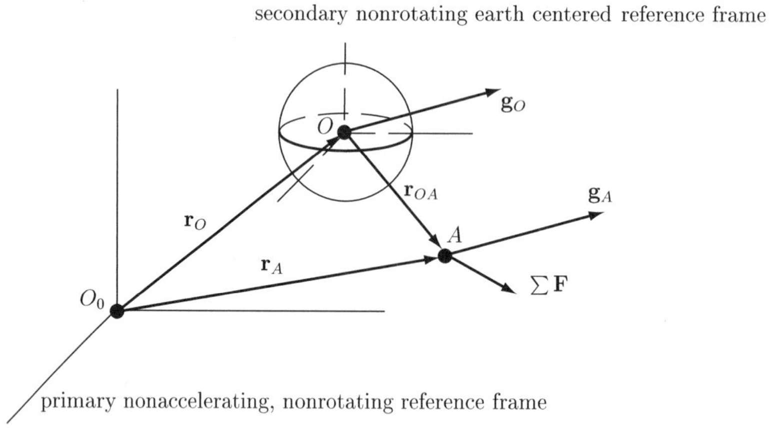

secondary nonrotating earth centered reference frame

primary nonaccelerating, nonrotating reference frame

FIGURE I.4.4 *Nonaccelerating, nonrotating reference frame with the origin at O_0, and a secondary nonrotating, earth-centered reference frame with the origin at O.*

Newton's second law for an object A of mass m, using the hypothetical nonaccelerating, nonrotating reference frame with the origin at O_0, can be written as

$$m\mathbf{a}_A = m\mathbf{g}_A + \sum \mathbf{F}, \qquad (\text{I}.4.36)$$

where $\mathbf{a}_A$ is the acceleration of A relative to O_0, $\mathbf{g}_A$ is the resulting gravitational acceleration, and $\sum \mathbf{F}$ is the sum of all other external forces acting on A. The acceleration of A relative to O_0 is

$$\mathbf{a}_A = \mathbf{a}_O + \mathbf{a}_{Arel},$$

where $\mathbf{a}_{Arel}$ is the acceleration of A relative to the earth-centered reference frame. The acceleration of the origin O is equal to the gravitational acceleration of the earth $\mathbf{a}_O = \mathbf{g}_O$. The earth-centered reference frame does not rotate ($\boldsymbol{\omega} = \mathbf{0}$). If the object A is on or near the earth, its gravitational acceleration $\mathbf{g}_A$ due to the attraction of the sun, etc., is nearly equal to the gravitational acceleration of the earth $\mathbf{g}_O$, and Eq. (I.4.36) becomes

$$\sum \mathbf{F} = m\mathbf{a}_{Arel}. \tag{I.4.37}$$

Newton's second law can be applied using a nonrotating, earth-centered reference frame if the object is near the earth. In most applications, Newton's second law can be applied using an earth-fixed reference frame. Figure I.4.5 shows a nonrotating reference frame with

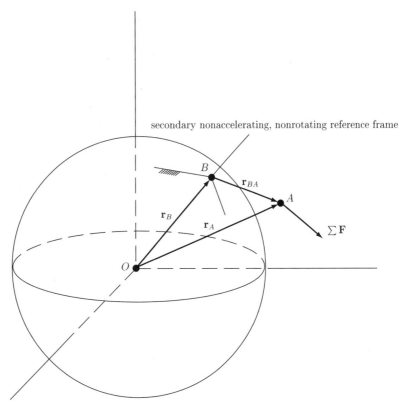

FIGURE I.4.5 *Nonrotating reference frame with the origin at the center of the earth O and a secondary earth-fixed reference frame with the origin at B.*

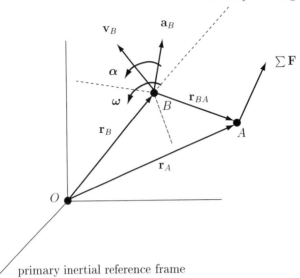

secondary rotating reference frame

$\mathbf{v}_B$ $\mathbf{a}_B$

$\sum \mathbf{F}$

α

ω B $\mathbf{r}_{BA}$

$\mathbf{r}_B$ A

$\mathbf{r}_A$

O

primary inertial reference frame

FIGURE I.4.6 *Primary inertial reference frame with the origin at O and a secondary reference frame with the origin at B.*

its origin at the center of the earth O and a secondary earth-fixed reference frame with its origin at a point B. The earth-fixed reference frame with the origin at B can be assumed to be an inertial reference and $\sum \mathbf{F} = m\mathbf{a}_{Arel}$, where $\mathbf{a}_{Arel}$ is the acceleration of A relative to the earth-fixed reference frame.

The motion of an object A can be analyzed using a primary inertial reference frame with its origin at the point O (Fig. I.4.6). A secondary reference frame with its origin at B undergoes an arbitrary motion with angular velocity ω and angular acceleration α. Newton's second law for the object A of mass m is $\sum \mathbf{F} = m\mathbf{a}_A$, where $\mathbf{a}_A$ is the acceleration of A relative to O. Newton's second law can be written in the form:

$$\sum \mathbf{F} - m[\mathbf{a}_B + 2\omega \times \mathbf{v}_{Arel} + \alpha \times \mathbf{r}_{BA} + \omega \times (\omega \times \mathbf{r}_{BA})] = m\mathbf{a}_{Arel}, \qquad (I.4.38)$$

where $\mathbf{a}_{Arel}$ is the acceleration of A relative to the secondary reference frame. The term $\mathbf{a}_B$ is the acceleration of the origin B of the secondary reference frame relative to the primary inertial reference. The term $2\omega \times \mathbf{v}_{Arel}$ is the Coriolis acceleration, and the term $-2m\omega \times \mathbf{v}_{Arel}$ is called the Coriolis force. This is Newton's second law expressed in terms of a secondary reference frame undergoing an arbitrary motion relative to an inertial primary reference frame.

The classical method for obtaining the velocities and/or accelerations of links and joints is to compute the derivatives of the positions and/or velocities with respect to time.

I.4.2 Driver Link

For a driver link in rotational motion [see Fig. I.3.3(a)], the following position relation can be written:

$$x_B(t) = x_A + L_{AB} \cos \phi(t),$$
$$y_B(t) = y_A + L_{AB} \sin \phi(t). \tag{I.4.39}$$

Differentiating Eq. (I.4.39) with respect to time, the following expressions are obtained:

$$v_{Bx} = \dot{x}_B = \frac{d\, x_B(t)}{d\, t} = -L_{AB}\dot{\phi} \sin \phi,$$
$$v_{By} = \dot{y}_B = \frac{d\, y_B(t)}{d\, t} = L_{AB}\dot{\phi} \cos \phi. \tag{I.4.40}$$

The angular velocity of the driver link is $\omega = \dot{\phi}$. The time derivative of Eq. (I.4.40) yields

$$a_{Bx} = \ddot{x}_B = \frac{d\, v_B(t)}{d\, t} = -L_{AB}\dot{\phi}^2 \cos \phi - L_{AB}\ddot{\phi} \sin \phi,$$
$$a_{By} = \ddot{y}_B = \frac{d\, v_B(t)}{d\, t} = -L_{AB}\dot{\phi}^2 \sin \phi + L_{AB}\ddot{\phi} \cos \phi, \tag{I.4.41}$$

where $\alpha = \ddot{\phi}$ is the angular acceleration of the driver link AB.

I.4.3 RRR Dyad

For the RRR dyad (see Fig. I.3.4) there are two quadratic equations of the form

$$[x_C(t) - x_A]^2 + [y_C(t) - y_A]^2 = L_{AC}^2 = L_2^2,$$
$$[x_C(t) - x_B]^2 + [y_C(t) - y_B]^2 = L_{BC}^2 = L_3^2. \tag{I.4.42}$$

Solving the above system of quadratic equations, the coordinates $x_C(t)$ and $y_C(t)$ are obtained.

The derivative of Eq. (I.4.42) with respect to time yields

$$(x_C - x_A)\,(\dot{x}_C - \dot{x}_A) + (y_C - y_A)\,(\dot{y}_C - \dot{y}_A) = 0,$$
$$(x_C - x_B)\,(\dot{x}_C - \dot{x}_B) + (y_C - y_B)\,(\dot{y}_C - \dot{y}_B) = 0, \tag{I.4.43}$$

From (Eq. I.4.43) the velocity vector of the joint C, $\mathbf{v}_C = [\dot{x}_C, \dot{y}_C]^T$, is written in matrix form:

$$\mathbf{v}_C = \mathbf{M}_1 \cdot \mathbf{v}, \tag{I.4.44}$$

where

$$\mathbf{v} = [\dot{x}_A, \dot{y}_A, \dot{x}_B, \dot{y}_B]^T,$$

$$\mathbf{M}_1 = \mathbf{A}_1^{-1} \cdot \mathbf{A}_2,$$

$$\mathbf{A}_1 = \begin{bmatrix} x_C - x_A & y_C - y_A \\ x_C - x_B & y_C - y_B \end{bmatrix},$$

$$\mathbf{A}_2 = \begin{bmatrix} x_C - x_A & y_C - y_A & 0 & 0 \\ 0 & 0 & x_C - x_B & y_C - y_B \end{bmatrix}.$$

Similarly, by differentiating Eq. (I.4.43), the following acceleration equations are obtained:

$$(\dot{x}_C - \dot{x}_A)^2 + (x_C - x_A)(\ddot{x}_C - \ddot{x}_A)$$
$$+ (\dot{y}_C - \dot{y}_A)^2 + (y_C - y_A)(\ddot{y}_C - \ddot{y}_A) = 0,$$
$$(\dot{x}_C - \dot{x}_B)^2 + (x_C - x_B)(\ddot{x}_C - \ddot{x}_B)$$
$$+ (\dot{y}_C - \dot{y}_B)^2 + (y_C - y_B)(\ddot{y}_C - \ddot{y}_B) = 0. \qquad (\text{I.4.45})$$

The acceleration vector of the joint C is obtained from the above system of equations:

$$\mathbf{a}_C = [\ddot{x}_C, \ddot{y}_C]^T = \mathbf{M}_1 \cdot \mathbf{a} + \mathbf{M}_2, \qquad (\text{I.4.46})$$

where

$$\mathbf{a} = [\ddot{x}_A, \ddot{y}_A, \ddot{x}_B, \ddot{y}_B]^T,$$

$$\mathbf{M}_2 = -\mathbf{A}_1^{-1} \cdot \mathbf{A}_3,$$

$$\mathbf{A}_3 = \begin{bmatrix} (\dot{x}_C - \dot{x}_A)^2 + (\dot{y}_C - \dot{y}_A)^2 \\ (\dot{x}_C - \dot{x}_B)^2 + (\dot{y}_C - \dot{y}_B)^2 \end{bmatrix}.$$

To compute the angular velocity and acceleration of the RRR dyad, the following equations are written for the angles $\phi_2(t)$ and $\phi_3(t)$:

$$y_C(t) - y_A + [x_C(t) - x_A] \tan \phi_2(t) = 0,$$
$$y_C(t) - y_B + [x_C(t) - x_B] \tan \phi_3(t) = 0. \qquad (\text{I.4.47})$$

The derivative with respect to time of Eq. (I.4.47) yields

$$\dot{y}_C - \dot{y}_A - (\dot{x}_C - \dot{x}_A) \tan \phi_2 - (x_C - x_A)\frac{1}{\cos^2 \phi_2} \dot{\phi}_2 = 0,$$

$$\dot{y}_C - \dot{y}_B - (\dot{x}_C - \dot{x}_B) \tan \phi_3 - (x_C - x_B)\frac{1}{\cos^2 \phi_3} \dot{\phi}_3 = 0. \qquad (\text{I.4.48})$$

The angular velocity vector is computed as

$$\boldsymbol{\omega} = [\dot{\phi}_2, \dot{\phi}_3]^T = [\omega_2, \omega_3]^T = \boldsymbol{\Omega}_1 \cdot \mathbf{v} + \boldsymbol{\Omega}_2 \cdot \mathbf{v}_C, \tag{I.4.49}$$

where

$$\boldsymbol{\Omega}_1 = \begin{bmatrix} \dfrac{x_C - x_A}{L_2^2} & -\dfrac{x_C - x_A}{L_2^2} & 0 & 0 \\[2ex] 0 & 0 & \dfrac{x_C - x_B}{L_3^2} & -\dfrac{x_C - x_B}{L_3^2} \end{bmatrix},$$

$$\boldsymbol{\Omega}_2 = \begin{bmatrix} -\dfrac{x_C - x_A}{L_2^2} & \dfrac{x_C - x_A}{L_2^2} \\[2ex] -\dfrac{x_C - x_B}{L_3^2} & \dfrac{x_C - x_B}{L_3^2} \end{bmatrix}.$$

Differentiating Eq. 4.49, the angular acceleration vector $\boldsymbol{\alpha} = \dot{\boldsymbol{\omega}}$ is

$$\boldsymbol{\alpha} = [\ddot{\phi}_2, \ddot{\phi}_3]^T = [\alpha_2, \alpha_3]^T = \dot{\boldsymbol{\Omega}}_1 \cdot \mathbf{v} + \dot{\boldsymbol{\Omega}}_2 \cdot \mathbf{v}_C + \boldsymbol{\Omega}_1 \cdot \mathbf{a} + \boldsymbol{\Omega}_2 \cdot \mathbf{a}_C. \tag{I.4.50}$$

I.4.4 RRT Dyad

For the RRT dyad [see Fig. I.3.5(a)], the following equations can be written for position analysis:

$$[x_C(t) - x_A]^2 + [y_C(t) - y_A]^2 = AC^2 = L_{AC}^2 = L_2^2,$$
$$[x_C(t) - x_B] \sin \alpha - [y_C(t) - y_B] \cos \alpha = \pm h. \tag{I.4.51}$$

From the above system of equations, $x_C(t)$ and $y_C(t)$ can be computed. The time derivative of Eq. (I.4.51) yields

$$(x_C - x_A)(\dot{x}_C - \dot{x}_A) + (y_C - y_A)(\dot{y}_C - \dot{y}_A) = 0,$$
$$(\dot{x}_C - \dot{x}_B) \sin \alpha - (\dot{y}_C - \dot{y}_B) \cos \alpha = 0. \tag{I.4.52}$$

The solution for the velocity vector of the joint C from Eq. (I.4.52) is

$$\mathbf{v}_C = [\dot{x}_C, \dot{y}_C]^T = \mathbf{M}_3 \cdot \mathbf{v}, \tag{I.4.53}$$

where

$$\mathbf{M}_3 = \mathbf{A}_4^{-1} \cdot \mathbf{A}_5,$$

$$\mathbf{A}_4 = \begin{bmatrix} x_C - x_A & y_C - y_A \\ \sin \alpha & -\cos \alpha \end{bmatrix},$$

$$\mathbf{A}_5 = \begin{bmatrix} x_C - x_A & y_C - y_A & 0 & 0 \\ 0 & 0 & \sin \alpha & -\cos \alpha \end{bmatrix}.$$

Differentiating Eq. (I.4.52) with respect to time

$$(\dot{x}_C - \dot{x}_A)^2 + (x_C - x_A)(\ddot{x}_C - \ddot{x}_A)$$
$$+ (\dot{y}_C - \dot{y}_A)^2 + (y_C - y_A)(\ddot{y}_C - \ddot{y}_A) = 0,$$
$$(\ddot{x}_C - \ddot{x}_B)\sin\alpha - (\ddot{y}_C - \ddot{y}_B)\cos\alpha = 0, \tag{I.4.54}$$

the acceleration vector $\mathbf{a}_C$ is obtained as

$$\mathbf{a}_C = [\ddot{x}_C, \ddot{y}_C]^T = \mathbf{M}_3 \cdot \mathbf{a} + \mathbf{M}_4, \tag{I.4.55}$$

where

$$\mathbf{M}_4 = -\mathbf{A}_4^{-1} \cdot \mathbf{A}_6,$$
$$\mathbf{A}_6 = \begin{bmatrix} (\dot{x}_C - \dot{x}_A)^2 + (\dot{y}_C - \dot{y}_A)^2 \\ 0 \end{bmatrix}. \tag{I.4.56}$$

The angular position of the element 2 is described by the following equation:

$$y_C(t) - y_A - [x_C(t) - x_A]\tan\phi_2(t) = 0. \tag{I.4.57}$$

The time derivative of Eq. (I.4.57) yields

$$\dot{y}_C - \dot{y}_A - (\dot{x}_C - \dot{x}_A)\tan\phi_2 - (x_C - x_A)\frac{1}{\cos^2\phi_2}\dot{\phi}_2 = 0, \tag{I.4.58}$$

and the angular velocity of the element 2 is

$$\omega_2 = \frac{x_C - x_A}{L_2^2}[(\dot{y}_C - \dot{y}_A) - (\dot{x}_C - \dot{x}_A)\tan\phi_2]. \tag{I.4.59}$$

The angular acceleration of the element 2 is $\alpha_2 = \dot{\omega}_2$.

I.4.5 RTR Dyad

For the RTR dyad [see Fig. I.3.6(a)] the position relations are

$$[x_C(t) - x_A]^2 + [y_C(t) - y_A]^2 = L_2^2,$$
$$\tan\alpha = \frac{\dfrac{y_C - y_B}{x_C - x_B} - \dfrac{y_C - y_A}{x_C - x_A}}{1 + \dfrac{y_C - y_B}{x_C - x_B} \cdot \dfrac{y_C - y_A}{x_C - x_A}}$$
$$= \frac{(y_C - y_B)(x_C - x_A) - (y_C - y_A)(x_C - x_B)}{(x_C - x_B)(x_C - x_A) + (y_C - y_B)(y_C - y_A)}. \tag{I.4.60}$$

The time derivative of Eq. (I.4.60) yields

$$(x_C - x_A)(\dot{x}_C - \dot{x}_A) + (y_C - y_A)(\dot{y}_C - \dot{y}_A) = 0,$$

$$\tan\alpha\,[(\dot{x}_C - \dot{x}_B)(x_C - x_A) + (x_C - x_B)(\dot{x}_C - \dot{x}_A)]$$

$$+ \tan\alpha\,[(\dot{y}_C - \dot{y}_A)(y_C - y_B) + (y_C - y_A)(\dot{y}_C - \dot{y}_B)]$$

$$+ (\dot{y}_C - \dot{y}_A)(x_C - x_B) + (y_C - y_A)(\dot{x}_C - \dot{x}_B) -$$

$$(\dot{y}_C - \dot{y}_B)(x_C - x_A) - (y_C - y_B)(\dot{x}_C - \dot{x}_A) = 0, \qquad (I.4.61)$$

or in a matrix form

$$\mathbf{A}_7 \cdot \mathbf{v}_C = \mathbf{A}_8 \cdot \mathbf{v}, \qquad (I.4.62)$$

where

$$\mathbf{A}_7 = \begin{bmatrix} x_C - x_A & y_C - y_A \\ \gamma_1 & \gamma_2 \end{bmatrix},$$

$$\mathbf{A}_8 = \begin{bmatrix} x_C - x_A & y_C - y_A & 0 & 0 \\ \gamma_3 & \gamma_4 & \gamma_5 & \gamma_6 \end{bmatrix}.$$

In addition,

$$\gamma_1 = [(x_C - x_B) + (x_C - x_A)]\tan\alpha - (y_C - y_B) + (y_C - y_A),$$

$$\gamma_2 = [(y_C - y_A) + (y_C - y_B)]\tan\alpha - (x_C - x_A) + (x_C - x_B),$$

$$\gamma_3 = (x_C - x_B)\tan\alpha + (y_C - y_B),$$

$$\gamma_4 = (x_C - x_A)\tan\alpha + (y_C - y_A),$$

$$\gamma_5 = (y_C - y_B)\tan\alpha + (x_C - x_B),$$

$$\gamma_6 = (y_C - y_A)\tan\alpha - (x_C - x_A).$$

The solution for the velocity vector, $\mathbf{v}_C$, of the joint C, from Eq. (I.4.62) is

$$\mathbf{v}_C = \mathbf{M}_5 \cdot \mathbf{v}, \qquad (I.4.63)$$

where

$$\mathbf{M}_5 = \mathbf{A}_7^{-1} \cdot \mathbf{A}_8.$$

Differentiating Eq. (I.4.62), the following relation is obtained:

$$\mathbf{A}_7 \cdot \mathbf{a}_C = \mathbf{A}_8 \cdot \mathbf{a} - \mathbf{A}_9, \qquad (I.4.64)$$

where

$$\mathbf{A}_9 = \left[\begin{array}{c} (\dot{x}_C - \dot{x}_A)^2 + (\dot{y}_C - \dot{y}_A)^2 \\ \gamma_7 \end{array} \right],$$

$$\gamma_7 = 2(\dot{x}_C - \dot{x}_B)(\dot{x}_C - \dot{x}_A) \tan \alpha + 2(\dot{y}_C - \dot{y}_B)(\dot{y}_C - \dot{y}_A) \tan \alpha$$
$$- 2(\dot{y}_C - \dot{y}_B)(\dot{x}_C - \dot{x}_A) + 2(\dot{y}_C - \dot{y}_A)(\dot{x}_C - \dot{x}_B).$$

The acceleration vector of the joint C is

$$\mathbf{a}_C = \mathbf{M}_5 \cdot \mathbf{a} - \mathbf{M}_6, \tag{I.4.65}$$

where

$$\mathbf{M}_6 = \mathbf{A}_7^{-1} \cdot \mathbf{A}_9.$$

To compute the angular velocities for the RTR dyad, the following equations can be written:

$$y_C(t) - y_A = [x_C(t) - x_A] \tan \phi_2$$
$$\phi_3 = \phi_2 + \alpha. \tag{I.4.66}$$

The time derivative of Eq. (I.4.66) yields

$$(\dot{y}_C - \dot{y}_A) = (\dot{x}_C - \dot{x}_A) \tan \phi_2 + (x_C - x_A) \frac{1}{\cos^2 \phi_2} \dot{\phi}_2$$
$$\dot{\phi}_3 = \dot{\phi}_2. \tag{I.4.67}$$

The angular velocities of the links 2 and 3 are

$$\omega_2 = \omega_3 = \frac{\cos^2 \phi_2}{x_C - x_A} [(\dot{y}_C - \dot{y}_A) - (\dot{x}_C - \dot{x}_A) \tan \phi_2]. \tag{I.4.68}$$

The angular accelerations are found to be

$$\alpha_2 = \alpha_3 = \dot{\omega}_2 = \dot{\omega}_3. \tag{I.4.69}$$

I.4.6 TRT Dyad

For the TRT dyad (see Fig. I.3.7) the two position equations are

$$[x_C(t) - x_A] \sin \alpha - [y_C(t) - y_A] \cos \alpha = \pm d,$$
$$[x_C(t) - x_B] \sin \beta - [y_C(t) - y_B] \cos \beta = \pm h. \tag{I.4.70}$$

The derivative, with respect to time of Eq. (I.4.70), yields

$$(\dot{x}_C - \dot{x}_A)\sin\alpha - (\dot{y}_C - \dot{y}_A)\cos\alpha$$
$$+ (x_C - x_A)\dot{\alpha}\cos\alpha + (y_C - y_A)\dot{\alpha}\sin\alpha = 0,$$
$$(\dot{x}_C - \dot{x}_B)\sin\beta - (\dot{y}_C - \dot{y}_B)\cos\beta$$
$$+ (x_C - x_B)\dot{\beta}\cos\beta + (y_C - y_B)\dot{\beta}\sin\beta = 0, \tag{I.4.71}$$

or in a matrix form

$$\mathbf{A}_{10} \cdot \mathbf{v}_C = \mathbf{A}_{11} \cdot \mathbf{v}_1, \tag{I.4.72}$$

where

$$\mathbf{v}_1 = [\dot{x}_A, \dot{y}_A, \dot{\alpha}, \dot{x}_B, \dot{y}_B, \dot{\beta}]^T,$$

$$\mathbf{A}_{10} = \begin{bmatrix} -\sin\alpha & -\cos\alpha \\ \sin\beta & -\cos\beta \end{bmatrix},$$

$$\mathbf{A}_{11} = \begin{bmatrix} \sin\alpha & -\cos\alpha & \xi_1 & 0 & 0 & 0 \\ 0 & 0 & 0 & \sin\beta & -\cos\beta & \xi_2 \end{bmatrix},$$

$$\xi_1 = (x_A - x_C)\cos\alpha + (y_A - y_C)\sin\alpha,$$
$$\xi_2 = (x_B - x_C)\cos\beta + (y_B - y_C)\sin\beta.$$

The solution of Eq. (I.4.72) gives the velocity of the joint C as

$$\mathbf{v}_C = \mathbf{M}_7 \cdot \mathbf{v}_1, \tag{I.4.73}$$

where

$$\mathbf{M}_7 = \mathbf{A}_{10}^{-1} \cdot \mathbf{A}_{11}.$$

Differentiating Eq. (I.4.72), with respect to time, gives

$$\mathbf{A}_{10} \cdot \mathbf{a}_C = \mathbf{A}_{11} \cdot \mathbf{a}_1 - \mathbf{A}_{12}, \tag{I.4.74}$$

where

$$\mathbf{a}_1 = [\ddot{x}_A, \ddot{y}_A, \ddot{\alpha}, \ddot{x}_B, \ddot{y}_B, \ddot{\beta}]^T,$$

$$\mathbf{A}_{12} = \begin{bmatrix} \xi_3 \\ \xi_4 \end{bmatrix},$$

$$\xi_3 = 2(\dot{x}_C - \dot{x}_A)\dot{\alpha}\cos\alpha + 2(\dot{y}_C - \dot{y}_A)\dot{\beta}\sin\alpha$$
$$- (x_C - x_A)\dot{\alpha}^2\sin\alpha + (y_C - y_A)\dot{\alpha}^2\cos\alpha,$$

$$\xi_4 = 2(\dot{x}_C - \dot{x}_B)\dot{\beta}\cos\beta + 2(\dot{y}_C - \dot{y}_B)\dot{\beta}\sin\beta$$
$$- (x_C - x_B)\dot{\beta}^2\sin\beta + (y_C - y_B)\dot{\beta}^2\cos\beta.$$

The solution of Eq. (I.4.74) gives the acceleration vector of joint C as

$$\mathbf{a}_C = \mathbf{M}_7 \cdot \mathbf{a} + \mathbf{M}_8. \qquad (I.4.75)$$

where

$$\mathbf{M}_8 = \mathbf{A}_{10}^{-1} \cdot \mathbf{A}_{12}.$$

I.4.7 Examples

EXAMPLE I.4.1: R-TRR mechanism

The following dimensions are given for the mechanism shown in Figure I.4.7: $AC = a = 0.100$ m and $BC = 0.300$ m. The angle of the driver link 1 with the horizontal axis is $\phi = \phi_1 = 45°$. The coordinates of joint B are $x_B = y_B = 0.256$ m. The driver link 1 rotates with a constant speed of $n_1 = 30$ rpm. Find the velocities and the accelerations of the mechanism.

Solution A Cartesian reference frame with the origin at A is selected. The coordinates of joint A are

$$x_A = y_A = 0,$$

the coordinates of joint C are

$$x_C = AC = 0.100\text{ m} \quad \text{and} \quad y_C = 0,$$

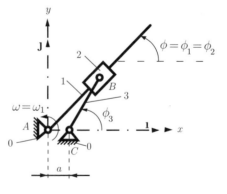

FIGURE I.4.7 *R-TRR mechanism for Example I.4.1.*

EXAMPLE I.4.1: R-TRR mechanism–*Cont'd*

and the coordinates of joint B are

$$x_B = 0.256 \text{ m} \quad \text{and} \quad y_B = 0.256 \text{ m}.$$

The position of joint B is calculated from the equations

$$\tan \phi(t) = \frac{y_B(t)}{x_B(t)} \quad \text{and} \quad [x_B(t) - x_C]^2 + [y_B(t) - y_C]^2 = BC^2,$$

or

$$x_B(t) \sin \phi(t) = y_B(t) \cos \phi(t),$$

$$[x_B(t) - x_C]^2 + [y_B(t) - y_C]^2 = BC^2. \tag{I.4.76}$$

The linear velocity of point B on link 3 or 2 is

$$\mathbf{v}_B = \mathbf{v}_{B_3} = \mathbf{v}_{B_2} = \dot{x}_B \mathbf{I} + \dot{y}_B \mathbf{J},$$

where

$$\dot{x}_B = \frac{dx_B}{dt} \quad \text{and} \quad \dot{y}_B = \frac{dy_B}{dt}.$$

The velocity analysis is carried out differentiating Eq. (I.4.76):

$$\dot{x}_B \sin \phi + x_B \dot{\phi} \cos \phi = \dot{y}_B \cos \phi - y_B \dot{\phi} \sin \phi,$$

$$\dot{x}_B(x_B - x_C) + \dot{y}_B(y_B - y_C) = 0,$$

or

$$\dot{x}_B \sin \phi + x_B \omega \cos \phi = \dot{y}_B \cos \phi - y_B \omega \sin \phi,$$

$$\dot{x}_B(x_B - x_C) + \dot{y}_B(y_B - y_C) = 0. \tag{I.4.77}$$

The magnitude of the angular velocity of the driver link 1 is

$$\omega = \omega_1 = \dot{\phi} = \frac{\pi n_1}{30} = \frac{\pi (30 \text{ rpm})}{30} = 3.141 \text{ rad/s}. \tag{I.4.78}$$

The angular velocity of link 1 is

$$\boldsymbol{\omega} = \boldsymbol{\omega}_1 = \omega \mathbf{k} = 3.141 \, \mathbf{k} \text{ rad/s}.$$

Continued

EXAMPLE I.4.1: R-TRR mechanism–*Cont'd*

The link 2 and the driver link 1 have the same angular velocity $\omega_1 = \omega_2$.
For the given numerical data Eq. (I.4.77) becomes

$$\dot{x}_B \sin 45° + 0.256\,(3.141)\cos 45° = \dot{y}_B \cos 45° - 0.256\,(3.141)\sin 45°,$$

$$\dot{x}_B(0.256 - 0.1) + \dot{y}_B(0.256 - 0) = 0. \qquad (I.4.79)$$

The solution of Eq. (I.4.79) gives

$$\dot{x}_B = -0.999 \text{ m/s} \quad \text{and} \quad \dot{y}_B = 0.609 \text{ m/s}.$$

The velocity of B is

$$\mathbf{v}_B = \mathbf{v}_{B_3} = \mathbf{v}_{B_2} = -0.999\,\mathbf{\imath} + 0.609\,\mathbf{\jmath} \text{ m/s},$$

$$|\mathbf{v}_B| = |\mathbf{v}_{B_3}| = |\mathbf{v}_{B_2}| = \sqrt{(-0.999)^2 + (0.609)^2} = 1.171 \text{ m/s}.$$

The acceleration analysis is obtained using the derivative of the velocities given by
Eq. (I.4.77):

$$\ddot{x}_B \sin\phi + \dot{x}_B\,\omega\cos\phi + \dot{x}_B\,\omega\cos\phi - x_B\,\omega^2\sin\phi$$

$$= \ddot{y}_B \cos\phi - \dot{y}_B\,\omega\sin\phi - \dot{y}_B\,\omega\sin\phi + y_B\,\omega^2\cos\phi,$$

$$\ddot{x}_B(x_B - x_C) + \dot{x}_B^2 + \ddot{y}_B(y_B - y_C) + \dot{y}_B^2 = 0. \qquad (I.4.80)$$

The magnitude of the angular acceleration of the driver link 1 is

$$\alpha = \dot{\omega} = \ddot{\phi} = 0.$$

Numerically, Eq. (I.4.80) gives

$$\ddot{x}_B \sin 45° + 2\,(-0.999)\,(3.141)\cos 45° - 0.256\,(3.141)^2\sin 45°$$

$$= \ddot{y}_B \cos 45° - 2\,(0.609)\,(3.141)\sin 45° + 0.256\,(3.141)^2\cos 45°,$$

$$\ddot{x}_B(0.256 - 0.1) + (-0.999)^2 + \ddot{y}_B(0.256) + 0.609^2 = 0. \qquad (I.4.81)$$

The solution of Eq. (I.4.81) is

$$\ddot{x}_B = -1.802 \text{ m/s}^2 \quad \text{and} \quad \ddot{y}_B = -4.255 \text{ m/s}^2.$$

The acceleration of B on link 3 or 2 is

$$\mathbf{a}_B = \mathbf{a}_{B_3} = \mathbf{a}_{B_2} = \ddot{x}_B\,\mathbf{\imath} + \ddot{y}_B\,\mathbf{\jmath} = -1.802\,\mathbf{\imath} - 4.255\,\mathbf{\jmath} \text{ m/s}^2,$$

$$|\mathbf{a}_B| = |\mathbf{a}_{B_3}| = |\mathbf{a}_{B_2}| = \sqrt{(-1.802)^2 + (-4.255)^2} = 4.620 \text{ m/s}^2.$$

EXAMPLE I.4.1: R-TRR mechanism–*Cont'd*

The slope of the link 3 (the points B and C are on the straight line BC) is

$$\tan\phi_3(t) = \frac{y_B(t) - y_C}{x_B(t) - x_C},$$

or

$$[x_B(t) - x_C]\sin\phi_3(t) = [y_B(t) - y_C]\cos\phi_3(t). \qquad (I.4.82)$$

The angle ϕ_3 is computed as follows:

$$\phi_3 = \arctan\frac{y_B - y_C}{x_B - x_C} = \arctan\frac{0.256}{0.256 - 0.1} = 1.023 \text{ rad} = 58.633°.$$

The derivative of Eq. (I.4.82) yields

$$\dot{x}_B \sin\phi_3 + (x_B - x_C)\dot{\phi}_3 \cos\phi_3 = \dot{y}_B \cos\phi_3 - (y_B - y_C)\dot{\phi}_3 \sin\phi_3,$$

or

$$\dot{x}_B \sin\phi_3 + (x_B - x_C)\omega_3 \cos\phi_3 = \dot{y}_B \cos\phi_3 - (y_B - y_C)\omega_3 \sin\phi_3, \qquad (I.4.83)$$

where $\omega_3 = \dot{\phi}_3$.

Numerically, Eq. (I.4.83) gives

$$-0.999 \sin 58.633° + (0.256 - 0.1)\omega_3 \cos 58.633°$$

$$= 0.609 \cos 58.633° - 0.256\,\omega_3 \sin 58.633°,$$

with the solution $\omega_3 = 3.903$ rad/s.

The angular velocity of link 3 is

$$\boldsymbol{\omega}_3 = \omega_3\,\mathbf{k} = 3.903\,\mathbf{k} \text{ rad/s}.$$

The angular acceleration of link 3, $\alpha_3 = \dot{\omega}_3 = \ddot{\phi}_3$, is obtained using the derivative of the Eq. (I.4.83):

$$\ddot{x}_B \sin\phi_3 + \dot{x}_B \omega_3 \cos\phi_3 + \dot{x}_B \omega_3 \cos\phi_3$$

$$+ (x_B - x_C)\dot{\omega}_3 \cos\phi_3 - (x_B - x_C)\omega_3^2 \sin\phi_3$$

$$= \ddot{y}_B \cos\phi_3 - \dot{y}_B \omega_3 \sin\phi_3 - \dot{y}_B \omega_3 \sin\phi_3$$

$$- (y_B - y_C)\dot{\omega}_3 \sin\phi_3 - (y_B - y_C)\omega_3^2 \cos\phi_3,$$

Continued

EXAMPLE I.4.1: R-TRR mechanism–*Cont'd*

or

$$\ddot{x}_B \sin\phi_3 + 2\,\dot{x}_B\,\omega_3\,\cos\phi_3 + (x_B - x_C)\,\alpha_3\,\cos\phi_3 - (x_B - x_C)\,\omega_3^2\,\sin\phi_3$$
$$= \ddot{y}_B \cos\phi_3 - 2\,\dot{y}_B\,\omega_3\,\sin\phi_3 - (y_B - y_C)\,\alpha_3\,\sin\phi_3 - (y_B - y_C)\,\omega_3^2\,\cos\phi_3.$$

Numerically, the previous equation becomes

$$- 1.802\,\sin 58.633° + 2\,(-0.999)\,(3.903)\,\cos 58.633°$$
$$+ (0.256 - 0.1)\,\alpha_3\,\cos 58.633° - (0.256 - 0.1)\,(3.903)^2\,\sin 58.633°$$
$$= -4.255\,\cos 58.633° - 2\,(0.609)\,(3.903)\,\sin 58.633°$$
$$- 0.256\,\alpha_3\,\sin 58.633° - 0.256\,(3.903)^2\,\cos 58.633°,$$

with the solution $\alpha_3 = -2.252$ rad/s^2. The angular acceleration of link 3 is

$$\alpha_3 = \alpha_3\,\mathbf{k} = -2.252\,\mathbf{k}\ \text{rad/s}^2.$$

The velocity of the point B_1 on link 1 is calculated with the expression of velocity field of two points (B_1 and A) on the same rigid body (link 1):

$$\mathbf{v}_{B_1} = \mathbf{v}_A + \omega_1 \times \mathbf{r}_{AB} = \omega_1 \times \mathbf{r}_{AB} = \begin{vmatrix} \mathbf{i} & \mathbf{j} & \mathbf{k} \\ 0 & 0 & \omega_1 \\ x_B & y_B & 0 \end{vmatrix}$$

$$= \begin{vmatrix} \mathbf{i} & \mathbf{j} & \mathbf{k} \\ 0 & 0 & 3.141 \\ 0.256 & 0.256 & 0 \end{vmatrix} = -0.804\,\mathbf{i} + 0.804\,\mathbf{j}\ \text{m/s}.$$

The velocity field of two points (B_1 and B_2) not situated on the same rigid body (B_1 is on link 1 and B_2 is on link 2) is calculated with

$$\mathbf{v}_{B_2} = \mathbf{v}_{B_1} + \mathbf{v}^r_{B_2 B_1} = \mathbf{v}_{B_1} + \mathbf{v}^r_{B_{21}},$$

where $\mathbf{v}^r_{B_{21}}$ is the relative velocity of the point B_2 on link 2 with respect to the point B_1 on link 1:

$$\mathbf{v}^r_{B_{21}} = \mathbf{v}_{B_2} - \mathbf{v}_{B_1} = -0.999\,\mathbf{i} + 0.609\,\mathbf{j} - (-0.804\,\mathbf{i} + 0.804\,\mathbf{j})$$
$$= -0.195\,\mathbf{i} - 0.195\,\mathbf{j}\ \text{m/s}.$$

The relation between the angular velocities of link 2 and link 3 is

$$\omega_2 = \omega_3 + \omega_{23},$$

EXAMPLE I.4.1: R-TRR mechanism–*Cont'd*

and the relative angular velocity of link 2 with respect to link 3 is

$$\omega_{23} = \omega_2 - \omega_3 = 3.141 \, \mathbf{k} - 3.903 \, \mathbf{k} = -0.762 \, \mathbf{k} \text{ rad/s}.$$

The acceleration of the point B_1 on link 1 is

$$\mathbf{a}_{B_1} = \mathbf{a}_A + \boldsymbol{\alpha}_1 \times \mathbf{r}_{AB} - \omega_1^2 \, \mathbf{r}_{AB} = -\omega_1^2 \, \mathbf{r}_{AB} = -\omega_1^2 (x_B \, \mathbf{i} + y_B \, \mathbf{j})$$
$$= -3.141^2 (0.256 \, \mathbf{i} + 0.256 \, \mathbf{j}) = -2.528 \, \mathbf{i} - 2.528 \, \mathbf{j} \text{ m/s}^2.$$

The acceleration of B_2 in terms of B_1 is

$$\mathbf{a}_{B_2} = \mathbf{a}_{B_1} + \mathbf{a}_{B_{21}}^r + 2 \, \boldsymbol{\omega}_1 \times \mathbf{v}_{B_{21}}^r,$$

where $\mathbf{a}_{B_{21}}^r$ is the relative acceleration of the point B_2 on link 2 with respect to the point B_1 on link 1 and $2 \, \boldsymbol{\omega}_1 \times \mathbf{v}_{B_{21}}^r$ is the Coriolis acceleration:

$$\mathbf{a}_{B_{21}}^c = 2 \, \boldsymbol{\omega}_1 \times \mathbf{v}_{B_{21}}^r = 2 \, \boldsymbol{\omega}_2 \times \mathbf{v}_{B_{21}}^r = \begin{vmatrix} \mathbf{i} & \mathbf{j} & \mathbf{k} \\ 0 & 0 & \omega_1 \\ v_{B_{21}x}^r & v_{B_{21}y}^r & 0 \end{vmatrix}$$

$$= \begin{vmatrix} \mathbf{i} & \mathbf{j} & \mathbf{k} \\ 0 & 0 & 3.141 \\ -0.195 & -0.195 & 0 \end{vmatrix} = 1.226 \, \mathbf{i} - 1.226 \, \mathbf{j} \text{ m/s}^2.$$

The relative acceleration of B_2 with respect to B_1 is

$$\mathbf{a}_{B_{21}}^r = \mathbf{a}_{B_2} - \mathbf{a}_{B_1} - \mathbf{a}_{B_{21}}^c$$
$$= -1.802 \, \mathbf{i} - 4.255 \, \mathbf{j} - (-2.528 \, \mathbf{i} - 2.528 \, \mathbf{j}) - (1.226 \, \mathbf{i} - 1.226 \, \mathbf{j})$$
$$= -0.5 \, \mathbf{i} - 0.5 \, \mathbf{j} \text{ m/s}^2.$$

The relative angular acceleration of link 2 with respect to link 3 is

$$\alpha_{23} = \alpha_2 - \alpha_3 = -\alpha_3 = 2.252 \, \mathbf{k} \text{ rad/s}^2,$$

where $\alpha_2 = \alpha_1 = \mathbf{0}$.

EXAMPLE I.4.2: R-RTR-RRT mechanism

The mechanism shown in Figure I.4.8 has the dimensions: $AB = 0.100$ m, $AC = 0.150$ m, $CD = 0.075$ m, and $DE = 0.200$ m. The angle of the driver link 1 with the horizontal axis is $\phi = \phi_1 = 45°$, and the angular speed of the driver link 1 is $\omega = \omega_1 = 4.712$ rad/s. Find the velocities and the accelerations of the mechanism.

Continued

EXAMPLE I.4.2: R-RTR-RRT mechanism–*Cont'd*

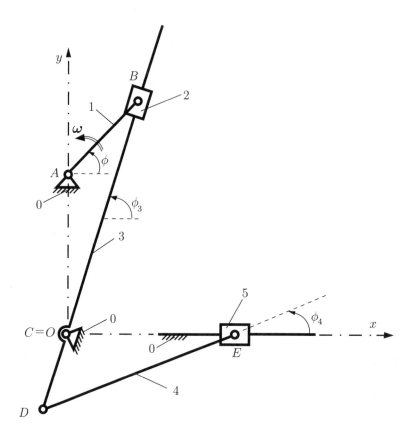

FIGURE I.4.8 *R-RTR-RRT mechanism for Example I.4.2.*

Solution The origin of the fixed reference frame is at $C \equiv 0$. The position of the fixed joint A is

$$x_A = 0, \quad y_A = AC = 0.150 \text{ m}.$$

The position of joint B is

$$x_B(t) = x_A + AB \cos \phi(t), \quad y_B(t) = y_A + AB \sin \phi(t),$$

and for $\phi = 45°$, the position is

$$x_B = 0 + 0.100 \cos 45° = 0.070 \text{ m}, \quad y_B = 0.150 + 0.100 \sin 45° = 0.220 \text{ m}.$$

The linear velocity vector of B is

$$\mathbf{v}_B = \dot{x}_B \mathbf{1} + \dot{y}_B \mathbf{J},$$

EXAMPLE I.4.2: R-RTR-RRT mechanism–*Cont'd*

where

$$\dot{x}_B = \frac{dx_B}{dt} = -AB\,\dot{\phi}\,\sin\phi, \quad \dot{y}_B = \frac{dy_B}{dt} = AB\,\dot{\phi}\,\cos\phi.$$

With $\phi = 45°$ and $\dot{\phi} = \omega = 4.712$ rad/s:

$$\dot{x}_B = -0.100\,(4.712)\,\sin 45° = -0.333 \text{ m/s},$$

$$\dot{y}_B = 0.100\,(4.712)\,\cos 45° = 0.333 \text{ m/s},$$

$$v_B = |\mathbf{v}_B| = \sqrt{\dot{x}_B^2 + \dot{y}_B^2} = \sqrt{(-0.333)^2 + 0.333^2} = 0.471 \text{ m/s}.$$

The linear acceleration vector of B is

$$\mathbf{a}_B = \ddot{x}_B \mathbf{\imath} + \ddot{y}_B \mathbf{J},$$

where

$$\ddot{x}_B = \frac{d\dot{x}_B}{dt} = -AB\dot{\phi}^2 \cos\phi - AB\ddot{\phi}\sin\phi,$$

$$\ddot{y}_B = \frac{d\dot{y}_B}{dt} = -AB\dot{\phi}^2 \sin\phi + AB\ddot{\phi}\cos\phi.$$

The angular acceleration of link 1 is $\ddot{\phi} = \dot{\omega} = 0$. The numerical values for the acceleration of B are

$$\ddot{x}_B = -0.100\,(4.712)^2 \cos 45° = -1.569 \text{ m/s}^2,$$

$$\ddot{y}_B = -0.100\,(4.712)^2 \sin 45° = -1.569 \text{ m/s}^2,$$

$$a_B = |\mathbf{a}_B| = \sqrt{\ddot{x}_B^2 + \ddot{y}_B^2} = \sqrt{(-1.569)^2 + (-1.569)^2} = 2.220 \text{ m/s}^2.$$

The velocity and acceleration of point B on link 1 (or on link 2) can also be calculated with the relations

$$\mathbf{v}_B = \mathbf{v}_{B_1} = \mathbf{v}_{B_2} = \mathbf{v}_A + \boldsymbol{\omega}_1 \times \mathbf{r}_{AB} = \begin{vmatrix} \mathbf{\imath} & \mathbf{J} & \mathbf{k} \\ 0 & 0 & \omega_1 \\ x_B - x_A & y_B - y_A & 0 \end{vmatrix}$$

$$= \begin{vmatrix} \mathbf{\imath} & \mathbf{J} & \mathbf{k} \\ 0 & 0 & 4.712 \\ 0.070 - 0.15 & 0.220 & 0 \end{vmatrix} = -0.333\,\mathbf{\imath} + 0.333\,\mathbf{J} \text{ m/s},$$

$$\mathbf{a}_B = \mathbf{a}_{B_1} = \mathbf{a}_{B_2} = \mathbf{a}_A + \boldsymbol{\alpha}_1 \times \mathbf{r}_{AB} - \omega_1^2 \mathbf{r}_{AB} = -\omega_1^2 \mathbf{r}_{AB}$$

$$= 4.712^2[(0.070 - 0.15)\,\mathbf{\imath} + 0.220\,\mathbf{J}] = -1.569\,\mathbf{\imath} - 1.569\,\mathbf{J} \text{ m/s}^2,$$

where $\boldsymbol{\omega}_1 = \omega_1\,\mathbf{k} = \omega\,\mathbf{k}$ and $\boldsymbol{\alpha}_1 = \dot{\omega}_1 = \mathbf{0}$.

Continued

EXAMPLE I.4.2: R-RTR-RRT mechanism–*Cont'd*

The points B and C are located on the same straight line BD:

$$y_B(t) - y_C - [x_B(t) - x_C] \tan \phi_3(t) = 0. \qquad (I.4.84)$$

The angle $\phi_3 = \phi_2$ is computed as follows:

$$\phi_3 = \phi_2 = \arctan \frac{y_B - y_C}{x_B - x_C},$$

and for $\phi = 45°$ is obtained by

$$\phi_3 = \arctan \frac{0.22}{0.07} = 72.235°.$$

The derivative of Eq. (I.4.84) yields

$$\dot{y}_B - \dot{y}_C - (\dot{x}_B - \dot{x}_C) \tan \phi_3 - (x_B - x_C)\frac{1}{\cos^2 \phi_3} \dot{\phi}_3 = 0. \qquad (I.4.85)$$

The angular velocity of link 3, $\omega_3 = \omega_2 = \dot{\phi}_3$, is computed as follows

$$\omega_3 = \omega_2 = \frac{\cos^2 \phi_3 [\dot{y}_B - \dot{y}_C - (\dot{x}_B - \dot{x}_C) \tan \phi_3]}{x_B - x_C},$$

and

$$\omega_3 = \frac{\cos^2 72.235°(0.333 + 0.333 \tan 72.235°)}{0.07} = 1.807 \text{ rad/s.}$$

The angular acceleration of link 3, $\alpha_3 = \alpha_2 = \ddot{\phi}_3$, is computed from the time derivative of Eq. (I.4.85)

$$\ddot{y}_B - \ddot{y}_C - (\ddot{x}_B - \ddot{x}_C)\tan \phi_3 - 2(\dot{x}_B - \dot{x}_C)\frac{1}{\cos^2 \phi_3}\dot{\phi}_3$$

$$- 2(x_B - x_C)\frac{\sin \phi_3}{\cos^3 \phi_3} \dot{\phi}_3^2 - (x_B - x_C)\frac{1}{\cos^2 \phi_3} \ddot{\phi}_3 = 0.$$

The solution of the previous equation is

$$\alpha_3 = \alpha_2 = [\ddot{y}_B - \ddot{y}_C - (\ddot{x}_B - \ddot{x}_C)\tan \phi_3 - 2(\dot{x}_B - \dot{x}_C)\frac{1}{\cos^2 \phi_3}\dot{\phi}_3$$

$$- 2(x_B - x_C)\frac{\sin \phi_3}{\cos^3 \phi_3} \dot{\phi}_3^2]\frac{\cos^2 \phi_3}{x_B - x_C},$$

EXAMPLE I.4.2: R-RTR-RRT mechanism–*Cont'd*

and for the given numerical data:

$$\alpha_3 = \alpha_2 = [-1.569 + 1.569 \tan 72.235° + 2(0.333)\frac{1}{\cos^2 72.235°}1.807$$

$$- 2(0.07)\frac{\sin 72.235°}{\cos^3 72.235°}(1.807)^2]\frac{\cos^2 72.235°}{0.07} = 1.020 \text{ rad/s}^2.$$

The links 2 and 3 have the same angular velocity $\boldsymbol{\omega}_3 = \boldsymbol{\omega}_2 = \omega_3 \, \mathbf{k}$ and the same angular acceleration $\boldsymbol{\alpha}_3 = \boldsymbol{\alpha}_2 = \alpha_3 \, \mathbf{k}$. The relative angular velocity of link 2 relative to link 1 is

$$\boldsymbol{\omega}_{21} = \boldsymbol{\omega}_2 - \boldsymbol{\omega}_1 = (1.807 - 4.712) \, \mathbf{k} = -2.905 \, \mathbf{k} \text{ rad/s},$$

and the relative angular acceleration of link 2 relative to link 1 is

$$\boldsymbol{\alpha}_{21} = \boldsymbol{\alpha}_2 - \boldsymbol{\alpha}_1 = \boldsymbol{\alpha}_2 = 1.020 \, \mathbf{k} \text{ rad/s}^2.$$

The velocity and acceleration of point B on link 3 are calculated with

$$\mathbf{v}_{B_3} = \mathbf{v}_C + \boldsymbol{\omega}_3 \times \mathbf{r}_{CB} = \begin{vmatrix} \mathbf{\imath} & \mathbf{\jmath} & \mathbf{k} \\ 0 & 0 & \omega_3 \\ x_B & y_B & 0 \end{vmatrix}$$

$$= \begin{vmatrix} \mathbf{\imath} & \mathbf{\jmath} & \mathbf{k} \\ 0 & 0 & 1.807 \\ 0.070 & 0.220 & 0 \end{vmatrix} = -0.398 \, \mathbf{\imath} + 0.127 \, \mathbf{\jmath} \text{ m/s},$$

$$\mathbf{a}_{B_3} = \mathbf{a}_C + \boldsymbol{\alpha}_3 \times \mathbf{r}_{CB} - \omega_3^2 \mathbf{r}_{CB}$$

$$= \begin{vmatrix} \mathbf{\imath} & \mathbf{\jmath} & \mathbf{k} \\ 0 & 0 & \alpha_3 \\ x_B & y_B & 0 \end{vmatrix} - \omega_3^2(x_B \, \mathbf{\imath} + y_B \, \mathbf{\jmath})$$

$$= \begin{vmatrix} \mathbf{\imath} & \mathbf{\jmath} & \mathbf{k} \\ 0 & 0 & 1.020 \\ 0.070 & 0.220 & 0 \end{vmatrix} - 1.807^2(0.070 \, \mathbf{\imath} + 0.220 \, \mathbf{\jmath})$$

$$= -0.456 \, \mathbf{\imath} - 0.649 \, \mathbf{\jmath} \text{ m/s}^2.$$

The velocity field of two points (B_2 and B_3) not situated on the same rigid body (B_2 is on link 2 and B_3 is on link 3) is expressed by

$$\mathbf{v}_{B_2} = \mathbf{v}_{B_3} + \mathbf{v}_{B_{23}}^r,$$

Continued

EXAMPLE I.4.2: R-RTR-RRT mechanism–*Cont'd*

and

$$\mathbf{v}^r_{B_{23}} = \mathbf{v}_{B_2} - \mathbf{v}_{B_3} = -0.333\ \mathbf{i} + 0.333\ \mathbf{j} - (-0.398\ \mathbf{i} + 0.127\ \mathbf{j})$$
$$= 0.065\ \mathbf{i} + 0.205\ \mathbf{j}\ \text{m/s}.$$

The expression for the Coriolis acceleration is

$$\mathbf{a}^c_{B_{23}} = 2\ \omega_2 \times \mathbf{v}^r_{B_{23}} = 2\ \omega_3 \times \mathbf{v}^r_{B_{23}}$$

$$= 2 \begin{vmatrix} \mathbf{i} & \mathbf{j} & \mathbf{k} \\ 0 & 0 & \omega_3 \\ v^r_{B_{23x}} & v^r_{B_{21y}} & 0 \end{vmatrix} = 2 \begin{vmatrix} \mathbf{i} & \mathbf{j} & \mathbf{k} \\ 0 & 0 & 1.807 \\ 0.065 & 0.205 & 0 \end{vmatrix}$$

$$= -0.742\ \mathbf{i} + 0.237\ \mathbf{j}\ \text{m/s}^2.$$

The relative acceleration of B_2 with respect to B_3 is

$$\mathbf{a}^r_{B_{23}} = \mathbf{a}_{B_2} - \mathbf{a}_{B_3} - \mathbf{a}^c_{B_{23}}$$
$$= -1.569\ \mathbf{i} - 1.569\ \mathbf{j} - (-0.456\,\mathbf{i} - 0.649\,\mathbf{j}) - (-0.742\ \mathbf{i} + 0.237\ \mathbf{j})$$
$$= -0.5\ \mathbf{i} - 0.5\ \mathbf{j}\ \text{m/s}^2.$$

The position of the joint D is given by the following quadratic equations:

$$[x_D(t) - x_C]^2 + [y_D(t) - y_C]^2 = CD^2,$$
$$[x_D(t) - x_C]\ \sin\phi_3(t) - [y_D(t) - y_C]\ \cos\phi_3(t) = 0,$$

The previous equations are rewritten as follows:

$$x_D^2(t) + y_D^2(t) = CD^2,$$
$$x_D(t)\ \sin\phi_3(t) - y_D(t)\ \cos\phi_3(t) = 0. \tag{I.4.86}$$

For $\phi = 45°$, the coordinates of joint D are

$$x_D = \pm \frac{CD}{\sqrt{1 + \tan^2\phi_3}} = \pm \frac{0.075}{\sqrt{1 + \tan^2 72.235°}} = -0.023\,\text{m},$$
$$y_D = x_D \tan\phi_3 = -0.023 \tan 72.235° = -0.071\,\text{m}.$$

The negative value for x_D was selected for this position of the mechanism. The velocity analysis is carried out differentiating Eq. (I.4.86)

$$x_D \dot{x}_D + y_D \dot{y}_D = 0,$$
$$\dot{x}_D \sin\phi_3 + x_D \cos\phi_3\ \dot\phi_3 - \dot{y}_D \cos\phi_3 + y_D \sin\phi_3\ \dot\phi_3 = 0. \tag{I.4.87}$$

EXAMPLE I.4.2: R-RTR-RRT mechanism–*Cont'd*

For the given data, Eq. (I.4.87) becomes

$$- 0.023\dot{x}_D - 0.071 y_D = 0,$$
$$0.952\,\dot{x}_D - 0.023(0.305)(1.807) - 0.305\,\dot{y}_D - 0.071(0.952)(1.807) = 0.$$

The solution is

$$\dot{x}_D = 0.129 \text{ m/s}, \quad \dot{y}_D = -0.041 \text{ m/s}.$$

The magnitude of the velocity of joint D is

$$v_D = |\mathbf{v}_D| = \sqrt{\dot{x}_D^2 + \dot{y}_D^2} = \sqrt{0.129^2 + (-0.041)^2} = 0.135 \text{ m/s}.$$

The acceleration analysis is obtained using the derivative of the velocity given by Eq. (I.4.87):

$$\dot{x}_D^2 + x_D\ddot{x}_D + \dot{y}_D^2 + y_D\ddot{y}_D = 0,$$
$$\ddot{x}_D \sin \phi_3 + 2\dot{x}_D\dot{\phi}_3 \cos \phi_3 - x_D\dot{\phi}_3^2 \sin \phi_3 + x_D\ddot{\phi}_3 \cos \phi_3$$
$$- \ddot{y}_D \cos \phi_3 + 2\dot{y}_D\dot{\phi}_3 \sin \phi_3 + y_D\dot{\phi}_3^2 \cos \phi_3 + y_D\ddot{\phi}_3^2 \sin \phi_3 = 0,$$

or

$$0.129^2 + (-0.022)\ddot{x}_D + (-0.041)^2 + (-0.071)\ddot{y}_D = 0,$$
$$\ddot{x}_D \sin 72.235° + 2(0.129)(1.807) \cos 72.235° - (-0.022)(1.807)^2 \sin 72.235°$$
$$+ (-0.022)(1.020) \cos 72.235° - \ddot{y}_D \cos 72.235° + 2(-0.041)(1.807) \sin 72.235°$$
$$+ (-0.071)(1.807)^2 \cos 72.235° + (-0.071)(1.020)^2 \sin 72.235° = 0.$$

The solution of the previous system is

$$\ddot{x}_D = 0.147 \text{ m/s}^2, \quad \ddot{y}_D = 0.210 \text{ m/s}^2.$$

The absolute acceleration of joint D is

$$a_D = |\mathbf{a}_D| = \sqrt{\ddot{x}_D^2 + \ddot{y}_D^2} = \sqrt{(0.150)^2 + (0.212)^2} = 0.256 \text{ m/s}^2.$$

The position of joint E is determined from the following equation:

$$[x_E(t) - x_D(t)]^2 + [y_E(t) - y_D(t)]^2 = DE^2,$$

Continued

EXAMPLE I.4.2: R-RTR-RRT mechanism–*Cont'd*

and with the coordinate $y_E = 0$:

$$[x_E(t) - x_D(t)]^2 + y_D^2(t) = DE^2. \tag{I.4.88}$$

With the given numerical values Eq. (I.4.88) becomes

$$(x_E + 0.023)^2 + (0.071)^2 = 0.2^2,$$

with the correct solution $x_E = 0.164$ m.

The velocity of joint E is determined by differentiating Eq. (I.4.88) as follows

$$2(\dot{x}_E - \dot{x}_D)(x_E - x_D) + 2y_D \dot{y}_D = 0, \tag{I.4.89}$$

or

$$\dot{x}_E - \dot{x}_D = -\frac{y_D \dot{y}_D}{x_E - x_D}.$$

The solution of the above equation is

$$\dot{x}_E = 0.129 - \frac{(-0.071)(-0.041)}{0.164 + 0.023} = 0.113 \text{ m/s}.$$

The derivative of Eq. (I.4.89) yields

$$(\ddot{x}_E - \ddot{x}_D)(x_E - x_D) + (\dot{x}_E - \dot{x}_D)^2 + \dot{y}_D^2 + y_D \ddot{y}_D = 0,$$

with the solution

$$\ddot{x}_E = \ddot{x}_D - \frac{\dot{y}_D^2 + y_D \ddot{y}_D + (\dot{x}_E - \dot{x}_D)^2}{x_E - x_D},$$

or with numerical values

$$\ddot{x}_E = 0.150 - \frac{(-0.041)^2 + (-0.071)(0.21) + (0.112 - 0.129)^2}{0.164 + 0.023} = 0.217 \text{ m/s}^2.$$

The angle ϕ_4 is determined from the following equation:

$$y_E - y_D(t) - [x_E(t) - x_D(t)] \tan \phi_4(t) = 0, \tag{I.4.90}$$

where $y_E = 0$. The above equation can be rewritten

$$-y_D(t) - [x_E(t) - x_D(t)] \tan \phi_4(t) = 0, \tag{I.4.91}$$

EXAMPLE I.4.3: R-RTR-RRT mechanism—*Cont'd*

and the solution is

$$\phi_4 = \arctan\left(\frac{-y_D}{x_E - x_D}\right) = \arctan\left(\frac{0.071}{0.164 + 0.023}\right) = 20.923°.$$

The derivative of Eq. (I.4.91) yields

$$-\dot{y}_D - (\dot{x}_E - \dot{x}_D)\tan\phi_4 - (x_E - x_D)\frac{1}{\cos^2\phi_4}\dot{\phi}_4 = 0. \qquad (I.4.92)$$

Hence,

$$\omega_4 = \dot{\phi}_4 = -\frac{\cos^2\phi_4[\dot{y}_D + (\dot{x}_E - \dot{x}_D)\tan\phi_4]}{x_E - x_D}$$

$$= -\frac{\cos^2 20.923°[-0.041 + (0.113 - 0.129)\tan 20.923°]}{0.164 - (-0.022)}$$

$$= 0.221 \text{ rad/s.}$$

The angular acceleration of link 4 is determined by differentiating Eq. (I.4.92) as follows:

$$-\ddot{y}_D - (\ddot{x}_E - \ddot{x}_D)\tan\phi_4 - 2(\dot{x}_E - \dot{x}_D)\frac{1}{\cos^2\phi_4}\dot{\phi}_4$$

$$-2(x_E - x_D)\frac{\sin\phi_4}{\cos^3\phi_4}\dot{\phi}_4^2 - (x_E - x_D)\frac{1}{\cos^2\phi_4}\ddot{\phi}_4 = 0,$$

or

$$-0.210 - (0.217 - 0.147)\tan 20.923° - 2(0.113 - 0.129)\frac{1}{\cos^2 20.923°}0.221$$

$$-2(0.164 + 0.022)\frac{\sin 20.923°}{\cos^3 20.923°}0.221^2 - (0.164 + 0.022)\frac{1}{\cos^2 20.923°}\ddot{\phi}_4 = 0,$$

The solution of the previous equation is

$$\alpha_4 = \ddot{\phi}_4 = -1.105 \text{ rad/s}^2.$$

I.4.8 Problems

I.4.1 The four-bar mechanism shown in Figure I.3.16 has the dimensions:
$AB = CD = 0.04$ m and $AD = BC = 0.09$ m. The driver link AB rotates with a
constant angular speed of 120 rpm. Find the velocities and the accelerations of the

four-bar mechanism for the case when the angle of the driver link AB with the horizontal axis is $\phi = 30°$.

I.4.2 The constant angular speed of the driver link 1 of the mechanism shown in Figure I.4.9 is $\omega = \omega_1 = 10$ rad/s. The distance from link 3 to the horizontal axis Ax is $a = 55$ mm. Find the velocity and the acceleration of point C on link 3 for $\phi = 45°$.

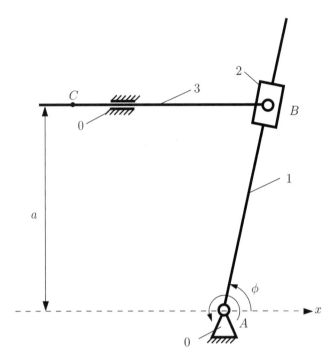

FIGURE I.4.9 *Mechanism for Problem I.4.2.*

I.4.3 The slider crank mechanism shown in Figure I.4.10 has the dimensions: $AB = 0.1$ m and $BC = 0.2$ m. The driver link 1 rotates with a constant angular speed of $n = 60$ rpm. Find the velocity and acceleration of the slider 3 when the angle of the driver link with the horizontal axis is $\phi = 45°$.

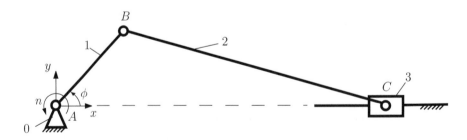

FIGURE I.4.10 *Mechanism for Problem I.4.3.*

I.4.4 The planar mechanism considered is shown in Figure I.3.19. The following data are given: $AB = 0.150$ m, $BC = 0.400$ m, $CD = 0.370$ m, $CE = 0.230$ m, $EF = CE$, $L_a = 0.300$ m, $L_b = 0.450$ m, and $L_c = CD$. The constant angular speed of the driver link 1 is 60 rpm. Find the velocities and the accelerations of the mechanism for $\phi = \phi_1 = 30°$.

I.4.5 The R-RRR-RTT mechanism is shown in Figure I.3.20. The following data are given: $AB = 0.080$ m, $BC = 0.350$ m, $CE = 0.200$ m, $CD = 0.150$ m, $L_a = 0.200$ m, $L_b = 0.350$ m, and $L_c = 0.040$ m. The driver link 1 rotates with a constant angular speed of $n = 300$ rpm. Find the velocities and the accelerations of the mechanism when the angle of the driver link with the horizontal axis is $\phi = 155°$.

I.4.6 The mechanism shown in Figure I.3.21 has the following dimensions: $AB = 60$ mm, $AD = 200$ mm, $BC = 140$ mm, $CE = 50$ mm, $EF = 170$ mm, and $a = 130$ mm. The constant angular speed of the driver link 1 is $n = 300$ rpm. Find the velocities and the accelerations of the mechanism when the angle of the driver link 1 with the horizontal axis is $\phi = \phi_1 = 30°$.

I.4.7 The dimensions for the mechanism shown in Figure I.3.22 are: $AB = 120$ mm, $BD = 320$ mm, $BC = 110$ mm, $CD = 300$ mm, $DE = 200$ mm, $CF = 400$ mm, $AE = 320$ mm, and $b = 80$ mm. The constant angular speed of the driver link 1 is $n = 30$ rpm. Find the velocities and the accelerations of the mechanism for $\phi = \phi_1 = 30°$.

I.4.8 The mechanism in Figure I.3.23 has the dimensions: $AB = 50$ mm, $AC = 25$ mm, $BD = 100$ mm, $DE = 140$ mm, $EF = 80$ mm, $L_a = 130$ mm, and $L_b = 30$ mm. Find the velocities and the accelerations of the mechanism if the constant angular speed of the driver link 1 is $n = 100$ rpm and for $\phi = \phi_1 = 150°$.

I.4.9 The dimensions for the mechanism shown in Figure I.3.24 are: $AB = 180$ mm, $BC = 470$ mm, $AD = 430$ mm, $CD = 270$ mm, $DE = 180$ mm, $EF = 400$ mm, and $L_a = 70$ mm. The constant angular speed of the driver link 1 is $n = 220$ rpm. Find the velocities and the accelerations of the mechanism for $\phi = \phi_1 = 45°$.

I.4.10 The mechanism in Figure I.3.25 has the dimensions: $AB = 200$ mm, $AC = 600$ mm, $BD = 1000$ mm, $L_a = 150$ mm, and $L_b = 250$ mm. The driver link 1 rotates with a constant angular speed of $n = 60$ rpm. Find the velocities and the accelerations of the mechanism for $\phi = \phi_1 = 120°$.

I.4.11 Figure I.3.26 shows a mechanism with the following dimensions: $AB = 250$ mm, $BD = 900$ mm, and $L_a = 300$ mm. The constant angular speed of the driver link 1 is $n = 500$ rpm. Find the velocities and the accelerations of the mechanism when the angle of the driver link 1 with the horizontal axis is $\phi = 240°$.

I.4.12 The mechanism in Figure I.3.27 has the dimensions: $AB = 150$ mm, $AC = 350$ mm, $BD = 530$ mm, $DE = 300$ mm, $EF = 200$ mm, $L_a = 55$ mm, and $L_b = 250$ mm. The constant angular speed of the driver link 1 is $n = 30$ rpm. Find the velocities and the accelerations of the mechanism for $\phi = \phi_1 = 120°$.

I.4.13 Figure I.3.28 shows a mechanism with the following dimensions: $AB = 150$ mm, $BC = 550$ mm, $CD = DE = 220$ mm, $EF = 400$ mm, $L_a = 530$ mm, and $L_b = L_c = 180$ mm. Find the velocities and the accelerations of the mechanism if the constant angular speed of the driver link 1 is $n = 1000$ rpm and for $\phi = \phi_1 = 150°$.

I.4.14 Figure I.3.29 shows a mechanism with the following dimensions: $AB = 250$ mm, $BC = 1200$ mm, $CE = 400$ mm, $CD = 800$ mm, $EF = 700$ mm, $L_a = 650$ mm, $L_b = 1000$ mm, and $L_c = 1200$ mm. The constant angular speed of the driver link 1 is $n = 70$ rpm. Find the velocities and the accelerations of the mechanism for $\phi = \phi_1 = 120°$.

I.4.15 Figure I.3.30 shows a mechanism with the following dimensions: $AB = 100$ mm, $BC = 270$ mm, $CF = 260$ mm, $CD = 90$ mm, $DE = 300$ mm, $L_a = 350$ mm, $L_b = 200$ mm, and $L_c = 120$ mm. The constant angular speed of the driver link 1 is $n = 100$ rpm. Find the velocities and the accelerations of the mechanism when the angle of the driver link 1 with the horizontal axis is $\phi = 60°$.

I.4.16 Figure I.3.31 shows a mechanism with the following dimensions: $AB = 40$ mm, $BC = 100$ mm, $AD = 50$ mm, and $BE = 110$ mm. The constant angular speed of the driver link 1 is $n = 250$ rpm. Find the velocities and the accelerations of the mechanism if the angle of the driver link 1 with the horizontal axis is $\phi = 30°$.

I.4.17 The dimensions of the mechanism shown in Figure I.3.32 are: $AB = 100$ mm, $BC = 200$ mm, $BE = 400$ mm, $CE = 600$ mm, $CD = 220$ mm, $EF = 800$ mm, $L_a = 250$ mm, $L_b = 150$ mm, and $L_c = 100$ mm. The constant angular speed of the driver link 1 is $n = 100$ rpm. Find the velocities and the accelerations of the mechanism for $\phi = \phi_1 = 150°$.

I.4.18 The dimensions of the mechanism shown in Figure I.3.33 are: $AB = 200$ mm, $AC = 300$ mm, $CD = 500$ mm, $DE = 250$ mm, and $L_a = 400$ mm. Find the positions of the joints and the angles of the links. The constant angular speed of the driver link 1 is $n = 40$ rpm. Find the velocities and the accelerations of the mechanism when the angle of the driver link 1 with the horizontal axis is $\phi = 60°$.

I.4.19 The dimensions of the mechanism shown in Figure I.3.34 are: $AB = 160$ mm, $AC = 90$ mm, $CD = 150$ mm, and $DE = 400$ mm. The constant angular speed of the driver link 1 is $n = 70$ rpm. Find the velocities and the accelerations of the mechanism for $\phi = \phi_1 = 45°$.

I.4.20 The dimensions of the mechanism shown in Figure I.3.35 are: $AB = 150$ mm, $AC = 250$ mm, and $CD = 450$ mm. For the distance b select a suitable value. The constant angular speed of the driver link 1 is $n = 80$ rpm. Find the velocities and the accelerations of the mechanism for $\phi = \phi_1 = 30°$.

I.4.21 The dimensions of the mechanism shown in Figure I.3.36 are: $AB = 180$ mm, $AC = 90$ mm, and $CD = 200$ mm. The constant angular speed of the driver link 1 is $n = 180$ rpm. Find the velocities and the accelerations of the mechanism for $\phi = \phi_1 = 60°$.

I.4.22 The dimensions of the mechanism shown in Figure I.3.37 are: $AB = 180$ mm, $AC = 500$ mm, $BD = L_a = 770$ mm, and $DE = 600$ mm. The constant angular speed of the driver link 1 is $n = n_1 = 700$ rpm. Find the velocities and the accelerations of the mechanism for $\phi = \phi_1 = 45°$.

I.4.23 The dimensions of the mechanism shown in Figure I.3.38 are: $AB = 220$ mm, $AD = 600$ mm, and $BC = 250$ mm. The constant angular speed of the driver link 1 is $n = 700$ rpm. Find the velocities and the accelerations of the mechanism for $\phi = \phi_1 = 120°$. Select a suitable value for the distance a.

I.4.24 Refer to Example I.3.1. The mechanism in Figure I.3.11(a) has the dimensions: $AB = 0.20$ m, $AD = 0.40$ m, $CD = 0.70$ m, $CE = 0.30$ m, and $y_E = 0.35$ m. The constant angular speed of the driver link 1 is $n = 1000$ rpm. Find the velocities and the accelerations of the mechanism for the given input angle $\phi = \phi_1 = 60°$.

I.4.25 Refer to Example I.3.2. The mechanism in Figure I.3.12 has the dimensions: $AB = 0.02$ m, $BC = 0.03$ m, $CD = 0.06$ m, $AE = 0.05$ m, and $L_a = 0.02$ m. The constant angular speed of the driver link 1 is $n = 600$ rpm. Find the velocities and the accelerations of the mechanism for the given input angle $\phi = \phi_1 = \pi/3$.

I.4.26 Refer to Example I.3.3. The mechanism in Figure I.3.15 has the dimensions: $AC = 0.100$ m, $BC = 0.300$ m, $BD = 0.900$ m, and $L_a = 0.100$ m. The constant angular speed of the driver link 1 is $n = 100$ rpm. Find the velocities and the accelerations of the mechanism for $\phi = 30°$.

I.5   Contour Equations

This chapter provides an algebraic method to compute the velocities and accelerations of any closed kinematic chain. The classical method for obtaining the velocities and accelerations involves the computation of the derivative with respect to time of the position vectors. The method of contour equations avoids this task and uses only algebraic equations [4, 56]. Using this approach, a numerical implementation is much more efficient. The method described here can be applied to planar and spatial mechanisms.

Two rigid links (j) and (k) are connected by a joint (kinematic pair) at A (Fig. I.5.1). The point A_j of the rigid body (j) is guided along a path prescribed in the body (k). The points A_j belonging to body (j) and the A_k belonging to body (k) are coincident at the instant of motion under consideration. The following relation exists between the velocity $\mathbf{v}_{A_j}$ of the point A_j and the velocity $\mathbf{v}_{A_k}$ of the point A_k:

$$\mathbf{v}_{A_j} = \mathbf{v}_{A_k} + \mathbf{v}^r_{A_{jk}}, \tag{I.5.1}$$

where $\mathbf{v}^r_{A_{jk}} = \mathbf{v}^r_{A_j A_k}$ indicates the velocity of A_j as seen by an observer at A_k attached to body k or the relative velocity of A_j with respect to A_k, allowed at the joint A. The direction of $\mathbf{v}^r_{A_{jk}}$ is obviously tangent to the path prescribed in the body (k).

From Eq. (I.5.1) the accelerations of A_j and A_k are expressed as

$$\mathbf{a}_{A_j} = \mathbf{a}_{A_k} + \mathbf{a}^r_{A_{jk}} + \mathbf{a}^c_{A_{jk}}, \tag{I.5.2}$$

where $\mathbf{a}^c_{A_{jk}} = \mathbf{a}^c_{A_j A_k}$ is known as the *Coriolis acceleration* and is given by

$$\mathbf{a}^c_{A_{jk}} = 2\,\omega_k \times \mathbf{v}^r_{A_{jk}}, \tag{I.5.3}$$

where ω_k is the angular velocity of the body (k).

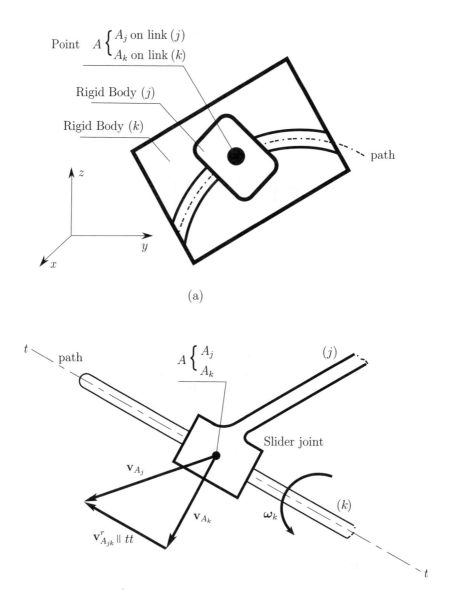

Point A $\begin{cases} A_j \text{ on link } (j) \\ A_k \text{ on link } (k) \end{cases}$

Rigid Body (j)

Rigid Body (k)

path

z

y

x

(a)

t

path

A $\begin{cases} A_j \\ A_k \end{cases}$

(j)

Slider joint

$\mathbf{v}_{A_j}$

$\mathbf{v}_{A_k}$

ω_k

(k)

$\mathbf{v}^r_{A_{jk}} \parallel tt$

t

(b)

FIGURE I.5.1 *Two rigid links (j) and (k) connected by a joint at A: (a) general case, (b) slider joint in general motion.*

Equations (I.5.1) and (I.5.2) are useful even for coincident points belonging to two links that may not be directly connected. A graphical representation of Eq. (I.5.1) is shown in Figure I.5.1(b) for a rotating slider joint.

Figure I.5.2 shows a monocontour closed kinematic chain with n rigid links. The joint A_i, $i = 0, 1, 2, \ldots, n$ is the connection between the links (i) and $(i-1)$. The last link n is

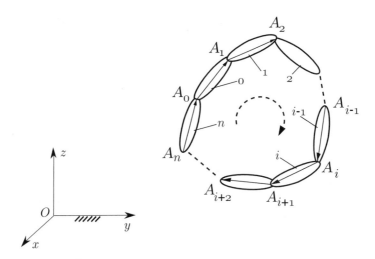

FIGURE I.5.2 *Monocontour closed kinematic chain with n rigid links.*

connected with the first link 0 of the chain. For the closed kinematic chain, a path is chosen from link 0 to link n. At the joint A_i there are two instantaneously coincident points: 1) the point $A_{i,i}$ belonging to link (i), $A_{i,i} \in (i)$, and 2) the point $A_{i,i-1}$ belonging to body $(i-1)$, $A_{i,i-1} \in (i-1)$.

I.5.1 Contour Velocity Equations

The absolute angular velocity, $\omega_i = \omega_{i,0}$, of the rigid body (i), or the angular velocity of the rigid body (i) with respect to the "fixed" reference frame $Oxyz$ is

$$\omega_i = \omega_{i-1} + \omega_{i,i-1}, \tag{I.5.4}$$

where $\omega_{i-1} = \omega_{i-1,0}$ is the absolute angular velocity of the rigid body $(i-1)$ [or the angular velocity of the rigid body $(i-1)$ with respect to the "fixed" reference frame $Oxyz$] and $\omega_{i,i-1}$ is the relative angular velocity of the rigid body (i) with respect to the rigid body $(i-1)$.

For the n link closed kinematic chain the following expressions are obtained for the angular velocities:

$$\omega_1 = \omega_0 + \omega_{1,0}$$
$$\omega_2 = \omega_1 + \omega_{2,1}$$
$$\dots\dots\dots\dots\dots\dots\dots\dots\dots\dots\dots\dots\dots\dots\dots\dots$$
$$\omega_i = \omega_{i-1} + \omega_{i,i-1}$$
$$\dots\dots\dots\dots\dots\dots\dots\dots\dots\dots\dots\dots\dots\dots\dots\dots$$
$$\omega_0 = \omega_n + \omega_{0,n}. \tag{I.5.5}$$

Summing the expressions given in Eq. (I.5.5), the following relation is obtained:

$$\omega_{1,0} + \omega_{2,1} + \cdots + \omega_{0,n} = \mathbf{0}, \tag{I.5.6}$$

which may be rewritten as

$$\sum_{(i)} \omega_{i,i-1} = \mathbf{0}. \tag{I.5.7}$$

Equation (I.5.7) represents the first vectorial equation for the angular velocities of a simple closed kinematic chain.

The following relation exists between the velocity $\mathbf{v}_{A_{i,i}}$ of the point $A_{i,i}$ and the velocity $\mathbf{v}_{A_{i,i-1}}$ of the point $A_{i,i-1}$

$$\mathbf{v}_{A_{i,i}} = \mathbf{v}_{A_{i,i-1}} + \mathbf{v}^r_{A_{i,i-1}}, \tag{I.5.8}$$

where $\mathbf{v}^r_{A_{i,i-1}} = \mathbf{v}^r_{A_{i,i}A_{i,i-1}}$ is the relative velocity of $A_{i,i}$ on link (i) with respect to $A_{i,i-1}$ on link $(i-1)$. Using the velocity relation for two particles on the rigid body (i), the following relation exists:

$$\mathbf{v}_{A_{i+1,i}} = \mathbf{v}_{A_{i,i}} + \omega_i \times \mathbf{r}_{A_i A_{i+1}}, \tag{I.5.9}$$

where ω_i is the absolute angular velocity of the link (i) in the reference frame $Oxyz$, and $\mathbf{r}_{A_i A_{i+1}}$ is the distance vector from A_i to A_{i+1}. Using Eqs. (I.5.8) and (I.5.9), the velocity of the point $A_{i+1,i} \in (i+1)$ is written as

$$\mathbf{v}_{A_{i+1,i}} = \mathbf{v}_{A_{i,i-1}} + \omega_i \times \mathbf{r}_{A_i A_{i+1}} + \mathbf{v}^r_{A_{i,i-1}}. \tag{I.5.10}$$

For the n link closed kinematic chain the following expressions are obtained:

$$\mathbf{v}_{A_{3,2}} = \mathbf{v}_{A_{2,1}} + \omega_2 \times \mathbf{r}_{A_2 A_3} + \mathbf{v}^r_{A_{2,1}}$$

$$\mathbf{v}_{A_{4,3}} = \mathbf{v}_{A_{3,2}} + \omega_3 \times \mathbf{r}_{A_3 A_4} + \mathbf{v}^r_{A_{3,2}}$$

$$\cdots\cdots\cdots\cdots\cdots\cdots\cdots\cdots\cdots\cdots\cdots\cdots\cdots$$

$$\mathbf{v}_{A_{i+1,i}} = \mathbf{v}_{A_{i,i-1}} + \omega_i \times \mathbf{r}_{A_i A_{i+1}} + \mathbf{v}^r_{A_{i,i-1}}$$

$$\cdots\cdots\cdots\cdots\cdots\cdots\cdots\cdots\cdots\cdots\cdots\cdots\cdots$$

$$\mathbf{v}_{A_{1,0}} = \mathbf{v}_{A_{0,n}} + \omega_0 \times \mathbf{r}_{A_0 A_1} + \mathbf{v}^r_{A_{0,n}}$$

$$\mathbf{v}_{A_{2,1}} = \mathbf{v}_{A_{1,0}} + \omega_1 \times \mathbf{r}_{A_1 A_2} + \mathbf{v}^r_{A_{1,0}}. \tag{I.5.11}$$

Summing the relations in Eq. (I.5.11):

$$\left[\omega_1 \times \mathbf{r}_{A_1 A_2} + \omega_2 \times \mathbf{r}_{A_2 A_3} + \cdots + \omega_i \times \mathbf{r}_{A_i A_{i+1}} + \cdots + \omega_0 \times \mathbf{r}_{A_0 A_1} \right]$$
$$+ \left[\mathbf{v}^r_{A_{2,1}} + \mathbf{v}^r_{A_{3,2}} + \cdots + \mathbf{v}^r_{A_{i,i-1}} + \cdots + \mathbf{v}^r_{A_{0,n}} + \mathbf{v}^r_{A_{1,0}} \right] = \mathbf{0}. \tag{I.5.12}$$

Because the reference system $Oxyz$ is considered "fixed", the vector $\mathbf{r}_{A_{i-1}A_i}$ is written in terms of the position vectors of the points A_{i-1} and A_i:

$$\mathbf{r}_{A_{i-1}A_i} = \mathbf{r}_{A_i} - \mathbf{r}_{A_{i-1}}, \tag{I.5.13}$$

where $\mathbf{r}_{A_i} = \mathbf{r}_{OA_i}$ and $\mathbf{r}_{A_{i-1}} = \mathbf{r}_{OA_{i-1}}$. Equation (I.5.12) becomes

$$\left[\mathbf{r}_{A_1} \times (\boldsymbol{\omega}_1 - \boldsymbol{\omega}_0) + \mathbf{r}_{A_2} \times (\boldsymbol{\omega}_2 - \boldsymbol{\omega}_1) + \cdots + \mathbf{r}_{A_0} \times (\boldsymbol{\omega}_0 - \boldsymbol{\omega}_n)\right]$$
$$+ \left[\mathbf{v}^r_{A_{1,0}} + \mathbf{v}^r_{A_{2,1}} + \cdots + \mathbf{v}^r_{A_{i,i-1}} + \cdots + \mathbf{v}^r_{A_{0,n}}\right] = \mathbf{0}. \tag{I.5.14}$$

Using Eq. (I.5.5), Eq. (I.5.14) becomes

$$\left[\mathbf{r}_{A_1} \times \boldsymbol{\omega}_{1,0} + \mathbf{r}_{A_2} \times \boldsymbol{\omega}_{2,1} + \cdots + \mathbf{r}_{A_0} \times \boldsymbol{\omega}_{0,n}\right]$$
$$+ \left[\mathbf{v}^r_{A_{1,0}} + \mathbf{v}^r_{A_{2,1}} + \cdots + \mathbf{v}^r_{A_{0,n}}\right] = \mathbf{0}. \tag{I.5.15}$$

The previous equation is written as

$$\sum_{(i)} \mathbf{r}_{A_i} \times \boldsymbol{\omega}_{i,i-1} + \sum_{(i)} \mathbf{v}^r_{A_{i,i-1}} = \mathbf{0}. \tag{I.5.16}$$

Equation (I.5.16) represents the second vectorial equation for the angular velocities of a simple closed kinematic chain.

Equations such as

$$\sum_{(i)} \boldsymbol{\omega}_{i,i-1} = \mathbf{0} \quad \text{and} \quad \sum_{(i)} \mathbf{r}_{A_i} \times \boldsymbol{\omega}_{i,i-1} + \sum_{(i)} \mathbf{v}^r_{A_{i,i-1}} = \mathbf{0}, \tag{I.5.17}$$

represent the velocity equations for a simple closed kinematic chain.

I.5.2 Contour Acceleration Equations

The absolute angular acceleration, $\boldsymbol{\alpha}_i = \boldsymbol{\alpha}_{i,0}$, of the rigid body (i) [or the angular acceleration of the rigid body (i) with respect to the "fixed" reference frame $Oxyz$] is

$$\boldsymbol{\alpha}_i = \boldsymbol{\alpha}_{i-1} + \boldsymbol{\alpha}_{i,i-1} + \boldsymbol{\omega}_i \times \boldsymbol{\omega}_{i,i-1}, \tag{I.5.18}$$

where $\boldsymbol{\alpha}_{i-1} = \boldsymbol{\alpha}_{i-1,0}$ is the absolute angular acceleration of the rigid body $(i-1)$ [or the angular acceleration of the rigid body $(i-1)$ with respect to the "fixed" reference frame $Oxyz$] and $\boldsymbol{\alpha}_{i,i-1}$ is the relative angular acceleration of the rigid body (i) with respect to the rigid body $(i-1)$.

For the n link closed kinematic chain the following expressions are obtained for the angular accelerations:

$$\alpha_2 = \alpha_1 + \alpha_{2,1} + \omega_2 \times \omega_{2,1}$$

$$\alpha_3 = \alpha_2 + \alpha_{3,2} + \omega_3 \times \omega_{3,2}$$

$$\cdots\cdots\cdots\cdots\cdots\cdots$$

$$\alpha_i = \alpha_{i-1} + \alpha_{i,i-1} + \omega_i \times \omega_{i,i-1}$$

$$\cdots\cdots\cdots\cdots\cdots\cdots$$

$$\alpha_1 = \alpha_0 + \alpha_{1,0} + \omega_1 \times \omega_{1,0}. \qquad (\text{I.5.19})$$

Summing all the expressions in Eq. (I.5.19):

$$\alpha_{2,1} + \alpha_{3,2} + \cdots + \alpha_{1,0} + \omega_2 \times \omega_{2,1} + \cdots + \omega_1 \times \omega_{1,0} = \mathbf{0}. \qquad (\text{I.5.20})$$

Equation (I.5.20) is rewritten as

$$\sum_{(i)} \alpha_{i,i-1} + \sum_{(i)} \omega_i \times \omega_{i,i-1} = \mathbf{0}. \qquad (\text{I.5.21})$$

Equation (I.5.21) represents the first vectorial equation for the angular accelerations of a simple closed kinematic chain.

Using the acceleration distributions of the relative motion of two rigid bodies (i) and $(i-1)$:

$$\mathbf{a}_{A_{i,i}} = \mathbf{a}_{A_{i,i-1}} + \mathbf{a}^r_{A_{i,i-1}} + \mathbf{a}^c_{A_{i,i-1}}, \qquad (\text{I.5.22})$$

where $\mathbf{a}_{A_{i,i}}$ and $\mathbf{a}_{A_{i,i-1}}$ are the linear accelerations of the points $A_{i,i}$ and $A_{i,i-1}$, and $\mathbf{a}^r_{A_{i,i-1}} = \mathbf{a}^r_{A_{i,i}A_{i,i-1}}$ is the relative acceleration between $A_{i,i}$ on link (i) and $A_{i,i-1}$ on link $(i-1)$. Finally, $\mathbf{a}^c_{A_{i,i-1}}$ is the Coriolis acceleration defined as

$$\mathbf{a}^c_{A_{i,i-1}} = 2\,\omega_{i-1} \times \mathbf{v}^r_{A_{i,i-1}}. \qquad (\text{I.5.23})$$

Using the acceleration distribution relations for two particles on a rigid body:

$$\mathbf{a}_{A_{i+1,i}} = \mathbf{a}_{A_{i,i}} + \alpha_i \times \mathbf{r}_{A_i A_{i+1}} + \omega_i \times (\omega_i \times \mathbf{r}_{A_i A_{i+1}}), \qquad (\text{I.5.24})$$

where α_i is the angular acceleration of the link (i). From Eqs. (I.5.22) and (I.5.24):

$$\mathbf{a}_{A_{i+1,i}} = \mathbf{a}_{A_{i,i-1}} + \mathbf{a}^r_{A_{i,i-1}} + \mathbf{a}^c_{A_{i,i-1}} + \alpha_i \times \mathbf{r}_{A_i A_{i+1}} + \omega_i \times (\omega_i \times \mathbf{r}_{A_i A_{i+1}}). \qquad (\text{I.5.25})$$

Writing similar equations for all the links of the kinematic chain, the following relations are obtained:

$$\mathbf{a}_{A_{3,2}} = \mathbf{a}_{A_{2,1}} + \mathbf{a}^r_{A_{2,1}} + \mathbf{a}^c_{A_{2,1}} + \boldsymbol{\alpha}_2 \times \mathbf{r}_{A_2 A_3} + \boldsymbol{\omega}_2 \times (\boldsymbol{\omega}_2 \times \mathbf{r}_{A_2 A_3}),$$

$$\mathbf{a}_{A_{4,3}} = \mathbf{a}_{A_{3,2}} + \mathbf{a}^r_{A_{3,2}} + \mathbf{a}^c_{A_{3,2}} + \boldsymbol{\alpha}_3 \times \mathbf{r}_{A_3 A_4} + \boldsymbol{\omega}_3 \times (\boldsymbol{\omega}_3 \times \mathbf{r}_{A_3 A_4}),$$

$$\dots \dots \dots \dots \dots \dots \dots \dots \dots \dots \dots \dots \dots \dots \dots \dots \dots \dots \dots \dots$$

$$\mathbf{a}_{A_{1,0}} = \mathbf{a}_{A_{0,n}} + \mathbf{a}^r_{A_{0,n}} + \mathbf{a}^c_{A_{0,n}} + \boldsymbol{\alpha}_0 \times \mathbf{r}_{A_0 A_1} + \boldsymbol{\omega}_0 \times (\boldsymbol{\omega}_0 \times \mathbf{r}_{A_0 A_1}),$$

$$\mathbf{a}_{A_{2,1}} = \mathbf{a}_{A_{1,0}} + \mathbf{a}^r_{A_{1,0}} + \mathbf{a}^c_{A_{1,0}} + \boldsymbol{\alpha}_1 \times \mathbf{r}_{A_1 A_2} + \boldsymbol{\omega}_1 \times (\boldsymbol{\omega}_1 \times \mathbf{r}_{A_1 A_2}). \tag{I.5.26}$$

Summing the expressions in Eq. (I.5.26):

$$\left[\mathbf{a}^r_{A_{1,0}} + \mathbf{a}^r_{A_{2,1}} + \cdots + \mathbf{a}^r_{A_{0,n}} \right] + \left[\mathbf{a}^c_{A_{1,0}} + \mathbf{a}^c_{A_{2,1}} + \cdots + \mathbf{a}^c_{A_{0,n}} \right]$$

$$+ \left[\boldsymbol{\alpha}_1 \times \mathbf{r}_{A_1 A_2} + \boldsymbol{\alpha}_2 \times \mathbf{r}_{A_2 A_3} + \cdots + \boldsymbol{\alpha}_0 \times \mathbf{r}_{A_0 A_1} \right] + \boldsymbol{\omega}_1 \times (\boldsymbol{\omega}_1 \times \mathbf{r}_{A_1 A_2})$$

$$+ \boldsymbol{\omega}_2 \times (\boldsymbol{\omega}_2 \times \mathbf{r}_{A_2 A_3}) + \cdots + \boldsymbol{\omega}_0 \times (\boldsymbol{\omega}_0 \times \mathbf{r}_{A_0 A1}) = \mathbf{0}. \tag{I.5.27}$$

Using the relation $\mathbf{r}_{A_{i-1} A_i} = \mathbf{r}_{A_i} - \mathbf{r}_{A_{i-1}}$ in Eq. (I.5.27):

$$\left[\mathbf{a}^r_{A_{1,0}} + \mathbf{a}^r_{A_{2,1}} + \cdots + \mathbf{a}^r_{A_{0,n}} \right] + \left[\mathbf{a}^c_{A_{1,0}} + \mathbf{a}^c_{A_{2,1}} + \cdots + \mathbf{a}^c_{A_{0,n}} \right]$$

$$+ \left[\mathbf{r}_{A_1} \times (\boldsymbol{\alpha}_{1,0} + \boldsymbol{\omega}_1 \times \boldsymbol{\omega}_{1,0}) + \cdots + \mathbf{r}_{A_0} \times (\boldsymbol{\alpha}_{0,n} + \boldsymbol{\omega}_0 \times \boldsymbol{\omega}_{0,n}) \right]$$

$$+ \boldsymbol{\omega}_1 \times (\boldsymbol{\omega}_1 \times \mathbf{r}_{A_1 A_2}) + \boldsymbol{\omega}_2 \times (\boldsymbol{\omega}_2 \times \mathbf{r}_{A_2 A_3}) + \cdots + \boldsymbol{\omega}_0 \times (\boldsymbol{\omega}_0 \times \mathbf{r}_{A_0 A_1})$$

$$= \mathbf{0}. \tag{I.5.28}$$

Equation (I.5.28) is rewritten as

$$\sum_{(i)} \mathbf{a}^r_{A_{i,i-1}} + \sum_{(i)} \mathbf{a}^c_{A_{i,i-1}} + \sum_{(i)} \mathbf{r}_{A_i} \times (\boldsymbol{\alpha}_{i,i-1} + \boldsymbol{\omega}_i \times \boldsymbol{\omega}_{i,i-1})$$

$$+ \sum_{(i)} \boldsymbol{\omega}_i \times (\boldsymbol{\omega}_i \times \mathbf{r}_{A_i A_{i+1}}) = \mathbf{0}. \tag{I.5.29}$$

Equation (I.5.29) represents the second vectorial equation for the angular accelerations of a simple closed kinematic chain. Thus,

$$\sum_{(i)} \boldsymbol{\alpha}_{i,i-1} + \sum_{(i)} \boldsymbol{\omega}_i \times \boldsymbol{\omega}_{i,i-1} = \mathbf{0} \quad \text{and}$$

$$\sum_{(i)} \mathbf{r}_{A_i} \times (\boldsymbol{\alpha}_{i,i-1} + \boldsymbol{\omega}_i \times \boldsymbol{\omega}_{i,i-1}) + \sum_{(i)} \mathbf{a}^r_{A_{i,i-1}} + \sum_{(i)} \mathbf{a}^c_{A_{i,i-1}}$$

$$+ \sum_{(i)} \boldsymbol{\omega}_i \times (\boldsymbol{\omega}_i \times \mathbf{r}_{A_iA_{i+1}}) = \mathbf{0}. \tag{I.5.30}$$

are the acceleration equations for the case of a simple closed kinematic chain.

Remarks

1. For a closed kinematic chain in planar motion, simplified relations are obtained because

$$\boldsymbol{\omega}_i \times (\boldsymbol{\omega}_i \times \mathbf{r}_{A_iA_{i+1}}) = -\omega_i^2 \mathbf{r}_{A_iA_{i+1}} \quad \text{and} \quad \boldsymbol{\omega}_i \times \boldsymbol{\omega}_{i,i-1} = \mathbf{0}. \tag{I.5.31}$$

Equations

$$\sum_{(i)} \boldsymbol{\alpha}_{i,i-1} = \mathbf{0} \quad \text{and}$$

$$\sum_{(i)} \mathbf{r}_{A_i} \times \boldsymbol{\alpha}_{i,i-1} + \sum_{(i)} \mathbf{a}^r_{A_{i,i-1}} + \sum_{(i)} \mathbf{a}^c_{A_{i,i-1}} - \omega_i^2 \mathbf{r}_{A_iA_{i+1}} = \mathbf{0}. \tag{I.5.32}$$

represent the acceleration equations for a simple closed kinematic chain in planar motion.

2. The Coriolis acceleration, given by the expression

$$\mathbf{a}^c_{A_{i,i-1}} = 2\boldsymbol{\omega}_{i-1} \times \mathbf{v}^r_{A_{i,i-1}} \tag{I.5.33}$$

vanishes when $\boldsymbol{\omega}_{i-1} = \mathbf{0}$, or $\mathbf{v}^r_{A_{i,i-1}} = \mathbf{0}$, or when $\boldsymbol{\omega}_{i-1}$ is parallel to $\mathbf{v}^r_{A_{i,i-1}}$.

I.5.3   Independent Contour Equations

A diagram is used to represent to a mechanism in the following way: the numbered links are the nodes of the diagram and are represented by circles, and the joints are represented by lines which connect the nodes.

Figure I.5.3 shows the diagram that represents a planar mechanism. The maximum number of independent contours is given by

$$N = c - n \quad \text{or} \quad n_c = N = c - p + 1, \tag{I.5.34}$$

where c is the number of joints, n is the number of moving links, and p is the number of links.

The equations for velocities and accelerations are written for any closed contour of the mechanism. However, it is best to write the contour equations only for the independent loops of the diagram representing the mechanism.

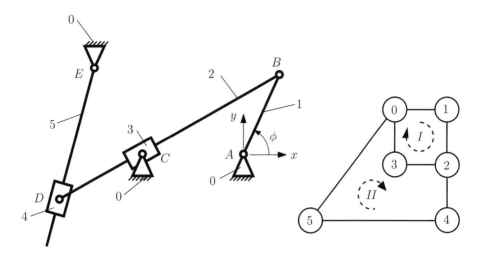

FIGURE I.5.3 *Planar mechanism and the diagram that represents the mechanism.*

Step 1. Determine the position analysis of the mechanism.

Step 2. Draw a diagram representing the mechanism and select the independent contours. Determine a path for each contour.

Step 3. For each closed loop write the contour velocity relations [Eq. (I.5.17)], and contour acceleration relations [Eq. (I.5.30)]. For a closed kinematic chain in planar motion the following equations will be used:

$$\sum_{(i)} \boldsymbol{\omega}_{i,i-1} = \mathbf{0},$$

$$\sum_{(i)} \mathbf{r}_{A_i} \times \boldsymbol{\omega}_{i,i-1} + \sum_{(i)} \mathbf{v}^r_{A_{i,i-1}} = \mathbf{0}. \qquad (I.5.35)$$

$$\sum_{(i)} \boldsymbol{\alpha}_{i,i-1} = \mathbf{0},$$

$$\sum_{(i)} \mathbf{r}_{A_i} \times \boldsymbol{\alpha}_{i,i-1} + \sum_{(i)} \mathbf{a}^r_{A_{i,i-1}} + \sum_{(i)} \mathbf{a}^c_{A_{i,i-1}} - \omega_i^2 \mathbf{r}_{A_i A_{i+1}} = \mathbf{0}. \qquad (I.5.36)$$

Step 4. Project on a Cartesian reference system the velocity and acceleration equations. Linear algebraic equations are obtained where the unknowns are

- the components of the relative angular velocities $\boldsymbol{\omega}_{j,j-1}$;
- the components of the relative angular accelerations $\boldsymbol{\alpha}_{j,j-1}$;
- the components of the relative linear velocities $\mathbf{v}^r_{Aj,j-1}$;
- the components of the relative linear accelerations $\mathbf{a}^r_{Aj,j-1}$.

Solve the algebraic system of equations and determine the unknown kinematic parameters.

Step 5. Determine the absolute angular velocities ω_j and the absolute angular accelerations α_j. Compute the velocities and accelerations of the characteristic points and joints.

In the following examples, the contour method is applied to determine the velocities and accelerations distribution for several planar mechanisms. The following notation will be used:

ω_{ij} is the relative angular velocity vector of the link i with respect to the link j. When the link j is the ground (denoted as link 0), then $\omega_i = \omega_{i0}$ also denotes the absolute angular velocity vector of the link i. The magnitude of ω_{ij}, is ω_{ij} i.e., $|\omega_{ij}| = \omega_{ij}$.

$v^r_{A_{ij}}$ is the relative linear velocity of the point A_i on link i with respect to the point A_j on link j. The point A_i belonging to link j, and the point A_j, belonging to link j, are coincident at the instant of motion under consideration.

α_{ij} is the relative angular acceleration vector of the link i with respect to the rigid body j. When the link j is the ground, then $\alpha_i = \alpha_{i0}$ also denotes the absolute angular acceleration vector of the rigid body i.

$a^r_{A_{ij}}$ is the relative linear acceleration vector of A_i on link i with respect to A_j on link j.

$a^c_{A_{ij}}$ is the Coriolis acceleration of A_i with respect to A_j.

r_{BC} denotes a vector from the joint B to the joint C.

x_B, y_B, z_B denote the coordinates of the point B with respect to the fixed reference frame.

v_B denotes the linear velocity vector of the point B with respect to the fixed reference frame.

a_B denotes the linear acceleration vector of the point B with respect to the fixed reference frame.

I.5.4 Example

The planar mechanism considered in this example is depicted in Figure I.5.4(a). The following data are given: $AC = 0.100$ m, $BC = 0.300$ m, $BD = 0.900$ m, and $L_a = 0.100$ m. The angle of the driver element (link AB) with the horizontal axis is $\phi = 45°$. A Cartesian reference frame with the origin at A ($x_A = y_A = 0$) is selected. The coordinates of joint C are $x_C = AC$, $y_C = 0$. The coordinates of joint B are $x_B = 0.256$ m, $y_B = 0.256$ m. The coordinates of joint D are $x_D = 1.142$ m, $y_D = 0.100$ m. The position vectors r_{AB}, r_{AC}, and r_{AD} are defined as follows:

$$r_{AB} = x_B \mathbf{1} + y_B \mathbf{J} = 0.256 \mathbf{1} + 0.256 \mathbf{J},$$

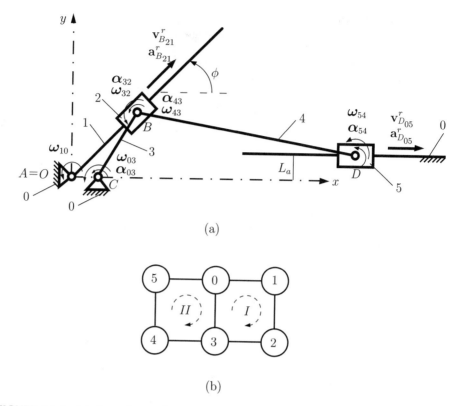

(a)

(b)

FIGURE I.5.4 *(a) Example mechanism (R-TRR-RRT), and (b) diagram that represents the mechanism.*

$$\mathbf{r}_{AC} = x_C \mathbf{\imath} + y_C \mathbf{J} = 0.100 \, \mathbf{\imath},$$

$$\mathbf{r}_{AD} = x_D \mathbf{\imath} + y_D \mathbf{J} = 1.142 \, \mathbf{\imath} + 0.100 \, \mathbf{J}.$$

The angular velocity of the driver link is $n_1 = 100$ rpm, or

$$\omega_{10} = \omega_1 = n\frac{\pi}{30} = 100\frac{\pi}{30} \text{ rad/s} = 10.472 \text{ rad/s}.$$

The mechanism has six links and seven full joints. Using Eq. (I.5.34), the number of independent loops is given by

$$n_c = l - p + 1 = 7 - 6 + 1 = 2.$$

This mechanism has two independent contours. The first contour *I* contains the links 0, 1, 2, and 3, while the second contour *II* contains the links 0, 3, 4, and 5. The diagram representing the mechanism is given in Figure I.5.4(b). Clockwise paths are chosen for each closed loop *I* and *II*.

First Contour

According to Figure I.5.5, the first contour has

- rotational joint R between links 0 and 1 (joint A);
- translational joint T between links 1 and 2 (joint B_T);
- rotational joint R between links 2 and 3 (joint B_R);
- rotational joint R between links 3 and 0 (joint C).

For the velocity analysis, the following equations are written using Eq. (I.5.35):

$$\boldsymbol{\omega}_{10} + \boldsymbol{\omega}_{32} + \boldsymbol{\omega}_{03} = \mathbf{0},$$

$$\mathbf{r}_{AB} \times \boldsymbol{\omega}_{32} + \mathbf{r}_{AC} \times \boldsymbol{\omega}_{03} + \mathbf{v}^r_{B_{21}} = \mathbf{0}, \qquad (I.5.37)$$

where $\boldsymbol{\omega}_{10} = \omega_{10}\,\mathbf{k} = 10.47\,\mathbf{k}$ rad/s, $\boldsymbol{\omega}_{32} = \omega_{32}\,\mathbf{k}$, and $\boldsymbol{\omega}_{03} = \omega_{03}\,\mathbf{k}$.

The relative velocity of B_2 on link 2 with respect to B_1 on link 1, $\mathbf{v}^r_{B_{21}}$, has $\mathbf{I}$ and $\mathbf{J}$ components:

$$\mathbf{v}^r_{B_{21}} = v^r_{B_{21x}}\,\mathbf{I} + v^r_{B_{21y}}\,\mathbf{J} = v^r_{B_{21}}\cos\phi\,\mathbf{I} + v^r_{B_{21}}\sin\phi\,\mathbf{J},$$

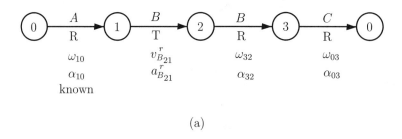

(a)

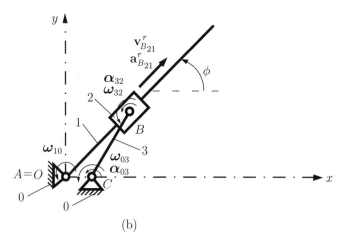

(b)

FIGURE I.5.5 *First contour RTRR: (a) diagram, and (b) mechanism.*

where $v^r_{B_{21}}$ is the magnitude of the vector $\mathbf{v}^r_{B_{21}}$ i.e., $|\mathbf{v}^r_{B+21}| = v^r_{B_{21}}$. The sign of the unknown relative velocities is selected as positive as shown in Figures I.5.4(a) and I.5.5(b). The numerical computation will then give the correct orientation of the unknown vectors. The unknowns in Eq. (I.5.37) are ω_{32}, ω_{03}, and $v^r_{B_{21}}$. Equation (I.5.37) becomes

$$\omega_{10}\,\mathbf{k} + \omega_{32}\,\mathbf{k} + \omega_{03}\,\mathbf{k} = \mathbf{0},$$

$$\begin{vmatrix} \mathbf{i} & \mathbf{j} & \mathbf{k} \\ x_B & y_B & 0 \\ 0 & 0 & \omega_{32} \end{vmatrix} + \begin{vmatrix} \mathbf{i} & \mathbf{j} & \mathbf{k} \\ x_C & y_C & 0 \\ 0 & 0 & \omega_{03} \end{vmatrix}$$

$$+ v^r_{B_{21}} \cos\phi\,\mathbf{i} + v^r_{B_{21}} \sin\phi\,\mathbf{j} = \mathbf{0}. \qquad (I.5.38)$$

Equation (I.5.38) is projected onto the "fixed" reference frame $Oxyz$:

$$\omega_{10} + \omega_{32} + \omega_{03} = 0,$$

$$y_B\,\omega_{32} + y_C\,\omega_{03} + v^r_{B_{21}} \cos\phi = 0,$$

$$-x_B\,\omega_{32} - x_C\,\omega_{03} + v^r_{B_{21}} \sin\phi = 0, \qquad (I.5.39)$$

or numerically as

$$10.472 + \omega_{32} + \omega_{03} = 0,$$

$$0.256\,\omega_{32} + v^r_{B_{21}} \cos 45° = 0,$$

$$-0.256\,\omega_{32} - 0.100\,\omega_{03} + v^r_{B_{21}} \sin 45° = 0. \qquad (I.5.40)$$

Equation (I.5.40) represents a system of three equations with three unknowns: ω_{32}, ω_{03}, and $v^r_{B_{21}}$. Solving the algebraic equations, the following numerical values are obtained: $\omega_{32} = 2.539$ rad/s, $\omega_{03} = -13.011$ rad/s, and $v^r_{B_{21}} = -0.920$ m/s.

The absolute angular velocity of link 3 is

$$\boldsymbol{\omega}_{30} = -\omega_{03} = 13.011\,\mathbf{k}\ \text{rad/s}. \qquad (I.5.41)$$

The velocity of point $B_2 = B_3$ is computed with the expression of velocity field of two points (B_3 and C) on the same rigid body (link 3):

$$\mathbf{v}_{B_2} = \mathbf{v}_{B_3} = \mathbf{v}_C + \boldsymbol{\omega}_{30} \times \mathbf{r}_{CB} = \begin{vmatrix} \mathbf{i} & \mathbf{j} & \mathbf{k} \\ 0 & 0 & \omega_{30} \\ x_B - x_C & y_B - y_C & 0 \end{vmatrix}$$

$$= \begin{vmatrix} \mathbf{i} & \mathbf{j} & \mathbf{k} \\ 0 & 0 & 13.011 \\ 0.256 - 0.100 & 0.256 & 0 \end{vmatrix} = -3.333\,\mathbf{i} + 2.032\,\mathbf{j}\ \text{m/s},$$

where $\mathbf{v}_C = \mathbf{0}$ because joint C is grounded.

Link 2 and driver link 1 have the same angular velocity:

$$\omega_{10} = \omega_{20} = \omega_{30} + \omega_{23} = 13.011\mathbf{k} - 2.539\mathbf{k} = 10.472\mathbf{k} \text{ rad/s.}$$

The velocity of point B_1 on link 1 is calculated with the expression of velocity field of two points (B_1 and A) on the same rigid body (link 1):

$$\mathbf{v}_{B_1} = \mathbf{v}_A + \omega_{10} \times \mathbf{r}_{AB} = \omega_{10} \times \mathbf{r}_{AB} = \begin{vmatrix} \mathbf{i} & \mathbf{j} & \mathbf{k} \\ 0 & 0 & \omega_{10} \\ x_B & y_B & 0 \end{vmatrix}$$

$$= \begin{vmatrix} \mathbf{i} & \mathbf{j} & \mathbf{k} \\ 0 & 0 & 10.472 \\ 0.256 & 0.256 & 0 \end{vmatrix} = -2.682\,\mathbf{i} + 2.682\,\mathbf{j} \text{ m/s.}$$

Another way of calculating the velocity of the point $B_2 = B_3$ is with the help of velocity field of two points (B_1 and B_2) not situated on the same rigid body (B_1 is on link 1 and B_2 is on link 2):

$$\mathbf{v}_{B_2} = \mathbf{v}_{B_1} + \mathbf{v}^r_{B_{21}},$$

where $\mathbf{v}^r_{B_{21}} = v^r_{B_{21}} \cos\phi\,\mathbf{i} + v^r_{B_{21}} \sin\phi\mathbf{j} = -0.651\,\mathbf{i} - 0.651\,\mathbf{j}$ m/s.
For the acceleration analysis, the following equations are written using Eq. (I.5.36):

$$\alpha_{10} + \alpha_{32} + \alpha_{03} = \mathbf{0},$$
$$\mathbf{r}_{AB} \times \alpha_{32} + \mathbf{r}_{AC} \times \alpha_{03} + \mathbf{a}_{B_{21}} + \mathbf{a}^c_{B_{21}} - \omega^2_{10}\mathbf{r}_{AB} - \omega^2_{30}\mathbf{r}_{BC} = \mathbf{0}, \qquad\text{(I.5.42)}$$

where $\alpha_{10} = \dot{\omega}_{10}\,\mathbf{k} = \mathbf{0}$, $\alpha_{32} = \alpha_{32}\,\mathbf{k}$, and $\alpha_{03} = \alpha_{03}\,\mathbf{k}$.
The relative acceleration of B_2 on link 2 with respect to B_1 on link 1, $\mathbf{a}^r_{B_{21}}$, has $\mathbf{i}$ and $\mathbf{j}$ components:

$$\mathbf{a}^r_{B_{21}} = a^r_{B_{21x}}\,\mathbf{i} + a^r_{B_{21y}}\mathbf{j} = a^r_{B_{21}} \cos\phi\,\mathbf{i} + a^r_{B_{21}} \sin\phi\,\mathbf{j}.$$

The sign of the unknown relative accelerations is selected positive and then the numerical computation will give the correct orientation of the unknown acceleration vectors. The expression for the Coriolis acceleration is

$$\mathbf{a}^c_{B_{21}} = 2\,\omega_{10} \times \mathbf{v}^r_{B_{21}} = 2\,\omega_{20} \times \mathbf{v}^r_{B_{21}}$$

$$= 2\begin{vmatrix} \mathbf{i} & \mathbf{j} & \mathbf{k} \\ 0 & 0 & \omega_{10} \\ v^r_{B_{21}} \cos\phi & v^r_{B_{21}} \sin\phi & 0 \end{vmatrix} = -2v^r_{B_{21}}\omega_{10} \sin\phi\,\mathbf{i} + 2v^r_{B_{21}}\omega_{10} \cos\phi\,\mathbf{j}$$

$$= -2(-0.920)(10.472) \sin 45°\,\mathbf{i} + 2(-0.920)(10.472) \cos 45°\,\mathbf{j}$$

$$= 13.629\mathbf{i} - 13.629\,\mathbf{j} \text{ m/s}^2.$$

The unknowns in Eq. (I.5.42) are α_{32}, α_{03}, and $a^r_{B_{21}}$. Equation (I.5.42) becomes

$$\alpha_{32}\,\mathbf{k} + \alpha_{03}\,\mathbf{k} = \mathbf{0},$$

$$\begin{vmatrix} \mathbf{I} & \mathbf{J} & \mathbf{k} \\ x_B & y_B & 0 \\ 0 & 0 & \alpha_{32} \end{vmatrix} + \begin{vmatrix} \mathbf{I} & \mathbf{J} & \mathbf{k} \\ x_C & y_C & 0 \\ 0 & 0 & \alpha_{03} \end{vmatrix} + a^r_{B_{21}} \cos\phi\,\mathbf{I} + a^r_{B_{21}} \sin\phi\,\mathbf{J}$$

$$+ \mathbf{a}^c_{B_{21}} - \omega^2_{10}(x_B\,\mathbf{I} + y_B\,\mathbf{J}) - \omega^2_{30}[(x_C - x_B)\,\mathbf{I} + (y_C - y_B)\,\mathbf{J}] = \mathbf{0}.$$

The previous equations are projected onto the "fixed" reference frame $Oxyz$:

$$\alpha_{32} + \alpha_{03} = 0,$$

$$y_B\,\alpha_{32} + y_C\,\alpha_{03} + a^r_{B_{21}}\cos\phi - 2v^r_{B_{21}}\omega_{10}\sin\phi - \omega^2_{10}x_B - \omega^2_{30}(x_C - x_B) = 0,$$

$$-x_B\,\alpha_{32} - x_C\,\alpha_{03} + a^r_{B_{21}}\sin\phi + 2v^r_{B_{21}}\omega_{10}\cos\phi - \omega^2_{10}y_B - \omega^2_{30}(y_C - y_B) = 0,$$

or numerically as

$$\alpha_{32} + \alpha_{03} = 0,$$

$$0.256\,\alpha_{32} + a^r_{B_{21}}\cos 45° + 13.626 - (10.472)^2(0.256)$$

$$- (13.011)^2(0.100 - 0.256) = 0,$$

$$- 0.256\,\alpha_{32} - 0.100\,\alpha_{03} + a^r_{B_{21}}\sin 45° - 13.626 - (10.472)^2(0.256)$$

$$- (13.011)^2(0 - 0.256) = 0. \tag{I.5.43}$$

Equation (I.5.43) represents a system of three equations with three unknowns: α_{32}, α_{03}, and $a^r_{B_{21}}$. Solving the algebraic equations, the following numerical values are obtained: $\alpha_{32} = -25.032$ rad/s^2, $\alpha_{03} = 25.032$ rad/s^2, and $a^r_{B_{21}} = -7.865$ m/s^2.

The absolute angular acceleration of link 3 is

$$\alpha_{30} = -\alpha_{03} = -25.032\,\mathbf{k}\ \text{rad/s}^2.$$

The velocity of the point $B_2 = B_3$ is computed with the expression of velocity field of two points (B_3 and C) on the same rigid body (link 3):

$$\mathbf{v}_{B_2} = \mathbf{v}_{B_3} = \mathbf{v}_C + \boldsymbol{\omega}_{30} \times \mathbf{r}_{CB} = \begin{vmatrix} \mathbf{I} & \mathbf{J} & \mathbf{k} \\ 0 & 0 & \omega_{30} \\ x_B - x_C & y_B - y_C & 0 \end{vmatrix}$$

$$= \begin{vmatrix} \mathbf{I} & \mathbf{J} & \mathbf{k} \\ 0 & 0 & 13.011 \\ 0.256 - 0.100 & 0.256 & 0 \end{vmatrix} = -3.333\,\mathbf{I} + 2.032\,\mathbf{J}\ \text{m/s},$$

where $\mathbf{v}_C = \mathbf{0}$ because joint C is grounded.

Link 2 and driver link 1 have the same angular velocity:

$$\omega_{10} = \omega_{20} = \omega_{30} + \omega_{23} = 13.011\mathbf{k} - 2.539\,\mathbf{k} = 10.472\,\mathbf{k}\ \text{rad/s}.$$

The velocity of point B_1 on link 1 is calculated with the expression of velocity field of two points (B_1 and A) on the same rigid body (link 1):

$$\mathbf{v}_{B_1} = \mathbf{v}_A + \omega_{10} \times \mathbf{r}_{AB} = \omega_{10} \times \mathbf{r}_{AB} = \begin{vmatrix} \mathbf{I} & \mathbf{J} & \mathbf{k} \\ 0 & 0 & \omega_{10} \\ x_B & y_B & 0 \end{vmatrix}$$

$$= \begin{vmatrix} \mathbf{I} & \mathbf{J} & \mathbf{k} \\ 0 & 0 & 10.472 \\ 0.256 & 0.256 & 0 \end{vmatrix} = -2.682\,\mathbf{I} + 2.682\,\mathbf{J}\ \text{m/s}.$$

Another way of calculating the velocity of the point $B_2 = B_3$ is with the help of velocity field of two points (B_1 and B_2) not situated on the same rigid body (B_1 is on link 1 and B_2 is on link 2):

$$\mathbf{v}_{B_2} = \mathbf{v}_{B_1} + \mathbf{v}_{B_{21}}^r,$$

where $\mathbf{v}_{B_{21}}^r = v_{B_{21}}^r \cos\phi\,\mathbf{I} + v_{B_{21}}^r \sin\phi\,\mathbf{J} = -0.651\,\mathbf{I} - 0.651\ \mathbf{J}$ m/s.
The angular acceleration of link 3 is:

$$\alpha_{30} = -\alpha_{03} = \alpha_{32} = -25.032\,\mathbf{k}\ \text{rad/s}^2.$$

The absolute linear acceleration of point B_3 is computed as follows:

$$\mathbf{a}_{B_3} = \mathbf{a}_C + \alpha_{30} \times \mathbf{r}_{CB} - \omega_{30}^2\mathbf{r}_{CB} = -20.026\,\mathbf{I} - 47.277\,\mathbf{J}\ \text{m/s}^2.$$

Second Contour Analysis

According to Figure I.5.6, the second contour is described as

- rotational joint R between links 0 and 3 (joint C);
- rotational joint R between links 3 and 4 (joint B) ;
- rotational joint R between links 4 and 5 (joint D_R);
- translational joint T between links 5 and 0 (joint D_T).

For the velocity analysis, the following equations are written:

$$\omega_{30} + \omega_{43} + \omega_{54} = \mathbf{0},$$

$$\mathbf{r}_{AC} \times \omega_{30} + \mathbf{r}_{AB} \times \omega_{43} + \mathbf{r}_{AD} \times \omega_{54} + \mathbf{v}_{D05}^r = \mathbf{0}. \tag{I.5.44}$$

The relative linear velocity $\mathbf{v}_{D05}^r$ has only one component, along the x-axis:

$$\mathbf{v}_{D05}^r = v_{D05}^r\,\mathbf{I}.$$

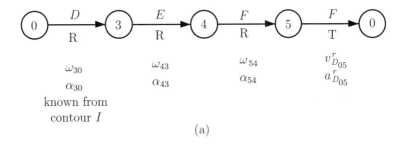

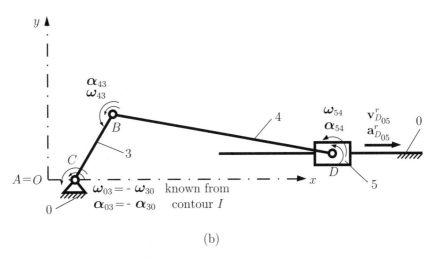

FIGURE I.5.6 *Second contour RRRT: (a) diagram, and (b) mechanism.*

The unknown parameters in Eq. (I.5.44) are ω_{43}, ω_{54}, and $v^r_{D_{05}}$. The following numerical values are obtained: $\omega_{43} = -15.304$ rad/s, $\omega_{54} = 2.292$ rad/s, and $v^r_{D_{05}} = 3.691$ m/s.

The angular velocity of the link BD is

$$\boldsymbol{\omega}_{40} = \boldsymbol{\omega}_{30} + \boldsymbol{\omega}_{43} = -\boldsymbol{\omega}_{54} = -2.292\,\mathbf{k} \text{ rad/s.}$$

The absolute linear velocity of the point $D_4 = D_5$ is computed as follows:

$$\mathbf{v}_{D_4} = \mathbf{v}_{D_5} = \mathbf{v}_{B_4} + \boldsymbol{\omega}_{40} \times \mathbf{r}_{BD} = -\mathbf{v}^r_{D_{05}} = -3.691\,\boldsymbol{\imath} \text{ m/s,}$$

where $\mathbf{v}_{B_4} = \mathbf{v}_{B_3}$.

For the acceleration analysis the following equations exist:

$$\boldsymbol{\alpha}_{30} + \boldsymbol{\alpha}_{43} + \boldsymbol{\alpha}_{54} = \mathbf{0},$$

$$\mathbf{a}^r_{D_{05}} + \mathbf{a}^c_{D_{05}} + \mathbf{r}_{AC} \times \boldsymbol{\alpha}_{30} + \mathbf{r}_{AB} \times \boldsymbol{\alpha}_{43} + \mathbf{r}_{AD} \times \boldsymbol{\alpha}_{54}$$

$$- \omega^2_{30}\mathbf{r}_{CB} - \omega^2_{40}\mathbf{r}_{BD} = \mathbf{0}. \tag{I.5.45}$$

Because the slider 5 does not rotate ($\omega_{50} = \mathbf{0}$), the Coriolis acceleration is

$$\mathbf{a}^c_{D_{05}} = 2\omega_{50} \times \mathbf{v}^r_{D_{05}} = \mathbf{0}.$$

The unknowns in Eq. (I.5.45) are α_{43}, α_{54}, and $a^r_{D_{05}}$. The following numerical results are obtained: $\alpha_{43} = 77.446$ rad/s^2, $\alpha_{54} = -52.414$ rad/s^2, and $a^r_{D_{05}} = 16.499$ m/s^2.

The absolute angular acceleration of the link BD is

$$\alpha_{40} = \alpha_{30} + \alpha_{43} = \alpha_{45} = 52.414 \, \mathbf{k} \text{ rad/s}^2,$$

and the linear acceleration of the point $D_4 = D_5$ is

$$\mathbf{a}_{D_4} = \mathbf{a}_{D_5} = \mathbf{a}_{B_4} + \alpha_{40} \times \mathbf{r}_{BD} - \omega^2_{40}\mathbf{r}_{BD} = -16.499 \, \mathbf{\imath} \text{ m/s}^2,$$

where $\mathbf{a}_{B_4} = \mathbf{a}_{B_3}$.

I.5.5 Problems

I.5.1 The four-bar mechanism shown in Figure I.3.10(a) has the dimensions: $AB = 80$ mm, $BC = 210$ mm, $CD = 120$ mm, and $AD = 190$ mm. The driver link AB rotates with a constant angular speed of 200 rpm. Find the velocities and the accelerations of the four-bar mechanism using the contour equations method for the case when the angle of the driver link AB with the horizontal axis is $\phi = 60°$.

I.5.2 The angular speed of the driver link 1, of the mechanism shown in Figure I.4.9, is $\omega = \omega_1 = 20$ rad/s. The distance from link 3 to the horizontal axis Ax is $a = 55$ mm. Using the contour equations, find the velocity and the acceleration of point C on link 3 for $\phi = 30°$.

I.5.3 The slider crank mechanism shown in Figure I.4.10 has the dimensions $AB = 0.4$ m and $BC = 1$ m. The driver link 1 rotates with a constant angular speed of $n = 160$ rpm. Find the velocity and acceleration of the slider 3 using the contour equations when the angle of the driver link with the horizontal axis is $\phi = 30°$.

I.5.4 The planar mechanism considered is shown in Figure I.3.19. The following data are given: $AB = 0.150$ m, $BC = 0.400$ m, $CD = 0.370$ m, $CE = 0.230$ m, $EF = CE$, $L_a = 0.300$ m, $L_b = 0.450$ m, and $L_c = CD$. The angular speed of the driver link 1 is constant and has the value 180 rpm. Using the contour equations method, find the velocities and the accelerations of the mechanism for $\phi = \phi_1 = 30°$.

I.5.5 The R-RRR-RTT mechanism is shown in Figure I.3.20. The following data are given: $AB = 0.080$ m, $BC = 0.350$ m, $CE = 0.200$ m, $CD = 0.150$ m, $L_a = 0.200$ m, $L_b = 0.350$ m, and $L_c = 0.040$ m. The driver link 1 rotates with a constant angular speed of $n = 1200$ rpm. For $\phi = 145°$ find the velocities and the accelerations of the mechanism with the contour equations.

I.5.6 The mechanism shown in Figure I.3.21 has the following dimensions: $AB = 80$ mm, $AD = 250$ mm, $BC = 180$ mm, $CE = 60$ mm, $EF = 200$ mm, and $a = 170$ mm. The constant angular speed of the driver link 1 is $n = 400$ rpm. Find the velocities and the accelerations of the mechanism using the contour equations when the angle of the driver link 1 with the horizontal axis is $\phi = \phi_1 = 300°$.

I.5.7 The dimensions for the mechanism shown in Figure I.3.22 are: $AB = 150$ mm, $BD = 400$ mm, $BC = 140$ mm, $CD = 400$ mm, $DE = 250$ mm, $CF = 500$ mm, $AE = 380$ mm, and $b = 100$ mm. The constant angular speed of the driver link 1 is $n = 40$ rpm. Find the velocities and the accelerations of the mechanism for $\phi = \phi_1 = 210°$. Use the contour equations method.

I.5.8 The mechanism in Figure I.3.23 has the dimensions: $AB = 200$ mm, $AC = 100$ mm, $BD = 400$ mm, $DE = 550$ mm, $EF = 300$ mm, $L_a = 500$ mm, and $L_b = 100$ mm. Using the contour equations method find the velocities and the accelerations of the mechanism if the constant angular speed of the driver link 1 is $n = 70$ rpm, and for $\phi = \phi_1 = 210°$.

I.5.9 The dimensions for the mechanism shown in Figure I.3.24 are: $AB = 150$ mm, $BC = 400$ mm, $AD = 360$ mm, $CD = 210$ mm, $DE = 130$ mm, $EF = 400$ mm, and $L_a = 40$ mm. The constant angular speed of the driver link 1 is $n = 250$ rpm. Using the contour equations find the velocities and the accelerations of the mechanism for $\phi = \phi_1 = 30°$.

I.5.10 The mechanism in Figure I.3.25 has the dimensions: $AB = 250$ mm, $AC = 800$ mm, $BD = 1200$ mm, $L_a = 180$ mm, and $L_b = 300$ mm. The driver link 1 rotates with a constant angular speed of $n = 50$ rpm. Find the velocities and the accelerations of the mechanism for $\phi = \phi_1 = 210°$. Use the contour equations method.

I.5.11 Figure I.3.26 shows a mechanism with the following dimensions: $AB = 120$ mm, $BD = 400$ mm, and $L_a = 150$ mm. The constant angular speed of the driver link 1 is $n = 600$ rpm. Find the velocities and the accelerations of the mechanism, using the contour equations method, when the angle of the driver link 1 with the horizontal axis is $\phi = 210°$.

I.5.12 The mechanism in Figure I.3.27 has the dimensions: $AB = 200$ mm, $AC = 500$ mm, $BD = 800$ mm, $DE = 400$ mm, $EF = 270$ mm, $L_a = 70$ mm, and $L_b = 300$ mm. The constant angular speed of the driver link 1 is $n = 40$ rpm. Using the contour equations, find the velocities and the accelerations of the mechanism for $\phi = \phi_1 = 300°$.

I.5.13 Figure I.3.28 shows a mechanism with the following dimensions: $AB = 200$ mm, $BC = 750$ mm, $CD = DE = 300$ mm, $EF = 500$ mm, $L_a = 750$ mm, and $L_b = L_c = 250$ mm. Find the velocities and the accelerations of the mechanism, using the contour equations method, if the constant angular speed of the driver link 1 is $n = 1100$ rpm and for $\phi = \phi_1 = 120°$.

I.5.14 Figure I.3.29 shows a mechanism with the following dimensions: $AB = 120$ mm, $BC = 550$ mm, $CE = 180$ mm, $CD = 350$ mm, $EF = 300$ mm, $L_a = 320$ mm, $L_b = 480$ mm, and $L_c = 600$ mm. The constant angular speed of the driver link 1 is $n = 100$ rpm. Find the velocities and the accelerations of the mechanism, using the contour equations, for $\phi = \phi_1 = 30°$.

I.5.15 Figure I.3.30 shows a mechanism with the following dimensions: $AB = 180$ mm, $BC = 520$ mm, $CF = 470$ mm, $CD = 165$ mm, $DE = 540$ mm, $L_a = 630$ mm, $L_b = 360$ mm, and $L_c = 210$ mm. The constant angular speed of the driver link 1 is

$n = 70$ rpm. Use the contour equations to calculate the velocities and the accelerations of the mechanism when the angle of the driver link 1 with the horizontal axis is $\phi = 210°$.

I.5.16 Figure I.3.31 shows a mechanism with the following dimensions: $AB = 60$ mm, $BC = 150$ mm, $AD = 70$ mm, and $BE = 170$ mm. The constant angular speed of the driver link 1 is $n = 300$ rpm. Find the velocities and the accelerations of the mechanism, using the contour equations, if the angle of the driver link 1 with the horizontal axis is $\phi = 210°$.

I.5.17 The dimensions of the mechanism shown in Figure I.3.32 are: $AB = 90$ mm, $BC = 240$ mm, $BE = 400$ mm, $CE = 600$ mm, $CD = 220$ mm, $EF = 900$ mm, $L_a = 250$ mm, $L_b = 150$ mm, and $L_c = 100$ mm. The constant angular speed of the driver link 1 is $n = 50$ rpm. Employing the contour equations find the velocities and the accelerations of the mechanism for $\phi = \phi_1 = 210°$.

I.5.18 The dimensions of the mechanism shown in Figure I.3.33 are: $AB = 180$ mm, $AC = 300$ mm, $CD = 400$ mm, $DE = 200$ mm, and $L_a = 360$ mm. The constant angular speed of the driver link 1 is $n = 90$ rpm. Use the contour equations to calculate the velocities and the accelerations of the mechanism when the angle of the driver link 1 with the horizontal axis is $\phi = 30°$.

I.5.19 The dimensions of the mechanism shown in Figure I.3.34 are: $AB = 80$ mm, $AC = 40$ mm, $CD = 100$ mm, and $DE = 300$ mm. The constant angular speed of the driver link 1 is $n = 60$ rpm. Use the contour equations to calculate the velocities and the accelerations of the mechanism for $\phi = \phi_1 = 210°$.

I.5.20 The dimensions of the mechanism shown in Figure I.3.35 are: $AB = 200$ mm, $AC = 350$ mm, and $CD = 600$ mm. For the distance b select a suitable value. The constant angular speed of the driver link 1 is $n = 90$ rpm. Find the velocities and the accelerations of the mechanism for $\phi = \phi_1 = 120°$.

I.5.21 The dimensions of the mechanism shown in Figure I.3.36 are: $AB = 80$ mm, $AC = 60$ mm, and $CD = 70$ mm. The constant angular speed of the driver link 1 is $n = 220$ rpm. Find the velocities and the accelerations of the mechanism, using the contour equations, for $\phi = \phi_1 = 240°$.

I.5.22 The dimensions of the mechanism shown in Figure I.3.37 are: $AB = 150$ mm, $AC = 420$ mm, $BD = L_a = 650$ mm, and $DE = 350$ mm. The constant angular speed of the driver link 1 is $n = n_1 = 650$ rpm. Find the velocities and the accelerations of the mechanism, using the contour equations, for $\phi = \phi_1 = 240°$.

I.5.23 The dimensions of the mechanism shown in Figure I.3.38 are: $AB = 200$ mm, $AD = 500$ mm, and $BC = 250$ mm. The constant angular speed of the driver link 1 is $n = 160$ rpm. Use the contour equations to calculate the velocities and the accelerations of the mechanism for $\phi = \phi_1 = 240°$. Select a suitable value for the distance a.

I.5.24 The mechanism in Figure I.3.11(a) has the dimensions: $AB = 0.20$ m, $AD = 0.40$ m, $CD = 0.70$ m, $CE = 0.30$ m, and $y_E = 0.35$ m. The constant angular speed of the driver link 1 is $n = 2600$ rpm. Using the contour equations find the velocities and the accelerations of the mechanism for the given input angle $\phi = \phi_1 = 210°$.

I.5.25 The mechanism in Figure I.3.12 has the dimensions: $AB = 0.03$ m, $BC = 0.05$ m, $CD = 0.08$ m, $AE = 0.07$ m, and $L_a = 0.025$ m. The constant angular speed of the driver link 1 is $n = 90$ rpm. Employing the contour equations, find the velocities and the accelerations of the mechanism for the given input angle $\phi = \phi_1 = \pi/3$.

I.5.26 The mechanism in Figure I.3.15 has the dimensions: $AC = 0.200$ m, $BC = 0.300$ m, $BD = 1.000$ m, and $L_a = 0.050$ m. The constant angular speed of the driver link 1 is $n = 1500$ rpm. Use the contour equations to calculate the velocities and the accelerations of the mechanism for $\phi = 330°$.

I.6   Dynamic Force Analysis

For a kinematic chain it is important to know how forces and moments are transmitted from the input to the output, so that the links can be properly designated. The friction effects are assumed to be negligible in the force analysis presented here.

I.6.1   Equation of Motion for the Mass Center

Consider a system of N particles. A particle is an object whose shape and geometrical dimensions are not significant to the investigation of its motion. An arbitrary collection of matter with total mass m can be divided into N particles, the ith particle having mass, m_i (Fig. I.6.1):

$$m = \sum_{i=1}^{N} m_i.$$

A rigid body can be considered as a collection of particles in which the number of particles approaches infinity and in which the distance between any two points remains constant. As N approaches infinity, each particle is treated as a differential mass element, $m_i \to dm$, and the summation is replaced by integration over the body:

$$m = \int_{\text{body}} dm.$$

The position of the mass center of a collection of particles is defined by

$$\mathbf{r}_C = \frac{1}{m} \sum_{i=1}^{N} m_i \mathbf{r}_i, \qquad (\text{I.6.1})$$

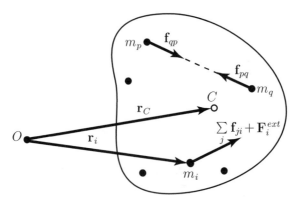

FIGURE I.6.1 *Rigid body as a collection of particles.*

where $\mathbf{r}_i$ is the position vector from the origin O to the ith particle. As $N \to \infty$, the summation is replaced by integration over the body:

$$\mathbf{r}_C = \frac{1}{m} \int\limits_{body} \mathbf{r}\, dm, \tag{I.6.2}$$

where $\mathbf{r}$ is the vector from the origin O to differential element dm.

The time derivative of Eq. (I.6.1) gives

$$\sum_{i=1}^{N} m_i \frac{d^2\mathbf{r}_i}{dt^2} = m\frac{d^2\mathbf{r}_C}{dt^2} = m\mathbf{a}_C, \tag{I.6.3}$$

where $\mathbf{a}_C$ is the acceleration of the mass center. The acceleration of the mass center can be related to the external forces acting on the system. This relationship is obtained by applying Newton's laws to each of the individual particles in the system. Any such particle is acted on by two types of forces. One type is exerted by other particles that are also part of the system. Such forces are called internal forces (internal to the system). Additionally, a particle can be acted on by a force that is exerted by a particle or object not included in the system. Such a force is known as an external force (external to the system). Let $\mathbf{f}_{ij}$ be the internal force exerted on the jth particle by the ith particle. Newton's third law (action and reaction) states that the jth particle exerts a force on the ith particle of equal magnitude, and opposite direction, and collinear with the force exerted by the ith particle on the jth particle (Fig. I.6.1):

$$\mathbf{f}_{ji} = -\mathbf{f}_{ij}, \quad j \neq i.$$

Newton's second law for the ith particle must include all of the internal forces exerted by all of the other particles in the system on the ith particle, plus the sum of any external forces exerted by particles, objects outside of the system on the ith particle:

$$\sum_j \mathbf{f}_{ji} + \mathbf{F}_i^{ext} = m_i \frac{d^2 \mathbf{r}_i}{dt^2}, \quad j \neq i, \tag{I.6.4}$$

where $\mathbf{F}_i^{ext}$ is the external force on the ith particle. Equation (I.6.4) is written for each particle in the collection of particles. Summing the resulting equations over all of the particles from $i = 1$ to N the following relation is obtained:

$$\sum_i \sum_j \mathbf{f}_{ji} + \sum_i \mathbf{F}_i^{ext} = m\mathbf{a}_C, \quad j \neq i. \tag{I.6.5}$$

The sum of the internal forces includes pairs of equal and opposite forces. The sum of any such pair must be zero. The sum of all of the internal forces on the collection of particles is zero (Newton's third law):

$$\sum_i \sum_j \mathbf{f}_{ji} = \mathbf{0}, \quad j \neq i.$$

The term $\sum_i \mathbf{F}_i^{ext}$ is the sum of the external forces on the collection of particles:

$$\sum_i \mathbf{F}_i^{ext} = \mathbf{F}.$$

One can conclude that the sum of the external forces acting on a closed system equals the product of the mass and the acceleration of the mass center:

$$m \, \mathbf{a}_C = \mathbf{F}. \tag{I.6.6}$$

Considering Figure I.6.2 for a rigid body and introducing the distance $\mathbf{q}$ in Eq. (I.6.2) gives

$$\mathbf{r}_C = \frac{1}{m} \int_{body} \mathbf{r} \, dm = \frac{1}{m} \int_{body} (\mathbf{r}_C + \mathbf{q}) \, dm = \mathbf{r}_C + \frac{1}{m} \int_{body} \mathbf{q} \, dm. \tag{I.6.7}$$

It results

$$\frac{1}{m} \int_{body} \mathbf{q} \, dm = \mathbf{0}, \tag{I.6.8}$$

that is the weighed average of the displacement vector about the mass center is zero. The equation of motion for the differential element dm is

$$\mathbf{a} \, dm = d\mathbf{F},$$

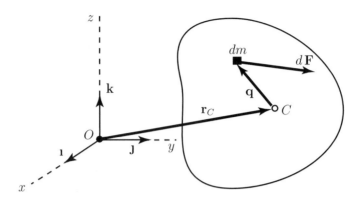

FIGURE I.6.2 *Rigid body with differential element dm.*

where $d\mathbf{F}$ is the total force acting on the differential element. For the entire body:

$$\int_{\text{body}} \mathbf{a}\, dm = \int_{\text{body}} d\mathbf{F} = \mathbf{F}, \qquad (\text{I}.6.9)$$

where $\mathbf{F}$ is the resultant of all forces. This resultant contains contributions only from the external forces, as the internal forces cancel each other. Introducing Eq. (I.6.7) into Eq. (I.6.9), the Newton's second law for a rigid body is obtained:

$$m\,\mathbf{a}_C = \mathbf{F}$$

The derivation of the equations of motion is valid for the general motion of a rigid body. These equations are equally applicable to planar and three-dimensional motions.

Resolving the sum of the external forces into Cartesian rectangular components

$$\mathbf{F} = F_x\,\mathbf{i} + F_y\,\mathbf{j} + F_z\,\mathbf{k},$$

and the position vector of the mass center

$$\mathbf{r}_C = x_C(t)\,\mathbf{i} + y_C(t)\,\mathbf{j} + z_C(t)\,\mathbf{k},$$

Newton's second law for the rigid body is

$$m\ddot{\mathbf{r}}_C = \mathbf{F}, \qquad (\text{I}.6.10)$$

or

$$m\ddot{x}_C = F_x, \quad m\ddot{y}_C = F_y, \quad m\ddot{z}_C = F_z. \qquad (\text{I}.6.11)$$

I.6.2 Angular Momentum Principle for a System of Particles

An arbitrary system with the mass m can be divided into N particles $P_1, P_2, \ldots, P_N$. The position vector of the ith particle relative to an origin point O is $\mathbf{r}_i = \mathbf{r}_{OP_i}$ and the mass of the ith particle is m_i (Fig. I.6.3). The position of the mass center C of the system is $\mathbf{r}_C = \sum_{i=1}^{N} m_i \mathbf{r}_i / m$. The position of the particle P_i of the system relative to O is

$$\mathbf{r}_i = \mathbf{r}_C + \mathbf{r}_{CP_i}. \tag{I.6.12}$$

Multiplying Eq. (I.6.12) by m_i, summing from 1 to N, the following relation is obtained:

$$\sum_{i=1}^{N} m_i \mathbf{r}_{CP_i} = \mathbf{0}. \tag{I.6.13}$$

The total angular momentum of the system about its mass center C is the sum of the angular momenta of the particles about C:

$$\mathbf{H}_C = \sum_{i=1}^{N} \mathbf{r}_{CP_i} \times m_i \mathbf{v}_i, \tag{I.6.14}$$

where $\mathbf{v}_i = \dfrac{d\mathbf{r}_i}{dt}$ is the velocity of the particle P_i.

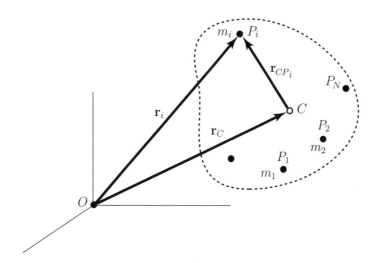

FIGURE I.6.3 *System of particles.*

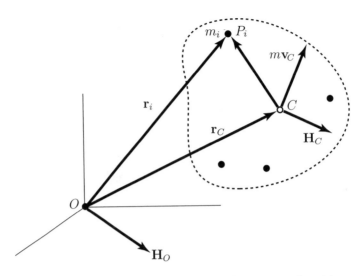

FIGURE I.6.4 *Angular momentum about the mass center for a system of particles.*

The total angular momentum of the system about O is the sum of the angular momenta of the particles

$$\mathbf{H}_O = \sum_{i=1}^{N} \mathbf{r}_i \times m_i \mathbf{v}_i = \sum_{i=1}^{N} (\mathbf{r}_C + \mathbf{r}_{CP_i}) \times m_i \mathbf{v}_i = \mathbf{r}_C \times m\mathbf{v}_C + \mathbf{H}_C, \qquad (\text{I.6.15})$$

or the total angular momentum about O is the sum of the angular momentum about O due to the velocity $\mathbf{v}_C$ of the mass center of the system and the total angular momentum about the mass center (Fig. I.6.4).

Newton's second law for the ith particle is

$$\sum_j \mathbf{f}_{ji} + \mathbf{F}_i^{ext} = m_i \frac{d\mathbf{v}_i}{dt}, \quad j \neq i,$$

and the cross product with the position vector $\mathbf{r}_i$, and sum from $i = 1$ to N gives

$$\sum_i \sum_j \mathbf{r}_i \times \mathbf{f}_{ji} + \sum_i \mathbf{r}_i \times \mathbf{F}_i^{ext} = \sum_i \mathbf{r}_i \times \frac{d}{dt}(m_i \mathbf{v}_i), \quad j \neq i. \qquad (\text{I.6.16})$$

The first term on the left side of Eq. (I.6.16) is the sum of the moments about O due to internal forces, and

$$\mathbf{r}_i \times \mathbf{f}_{ji} + \mathbf{r}_i \times \mathbf{f}_{ij} = \mathbf{r}_i \times (\mathbf{f}_{ji} + \mathbf{f}_{ij}) = \mathbf{0}, \quad j \neq i.$$

The term vanishes because the internal forces between each pair of particles are equal, opposite, and directed along the straight line between the two particles (Fig. I.6.1.) The second term on the left side of Eq. (I.6.16),

$$\sum_i \mathbf{r}_i \times \mathbf{F}_i^{ext} = \sum \mathbf{M}_O,$$

represents the sum of the moments about O due to external forces and couples. The term on the right side of Eq. (I.6.16) is

$$\sum_i \mathbf{r}_i \times \frac{d}{dt}(m_i \mathbf{v}_i) = \sum_i \left[\frac{d}{dt}(\mathbf{r}_i \times m_i \mathbf{v}_i) - \mathbf{v}_i \times m_i \mathbf{v}_i \right] = \frac{d\mathbf{H}_O}{dt}, \qquad (I.6.17)$$

which represents the rate of change of the total angular momentum of the system about the point O.

Equation (I.6.16) is rewritten as

$$\frac{d\mathbf{H}_O}{dt} = \sum \mathbf{M}_O. \qquad (I.6.18)$$

The rate of change of the angular momentum about O equals the sum of the moments about O due to external forces and couples.

Using Eqs. (I.6.15) and (I.6.18), the following result is obtained:

$$\sum \mathbf{M}_O = \frac{d}{dt}(\mathbf{r}_C \times m\mathbf{v}_C + \mathbf{H}_C) = \mathbf{r}_C \times m\mathbf{a}_C + \frac{d\mathbf{H}_C}{dt}, \qquad (I.6.19)$$

where $\mathbf{a}_C$ is the acceleration of the mass center.

With the relation

$$\sum \mathbf{M}_O = \sum \mathbf{M}_C + \mathbf{r}_C \times \mathbf{F} = \sum \mathbf{M}_C + \mathbf{r}_C \times m\mathbf{a}_C,$$

Eq. (I.6.19) becomes

$$\frac{d\mathbf{H}_C}{dt} = \sum \mathbf{M}_C. \qquad (I.6.20)$$

The rate of change of the angular momentum about the mass center equals the sum of the moments about the mass center.

I.6.3 Equations of Motion for General Plane Motion

An arbitrary rigid body with the mass m can be divided into N particles P_i, $i = 1, 2, \ldots, N$. The position vector of the P_i particle is $\mathbf{r}_i = OP_i$ and the mass of the particle is m_i. Figure I.6.5(a) represents the rigid body moving with general planar motion in the (X, Y) plane. The origin of the Cartesian reference frame is O. The mass center C of the rigid body is located in the plane of the motion, $C \in (X, Y)$.

Let $d_O = OZ$ be the axis through the fixed origin point O that is perpendicular to the plane of motion of the rigid body (X, Y), $d_O \perp (X, Y)$. Let $d_C = Czz$ be the parallel axis

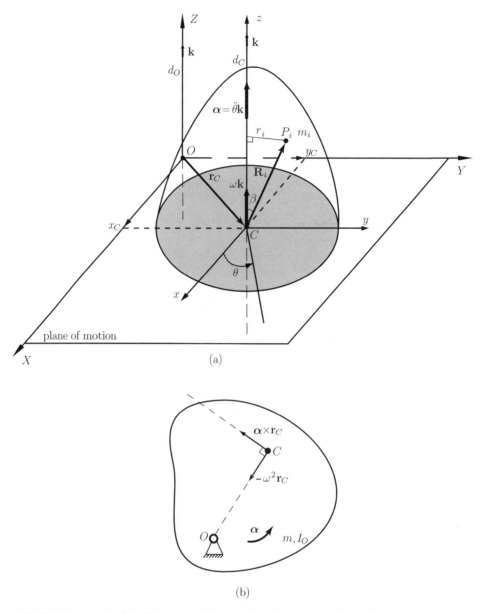

FIGURE I.6.5 *(a) Rigid body in general plane motion; (b) rotation about a fixed point.*

through the mass center C, $d_C \| d_O$. The rigid body has a general planar motion and the angular velocity vector is $\boldsymbol{\omega} = \omega \mathbf{k}$. The unit vector of the $d_C = Czz$ axis is $\mathbf{k}$.

The velocity of the P_i particle relative to the mass center is

$$\frac{d\mathbf{R}_i}{dt} = \omega \mathbf{k} \times \mathbf{R}_i,$$

where $\mathbf{R}_i = \mathbf{r}_{CP_i}$. The sum of the moments about O due to external forces and couples is

$$\sum \mathbf{M}_O = \frac{d\mathbf{H}_O}{dt} = \frac{d}{dt}[(\mathbf{r}_C \times m\mathbf{v}_C) + \mathbf{H}_C], \qquad (\text{I.6.21})$$

where

$$\mathbf{H}_C = \sum_i [\mathbf{R}_i \times m_i(\omega \mathbf{k} \times \mathbf{R}_i)],$$

is the angular momentum about d_C. The magnitude of the angular momentum about d_C is

$$
\begin{aligned}
H_C = \mathbf{H}_C \cdot \mathbf{k} &= \sum_i [\mathbf{R}_i \times m_i(\omega \mathbf{k} \times \mathbf{R}_i)] \cdot \mathbf{k} \\
&= \sum_i m_i [(\mathbf{R}_i \times \mathbf{k}) \times \mathbf{R}_i)] \cdot \mathbf{k}\,\omega = \sum_i m_i [(\mathbf{R}_i \times \mathbf{k}) \cdot (\mathbf{R}_i \times \mathbf{k})]\,\omega \\
&= \sum_i m_i \, |\mathbf{R}_i \times \mathbf{k}|^2 \, \omega = \sum_i m_i r_i^2 \omega, \qquad (\text{I.6.22})
\end{aligned}
$$

where the term $|\mathbf{k} \times \mathbf{R}_i| = r_i$ is the perpendicular distance from d_C to the P_i particle. The identity

$$(\mathbf{a} \times \mathbf{b}) \cdot \mathbf{c} = \mathbf{a} \cdot (\mathbf{b} \times \mathbf{c}).$$

has been used.

The summation $\sum_i m_i r_i^2$ is replaced by integration over the body $\int r^2 \, dm$ and is defined as mass moment of inertia I_{Czz} of the body about the z-axis through C:

$$I_{Czz} = \sum_i m_i r_i^2.$$

The mass moment of inertia I_{Czz} is a constant property of the body and is a measure of the rotational inertia or resistance to change in angular velocity due to the radial distribution of the rigid body mass around z-axis through C.

Equation (I.6.22) defines the angular momentum of the rigid body about d_C (z-axis through C):

$$H_C = I_{Czz}\,\omega \quad \text{or} \quad \mathbf{H}_C = I_{Czz}\,\omega\,\mathbf{k} = I_{Czz}\,\boldsymbol{\omega}.$$

Substituting this expression into Eq. (I.6.21) gives

$$\sum \mathbf{M}_O = \frac{d}{dt}[(\mathbf{r}_C \times m\mathbf{v}_C) + I_{Czz}\boldsymbol{\omega}] = (\mathbf{r}_C \times m\mathbf{a}_C) + I_{Czz}\boldsymbol{\alpha}. \qquad (\text{I.6.23})$$

The rotational equation of motion for the rigid body is

$$I_{Czz}\,\boldsymbol{\alpha} = \sum \mathbf{M}_C \quad \text{or} \quad I_{Czz}\,\alpha\,\mathbf{k} = \sum M_C\,\mathbf{k}. \qquad (\text{I.6.24})$$

For general planar motion the angular acceleration is

$$\boldsymbol{\alpha} = \dot{\boldsymbol{\omega}} = \ddot{\theta}\mathbf{k}, \tag{I.6.25}$$

where the angle θ describes the position, or orientation, of the rigid body about a fixed axis.

If the rigid body is a plate moving in the plane of motion (X, Y), the mass moment of inertia of the rigid body about z-axis through C becomes the polar mass moment of inertia of the rigid body about C, $I_{Czz} = I_C$. For this case the Eq. (I.6.24) gives

$$I_C\,\boldsymbol{\alpha} = \sum \mathbf{M}_C. \tag{I.6.26}$$

A special application of Eq. (6.26) is for rotation about a fixed point. Consider the special case when the rigid body rotates about the fixed point O as shown in Figure I.6.5(b). It follows that the acceleration of the mass center is expressed as

$$\mathbf{a}_C = \boldsymbol{\alpha} \times \mathbf{r}_C - \omega^2 \mathbf{r}_C. \tag{I.6.27}$$

The relation between the sum of the moments of the external forces about the fixed point O and the product $I_{Czz}\,\boldsymbol{\alpha}$ is given by Eq. (I.6.23):

$$\sum \mathbf{M}_O = \mathbf{r}_C \times m\mathbf{a}_C + I_{Czz}\boldsymbol{\alpha}. \tag{I.6.28}$$

Equations (I.6.27) and (I.6.28) give

$$
\begin{aligned}
\sum \mathbf{M}_O &= \mathbf{r}_C \times m\,(\boldsymbol{\alpha} \times \mathbf{r}_C - \omega^2 \mathbf{r}_C) + I_{Czz}\boldsymbol{\alpha} \\
&= m\,\mathbf{r}_C \times (\boldsymbol{\alpha} \times \mathbf{r}_C) + I_{Czz}\boldsymbol{\alpha} \\
&= m\,[(\mathbf{r}_C \cdot \mathbf{r}_C)\boldsymbol{\alpha} - (\mathbf{r}_C \cdot \boldsymbol{\alpha})\mathbf{r}_C] + I_{Czz}\boldsymbol{\alpha} \\
&= m\,r_C^2\boldsymbol{\alpha} + I_{Czz}\boldsymbol{\alpha} = (m\,r_C^2 + I_{Czz})\boldsymbol{\alpha}. \tag{I.6.29}
\end{aligned}
$$

According to parallel-axis theorem

$$I_{Ozz} = I_{Czz} + m\,r_C^2$$

where I_{Ozz} denotes the mass moment of inertia of the rigid body about z-axis through O. For the special case of rotation about a fixed point O one can use the formula

$$I_{Ozz}\boldsymbol{\alpha} = \sum \mathbf{M}_O. \tag{I.6.30}$$

The general equations of motion for a rigid body in plane motion are (Fig. I.6.6):

$$\mathbf{F} = m\mathbf{a}_C \quad \text{or} \quad \mathbf{F} = m\ddot{\mathbf{r}}_C, \tag{I.6.31}$$

$$\sum \mathbf{M}_C = I_{Czz}\,\boldsymbol{\alpha}, \tag{I.6.32}$$

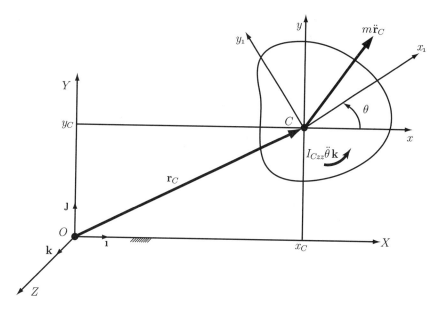

FIGURE I.6.6 *Rigid body in plane motion.*

or using the Cartesian components:

$$m\ddot{x}_C = \sum F_x, \quad m\ddot{y}_C = \sum F_y, \quad I_{Czz}\ddot{\theta} = \sum M_C. \qquad (\text{I.6.33})$$

Equations (I.6.31) and (I.6.32) are interpreted in two ways:

1. The forces and moments are known and the equations are solved for the motion of the rigid body (direct dynamics).
2. The motion of the RB is known and the equations are solved for the force and moments (inverse dynamics).

The dynamic force analysis in this chapter is based on the known motion of the mechanism.

I.6.4 D'Alembert's Principle

Newton's second law can be writen as

$$\mathbf{F} + (-m\mathbf{a}_C) = \mathbf{0}, \quad \text{or} \quad \mathbf{F} + \mathbf{F}_{in} = \mathbf{0}, \qquad (\text{I.6.34})$$

where the term $\mathbf{F}_{in} = -m\mathbf{a}_C$ is the *inertia force*. Newton's second law can be regarded as an "equilibrium" equation.

Equation (I.6.23) relates the total moment about a fixed point O to the acceleration of the mass center and the angular acceleration

$$\sum \mathbf{M}_O = (\mathbf{r}_C \times m\mathbf{a}_C) + I_{Czz}\boldsymbol{\alpha},$$

or

$$\sum \mathbf{M}_O + [\mathbf{r}_C \times (-m\mathbf{a}_C)] + (-I_{Czz}\boldsymbol{\alpha}) = \mathbf{0}. \qquad (I.6.35)$$

The term $\mathbf{M}_{in} = -I_{Czz}\boldsymbol{\alpha}$ is the *inertia moment*. The sum of the moments about any point, including the moment due to the inertial force $-m\mathbf{a}$ acting at mass center and the inertial moment, equals zero.

The equations of motion for a rigid body are analogous to the equations for static equilibrium: The sum of the forces equals zero and the sum of the moments about any point equals zero when the inertial forces and moments are taken into account. This is called *D'Alembert's principle*.

The dynamic force analysis is expressed in a form similar to static force analysis:

$$\sum \mathbf{R} = \sum \mathbf{F} + \mathbf{F}_{in} = \mathbf{0}, \qquad (I.6.36)$$

$$\sum \mathbf{T}_C = \sum \mathbf{M}_C + \mathbf{M}_{in} = \mathbf{0}, \qquad (I.6.37)$$

where $\sum \mathbf{F}$ is the vector sum of all external forces (resultant of external force), and $\sum \mathbf{M}_C$ is the sum of all external moments about the center of mass C (resultant external moment).

For a rigid body in plane motion in the xy plane,

$$\mathbf{a}_C = \ddot{x}_C \mathbf{\imath} + \ddot{y}_C \mathbf{\jmath}, \quad \boldsymbol{\alpha} = \alpha\, \mathbf{k},$$

with all external forces in that plane, Eqs. (I.6.36) and (I.6.37) become

$$\sum R_x = \sum F_x + F_{inx} = \sum F_x + (-m\ddot{x}_C) = 0, \qquad (I.6.38)$$

$$\sum R_y = \sum F_y + F_{iny} = \sum F_y + (-m\ddot{y}_C) = 0, \qquad (I.6.39)$$

$$\sum T_C = \sum M_C + M_{in} = \sum M_C + (-I_C\alpha) = 0. \qquad (I.6.40)$$

With d'Alembert's principle the moment summation can be about any arbitrary point P:

$$\sum \mathbf{T}_P = \sum \mathbf{M}_P + \mathbf{M}_{in} + \mathbf{r}_{PC} \times \mathbf{F}_{in} = \mathbf{0}, \qquad (I.6.41)$$

where

- $\sum \mathbf{M}_P$ is the sum of all external moments about P,
- $\mathbf{M}_{in}$ is the inertia moment,
- $\mathbf{F}_{in}$ is the inertia force, and
- $\mathbf{r}_{PC}$ is a vector from P to C.

The dynamic analysis problem is reduced to a static force and moment balance problem where the inertia forces and moments are treated in the same way as external forces and moments.

I.6.5 Free-Body Diagrams

A free-body diagram is a drawing of a part of a complete system, isolated in order to determine the forces acting on that rigid body.

The following force convention is defined: $\mathbf{F}_{ij}$ represents the force exerted by link i on link j. Figure I.6.7 shows various free-body diagrams that are considered in the analysis of a crank slider mechanism Fig. I.6.7(a). In Figure I.6.7(b), the free body consists of the three moving links isolated from the frame 0. The forces acting on the system include a driving moment $\mathbf{M}$, external driven force $\mathbf{F}$, and the forces transmitted from the frame at joint A, $\mathbf{F}_{01}$, and at joint C, $\mathbf{F}_{03}$. Figure I.6.7(c) is a free-body diagram of the two links 1 and 2. Figure I.6.7(d) is a free-body diagram of a single link.

The force analysis can be accomplished by examining individual links or a subsystem of links. In this way the joint forces between links as well as the required input force or moment for a given output load are computed.

I.6.6 Joint Forces Analysis Using Individual Links

Figure I.6.8(a) is a schematic diagram of a crank slider mechanism comprised of a crank 1, a connecting rod 2, and a slider 3. The center of mass of link 1 is C_1, the center of mass of link 2 is C_2, and the center of mass of slider 3 is C. The mass of the crank is m_1, the mass of the connecting rod is m_2, and the mass of the slider is m_3. The moment of inertia of link i is I_{Ci}, $i = 1, 2, 3$.

The gravitational force is $\mathbf{G}_i = -m_i g \mathbf{J}$, $i = 1, 2, 3$, where $g = 9.81$ m/s^2 is the acceleration of gravity.

For a given value of the crank angle ϕ and a known driven force $\mathbf{F}_{ext}$ the joint reactions and the drive moment $\mathbf{M}$ on the crank are computed using free-body diagrams of the individual links.

Figures I.6.8(b), (c), and (d) show free-body diagrams of the crank 1, the connecting rod 2, and the slider 3. For each moving link the dynamic equilibrium equations are applied.

For the slider 3 the vector sum of the all the forces (external forces $\mathbf{F}_{ext}$, gravitational force $\mathbf{G}_3$, inertia forces $\mathbf{F}_{in\,3}$, joint forces $\mathbf{F}_{23}$, $\mathbf{F}_{03}$) is zero [Fig. I.6.8(d)]:

$$\sum \mathbf{F}^{(3)} = \mathbf{F}_{23} + \mathbf{F}_{in\,3} + \mathbf{G}_3 + \mathbf{F}_{ext} + \mathbf{F}_{03} = \mathbf{0}.$$

Projecting this force onto x- and y-axes gives

$$\sum \mathbf{F}^{(3)} \cdot \mathbf{I} = F_{23x} + (-m_3 \ddot{x}_C) + F_{ext} = 0, \tag{I.6.42}$$

$$\sum \mathbf{F}^{(3)} \cdot \mathbf{J} = F_{23y} - m_3 g + F_{03y} = 0. \tag{I.6.43}$$

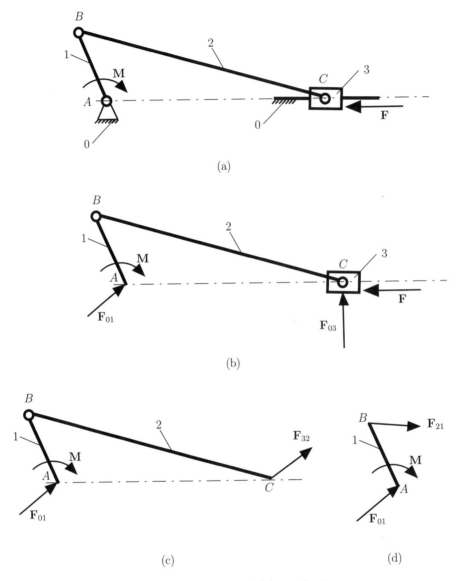

FIGURE I.6.7 Free-body diagrams for a crank slider mechanism.

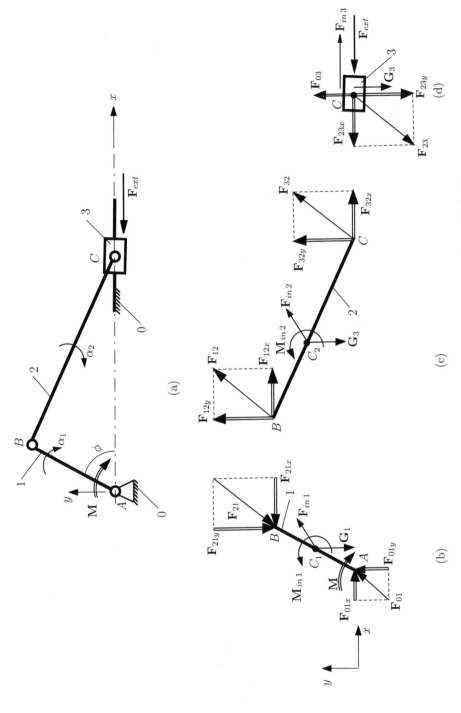

FIGURE I.6.8 (a) Crank slider mechanism; free-body diagrams: (b) crank 1, (c) connecting rod 2, and (d) slider 3.

For the connecting rod 2 [Fig. I.6.8(c)], two vectorial equations can be written:

$$\sum \mathbf{F}^{(2)} = \mathbf{F}_{32} + \mathbf{F}_{in\,2} + \mathbf{G}_2 + \mathbf{F}_{12} = \mathbf{0},$$

$$\sum \mathbf{M}_B^{(2)} = (\mathbf{r}_C - \mathbf{r}_B) \times \mathbf{F}_{32} + (\mathbf{r}_{C2} - \mathbf{r}_B) \times (\mathbf{F}_{in\,2} + \mathbf{G}_2) + \mathbf{M}_{in\,2} = \mathbf{0},$$

or

$$\sum \mathbf{F}^{(2)} \cdot \mathbf{\imath} = F_{32x} + (-m_2 \ddot{x}_{C2}) + F_{12x} = 0, \tag{I.6.44}$$

$$\sum \mathbf{F}^{(2)} \cdot \mathbf{\jmath} = F_{32y} + (-m_2 \ddot{y}_{C2}) - m_2\, g + F_{12y} = 0, \tag{I.6.45}$$

$$\begin{vmatrix} \mathbf{\imath} & \mathbf{\jmath} & \mathbf{k} \\ x_C - x_B & y_C - y_B & 0 \\ F_{32x} & F_{32y} & 0 \end{vmatrix} + \begin{vmatrix} \mathbf{\imath} & \mathbf{\jmath} & \mathbf{k} \\ x_{C2} - x_B & y_{C2} - y_B & 0 \\ -m_2 \ddot{x}_{C2} & -m_2 \ddot{y}_{C2} - m_2\, g & 0 \end{vmatrix} - I_{C2}\, \alpha_2\, \mathbf{k} = \mathbf{0}. \tag{I.6.46}$$

For the crank 1 [Fig. I.6.8(b)], there are two vectorial equations:

$$\sum \mathbf{F}^{(1)} = \mathbf{F}_{21} + \mathbf{F}_{in\,1} + \mathbf{G}_1 + \mathbf{F}_{01} = \mathbf{0},$$

$$\sum \mathbf{M}_A^{(1)} = \mathbf{r}_B \times \mathbf{F}_{21} + \mathbf{r}_{C1} \times (\mathbf{F}_{in\,1} + \mathbf{G}_1) + \mathbf{M}_{in\,1} + \mathbf{M} = \mathbf{0},$$

or

$$\sum \mathbf{F}^{(1)} \cdot \mathbf{\imath} = F_{21x} + (-m_1 \ddot{x}_{C1}) + F_{01x} = 0, \tag{I.6.47}$$

$$\sum \mathbf{F}^{(1)} \cdot \mathbf{\jmath} = F_{21y} + (-m_1 \ddot{y}_{C1}) - m_1\, g + F_{01y} = 0, \tag{I.6.48}$$

$$\begin{vmatrix} \mathbf{\imath} & \mathbf{\jmath} & \mathbf{k} \\ x_B & y_B & 0 \\ F_{21x} & F_{21y} & 0 \end{vmatrix} + \begin{vmatrix} \mathbf{\imath} & \mathbf{\jmath} & \mathbf{k} \\ x_{C1} & y_{C1} & 0 \\ -m_1 \ddot{x}_{C1} & -m_1 \ddot{y}_{C1} - m_1\, g & 0 \end{vmatrix} - I_{C1}\, \alpha_1\, \mathbf{k} + M\, \mathbf{k} = \mathbf{0}, \tag{I.6.49}$$

where $M = |\mathbf{M}|$ is the magnitude of the input moment on the crank.

The eight scalar unknowns F_{03y}, $F_{23x} = -F_{32x}$, $F_{23y} = -F_{32y}$, $F_{12x} = -F_{21x}$, $F_{12y} = -F_{21y}$, F_{01x}, F_{01y}, and M are computed from the set of eight equations (I.6.42), through (I.6.49).

I.6.7 Joint Forces Analysis Using Contour Method

An analytical method to compute joint forces that can be applied for both planar and spatial mechanisms will be presented. The method is based on the decoupling of a closed kinematic chain and writing the dynamic equilibrium equations. The kinematic links are loaded with external forces and inertia forces and moments.

A general monocontour closed kinematic chain is considered in Figure I.6.9. The joint force between the links $i-1$ and i (joint A_i) will be determined. When these two links $i-1$ and i are separated [Fig. I.6.9(b)] the joint forces $\mathbf{F}_{i-1,i}$ and $\mathbf{F}_{i,i-1}$ are introduced and

$$\mathbf{F}_{i-1,i} + \mathbf{F}_{i,i-1} = \mathbf{0}. \qquad (I.6.50)$$

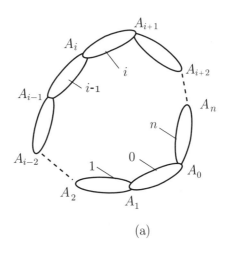

(a)

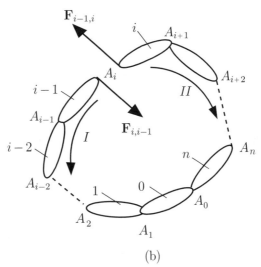

(b)

FIGURE I.6.9 (a) Monocontour closed kinematic chain, (b) joint at A_i replaced by the joint forces $\mathbf{F}_{i-1,i}$ and $\mathbf{F}_{i,i-1}$: $\mathbf{F}_{i-1,i} + \mathbf{F}_{i,i-1} = \mathbf{0}$.

TABLE I.6.1 Joint forces for several joints

Type of the kinematic pair	Joint force or moment	Unknowns	Equilibrium condition
 rotational joint	$\mathbf{F}_x + \mathbf{F}_y = \mathbf{F}$ $\mathbf{F} \perp \Delta\Delta$	$\|\mathbf{F}_x\| = F_x$ $\|\mathbf{F}_y\| = F_y$	$\mathbf{M}_\Delta = 0$
 translatioal joint	$\mathbf{F} \perp \Delta\Delta$	$\|\mathbf{F}\| = F_x$ x	$\mathbf{F}_\Delta = 0$
 cylindrical joint	$\mathbf{F}_x + \mathbf{F}_y = \mathbf{F}$ $\mathbf{F} \perp \Delta\Delta$	$\|\mathbf{F}_x\| = F_x$ $\|\mathbf{F}_y\| = F_y$ x	$\mathbf{F}_\Delta = 0$ $\mathbf{M}_\Delta = 0$
 roll-slide joint	$\mathbf{F} \perp \Delta\Delta$ $\mathbf{F} \parallel \mathbf{n}$	$\|\mathbf{F}\| = F$ x	$\mathbf{F}_\Delta = 0$ $\mathbf{M}_\Delta = 0$
 sphere joint	$\mathbf{F}_x + \mathbf{F}_y + \mathbf{F}_z = \mathbf{F}$	$\|\mathbf{F}_x\| = F_x$ $\|\mathbf{F}_y\| = F_y$ $\|\mathbf{F}_z\| = F_z$	$\mathbf{M}_{\Delta_1} = 0$ $\mathbf{M}_{\Delta_2} = 0$ $\mathbf{M}_{\Delta_3} = 0$

Table I.6.1 shows the joint forces for several joints. The following notations have been used: $\mathbf{M}_\Delta$ is the moment with respect to the axis Δ, and F_Δ is the projection of the force vector $\mathbf{F}$ onto the axis Δ.

It is helpful to "mentally disconnect" the two links $(i-1)$ and i, which create the joint A_i, from the rest of the mechanism. The joint at A_i will be replaced by the joint forces $\mathbf{F}_{i-1,i}$ and

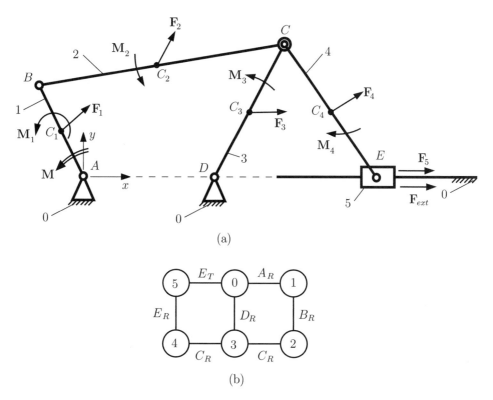

(a)

(b)

FIGURE I.6.10   (a) Mechanism, and (b) diagram representing the mechanism.

$F_{i,i-1}$. The closed kinematic chain has been transformed into two open kinematic chains, and two paths *I* and *II* are associated. The two paths start from A_i.

For the path *I* (counterclockwise), starting at A_i and following *I* the first joint encountered is A_{i-1}. For the link $i-1$ left behind, dynamic equilibrium equations are written according to the type of joint at A_{i-1}. Following the same path *I*, the next joint encountered is A_{i-2}. For the subsystem ($i-1$ and $i-2$) equilibrium conditions corresponding to the type of joint at A_{i-2} can be specified, and so on. A similar analysis is performed for the path *II* of the open kinematic chain. The number of equilibrium equations written is equal to the number of unknown scalars introduced by joint A_i (joint forces at this joint). For a joint, the number of equilibrium conditions is equal to the number of relative mobilities of the joint.

The five moving link ($j = 1, 2, 3, 4, 5$) mechanism shown in Figure I.6.10(a) has the center of mass locations designated by $C_j(x_{Cj}, y_{Cj}, 0)$. The following analysis will consider the relationships of the inertia forces $\mathbf{F}_{inj}$, the inertia moments $\mathbf{M}_{inj}$, the gravitational force $\mathbf{G}_j$, the driven force, $\mathbf{F}_{ext}$, to the joint reactions $\mathbf{F}_{ij}$ and the drive moment $\mathbf{M}$ on the crank 1.

To simplify the notation the total vector force at C_j is written as $\mathbf{F}_j = \mathbf{F}_{inj} + \mathbf{G}_j$ and the inertia moment of link j is written as $\mathbf{M}_j = \mathbf{M}_{inj}$. The diagram representing the mechanism is depicted in Figure I.6.10(b) and has two contours 0-1-2-3-0 and 0-3-4-5-0.

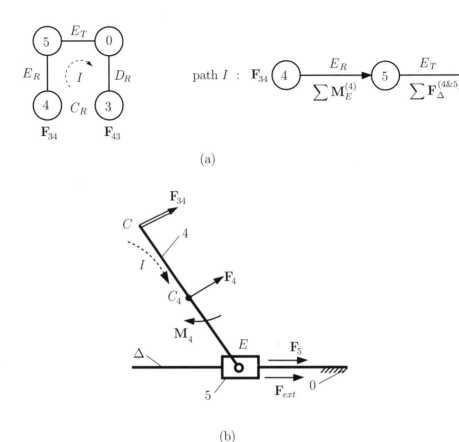

(a)

(b)

FIGURE I.6.11 *Joint force* $\mathbf{F}_{34}$ *(a) calculation diagram, and (b) force diagram.*

Remark

The joint at C represents a ramification point for the mechanism and the diagram, and the dynamic force analysis will start with this joint. The force computation starts with the contour 0-3-4-5-0 because the driven load $\mathbf{F}_{ext}$ on link 5 is given.

Contour 0-3-4-5-0

Reaction $\mathbf{F}_{34}$

The rotation joint at C (or C_R, where the subscript R means rotation), between 3 and 4, is replaced with the unknown reaction (Fig. I.6.11):

$$\mathbf{F}_{34} = -\mathbf{F}_{43} = F_{34x}\,\mathbf{i} + F_{34y}\,\mathbf{J}.$$

If the path I is followed [Fig. I.6.11(a)] for the rotation joint at E (E_R), a moment equation is written:

$$\sum\mathbf{M}_E^{(4)} = (\mathbf{r}_C - \mathbf{r}_E) \times \mathbf{F}_{32} + (\mathbf{r}_{C4} - \mathbf{r}_E) \times \mathbf{F}_4 + \mathbf{M}_4 = \mathbf{0},$$

or

$$\begin{vmatrix} \mathbf{I} & \mathbf{J} & \mathbf{k} \\ x_C - x_E & y_C - y_E & 0 \\ F_{34x} & F_{34y} & 0 \end{vmatrix} + \begin{vmatrix} \mathbf{I} & \mathbf{J} & \mathbf{k} \\ x_{C4} - x_E & y_{C4} - y_E & 0 \\ F_{4x} & F_{4y} & 0 \end{vmatrix} + M_4 \mathbf{k} = \mathbf{0}. \quad (I.6.51)$$

Continuing on path I, the next joint is the translational joint at D (D_T). The projection of all the forces that act on 4 and 5 onto the sliding direction Δ (x-axis) should be zero.

$$\sum \mathbf{F}_\Delta^{(4\&5)} = \sum \mathbf{F}^{(4\&5)} \cdot \mathbf{I} = (\mathbf{F}_{34} + \mathbf{F}_4 + \mathbf{F}_5 + \mathbf{F}_{ext}) \cdot \mathbf{I}$$
$$= F_{34x} + F_{4x} + F_{5x} + F_{ext} = 0. \quad (I.6.52)$$

The system of Eqs. (I.6.51) and (I.6.52) is solved and the two unknowns F_{34x} and F_{34y} are obtained.

Reaction $\mathbf{F}_{45}$

The rotation joint at E (E_R), between 4 and 5, is replaced with the unknown reaction (Fig. I.6.12):

$$\mathbf{F}_{45} = -\mathbf{F}_{54} = F_{45x} \mathbf{I} + F_{45y} \mathbf{J}.$$

If the path I is traced [Fig. I.6.12(a)] for the pin joint at C (C_R), a moment equation is written:

$$\sum \mathbf{M}_C^{(4)} = (\mathbf{r}_E - \mathbf{r}_C) \times \mathbf{F}_{54} + (\mathbf{r}_{C4} - \mathbf{r}_C) \times \mathbf{F}_4 + M_4 = \mathbf{0},$$

or

$$\begin{vmatrix} \mathbf{I} & \mathbf{J} & \mathbf{k} \\ x_E - x_C & y_E - y_C & 0 \\ -F_{45x} & -F_{45y} & 0 \end{vmatrix} + \begin{vmatrix} \mathbf{I} & \mathbf{J} & \mathbf{k} \\ x_{C4} - x_C & y_{C4} - y_C & 0 \\ F_{4x} & F_{4y} & 0 \end{vmatrix} + M_4 \mathbf{k} = \mathbf{0}. \quad (I.6.53)$$

For the path II, the slider joint at E (E_T) is encountered. The projection of all the forces that act on 5 onto the sliding direction Δ (x-axis) should be zero.

$$\sum \mathbf{F}_\Delta^{(5)} = \sum \mathbf{F}^{(5)} \cdot \mathbf{I} = (\mathbf{F}_{45} + \mathbf{F}_5 + \mathbf{F}_{ext}) \cdot \mathbf{I}$$
$$= F_{45x} + F_{5x} + F_{ext} = 0. \quad (I.6.54)$$

The unknown force components F_{45x} and F_{45y} are calculated from Eqs. (I.6.53) and (I.6.54).

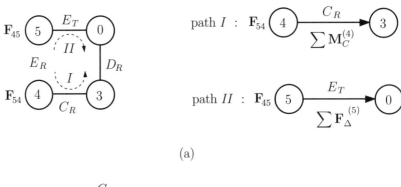

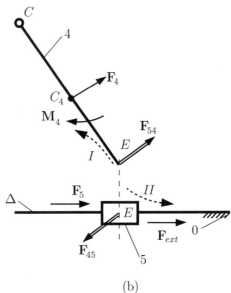

(b)

FIGURE I.6.12 *Joint force* $\mathbf{F}_{45}$ *(a) calculation diagram, and (b) force diagram.*

Reaction $\mathbf{F}_{05}$

The slider joint at E (E_T), between 0 and 5, is replaced with the unknown reaction (Fig. I.6.13):

$$\mathbf{F}_{05} = F_{05y}\,\mathbf{J}.$$

The reaction joint introduced by the translational joint is perpendicular on the sliding direction $\mathbf{F}_{05} \perp \Delta$. The application point P of force $\mathbf{F}_{05}$ is unknown.

If the path I is followed [Fig. I.6.13(a)] for the pin joint at E (E_R), a moment equation is written for link 5:

$$\sum \mathbf{M}_E^{(5)} = (\mathbf{r}_P - \mathbf{r}_E) \times \mathbf{F}_{05} = \mathbf{0},$$

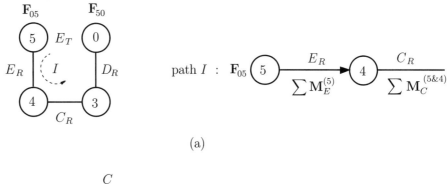

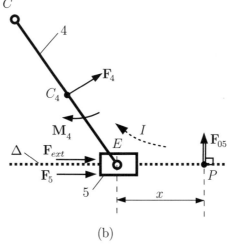

FIGURE I.6.13 *Joint force* $\mathbf{F}_{05}$ *(a) calculation diagram, and (b) force diagram.*

or

$$x F_{05y} = 0 \quad \Rightarrow x = 0. \tag{I.6.55}$$

The application point is at E ($P \equiv E$).

Continuing on path I, the next joint is the pin joint C (C_R).

$$\sum \mathbf{M}_C^{(4\&5)} = (\mathbf{r}_E - \mathbf{r}_C) \times (\mathbf{F}_{05} + \mathbf{F}_5 + \mathbf{F}_{ext}) + (\mathbf{r}_{C4} - \mathbf{r}_C) \times \mathbf{F}_4 + \mathbf{M}_4 = \mathbf{0},$$

or

$$\begin{vmatrix} \mathbf{i} & \mathbf{j} & \mathbf{k} \\ x_E - x_C & y_E - y_C & 0 \\ F_{5x} + F_{ext} & F_{05y} & 0 \end{vmatrix} + \begin{vmatrix} \mathbf{i} & \mathbf{j} & \mathbf{k} \\ x_{C4} - x_C & y_{C4} - y_C & 0 \\ F_{4x} & F_{4y} & 0 \end{vmatrix} + M_4 \mathbf{k} = \mathbf{0}. \tag{I.6.56}$$

The joint reaction force F_{05y} is computed from Eq. (I.6.56).

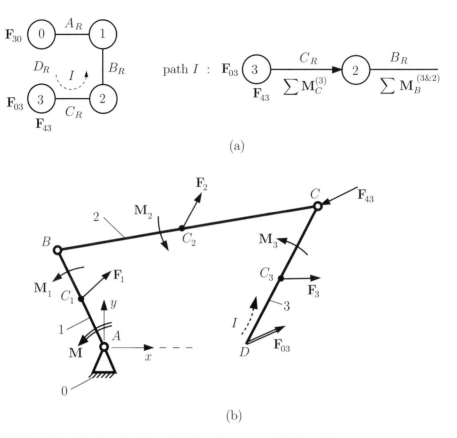

(a)

(b)

FIGURE I.6.14 *Joint force* $\mathbf{F}_{03}$ *(a) calculation diagram, and (b) force diagram.*

Contour 0-1-2-3-0

For this contour the joint force $\mathbf{F}_{43} = -\mathbf{F}_{34}$ at the ramification point C is considered as a known external force.

Reaction $\mathbf{F}_{03}$

The pin joint D_R, between 0 and 3, is replaced with the unknown reaction force (Fig. I.6.14):

$$\mathbf{F}_{03} = F_{03x}\,\mathbf{i} + F_{03y}\,\mathbf{J}.$$

If the path I is followed [Fig. I.6.14(a)], a moment equation is written for the pin joint C_R for link 3:

$$\sum \mathbf{M}_C^{(3)} = (\mathbf{r}_D - \mathbf{r}_C) \times \mathbf{F}_{03} + (\mathbf{r}_{C3} - \mathbf{r}_C) \times \mathbf{F}_3 + \mathbf{M}_3 = \mathbf{0},$$

or

$$\begin{vmatrix} \mathbf{1} & \mathbf{J} & \mathbf{k} \\ x_D - x_C & y_D - y_C & 0 \\ F_{03x} & F_{03y} & 0 \end{vmatrix} + \begin{vmatrix} \mathbf{1} & \mathbf{J} & \mathbf{k} \\ x_{C3} - x_C & y_{C3} - y_C & 0 \\ F_{3x} & F_{3y} & 0 \end{vmatrix} + M_3 \mathbf{k} = \mathbf{0}. \qquad (\text{I.6.57})$$

Continuing on path I, the next joint is the pin joint B_R and a moment equation is written for links 3 and 2:

$$\sum \mathbf{M}_B^{(3\&2)} = (\mathbf{r}_D - \mathbf{r}_B) \times \mathbf{F}_{03} + (\mathbf{r}_{C3} - \mathbf{r}_B) \times \mathbf{F}_3 + \mathbf{M}_3 + (\mathbf{r}_C - \mathbf{r}_B) \times \mathbf{F}_{43}$$

$$+ (\mathbf{r}_{C2} - \mathbf{r}_B) \times \mathbf{F}_2 + \mathbf{M}_2 = \mathbf{0},$$

or

$$\begin{vmatrix} \mathbf{1} & \mathbf{J} & \mathbf{k} \\ x_D - x_B & y_D - y_B & 0 \\ F_{03x} & F_{03y} & 0 \end{vmatrix} + \begin{vmatrix} \mathbf{1} & \mathbf{J} & \mathbf{k} \\ x_{C3} - x_B & y_{C3} - y_B & 0 \\ F_{3x} & F_{3y} & 0 \end{vmatrix}$$

$$+ M_3 \mathbf{k} + \begin{vmatrix} \mathbf{1} & \mathbf{J} & \mathbf{k} \\ x_C - x_B & y_C - y_B & 0 \\ F_{43x} & F_{43y} & 0 \end{vmatrix}$$

$$+ \begin{vmatrix} \mathbf{1} & \mathbf{J} & \mathbf{k} \\ x_{C2} - x_B & y_{C2} - y_B & 0 \\ F_{2x} & F_{2y} & 0 \end{vmatrix} + M_2 \mathbf{k} = \mathbf{0}. \qquad (\text{I.6.58})$$

The two components F_{03x} and F_{03y} of the joint force are obtained from Eqs. (I.6.57) and (I.6.58).

Reaction $\mathbf{F}_{23}$

The pin joint C_R, between 2 and 3, is replaced with the unknown reaction force (Fig. I.6.15)

$$\mathbf{F}_{23} = F_{23x} \mathbf{1} + F_{23y} \mathbf{J}.$$

If the path I is followed, as in Figure I.6.15(a), a moment equation is written for the pin joint D_R for link 3:

$$\sum \mathbf{M}_D^{(3)} = (\mathbf{r}_C - \mathbf{r}_D) \times (\mathbf{F}_{23} + \mathbf{F}_{43}) + (\mathbf{r}_{C3} - \mathbf{r}_D) \times \mathbf{F}_3 + \mathbf{M}_3 = \mathbf{0},$$

or

$$\begin{vmatrix} \mathbf{1} & \mathbf{J} & \mathbf{k} \\ x_C - x_D & y_C - y_D & 0 \\ F_{23x} + F_{43x} & F_{23y} + F_{43y} & 0 \end{vmatrix} + \begin{vmatrix} \mathbf{1} & \mathbf{J} & \mathbf{k} \\ x_{C3} - x_D & y_{C3} - y_D & 0 \\ F_{3x} & F_{3y} & 0 \end{vmatrix} + M_3 \mathbf{k} = \mathbf{0}. \qquad (\text{I.6.59})$$

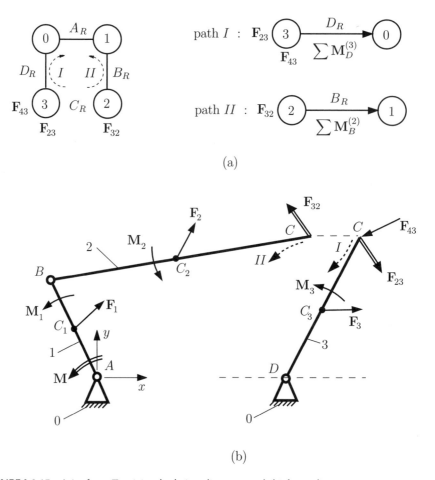

(a)

(b)

FIGURE I.6.15 *Joint force $\mathbf{F}_{23}$ (a) calculation diagram, and (b) force diagram.*

For the path II, the first joint encountered is the pin joint B_R and a moment equation is written for link 2:

$$\sum \mathbf{M}_B^{(2)} = (\mathbf{r}_C - \mathbf{r}_B) \times (-\mathbf{F}_{23}) + (\mathbf{r}_{C2} - \mathbf{r}_B) \times \mathbf{F}_2 + \mathbf{M}_2 = \mathbf{0},$$

or

$$\begin{vmatrix} \mathbf{I} & \mathbf{J} & \mathbf{k} \\ x_C - x_B & y_C - y_B & 0 \\ -F_{23x} & -F_{23y} & 0 \end{vmatrix} + \begin{vmatrix} \mathbf{I} & \mathbf{J} & \mathbf{k} \\ x_{C2} - x_B & y_{C2} - y_B & 0 \\ F_{2x} & F_{2y} & 0 \end{vmatrix} + M_2 \mathbf{k} = \mathbf{0}. \qquad (I.6.60)$$

The two force components F_{23x} and F_{23y} of the joint force are obtained from Eqs. (I.6.59) and (I.6.60).

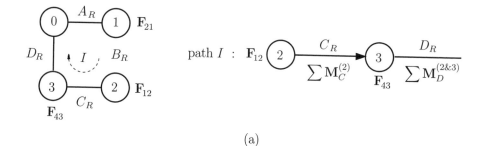

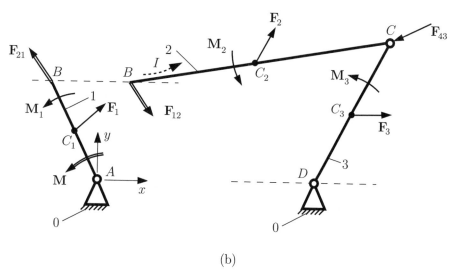

(b)

FIGURE I.6.16 *Joint force* $\mathbf{F}_{12}$ *(a) calculation diagram, and (b) force diagram.*

Reaction $\mathbf{F}_{12}$

The pin joint B_R, between 1 and 2, is replaced with the unknown reaction force (Fig. I.6.16):

$$\mathbf{F}_{12} = F_{12x}\,\mathbf{I} + F_{12y}\,\mathbf{J}.$$

If the path I is followed, as in Figure I.6.16(a), a moment equation is written for the pin joint C_R for link 2:

$$\sum \mathbf{M}_C^{(2)} = (\mathbf{r}_B - \mathbf{r}_C) \times \mathbf{F}_{12} + (\mathbf{r}_{C2} - \mathbf{r}_C) \times \mathbf{F}_2 + \mathbf{M}_2 = \mathbf{0},$$

or

$$\begin{vmatrix} \mathbf{I} & \mathbf{J} & \mathbf{k} \\ x_B - x_C & y_B - y_C & 0 \\ F_{12x} & F_{12y} & 0 \end{vmatrix} + \begin{vmatrix} \mathbf{I} & \mathbf{J} & \mathbf{k} \\ x_{C2} - x_C & y_{C2} - y_C & 0 \\ F_{2x} & F_{2y} & 0 \end{vmatrix} + M_2\mathbf{k} = \mathbf{0}. \qquad (\text{I.6.61})$$

Continuing on path I the next joint encountered is the pin joint D_R, and a moment equation is written for links 2 and 3:

$$\sum \mathbf{M}_D^{(2\&3)} = (\mathbf{r}_B - \mathbf{r}_D) \times \mathbf{F}_{12} + (\mathbf{r}_{C2} - \mathbf{r}_D) \times \mathbf{F}_2 + \mathbf{M}_2 +$$

$$(\mathbf{r}_C - \mathbf{r}_D) \times \mathbf{F}_{43} + (\mathbf{r}_{C3} - \mathbf{r}_D) \times \mathbf{F}_3 + \mathbf{M}_3 = \mathbf{0},$$

or

$$\begin{vmatrix} \mathbf{i} & \mathbf{j} & \mathbf{k} \\ x_B - x_D & y_B - y_D & 0 \\ F_{12x} & F_{12y} & 0 \end{vmatrix} + \begin{vmatrix} \mathbf{i} & \mathbf{j} & \mathbf{k} \\ x_{C2} - x_D & y_{C2} - y_D & 0 \\ F_{2x} & F_{2y} & 0 \end{vmatrix} + M_2 \mathbf{k}$$

$$+ \begin{vmatrix} \mathbf{i} & \mathbf{j} & \mathbf{k} \\ x_C - x_D & y_C - y_D & 0 \\ F_{43x} & F_{43y} & 0 \end{vmatrix} + \begin{vmatrix} \mathbf{i} & \mathbf{j} & \mathbf{k} \\ x_{C3} - x_D & y_{C3} - y_D & 0 \\ F_{3x} & F_{3y} & 0 \end{vmatrix} + M_3 \mathbf{k} = \mathbf{0}. \qquad (I.6.62)$$

The two components F_{12x} and F_{12y} of the joint force are computed from Eqs. (I.6.61) and (I.6.62).

Reaction $\mathbf{F}_{01}$ and driver moment $\mathbf{M}$

The pin joint A_R, between 0 and 1, is replaced with the unknown reaction force (Fig. I.6.17):

$$\mathbf{F}_{01} = F_{01x} \mathbf{i} + F_{01y} \mathbf{j}.$$

The unknown driver moment is $\mathbf{M} = M \mathbf{k}$. If the path I is followed [Fig. I.6.17(a)], a moment equation is written for the pin joint B_R for link 1:

$$\sum \mathbf{M}_B^{(1)} = (\mathbf{r}_A - \mathbf{r}_B) \times \mathbf{F}_{01} + (\mathbf{r}_{C1} - \mathbf{r}_B) \times \mathbf{F}_1 + \mathbf{M}_1 + \mathbf{M} = \mathbf{0},$$

or

$$\begin{vmatrix} \mathbf{i} & \mathbf{j} & \mathbf{k} \\ x_A - x_B & y_A - y_B & 0 \\ F_{01x} & F_{01y} & 0 \end{vmatrix} + \begin{vmatrix} \mathbf{i} & \mathbf{j} & \mathbf{k} \\ x_{C1} - x_B & y_{C1} - y_B & 0 \\ F_{1x} & F_{1y} & 0 \end{vmatrix} + M_1 \mathbf{k} + M \mathbf{k} = \mathbf{0}. \qquad (I.6.63)$$

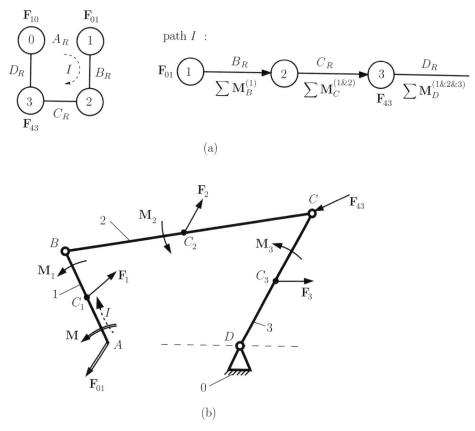

FIGURE I.6.17 *Joint force* $\mathbf{F}_{01}$ *(a) calculation diagram, and (b) force diagram.*

Continuing on path I, the next joint encountered is the pin joint C_R and a moment equation is written for links 1 and 2:

$$\sum \mathbf{M}_C^{(1\&2)} = (\mathbf{r}_A - \mathbf{r}_C) \times \mathbf{F}_{01} + (\mathbf{r}_{C1} - \mathbf{r}_C) \times \mathbf{F}_1 + \mathbf{M}_1 + \mathbf{M}$$

$$+ (\mathbf{r}_{C2} - \mathbf{r}_C) \times \mathbf{F}_2 + \mathbf{M}_2 = \mathbf{0}. \qquad (I.6.64)$$

Equation (I.6.64) is the vector sum of the moments about D_R of all forces and moments that act on links 1, 2, and 3.

$$\sum \mathbf{M}_D^{(1\&2\&3)} = (\mathbf{r}_A - \mathbf{r}_D) \times \mathbf{F}_{01} + (\mathbf{r}_{C1} - \mathbf{r}_D) \times \mathbf{F}_1 + \mathbf{M}_1 + \mathbf{M} + (\mathbf{r}_{C2} - \mathbf{r}_D)$$

$$\times \mathbf{F}_2 + \mathbf{M}_2 + (\mathbf{r}_C - \mathbf{r}_D) \times \mathbf{F}_{43} + (\mathbf{r}_{C3} - \mathbf{r}_D) \times \mathbf{F}_3 + \mathbf{M}_3 = \mathbf{0}. \quad (I.6.65)$$

The components F_{01x}, F_{01y} and M are computed from Eqs. (I.6.63), (I.6.64), and (I.6.65).

I.6.8 Joint Force Analysis Using Dyads

RRR Dyad

Figure I.6.18 shows an RRR dyad with two links, 2 and 3, and three pin joints, B, C, and D. The unknowns are the joint reaction forces:

$$\mathbf{F}_{12} = F_{12x}\mathbf{i} + F_{12y}\mathbf{J},$$
$$\mathbf{F}_{43} = F_{43x}\mathbf{i} + F_{43y}\mathbf{J},$$
$$\mathbf{F}_{23} = -\mathbf{F}_{32} = F_{23x}\mathbf{i} + F_{23y}\mathbf{J}. \tag{I.6.66}$$

The inertia forces and external forces $\mathbf{F}_j = F_{ji}\mathbf{i} + F_{jj}\mathbf{J}$, inertia moments and external moments $\mathbf{M}_j = M_j\mathbf{k}$, ($j = 2,3$) are given.

To determine $\mathbf{F}_{12}$ and $\mathbf{F}_{43}$, the following equations are written:

• sum of all forces on links 2 and 3 is zero:

$$\sum \mathbf{F}^{(2\&3)} = \mathbf{F}_{12} + \mathbf{F}_2 + \mathbf{F}_3 + \mathbf{F}_{43} = \mathbf{0},$$

or

$$\sum \mathbf{F}^{(2\&3)} \cdot \mathbf{i} = F_{12x} + F_{2x} + F_{3x} + F_{43x} = 0,$$
$$\sum \mathbf{F}^{(2\&3)} \cdot \mathbf{J} = F_{12y} + F_{2y} + F_{3y} + F_{43y} = 0. \tag{I.6.67}$$

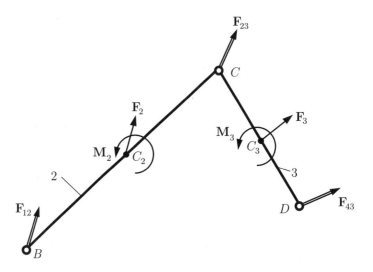

FIGURE I.6.18   *Joint forces for RRR dyad.*

- sum of moments of all forces and moments on link 2 about C is zero:

$$\sum \mathbf{M}_C^{(2)} = (\mathbf{r}_B - \mathbf{r}_C) \times \mathbf{F}_{12} + (\mathbf{r}_{C2} - \mathbf{r}_C) \times \mathbf{F}_2 + \mathbf{M}_2 = \mathbf{0}. \qquad (\text{I.6.68})$$

- sum of moments of all forces and moments on link 3 about C is zero:

$$\sum \mathbf{M}_C^{(3)} = (\mathbf{r}_D - \mathbf{r}_C) \times \mathbf{F}_{43} + (\mathbf{r}_{C3} - \mathbf{r}_C) \times \mathbf{F}_3 + \mathbf{M}_3 = \mathbf{0}. \qquad (\text{I.6.69})$$

The components F_{12x}, F_{12y}, F_{43x}, and F_{43y} are calculated from Eqs. (I.6.67), (I.6.68), and (I.6.69). The reaction force $\mathbf{F}_{32} = -\mathbf{F}_{23}$ is computed from the sum of all forces on link 2:

$$\sum \mathbf{F}^{(2)} = \mathbf{F}_{12} + \mathbf{F}_2 + \mathbf{F}_{32} = \mathbf{0},$$

or

$$\sum \mathbf{F}^{(2)} \cdot \mathbf{I} = F_{12x} + F_{2x} + F_{32x} = 0,$$

$$\sum \mathbf{F}^{(2)} \cdot \mathbf{J} = F_{12y} + F_{2y} + F_{32y} = 0. \qquad (\text{I.6.70})$$

RRT Dyad

Figure I.6.19 shows an RRT dyad with the unknown joint reaction forces $\mathbf{F}_{12}$, $\mathbf{F}_{43}$, and $\mathbf{F}_{23} = -\mathbf{F}_{32}$. The joint reaction force $\mathbf{F}_{43}$ is perpendicular to the sliding direction $\mathbf{F}_{43} \perp \Delta$ or

$$\mathbf{F}_{43} \cdot \Delta = (F_{43x}\mathbf{I} + F_{43y}\mathbf{J}) \cdot (\cos\theta\mathbf{I} + \sin\theta\mathbf{J}) = 0. \qquad (\text{I.6.71})$$

In order to determine $\mathbf{F}_{12}$ and $\mathbf{F}_{43}$ the following equations are written:

- sum of all the forces on links 2 and 3 is zero:

$$\sum \mathbf{F}^{(2\&3)} = \mathbf{F}_{12} + \mathbf{F}_2 + \mathbf{F}_3 + \mathbf{F}_{43} = \mathbf{0},$$

or

$$\sum \mathbf{F}^{(2\&3)} \cdot \mathbf{I} = F_{12x} + F_{2x} + F_{3x} + F_{43x} = 0,$$

$$\sum \mathbf{F}^{(2\&3)} \cdot \mathbf{J} = F_{12y} + F_{2y} + F_{3y} + F_{43y} = 0. \qquad (\text{I.6.72})$$

- sum of moments of all the forces and the moments on link 2 about C is zero:

$$\sum \mathbf{M}_C^{(2)} = (\mathbf{r}_B - \mathbf{r}_C) \times \mathbf{F}_{12} + (\mathbf{r}_{C2} - \mathbf{r}_C) \times \mathbf{F}_2 + \mathbf{M}_2 = \mathbf{0}. \qquad (\text{I.6.73})$$

The components F_{12x}, F_{12y}, F_{43x}, and F_{43y} are calculated from Eqs. (I.6.71), (I.6.72), and (I.6.73). The reaction force components F_{32x} and F_{32y} are computed from the sum of all the forces on link 2:

$$\sum \mathbf{F}^{(2)} = \mathbf{F}_{12} + \mathbf{F}_2 + \mathbf{F}_{32} = \mathbf{0},$$

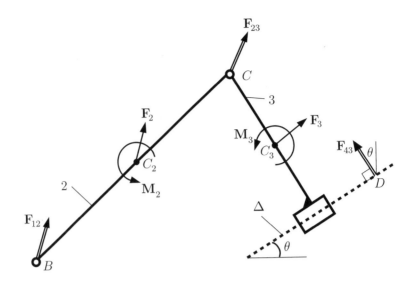

FIGURE I.6.19 *Joint forces for RRT dyad.*

or

$$\sum \mathbf{F}^{(2)} \cdot \mathbf{1} = F_{12x} + F_{2x} + F_{32x} = 0,$$

$$\sum \mathbf{F}^{(2)} \cdot \mathbf{J} = F_{12y} + F_{2y} + F_{32y} = 0. \qquad (I.6.74)$$

RTR Dyad

The unknown joint reaction forces $\mathbf{F}_{12}$ and $\mathbf{F}_{43}$ are calculated from the relations (Fig. I.6.20):

- sum of all the forces on links 2 and 3 is zero:

$$\sum \mathbf{F}^{(2\&3)} = \mathbf{F}_{12} + \mathbf{F}_2 + \mathbf{F}_3 + \mathbf{F}_{43} = \mathbf{0},$$

or

$$\sum \mathbf{F}^{(2\&3)} \cdot \mathbf{1} = F_{12x} + F_{2x} + F_{3x} + F_{43x} = 0,$$

$$\sum \mathbf{F}^{(2\&3)} \cdot \mathbf{J} = F_{12y} + F_{2y} + F_{3y} + F_{43y} = 0. \qquad (I.6.75)$$

- sum of the moments of all the forces and moments on links 2 and 3 about B is zero:

$$\sum \mathbf{M}_B^{(2\&3)} = (\mathbf{r}_D - \mathbf{r}_B) \times \mathbf{F}_{43} + (\mathbf{r}_{C3} - \mathbf{r}_B) \times \mathbf{F}_3 + \mathbf{M}_3$$

$$+ (\mathbf{r}_{C2} - \mathbf{r}_B) \times \mathbf{F}_2 + \mathbf{M}_2 = \mathbf{0}. \qquad (I.6.76)$$

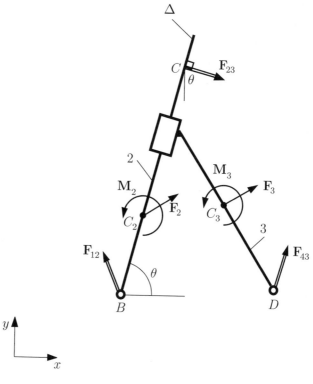

FIGURE I.6.20 *Joint forces for RTR dyad.*

- sum of all the forces on link 2 projected onto the sliding direction $\Delta = \cos\theta\mathbf{i} + \sin\theta\mathbf{J}$ is zero:

$$\sum \mathbf{F}^{(2)} \cdot \Delta = (\mathbf{F}_{12} + \mathbf{F}_2) \cdot (\cos\theta\mathbf{i} + \sin\theta\mathbf{J}) = 0. \qquad (I.6.77)$$

The components F_{12x}, F_{12y}, F_{43x}, and F_{43y} are calculated from Eqs. (I.6.75), (I.6.76), and (I.6.77).

The force components F_{32x} and F_{32y} are computed from the sum of all the forces on link 2:

$$\sum \mathbf{F}^{(2)} = \mathbf{F}_{12} + \mathbf{F}_2 + \mathbf{F}_{32} = \mathbf{0},$$

or

$$\sum \mathbf{F}^{(2)} \cdot \mathbf{i} = F_{12x} + F_{2x} + F_{32x} = 0,$$

$$\sum \mathbf{F}^{(2)} \cdot \mathbf{J} = F_{12y} + F_{2y} + F_{32y} = 0. \qquad (I.6.78)$$

I.6.9 Examples

EXAMPLE I.6.1:

The R-RTR mechanism shown in Figure I.6.21(a) has the dimensions: $AB = 0.14$ m, $AC = 0.06$ m, and $CF = 0.2$ m. The driver link 1 makes an angle $\phi = \phi_1 = \frac{\pi}{3}$ rad with the horizontal axis and rotates with a constant speed of $n = n_1 = 30\pi$ rpm. The position vectors of the points A, B, C, and F are

$$\mathbf{r}_A = 0\mathbf{\imath} + 0\mathbf{\jmath} \text{ m},$$

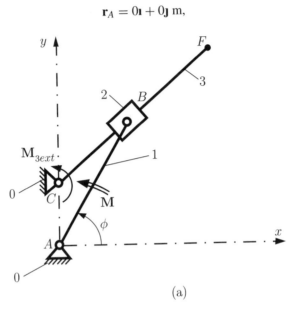

(a)

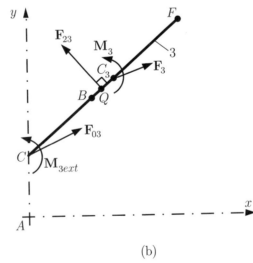

(b)

FIGURE I.6.21 *Joint forces for R-RTR mechanism (Example I.6.1).*

EXAMPLE I.6.1: *Cont'd*

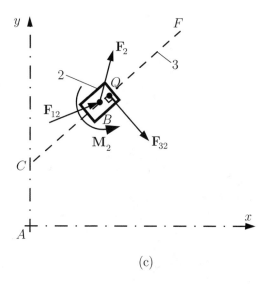

(c)

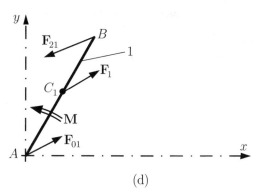

(d)

FIGURE I.6.21 *Continued*

$$\mathbf{r}_B = \mathbf{r}_{C_2} = x_B\mathbf{I} + y_B\mathbf{J} = 0.07\mathbf{I} + 0.121\mathbf{J} \text{ m},$$

$$\mathbf{r}_C = x_C\mathbf{I} + y_C\mathbf{J} = 0\mathbf{I} + 0.06\mathbf{J} \text{ m},$$

$$\mathbf{r}_F = x_F\mathbf{I} + y_F\mathbf{J} = 0.150\mathbf{I} + 0.191\mathbf{J} \text{ m},$$

where the mass center of the slider 2 is at B ($B = C_2$). The position vectors of the mass centers of links 1 and 3 are

$$\mathbf{r}_{C_1} = x_{C_1}\mathbf{I} + y_{C_1}\mathbf{J} = \frac{x_B}{2}\mathbf{I} + \frac{y_B}{2}\mathbf{J} = 0.035\mathbf{I} + 0.06\mathbf{J} \text{ m},$$

$$\mathbf{r}_{C_3} = x_{C_3}\mathbf{I} + y_{C_3}\mathbf{J} = \frac{x_C + x_F}{2}\mathbf{I} + \frac{y_C + y_F}{2}\mathbf{J} = 0.075\mathbf{I} + 0.125\mathbf{J} \text{ m}.$$

Continued

EXAMPLE I.6.1: *Cont'd*

The total forces and moments at C_j, $j = 1, 2, 3$ are $\mathbf{F}_j = \mathbf{F}_{inj} + \mathbf{G}_j$ and $\mathbf{M}_j = \mathbf{M}_{inj}$, where $\mathbf{F}_{inj}$ is the inertia force, $\mathbf{M}_j$ is the inertia moment, and $\mathbf{G}_j = -m_j g \mathbf{J}$ is the gravity force with gravity acceleration $g = 9.81$ m/s^2.

$$\mathbf{F}_1 = 0.381\mathbf{i} + 0.437\mathbf{J} \text{ N}, \quad \mathbf{M}_1 = 0\mathbf{k} \quad \text{N} \cdot \text{m},$$

$$\mathbf{F}_2 = 0.545\mathbf{i} + 0.160\mathbf{J} \text{ N}, \quad \mathbf{M}_2 = -0.001\mathbf{k} \text{ N} \cdot \text{m},$$

$$\mathbf{F}_3 = 3.302\mathbf{i} - 0.539\mathbf{J} \text{ N}, \quad \mathbf{M}_3 = -0.046\mathbf{k} \text{ N} \cdot \text{m}.$$

The external moment on link 3 is $\mathbf{M}_{3ext} = -1000\mathbf{k}$ N·m. Determine the moment $\mathbf{M}$ required for dynamic equilibrium and the joint forces for the mechanism using the free-body diagrams of the individual links.

Solution For each link two vectorial equations are written:

$$\sum \mathbf{F}_j + \mathbf{F}_{inj} = 0 \quad \text{and} \quad \sum \mathbf{M}_{C_j} + \mathbf{M}_{inj} = 0, \qquad (\text{I.6.79})$$

where $\sum \mathbf{F}_j$ is the vector sum of all external forces (resultant of external force) on link j, and $\sum \mathbf{M}_{C_j}$ is the sum of all external moments on link j about the mass center C_j. The force analysis will start with link 3 because the moment $\mathbf{M}_{3ext}$ is known.

Link 3

For the free-body diagram of link 3 shown in Figure I.6.21(b), Eq. (I.6.79) gives

$$\mathbf{F}_{03} + \mathbf{F}_{in3} + \mathbf{G}_3 + \mathbf{F}_{23} = 0,$$

$$\mathbf{r}_{C_3C} \times \mathbf{F}_{03} + \mathbf{r}_{C_3Q} \times \mathbf{F}_{23} + \mathbf{M}_{in3} + \mathbf{M}_{3ext} = 0,$$

or

$$\mathbf{F}_{03} + \mathbf{F}_3 + \mathbf{F}_{23} = 0,$$

$$\mathbf{r}_{C_3C} \times \mathbf{F}_{03} + \mathbf{r}_{C_3Q} \times \mathbf{F}_{23} + \mathbf{M}_3 + \mathbf{M}_{3ext} = 0, \qquad (\text{I.6.80})$$

where the unknowns are

$$\mathbf{F}_{03} = F_{03x} \mathbf{i} + F_{03y} \mathbf{J}, \quad \mathbf{F}_{23} = F_{23x} \mathbf{i} + F_{23y} \mathbf{J},$$

and the position vector $\mathbf{r}_Q = x_Q \mathbf{i} + y_Q \mathbf{J}$ of the application point of the joint force $\mathbf{F}_{23}$. Numerically, Eq. (I.6.80) becomes

$$3.302 + F_{03x} + F_{23x} = 0, \qquad (\text{I.6.81})$$

$$-0.539 + F_{03y} + F_{23y} = 0, \qquad (\text{I.6.82})$$

$$- 1000.05 + 0.065F_{03x} - 0.075F_{03y} + 0.125F_{23x} - 0.075F_{23y}$$
$$+ F_{23y}x_Q - F_{23x}y_Q = 0. \tag{I.6.83}$$

The application point Q of the joint force $\mathbf{F}_{23}$ is on the line BC:

$$\frac{y_B - y_C}{x_B - x_C} = \frac{y_Q - y_C}{x_Q - x_C} \quad \text{or} \quad 0.874 - \frac{y_Q - 0.06}{x_Q} = 0. \tag{I.6.84}$$

The joint force $\mathbf{F}_{23}$ is perpendicular to the sliding direction BC:

$$\mathbf{F}_{23} \cdot \mathbf{r}_{BC} = 0 \quad \text{or} \quad - 0.07F_{23x} - 0.061F_{23y} = 0. \tag{I.6.85}$$

There are five scalar equations, Eqs. (I.6.81) through (I.6.85), and six unknowns, F_{03x}, F_{03y}, F_{23x}, F_{23y}, x_Q, y_Q. The force analysis will continue with link 2.

Link 2
Figure I.6.21(c) shows the free-body diagram of link 2 and Eq. (I.6.79) gives

$$\mathbf{F}_{12} + \mathbf{F}_{in\,2} + \mathbf{G}_2 + \mathbf{F}_{32} = \mathbf{0},$$
$$\mathbf{r}_{BQ} \times \mathbf{F}_{32} + \mathbf{M}_{in\,2} = \mathbf{0},$$

or

$$\mathbf{F}_{12} + \mathbf{F}_2 - \mathbf{F}_{23} = \mathbf{0},$$
$$\mathbf{r}_{BQ} \times (-\mathbf{F}_{23}) + \mathbf{M}_2 = \mathbf{0},$$

where the new unknown is introduced (the reaction of link 1 on link 2):

$$\mathbf{F}_{12} = F_{12x}\,\mathbf{\imath} + F_{12y}\,\mathbf{J}.$$

Numerically, the previous equations becomes

$$0.545 + F_{12x} - F_{23x} = 0, \tag{I.6.86}$$
$$0.160 + F_{12y} - F_{23y} = 0, \tag{I.6.87}$$
$$- 0.001 - 0.121\,F_{23x} + 0.07\,F_{23y} - x_Q\,F_{23y} + y_Q\,F_{23x} = 0. \tag{I.6.88}$$

Now there is a system of eight scalar equations, Eqs. (I.6.81) through (I.6.88), eight unknowns, and the solution is

$$\mathbf{F}_{03} = F_{03x}\,\mathbf{\imath} + F_{03y}\,\mathbf{J} = 7078.41\,\mathbf{\imath} - 8093.7\,\mathbf{J} \quad \text{N},$$
$$\mathbf{F}_{23} = F_{23x}\,\mathbf{\imath} + F_{23y}\,\mathbf{J} = -7081.72\,\mathbf{\imath} + 8094.24\,\mathbf{J} \quad \text{N},$$
$$\mathbf{F}_{12} = F_{12x}\,\mathbf{\imath} + F_{12y}\,\mathbf{J} = -7082.26\,\mathbf{\imath} + 8094.08\,\mathbf{J} \quad \text{N},$$
$$\mathbf{r}_Q = x_Q\,\mathbf{\imath} + y_Q\,\mathbf{J} = 0.069\,\mathbf{\imath} + 0.121\,\mathbf{J} \quad \text{m}.$$

Link 1

Figure I.6.21(d) shows the free-body diagram of link 1. The sum of all the forces for the driver link 1 gives

$$\mathbf{F}_{21} + \mathbf{F}_{in\,1} + \mathbf{G}_1 + \mathbf{F}_{01} = \mathbf{0}, \quad \text{or} \quad -\mathbf{F}_{12} + \mathbf{F}_1 + \mathbf{F}_{01} = \mathbf{0}.$$

The reaction of the ground 0 on link 1 is

$$\mathbf{F}_{01} = \mathbf{F}_{12} - \mathbf{F}_1 = -7082.26\,\mathbf{i} + 8094.08\,\mathbf{j} - (0.381\mathbf{i} + 0.437\mathbf{j})$$
$$= -7082.64\,\mathbf{i} + 8094.52\,\mathbf{j} \quad \text{N.}$$

The sum of the moments about the mass center C_1 for link 1 gives the equilibrium moment

$$\mathbf{r}_{C_1 B} \times \mathbf{F}_{21} + \mathbf{r}_{C_1 A} \times \mathbf{F}_{01} + \mathbf{M} = \mathbf{0},$$

or

$$\mathbf{M} = \mathbf{r}_{C_1 B} \times \mathbf{F}_{12} - \mathbf{r}_{C_1 A} \times \mathbf{F}_{01}$$
$$= 712.632\,\mathbf{k} + 712.671\,\mathbf{k} = 1425.303\,\mathbf{k} \quad \text{N} \cdot \text{m.}$$

EXAMPLE I.6.2:

Calculate the moment $\mathbf{M}$ required for dynamic equilibrium and the joint forces for the mechanism shown in Figure I.6.22 using the contour method. The position of the crank angle is $\phi = \frac{\pi}{4}$ rad. The dimensions are $AC = 0.10$ m, $BC = 0.30$ m, $BD = 0.90$ m, and $L_a = 0.10$ m, and the external force on slider 5 is $F_{ext} = 100$ N. The angular speed of crank 1 is $n_1 = 100$ rpm, or $\omega_1 = 100\,\frac{\pi}{30}$ rad/s. The center of mass locations of links $j = 1, 2, \ldots, 5$ (with the masses m_j) are designated by $C_j(x_{Cj}, y_{Cj}, 0)$. The position vectors of the joints and the centers of mass are

$$\mathbf{r}_A = 0\mathbf{i} + 0\mathbf{j} \text{ m},$$
$$\mathbf{r}_{C1} = 0.212\mathbf{i} + 0.212\mathbf{j} \text{ m},$$
$$\mathbf{r}_B = \mathbf{r}_{C2} = 0.256\mathbf{i} + 0.256\mathbf{j} \text{ m},$$
$$\mathbf{r}_{C3} = 0.178\mathbf{i} + 0.128\mathbf{j} \text{ m},$$
$$\mathbf{r}_C = 0.100\mathbf{i} + 0.000\mathbf{j} \text{ m},$$
$$\mathbf{r}_{C4} = 0.699\mathbf{i} + 0.178\mathbf{j} \text{ m},$$
$$\mathbf{r}_D = \mathbf{r}_{C5} = 1.142\mathbf{i} + 0.100\mathbf{j} \text{ m.}$$

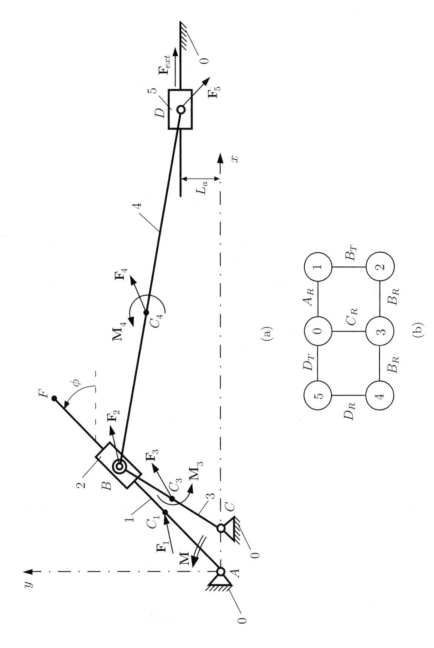

FIGURE I.6.22 (a) Mechanism, and (b) diagram representing the mechanism with two contours.

EXAMPLE I.6.2: *Cont'd*

The total forces and moments at C_j are $\mathbf{F}_j = \mathbf{F}_{inj} + \mathbf{G}_j$ and $\mathbf{M}_j = \mathbf{M}_{inj}$, where $\mathbf{F}_{inj}$ is the inertia force, $\mathbf{M}_j$ is the inertia moment, and $\mathbf{G}_j = -m_j\, g\, \mathbf{J}$ is the gravity force with gravity acceleration $g = 9.81$ m/s^2.

$$\mathbf{F}_1 = 5.514\mathbf{\imath} + 3.189\mathbf{J} \ \text{N},$$

$$\mathbf{F}_2 = 0.781\mathbf{\imath} + 1.843\mathbf{J} \ \text{N},$$

$$\mathbf{F}_3 = 1.202\mathbf{\imath} + 1.660\mathbf{J} \ \text{N},$$

$$\mathbf{F}_4 = 6.466\mathbf{\imath} + 4.896\mathbf{J} \ \text{N},$$

$$\mathbf{F}_5 = 0.643\mathbf{\imath} - 0.382\mathbf{J} \ \text{N},$$

$$\mathbf{M}_1 = \mathbf{M}_2 = \mathbf{M}_5 = 0\mathbf{k} \ \text{N} \cdot \text{m},$$

$$\mathbf{M}_3 = 0.023\mathbf{k} \ \text{N} \cdot \text{m},$$

$$\mathbf{M}_4 = -1.274\mathbf{k} \ \text{N} \cdot \text{m}.$$

Solution The diagram representing the mechanism is shown in Figure I.6.22(b) and has two contours, 0-1-2-3-0 and 0-3-4-5-0.

Contour 0-3-4-5-0

The joint at B represents a ramification point, and the dynamic force analysis will start with this joint.

Reaction F_{34}

The rotation joint at B_R, between 3 and 4, is replaced with the unknown reaction (Fig. I.6.23):

$$\mathbf{F}_{34} = -\mathbf{F}_{43} = F_{34x}\,\mathbf{\imath} + F_{34y}\,\mathbf{J}.$$

If the path I is followed [Fig. I.6.23(a)], a moment equation is written for the rotation joint D_R:

$$\sum \mathbf{M}_D^{(4)} = (\mathbf{r}_B - \mathbf{r}_D) \times \mathbf{F}_{34} + (\mathbf{r}_{C4} - \mathbf{r}_D) \times \mathbf{F}_4 + \mathbf{M}_4 = \mathbf{0}. \tag{I.6.89}$$

Continuing on path I the next joint is the slider joint D_T, and a force equation is written. The projection of all the forces that act on 4 and 5 onto the sliding direction x is zero:

$$\sum \mathbf{F}^{(4\&5)} \cdot \mathbf{\imath} = (\mathbf{F}_{34} + \mathbf{F}_4 + \mathbf{F}_5 + \mathbf{F}_{ext}) \cdot \mathbf{\imath}$$

$$= F_{34x} + F_{4x} + F_{5x} + F_{ext} = 0. \tag{I.6.90}$$

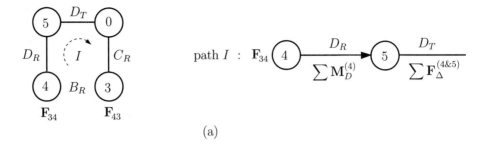

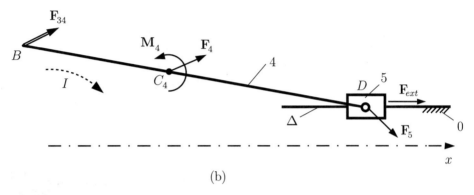

(b)

FIGURE I.6.23 *Joint force* $\mathbf{F}_{34}$ *(a) calculation diagram, and (b) force diagram.*

Solving the system of Eqs. (I.6.89) and (I.6.90):

$$F_{34x} = -107.110 \text{ N} \quad \text{and} \quad F_{34y} = 14.415 \text{ N}.$$

Reaction F₄₅

The pin joint at D_R, between 4 and 7, is replaced with the reaction force (Fig. I.6.24):

$$\mathbf{F}_{45} = -\mathbf{F}_{54} = F_{45x}\,\mathbf{1} + F_{45y}\,\mathbf{J}.$$

For the path I, shown Figure I.6.24(a), a moment equation about B_R is written for link 4:

$$\sum \mathbf{M}_B^{(4)} = (\mathbf{r}_D - \mathbf{r}_B) \times \mathbf{F}_{54} + (\mathbf{r}_{C4} - \mathbf{r}_B) \times \mathbf{F}_4 + \mathbf{M}_4 = \mathbf{0}. \tag{I.6.91}$$

For the path II, an equation for the forces projected onto the sliding direction of the joint D_T is written for link 5:

$$\sum \mathbf{F}^{(5)} \cdot \mathbf{1} = (\mathbf{F}_{45} + \mathbf{F}_5 + \mathbf{F}_{ext}) \cdot \mathbf{1}$$

$$= F_{45x} + F_{5x} + F_{ext} = 0. \tag{I.6.92}$$

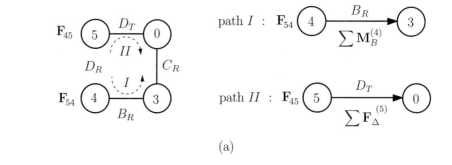

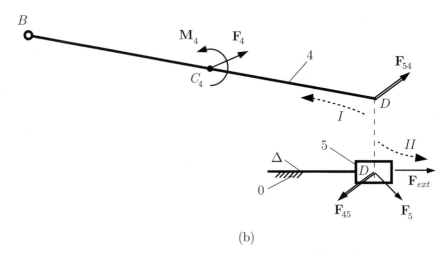

FIGURE I.6.24 *Joint force* $\mathbf{F}_{45}$ *(a) calculation diagram, and (b) force diagram.*

The joint force $\mathbf{F}_{45}$ is obtained from the system of Eqs. (I.6.91) and (I.6.92):

$$F_{45x} = -100.643 \text{ N} \quad \text{and} \quad F_{45y} = 19.310 \text{ N}.$$

Reaction $\mathbf{F}_{05}$

The reaction force $\mathbf{F}_{05}$ is perpendicular to the sliding direction of joint D_T (Fig. I.6.25):

$$\mathbf{F}_{05} = F_{05y} \, \mathbf{J}.$$

The application point of the unknown reaction force $\mathbf{F}_{05}$ is computed from a moment equation about D_R for link 5 (path I) [Fig. I.6.25(a)]:

$$\sum \mathbf{M}_D^{(5)} = (\mathbf{r}_P - \mathbf{r}_D) \times \mathbf{F}_{05} = \mathbf{0}, \tag{I.6.93}$$

or

$$x \, F_{05y} = 0 \quad \Rightarrow x = 0. \tag{I.6.94}$$

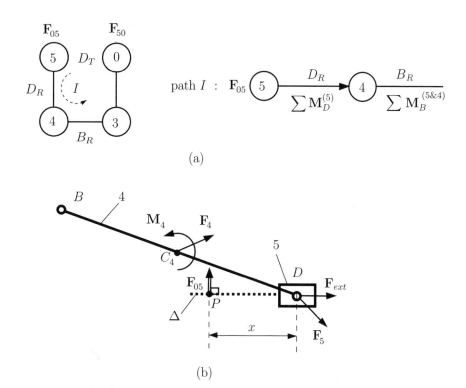

(a)

(b)

FIGURE I.6.25 *Joint force $\mathbf{F}_{05}$ (a) calculation diagram, and (b) force diagram.*

The application point of the reaction force $\mathbf{F}_{05}$ is at D ($P \equiv D$). The magnitude of the reaction force F_{05y} is obtained from a moment equation about B_R for links 5 and 4 (path I):

$$\sum \mathbf{M}_B^{(5\&4)} = (\mathbf{r}_D - \mathbf{r}_B) \times (\mathbf{F}_{05} + \mathbf{F}_5 + \mathbf{F}_{ext}) +$$

$$(\mathbf{r}_{C4} - \mathbf{r}_B) \times \mathbf{F}_4 + \mathbf{M}_4 = \mathbf{0}. \tag{I.6.95}$$

Solving the above equation:

$$F_{05y} = -18.928 \text{ N}.$$

Contour 0-1-2-3-0

The reaction force $\mathbf{F}_{43} = 107.110\,\mathbf{i} - 14.415\,\mathbf{j}$ N is considered as an external force for this contour at B.

Reaction $\mathbf{F}_{23}$

The rotation joint at B_R, between 2 and 3, is replaced with the unknown reaction force (Fig. I.6.26):

$$\mathbf{F}_{23} = -\mathbf{F}_{32} = F_{23x}\,\mathbf{i} + F_{23y}\,\mathbf{j}.$$

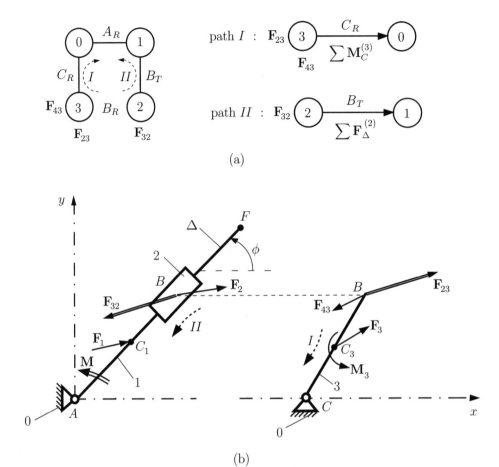

FIGURE I.6.26   *Joint force* $\mathbf{F}_{23}$ *(a) calculation diagram, and (b) force diagram.*

If the path *I* is followed, as in Figure I.6.26(a), a moment equation is written for the pin joint C_R for link 3:

$$\sum \mathbf{M}_C^{(3)} = (\mathbf{r}_B - \mathbf{r}_C) \times (\mathbf{F}_{23} + \mathbf{F}_{43}) + (\mathbf{r}_{C3} - \mathbf{r}_C) \times \mathbf{F}_3 + \mathbf{M}_3 = \mathbf{0}. \qquad (I.6.96)$$

For the path *II*, an equation for the forces projected in the direction Δ, the sliding direction of the joint B_T is written for the link 2:

$$\sum \mathbf{F}^{(2)} \cdot \Delta = (\mathbf{F}_{32} + \mathbf{F}_2) \cdot (\cos \phi \mathbf{i} + \sin \phi \mathbf{j}) = 0. \qquad (I.6.97)$$

The joint force $\mathbf{F}_{23}$ is calculated from Eqs. (I.6.96) and (I.6.97):

$$F_{23x} = -71.155 \text{ N} \quad \text{and} \quad F_{23y} = 73.397 \text{ N}.$$

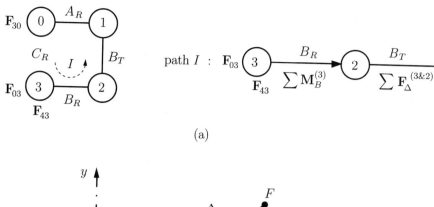

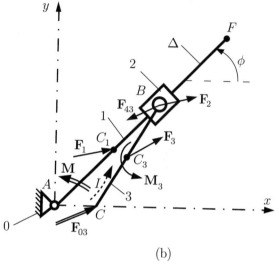

(a)

(b)

FIGURE I.6.27 *Joint force* $\mathbf{F}_{03}$ *(a) calculation diagram, and (b) force diagram.*

Reaction $\mathbf{F}_{03}$

For the joint reaction force $\mathbf{F}_{03}$ at C_R, there is only path I. For the pin joint B_R one moment equation is written for link 3 (Fig. I.6.27):

$$\sum \mathbf{M}_B^{(3)} = (\mathbf{r}_C - \mathbf{r}_B) \times \mathbf{F}_{03} + (\mathbf{r}_{C3} - \mathbf{r}_B) \times \mathbf{F}_3 + \mathbf{M}_3 = \mathbf{0}. \qquad (I.6.98)$$

A force equation is written for links 3 and 2 for the slider joint B_T:

$$\sum \mathbf{F}^{(3\&2)} \cdot \Delta = (\mathbf{F}_{03} + \mathbf{F}_3 + \mathbf{F}_{43} + \mathbf{F}_2) \cdot (\cos \phi \mathbf{i} + \sin \phi \mathbf{j}) = 0. \qquad (I.6.99)$$

The components of the unknown force are obtained by solving the system of Eqs. (I.6.98) and (I.6.99):

$$F_{03x} = -37.156 \text{ N} \quad \text{and} \quad F_{03y} = -60.643 \text{ N}.$$

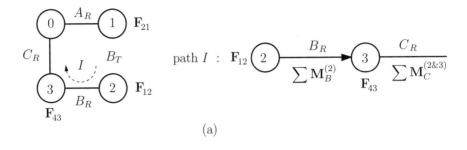

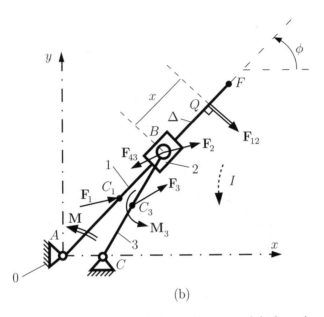

FIGURE I.6.28 *Joint force* $\mathbf{F}_{12}$ *(a) calculation diagram, and (b) force diagram.*

Reaction F_{12}

The slider joint at B_T, between 1 and 2, is replaced with the reaction force (Fig I.6.28):

$$\mathbf{F}_{12} = -\mathbf{F}_{21} = F_{12x}\,\mathbf{ı} + F_{12y}\,\mathbf{J}.$$

The reaction force $\mathbf{F}_{12}$ is perpendicular to the sliding direction Δ:

$$\mathbf{F}_{12} \cdot \Delta = (F_{12x}\,\mathbf{ı} + F_{12y}\,\mathbf{J}) \cdot (\cos\phi\,\mathbf{ı} + \sin\phi\,\mathbf{J}) =$$
$$F_{12x}\,\cos\phi + F_{12y}\,\sin\phi = 0. \tag{I.6.100}$$

The point of application of force $\mathbf{F}_{12}$ is determined from the equation (path I)

$$\sum \mathbf{M}_B^{(2)} = (\mathbf{r}_Q - \mathbf{r}_B) \times \mathbf{F}_{12} = \mathbf{0}, \tag{I.6.101}$$

or

$$x F_{12} = 0 \quad \Rightarrow x = 0, \tag{I.6.102}$$

and the force $\mathbf{F}_{12}$ acts at B.

Continuing on path I, a moment equation is written for links 2 and 3 with respect to the pin joint C_R:

$$\sum \mathbf{M}_C^{(2\&3)} = (\mathbf{r}_B - \mathbf{r}_C) \times (\mathbf{F}_{12} + \mathbf{F}_2 + \mathbf{F}_{43})$$
$$+ (\mathbf{r}_{C3} - \mathbf{r}_C) \times \mathbf{F}_3 + \mathbf{M}_3 = \mathbf{0}. \tag{I.6.103}$$

The two components of the joint force $\mathbf{F}_{12}$ are computed from Eqs. (I.6.100) and (I.6.103):

$$F_{12x} = -71.936 \text{ N} \quad \text{and} \quad F_{12y} = 71.936 \text{ N}.$$

Reaction $\mathbf{F}_{01}$ and Equilibrium Moment $\mathbf{M}$

The pin joint A_R, between 0 and 1, is replaced with the unknown reaction (Fig. I.6.29):

$$\mathbf{F}_{01} = F_{01x} \mathbf{I} + F_{01y} \mathbf{J}.$$

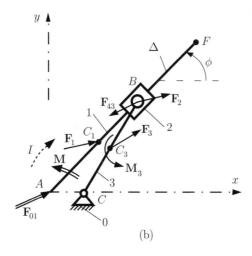

(a)

(b)

FIGURE I.6.29 *Joint force $\mathbf{F}_{01}$ (a) calculation diagram, and (b) force diagram.*

The unknown equilibrium moment is $\mathbf{M} = M\,\mathbf{k}$. If the path I is followed [Fig. I.6.29(a)] for the slider joint B_T, a force equation is written for link 1:

$$\sum \mathbf{F}^{(1)} \cdot \Delta = (\mathbf{F}_{01} + \mathbf{F}_1) \cdot (\cos \phi \mathbf{i} + \sin \phi \mathbf{j}) = 0. \qquad (I.6.104)$$

Continuing on path I the next joint encountered is the pin joint B_R, and a moment equation is written for links 1 and 2:

$$\sum \mathbf{M}_B^{(1\&2)} = -\mathbf{r}_B \times \mathbf{F}_{01} + (\mathbf{r}_{C1} - \mathbf{r}_B) \times \mathbf{F}_1 + \mathbf{M} = \mathbf{0}. \qquad (I.6.105)$$

Equation (I.6.105) is the vector sum of the moments about C_R of all forces and moments that acts on links 1, 2, and 3:

$$\sum \mathbf{M}_C^{(1\&2\&3)} = -\mathbf{r}_C \times \mathbf{F}_{01} + (\mathbf{r}_{C1} - \mathbf{r}_C) \times \mathbf{F}_1 + \mathbf{M}$$

$$+ (\mathbf{r}_B - \mathbf{r}_C) \times (\mathbf{F}_2 + \mathbf{F}_{43}) + \mathbf{M}_3 + (\mathbf{r}_{C3} - \mathbf{r}_C) \times \mathbf{F}_3 = \mathbf{0}. \qquad (I.6.106)$$

From Eqs. (I.6.104), (I.6.105), and (I.6.106) the components F_{01x}, F_{01y}, and M are computed:

$$F_{01x} = -77.451 \text{ N}, \quad F_{01y} = 68.747 \text{ N}, \quad \text{and} \quad M = 37.347 \text{ N} \cdot \text{m}.$$

EXAMPLE I.6.3:

For the R-TRR-RRT mechanism in Example I.6.2, calculate the moment $\mathbf{M}$ required for dynamic equilibrium of the mechanism and the joint forces using the dyad method.

Solution $B_R\, D_R\, D_T$ dyad

Figure I.6.30(a) shows the last dyad $B_R\, D_R\, D_T$ with the unknown joint reactions $\mathbf{F}_{34}$, $\mathbf{F}_{05}$, and $\mathbf{F}_{45} = -\mathbf{F}_{54}$. The joint reaction $\mathbf{F}_{05}$ is perpendicular to the sliding direction $\mathbf{F}_{05} \perp \Delta = \mathbf{i}$ or

$$\mathbf{F}_{05} = F_{05y}\mathbf{j}. \qquad (I.6.107)$$

The following equations are written to determine $\mathbf{F}_{34}$ and $\mathbf{F}_{05}$:

• sum of all the forces on links 4 and 5 is zero:

$$\sum \mathbf{F}^{(4\&5)} = \mathbf{F}_{34} + \mathbf{F}_4 + \mathbf{F}_5 + \mathbf{F}_{ext} + \mathbf{F}_{05} = \mathbf{0},$$

or

$$\sum \mathbf{F}^{(2\&3)} \cdot \mathbf{i} = F_{43x} + F_{4x} + F_{5x} + F_{ext} = 0,$$

$$\sum \mathbf{F}^{(2\&3)} \cdot \mathbf{j} = F_{43y} + F_{4y} + F_{5y} + F_{05y} = 0. \qquad (I.6.108)$$

EXAMPLE I.6.3: *Cont'd*

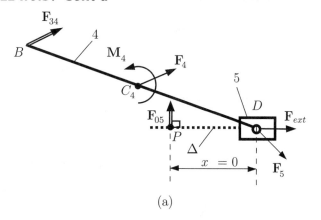

(a)

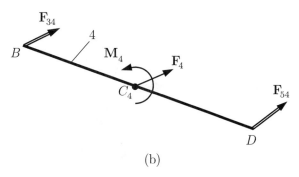

(b)

FIGURE I.6.30   *Joint reactions for the dyad $B_R D_R D_T$.*

- sum of moments of all the forces and moments on link 4 about D_R is zero

$$\sum \mathbf{M}_D^{(4)} = (\mathbf{r}_B - \mathbf{r}_D) \times \mathbf{F}_{43} + (\mathbf{r}_{C4} - \mathbf{r}_D) \times \mathbf{F}_4 + \mathbf{M}_4 = \mathbf{0}. \qquad (\text{I.6.109})$$

From Eqs. (I.6.108) and (I.6.109) the unknown components are calculated:

$$F_{34x} = -107.110 \text{ N}, \quad F_{34y} = 14.415 \text{ N}, \quad \text{and} \quad F_{05y} = -18.928 \text{ N}.$$

The reaction components F_{54x} and F_{54y} are computed from the sum of all the forces on link 4 [Fig. I.6.30(b)]:

$$\sum \mathbf{F}^{(4)} = \mathbf{F}_{34} + \mathbf{F}_4 + \mathbf{F}_{54} = \mathbf{0},$$

or

$$\sum \mathbf{F}^{(4)} \cdot \mathbf{\imath} = F_{34x} + F_{4x} + F_{54x} = 0,$$

$$\sum \mathbf{F}^{(4)} \cdot \mathbf{J} = F_{34y} + F_{5y} + F_{54y} = 0, \qquad (\text{I.6.110})$$

Continued

EXAMPLE I.6.3: *Cont'd*

and

$$F_{54x} = 100.643 \text{ N} \quad \text{and} \quad F_{54y} = -19.310 \text{ N}.$$

$B_T B_R C_R$ dyad

Figure I.6.31(a) shows the first dyad $B_T B_R C_R$ with the unknown joint reaction forces $\mathbf{F}_{12}$, $\mathbf{F}_{03}$, and $\mathbf{F}_{23} = -\mathbf{F}_{32}$. The joint reaction force $\mathbf{F}_{12}$ is perpendicular to the sliding direction $\mathbf{F}_{12} \perp \Delta$ or

$$\mathbf{F}_{12} \cdot \Delta = (F_{12x}\mathbf{I} + F_{12y}\mathbf{J}) \cdot (\cos \phi \mathbf{I} + \sin \phi \mathbf{J}) = 0. \qquad (I.6.111)$$

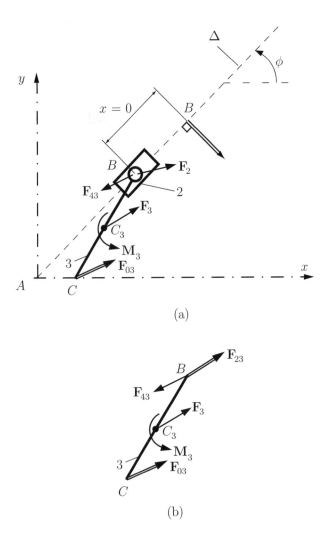

(a)

(b)

FIGURE I.6.31 *Joint reactions for the dyad $B_T B_R C_R$.*

EXAMPLE I.6.3: *Cont'd*

The following equations are written in order to determine the forces $\mathbf{F}_{12}$ and $\mathbf{F}_{03}$:

- sum of all forces on links 2 and 3 is zero:

$$\sum \mathbf{F}^{(2\&3)} = \mathbf{F}_{12} + \mathbf{F}_2 + \mathbf{F}_3 + \mathbf{F}_{43} + \mathbf{F}_{03} = \mathbf{0},$$

or

$$\sum \mathbf{F}^{(2\&3)} \cdot \mathbf{\imath} = F_{12x} + F_{2x} + F_{3x} + F_{43x} + F_{03x} = 0,$$

$$\sum \mathbf{F}^{(2\&3)} \cdot \mathbf{\jmath} = F_{12y} + F_{2y} + F_{3y} + F_{43y} + F_{03x} = 0. \qquad (I.6.112)$$

- sum of moments of all the forces and the moments on link 3 about B_R is zero:

$$\sum \mathbf{M}_B^{(3)} = (\mathbf{r}_C - \mathbf{r}_B) \times \mathbf{F}_{03} + (\mathbf{r}_{C3} - \mathbf{r}_B) \times \mathbf{F}_3 + \mathbf{M}_3 = \mathbf{0}. \qquad (I.6.113)$$

From Eqs. (I.6.111), (I.6.112), and (I.6.113) the following components are obtained:

$$F_{12x} = -71.936 \text{ N} \quad \text{and} \quad F_{12y} = 71.936 \text{ N},$$

$$F_{03x} = -37.156 \text{ N} \quad \text{and} \quad F_{03y} = -60.643 \text{ N}.$$

The reaction components F_{23x} and F_{23y} are computed from the sum of all the forces on link 3 [Fig. I.6.31(b)]:

$$\sum \mathbf{F}^{(3)} = \mathbf{F}_{23} + \mathbf{F}_3 + \mathbf{F}_{43} + \mathbf{F}_{03} = \mathbf{0},$$

or

$$\sum \mathbf{F}^{(3)} \cdot \mathbf{\imath} = F_{23x} + F_{3x} + F_{43x} + F_{03x} = 0,$$

$$\sum \mathbf{F}^{(2)} \cdot \mathbf{\jmath} = F_{23y} + F_{3y} + F_{43y} + F_{03y} = 0, \qquad (I.6.114)$$

and solving the equations

$$F_{23x} = -71.155 \text{ N} \quad \text{and} \quad F_{23y} = 73.397 \text{ N}.$$

Driver link

A force equation for the driver can be written to determine the joint reaction $\mathbf{F}_{01}$ (Fig. I.6.32):

$$\sum \mathbf{F}^{(1)} = \mathbf{F}_{01} + \mathbf{F}_1 + \mathbf{F}_{21} = \mathbf{0},$$

or

$$\sum \mathbf{F}^{(1)} \cdot \mathbf{\imath} = F_{01x} + F_{1x} + F_{21x} = 0,$$

$$\sum \mathbf{F}^{(1)} \cdot \mathbf{\jmath} = F_{01y} + F_{1y} + F_{21y} = 0, \qquad (I.6.115)$$

Continued

EXAMPLE I.6.3: *Cont'd*

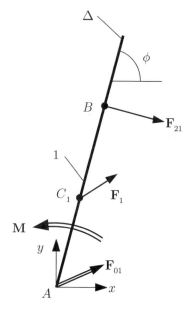

FIGURE I.6.32 *Joint reactions for the driver link.*

Solving the above equations gives

$$F_{01x} = -77.451 \text{ N} \quad \text{and} \quad F_{01y} = 68.747 \text{ N}.$$

Sum of the moments about A_R for link 1 gives the equilibrium moment

$$\sum \mathbf{M}_A^{(1)} = \mathbf{r}_B \times \mathbf{F}_{21} + \mathbf{r}_{C1} \times \mathbf{F}_1 + \mathbf{M} = \mathbf{0}, \qquad (I.6.116)$$

and $M = 37.347 \text{ N·m}$.

I.6.10 Problems

I.6.1 Figure I.6.33 shows a uniform rod of mass m and length L. The rod is free to swing in a vertical plane. The rod is connected to the ground by a pin joint at the distance D from one end of the rod. The rod makes an angle $\theta(t)$ with the horizontal axis. The local acceleration of gravity is g. a). Find the differential equation or equations describing the motion of the rod. b). Determine the axial and shear components of the force exerted by the pin on the rod as the rod swings by any arbitrary position. c). When the rod is released from rest in the horizontal position, the initial value of the angular velocity is zero. Find the initial angular acceleration and the initial pin force components.

I.6.2 The four-bar mechanism shown in Figure I.3.10(a) has the dimensions: $AB = 80$ mm, $BC = 210$ mm, $CD = 120$ mm, and $AD = 190$ mm. The driver link AB rotates with a constant angular speed of 2400 rpm. The links are homogeneous

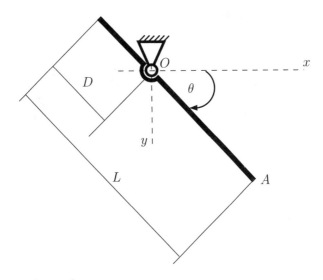

FIGURE I.6.33 *Uniform rod.*

rectangular prisms made of steel with the width $h = 0.010$ m and the depth $d = 0.001$ m. The external moment applied on the link CD is opposed to the motion of the link and has the value $|\mathbf{M}_{ext}| = 600$ N·m. The density of the material is $\rho_{Steel} = 8000$ kg/m^3 and the gravitational acceleration is $g = 9.807$ m/s^2. Find the equilibrium moment on link AB and the joint forces for $\phi = 120°$ using: a). free-body diagram of individual links; b). contour method; and c). dyads.

I.6.3 The slider crank mechanism shown in Figure I.4.10 has the dimensions $AB = 0.4$ m and $BC = 1$ m. The driver link 1 rotates with a constant angular speed of $n = 1600$ rpm. The links 1 and 2 have a rectangular shape made of steel with the width $h = 0.010$ m and the depth $d = 0.001$ m. The steel slider 3 has the width $w_{Slider} = 0.050$ m, the height $h_{Slider} = 0.020$ m, and the depth $d = 0.001$ m. The external force applied on the slider 3 is opposed to the motion of the slider and has the value $|\mathbf{F}_{ext}| = 800$ N. The density of the material is $\rho_{Steel} = 8000$ kg/m^3 and the gravitational acceleration is $g = 9.807$ m/s^2. Find the equilibrium moment on the driver link 1 and the joint forces for $\phi = 30°$ using: a). free-body diagram of individual links; b). contour method; and c). dyads.

I.6.4 The planar mechanism considered is shown in Figure I.3.19 and has the following data: $AB = 0.150$ m, $BC = 0.400$ m, $CD = 0.370$ m, $CE = 0.230$ m, $EF = CE$, $L_a = 0.300$ m, $L_b = 0.450$ m, and $L_c = CD$. The constant angular speed of the driver link 1 is 1800 rpm. The links 1, 2, 3, and 4 are homogeneous rectangular prisms with the width $h = 0.010$ m and the depth $d = 0.001$ m. The slider 5 has the width $w_{Slider} = 0.050$ m, the height $h_{Slider} = 0.020$ m, and the depth $d = 0.001$ m. The external force applied on the slider 5 is opposed to the motion of the slider and has the value $|\mathbf{F}_{ext}| = 500$ N. The density of the material is $\rho_{Steel} = 8000$ kg/m^3 and the gravitational acceleration is $g = 9.807$ m/s^2. Find the equilibrium moment on the driver link 1 and the joint forces for $\phi = \phi_1 = 60°$.

I.6.5 The R-RRR-RTT mechanism is shown in Figure I.3.20. The following data are given: $AB = 0.080$ m, $BC = 0.350$ m, $CE = 0.200$ m, $CD = 0.150$ m, $L_a = 0.200$ m, $L_b = 0.350$ m, and $L_c = 0.040$ m. The driver link 1 rotates with a constant angular speed of $n = 2200$ rpm. The links 1, 2, 3, and 5 are homogeneous rectangular prisms made of aluminum with the width $h = 0.010$ m and the depth $d = 0.001$ m. The aluminum slider 4 has the width $w_{Slider} = 0.050$ m, the height $h_{Slider} = 0.020$ m, and the depth $d = 0.001$ m. The external force applied on 5 is opposed to the motion of the link and has the value $|\mathbf{F}_{ext}| = 1000$ N. The density of the material is $\rho_{Al} = 2.8$ Mg/m^3 and the gravitational acceleration is $g = 9.807$ m/s^2. For $\phi = 145°$ find the equilibrium moment on the driver link 1 and the joint forces. Select suitable dimensions for the link 5.

I.6.6 The mechanism shown in Figure I.3.21 has the following dimensions: $AB = 100$ mm, $AD = 350$ mm, $BC = 240$ mm, $CE = 70$ mm, $EF = 300$ mm, and $a = 240$ mm. The constant angular speed of the driver link 1 is $n = 1400$ rpm. The links 1 and 4 are homogeneous rectangular prisms with the width $h = 0.010$ m and the depth $d = 0.001$ m. The link 2 has the width $h = 0.010$ m and the depth $d = 0.001$ m. The sliders 3 and 5 have the width $w_{Slider} = 0.050$ m, the height $h_{Slider} = 0.020$ m, and the depth $d = 0.001$ m. The external force applied on 5 is opposed to the motion of the link and has the value $|\mathbf{F}_{ext}| = 1200$ N. The density of the material is $\rho_{Iron} = 7.2$ Mg/m^3 and the gravitational acceleration is $g = 9.807$ m/s^2. Find the equilibrium moment on the driver link 1 and the joint forces for $\phi = \phi_1 = 30°$. Select a suitable dimension for link 2.

I.6.7 The dimensions for the mechanism shown in Figure I.3.22 are: $AB = 60$ mm, $BD = 160$ mm, $BC = 55$ mm, $CD = 150$ mm, $DE = 100$ mm, $CF = 250$ mm, $AE = 150$ mm, and $b = 40$ mm. The constant angular speed of the driver link 1 is $n = 1400$ rpm. The links 1, 3, and 4 are homogeneous rectangular prisms with the width $h = 0.010$ m and the depth $d = 0.001$ m. The slider 5 has the width $w_{Slider} = 0.050$ m, the height $h_{Slider} = 0.020$ m, and the depth $d = 0.001$ m. The plate 2 has the width $h = 0.010$ m and the depth $d = 0.001$ m. The external force applied on 5 is opposed to the motion of the link and has the value $|\mathbf{F}_{ext}| = 1500$ N. The density of the material is $\rho_{Bronze} = 8.7$ Mg/m^3 and the gravitational acceleration is $g = 9.807$ m/s^2. Find the equilibrium moment on the driver link 1 and the joint forces for $\phi = \phi_1 = 60°$.

I.6.8 The mechanism in Figure I.3.23 has the dimensions: $AB = 110$ mm, $AC = 55$ mm, $BD = 220$ mm, $DE = 300$ mm, $EF = 175$ mm, $L_a = 275$ mm, and $L_b = 65$ mm. The links 1, 2, 4, and 5 are homogeneous rectangular prisms with the width $h = 0.010$ m and the depth $d = 0.001$ m. The slider 3 has the width $w_{Slider} = 0.050$ m, the height $h_{Slider} = 0.020$ m, and the depth $d = 0.001$ m. The constant angular speed of the driver link 1 is $n = 2400$ rpm. The external moment on 5 is opposed to the motion of the link $\mathbf{M}_{ext} = -|\mathbf{M}_{ext}|\dfrac{\omega_5}{|\omega_5|}$, where

$|\mathbf{M}_{ext}| = 600$ N·m. The density of the material is $\rho_{Steel} = 8000$ kg/m^3 and the gravitational acceleration is $g = 9.807$ m/s^2. Find the equilibrium moment on the driver link 1 and the joint forces for $\phi = \phi_1 = 150°$.

I.6.9 The dimensions for the mechanism shown in Figure I.3.24 are: $AB = 250$ mm, $BC = 650$ mm, $AD = 600$ mm, $CD = 350$ mm, $DE = 200$ mm, $EF = 600$ mm, and $L_a = 100$ mm. The constant angular speed of the driver link 1 is $n = 2500$ rpm. The links 1, 2, 3, and 4 are homogeneous rectangular prisms with the width

$h = 0.010$ m and the depth $d = 0.001$ m. The slider 5 has the width $w_{Slider} = 0.050$ m, the height $h_{Slider} = 0.020$ m, and the depth $d = 0.001$ m. The external force on 5 is opposed to the motion of the link $\mathbf{F}_{ext} = -|\mathbf{F}_{ext}|\dfrac{\mathbf{v}_F}{|\mathbf{v}_F|}$, where $|\mathbf{F}_{ext}| = 1600$ N. The density of the material is $\rho_{Steel} = 8000$ kg/m^3 and the gravitational acceleration is $g = 9.807$ m/s^2. Find the equilibrium moment on the driver link 1 and the joint forces for $\phi = \phi_1 = 60°$.

I.6.10 The mechanism in Figure I.3.25 has the dimensions: $AB = 50$ mm, $AC = 160$ mm, $BD = 250$ mm, $L_a = 30$ mm, and $L_b = 60$ mm. The driver link 1 rotates with a constant angular speed of $n = 1500$ rpm. The links 1, 2, and 5 are homogeneous rectangular prisms with the width $h = 0.010$ m and the depth $d = 0.001$ m. The sliders 3 and 4 have the width $w_{Slider} = 0.050$ m, the height $h_{Slider} = 0.020$ m, and the depth $d = 0.001$ m. The external moment on 5 is opposed to the motion of the link $\mathbf{M}_{ext} = -|\mathbf{M}_{ext}|\dfrac{\omega_5}{|\omega_5|}$, where $|\mathbf{M}_{ext}| = 900$ N·m. The density of the material is $\rho_{Steel} = 8000$ kg/m^3 and the gravitational acceleration is $g = 9.807$ m/s^2. Find the equilibrium moment on the driver link 1 and the joint forces for $\phi = \phi_1 = 130°$. Select a suitable dimension for link 5.

I.6.11 Figure I.3.26 shows a mechanism with the following dimensions: $AB = 150$ mm, $BD = 500$ mm, and $L_a = 180$ mm. The constant angular speed of the driver link 1 is $n = 1600$ rpm. The links 1, 2, and 4 are homogeneous rectangular prisms with the width $h = 0.010$ m and the depth $d = 0.001$ m. The sliders 3 and 5 have the width $w_{Slider} = 0.050$ m, the height $h_{Slider} = 0.020$ m, and the depth $d = 0.001$ m. The external force on 5 is opposed to the motion of the link $\mathbf{F}_{ext} = -|\mathbf{F}_{ext}|\dfrac{\mathbf{v}_D}{|\mathbf{v}_D|}$, where $|\mathbf{F}_{ext}| = 2000$ N. The density of the material is $\rho_{Steel} = 8000$ kg/m^3 and the gravitational acceleration is $g = 9.807$ m/s^2. Find the equilibrium moment on the driver link 1 and the joint forces for $\phi = 210°$. Select a suitable dimension for link 4.

I.6.12 The mechanism in Figure I.3.27 has the dimensions: $AB = 20$ mm, $AC = 50$ mm, $BD = 150$ mm, $DE = 40$ mm, $EF = 27$ mm, $L_a = 7$ mm, and $L_b = 30$ mm. The constant angular speed of the driver link 1 is $n = 1400$ rpm. The links 1, 2, 4, and 5 are homogeneous rectangular prisms with the width $h = 0.010$ m and the depth $d = 0.001$ m. The slider 3 has the width $w_{Slider} = 0.050$ m, the height $h_{Slider} = 0.020$ m, and the depth $d = 0.001$ m. The external moment on 5 is opposed to the motion of the link $\mathbf{M}_{ext} = -|\mathbf{M}_{ext}|\dfrac{\omega_5}{|\omega_5|}$, where $|\mathbf{M}_{ext}| = 1500$ N·m. The density of the material is $\rho_{Steel} = 8000$ kg/m^3 and the gravitational acceleration is $g = 9.807$ m/s^2. Find the equilibrium moment on the driver link 1 and the joint forces for $\phi = \phi_1 = 120°$.

I.6.13 Figure I.3.28 shows a mechanism with the following dimensions: $AB = 250$ mm, $BC = 940$ mm, $CD = DE = 380$ mm, $EF = 700$ mm, $L_a = 930$ mm, and $L_b = L_c = 310$ mm. The driver link 1 rotates with a constant angular speed of $n = 1500$ rpm. The links 1, 2, 3, and 4 are homogeneous rectangular prisms with the width $h = 0.010$ m and the depth $d = 0.001$ m. The slider 5 has the width $w_{Slider} = 0.050$ m, the height $h_{Slider} = 0.020$ m, and the depth $d = 0.001$ m. The external force on 5 is opposed to the motion of the link $\mathbf{F}_{ext} = -|\mathbf{F}_{ext}|\dfrac{\mathbf{v}_F}{|\mathbf{v}_F|}$, where

$|\mathbf{F}_{ext}| = 2000$ N. The density of the material is $\rho_{Steel} = 8000$ kg/m^3 and the gravitational acceleration is $g = 9.807$ m/s^2. Find the equilibrium moment on the driver link 1 and the joint forces for $\phi = \phi_1 = 120°$.

I.6.14 Figure I.3.29 shows a mechanism with the following dimensions: $AB = 200$ mm, $BC = 900$ mm, $CE = 300$ mm, $CD = 600$ mm, $EF = 600$ mm, $L_a = 500$ mm, $L_b = 800$ mm, and $L_c = 1100$ mm. The constant angular speed of the driver link 1 is $n = 1000$ rpm. The links 1, 2, 3, and 4 are homogeneous rectangular prisms with the width $h = 0.010$ m and the depth $d = 0.001$ m. The slider 5 has the width $w_{Slider} = 0.050$ m, the height $h_{Slider} = 0.020$ m, and the depth $d = 0.001$ m. The external force on 5 is opposed to the motion of the link $\mathbf{F}_{ext} = -|\mathbf{F}_{ext}|\dfrac{\mathbf{v}_F}{|\mathbf{v}_F|}$, where

$|\mathbf{F}_{ext}| = 3000$ N. The density of the material is $\rho_{Steel} = 8000$ kg/m^3 and the gravitational acceleration is $g = 9.807$ m/s^2. Find the equilibrium moment on the driver link 1 and the joint forces for $\phi = \phi_1 = 150°$.

I.6.15 Figure I.3.30 shows a mechanism with the following dimensions: $AB = 200$ mm, $BC = 540$ mm, $CF = 520$ mm, $CD = 190$ mm, $DE = 600$ mm, $L_a = 700$ mm, $L_b = 400$ mm, and $L_c = 240$ mm. The constant angular speed of the driver link 1 is $n = 1200$ rpm. The links 1, 2, 3, and 4 are homogeneous rectangular prisms with the width $h = 0.010$ m and the depth $d = 0.001$ m. The slider 5 has the width $w_{Slider} = 0.050$ m, the height $h_{Slider} = 0.020$ m, and the depth $d = 0.001$ m. The external force on 5 is opposed to the motion of the link $\mathbf{F}_{ext} = -|\mathbf{F}_{ext}|\dfrac{\mathbf{v}_E}{|\mathbf{v}_E|}$, where

$|\mathbf{F}_{ext}| = 900$ N. The density of the material is $\rho_{Steel} = 8000$ kg/m^3 and the gravitational acceleration is $g = 9.807$ m/s^2. For $\phi = 30°$ find the equilibrium moment on the driver link 1 and the joint forces.

I.6.16 Figure I.3.31 shows a mechanism with the following dimensions: $AB = 80$ mm, $BC = 200$ mm, $AD = 90$ mm, and $BE = 220$ mm. The constant angular speed of the driver link 1 is $n = 1300$ rpm. The links 1, 2, and 4 are homogeneous rectangular prisms with the width $h = 0.010$ m and the depth $d = 0.001$ m. The sliders 3 and 5 have the width $w_{Slider} = 0.050$ m, the height $h_{Slider} = 0.020$ m, and the depth $d = 0.001$ m. The external force on 3 is opposed to the motion of the link $\mathbf{F}_{ext} = -|\mathbf{F}_{ext}|\dfrac{\mathbf{v}_C}{|\mathbf{v}_C|}$, where $|\mathbf{F}_{ext}| = 1900$ N. The density of the material is $\rho_{Steel} = 8000$ kg/m^3 and the gravitational acceleration is $g = 9.807$ m/s^2. Find the equilibrium moment on the driver link 1 and the joint forces for $\phi = 60°$.

I.6.17 The dimensions of the mechanism shown in Figure I.3.32 are: $AB = 80$ mm, $BC = 150$ mm, $BE = 300$ mm, $CE = 450$ mm, $CD = 170$ mm, $EF = 600$ mm, $L_a = 200$ mm, $L_b = 150$ mm, and $L_c = 50$ mm. The constant angular speed of the driver link 1 is $n = 1500$ rpm. The links 1, 3, and 4 are homogeneous rectangular prisms with the width $h = 0.010$ m and the depth $d = 0.001$ m. The slider 5 has the width $w_{Slider} = 0.050$ m, the height $h_{Slider} = 0.020$ m, and the depth $d = 0.001$ m. The plate 2 has the width $h = 0.010$ m and the depth $d = 0.001$ m. The external force applied on 5 is opposed to the motion of the link and has the value $|\mathbf{F}_{ext}| = 2000$ N. The density of the material is $\rho_{Bronze} = 8.7$ Mg/m^3 and the gravitational acceleration is $g = 9.807$ m/s^2. Find the equilibrium moment on the driver link 1 and the joint forces for $\phi = \phi_1 = 210°$.

I.6.18 The dimensions of the mechanism shown in Figure I.3.33 are: $AB = 140$ mm, $AC = 200$ mm, $CD = 350$ mm, $DE = 180$ mm, and $L_a = 300$ mm. The constant angular speed of the driver link 1 is $n = 900$ rpm. The links 1, 3, and 4 are homogeneous rectangular prisms with the width $h = 0.010$ m and the depth $d = 0.001$ m. The sliders 2 and 5 have the width $w_{Slider} = 0.050$ m, the height $h_{Slider} = 0.020$ m, and the depth $d = 0.001$ m. The external force on 5 is opposed to the motion of the link $\mathbf{F}_{ext} = -|\mathbf{F}_{ext}|\dfrac{\mathbf{v}_E}{|\mathbf{v}_E|}$, where $|\mathbf{F}_{ext}| = 1000$ N. The density of the material is $\rho_{Steel} = 8000$ kg/m^3 and the gravitational acceleration is $g = 9.807$ m/s^2. Find the equilibrium moment on the driver link 1 and the joint forces for $\phi = 60°$.

I.6.19 The dimensions of the mechanism shown in Figure I.3.34 are: $AB = 250$ mm, $AC = 100$ mm, $CD = 280$ mm, and $DE = 800$ mm. The constant angular speed of the driver link 1 is $n = 1600$ rpm. The links 1, 3, and 4 are homogeneous rectangular prisms with the width $h = 0.010$ m and the depth $d = 0.001$ m. The sliders 2 and 5 have the width $w_{Slider} = 0.050$ m, the height $h_{Slider} = 0.020$ m, and the depth $d = 0.001$ m. The external force on 5 is opposed to the motion of the link $\mathbf{F}_{ext} = -|\mathbf{F}_{ext}|\dfrac{\mathbf{v}_E}{|\mathbf{v}_E|}$, where $|\mathbf{F}_{ext}| = 900$ N. The density of the material is $\rho_{Steel} = 8000$ kg/m^3 and the gravitational acceleration is $g = 9.807$ m/s^2. For $\phi = \phi_1 = 210°$ find the equilibrium moment on the driver link 1 and the joint forces.

I.6.20 The dimensions of the mechanism shown in Figure I.3.35 are: $AB = 100$ mm, $AC = 200$ mm, and $CD = 350$ mm. The constant angular speed of the driver link 1 is $n = 900$ rpm. The links 1, 3, and 5 are homogeneous rectangular prisms with the width $h = 0.010$ m and the depth $d = 0.001$ m. The sliders 2 and 4 have the width $w_{Slider} = 0.050$ m, the height $h_{Slider} = 0.020$ m, and the depth $d = 0.001$ m. The external force on 5 is opposed to the motion of the link $\mathbf{F}_{ext} = -|\mathbf{F}_{ext}|\dfrac{\mathbf{v}_G}{|\mathbf{v}_G|}$, where $|\mathbf{F}_{ext}| = 2500$ N. The density of the material is $\rho_{Steel} = 8000$ kg/m^3 and the gravitational acceleration is $g = 9.807$ m/s^2. Find the equilibrium moment on the driver link 1 and the joint reaction forces for $\phi = \phi_1 = 45°$. Select suitable dimensions for the link 5 and the distance b.

I.6.21 The dimensions of the mechanism shown in Figure I.3.36 are: $AB = 140$ mm, $AC = 60$ mm, and $CD = 140$ mm. The constant angular speed of the driver link 1 is $n = 2200$ rpm. The links 1, 3, and 5 are homogeneous rectangular prisms with the width $h = 0.010$ m and the depth $d = 0.001$ m. The sliders 2 and 4 have the width $w_{Slider} = 0.050$ m, the height $h_{Slider} = 0.020$ m, and the depth $d = 0.001$ m. The external moment on 5 is opposed to the motion of the link $\mathbf{M}_{ext} = -|\mathbf{M}_{ext}|\dfrac{\omega_5}{|\omega_5|}$, where $|\mathbf{M}_{ext}| = 1500$ N·m. The density of the material is $\rho_{Steel} = 8000$ kg/m^3 and the gravitational acceleration is $g = 9.807$ m/s^2. Find the equilibrium moment on the driver link 1 and the joint forces for $\phi = \phi_1 = 60°$. Select suitable lengths for links 3 and 5.

I.6.22 The dimensions of the mechanism shown in Figure I.3.37 are: $AB = 110$ mm, $AC = 260$ mm, $BD = L_a = 400$ mm, and $DE = 270$ mm. The constant angular speed of the driver link 1 is $n = n_1 = 1000$ rpm. The links 1, 2, and 4 are homogeneous rectangular prisms with the width $h = 0.010$ m and the depth $d = 0.001$ m.

The sliders 3 and 5 have the width $w_{Slider} = 0.050$ m, the height $h_{Slider} = 0.020$ m, and the depth $d = 0.001$ m. The external force on 5 is opposed to the motion of the link $\mathbf{F}_{ext} = -|\mathbf{F}_{ext}|\dfrac{\mathbf{v}_E}{|\mathbf{v}_E|}$, where $|\mathbf{F}_{ext}| = 900$ N. The density of the material is $\rho_{Steel} = 8000$ kg/m^3 and the gravitational acceleration is $g = 9.807$ m/s^2. Find the equilibrium moment on the driver link 1 and the joint forces for $\phi = \phi_1 = 45°$.

I.6.23 The dimensions of the mechanism shown in Figure I.3.38 are: $AB = 180$ mm, $AD = 450$ mm, and $BC = 200$ mm. The constant angular speed of the driver link 1 is $n = 1600$ rpm. The links 1, 2, and 5 are homogeneous rectangular prisms with the width $h = 0.010$ m and the depth $d = 0.001$ m. The sliders 3 and 4 have the width $w_{Slider} = 0.050$ m, the height $h_{Slider} = 0.020$ m, and the depth $d = 0.001$ m. The external force on 5 is opposed to the motion of the link $\mathbf{F}_{ext} = -|\mathbf{F}_{ext}|\dfrac{\mathbf{v}_G}{|\mathbf{v}_G|}$, where $|\mathbf{F}_{ext}| = 1500$ N. The density of the material is $\rho_{Steel} = 8000$ kg/m^3 and the gravitational acceleration is $g = 9.807$ m/s^2. Find the equilibrium moment on the driver link 1 and the joint forces. Select suitable lengths for the link 5 for $\phi = \phi_1 = 135°$.

I.6.24 The mechanism in Figure I.3.11(a) has the dimensions: $AB = 0.20$ m, $AD = 0.40$ m, $CD = 0.70$ m, $CE = 0.30$ m, and $y_E = 0.35$ m. The constant angular speed of the driver link 1 is $n = 2600$ rpm. The links 1, 3, and 4 are homogeneous rectangular prisms with the width $h = 0.010$ m and the depth $d = 0.001$ m. The sliders 2 and 5 have the width $w_{Slider} = 0.050$ m, the height $h_{Slider} = 0.020$ m, and the depth $d = 0.001$ m. The external force on 5 is opposed to the motion of the link $\mathbf{F}_{ext} = -|\mathbf{F}_{ext}|\dfrac{\mathbf{v}_E}{|\mathbf{v}_E|}$, where $|\mathbf{F}_{ext}| = 1500$ N. The density of the material is $\rho_{Steel} = 8000$ kg/m^3 and the gravitational acceleration is $g = 9.807$ m/s^2. Find the equilibrium moment on the driver link 1 and the joint forces for $\phi = \phi_1 = 30°$.

I.6.25 The mechanism in Figure I.3.12 has the dimensions: $AB = 0.04$ m, $BC = 0.07$ m, $CD = 0.12$ m, $AE = 0.10$ m, and $L_a = 0.035$ m. The constant angular speed of the driver link 1 is $n = 900$ rpm. The links 1, 2, and 4 are homogeneous rectangular prisms with the width $h = 0.010$ m and the depth $d = 0.001$ m. The sliders 3 and 5 have the width $w_{Slider} = 0.050$ m, the height $h_{Slider} = 0.020$ m, and the depth $d = 0.001$ m. The external force on 5 is opposed to the motion of the link $\mathbf{F}_{ext} = -|\mathbf{F}_{ext}|\dfrac{\mathbf{v}_D}{|\mathbf{v}_D|}$, where $|\mathbf{F}_{ext}| = 1250$ N. The density of the material is $\rho_{Steel} = 8000$ kg/m^3 and the gravitational acceleration is $g = 9.807$ m/s^2. Find the equilibrium moment on the driver link 1 and the joint forces for $\phi = \phi_1 = 60°$.

I.6.26 The mechanism in Figure I.3.15 has the dimensions: $AC = 0.080$ m, $BC = 0.150$ m, $BD = 0.400$ m, and $L_a = 0.020$ m. The constant angular speed of the driver link 1 is $n = 1500$ rpm. The links 1, 3, and 4 are homogeneous rectangular prisms with the width $h = 0.010$ m and the depth $d = 0.001$ m. The sliders 2 and 5 have the width $w_{Slider} = 0.050$ m, the height $h_{Slider} = 0.020$ m, and the depth $d = 0.001$ m. The external force on 5 is opposed to the motion of the link $\mathbf{F}_{ext} = -|\mathbf{F}_{ext}|\dfrac{\mathbf{v}_D}{|\mathbf{v}_D|}$, where $|\mathbf{F}_{ext}| = 2000$ N. The density of the material is $\rho_{Steel} = 8000$ kg/m^3 and the gravitational acceleration is $g = 9.807$ m/s^2. Find the equilibrium moment on the driver link 1 and the joint forces for $\phi = \phi_1 = 60°$. Select a suitable length for link 1.

I.7 Simulation of Kinematic Chains with *Mathematica*^{TM}

A planar mechanism will be analyzed and simulated using the *Mathematica*^{TM} software. The planar R-RTR-RTR mechanism considered is shown in Figure I.7.1. The driver link is the rigid link 1 (the link *AB*). The following numerical data are given: $AB = 0.140$ m, $AC = 0.060$ m, $AE = 0.250$ m, $CD = 0.150$ m. The angle of the driver link 1 with the horizontal axis is ϕ.

I.7.1 Position Analysis

The *Mathematica*^{TM} commands for the input data are

```
AB=0.140;    AC=0.060;    AE=0.250;    CD=0.150;
```

Position Analysis for an Input Angle

The angle of the driver link 1 with the horizontal axis $\phi = 30°$. The *Mathematica*^{TM} command for the input angle is

```
phi=N[Pi]/6;
```

where **N[expr]** gives the numerical value of **expr** and **Pi** is the constant π, with numerical value approximately equal to 3.14159.

Position of joint A
A Cartesian reference frame xOy is selected. The joint A is in the origin of the reference frame, that is, $A \equiv O$,

$$x_A = 0, \ y_A = 0. \tag{I.7.1}$$

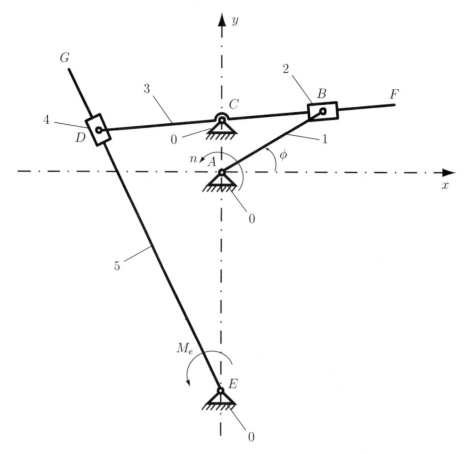

FIGURE I.7.1 *R-RTR-RTR mechanism.*

Position of joint C

The coordinates of the joint C are

$$x_C = 0, \quad y_C = AC = 0.060 \text{ m}. \tag{I.7.2}$$

Position of joint E

The coordinates of the joint E are

$$x_E = 0, \quad y_E = -AE = -0.250 \text{ m}. \tag{I.7.3}$$

The *Mathematica*^{TM} commands for Eqs. (I.7.1), (I.7.2), and (I.7.3) are

```
xA=0;  yA=0;
xC=0;  yC=AC;
xE=0;  yE=-AE;
```

Position of joint B

The unknowns are the coordinates of the joint B, x_B and y_B. Because the joint A is fixed and the angle ϕ is known, the coordinates of the joint B are computed from the following expressions

$$x_B = AB \cos \phi = 0.140 \cos 30° = 0.121 \text{ m},$$
$$y_B = AB \sin \phi = 0.140 \sin 30° = 0.070 \text{ m}. \tag{I.7.4}$$

The *Mathematica*^{TM} commands for Eq. (I.7.4) are

```
xB=AB Cos[phi];
yB=AB Sin[phi];
```

where **phi** is the angle ϕ in radians.

Position of joint D

The unknowns are the coordinates of the joint D, x_D and y_D. Knowing the positions of the joints B and C, one can compute the slope m and the intercept b of the line BC

$$m = \frac{(y_B - y_C)}{(x_B - x_C)},$$
$$b = y_B - m \, x_B. \tag{I.7.5}$$

The *Mathematica*^{TM} commands for Eq. (I.7.5) are

```
m=(yB-yC)/(xB-xC);
b=yB-m xB;
```

The joint D is located on the line BC:

$$y_D - m \, x_D - b = 0. \tag{I.7.6}$$

Furthermore, the length of the segment CD is constant:

$$(x_C - x_D)^2 + (y_C - y_D)^2 = CD^2. \tag{I.7.7}$$

The Eqs. (I.7.6) and (I.7.7) with *Mathematica*^{TM} commands are

```
eqnD1=(xDsol-xC)^2+(yDsol-yC)^2-CD^2==0;
eqnD2=yDsol-m xDsol-b==0;
```

The Eqs. (I.7.6) and (I.7.7) form a system from which the coordinates of the joint D can be computed. To solve the system of equations, a specific *Mathematica*^{TM} command will be used. The command **Solve[eqns, vars]** attempts to solve an equation or set of equations

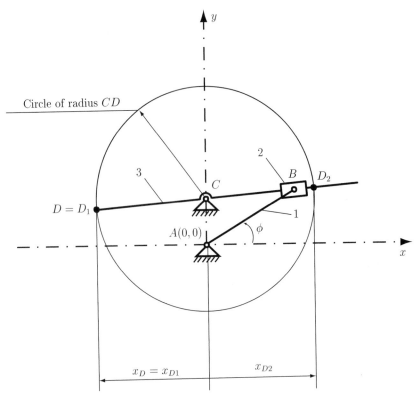

FIGURE I.7.2 *Solutions for the position of the joint D for $0 \leq \phi \leq 90°$ ($x_D \leq x_C$).*

eqns for the variables **vars**. For the mechanism

```
solutionD=Solve[{eqnD1,eqnD2},{xDsol,yDsol}];
```

two sets of solutions are found for the position of the joint D:

```
xD1=xDsol/.solutionD[[1]];
yD1=yDsol/.solutionD[[1]];
xD2=xDsol/.solutionD[[2]];
yD2=yDsol/.solutionD[[2]];
```

These solutions are located at the intersection of the line BC with the circle centered in C and radius CD (Fig. I.7.2), and they have the following numerical values:

$$x_{D1} = -0.149 \text{ m}, \quad y_{D1} = 0.047 \text{ m},$$
$$x_{D2} = 0.149 \text{ m}, \quad y_{D2} = 0.072 \text{ m}.$$

To determine the correct position of the joint D for the mechanism, an additional condition is needed. For the first quadrant, $0 \leq \phi \leq 90°$, the condition is $x_D \leq x_C$.

This condition with *Mathematica*[TM] is `If[condition, t, f]`, that gives `t` if `condition` evaluates to True, and `f` if it evaluates to False. For the considered mechanism, the following applies:

```
If[xD1<=xC, xD=xD1;yD=yD1, xD=xD2;yD=yD2];
```

Because $x_C = 0$ m, the coordinates of the joint D are

$$x_D = x_{D1} = -0.149 \text{ m},$$

$$y_D = y_{D1} = 0.047 \text{ m}.$$

The numerical solutions for B and D are printed using *Mathematica*[TM]:

```
Print["xB = ", xB, " m" ];
Print["yB = ", yB, " m" ];
Print["xD = ", xD, " m" ];
Print["yD = ", yD, " m" ];
```

The *Mathematica*[TM] program for the input angle $\phi = 30°$ is given in Program I.7.1. At the end of the program there are commands to draw the mechanism.

Position Analysis for a Complete Rotation

For a complete rotation of the driver link AB, $0 \le \phi \le 360°$, a step angle of $\phi = 60°$ is selected.

To calculate the position analysis for a complete cycle one can use the *Mathematica*[TM] command `For[start, test, incr, body]`. It executes `start`, then repeatedly evaluates `body` and `incr` until `test` fails to give True. In the case of the mechanism the following applies

```
For[phi=0, phi<=2*N[Pi], phi+=N[Pi]/3, Program block];
```

Method I

Method I uses constraint conditions for the mechanism for each quadrant. For the mechanism, there are several conditions for the position of the joint D.

For the angle ϕ located in the first quadrant $0° \le \phi \le 90°$ (Fig. I.7.2), and the fourth quadrant $270° \le \phi \le 360°$ (Fig. I.7.5), the following relation exists between x_D and x_C:

$$x_D \le x_C.$$

For the angle ϕ located in the second quadrant $90° < \phi \le 180°$ (Fig. I.7.3), and the third quadrant $180° < \phi < 270°$ (Fig. I.7.4), the following relation exists between x_D and x_C:

$$x_D \ge x_C.$$

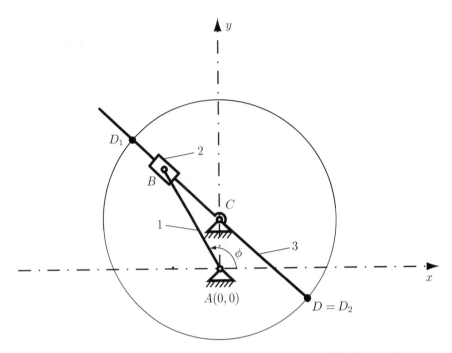

FIGURE I.7.3 *Solutions for the position of the joint D for 90° < φ ≤ 180° ($x_D \geq x_C$).*

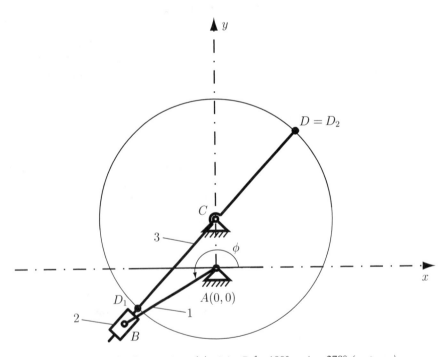

FIGURE I.7.4 *Solutions for the position of the joint D for 180° < φ < 270° ($x_D \geq x_C$).*

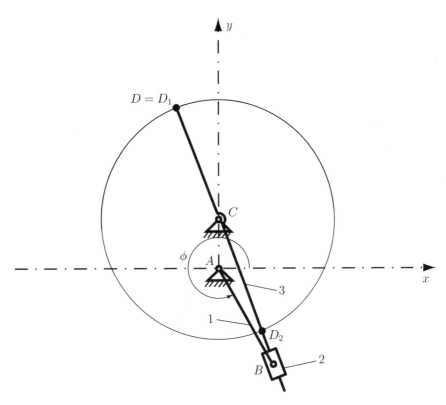

FIGURE I.7.5 *Solutions for the position of the joint D for* $270° \leq \phi \leq 360°$ $(x_D \leq x_C)$.

The following *Mathematica*^{TM} commands are used to determine the correct position of the joint D for all four quadrants:

```
If[0 <= phi <= N[Pi]/2 || 3 N[Pi]/2 <= phi <= 2 N[Pi],
If[xD1<=xC, xD=xD1;yD=yD1, xD=xD2;yD=yD2],
If[xD1>=xC, xD=xD1;yD=yD1, xD=xD2;yD=yD2]
];
```

where || is the logical OR function.

The *Mathematica*^{TM} program for a complete rotation of the driver link using method I is given in Program I.7.2. The graph of the mechanism for a complete rotation of the driver link is given in Figure I.7.6.

Method II

Another position analysis method for a complete rotation of the driver link uses constraint conditions for the initial value of the angle ϕ. For the mechanism, the correct position of the joint D is calculated using a simple function, the Euclidian distance between two points P and Q:

$$d = \sqrt{(x_P - x_Q)^2 + (y_P - y_Q)^2}.$$ (I.7.8)

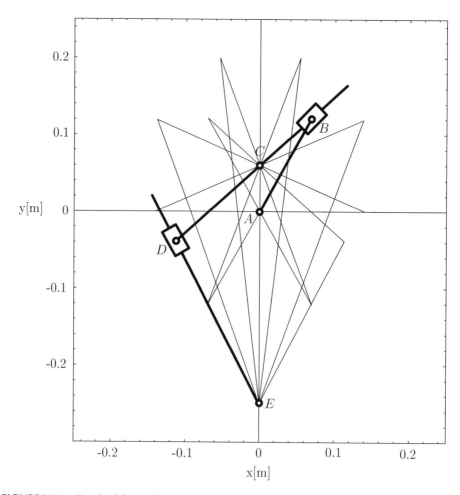

FIGURE I.7.6 *Graph of the mechanism for a complete rotation, $0 \le \phi \le 360°$.*

In *Mathematica*^{TM}, the following function is introduced:

```
Dist[xP_,yP_,xQ_,yQ_]:=Sqrt[(xP-xQ)^2+(yP-yQ)^2];
```

For the initial angle $\phi = 0°$, the constraint is $x_D \le x_C$, so the first position of the joint D, that is, D_0, is calculated for the first step $D = D_0 = D_k$, $k = 0$. For the next position of the joint, D_{k+1}, there are two solutions D_{k+1}^I and D_{k+1}^{II}, $k = 0$, 1, 2,.... In order to choose the correct solution of the joint, D_{k+1}, it is compared to the distances between the old position, D_k, and each new calculated positions D_{k+1}^I and D_{k+1}^{II}. The distances between the known solution D_k and the new solutions D_{k+1}^I and D_{k+1}^{II} are d_k^I and d_k^{II}. If the distance to the first solution is less than the distance to the second solution, $d_k^I < d_k^{II}$, then the correct answer is $D_{k+1} = D_{k+1}^I$, or else $D_{k+1} = D_{k+1}^{II}$ (Fig. I.7.7).

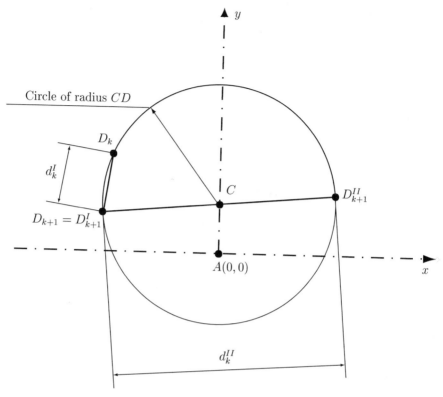

FIGURE I.7.7 *Distance condition for position analysis:* $d_k^I < d_k^{II} \Rightarrow D_{k+1} = D_{k+1}^I$.

The following *Mathematica*^{TM} commands are used to determine the correct position of the joint D using a single condition for all four quadrants:

```
increment=0;
For[phi=0, phi<=2*N[Pi], phi+=N[Pi]/3,
...
If[increment==0, If[xD1<xC, xD=xD1;yD=yD1, xD=xD2;yD=yD2],
dist1=Dist[xD1,yD1,xDold,yDold];
dist2=Dist[xD2,yD2,xDold,yDold];
If[dist1<dist2, xD=xD1;yD=yD1, xD=xD2;yD=yD2 ]
];
xDold=xD;
yDold=yD;
increment++;
...
];
```

With this algorithm the correct solution is selected using just one constraint relation for the initial step and then, automatically, the problem is solved. In this way it is not necessary to have different constraints for different quadrants.

The *Mathematica*^{TM} program for a complete rotation of the driver link using the second method is given in Program I.7.3.

I.7.2 Velocity and Acceleration Analysis

For the considered mechanism (Fig. I.7.1) the driver link 1 is rotating with a constant speed of $n = 50$ rpm. A *Mathematica*^{TM} program for velocity and acceleration analysis is presented here.

The *Mathematica*^{TM} commands for the angular speed, in rad/s, are

```
n=50; (* rpm *)
omega=n*N[Pi]/30; (* rad/s *)
```

The *Mathematica*^{TM} commands for coordinates of the joints A, C, and E are

```
xA=0; yA=0;
xC=0; yC=AC;
xE=0; yE=-AE;
```

The coordinates of the joint B ($B = B_1 = B_2$ on the link 1 or 2) are

$$x_B(t) = AB \cos \phi(t) \quad \text{and} \quad y_B(t) = AB \sin \phi(t).$$

To calculate symbolically the position of the joint B, the following *Mathematica*^{TM} commands are used:

```
xB=AB Cos[phi[t]];
yB=AB Sin[phi[t]];
```

where **phi[t]** represents the mathematical function $\phi(t)$. The function name is **phi** and it has one argument, the time **t**.

To calculate numerically the position of the joint B, the symbolic variables need to be substituted with the input data. To apply a transformation rule to a particular expression **expr**, type **expr/.lhs->rhs**. To apply a sequence of rules on each part of the expression **expr**, type **expr/.{lhs1->rhs1, lhs2->rhs2, ...}**.

For the mechanism, the transformation rule represents the initial data:

```
initdata={AB->0.14, AC->0.06, AE->0.25, CD->0.15,
phi[t]->N[Pi]/6, phi'[t]->omega, phi"[t]->0};
```

where **phi'[t]** is the first derivative of **phi** with respect to **t**, and **phi"[t]** is the second derivative of the function.

The command **Print[expr1, expr2, ...]** prints the **expr1, expr2, ...**, followed by a new line. To print the solutions of the position vector, the following commands are used:

```
Print["xB = ", xB," = ", xB/.initdata, " m" ];
Print["yB = ", yB," = ", yB/.initdata, " m" ];
```

The linear velocity vector of the joint B ($B = B_1 = B_2$) is

$$\mathbf{v}_B = \mathbf{v}_{B_1} = \mathbf{v}_{B_2} = \dot{x}_B\mathbf{I} + \dot{y}_B\mathbf{J},$$

where

$$\dot{x}_B = \frac{dx_B}{dt} = -AB\dot{\phi}\sin\phi \quad \text{and} \quad \dot{y}_B = \frac{dy_B}{dt} = AB\dot{\phi}\cos\phi,$$

are the components of the velocity vector of B. To calculate symbolically the components of the velocity vector using the *Mathematica*^{TM} the command **D[f, t]** is used, which gives the derivative of **f** with respect to **t**:

```
vBx=D[xB,t];
vBy=D[yB,t];
```

For the mechanism $\dot{\phi} = \omega = \dfrac{\pi n}{30} = \dfrac{\pi(50)}{30}$ rad/s = 5.235 rad/s, the numerical values are

$$\dot{x}_B = -0.140\,(5.235)\sin 30° = -0.366 \text{ m/s},$$
$$\dot{y}_B = 0.140\,(5.235)\cos 30° = 0.634 \text{ m/s}.$$

The solutions can be printed using *Mathematica*^{TM}:

```
Print["vBx = ", vBx," = ", vBx/.initdata, " m/s" ];
Print["vBy = ", vBy," = ", vBy/.initdata, " m/s" ];
```

The linear acceleration vector of the joint B ($B = B_1 = B_2$) is

$$\mathbf{a}_B = \ddot{x}_B\mathbf{I} + \ddot{y}_B\mathbf{J},$$

where

$$\ddot{x}_B = \frac{d\dot{x}_B}{dt} = -AB\dot{\phi}^2\cos\phi - AB\ddot{\phi}\sin\phi,$$
$$\ddot{y}_B = \frac{d\dot{y}_B}{dt} = -AB\dot{\phi}^2\sin\phi + AB\ddot{\phi}\cos\phi,$$

are the components of the acceleration vector of the joint B.

The *Mathematica*^{TM} commands used to calculate symbolically the components of the acceleration vector are

```
aBx=D[vBx,t];
aBy=D[vBy,t];
```

For the considered mechanism the angular acceleration of the link 1 is $\ddot{\phi} = \dot{\omega} = 0$. The numerical values of the acceleration of B are

$$\ddot{x}_B = -0.140\,(5.235)^2 \cos 30° = -3.323 \text{ m/s}^2,$$

$$\ddot{y}_B = -0.140\,(5.235)^2 \sin 30° = -1.919 \text{ m/s}^2.$$

The solutions printed with $Mathematica^{TM}$ are

```
Print["aBx = ", aBx," = ", aBx/.initdata, " m/s^2" ];
Print["aBy = ", aBy," = ", aBy/.initdata, " m/s^2" ];
```

The coordinates of the joint D are x_D and y_D. The $Mathematica^{TM}$ commands used to calculate the position of D are

```
mBC=(yB-yC)/(xB-xC);
bBC=yB-mBC xB;
eqnD1=(xDsol-xC)^2+(yDsol-yC)^2-CD^2==0;
eqnD2=yDsol-mBC xDsol-nBC==0;
solutionD=Solve[{eqnD1,eqnD2},{xDsol,yDsol}];
```

where **mBC** is the slope and **bBC** is the y-intercept of the line BC.

Two sets of solutions are found for the position of the joint D that are functions of the angle $\phi(t)$ (i.e., functions of time):

```
xD1=xDsol/.solutionD[[1]];
yD1=yDsol/.solutionD[[1]];
xD2=xDsol/.solutionD[[2]];
yD2=yDsol/.solutionD[[2]];
```

To determine the correct position of the joint D for the mechanism, an additional condition is needed. For the first quadrant, $0 \le \phi \le 90°$, the condition is $x_D \le x_C$. This condition using the $Mathematica^{TM}$ command is

```
If[xD1/.initdata<=xC, xD=xD1;yD=yD1, xD=xD2;yD=yD2];
```

The numerical solutions are printed using $Mathematica^{TM}$:

```
Print["xD = ", xD/.initdata, " m" ];
Print["yD = ", yD/.initdata, " m" ];
```

The linear velocity vector of the joint D ($D = D_3 = D_4$ on link 3 or link 4) is

$$\mathbf{v}_D = \mathbf{v}_{D_3} = \mathbf{v}_{D_4} = \dot{x}_D \mathbf{\imath} + \dot{y}_D \mathbf{\jmath},$$

where

$$\dot{x}_D = \frac{dx_D}{dt} \quad \text{and} \quad \dot{y}_D = \frac{dy_D}{dt},$$

are the components of the velocity vector of the joint D, respectively, on the x-axis and the y-axis.

To calculate symbolically the components of this velocity vector the following *Mathematica* ^{TM} commands are used:

```
vDx=D[xD,t];
vDy=D[yD,t];
```

For the considered mechanism the numerical values are

$$\dot{x}_D = 0.067 \text{ m/s} \quad \text{and} \quad \dot{y}_D = -0.814 \text{ m/s}.$$

The numerical solutions are printed using *Mathematica* ^{TM}:

```
Print["vDx = ", vDx/.initdata, " m/s" ];
Print["vDy = ", vDy/.initdata, " m/s" ];
```

The linear acceleration vector of $D = D_3 = D_4$ is

$$\mathbf{a}_D = \ddot{x}_D \mathbf{1} + \ddot{y}_D \mathbf{J},$$

where

$$\ddot{x}_D = \frac{d\dot{x}_D}{dt} \quad \text{and} \quad \ddot{y}_B = \frac{d\dot{y}_B}{dt}.$$

To calculate symbolically the components of the acceleration vector the following *Mathematica* ^{TM} commands are used:

```
aDx=D[vDx,t];
aDy=D[vDy,t];
```

The numerical values of the acceleration of D are

$$\ddot{x}_D = 4.617 \text{ m/s}^2 \quad \text{and} \quad \ddot{y}_D = -1.811 \text{ m/s}^2,$$

and can be printed using *Mathematica* ^{TM}:

```
Print["aDx = ", aDx/.initdata, " m/s^2" ];
Print["vDy = ", vDy/.initdata, " m/s^2" ];
```

The angle $\phi_3(t)$ is determined as a function of time t from the equation of the slope of the line BC:

$$\tan \phi_3(t) = m_{BC}(t).$$

The *Mathematica* ^{TM} function **ArcTan[z]** gives the arc tangent of the number **z**. To calculate symbolically the angle ϕ_3,

```
phi3=ArcTan[mBC];
```

The angular velocity $\omega_3(t)$ is the derivative with respect to time of the angle $\phi_3(t)$:

$$\omega_3 = \frac{d\phi_3(t)}{dt}.$$

Symbolically, the angular velocity ω_3 is calculated using *Mathematica*[TM]:

```
omega3=D[phi3,t];
```

The angular acceleration $\alpha_3(t)$ is the derivative with respect to time of the angular velocity $\omega_3(t)$:

$$\alpha_3(t) = \frac{d\omega_3(t)}{dt}.$$

Symbolically, using *Mathematica*[TM], the angular acceleration α_3 is

```
alpha3=D[omega3,t];
```

The numerical values of the angles, angular velocities, and angular accelerations for the links 2 and 3 are

$$\phi_3 = \phi_2 = 0.082 \text{ rad}, \quad \omega_3 = \omega_2 = 5.448 \text{ rad/s}, \quad \alpha_3 = \alpha_2 = 14.568 \text{ rad/s}^2.$$

The numerical solutions are printed using *Mathematica*[TM]:

```
Print["phi3=phi2= ",phi3/.initdata," rad "];
Print["omega3=omega2= ",omega3/.initdata," rad/s"];
Print["alpha3=alpha2= ",alpha3/.initdata," rad/s^2"];
```

The angle $\phi_5(t)$ is determined as a function of time t from the following equation:

$$\tan \phi_5(t) = \frac{y_D(t) - y_E}{x_D(t) - x_E},$$

and symbolically using *Mathematica*[TM]:

```
phi5=ArcTan[(yD-yE)/(xD-xE)];
```

The angular velocity $\omega_5(t)$ is the derivative with respect to time of the angle $\phi_5(t)$

$$\omega_5 = \frac{d\phi_5(t)}{dt}.$$

To calculate symbolically the angular velocity ω_5 using *Mathematica*[TM], the following command is used:

```
omega5=D[phi5,t];
```

The angular acceleration $\alpha_5(t)$ is the derivative with respect to time of the angular velocity $\omega_5(t)$:

$$\alpha_5(t) = \frac{d\omega_5(t)}{dt},$$

and it is calculated symbolically with *Mathematica*[TM]:

```
alpha5=D[omega5,t];
```

The numerical values of the angles, angular velocities, and angular accelerations for the link 5 and 4 are

$$\phi_5 = \phi_4 = 2.036 \text{ rad}, \quad \omega_5 = \omega_4 = 0.917 \text{ rad/s}, \quad \alpha_5 = \alpha_4 = -5.771 \text{ rad/s}^2.$$

The numerical solutions printed with *Mathematica*[TM] are

```
Print["phi5=phi4= ",phi5/.initdata," rad "];
Print["omega5=omega4= ",omega5/.initdata," rad/s"];
Print["alpha5=alpha4= ",alpha5/.initdata," rad/s^2"];
```

The *Mathematica*[TM] program for velocity and acceleration analysis is given in Program I.7.4.

I.7.3 Contour Equations for Velocities and Accelerations

The same planar R-RTR-RTR mechanism is considered in Figure I.7.8(a). The driver link 1 is rotating with a constant speed of $n = 50$ rpm. A *Mathematica*[TM] program for velocity and acceleration analysis using the contour equations is presented here.

The mechanism has five moving links and seven full joints. The number of independent contours is

$$n_c = c - n = 7 - 5 = 2,$$

where c is the number of joints and n is the number of moving links.

The mechanism has two independent contours. The first contour I contains the links 0, 1, 2, and 3, while the second contour II contains the links 0, 3, 4, and 5. The diagram of the mechanism is represented in Figure I.7.8(b). Clockwise paths are chosen for each closed contours I and II.

First Contour Analysis

Figure I.7.9(a) shows the first independent contour I with

- rotational joint R between the links 0 and 1 (joint A);
- rotational joint R between the links 1 and 2 (joint B);
- translational joint T between the links 2 and 3 (joint B);
- rotational joint R between the links 3 and 0 (joint C).

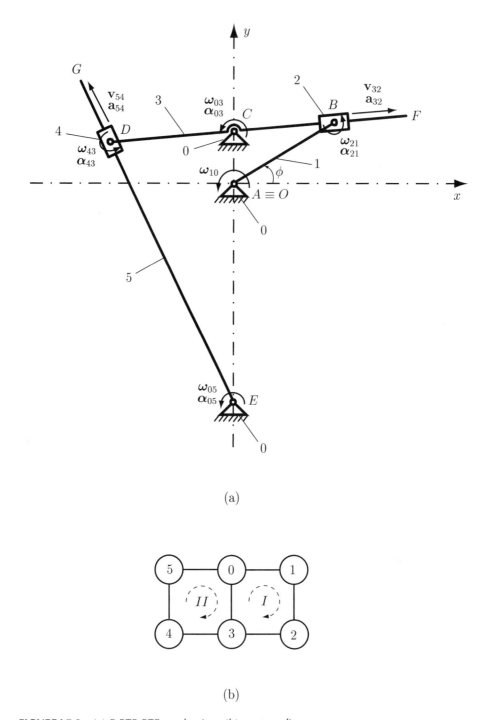

(a)

(b)

FIGURE I.7.8 *(a) R-RTR-RTR mechanism; (b) contour diagram.*

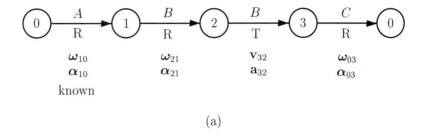

(a)

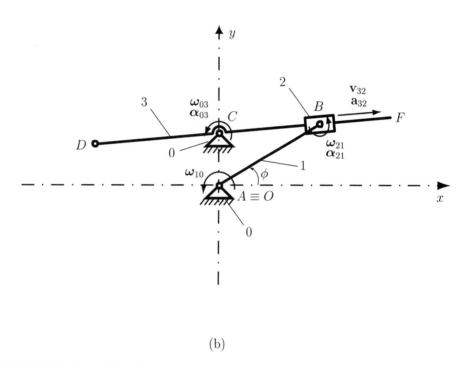

(b)

FIGURE I.7.9 *First independent contour.*

The angular velocity ω_{10} of the driver link is known:

$$\omega_{10} = \omega_1 = \omega = \frac{50\pi}{30} \text{ rad/s} = 5.235 \text{ rad/s}.$$

The origin of the reference frame is the point $A(0, 0, 0)$.

For the velocity analysis, the following vectorial equations are used:

$$\omega_{10} + \omega_{21} + \omega_{03} = \mathbf{0},$$

$$\mathbf{r}_{AB} \times \omega_{21} + \mathbf{r}_{AC} \times \omega_{03} + \mathbf{v}^r_{B32} = \mathbf{0}, \qquad (\text{I.7.9})$$

where $\mathbf{r}_{AB} = x_B\mathbf{I} + y_B\mathbf{J}$, $\mathbf{r}_{AC} = x_C\mathbf{I} + y_C\mathbf{J}$, and

$$\boldsymbol{\omega}_{10} = \omega_{10}\mathbf{k}, \ \boldsymbol{\omega}_{21} = \omega_{21}\mathbf{k}, \ \boldsymbol{\omega}_{03} = \omega_{03}\mathbf{k},$$

$$\mathbf{v}^r_{B32} = \mathbf{v}_{32} = v_{32}\cos\phi_{21}\mathbf{I} + v_{32}\sin\phi_2\mathbf{J}.$$

The sign of the relative angular velocities is selected as positive as shown in Figures I.7.8(a) and I.7.9(a). The numerical computation will then give the correct orientation of the unknown vectors. The components of the vectors $\mathbf{r}_{AB}$ and $\mathbf{r}_{AC}$, and the angle ϕ_2 are already known from the position analysis of the mechanism. Equation (I.7.9) becomes

$$\omega_{10}\mathbf{k} + \omega_{21}\mathbf{k} + \omega_{03}\mathbf{k} = \mathbf{0},$$

$$\begin{vmatrix} \mathbf{I} & \mathbf{J} & \mathbf{k} \\ x_B & y_B & 0 \\ 0 & 0 & \omega_{21} \end{vmatrix} + \begin{vmatrix} \mathbf{I} & \mathbf{J} & \mathbf{k} \\ x_C & y_C & 0 \\ 0 & 0 & \omega_{03} \end{vmatrix} + v_{32}\cos\phi_{21}\mathbf{I} + v_{32}\sin\phi_2\mathbf{J} = \mathbf{0}. \qquad (I.7.10)$$

In the *Mathematica*^{TM} environment, a three-dimensional vector **v** is written as a list of variables **v={x, y, z}**, where **x**, **y**, and **z** are the spatial coordinates of the vector **v**. The first component of the vector **v** is **x=v[[1]]**, the second component is **y=v[[2]]**, and the third component is **z=v[[3]]**. For the considered mechanism with *Mathematica*^{TM}, the following applies:

```
rB={xB, yB, 0};
rC={xC, yC, 0};
omega10v={0,0,omega};
omega21vSol={0,0,omega21Sol};
omega03vSol={0,0,omega03Sol};
v32vSol={v32Sol Cos[phi2],v32Sol Sin[phi2],0};
```

Equation (I.7.10) represents a system of three equations and with *Mathematica*^{TM} commands gives

```
eqIkv=(omega10v+omega21vSol+omega03vSol)[[3]]==0;
eqIiv=(Cross[rB,omega21vSol]+Cross[rC,omega03vSol]+v32vSol)[[1]]==0;
eqIjv=(Cross[rB,omega21vSol]+Cross[rC,omega03vSol]+v32vSol)[[2]]==0;
```

where the command **Cross[a,b]** gives the vector cross product of the vectors **a** and **b**.

The system of equations can be solved using the *Mathematica*^{TM} commands

```
solIvel=Solve[{eqIkv,eqIiv,eqIjv}, {omega21Sol,
omega03Sol,v32Sol}];
```

and the following numerical solutions are obtained:

$$\omega_{21} = 0.212 \text{ rad/s}, \ \omega_{03} = -5.448 \text{ rad/s}, \ \text{and} \ v_{32} = 0.313 \text{ m/s}.$$

To print the numerical values, the following *Mathematica*[TM] commands are used:

```
omega21v=omega21vSol/.solIvel[[1]];
omega03v=omega03vSol/.solIvel[[1]];
v32v=v32vSol/.solIvel[[1]];
Print["omega21 = ",omega21v];
Print["omega03 = ",omega03v];
Print["v32 = ",v32v];
Print["v32r = ",v32Sol/.solIvel[[1]] ];
```

The absolute angular velocities of the links 2 and 3 are

$$\omega_{20} = \omega_{30} = -\omega_{03} = 5.448 \mathbf{k} \text{ rad/s}.$$

The absolute linear velocities of the joints B and D are

$$\mathbf{v}_B = \mathbf{v}_{B_1} = \mathbf{v}_{B_2} = \mathbf{v}_A + \omega_{10} \times \mathbf{r}_{AB} = -0.366 \, \mathbf{\imath} + 0.634 \, \mathbf{\jmath} \text{ m/s},$$

$$\mathbf{v}_D = \mathbf{v}_{D_3} = \mathbf{v}_{D_4} = \mathbf{v}_C + \omega_{30} \times \mathbf{r}_{CD} = 0.067 \, \mathbf{\imath} - 0.814 \, \mathbf{\jmath} \text{ m/s},$$

where $\mathbf{v}_A = \mathbf{0}$ and $\mathbf{v}_C = \mathbf{0}$, because the joints A and C are grounded and

$$\mathbf{r}_{CD} = \mathbf{r}_{AD} - \mathbf{r}_{AC}.$$

The *Mathematica*[TM] commands for the absolute velocities are

```
omega20v=omega30v=-omega03v;
vBv=Cross[omega10v,rB];
vDv=Cross[omega30v,(rD-rC)];
Print["omega20 = omega30 = ",omega20v];
Print["vB = ",vBv];
Print["vD = ",vDv];
```

For the acceleration analysis, the following vectorial equations are used:

$$\alpha_{10} + \alpha_{21} + \alpha_{03} = \mathbf{0},$$

$$\mathbf{r}_{AB} \times \alpha_{21} + \mathbf{r}_{AC} \times \alpha_{03} + \mathbf{a}_{B32}^r + \mathbf{a}_{B32}^c - \omega_{10}^2 \mathbf{r}_{AB} - \omega_{20}^2 \mathbf{r}_{BC} = \mathbf{0}. \qquad (I.7.11)$$

where

$$\alpha_{10} = \alpha_{10}\mathbf{k}, \ \alpha_{21} = \alpha_{21}\mathbf{k}, \ \alpha_{03} = \alpha_{03}\mathbf{k},$$

$$\mathbf{a}_{B32}^r = \mathbf{a}_{32} = a_{32}\cos\phi_2\mathbf{\imath} + a_{32}\sin\phi_2\mathbf{\jmath},$$

$$\mathbf{a}_{B32}^c = \mathbf{a}_{32}^c = 2\omega_{20} \times \mathbf{v}_{32}.$$

The driver link has a constant angular velocity and $\alpha_{10} = \dot{\omega}_{10} = 0$. The acceleration vectors using the *Mathematica*[TM] commands are:

```
alpha10v={0,0,0};
alpha21vSol={0,0,alpha21Sol};
```

```
alpha03vSol={0,0,alpha03Sol};
a32vSol={a32Sol Cos[phi2],a32Sol Sin[phi2],0};
```

Equation (I.7.11) represents a system of three equations and using *Mathematica*[TM] commands gives

```
eqIka=(alpha10v+alpha21vSol+alpha03vSol)[[3]]==0;
```

```
eqIia=( Cross[rB,alpha21vSol]+Cross[rC,alpha03vSol]+
a32vSol+2Cross[omega20v,v32v]-(omega10v.omega10v)rB-
(omega20v.omega20v)(rC-rB) )[[1]]==0;
```

```
eqIja=( Cross[rB,alpha21vSol]+Cross[rC,alpha03vSol]+
a32vSol+2Cross[omega20v,v32v]-(omega10v.omega10v)rB-
(omega20v.omega20v)(rC-rB) )[[2]]==0;
```

The unknowns in the Eq. (I.7.18) are α_{21}, α_{03}, and a_{32}. The system of equations is solved using the *Mathematica*[TM] commands

```
solIacc=Solve[{eqIka,eqIia,eqIja}, {alpha21Sol, alpha03Sol,a32Sol}];
```

The following numerical solutions are then obtained

$$\alpha_{21} = 14.568 \text{ rad/s}^2, \quad \alpha_{03} = -14.568 \text{ rad/s}^2, \quad \text{and} \quad a_{32} = -0.140 \text{ m/s}^2.$$

To print the numerical values, the following *Mathematica*[TM] commands are used:

```
alpha21v=alpha21vSol/.solIacc[[1]];
alpha03v=alpha03vSol/.solIacc[[1]];
a32v=a32vSol/.solIacc[[1]];
Print["alpha21 = ",alpha21v];
Print["alpha03 = ",alpha03v];
Print["a32 = ",a32v];
Print["a32r = ",a32Sol/.solIacc[[1]]];
```

The absolute angular accelerations of the links 2 and 3 are

$$\alpha_{20} = \alpha_{30} = -\alpha_{03} = 14.568 \text{ } \mathbf{k} \text{ rad/s}^2.$$

The absolute linear accelerations of the joints B and D are obtained from the following equation:

$$\mathbf{a}_B = \mathbf{a}_A + \boldsymbol{\alpha}_{10} \times \mathbf{r}_{AB} - \omega_{10}^2 \mathbf{AB} = -3.323 \text{ } \mathbf{\imath} - 1.919 \text{ } \mathbf{j} \text{ m/s}^2,$$

$$\mathbf{a}_D = \mathbf{a}_C + \boldsymbol{\alpha}_{30} \times \mathbf{r}_{CD} - \omega_{30}^2 \mathbf{r}_{CD} = 4.617 \text{ } \mathbf{\imath} - 1.811 \text{ } \mathbf{j} \text{ m/s}^2,$$

where $\mathbf{a}_A = \mathbf{0}$ and $\mathbf{a}_C = \mathbf{0}$, because the joints A and C are grounded.

To print the absolute accelerations with *Mathematica*[TM], the following relations are used

```
alpha20v=alpha30v=-alpha03v;
aBv=-(omega10v.omega10v)rB;
aDv=Cross[alpha30v,(rD-rC)]-(omega20v.omega20v)(rD-rC);
Print["alpha20 = alpha30 = ",alpha30v];
Print["aB = ",aBv];
Print["aD = ",aDv];
```

Second Contour Analysis

Figure I.7.10(a) depicts the second independent contour *II*

- rotational joint R between the links 0 and 3 (joint C);
- rotational joint R between the links 3 and 4 (joint D);
- translational joint T between the links 4 and 5 (joint D);
- rotational joint R between the links 5 and 0 (joint E).

For the velocity analysis, the following vectorial equations are used:

$$\omega_{30} + \omega_{43} + \omega_{05} = \mathbf{0},$$

$$\mathbf{r}_{AC} \times \omega_{30} + \mathbf{r}_{AD} \times \omega_{43} + \mathbf{r}_{AE} \times \omega_{05} + \mathbf{v}_{D54}^r = \mathbf{0}, \qquad (\text{I.7.12})$$

where $\mathbf{r}_{AD} = x_D\mathbf{I} + y_D\mathbf{J}$, $\mathbf{r}_{AE} = x_E\mathbf{I} + y_E\mathbf{J}$, and

$$\omega_{30} = \omega_{30}\mathbf{k}, \quad \omega_{43} = \omega_{43}\mathbf{k}, \quad \omega_{05} = \omega_{05}\mathbf{k},$$

$$\mathbf{v}_{D54}^r = \mathbf{v}_{54} = v_{54}\cos\phi_4\mathbf{I} + v_{54}\sin\phi_4\mathbf{J}.$$

The sign of the relative angular velocities is selected as positive as shown in Figures I.7.8(a) and I.7.10(a). The numerical computation will then give the correct orientation of the unknown vectors. The components of the vectors $\mathbf{r}_{AD}$ and $\mathbf{r}_{AE}$, and the angle ϕ_4 are already known from the position analysis of the mechanism.

The unknown vectors with *Mathematica*[TM] commands are

```
omega43vSol={0,0,omega43Sol};
omega05vSol={0,0,omega05Sol};
v54vSol={v54Sol Cos[phi4],v54Sol Sin[phi4],0};
```

Equation (I.7.12) becomes

$$\omega_{30}\mathbf{k} + \omega_{43}\mathbf{k} + \omega_{05}\mathbf{k} = \mathbf{0},$$

$$\begin{vmatrix} \mathbf{I} & \mathbf{J} & \mathbf{k} \\ x_C & y_C & 0 \\ 0 & 0 & \omega_{30} \end{vmatrix} + \begin{vmatrix} \mathbf{I} & \mathbf{J} & \mathbf{k} \\ x_D & y_D & 0 \\ 0 & 0 & \omega_{43} \end{vmatrix} + \begin{vmatrix} \mathbf{I} & \mathbf{J} & \mathbf{k} \\ x_E & y_E & 0 \\ 0 & 0 & \omega_{05} \end{vmatrix}$$

$$+ v_{32}\cos\phi_4\mathbf{I} + v_{32}\sin\phi_4\mathbf{J} = \mathbf{0}. \qquad (\text{I.7.13})$$

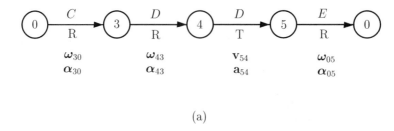

(a)

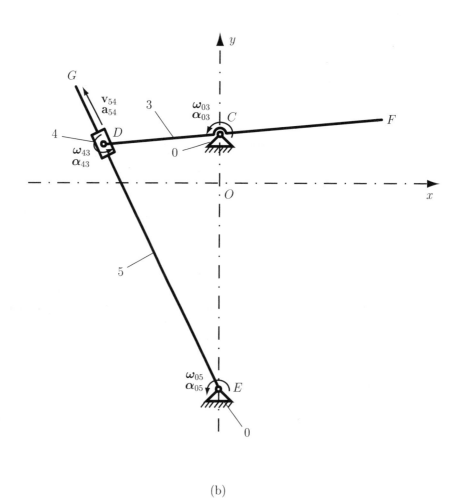

(b)

FIGURE I.7.10 *Second independent contour.*

Equation (I.7.13) projected onto the "fixed" reference frame $Oxyz$ gives

$$\omega_{30} + \omega_{43} + \omega_{05} = 0,$$

$$y_C\omega_{30} + y_D\omega_{43} + y_E\omega_{05} + v_{54}\cos\phi_4 = 0,$$

$$-x_C\omega_{30} - x_D\omega_{43} - x_E\omega_{05} + v_{54}\sin\phi_4 = 0. \qquad (I.7.14)$$

The above system of equations using the following *Mathematica*^{TM} commands becomes

```
  eqIIkv=(omega30v+omega43vSol+omega05vSol)[[3]]==0;
  eqIIiv=(Cross[rC,omega30v]+Cross[rD,omega43vSol]+
Cross[rE,omega05vSol]+v54vSol)[[1]]==0;
  eqIIjv=(Cross[rC,omega30v]+Cross[rD,omega43vSol]+
Cross[rE,omega05vSol]+v54vSol)[[2]]==0;
```

Equation (I.7.14) represents an algebraic system of three equations with three unknowns: ω_{43}, ω_{05}, and v_{54}. The system is solved using the *Mathematica*^{TM} commands

```
  solIIvel=Solve[{eqIIkv,eqIIiv,eqIIjv}, {omega43Sol,
omega05Sol,v54Sol}];
```

The following numerical solutions are obtained:

$$\omega_{43} = -4.531 \text{ rad/s}, \quad \omega_{05} = -0.917 \text{ rad/s}, \quad \text{and} \quad v_{54} = 0.757 \text{ m/s}.$$

To print the numerical values with *Mathematica*^{TM}, the following commands are used:

```
  omega43v=omega43vSol/.solIIvel[[1]];
  omega05v=omega05vSol/.solIIvel[[1]];
  v54v=v54vSol/.solIIvel[[1]];
  Print["omega43 = ",omega43v];
  Print["omega05 = ",omega05v];
  Print["v54 = ",v54v];
  Print["v54r=",v54Sol/.solIIvel[[1]] ];
```

The absolute angular velocities of the links 4 and 5 are

$$\omega_{40} = \omega_{50} = -\omega_{05} = 0.917 \, \mathbf{k} \text{ rad/s}, \qquad (I.7.15)$$

and with *Mathematica*^{TM} commands, they are

```
  omega40v=omega50v=-omega05v;
  Print["omega40 = omega50 = ",omega50v];
```

For the acceleration analysis, the following vectorial equations are used:

$$\alpha_{30} + \alpha_{43} + \alpha_{05} = \mathbf{0}, \qquad (I.7.16)$$

$$\mathbf{r}_{AC} \times \alpha_{30} + \mathbf{r}_{AD} \times \alpha_{43} + \mathbf{r}_{AE} \times \alpha_{05} + \mathbf{a}^r_{D54} + \mathbf{a}^c_{B54} - \omega^2_{30}\mathbf{r}_{CD} - \omega^2_{40}\mathbf{r}_{DE} = \mathbf{0}.$$

where

$$\boldsymbol{\alpha}_{30} = \alpha_{30}\mathbf{k}, \ \boldsymbol{\alpha}_{43} = \alpha_{43}\mathbf{k}, \ \boldsymbol{\alpha}_{05} = \alpha_{05}\mathbf{k},$$

$$\mathbf{a}_{B54}^r = \mathbf{a}_{54} = a_{54}\cos\phi_4\mathbf{I} + a_{54}\sin\phi_4\mathbf{J},$$

$$\mathbf{a}_{B54}^c = 2\boldsymbol{\omega}_{40} \times \mathbf{v}_{54}.$$

The unknown acceleration vectors using the *Mathematica*^{TM} commands are

```
alpha43vSol={0,0,alpha43Sol};
alpha05vSol={0,0,alpha05Sol};
a54vSol={a54Sol Cos[phi4],a54Sol Sin[phi4],0};
```

Equation (I.7.16) becomes

$$\alpha_{30}\mathbf{k} + \alpha_{43}\mathbf{k} + \alpha_{05}\mathbf{k} = \mathbf{0},$$

$$\begin{vmatrix} \mathbf{I} & \mathbf{J} & \mathbf{k} \\ x_C & y_C & 0 \\ 0 & 0 & \alpha_{30} \end{vmatrix} + \begin{vmatrix} \mathbf{I} & \mathbf{J} & \mathbf{k} \\ x_D & y_D & 0 \\ 0 & 0 & \alpha_{43} \end{vmatrix} + \begin{vmatrix} \mathbf{I} & \mathbf{J} & \mathbf{k} \\ x_E & y_E & 0 \\ 0 & 0 & \alpha_{05} \end{vmatrix}$$

$$+ a_{54}\cos\phi_4\mathbf{I} + a_{54}\sin\phi_4\mathbf{J} + \begin{vmatrix} \mathbf{I} & \mathbf{J} & \mathbf{k} \\ 0 & 0 & \omega_{40} \\ v_{54}\cos\phi_4 & v_{54}\sin\phi_4 & 0 \end{vmatrix}$$

$$- \omega_{30}^2[(x_D - x_C)\mathbf{I} + (y_D - y_C)\mathbf{J}]$$

$$- \omega_{40}^2[(x_E - x_D)\mathbf{I} + (y_E - y_D)\mathbf{J}] = \mathbf{0}. \tag{I.7.17}$$

Equation (I.7.17) can be rewritten as

$$\alpha_{30} + \alpha_{43} + \alpha_{05} = 0,$$

$$y_C\alpha_{30} + y_D\alpha_{43} + y_E\alpha_{05} + a_{54}\cos\phi_4 - 2\omega_{40}v_{54}\sin\phi_4$$

$$- \omega_{30}^2(x_D - x_C) - \omega_{40}^2(x_E - x_D) = 0,$$

$$-x_C\alpha_{30} - x_D\alpha_{43} - x_E\alpha_{05} + a_{54}\sin\phi_4 + 2\omega_{40}v_{54}\cos\phi_4$$

$$- \omega_{30}^2(y_D - y_C) - \omega_{40}^2(y_E - y_D) = 0. \tag{I.7.18}$$

The contour acceleration equations using *Mathematica*^{TM} commands are

```
eqIIka=(alpha30v+alpha43vSol+alpha05vSol)[[3]]==0;

eqIIia=( Cross[rC,alpha30v]+Cross[rD,alpha43vSol]+
Cross[rE,alpha05vSol]+a54vSol+2Cross[omega40v,v54v]-
(omega30v.omega30v)(rD-rC)-
(omega40v.omega40v)(rE-rD) )[[1]]==0;
```

```
eqIIja=( Cross[rC,alpha30v]+Cross[rD,alpha43vSol]+
Cross[rE,alpha05vSol]+a54vSol+2Cross[omega40v,v54v]-
(omega30v.omega30v)(rD-rC)-
(omega40v.omega40v)(rE-rD) )[[2]]==0;
```

The unknowns in Eq. (I.7.18) are α_{43}, α_{05}, and a_{54}. To solve the system, the following *Mathematica*[TM] command is used:

```
solIIacc=Solve[{eqIIka,eqIIia,eqIIja}, {alpha43Sol,
alpha05Sol,a54Sol}];
```

The following numerical solutions are obtained:

$$\alpha_{43} = -20.339 \text{ rad/s}^2, \ \alpha_{05} = 5.771 \text{ rad/s}^2, \ \text{and} \ a_{54} = 3.411 \text{ m/s}^2.$$

The *Mathematica*[TM] commands are

```
alpha43v=alpha43vSol/.solIIacc[[1]];
alpha05v=alpha05vSol/.solIIacc[[1]];
a54v=a54vSol/.solIIacc[[1]];
Print["alpha43 = ",alpha43v];
Print["alpha05 = ",alpha05v ];
Print["a54 = ",a54v];
Print["a54r=",a54Sol/.solIIacc[[1]]];
```

The absolute angular accelerations of the links 4 and 5 are

$$\alpha_{40} = \alpha_{50} = -\alpha_{05} = -5.771 \ \mathbf{k} \text{ rad/s}^2,$$

and with *Mathematica*[TM] they are

```
alpha40v=alpha50v=-alpha05v;
Print["alpha40 = alpha50 = ",alpha50v];
```

The *Mathematica*[TM] program for the velocity and acceleration analysis using the contour method is given in Program I.7.5.

I.7.4 Dynamic Force Analysis

In this section the motor moment $\mathbf{M}_m$ required for the dynamic equilibrium of the considered mechanism, shown in Figure I.7.11(a), is calculated. The joint reaction forces are also calculated. The widths of the links 1, 3, and 5 are $AB = 0.140$ m, $FD = 0.400$ m, and respectively, $EG = 0.500$ m. The height of the links 1, 3, and 5 is $h = 0.010$ m. The width of the links 2 and 4 is $w_{Slider} = 0.050$ m, and the height is $h_{Slider} = 0.020$ m. All five moving links are rectangular prisms with the depth $d = 0.001$ m. The angle of the driver link is $\phi = \frac{\pi}{6}$ rad and the angular velocity is $n = 50$ rpm. The external moment applied on link 5

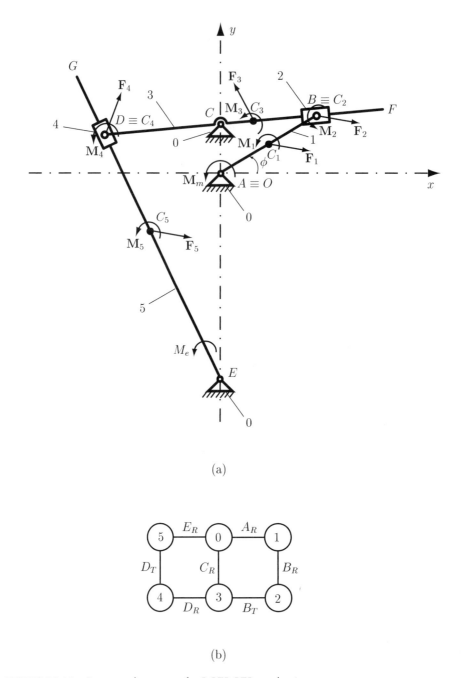

(a)

(b)

FIGURE I.7.11 *Forces and moments for R-RTR-RTR mechanism.*

is opposed to the motion of the link. Because $\omega_5 = 0.917\,\mathbf{k}$ rad/s, the external moment vector will be $\mathbf{M}_e = -100\,\mathbf{k}$ N·m. The density of the material is $\rho_{Steel} = \rho = 8000$ kg/m^3. The gravitational acceleration is $g = 9.807$ m/s^2. The center of mass locations of the links $i = 1, 2, \ldots, 5$ are designated by $C_i(x_{C_i}, y_{C_i}, 0)$.

The input data are introduced using a *Mathematica*[TM] rule:

```
rule={AB->0.14,AC->0.06,AE->0.25,CD->0.15,FD->0.4,
EG->0.5,h->0.01,d->0.001,hSlider->0.02,wSlider->0.05,
rho->8000,g->9.807,Me->-100.,
  phi[t]->N[Pi]/6,phi'[t]->omega,phi"[t]->0};
```

where **omega=n*N[Pi]/30**.

Inertia Forces and Moments

To calculate the inertia moment $\mathbf{M}_i$ and the total force $\mathbf{F}_i$ for the link $i = 1, 2, \ldots, 5$, the mass m_i, the acceleration of the center of mass $\mathbf{a}_{C_i}$, the gravity force $\mathbf{G}_i$, and the mass moment of inertia I_{C_i} are needed.

Link 1

The mass of the link is

$$m_1 = \rho \, AB \, h \, d.$$

The position, velocity, and acceleration for the center of mass C_1 are

$$\mathbf{r}_{C_1} = \mathbf{r}_B/2, \; \mathbf{v}_{C_1} = \mathbf{v}_B/2, \; \text{and} \; \mathbf{a}_{C_1} = \mathbf{a}_B/2.$$

The inertia force is

$$\mathbf{F}_{in1} = -m_1 \, \mathbf{a}_{C_1}.$$

The gravitational force is

$$\mathbf{G}_1 = -m_1 \, g \, \mathbf{k}.$$

The total force on link 1 at the mass center C_1 is

$$\mathbf{F}_1 = \mathbf{F}_{in1} + \mathbf{G}_1.$$

The mass moment of inertia is

$$I_{C_1} = m_1 \, (AB^2 + h^2)/12.$$

The moment of inertia is

$$\mathbf{M}_1 = \mathbf{M}_{in1} = -I_{C_1} \, \boldsymbol{\alpha}_1.$$

To calculate and print the numerical values of the total force $\mathbf{F}_1$ and the moment $\mathbf{M}_1$, the following *Mathematica*^{TM} commands are used:

```
m1=rho AB h d /.rule;
rC1=rB/2; vC1=vB/2; aC1=aB/2;
Fin1=-m1 aC1 /.rule;
G1={0,-m1*g,0} /.rule;
F1=(Fin1+G1) /.rule;
IC1=m1 (AB^2+h^2)/12 /.rule;
M1=Min1=-IC1 alpha1 /.rule;
Print["F1 = ",F1];
Print["M1 = ",M1];
```

Link 2

The mass of the link is

$$m_2 = \rho \, h_{Slider} \, w_{Slider} \, d.$$

The position, velocity, and acceleration for the center of mass C_2 are

$$\mathbf{r}_{C_2} = \mathbf{r}_B, \ \mathbf{v}_{C_2} = \mathbf{v}_B, \ \text{and} \ \mathbf{a}_{C_2} = \mathbf{a}_B.$$

The inertia force is

$$\mathbf{F}_{in2} = -m_2 \, \mathbf{a}_{C_2}.$$

The gravitational force is

$$\mathbf{G}_2 = -m_2 \, g \, \mathbf{k}.$$

The total force on slider 2 at B is

$$\mathbf{F}_2 = \mathbf{F}_{in2} + \mathbf{G}_2.$$

The mass moment of inertia is

$$I_{C_2} = m_2 \, (h_{Slider}^2 + w_{Slider}^2)/12.$$

The moment of inertia is

$$\mathbf{M}_2 = \mathbf{M}_{in2} = -I_{C_2}\alpha_2.$$

The *Mathematica*^{TM} commands for the total force $\mathbf{F}_2$ and the moment $\mathbf{M}_2$ are

```
m2=rho hSlider wSlider d /.rule;
rC2=rB; vC2=vB; aC2=aB;
Fin2=-m2 aC2 /.rule;
```

```
G2={0,-m2*g,0} /.rule;
F2=(Fin2+G2) /.rule;
IC2=m2 (hSlider^2+wSlider^2)/12 /.rule;
M2=Min2=-IC2 alpha2 /.rule;
Print["F2 = ",F2];
Print["M2 = ",M2];
```

Link 3

The mass of the link is

$$m_3 = \rho \, FD \, h \, d.$$

The position, velocity, and acceleration for the center of mass C_3 are

$$x_{C_3} = x_C + (FD/2 - CD) \, \cos\phi_3, \quad y_{C_3} = y_C + (FD/2 - CD) \, \sin\phi_3,$$

$$\mathbf{r}_{C_3} = x_{C_3}\mathbf{I} + y_{C_3}\mathbf{J}, \quad \mathbf{v}_{C_3} = \dot{x}_{C_3}\mathbf{I} + \dot{y}_{C_3}\mathbf{J}, \quad \text{and} \quad \mathbf{a}_{C_3} = \ddot{x}_{C_3}\mathbf{I} + \ddot{y}_{C_3}\mathbf{J}.$$

The inertia force is

$$\mathbf{F}_{in3} = -m_3 \, \mathbf{a}_{C_3}.$$

The gravitational force is

$$\mathbf{G}_3 = -m_3 \, g \, \mathbf{k}.$$

The total force at C_3 is

$$\mathbf{F}_3 = \mathbf{F}_{in3} + \mathbf{G}_3.$$

The mass moment of inertia is

$$I_{C_3} = m_3 \, (FD^2 + h^2)/12.$$

The total moment on link 3 is

$$\mathbf{M}_3 = \mathbf{M}_{in3} = -I_{C_3}\boldsymbol{\alpha}_3.$$

The force $\mathbf{F}_3$ and the moment $\mathbf{M}_3$ with *Mathematica*^{TM} are

```
m3=rho FD h d /.rule;
xC3=xC+(FD/2-CD) Cos[phi3];
yC3=yC+(FD/2-CD) Sin[phi3];
rC3={xC3,yC3,0};
vC3=D[rC3,t];
aC3=D[D[rC3,t],t];
Fin3=-m3 aC3 /.rule;
G3={0,-m3*g,0} /.rule;
```

```
F3=(Fin3+G3) /.rule;
IC3=m3 (FD^2+h^2)/12 /.rule;
M3=Min3=-IC3 alpha3 /.rule;
Print["F3 = ",F3];
Print["M3 = ",M3];
```

Link 4

The mass of the link is

$$m_4 = \rho \, h_{Slider} \, w_{Slider} \, d.$$

The position, velocity, and acceleration for the center of mass C_4 are

$$\mathbf{r}_{C_4} = \mathbf{r}_D, \; \mathbf{v}_{C_4} = \mathbf{v}_D, \text{ and } \mathbf{a}_{C_4} = \mathbf{a}_D.$$

The inertia force is

$$\mathbf{F}_{in4} = -m_4 \, \mathbf{a}_{C_4}.$$

The gravitational force is

$$\mathbf{G}_4 = -m_4 \, g \, \mathbf{k}.$$

The total force on slider 4 at D is

$$\mathbf{F}_4 = \mathbf{F}_{in4} + \mathbf{G}_4.$$

The mass moment of inertia is

$$I_{C_4} = m_4(h_{Slider}^2 + w_{Slider}^2)/12.$$

The moment of inertia is

$$\mathbf{M}_4 = \mathbf{M}_{in4} = -I_{C_4} \, \alpha_4.$$

To calculate and print the numerical values of the total force $\mathbf{F}_4$ and the moment $\mathbf{M}_4$, the following $Mathematica^{TM}$ commands are used:

```
m4=rho hSlider wSlider d /.rule;
rC4=rD; vC4=vD; aC4=aD;
Fin4=-m4 aC4 /.rule;
G4={0,-m4*g,0} /.rule;
F4=(Fin4+G4) /.rule;
IC4=m4 (hSlider^2+wSlider^2)/12 /.rule;
M4=Min4=-IC4 alpha4 /.rule;
Print["F4 = ",F4];
Print["M4 = ",M4];
```

Link 5

The mass of the link is

$$m_5 = \rho \, EG \, h \, d.$$

The position, velocity, and acceleration for the center of mass C_5 are

$$x_{C_5} = (EG/2) \cos \phi_5, \quad y_{C_5} = (EG/2) \sin \phi_5,$$

$$\mathbf{r}_{C_5} = x_{C_5}\mathbf{I} + y_{C_5}\mathbf{J}, \quad \mathbf{v}_{C_5} = \dot{x}_{C_5}\mathbf{I} + \dot{y}_{C_5}\mathbf{J}, \quad \text{and} \quad \mathbf{a}_{C_5} = \ddot{x}_{C_5}\mathbf{I} + \ddot{y}_{C_5}\mathbf{J}.$$

The inertia force is

$$\mathbf{F}_{in5} = -m_5 \, \mathbf{a}_{C_5}.$$

The gravitational force is

$$\mathbf{G}_5 = -m_5 \, g \, \mathbf{k}.$$

The total force on link 5 at C_5 is

$$\mathbf{F}_5 = \mathbf{F}_{in5} + \mathbf{G}_5.$$

The mass moment of inertia is

$$I_{C_5} = m_5 \, (EG^2 + h^2)/12.$$

The moment of inertia is

$$\mathbf{M}_5 = \mathbf{M}_{in5} = -I_{C_5} \, \boldsymbol{\alpha}_5.$$

The total force $\mathbf{F}_5$ and the moment $\mathbf{M}_5$ with *Mathematica*^{TM} are

```
m5=rho EG h d /.rule;
xC5=EG/2 Cos[phi5];
yC5=EG/2 Sin[phi5];
rC5={xC5,yC5,0};
vC5=D[rC5,t];
aC5=D[D[rC5,t],t];
Fin5=-m5 aC5 /.rule;
G5={0,-m5*g,0} /.rule;
F5=(Fin5+G5) /.rule;
IC5=m5 (EG^2+h^2)/12 /.rule;
M5=Min5=-IC5 alpha5 /.rule;
M5e={0,0,Me} /.rule;
Print["F5 = ",F5];
Print["M5 = ",M5];
```

The numerical values are

$$\mathbf{F}_1 = 0.018\mathbf{i} - 0.099\mathbf{j} \text{ N}, \quad \mathbf{M}_1 = \mathbf{0} \text{ N} \cdot \text{m},$$

$$\mathbf{F}_2 = 0.026\mathbf{i} - 0.063\mathbf{j} \text{ N}, \quad \mathbf{M}_2 = -0.00002\mathbf{k} \text{ N} \cdot \text{m},$$

$$\mathbf{F}_3 = 0.049\mathbf{i} - 0.333\mathbf{j} \text{ N}, \quad \mathbf{M}_3 = -0.00621\mathbf{k} \text{ N} \cdot \text{m},$$

$$\mathbf{F}_4 = -0.036\mathbf{i} - 0.063\mathbf{j} \text{ N}, \quad \mathbf{M}_4 = 0.00001\mathbf{k} \text{ N} \cdot \text{m},$$

$$\mathbf{F}_5 = -0.055\mathbf{i} - 0.410\mathbf{j} \text{ N}, \quad \mathbf{M}_5 = 0.00481\mathbf{k} \text{ N} \cdot \text{m}.$$

Joint Reaction Forces

The diagram representing the mechanism is shown in Figure I.7.11(b). It has two contours: 0-1-2-3-0 and 0-3-4-5-0.

Reaction force $\mathbf{F}_{05}$

The rotation joint E_R between the links 0 and 5 is replaced with the unknown reaction force $\mathbf{F}_{05}$ (Fig. I.7.12):

$$\mathbf{F}_{05} = F_{05x}\mathbf{i} + F_{05y}\mathbf{j}.$$

With *Mathematica*^{TM}, the force $\mathbf{F}_{05}$ is written as

```
F05Sol={F05xSol,F05ySol,0};
```

Following the path I, as shown in Figure I.7.12, a force equation is written for the translation joint D_T. The projection of all forces, that act on the link 5, onto the sliding direction $\mathbf{r}_{DE}$ is zero:

$$\sum \mathbf{F}^{(5)} \cdot \mathbf{r}_{DE} = (\mathbf{F}_5 + \mathbf{F}_{05}) \cdot \mathbf{r}_{DE} = 0, \tag{I.7.19}$$

where $\mathbf{r}_{DE} = \mathbf{r}_{AE} - \mathbf{r}_{AD}$.

Equation (I.7.19) with *Mathematica*^{TM} becomes

```
rDE=(rE-rD)/.rule;
eqER1=(F5+F05Sol).rDE==0;
```

where the command **a.b** gives the scalar product of the vectors **a** and **b**.

Continuing on the path I, a moment equation is written for the rotation joint D_R:

$$\sum \mathbf{M}_D^{(4\&5)} = \mathbf{r}_{DE} \times \mathbf{F}_{05} + \mathbf{r}_{DC_5} \times \mathbf{F}_5 + \mathbf{M}_4 + \mathbf{M}_5 + \mathbf{M}_e = \mathbf{0}, \tag{I.7.20}$$

where $\mathbf{r}_{DC_5} = \mathbf{r}_{AC_5} - \mathbf{r}_{AD}$.

Equation (I.7.20) with *Mathematica*^{TM} gives

```
rDC5=(rC5-rD)/.rule;
eqER2=(Cross[rDE,F05Sol]+Cross[rDC5,F5]+
M4+M5+M5e)[[3]]==0;
```

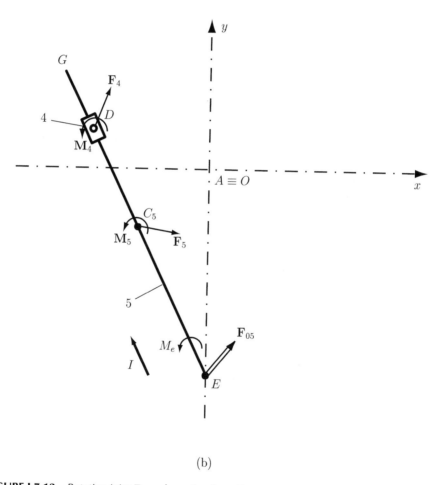

I

$\mathbf{F}_{05}$ $\quad 5 \quad$ D_T $\quad 4 \quad$ D_R

$\qquad \sum \mathbf{F}_{\Delta}^{(5)} \qquad \sum \mathbf{M}_{D}^{(4\&5)}$

(a)

(b)

FIGURE I.7.12 *Rotation joint E_R and reaction force $\mathbf{F}_{05}$.*

The system of two equations is solved using *Mathematica*^{TM} command

```
solF05=Solve[{eqER1,eqER2}, {F05xSol,F05ySol}];
```

The following numerical solution is obtained:

$$\mathbf{F}_{05} = 268.127\mathbf{I} + 135.039\mathbf{J} \ \text{N}.$$

Reaction force $\mathbf{F}_{45}$

The translation joint D_T between the links 4 and 5 is replaced with the unknown reaction force $\mathbf{F}_{45}$ (Fig. I.7.13):

$$\mathbf{F}_{45} = -\mathbf{F}_{54} = F_{45x}\mathbf{I} + F_{45y}\mathbf{J}.$$

The position of the application point P of the force $\mathbf{F}_{45}$ is unknown:

$$\mathbf{r}_{AP} = x_P\mathbf{I} + y_P\mathbf{J},$$

where x_P and y_P are the plane coordinates of the point P.

The force $\mathbf{F}_{45}$ and its point of application P with *Mathematica*^{TM} is written as

```
F45Sol={F45xSol,F45ySol,0};
rPSol={xPSol,yPSol,0};
```

Following the path *I* (Fig. I.7.13), a moment equation is written for the rotation joint E_R:

$$\sum \mathbf{M}_E^{(5)} = \mathbf{r}_{EP} \times \mathbf{F}_{45} + \mathbf{r}_{EC_5} \times \mathbf{F}_5 + \mathbf{M}_5 + \mathbf{M}_e = \mathbf{0}, \qquad (I.7.21)$$

where $\mathbf{r}_{EP} = \mathbf{r}_{AP} - \mathbf{r}_{AE}$, and $\mathbf{r}_{EC_5} = \mathbf{r}_{AC_5} - \mathbf{r}_{AE}$.

One can write Eq. (I.7.21) using the *Mathematica*^{TM} commands

```
rEP=(rPSol-rE)/.rule;
rEC5=(rC5-rE)/.rule;
eqDT1=(Cross[rEP,F45Sol]+Cross[rEC5,F5]+
M5+M5e) [[3]]==0;
```

Following the path *II* (Fig. I.7.13), a moment equation is written for the rotation joint D_R:

$$\sum \mathbf{M}_D^{(4)} = \mathbf{r}_{DP} \times \mathbf{F}_{54} + \mathbf{M}_4 = \mathbf{0}, \qquad (I.7.22)$$

where $\mathbf{r}_{DP} = \mathbf{r}_{AP} - \mathbf{r}_{AD}$ and $\mathbf{F}_{54} = -\mathbf{F}_{45}$.

Equation (I.7.22) with *Mathematica*^{TM} is

```
rDP=(rPSol-rD)/.rule;
eqDT2=(Cross[rDP,F54Sol]+M4) [[3]]==0;
```

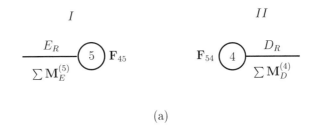

(a)

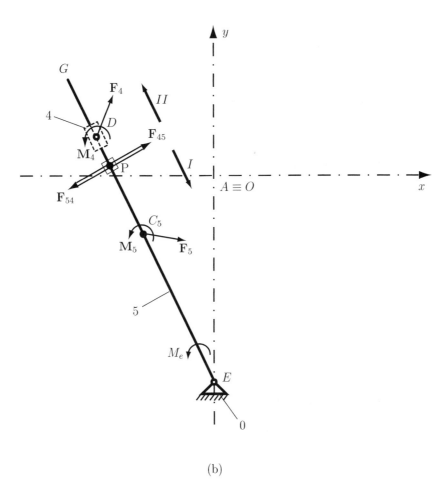

(b)

FIGURE I.7.13 *Translation joint D_T and reaction force $\mathbf{F}_{45}$.*

The direction of the unknown joint force $\mathbf{F}_{45}$ is perpendicular to the sliding direction $\mathbf{r}_{DE}$

$$\mathbf{F}_{45} \cdot \mathbf{r}_{DE} = 0, \qquad (\text{I.7.23})$$

and using *Mathematica*^{TM} command

```
eqDT3=F45Sol.rDE==0;
```

the application point P of the force $\mathbf{F}_{45}$ is located on the direction DE, that is

$$\frac{y_D - y_E}{x_D - x_E} = \frac{y_P - y_E}{x_P - x_E}. \qquad (\text{I.7.24})$$

One can write Eq. (I.7.24) using the *Mathematica*^{TM} commands

```
eqDT4=((yD-yE)/(xD-xE)/.rule)==
((yPSol-yE)/(xPSol-xE)/.rule);
```

The system of four equations is solved using the *Mathematica*^{TM} command

```
solF45=Solve[{eqDT1,eqDT2,eqDT3,eqDT4},
{F45xSol,F45ySol,xPSol,yPSol}];
```

The following numerical solutions are obtained:

$$\mathbf{F}_{45} = -268.072\mathbf{i} - 134.628\mathbf{j} \text{ N} \quad \text{and} \quad \mathbf{r}_{AP} = -0.149\mathbf{i} + 0.047\mathbf{j} \text{ m}.$$

Reaction force $\mathbf{F}_{34}$

The rotation joint D_R between the links 3 and 4 is replaced with the unknown reaction force $\mathbf{F}_{34}$ (Fig. I.7.14):

$$\mathbf{F}_{34} = -\mathbf{F}_{34} = F_{34x}\mathbf{i} + F_{34y}\mathbf{j},$$

and with *Mathematica*^{TM}

```
F34Sol={F34xSol,F34ySol,0};
```

Following the path I, a force equation can be written for the translation joint D_T. The projection of all forces, that act on the link 4, onto the sliding direction ED is zero:

$$\sum \mathbf{F}^{(4)} \cdot \mathbf{ED} = (\mathbf{F}_4 + \mathbf{F}_{34}) \cdot \mathbf{r}_{ED} = 0, \qquad (\text{I.7.25})$$

where $\mathbf{r}_{ED} = \mathbf{r}_{AD} - \mathbf{r}_{AE}$.

Equation (I.7.25) using *Mathematica*^{TM} gives

```
rED=(rD-rE)/.rule;
eqDR1=(F4+F34Sol).rED==0;
```

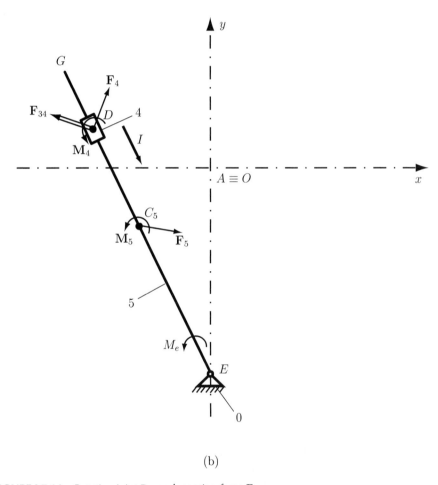

I

$\mathbf{F}_{34}$ ⑤ $\dfrac{D_T}{\sum \mathbf{F}_{\Delta}^{(4)}}$ ④ $\dfrac{E_R}{\sum \mathbf{M}_{E}^{(4\&5)}}$

(a)

(b)

FIGURE I.7.14 *Rotation joint D_R and reaction force $\mathbf{F}_{34}$.*

Continuing on the path I (Fig. I.7.14), a moment equation is written for the rotation joint E_R:

$$\sum \mathbf{M}_E^{(4\&5)} = \mathbf{r}_{ED} \times \mathbf{F}_{34} + \mathbf{r}_{EC_4} \times \mathbf{F}_4 + \mathbf{r}_{EC_5} \times \mathbf{F}_5 + \mathbf{M}_4 + \mathbf{M}_5 + \mathbf{M}_e = \mathbf{0}. \quad (I.7.26)$$

where $\mathbf{r}_{EC_5} = \mathbf{r}_{AC_5} - \mathbf{r}_{AE}$, and $\mathbf{r}_{EC_4} = \mathbf{r}_{AC_4} - \mathbf{r}_{AE}$.

Equation (I.7.26) with *Mathematica*^{TM} becomes

```
rEC5=(rC5-rE)/.rule;
rEC4=(rC4-rE)/.rule;
eqDR2=(Cross[rEC4,F4]+Cross[rEC5,F5]+
Cross[rED,F34Sol]+M4+M5+M5e)[[3]]==0;
```

The system of two equations is solved using the *Mathematica*^{TM} commands

```
solF34=Solve[{eqDR1,eqDR2}, {F34xSol,F34ySol}];
```

The following numerical solution is obtained:

$$\mathbf{F}_{34} = -268.035\mathbf{i} - 134.564\mathbf{j} \text{ N}.$$

Reaction force $\mathbf{F}_{03}$

The rotation joint C_R between the links 0 and 3 is replaced with the unknown reaction force $\mathbf{F}_{03}$ (Fig. I.7.15):

$$\mathbf{F}_{03} = F_{03x}\mathbf{i} + F_{03y}\mathbf{j}.$$

With *Mathematica*^{TM} the force $\mathbf{F}_{03}$ is written as

```
F03Sol={F03xSol,F03ySol,0};
```

Following the path I (Fig. I.7.15), a force equation is written for the translation joint B_T. The projection of all forces, that act on the link 3, onto the sliding direction CD is zero:

$$\sum \mathbf{F}^{(3)} \cdot \mathbf{r}_{CD} = (\mathbf{F}_{03} + \mathbf{F}_{43} + \mathbf{F}_3) \cdot \mathbf{r}_{CD} = 0, \quad (I.7.27)$$

where $\mathbf{r}_{CD} = \mathbf{r}_{AD} - \mathbf{r}_{AC}$.

Equation (I.7.27) with *Mathematica*^{TM} commands is

```
rCD=(rD-rC)/.rule;
eqCR1=(F03Sol+F43+F3).rCD==0;
```

Continuing on the path II (Fig. I.7.15), a moment equation is written for the rotation joint B_R:

$$\sum \mathbf{M}_B^{(3\&2)} = \mathbf{r}_{BC_3} \times \mathbf{F}_3 + \mathbf{r}_{BC} \times \mathbf{F}_{03} + \mathbf{r}_{BD} \times \mathbf{F}_{43} + \mathbf{M}_3 + \mathbf{M}_2 = \mathbf{0}, \quad (I.7.28)$$

where $\mathbf{r}_{BC_3} = \mathbf{r}_{AC_3} - \mathbf{r}_{AB}$, $\mathbf{r}_{BC} = \mathbf{r}_{AC} - \mathbf{r}_{AB}$, and $\mathbf{r}_{BD} = \mathbf{r}_{AD} - \mathbf{r}_{AB}$.

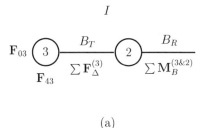

$$I$$

(a)

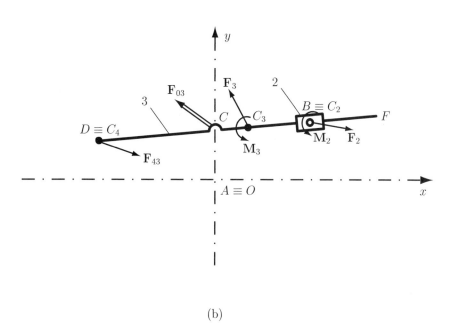

(b)

FIGURE I.7.15 *Rotation joint C_R and reaction force $\mathbf{F}_{03}$.*

With *Mathematica^{TM}* Eq. (I.7.28) gives

```
rBC3=(rC3-rB)/.rule;
rBC=(rC-rB)/.rule;
rBD=(rD-rB)/.rule;
eqCR2=(Cross[rBC3,F3]+Cross[rBC,F03Sol]+
Cross[rBD,F43]+M2+M3)[[3]]==0;
```

To solve the system of two equations the *Mathematica^{TM}* command is used:

```
solF03=Solve[{eqCR1,eqCR2}, {F03xSol,F03ySol}];
```

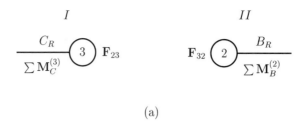

(a)

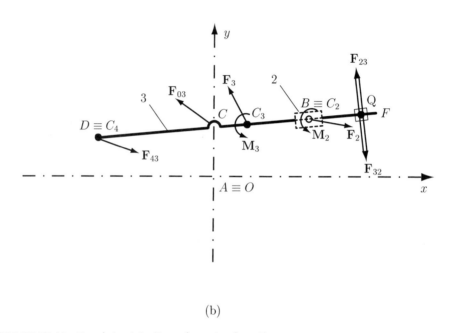

(b)

FIGURE I.7.16 *Translation joint B_T and reaction force $\mathbf{F}_{23}$.*

The following numerical solution is obtained:

$$\mathbf{F}_{03} = -256.71\mathbf{i} - 272.141\mathbf{j} \text{ N.}$$

Reaction force $\mathbf{F}_{23}$

The translation joint B_T between the links 2 and 3 is replaced with the unknown reaction force $\mathbf{F}_{23}$ (Fig. I.7.16):

$$\mathbf{F}_{23} = -\mathbf{F}_{32} = F_{23x}\mathbf{i} + F_{23y}\mathbf{j}.$$

The position of the application point Q of the force $\mathbf{F}_{23}$ is unknown:

$$\mathbf{r}_{AQ} = x_Q \mathbf{\imath} + y_Q \mathbf{J},$$

where x_Q and y_Q are the plane coordinates of the point Q.

The force $\mathbf{F}_{23}$ and its point of application Q are written in *Mathematica*^{TM} as

```
F34Sol={F34xSol,F34ySol,0};
rQSol={xQSol,yQSol,0};
```

Following the path I (Fig. I.7.16), a moment equation is written for the rotation joint C_R:

$$\sum \mathbf{M}_C^{(3)} = \mathbf{r}_{CQ} \times \mathbf{F}_{23} + \mathbf{r}_{CC_3} \times \mathbf{F}_3 + \mathbf{r}_{CD} \times \mathbf{F}_{43} + \mathbf{M}_3 = \mathbf{0}, \qquad (I.7.29)$$

where $\mathbf{r}_{CQ} = \mathbf{r}_{AQ} - \mathbf{r}_{AC}$, $\mathbf{r}_{CC_3} = \mathbf{r}_{AC_3} - \mathbf{r}_{AC}$, and $\mathbf{r}_{CD} = \mathbf{r}_{AD} - \mathbf{r}_{AC}$.

Using *Mathematica*^{TM}, Eq. (I.7.29) is written as

```
rCQ=(rQSol-rC)/.rule;
rCC3=(rC3-rC)/.rule;
rCD=(rD-rC)/.rule;
 eqBT1=(Cross[rCQ,F23Sol]+Cross[rCC3,F3]+
Cross[rCD,F43]+M3)[[3]]==0;
```

Following the path II (Fig. I.7.16), a moment equation is written for the rotation joint B_R:

$$\sum \mathbf{M}_B^{(2)} = \mathbf{r}_{BQ} \times \mathbf{F}_{32} + \mathbf{M}_2 = \mathbf{0}, \qquad (I.7.30)$$

where $\mathbf{r}_{BQ} = \mathbf{r}_{AQ} - \mathbf{r}_{AB}$.

Equation (I.7.30) with *Mathematica*^{TM} becomes

```
rBQ=(rQSol-rB)/.rule;
eqBT2=(Cross[rBQ,F32Sol]+M2)[[3]]==0;
```

The direction of the unknown joint force $\mathbf{F}_{23}$ is perpendicular to the sliding direction BC. The following relation is written:

$$\mathbf{F}_{23} \cdot \mathbf{r}_{BC} = 0,$$

or with *Mathematica*^{TM}, it is

```
eqBT3=F23Sol.rBC==0;
```

The application point Q of the force $\mathbf{F}_{23}$ is located on the direction BC, that is

$$\frac{y_C - y_B}{x_C - x_B} = \frac{y_C - y_Q}{x_C - x_Q}. \qquad (I.7.31)$$

Equation (I.7.31) with *Mathematica*^{TM} gives

```
eqBT4=((yC-yB)/(xC-xB)/.rule)==
((yC-yQSol)/(xC-xQSol)/.rule);
```

The system of four equations is solved using the *Mathematica*^{TM} command

```
solF23=Solve[{eqBT1,eqBT2,eqBT3,eqBT4},
{F23xSol,F23ySol,xQSol,yQSol}];
```

The following numerical solutions are obtained:

$$\mathbf{F}_{23} = -11.374\mathbf{\iota} + 137.91\mathbf{J} \text{ N} \quad \text{and} \quad \mathbf{r}_{AQ} = 0.121\mathbf{\iota} + 0.070\mathbf{J} \text{ m}.$$

Reaction force $\mathbf{F}_{12}$

The rotation joint B_R between the links 1 and 2 is replaced with the unknown reaction force $\mathbf{F}_{12}$ (Fig. I.7.17):

$$\mathbf{F}_{12} = -\mathbf{F}_{21} = F_{12x}\mathbf{\iota} + F_{12y}\mathbf{J}.$$

With *Mathematica*^{TM} it is written as:

```
F12Sol={F12xSol,F12ySol,0};
```

Following the path *I* (Fig. I.7.17), a force equation is written for the translation joint B_T. The projection of all forces that act on the link 2 onto the sliding direction BC is zero:

$$\sum \mathbf{F}^{(2)} \cdot \mathbf{r}_{BC} = (\mathbf{F}_{12} + \mathbf{F}_2) \cdot \mathbf{r}_{BC} = 0. \qquad (I.7.32)$$

Using *Mathematica*^{TM} it is written as:

```
rBC=(rC-rB)/.rule;
eqBR1=(F12Sol+F2).rBC==0;
```

Continuing on the path *I*, a moment equation is written for the rotation joint C_R:

$$\sum \mathbf{M}_C^{(2\&3)} = \mathbf{r}_{CB} \times \mathbf{F}_{12} + \mathbf{r}_{CC_2} \times \mathbf{F}_2 + \mathbf{r}_{CC_3} \times \mathbf{F}_3$$

$$+ \mathbf{r}_{CD} \times \mathbf{F}_{43} + \mathbf{M}_2 + \mathbf{M}_3 = \mathbf{0}, \qquad (I.7.33)$$

where $\mathbf{r}_{CB} = \mathbf{r}_{AB} - \mathbf{r}_{AC}$, $\mathbf{r}_{CC_2} = \mathbf{r}_{AC_2} - \mathbf{r}_{AC}$, $\mathbf{r}_{CC_3} = \mathbf{r}_{AC_3} - \mathbf{r}_{AC}$, and $\mathbf{r}_{CD} = \mathbf{r}_{AD} - \mathbf{r}_{AC}$. Using the *Mathematica*^{TM} commands, Eq. (I.7.33) gives

```
rCB=(rB-rC)/.rule;
rCC2=(rC2-rC)/.rule;
rCC3=(rC3-rC)/.rule;
rCD=(rD-rC)/.rule;
eqBR2=(Cross[rCB,F12Sol]+Cross[rCC2,F2]+
Cross[rCC3,F3]+Cross[rCD,F43]+M2+M3)[[3]]==0;
```

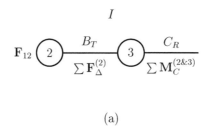

(a)

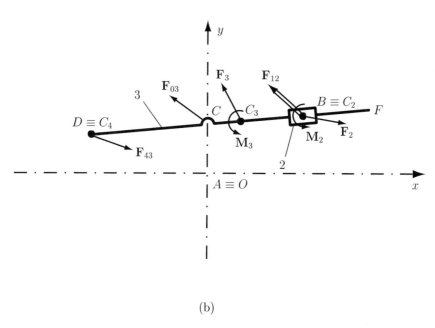

(b)

FIGURE I.7.17 *Rotation joint B_R and reaction force $\mathbf{F}_{12}$.*

The system of two equations is solved using the *Mathematica*^{TM} command:

```
solF12=Solve[{eqBR1,eqBR2},{F12xSol,F12ySol}];
```

and the following numerical solution is obtained:

$$\mathbf{F}_{12} = -11.401\mathbf{\imath} + 137.974\mathbf{\jmath} \text{ N.}$$

The motor moment M_m

The motor moment needed for the dynamic equilibrium of the mechanism is $\mathbf{M}_m = M_m \mathbf{k}$ (Fig. I.7.18) and with *Mathematica*^{TM} it is

```
M1mSol={0,0,MmSol};
```

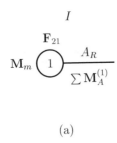

(a)

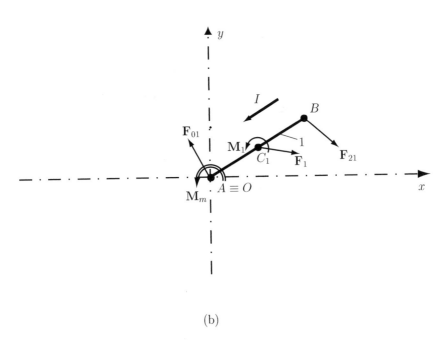

(b)

FIGURE I.7.18 *Dynamic equilibrium moment* $\mathbf{M}_m$.

Following the path I (Fig. I.7.18), a moment equation is written for the rotation joint A_R:

$$\sum \mathbf{M}_A^{(1)} = \mathbf{r}_{AB} \times \mathbf{F}_{21} + \mathbf{r}_{AC_1} \times \mathbf{F}_1 + \mathbf{M}_1 + \mathbf{M}_m = \mathbf{0}. \tag{I.7.34}$$

Equation (I.7.34) is solved using the *Mathematica*^{TM} commands:

```
eqMA=(Cross[rAB,F21]+Cross[rAC1,F1]+M1+
M1mSol)[[3]]==0;
   solMm=Solve[eqMA,MmSol];
```

```
Mm=MmSol/.solMm[[1]];
M1m={0,0,Mm};
Print["Mm = ",Mm];
```

The numerical solution is

$$\mathbf{M}_m = 17.533 \ \mathbf{k} \quad \text{N} \cdot \text{m}.$$

Reaction force $\mathbf{F}_{01}$

The rotation joint A_R between the links 0 and 1 is replaced with the unknown reaction force $\mathbf{F}_{01}$ (Fig. I.7.19):

$$\mathbf{F}_{01} = -\mathbf{F}_{10} = F_{01x}\mathbf{i} + F_{01y}\mathbf{J},$$

With *Mathematica*^{TM} it is written as:

```
F01Sol={F01xSol,F01ySol,0};
```

Following the path I (Fig. I.7.19), a moment equation is written for the rotation joint B_R:

$$\sum \mathbf{M}_B^{(1)} = \mathbf{r}_{BA} \times \mathbf{F}_{01} + \mathbf{r}_{BC_1} \times \mathbf{F}_1 + \mathbf{M}_1 + \mathbf{M}_m = \mathbf{0}, \tag{I.7.35}$$

where $\mathbf{r}_{BA} = -\mathbf{r}_{AB}$, and $\mathbf{r}_{BC_1} = \mathbf{r}_{AC_1} - \mathbf{r}_{AB}$.

Equation (I.7.35) using the *Mathematica*^{TM} commands gives

```
rBA=-rB/.rule;
rBC1=(rC1-rB)/.rule;
eqAR1=(Cross[rBA,F01Sol]+Cross[rBC1,F1]+M1+
M1m)[[3]]==0;
```

Continuing on the path I (Fig. I.7.19), a force equation is written for the translation joint B_T. The projection of all forces, that act on the links 1 and 2, onto the sliding direction BC is zero:

$$\sum \mathbf{F}^{(1\&2)} \cdot \mathbf{r}_{BC} = (\mathbf{F}_{01} + \mathbf{F}_1 + \mathbf{F}_2) \cdot \mathbf{r}_{BC} = 0, \tag{I.7.36}$$

or with *Mathematica*^{TM} it is

```
eqAR2=(F01Sol+F1+F2).rBC==0;
```

The system of two equations is solved using the *Mathematica*^{TM} command

```
solF01=Solve[{eqAR1,eqAR2},{F01xSol,F01ySol}];
```

The following numerical solution is obtained:

$$\mathbf{F}_{01} = -11.419\mathbf{i} + 138.073\mathbf{J} \quad \text{N}.$$

The *Mathematica*^{TM} program for the dynamic force analysis is presented in Program I.7.6.

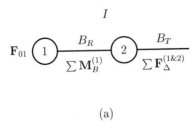

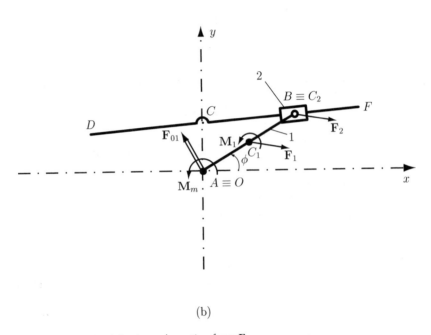

(b)

FIGURE I.7.19 *Rotation joint A_R and reaction force $\mathbf{F}_{01}$.*

I.7.5 Problems

I.7.1 Referring to Example I.3.1 (Fig. I.3.11), write a *Mathematica*^{TM} program for the position analysis of the mechanism.

I.7.2 Referring to Example I.3.2 (Fig. I.3.12), write a *Mathematica*^{TM} program for the position analysis of the mechanism.

I.7.3 Referring to Example I.3.3 (Fig. I.3.15), write a *Mathematica*^{TM} program for the position analysis of the mechanism.

I.7.4 Referring to Problem I.3.4 (Fig. I.3.19), write a *Mathematica*^{TM} program for the position analysis of the mechanism.

I.7.5 Referring to Problem I.3.5 (Fig. I.3.20), write a *Mathematica*^{TM} program for the position analysis of the mechanism.

I.7.6 Referring to Problem I.3.11 (Fig. I.3.26), write a *Mathematica*^{TM} program for the position analysis of the mechanism.

I.7.7 Referring to Example I.4.1 (Fig. I.4.7), write a *Mathematica*^{TM} program for the velocity and acceleration analysis of the mechanism.

I.7.8 Referring to Example I.4.2 (Fig. I.4.8), write a *Mathematica*^{TM} program for the velocity and acceleration analysis of the mechanism.

I.7.9 Referring to Problem I.4.1 [Fig. I.3.16(a)], write a *Mathematica*^{TM} program for the velocity and acceleration analysis of the mechanism.

I.7.10 Referring to Problem I.5.1 [Fig. I.3.16(a)], write a *Mathematica*^{TM} program for the velocity and acceleration analysis of the mechanism using the contour equations method.

I.7.11 Referring to Problem I.4.3 (Fig. I.4.10), write a *Mathematica*^{TM} program for the velocity and acceleration analysis of the mechanism.

I.7.12 Referring to Problem I.5.3 (Fig. I.3.10), write a *Mathematica*^{TM} program for the velocity and acceleration analysis of the mechanism using the contour equations method.

I.7.13 Referring to Problem I.4.4 (Fig. I.3.19), write a *Mathematica*^{TM} program for the velocity and acceleration analysis of the mechanism.

I.7.14 Referring to Problem I.5.4 (Fig. I.3.19), write a *Mathematica*^{TM} program for the velocity and acceleration analysis of the mechanism using the contour equations method.

I.7.15 Referring to Problem I.6.3 (Fig. I.4.10), write a *Mathematica*^{TM} program for the equilibrium moment and the joint forces of the mechanism.

I.7.16 Referring to Problem I.6.16 (Fig. I.3.31), write a *Mathematica*^{TM} program for the equilibrium moment and the joint forces of the mechanism.

I.7.17 Referring to Problem I.6.18 (Fig. I.3.33), write a *Mathematica*^{TM} program for the equilibrium moment and the joint forces of the mechanism.

I.7.18 Referring to Problem I.6.24 (Fig. I.3.11), write a *Mathematica*^{TM} program for the equilibrium moment and the joint forces of the mechanism.

I.7.19 Referring to Problem I.6.25 (Fig. I.3.12), write a *Mathematica*^{TM} program for the equilibrium moment and the joint forces of the mechanism.

I.7.20 Referring to Problem I.6.26 (Fig. I.3.15), write a *Mathematica*^{TM} program for the equilibrium moment and the joint forces of the mechanism.

I.7.6 Programs

PROGRAM I.7.1
```
(* POSITION ANALYSIS - input angle phi *)

Apply[Clear,Names["Global`*"]];
Off[General::spell];;
Off[General::spell1]
```

```
(* Input data *)
AB = 0.14 ;
AC = 0.06 ;
AE = 0.25 ;
CD = 0.15 ;

(* Input angle *)
phi = N[Pi]/6 ;

(* Position of joint A *)
xA = yA = 0;

(* Position of joint C *)
xC = 0 ;
yC = AC ;

(* Position of joint E *)
xE = 0 ;
yE = -AE ;

(* Position of joint B *)
xB = AB Cos[phi] ;
yB = AB Sin[phi] ;

(* Position of joint D *)

(* Parameters m and n of line BC: y = m x + b *)
m = ( yB - yC ) / ( xB - xC ) ;
b = yB - m xB ;
eqnD1 = ( xDsol - xC )^2 + ( yDsol - yC )^2 - CD^2 == 0 ;
eqnD2 = yDsol - m xDsol - b == 0 ;
solutionD = Solve [ { eqnD1 , eqnD2 } , { xDsol , yDsol } ];

(* Two solutions for D *)
xD1 = xDsol /. solutionD[[1]];
yD1 = yDsol /. solutionD[[1]];
xD2 = xDsol /. solutionD[[2]];
yD2 = yDsol /. solutionD[[2]];

(* Select the correct position for D *)
If [ xD1 <= xC , xD = xD1 ; yD = yD1 , xD = xD2 ; yD=yD2 ] ;

(* Print the solutions for B and D *)
Print["xB = ",xB," m"];
Print["yB = ",yB," m"];
Print["xD = ",xD," m"];
```

```
Print["yD = ",yD," m"];

(* Graph of the mecanism *)
markers = Table [ {
          Point [ { xA , yA } ] ,
          Point [ { xB , yB } ] ,
          Point [ { xC , yC } ] ,
          Point [ { xD , yD } ] ,
          Point [ { xE , yE } ]
          } ] ;

name = Table [ {
          Text [ "A" ,{0   , 0  },{-1 , 1 } ] ,
          Text [ "B" ,{xB , yB },{ 0 ,-1 } ] ,
          Text [ "C" ,{xC , yC },{ -1,-1 } ] ,
          Text [ "D" ,{xD , yD },{ 0 ,-1 } ] ,
          Text [ "E" ,{xE , yE },{ -1, 1 } ]
          } ] ;

graph = Graphics [
       { { RGBColor [ 1 , 0 , 0 ] ,
          Line [ { {xA,yA},{xB,yB} } ] } ,
        { RGBColor [ 0 , 1 , 0 ] ,
          Line [ { {xB,yB} , {xD,yD}} ] } ,
        { RGBColor [ 0 , 0 , 1 ] ,
          Line [ { {xD,yD}, {xE,yE}} ] } ,
        { RGBColor [ 1 , 1 , 1 ] ,
          PointSize [ 0.01 ] , markers } ,
        { name } } ] ;

Show [ Graphics [ graph ] ,
      PlotRange -> { { -.25 , .25 } ,
                     { -.3 , .25 } } ,
      Frame -> True,
      AxesOrigin -> {xA,yA},
      FrameLabel -> {"x","y"},
      Axes -> {True,True},
      AspectRatio -> Automatic ] ;

xB = 0.121244 m

yB = 0.07 m

xD = -0.149492 m

yD = 0.0476701 m
```

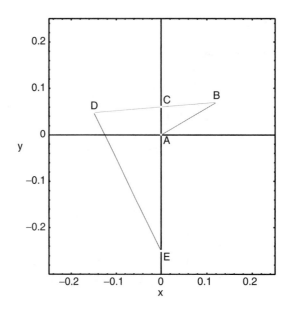

PROGRAM I.7.2
```
(* POSITION ANALYSIS - Complete rotation ( Method I ) *)

Apply[Clear,Names["Global`*"]];
Off[General::spell];
Off[General::spell1];

(* Input data *)
AB = 0.14 ;
AC = 0.06 ;
AE = 0.25 ;
CD = 0.15 ;

(* Position of joint A *)
xA = yA = 0;

(* Position of joint C *)
xC = 0 ;
yC = AC ;

(* Position of joint E *)
xE = 0 ;
yE = -AE ;

increment = 0 ;

For [ phi = 0 , phi <= 2*N[Pi] , phi += N[Pi]/3 ,
```

```
(* Position of joint B *)
xB = AB Cos[phi] ;
yB = AB Sin[phi] ;

(* Position of joint D *)

(* Parameters m and n of line BC: y = m x + b *)
m = ( yB - yC ) / ( xB - xC ) ;
b = yB - m xB ;
eqnD1 = ( xDsol - xC )^2 + ( yDsol - yC )^2 - CD^2 == 0 ;
eqnD2 = yDsol - m xDsol - b == 0 ;
solutionD = Solve [ { eqnD1 , eqnD2 } , { xDsol , yDsol } ];
(* Two solutions for D *)
xD1 = xDsol /. solutionD[[1]];
yD1 = yDsol /. solutionD[[1]];
xD2 = xDsol /. solutionD[[2]];
yD2 = yDsol /. solutionD[[2]];

(* Select the correct position for D *)
If [ 0 <= phi <= N[Pi]/2 || 3 N[Pi]/2 <= phi <= 2 N[Pi],
    If [ xD1 <= xC , xD = xD1 ; yD = yD1 , xD = xD2 ; yD=yD2] ,
    If [ xD1 >= xC , xD = xD1 ; yD = yD1 , xD = xD2 ; yD=yD2]
  ] ;

(* Print phi and the solutions for B and D *)

Print["phi = ",phi," rad = ",phi 180/N[Pi]," deg"];
Print["xB = ",xB," m"];
Print["yB = ",yB," m"];
Print["xD = ",xD," m"];
Print["yD = ",yD," m"];

(* Graph of the mechanism *)
markers = Table [ {
          Point [ { xA , yA } ] ,
          Point [ { xB , yB } ] ,
          Point [ { xC , yC } ] ,
          Point [ { xD , yD } ] ,
          Point [ { xE , yE } ]
          } ] ;

name = Table [ {
          Text [ "A" ,{0  , 0 },{-1 , 1 } ] ,
          Text [ "B" ,{xB , yB },{ 0 ,-1 } ] ,
          Text [ "C" ,{xC , yC },{ -1,-1 } ] ,
          Text [ "D" ,{xD , yD },{ 0 ,-1 } ] ,
          Text [ "E" ,{xE , yE },{ -1, 1 } ]
          } ] ;
```

```
graph [ increment ] = Graphics [
        { { RGBColor [ 1 , 0 , 0 ] ,
            Line [ { {xA,yA},{xB,yB} } ] } } ,
          { RGBColor [ 0 , 1 , 0 ] ,
            Line [ { {xB,yB} , {xD,yD}} ] } } ,
          { RGBColor [ 0 , 0 , 1 ] ,
            Line [ { {xD,yD}, {xE,yE}} ] } } ,
          { RGBColor [ 1 , 1 , 1 ] ,
            PointSize [ 0.01 ] , markers } ,
          { name } } ] ;

Show [ Graphics [ graph [ increment ] ] ,
        PlotRange -> { { -.25 , .25 } ,
                       { -.3 , .25 } } ,
        Frame -> True,
        AxesOrigin -> {xA,yA},
        FrameLabel -> {"x","y"},
        Axes -> {True,True},
        AspectRatio -> Automatic ] ;

increment++ ;

] ; (* End of FOR loop *)

phi = 0 rad = 0 deg

xB = 0.14 m

yB = 0 m

xD = -0.137872 m

yD = -0.119088 m
```

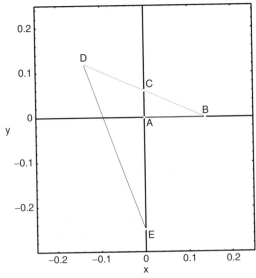

---

phi = 1.0472 rad = 60. deg

xB = 0.07 m

yB = 0.121244 m

xD = -0.112892 m

yD = -0.0387698 m

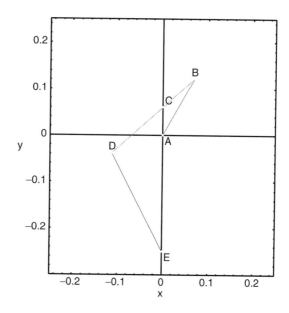

phi = 2.0944 rad = 120. deg

xB = -0.07 m

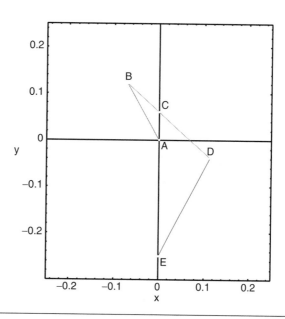

yB = 0.121244 m

xD = 0.112892 m

yD = -0.0387698 m

phi = 3.14159 rad = 180. deg

xB = -0.14 m

yB = 1.71451 × 10⁻¹⁷ m

xD = 0.137872 m

yD = 0.119088 m

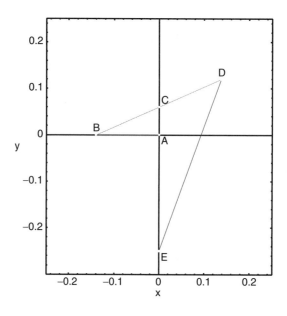

phi = 4.18879 rad = 240. deg

xB = -0.07 m

yB = -0.121244 m

xD = 0.0540425 m

yD = 0.199926 m

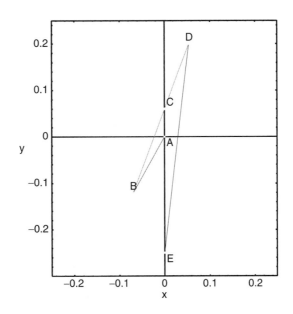

phi = 5.23599 rad = 300. deg

xB = 0.07 m

yB = -0.121244 m

xD = -0.0540425 m

yD = 0.199926 m

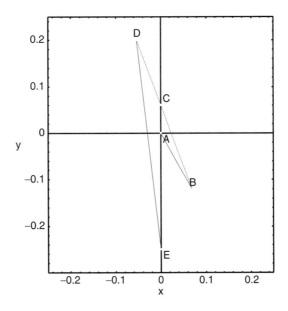

phi = 6.28319 rad = 360. deg

xB = 0.14 m

yB = -1.58635 × 10⁻¹⁶ m

xD = -0.137872 m

yD = 0.119088 m

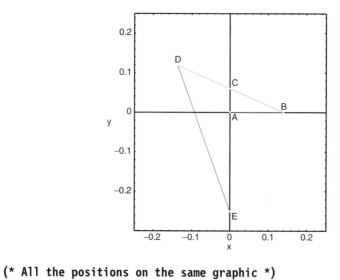

(* All the positions on the same graphic *)

```
Show [Table[graph [i] , { i , increment-1 } ] ,
      PlotRange -> { { -.25 , .25 } ,
                     { -.3 , .25 } } ,
      Frame -> True,
      AxesOrigin -> {xA,yA},
      FrameLabel -> {"x","y"},
      Axes -> {True,True},
      AspectRatio -> Automatic ] ;
```

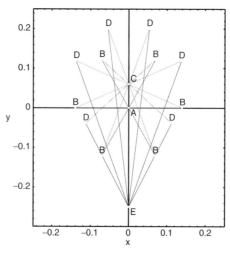

PROGRAM I.7.3

```
(* POSITION ANALYSIS - Complete rotation ( Method II ) *)

Apply[Clear,Names["Global`*"]];
Off[General::spell];
Off[General::spell1];

(* Euclidian distance function *)
Dist[xP_,yP_,xQ_,yQ_]:=Sqrt[(xP-xQ)^2+(yP-yQ)^2] ;

(* Input data *)
AB = 0.14 ;
AC = 0.06 ;
AE = 0.25 ;
CD = 0.15 ;

(* Position of joint A *)
xA = yA = 0 ;

(* Position of joint C *)
xC = 0 ;
yC = AC ;

(* Position of joint E *)
xE = 0 ;
yE = -AE ;

increment = 0 ;

For [ phi = 0 , phi <= 2*N[Pi] , phi += N[Pi]/6 ,

(* Position of joint B *)
xB = AB Cos [ phi ] ;
yB = AB Sin [ phi ] ;

(* Position of joint D *)

(* Parameters m and n of line BC: y = m x + b *)
m = ( yB - yC ) / ( xB - xC ) ;
b = yB - m xB ;
eqnD1 = ( xDsol - xC )^2 + ( yDsol - yC )^2 - CD^2 == 0 ;
eqnD2 = yDsol - m * xDsol - b == 0 ;
solutionD = Solve [ { eqnD1 , eqnD2 } , { xDsol , yDsol } ];
(* Two solutions for D *)
xD1 = xDsol /. solutionD[[1]];
yD1 = yDsol /. solutionD[[1]];
xD2 = xDsol /. solutionD[[2]];
yD2 = yDsol /. solutionD[[2]];
```

```
(* Select the correct position for D *)
If[increment==0, If[xD1<xC, xD=xD1;yD=yD1, xD=xD2;yD=yD2],
    dist1 = Dist[xD1,yD1,xDold,yDold];
    dist2 = Dist[xD2,yD2,xDold,yDold];
    If[dist1<dist2, xD=xD1;yD=yD1, xD=xD2;yD=yD2]
  ];
xDold = xD;
yDold = yD;

increment++;

Print["phi = ",phi," rad = ",phi 180/N[Pi]," deg"];
Print["xB = ",xB," m"];
Print["yB = ",yB," m"];
Print["xD = ",xD," m"];
Print["yD = ",yD," m"];

markers = Table [ {
            Point [ { xA , yA } ] ,
            Point [ { xB , yB } ] ,
            Point [ { xC , yC } ] ,
            Point [ { xD , yD } ] ,
            Point [ { xE , yE } ]
            } ] ;

name = Table [ {
            Text [ "A" ,{0 , 0 },{-1 , 1 } ] ,
            Text [ "B" ,{xB , yB },{ 0 ,-1 } ] ,
            Text [ "C" ,{xC , yC },{ -1,-1 } ] ,
            Text [ "D" ,{xD , yD },{ 0 ,-1 } ] ,
            Text [ "E" ,{xE , yE },{ -1, 1 } ]
            } ] ;

graph [ increment ] = Graphics [
        { { RGBColor [ 1 , 0 , 0 ] ,
            Line [ { {xA,yA},{xB,yB} } ] } ,
          { RGBColor [ 0 , 1 , 0 ] ,
            Line [ { {xB,yB} , {xD,yD}} ] } ,
          { RGBColor [ 0 , 0 , 1 ] ,
            Line [ { {xD,yD}, {xE,yE}} ] } ,
          { RGBColor [ 1 , 1 , 1 ] ,
            PointSize [ 0.01 ] , markers } ,
          { name } } ] ;

Show [ Graphics [ graph [ increment ] ] ,
      PlotRange -> { { -.25 , .25 } ,
                     { -.3 , .25 } } ,
```

```
        Frame -> True,
        AxesOrigin -> {xA,yA},
        FrameLabel -> {"x","y"},
        Axes -> {True,True},
        AspectRatio -> Automatic ] ;

] ; (* End of FOR loop *)

(* All positions on the same graphic *)

Show [Table[graph [i] , { i , increment } ] ,
        PlotRange -> { { -.25 , .25 } ,
                       { -.3 , .25 } } ,
        Frame -> True,
        AxesOrigin -> {xA,yA},
        FrameLabel -> {"x","y"},
        Axes -> {True,True},
        AspectRatio -> Automatic ] ;
```

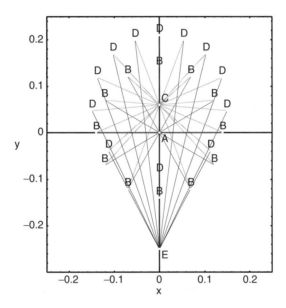

PROGRAM I.7.4
```
(* VELOCITY AND ACCELERATION ANALYSIS *)

Apply [Clear, Names["Global`*"] ] ;
Off[General::spell];
Off[General::spell];

n = 50 ; (* rpm *)
omega = n*N[Pi]/30 ; (* rad/s *)
```

```
(* Input data *)
initdata = {AB->0.14, AC->0.06, AE->0.25, CD->0.15, phi[t]->N[Pi]/6,
phi'[t]->omega, phi"[t]->0};

(* Position of joint A *)
xA = yA = 0;

(* Position of joint C *)
xC = 0 ;
yC = AC ;

(* Position of joint E *)
xE = 0 ;
yE = -AE ;

(* Position of joint B *)
xB = AB Cos[ phi[t] ] ;
yB = AB Sin[ phi[t] ] ;
Print["xB = ", xB ," = ", xB/.initdata, " m" ];
Print["yB = ", yB ," = ", yB/.initdata, " m" ];

(* Linear velocity of joint B *)
vBx = D[xB,t];
vBy = D[yB,t];
Print["vBx = ", vBx ," = ", vBx/.initdata, " m/s" ];
Print["vBy = ", vBy ," = ", vBy/.initdata, " m/s" ];

(* Linear acceleration of joint B *)
aBx = D[vBx,t];
aBy = D[vBy,t];
Print["aBx = ", aBx ," = ", aBx/.initdata, " m/s^2" ];
Print["aBy = ", aBy ," = ", aBy/.initdata, " m/s^2" ];

(* Position of joint D *)

(*Parameters m and n of line BC: y = m x + b *)
mBC = ( yB - YC ) / ( xB - xC ) ;
bBC = yB - mBC xB ;
eqn41 = ( xDsol - xC )^2 + ( yDsol - yC )^2 - CD^2 == 0 ;
eqn42 = yDsol - mBC xDsol - bBC == 0 ;
solutionD = Solve [ { eqn41 , eqn42 } , { xDsol , yDsol } ] ;
(* Two solutions for D *)
xD1 = xDsol /. solutionD[[1]];
yD1 = yDsol /. solutionD[[1]];
xD2 = xDsol /. solutionD[[2]];
yD2 = yDsol /. solutionD[[2]];
(* Select the correct position for D *)
```

```
If [ (xD1/.initdata)<=xC , xD=xD1 ; yD=yD1, xD=xD2; yD=yD2 ];
Print["xD = ", xD/.initdata, " m" ];
Print["yD = ", yD/.initdata, " m" ];

(* Linear velocity of joint D *)
vDx = D[xD,t];
vDy = D[yD,t];
Print["vDx = ", vDx/.initdata, " m/s" ];
Print["vDy = ", vDy/.initdata, " m/s" ];

(* Linear acceleration of joint D *)
aDx = D[vDx,t];
aDy = D[vDy,t];
Print["aDx = ", aDx/.initdata, " m/s^2" ];
Print["aDy = ", aDy/.initdata, " m/s^2" ];

(* Angular velocity and acceleration of the link BD *)

phi3 = ArcTan[ mBC ] ;
omega3 = D[ phi3 , t ] ;
alpha3 = D[ omega3, t ] ;
Print["phi3 = phi2 = ", phi3/.initdata , " rad " ];
Print["omega3 = omega2 = ", omega3/.initdata , " rad/s" ];
Print["alpha3 = alpha2 = ", alpha3/.initdata , " rad/s^2" ];

(* Angular velocity and acceleration of the link DE *)

phi5 = ArcTan[(yD-yE)/(xD-xE)] + N[Pi];
omega5 = D[ phi5, t ];
alpha5 = D[ omega5, t ];
Print["phi5 = phi4 = ", phi5/.initdata , " rad " ];
Print["omega5 = omega4 = ", omega5/.initdata , " rad/s" ];
Print["alpha5 = alpha4 = ", alpha5/.initdata , " rad/s^2" ];
```

$xB = AB \, Cos[phi[t]] = 0.121244$ m

$yB = AB \, Sin[phi[t]] = 0.07$ m

$vBx = -AB \, Sin[phi[t]]phi'[t] = -0.366519$ m/s

$vBy = AB \, Cos[phi[t]]phi'[t] = 0.63483$ m/s

$aBx = -AB \, Cos[phi[t]]phi'[t]^2 - AB \, Sin[phi[t]]phi''[t] = -3.32396$ m/s^2

$aBy = -AB \, Sin[phi[t]]phi'[t]^2 + AB \, Cos[phi[t]]phi''[t] = -1.91909$ m/s^2

$xD = -0.149492$ m

yD = 0.0476701 m

vDx = 0.0671766 m/s

vDy = -0.814473 m/s

aDx = 4.61708 m/s^2

aDy = -1.81183 m/s^2

phi3 = phi2 = 0.0822923 rad

omega3 = omega2 = 5.44826 rad/s

alpha3 = alpha2 = 14.5681 rad/s^2

phi5 = phi4 = 2.03621 rad

omega5 = omega4 = 0.917134 rad/s

alpha5 = alpha4 = -5.77155 rad/s^2

PROGRAM I.7.5

```
(* CONTOUR METHOD *)

Apply [Clear, Names["Global`*"] ] ;
Off[General::spell];
Off[General::spell1];

(* Input data *)

n = 50 ; (* rpm *)
omega = n N[Pi]/30 ; (* rad/s *)

AB = 0.14 ;
AC = 0.06 ;
AE = 0.25 ;
CD = 0.15 ;
phi = N[Pi]/6 ;

(* Position analysis *)

(* Position of joint A *)
xA = yA = 0;
rA = {xA, yA, 0};

(* Position of joint C *)
xC = 0 ;
yC = AC ;
```

```
rC = {xC, yC, 0};

(* Position of joint E *)
xE = 0 ;
yE = -AE ;
rE = {xE, yE, 0};

(* Position, velocity and acceleration of joint B *)
xB = AB Cos[ phi ] ;
yB = AB Sin[ phi ] ;
rB = {xB, yB, 0} ;

(* Position, velocity and acceleration of joint D *)

(* Parameters m and n of line BC: y = m x + b *)
mBC = ( yB - yC ) / ( xB - xC ) ;
bBC = yB - mBC xB ;
eqn41 = ( xDsol - xC )^2 + ( yDsol - yC )^2 - CD^2 == 0 ;
eqn42 = yDsol - mBC xDsol - bBC == 0 ;
solutionD = Solve [ { eqn41 , eqn42 } , { xDsol , yDsol } ] ;
(* Two solutions for D *)
xD1 = xDsol /. solutionD[[1]];
yD1 = yDsol /. solutionD[[1]];
xD2 = xDsol /. solutionD[[2]];
yD2 = yDsol /. solutionD[[2]];
(* Select the correct position for D *)
If [ xB >= xC , xD = xD1 ; yD = yD1 , xD = xD2 ; yD = yD2 ] ;
rD = {xD, yD, 0} ;

phi2 = ArcTan[ mBC ] ;
phi3 = phi2 ;

phi4 = ArcTan[(yD-yE)/(xD-xE)] + N[Pi] ;
phi5 = phi4 ;

(* ---------- *)
(* Velocities *)
(* ---------- *)

(* Contour I *)
Print["Contour I"];

(* Relative velocities *)

omega10v = { 0, 0, omega } ;
omega21vSol = { 0, 0, omega21Sol } ;
omega03vSol = { 0, 0, omega03Sol } ;
```

```
v32vSol = { v32Sol Cos[phi2], v32Sol Sin[phi2], 0 } ;

eqIkv = ( omega10v + omega21vSol + omega03vSol )[[3]] == 0 ;
eqIiv = ( Cross[rB,omega21vSol] + Cross[rC,omega03vSol]
        + v32vSol )[[1]] == 0 ;
eqIjv = ( Cross[rB,omega21vSol] + Cross[rC,omega03vSol]
        + v32vSol )[[2]] == 0 ;

solIvel = Solve[ { eqIkv, eqIiv, eqIjv }, { omega21Sol, omega03Sol,
        v32Sol } ] ;
omega21v = omega21vSol /.solIvel[[1]] ;
omega03v = omega03vSol /.solIvel[[1]] ;
v32v = v32vSol /.solIvel[[1]] ;

Print[ "omega21 = ", omega21v ] ;
Print[ "omega03 = ", omega03v ] ;
Print[ "v32 = ", v32v ] ;
Print[ "v32r = ", v32Sol/.solIvel[[1]] ] ;

(* Absolute velocities *)

omega20v = omega30v = - omega03v ;
vBv = Cross[omega10v,rB] ;
vDv = Cross[omega30v,(rD-rC)] ;

Print[ "omega20 = omega30 = ", omega30v ] ;
Print[ "vB = ", vBv ] ;
Print[ "vD = ", vDv ] ;

(* Relative accelerations *)

alpha10v = { 0, 0, 0 } ;
alpha21vSol = { 0, 0, alpha21Sol } ;
alpha03vSol = { 0, 0, alpha03Sol } ;
a32vSol = { a32Sol Cos[phi2], a32Sol Sin[phi2], 0 } ;

eqIka = ( alpha10v + alpha21vSol + alpha03vSol )[[3]] == 0 ;
eqIia = ( Cross[rB,alpha21vSol] + Cross[rC,alpha03vSol] + a32vSol + 2
Cross[omega20v,v32v] -
(omega10v.omega10v)rB-(omega20v.omega20v)(rC-rB) )[[1]] == 0 ;
eqIja = ( Cross[rB,alpha21vSol] + Cross[rC,alpha03vSol] + a32vSol + 2
Cross[omega20v,v32v] - (omega10v.omega10v)rB-(omega20v.omega20v)(rC-rB) )
[[2]] == 0 ;

solIacc = Solve[ { eqIka, eqIia, eqIja }, { alpha21Sol, alpha03Sol,
        a32Sol } ] ;
alpha21v = alpha21vSol /.solIacc[[1]] ;
```

```
alpha03v = alpha03vSol /.solIacc[[1]] ;
a32v = a32vSol /.solIacc[[1]] ;

Print[ "alpha21 = ", alpha21v ] ;
Print[ "alpha03 = ", alpha03v ] ;
Print[ "a32 = ", a32v ] ;
Print[ "a32r = ", a32Sol/.solIacc[[1]] ] ;

(* Absolute accelerations *)

alpha20v = alpha30v = - alpha03v ;
aBv = -(omega10v.omega10v) rB ;
aDv = Cross[alpha30v,(rD-rC)]-(omega20v.omega20v)(rD-rC) ;

Print[ "alpha20 = alpha30 = ", alpha30v ] ;
Print[ "aB = ", aBv ] ;
Print[ "aD = ", aDv ] ;

(* Contour II *)
Print["Contour II"];

(* Relative velocities *)

omega43vSol = { 0, 0, omega43Sol } ;
omega05vSol = { 0, 0, omega05Sol } ;
v54vSol = { v54Sol Cos[phi4], v54Sol Sin[phi4], 0 } ;

eqIIkv = ( omega30v + omega43vSol + omega05vSol )[[3]] == 0 ;

eqIIiv = ( Cross[rC,omega30v] + Cross[rD,omega43vSol] +
           Cross[rE,omega05vSol] + v54vSol )[[1]] == 0 ;

eqIIjv = ( Cross[rC,omega30v] + Cross[rD,omega43vSol] +
           Cross[rE,omega05vSol] + v54vSol )[[2]] == 0 ;

solIIvel = Solve[ { eqIIkv, eqIIiv, eqIIjv }, { omega43Sol,
           omega05Sol, v54Sol } ] ;
omega43v = omega43vSol /.solIIvel[[1]] ;
omega05v = omega05vSol /.solIIvel[[1]] ;
v54v = v54vSol /.solIIvel[[1]] ;

Print[ "omega43 = ", omega43v ] ;
Print[ "omega05 = ", omega05v ] ;
Print[ "v54 = ", v54v ] ;
Print[ "v54r = ", v54Sol/.solIIvel[[1]] ] ;

(* Absolute velocities *)
```

```
omega40v = omega50v = - omega05v ;
Print[ "omega40 = omega50 = ", omega50v ] ;

(* Relative accelerations *)

alpha43vSol = { 0, 0, alpha43Sol } ;
alpha05vSol = { 0, 0, alpha05Sol } ;
a54vSol = { a54Sol Cos[phi4], a54Sol Sin[phi4], 0 } ;

eqIIka = ( alpha30v + alpha43vSol + alpha05vSol )[[3]] == 0 ;

eqIIia = ( Cross[rC,alpha30v] + Cross[rD,alpha43vSol] +
          Cross[rE,alpha05vSol] + a54vSol + 2 Cross[omega40v,v54v] -
          (omega30v.omega30v)(rD-rC) -
          (omega40v.omega40v)(rE-rD) )[[1]] == 0 ;

eqIIja = ( Cross[rC,alpha30v] + Cross[rD,alpha43vSol] +
          Cross[rE,alpha05vSol] + a54vSol + 2 Cross[omega40v,v54v] -
          (omega30v.omega30v)(rD-rC) -
          (omega40v.omega40v)(rE-rD) )[[2]] == 0 ;

solIIacc = Solve[ { eqIIka, eqIIia, eqIIja }, { alpha43Sol, alpha05Sol,
            a54Sol } ] ;
alpha43v = alpha43vSol /.solIIacc[[1]] ;
alpha05v = alpha05vSol /.solIIacc[[1]] ;
a54v = a54vSol /.solIIacc[[1]] ;

Print[ "alpha43 = ", alpha43v ] ;
Print[ "alpha05 = ", alpha05v ] ;
Print[ "a54 = ", a54v ] ;
Print[ "a54r = ", a54Sol/.solIIacc[[1]] ] ;

(* Absolute accelerations *)

alpha40v = alpha50v = - alpha05v ;
Print[ "alpha40 = alpha50 = ", alpha50v ] ;

Contour I

omega21 = {0, 0, 0.21227}

omega03 = {0, 0, -5.44826}

v32 = {0.312037, 0.0257363, 0}

v32r = 0.313096
```

omega20 = omega30 = {0, 0, 5.44826}

vB = {-0.366519, 0.63483, 0.}

vD = {0.0671766, -0.814473, 0.}

alpha21 = {0, 0, 14.5681}

alpha03 = {0, 0, -14.5681}

a32 = {-0.140218, -0.11565, 0}

a32r = -0.140694

alpha20 = alpha30 = {0, 0, 14.5681}

aB = {-3.32396, -1.91909, 0}

aD = {4.61708, -1,81183, 0.}

Contour II

omega43 = {0, 0, -4.53112}

omega05 = {0, 0, -0.917134}

v54 = {-0.34018, 0.677368, 0}

v54r = 0.757991

omega40 = omega50 = {0, 0, 0.917134}

alpha43 = {0, 0, -20.3397}

alpha05 = {0, 0, 5.77155}

a54 = {-1.53085, 3.04823, 0}

a54r = 3.41104

alpha40 = alpha50 = {0, 0, -5.77155}

PROGRAM I.7.6
(* DYNAMIC FORCE ANALYSIS *)

Apply [Clear, Names["Global`*"]] ;

```
Off[General::spell];
Off[General::spell1];

(* Input data *)
n = 50 ; (* rpm *)
omega = n N[Pi]/30 ; (* rad/s *)

rule = {AB->0.14, AC->0.06, AE->0.25, CD->0.15, FD->0.4, EG->0.5,
h->0.01, d->0.001,
   hSlider->0.02, wSlider->0.05, rho->8000, g->9.807, Me->-100.,
phi[t]->N[Pi]/6,
   phi'[t]->omega, phi"[t]->0} ;

(* Position analysis *)

(* Position of joint A *)
xA = yA = 0;
rA = { xA, yA, 0};

(* Position of joint C *)
xC = 0 ;
yC = AC ;
rC = { xC, yC, 0} ;

(* Position of joint E *)
xE = 0 ;
yE = -AE ;
rE = { xE, yE, 0};

(* Position, velocity and acceleration of joint B *)
xB = AB Cos [ phi[t] ] ;
yB = AB Sin [ phi[t] ] ;
rB = { xB, yB, 0} ;
vB = D[rB,t] ;
aB = D[D[rB,t],t] ;

(* Position, velocity and acceleration of joint D *)

(* Parameters m and n of line BC: y = m x + b *)
mBC = ( yB - yC ) / ( xB - xC ) ;
bBC = yB - mBC xB ;
eqn41 = ( xDsol - xC )^2 + ( yDsol - yC )^2 - CD^2 == 0 ;
eqn42 = yDsol - mBC xDsol - bBC == 0 ;
solutionD = Solve [ { eqn41 , eqn42 } , { xDsol , yDsol } ];
(* Two solutions for D *)
xD1 = xDsol /. solutionD[[1]];
yD1 = yDsol /. solutionD[[1]];
xD2 = xDsol /. solutionD[[2]];
yD2 = yDsol /. solutionD[[2]];
```

```
(* Select the correct position for D *)
If[ (xD1/.rule)<=xC, xD=xD1; yD=yD1, xD=xD2; yD=yD2 ] ;
rD = { xD, yD, 0} ;
vD = D[rD,t] ;
aD = D[D[rD,t],t] ;

(* Angular velocity and acceleration of the link 1 *)
alpha1 = {0, 0, phi"[t]} ;

(* Angular velocity and acceleration of the link 2 and link 3 *)
phi2 = ArcTan[ mBC ] ;
alpha2 = {0, 0, D[D[phi2,t],t]} ;
phi3 = phi2 ;
alpha3 = alpha2 ;

(* Angular velocity and acceleration of the link 4 and link 5 *)

phi4 = ArcTan[(yD-yE)/(xD-xE)] + N[Pi] ;
alpha4 = {0, 0, D[D[phi4,t],t]} ;
phi5 = phi4 ;
alpha5 = alpha4 ;

(* ------------------------- *)
(* Inertia forces and moments *)
(* ------------------------- *)

(* Link 1 *)
m1 = rho AB h d /.rule ;
rC1 = rB/2 ;
vC1 = vB/2 ;
aC1 = aB/2 ;
Fin1 = - m1 aC1 /.rule ;
G1 = {0, -m1*g, 0} /.rule ;
F1 = ( Fin1 + G1 ) /.rule ;
IC1 = m1 (AB^2+h^2)/12 /.rule ;
M1 = Min1 = - IC1 alpha1 /.rule ;
Print["F1 = ", F1] ;
Print["M1 = ", M1] ;

(* Link 2 *)
m2 = rho hSlider wSlider d /.rule ;
rC2 = rB ;
vC2 = vB ;
aC2 = aB ;
Fin2 = - m2 aC2 /.rule ;
G2 = {0, -m2*g, 0} /.rule ;
F2 = ( Fin2 + G2 ) /.rule ;
IC2 = m2 (hSlider^2+wSlider^2)/12 /.rule ;
M2 = Min2 = - IC2 alpha2 /.rule ;
```

```
Print["F2 = ", F2] ;
Print["M2 = ", M2] ;


(* Link 3 *)
m3 = rho FD h d /.rule ;
xC3 = xC + (FD/2-CD) Cos [ phi3 ] ;
yC3 = yC + (FD/2-CD) Sin [ phi3 ] ;
rC3 = { xC3, yC3, 0 } ;
vC3 = D[rC3,t] ;
aC3 = D[D[rC3,t],t] ;
Fin3 = - m3 aC3 /.rule ;
G3 = {0, -m3*g, 0} /.rule ;
F3 = ( Fin3 + G3 ) /.rule ;
IC3 = m3 (FD^2+h^2)/12 /.rule ;
M3 = Min3 = - IC3 alpha3 /.rule ;
Print["F3 = ", F3] ;
Print["M3 = ", M3] ;


(* Link 4 *)
m4 = rho hSlider wSlider d /.rule ;
rC4 = rD ;
vC4 = vD ;
aC4 = aD ;
Fin4 = - m4 aC4 /.rule ;
G4 = {0, -m4*g, 0} /.rule ;
F4 = ( Fin4 + G4 ) /.rule ;
IC4 = m4 (hSlider^2+wSlider^2)/12 /.rule ;
M4 = Min4 = - IC4 alpha4 /.rule ;
Print["F4 = ", F4] ;
Print["M4 = ", M4] ;


(* Link 5 *)
m5 = rho EG h d /.rule ;
xC5 = EG/2 Cos [ phi5 ] ;
yC5 = EG/2 Sin [ phi5 ] ;
rC5 = { xC5, yC5, 0} ;
vC5 = D[rC5,t] ;
aC5 = D[D[rC5,t],t] ;
Fin5 = - m5 aC5 /.rule ;
G5 = {0, -m5*g, 0} /.rule ;
F5 = ( Fin5 + G5 ) /.rule ;
IC5 = m5 (EG^2+h^2)/12 /.rule ;
M5 = Min5 = - IC5 alpha5 /.rule ;
M5e = { 0, 0, Me } /.rule ;
Print["F5 = ", F5] ;
Print["M5 = ", M5] ;
```

```
(* --------------- *)
(* Joint reactions *)
(* --------------- *)

(*** Contour 0-3-4-5-0 ***)

(* Joint E_R *)

F05Sol = { F05xSol, F05ySol, 0 } ;

(* ∑ F for 5 *)
rDE = ( rE - rD ) /.rule ;
eqER1 = (F5+F05Sol).rDE == 0 ;

(* ∑ M_D for 4&5 *)
rDC5 = ( rC5 - rD ) /.rule ;
eqER2 = (Cross[rDE,F05Sol]+Cross[rDC5,F5]+M4+M5+M5e)[[3]] == 0 ;

solF05 = Solve[{eqER1,eqER2},{F05xSol,F05ySol}] ;
F05x = F05xSol /.solF05[[1]] ;
F05y = F05ySol /.solF05[[1]] ;
F05 = { F05x, F05y, 0 } ;
Print["F05 = ", F05] ;

(* Joint D_T *)

F45Sol = { F45xSol, F45ySol, 0 } ;
F54Sol = - F45Sol ;
rPSol = { xPSol, yPSol, 0 } ;

(* ∑ M_E for 5 *)
rEP = ( rPSol - rE ) /.rule ;
rEC5 = ( rC5 - rE ) /.rule ;
eqDT1 = (Cross[rEP,F45Sol]+Cross[rEC5,F5]+M5+M5e)[[3]] == 0 ;

(* ∑ M_D for 4 *)
rDP = ( rPSol - rD ) /.rule ;
eqDT2 = (Cross[rDP,F54Sol]+M4)[[3]] == 0 ;

eqDT3 = F45Sol.rDE == 0 ;
eqDT4 = ( (yD-yE)/(xD-xE)/.rule ) == ( (yPSol-yE)/(xPSol-xE)/.rule ) ;

solF45 = Solve[{eqDT1,eqDT2,eqDT3,eqDT4},{F45xSol,F45ySol,xPSol,
yPSol}] ;
F45x = F45xSol/.solF45[[1]] ;
F45y = F45ySol/.solF45[[1]] ;
F45 = { F45x, F45y, 0 } ;
```

```
    F54 = - F45 ;
    xP = xPSol/.solF45[[1]] ;
    yP = yPSol/.solF45[[1]] ;
    rP = { xP, yP, 0 } ;
    Print["F45 = ", F45] ;
    Print["rP = ", rP] ;

    (* Joint D_R *)

    F34Sol = { F34xSol, F34ySol, 0 } ;
    F43Sol = - F34Sol ;

    (* ∑ F for 4 *)
    rED = ( rD - rE ) /.rule ;
    eqDR1 = (F4+F34Sol).rED == 0 ;

    (* ∑ M_E for 4&5 *)
    rEC5 = ( rC5 - rE ) /.rule ;
    rEC4 = ( rC4 - rE ) /.rule ;
    eqDR2 = (Cross[rEC4,F4]+Cross[rEC5,F5]+Cross[rED,F34Sol]+M4+M5+M5e)[[3]]
== 0 ;

    solF34 = Solve[{eqDR1,eqDR2},{F34xSol,F34ySol}] ;
    F34x = F34xSol/.solF34[[1]] ;
    F34y = F34ySol/.solF34[[1]] ;
    F34 = { F34x, F34y, 0 } ;
    F43 = - F34 ;
    Print["F34 = ", F34] ;

    (*** Contour 0-1-2-3-0 ***)

    (* Joint C_R *)

    F03Sol = { F03xSol, F03ySol, 0 } ;

    (* ∑ F for 3 *)
    rCD = ( rD - rC ) /.rule ;
    eqCR1 = (F03Sol+F43+F3).rCD == 0 ;

    (* ∑ M_B for 3&2 *)
    rBC3 = ( rC3 - rB ) /.rule ;
    rBC = ( rC - rB ) /.rule ;
    rBD = ( rD - rB ) /.rule ;
    eqCR2 = (Cross[rBC3,F3]+Cross[rBC,F03Sol]+Cross[rBD,F43]+M2+M3)[[3]]
== 0 ;

    solF03 = Solve[{eqCR1,eqCR2},{F03xSol,F03ySol}] ;
```

```
    F03x = F03xSol/.solF03[[1]] ;
    F03y = F03ySol/.solF03[[1]] ;
    F03 = { F03x, F03y, 0 } ;
    Print["F03 = ", F03] ;

    (* Joint B_T *)

    F23Sol = { F23xSol, F23ySol, 0 } ;
    F32Sol = - F23Sol ;
    rQSol = { xQSol, yQSol, 0 } ;

    (* ∑ M_C for 3 *)
    rCQ = ( rQSol - rC ) /.rule ;
    rCC3 = ( rC3 - rC ) /.rule ;
    rCD = ( rD - rC ) /.rule ;
    eqBT1 = (Cross[rCQ,F23Sol]+Cross[rCC3,F3]+Cross[rCD,F43]+M3)[[3]]
== 0 ;

    (* ∑ M_B for 2 *)
    rBQ = ( rQSol - rB ) /.rule ;
    eqBT2 = (Cross[rBQ,F32Sol]+M2)[[3]] == 0 ;
    eqBT3 = F23Sol.rBC == 0 ;
    eqBT4 = ( (yC-yB)/(xC-xB)/.rule ) == ( (yC-yQSol)/(xC-xQSol)/.rule ) ;

    solF23 = Solve[{eqBT1,eqBT2,eqBT3,eqBT4},{F23xSol,F23ySol,xQSol,
yQSol}] ;
    F23x = F23xSol/.solF23[[1]] ;
    F23y = F23ySol/.solF23[[1]] ;
    F23 = { F23x, F23y, 0 } ;
    F32 = - F23 ;
    xQ = xQSol/.solF23[[1]] ;
    yQ = yQSol/.solF23[[1]] ;
    rQ = { xQ, yQ, 0 } ;
    Print["F23 = ", F23] ;
    Print["rQ = ", rQ] ;

    (* Joint B_R *)

    F12Sol = { F12xSol, F12ySol, 0 } ;
    F21Sol = - F12Sol ;

    (* ∑ F for 2 *)
    rBC = ( rC - rB ) /.rule ;
    eqBR1 = (F12Sol+F2).rBC == 0 ;

    (* ∑ M_C for 2&3 *)
    rCB = ( rB - rC ) /.rule ;
```

```
  rCC2 = ( rC2 - rC ) /.rule ;
  rCC3 = ( rC3 - rC ) /.rule ;
  rCD = ( rD - rC ) /.rule ;
  eqBR2 = (Cross[rCB,F12Sol]+Cross[rCC2,F2]+Cross[rCC3,F3]+
Cross[rCD,F43]+M2+M3)[[3]] == 0 ;

  solF12 = Solve[{eqBR1,eqBR2},{F12xSol,F12ySol}] ;
  F12x = F12xSol/.solF12[[1]] ;
  F12y = F12ySol/.solF12[[1]] ;
  F12 = { F12x, F12y, 0 } ;
  F21 = - F12 ;
  Print["F12 = ", F12] ;

  (* ∑ M_A for 1 *)

  M1mSol = { 0, 0, MmSol };

  rAB = rB /.rule ;
  rAC1 = rC1 /.rule ;
  eqMA = (Cross[rAB,F21]+Cross[rAC1,F1]+M1+M1mSol)[[3]] == 0 ;
  solMm = Solve[eqMA,MmSol] ;
  Mm = MmSol/.solMm[[1]] ;
  M1m = { 0, 0, Mm } ;
  Print["Mm = ", M1m] ;

  (* Joint A_R *)

  F01Sol = { F01xSol, F01ySol, 0 } ;

  (* ∑ M_B for 1 *)

  rBA = -rB /.rule ;
  rBC1 = ( rC1 - rB ) /.rule ;
  eqAR1 = (Cross[rBA,F01Sol]+Cross[rBC1,F1]+M1+M1m)[[3]] == 0 ;

  (* ∑ F for 1&2 *)

  eqAR2 = (F01Sol+F1+F2).rBC == 0 ;
  solF01 = Solve[{eqAR1,eqAR2},{F01xSol,F01ySol}] ;
  F01x = F01xSol/.solF01[[1]] ;
  F01y = F01ySol/.solF01[[1]] ;
  F01 = { F01x, F01y, 0 } ;
  Print["F01 = ", F01] ;

F1 = {0.0186142, -0.0990915, 0}

M1 = {0, 0, 0}
```

```
F2 = {0.0265917, -0.0631033, 0}

M2 = {0, 0, -0.000028165}

F3 = {0.0492489, -0.033315, 0}

M3 = {0, 0, -0.00621962}

F4 = {-0.0369367, -0.0639614, 0}

M4 = {0, 0, 0.0000111583}

F5 = {-0.0553516, -0.410666, 0}

M5 = {0, 0, 0.00481155}

F05 = {268.127, 135.039, 0}

F45 = {-268.072, -134.628, 0}

rP = {-0.149492, 0.0476701, 0}

F34 = {-268.035, -134.564, 0}

F03 = {-256.71, -272.141, 0}

F23 = {-11.3747, 137.91, 0}

rQ = {0.121243, 0.07, 0}

F12 = {-11.4013, 137.974, 0}

Mm = {0, 0, 17.5332}

F01 = {-11.4199, 138.073, 0}
```

I.8   Packages for Kinematic Chains

I.8.1   Driver Link

Packages can be used to calculate the position, velocity, and acceleration of a driver link in rotational motion [Fig. I.8.1.(a)]. For the position analysis, the input data are the coordinates (x_A, y_A) of the start joint A with respect to the reference frame $xOyz$, the length of the link AB, and the angle ϕ with the horizontal axis. For the velocity and the acceleration analysis, the angular velocity $\omega = \dot{\phi}$ and the angular acceleration $\alpha = \ddot{\phi}$ are considered. The output data are the position, velocity, and acceleration of the end point B.

The position equations for the driver link are

$$x_B = x_A + AB \cos \phi,$$
$$y_B = y_A + AB \sin \phi, \tag{I.8.1}$$

where x_B and y_B are the coordinates of the point B.

The velocity equations for the driver link are

$$v_{Bx} = -AB\omega \sin \phi,$$
$$v_{By} = AB\omega \cos \phi, \tag{I.8.2}$$

where v_{Bx} and v_{By} are the velocity components of the point B on the x- and y-axes.

The acceleration equations for the driver link are

$$a_{Bx} = -AB\omega^2 \cos \phi - AB\alpha \sin \phi,$$
$$a_{By} = -AB\omega^2 \sin \phi + AB\alpha \cos \phi, \tag{I.8.3}$$

where a_{Bx} and a_{By} are the acceleration components of the point B on the x- and y-axes.

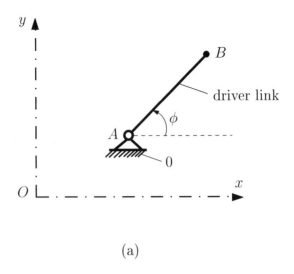

(a)

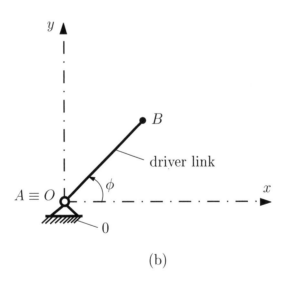

(b)

FIGURE I.8.1 *Driver link.*

In order to compute the position, velocity, and acceleration of the joint B, using *Mathematica*^{TM}, the necessary commands can be can collected in a function. The name of the function is **Driver**.

```
Driver[xA_,yA_,AB_,phi_,omega_,alpha_]:=
Block[{ xB, yB, vBx, vBy, aBx, aBy },
```

```
xB = xA + AB Cos[phi] ;
yB = yA + AB Sin[phi] ;
vBx = - AB omega Sin[phi] ;
vBy = AB omega Cos[phi] ;
aBx = - AB omega^2 Cos[phi] - AB alpha Sin[phi] ;
aBy = - AB omega^2 Sin[phi] + AB alpha Cos[phi] ;
Return[{ xB, yB, vBx, vBy, aBx, aBy } ] ;
] ;
```

The input data, the variable parts of the computation, and the output data are defined as parameters to this function.

All the variables local to **Driver[]** are declared in the **Block[]** statement to isolate them from any values they might have globally. The *Mathematica*^{TM} command **Block[{x, y, ...}, expr]** specifies that **expr** is to be evaluated with local values for the symbols **x, y, ...** . In our case, the local variables are **xB, yB, vBx, vBy, aBx, aBy**, and **expr** is the body of the function.

The *Mathematica*^{TM} command **Return[expr]** returns the value **expr** from a function. For the driver, **expr** represents the output data and it is a vector that contains the elements **xB, yB, vBx, vBy, aBx**, and **aBy**.

The mechanism that *Mathematica*^{TM} provides for keeping the variables used in a package different from those used in the main session is called *context*. As each symbol is read from the terminal or from a file, *Mathematica*^{TM} checks to see whether this symbol has already been used before. If it has been encountered before, the new instance is made to refer to that previously read symbol. If the symbol has not been encountered before, a new entry in the symbol table is created. Each symbol belongs to a certain context. Within one context the names of the symbols are unique, but the same name can occur in two different contexts. For the driver the proper context is

```
Driver::usage = "Driver[xA,yA,AB,phi,omega,alpha]
computes the driver link position,
velocity and acceleration vectors."
  Begin["Private`"]
  Driver[xA_,yA_,AB_,phi_,omega_,alpha_]:=
  Block[ { xB, yB, vBx, vBy, aBx, aBy },
  xB = xA + AB Cos[phi] ;
  yB = yA + AB Sin[phi] ;
  vBx = - AB omega Sin[phi] ;
  vBy = AB omega Cos[phi] ;
  aBx = - AB omega^2 Cos[phi] - AB alpha Sin[phi] ;
  aBy = - AB omega^2 Sin[phi] + AB alpha Cos[phi] ;
  Return[ { xB, yB, vBx, vBy, aBx, aBy } ] ;
  ]
  End[ ]
```

The local variables **xB, yB, vBx, vBy, aBx**, and **aBy** are now created in the context **Private`** which is not searched when one types a variable name later on.

The usage message defined for the symbol **Driver** is there to provide documentation for the function and to make sure that **Driver** is defined in the current context. If it had not been defined before entering the context **Private´**, it would not be found later on.

The *Mathematica*[TM] command **End[]** returns the present context and reverts to the previous one.

The functions that the package provides are put into a separate context which must be visible to be able to use the functions later on. This can be done using the pair of *Mathematica*[TM] commands **BeginPackage[]** and **EndPackage[]**. Thus, the following *Mathematica*[TM] package is introduced:

```
BeginPackage["Driver´"]
Driver::usage = "Driver[xA_,yA_,AB_,phi_,omega_,alpha_]
computes the driver position, velocity and acceleration vectors."
  Begin["´Private´"]
Driver[xA_,yA_,AB_,phi_,omega_,alpha_]:=
Block[ { xB, yB, vBx, vBy, aBx, aBy },
xB = xA + AB Cos[phi] ;
yB = yA + AB Sin[phi] ;
vBx = - AB omega Sin[phi] ;
vBy = AB omega Cos[phi] ;
aBx = - AB omega^2 Cos[phi] - AB alpha Sin[phi] ;
aBy = - AB omega^2 Sin[phi] + AB alpha Cos[phi] ;
Return[ { xB, yB, vBx, vBy, aBx, aBy } ] ;
]
End[ ]
EndPackage[ ]
```

The command **BeginPackage["Driver´"]** sets **Driver´** to be the current context, and the command **EndPackage[]** ends the package, prepending **Driver´** to the context search path.

Note the initial backquote in the context name inside the command **Begin["´Private´"]**. This establishes **´Private´** as a subcontext of the context **Driver´** (so its full name is **Driver´Private´**).

The name of the source file for the *Mathematica*[TM] package **Driver** is **Driver.m**, as shown in Program I.8.1.

Example

A driver link is shown in Figure I.8.1.(b). The input data are $AB = 0.20$ m, the angle between the driver link AB and the horizontal axis, $\phi = 30°$, and the angular velocity, $\omega = 5$ rad/s. Calculate the position, velocity, and acceleration components of the joint B. The Cartesian reference frame $xOyz$ is chosen with $A \equiv O$.

The *Mathematica*[TM] package **Driver** is loaded in the main *Mathematica*[TM] session using the command

```
<<Driver.m ;
```

To compute the numerical values of the position, velocity, and acceleration components for the joint B, the *Mathematica*[TM] function **Driver** is used.

Position Analysis

Since the joint A is the origin of the reference frame $xAyz$, the coordinates of the joint A are

$$x_A = y_A = 0.$$

The coordinates of the joint B are

$$x_B = x_A + AB \cos \phi = 0.173 \text{ m},$$
$$y_B = y_A + AB \sin \phi = 0.10 \text{ m}. \tag{I.8.4}$$

The coordinates x_B and y_B are the first and, respectively, the second component of the vector returned by the function **Driver**:

```
xB = Driver[xA,yA,AB,phi,omega,alpha][[1]] ;
yB = Driver[xA,yA,AB,phi,omega,alpha][[2]] ;
```

Velocity Analysis

The components v_{Bx} on the x-axis and v_{By} on the y-axis of the velocity for the joint B are

$$v_{Bx} = -AB\omega \sin \phi = -0.50 \text{ m/s},$$
$$v_{By} = AB\omega \cos \phi = 0.866 \text{ m/s}. \tag{I.8.5}$$

The components v_{Bx} and v_{By} are the third and, respectively, the fourth component of the vector returned by the function **Driver**:

```
vBx = Driver[xA,yA,AB,phi,omega,alpha][[3]] ;
vBy = Driver[xA,yA,AB,phi,omega,alpha][[4]] ;
```

Acceleration Analysis

The components a_{Bx} on the x-axis and a_{By} on the y-axis of the velocity for the joint B are

$$a_{Bx} = -AB\omega^2 \cos \phi - AB\alpha \sin \phi = -4.330 \text{ m/s}^2,$$
$$a_{By} = -AB\omega^2 \sin \phi + AB\alpha \cos \phi = -2.50 \text{ m/s}^2. \tag{I.8.6}$$

The components a_{Bx} and a_{By} are the fifth and, respectively, the sixth component of the vector returned by the function **Driver**:

```
aBx = Driver[xA,yA,AB,phi,omega,alpha][[5]] ;
aBy = Driver[xA,yA,AB,phi,omega,alpha][[6]] ;
```

The *Mathematica*^{TM} program and the numerical results are shown in Program I.8.2.

I.8.2   Position Analysis

RRR Dyad

The RRR dyad is shown in Figure I.8.2(a). The input data are the coordinates of the joint $M(x_M, y_M)$, the coordinates of the joint $N(x_N, y_N)$, and the lengths of the segments MP and NP. The output data are the coordinates of the joint $P(x_P, y_P)$.

The position equations for the RRR dyad are

$$(x_M - x_P)^2 + (y_M - y_P)^2 = MP^2,$$

$$(x_N - x_P)^2 + (y_N - y_P)^2 = NP^2, \tag{I.8.7}$$

where the unknowns are the coordinates x_P and y_P of the joint P. There are two solutions for the position of the joint P: (x_{P1}, y_{P1}) and (x_{P2}, y_{P2}).

The *Mathematica*^{TM} function for the positions of x_{P_1}, y_{P_1}, x_{P_2}, y_{P_2} is

```
PosRRR::usage = "PosRRR[xM,yM,xN,yN,MP,NP]
Computes the position vectors for RRR dyad"
  Begin["`Private`"]
  PosRRR[xM_,yM_,xN_,yN_,MP_,NP_]:=
  Block[
  {xPSol,yPSol,xP1,yP1,xP2,yP2,eqRRR1,eqRRR2,solRRR},
  eqRRR1 = (xM-xPSol)^2 + (yM-yPSol)^2 == MP^2 ;
  eqRRR2 = (xN-xPSol)^2 + (yN-yPSol)^2 == NP^2 ;
  solRRR = Solve[{eqRRR1,eqRRR2},{xPSol,yPSol}];
  xP1 = xPSol/.solRRR[[1]] ;
  yP1 = yPSol/.solRRR[[1]] ;
  xP2 = xPSol/.solRRR[[2]] ;
  yP2 = yPSol/.solRRR[[2]] ;
  Return[xP1, yP1, xP2, yP2] ;
  ]
  End[ ]
```

RRT Dyad

The RRT dyad is shown in Figure I.8.2(b). The input data are the coordinates of the joint $M(x_M, y_M)$, the coordinates of the point $N(x_N, y_N)$ on the sliding direction, the length of the segment MP, and the value of the angle θ. The output data are the coordinates of the joint $P(x_P, y_P)$.

The position equations for the RRT dyad are

$$(x_M - x_P)^2 + (y_M - y_P)^2 = MP^2,$$

$$\tan \theta = \frac{y_P - y_N}{x_P - x_N}, \tag{I.8.8}$$

where the unknowns are the coordinates x_P and y_P of the joint P. There are two solutions for the position of the joint P, those are (x_{P1}, y_{P1}) and (x_{P2}, y_{P2}).

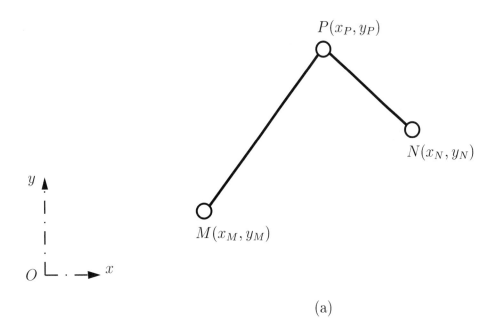

(a)

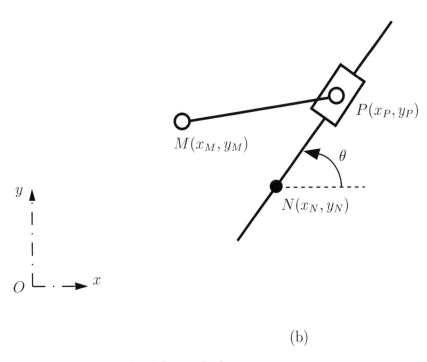

(b)

FIGURE I.8.2 *(a) RRR dyad and (b) RRT dyad.*

If the value of the angle θ is 90° or 180°, then $x_P = x_N$ and the following equation is used to find the coordinate y_P of the point P:

$$(x_M - x_N)^2 + (y_M - y_P)^2 = MP^2.$$ (I.8.9)

The *Mathematica*[TM] function for the position analysis is

```
PosRRT::usage = "PosRRT[xM,yM,xN,yN,MP,theta]
Computes the position vectors for RRT dyad"
  Begin["`Private`"]
  PosRRT[xM_,yM_,xN_,yN_,MP_,theta_]:=
  Block[
  {xPSol, yPSol, xP1, yP1,xP2, yP2, eqRRT, solRRT, eqRRT1, eqRRT2 },
  If[ (theta==Pi/2)||(theta==3*Pi/2),
  xP1 = xP2 = xN ;
  eqRRT = (xM-xN)^2 + (yM-yPSol)^2 == MP^2 ;
  solRRT = Solve[ eqRRT, yPSol ] ;
  yP1 = yPSol/.solRRT[[1]] ;
  yP2 = yPSol/.solRRT[[2]] ,
  eqRRT1 = (xM-xPSol)^2 + (yM-yPSol)^2 == MP^2 ;
  eqRRT2 = Tan[theta] == (yPSol-yN)/(xPSol-xN) ;
  solRRT = Solve[{eqRRT1,eqRRT2},{xPSol,yPSol}] ;
  xP1 = xPSol/.solRRT[[1]] ;
  yP1 = yPSol/.solRRT[[1]] ;
  xP2 = xPSol/.solRRT[[2]] ;
  yP2 = yPSol/.solRRT[[2]] ;
  ] ;
  Return[ { xP1, yP1, xP2, yP2 } ] ;
  ]
  End[ ]
```

The functions **PosRRR** and **PosRRT** are included in the *Mathematica*[TM] package **Position**. The name of the source file for the package is **Position.m** and is given in Program I.8.3.

R-RTR-RRT Mechanism

The planar R-RTR-RRT mechanism considered is shown in Figure I.8.3(a). Given the input data $AB = 0.20$ m, $AD = 0.40$ m, $CD = 0.70$ m, $CE = 0.30$ m, the angle of the driver link AB with the horizontal axis, $\phi = 45°$, and the angular velocity, $\omega = 5$ rad/s, calculate the positions of the joints. The distance from the slider 5 to the horizontal axis Ox is $y_E = 0.35$ m. The Cartesian reference frame $xOyz$ is chosen with $A \equiv O$.

The *Mathematica*[TM] packages **Driver** and **Position** are loaded in the main program using the commands

```
<<Driver.m ;
<<Position.m ;
```

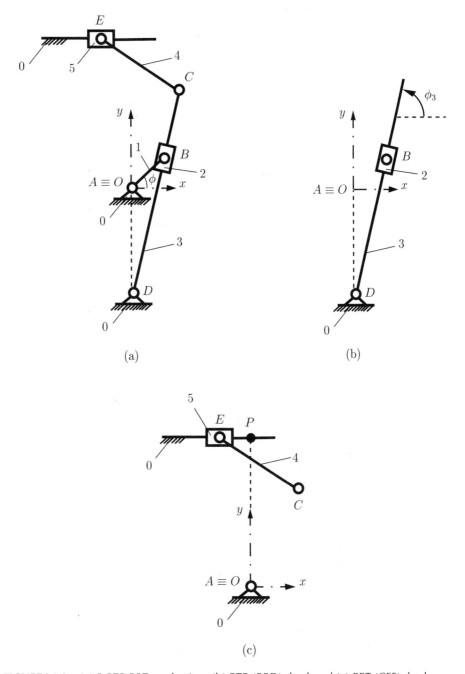

FIGURE I.8.3 (a) R-RTR-RRT mechanism, (b) RTR (BBD) dyad, and (c) RRT (CEE) dyad.

Position of the joint A

Since the joint A is the origin of the reference frame $xAyz$, the coordinates of the joint A are

$$x_A = y_A = 0.$$

Position of the joint B

The coordinates of the joint B are

$$x_B = x_A + AB \cos \phi = 0.141 \text{ m},$$
$$y_B = y_A + AB \sin \phi = 0.141 \text{ m}. \tag{I.8.10}$$

The numerical values for the coordinates of the joint B are obtained using the *Mathematica*^{TM} function **Driver**:

```
xB = Driver[xA,yA,AB,phi,omega,alpha][[1]] ;
yB = Driver[xA,xB,AB,phi,omega,alpha][[2]] ;
```

The RTR (BBD) dyad is represented in Figure I.8.3(b).

Position of the joint D

The coordinates of the joint D are

$$x_D = 0, \quad y_D = -AD = -0.400 \text{ m}.$$

The angle ϕ_3 is

$$\phi_3 = \arctan \frac{y_B - y_D}{x_B - x_D} = 75.36°.$$

The next dyad RRT (CEE) is represented in Figure I.8.3(c).

Position of the joint C

The coordinates of the joint C are

$$x_C = x_D + CD \cos \phi_3 = 0.176 \text{ m},$$
$$y_C = y_D + CD \sin \phi_3 = 0.277 \text{ m}.$$

Position of the joint E

In this particular case, the coordinate y_E of the joint E is constant:

$$y_E = 0.350\text{m}.$$

The coordinate x_E of the joint E is calculated using the equation

$$(x_C - x_E)^2 + (y_C - y_E)^2 = CE^2. \tag{I.8.11}$$

There are two solutions for x_E:

$$x_{E1} = -0.114 \text{ m, and } x_{E2} = 0.467 \text{ m}.$$

The correct solution for x_E is selected using the condition $x_E < x_C$:

$$x_E = -0.114 \text{ m}.$$

The numerical solution for x_E is obtained using the *Mathematica*[TM] function **PosRRT**

```
xE1 = PosRRT[xC,yC,0,yE,CE,phi5][[1]] ;
xE2 = PosRRT[xC,yC,0,yE,CE,phi5][[3]] ;
(* Choose the correct solution *)
If[ (xE1<xC), xE=xE1, xE=xE2 ] ;
```

The input data are the coordinates of the joint $C(x_C, y_C)$, the coordinates of the point $P(0, y_E)$ located on the sliding direction, the length of the link CE, and the angle between the sliding direction and the horizontal axis Ox, **phi5** $= 180°$.

The output data are the first and the third element of the vector returned by the function **PosRRT**, which are the x-coordinates of the joint E. The second and the fourth element are constant and equal to the y-coordinate of the joint E.

The numerical values are printed using the *Mathematica*[TM] commands:

```
Print["rB = ",{xB,yB,0}," [m]" ];
Print["rC = ",{xC,yC,0}," [m]" ];
Print["rE = ",{xE,yE,0}," [m]" ];
```

The *Mathematica*[TM] program and the numerical results are shown in Program I.8.4.

R-RRR-RRT Mechanism

The planar R-RRR-RRT mechanism considered is shown in Figure I.8.4(a). Given the input data $AB = 0.15$ m, $BC = 0.40$ m, $CD = 0.37$ m, $CE = 0.23$ m, $EF = CE$, $La = 0.30$ m, $Lb = 0.45$ m, $Lc = CD$, and the angle of the driver link AB with the horizontal axis, $\phi = 45°$, calculate the positions of the joints. The distance from the slider 5 to the horizontal axis Ox is $y_E = 0.35$ m. The Cartesian reference frame $xOyz$ is chosen, with $A \equiv O$.

The *Mathematica*[TM] packages **Driver** and **Position** are loaded in the main program using the commands:

```
<<Driver.m ;
<<Position.m ;
```

Position of the joint A
Since the joint A is the origin of the reference frame $xAyz$, the coordinates of the joint A are

$$x_A = y_A = 0.$$

Position of the joint B
The coordinates of the joint B are

$$x_B = x_A + AB \cos \phi = 0.106 \text{ m},$$

$$y_B = y_A + AB \sin \phi = 0.106 \text{ m}. \qquad \text{(I.8.12)}$$

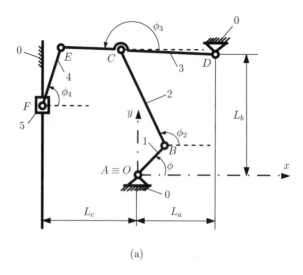

(a)

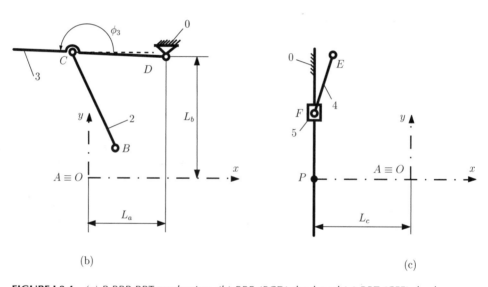

(b)   (c)

FIGURE I.8.4 *(a) R-RRR-RRT mechanism, (b) RRR (BCD) dyad, and (c) RRT (CEE) dyad.*

The numerical values for the coordinates of the joint B are obtained using the *Mathematica*^{TM} function **Driver**:

```
xB = Driver[xA,yA,AB,phi,omega,alpha][[1]] ;
yB = Driver[xA,xB,AB,phi,omega,alpha][[2]] ;
```

The RRR (*BCD*) dyad is represented in Figure I.8.4(b).

Position of the joint D

The coordinates of the joint D are

$$x_D = L_a = 0.300 \text{ m}, \ y_D = L_b = 0.450 \text{ m}.$$

Position of the joint C

The coordinates x_C and y_C of the joint C are calculated using the equations

$$(x_B - x_C)^2 + (y_B - y_C)^2 = BC^2,$$
$$(x_D - x_C)^2 + (y_D - y_C)^2 = CD^2. \tag{I.8.13}$$

There are two solutions for the coordinates of the joint C:

$$x_{C1} = -0.069 \text{ m}, \ y_{C1} = 0.465 \text{ m},$$
$$x_{C2} = 0.504 \text{ m}, \ y_{C2} = 0.141 \text{ m}.$$

The correct solution is selected using the condition $y_C > y_B$:

$$x_{C1} = -0.069 \text{ m}, \ y_{C1} = 0.465 \text{ m}.$$

The numerical solutions for the coordinates of the joint C using the *Mathematica*™ commands are

```
xC1 = PosRRR[xB,yB,xD,yD,BC,CD][[1]] ;
yC1 = PosRRR[xB,yB,xD,yD,BC,CD][[2]] ;
xC2 = PosRRR[xB,yB,xD,yD,BC,CD][[3]] ;
yC2 = PosRRR[xB,yB,xD,yD,BC,CD][[4]] ;
(* Choose the correct solution *)
If[ (yC1>yB), xC=xC1;yC=yC1, xC=xC2;yC=yC2 ] ;
```

The input data are the coordinates of the joint $B(x_B, y_B)$, the coordinates of the joint $D(x_D, y_D)$, and the lengths of the links BC and CD. The output data are the four elements of the vector returned by the function **PosRRR**, which are the coordinates of the joint C.

The angle ϕ_3 between the link 3 and the horizontal axis Ox is

$$\phi_3 = \arctan \frac{y_C - y_D}{x_C - x_D} + \pi = 3.099 \text{ rad}.$$

Position of the joint E

The coordinates of the joint E are

$$x_E = x_C + CE \cos \phi_3 = -0.299 \text{ m},$$
$$y_E = y_C + CE \sin \phi_3 = 0.474 \text{ m}.$$

The next dyad RRT (CEE) is represented in Figure I.8.4(c).

Position of the joint F

In this particular case, the x-coordinate x_F of the joint F is constant:

$$x_F = -L_c = -0.370 \text{ m}.$$

The y-coordinate y_F of the joint F is calculated using the equation

$$(x_E - x_F)^2 + (y_E - y_F)^2 = EF^2. \qquad (I.8.14)$$

There are two solutions for y_F:

$$y_{F1} = 0.256 \text{ m, and } y_{F2} = 0.693 \text{ m}.$$

The correct solution for y_F is chosen using the condition $y_F < y_E$:

$$y_F = 0.256 \text{ m}.$$

The numerical solutions for y_F using the *Mathematica*^{TM} commands are

```
yF1 = PosRRT[xE,yE,-Lc,0,EF,phi5][[2]] ;
yF2 = PosRRT[xE,yE,-Lc,0,EF,phi5][[4]] ;
(* Choose the correct solution *)
If[ (yF1<yE), yF=yF1, yF=yF2 ] ;
```

The input data are the coordinates of the joint $E(x_E, y_E)$, the coordinates of the point $P(-L_c, 0)$ on the sliding direction, the length of the link EF, and the angle between the sliding direction and the horizontal axis Ox, **phi5**=90°. The output data are the second and the fourth element of the vector returned by the function **PosRRT**, which are the y-coordinates of the joint F. The first and third elements are constant and equal to the x-coordinate of the joint F. The numerical values are printed using the *Mathematica*^{TM} commands:

```
Print["rB = ",{xB,yB,0}," [m]" ];
Print["rC = ",{xC,yC,0}," [m]" ];
Print["rE = ",{xE,yE,0}," [m]" ];
Print["rF = ",{xF,yF,0}," [m]" ];
```

The angle ϕ_2 between the link 2 and the horizontal axis Ox is

$$\phi_2 = \arctan \frac{y_C - y_B}{x_C - x_B} + \pi = 2.025 \text{ rad}.$$

The angle ϕ_4 between the link 4 and the horizontal axis Ox is

$$\phi_4 = \arctan \frac{y_E - y_F}{x_E - x_F} = 1.259 \text{ rad}.$$

The *Mathematica*^{TM} program and the numerical results are shown in Program I.8.5.

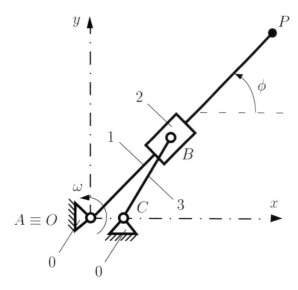

FIGURE I.8.5 *R-RRT mechanism.*

R-RRT Mechanism

The planar R-RRT mechanism considered is shown in Figure I.8.5. Given the input data $AC = 0.10$ m, $BC = 0.30$ m, $AP = 0.50$ m, and the angle of the driver link AB with the horizontal axis, $\phi = 45°$, calculate the positions of the joints. The Cartesian reference frame $xOyz$ is chosen with $A \equiv O$.

The *Mathematica*^{TM} packages **Driver** and **Position** are loaded in the main program using the commands

```
<<Driver.m ;
<<Position.m ;
```

Position of the joint A

Since the joint A is the origin of the reference frame $xAyz$, the coordinates of the joint A are

$$x_A = y_A = 0.$$

Position of the joint B

In order to calculate the position of the point B, the position of the point P located on the driver link AB is calculated with

$$x_P = x_A + AP \cos \phi = 0.353 \text{ m},$$

$$y_P = y_A + AP \sin \phi = 0.353 \text{ m}. \tag{I.8.15}$$

The numerical values for the coordinates of the point P are obtained using the *Mathematica*^{TM} function `Driver`:

```
xP = Driver[xA,yA,AP,phi,omega,alpha][[1]] ;
yP = Driver[xA,xB,AP,phi,omega,alpha][[2]] ;
```

The coordinates of the point B are calculated using the equations

$$(x_C - x_B)^2 + (y_C - y_B)^2 = BC^2,$$

$$\tan \theta = \frac{y_B - y_P}{x_B - x_P}.$$

There are two solutions for the coordinates of the point B:

$$x_{B1} = -0.156 \text{ m}, \ y_{B1} = -0.156 \text{ m},$$

$$x_{B2} = 0.256 \text{ m}, \ y_{B2} = 0.256 \text{ m}.$$

The correct solution is selected using the condition $y_B > y_C = 0$:

$$y_B = y_{B_2} = 0.256 \text{ m}.$$

The numerical solutions for the coordinates of the point B are obtained using the *Mathematica*^{TM} commands

```
xB1 = PosRRT[xC,yC,xP,yP,BC,phi][[1]] ;
yB1 = PosRRT[xC,yC,xP,yP,BC,phi][[2]] ;
xB2 = PosRRT[xC,yC,xP,yP,BC,phi][[3]] ;
yB2 = PosRRT[xC,yC,xP,yP,BC,phi][[4]] ;
(* Choose the correct solution *)
If[ (yB1>yC), xB=xB1;yB=yB1, xB=xB2;yB=yB2 ] ;
```

The numerical values are printed using the *Mathematica*^{TM} command

```
Print["rB = ",{xB,yB,0}," [m]" ];
```

The *Mathematica*^{TM} program and the numerical results are shown in Program I.8.6.

I.8.3 Velocity and Acceleration Analysis

RRR Dyad

The input data are the coordinates $x_M, y_M, x_N, y_N, x_P, y_P$ of the joints M, N, and P, the velocities $\dot{x}_M, \dot{y}_M, \dot{x}_N, \dot{y}_N$, and the accelerations $\ddot{x}_M, \ddot{y}_M, \ddot{x}_N, \ddot{y}_N$ of the joints M and N. The output data are the velocities $\dot{x}_P, \dot{y}_P$ and acceleration components $\ddot{x}_P$ and $\ddot{y}_P$ of the joint P.

The velocity equations for the RRR dyad are obtained taking the derivative of the position equations

$$(x_M - x_P)(\dot{x}_M - \dot{x}_P) + (y_M - y_P)(\dot{y}_M - \dot{y}_P) = 0,$$
$$(x_N - x_P)(\dot{x}_N - \dot{x}_P) + (y_N - y_P)(\dot{y}_N - \dot{y}_P) = 0, \tag{I.8.16}$$

where the unknowns are the velocity components $\dot{x}_P$ and $\dot{y}_P$ of the joint P.

The acceleration equations for the RRR dyad are obtained taking the derivative of the velocity equations

$$(x_M - x_P)(\ddot{x}_M - \ddot{x}_P) + (\dot{x}_M - \dot{x}_P)^2 + (y_M - y_P)(\ddot{y}_M - \ddot{y}_P) + (\dot{y}_M - \dot{y}_P)^2 = 0,$$
$$(x_N - x_P)(\ddot{x}_N - \ddot{x}_P) + (\dot{x}_N - \dot{x}_P)^2 + (y_N - y_P)(\ddot{y}_N - \ddot{y}_P) + (\dot{y}_N - \dot{y}_P)^2 = 0, \tag{I.8.17}$$

where the unknowns are the acceleration components $\ddot{x}_P$ and $\ddot{y}_P$ of the joint P.

The *Mathematica*^{TM} function for the velocity and acceleration analysis is

```
VelAccRRR::usage = "VelAccRRR[xM,yM,xN,yN,xP,yP,
vMx,vMy,vNx,vNy,aMx,aMy,aNx,aNy] computes the velocity
and acceleration vectors for RRR dyad"
  Begin["`Private`"]
  VelAccRRR[xM_,yM_,xN_,yN_,xP_,yP_,vMx_,vMy_,vNx_,vNy_,
aMx_,aMy_, aNx_,aNy_]:=
  Block[
  { vPxSol, vPySol, aPxSol, aPySol, vPx, vPy, aPx, aPy,
eqRRR1v, eqRRR2v, solRRRv, eqRRR1a, eqRRR2a, solRRRa },
  (* Velocities *)
  eqRRR1v=(xM-xP) (vMx-vPxSol)+(yM-yP) (vMy-vPySol)==0;
  eqRRR2v=(xN-xP) (vNx-vPxSol)+(yN-yP) (vNy-vPySol)==0;
  solRRRv=Solve[{eqRRR1v, eqRRR2v},{vPxSol, vPySol}];
  vPx = vPxSol/.solRRRv[[1]] ;
  vPy = vPySol/.solRRRv[[1]] ;
  (* Accelerations *)
  eqRRR1a = (xM-xP) (aMx-aPxSol) + (vMx-vPx)^2 +
(yM-yP) (aMy-aPySol) + (vMy-vPy)^2 == 0 ;
  eqRRR2a = (xN-xP) (aNx-aPxSol) + (vNx-vPx)^2 +
(yN-yP) (aNy-aPySol) + (vNy-vPy)^2 == 0 ;
  solRRRa=Solve[{eqRRR1a, eqRRR2a},{aPxSol, aPySol}];
  aPx = aPxSol/.solRRRa[[1]] ;
  aPy = aPySol/.solRRRa[[1]] ;
  Return[ { vPx, vPy, aPx, aPy } ] ; ]
  End[ ]
```

RRT Dyad

The input data are the coordinates $x_M, y_M, x_N, y_N, x_P, y_P$ of the joints M, N and P, the velocities $\dot{x}_M, \dot{y}_M, \dot{x}_N, \dot{y}_N$, the accelerations $\ddot{x}_M, \ddot{y}_M, \ddot{x}_N, \ddot{y}_N$ of the joints M and N, the angle θ, the angular velocity and acceleration $\dot{\theta}$ and $\ddot{\theta}$. The output data are the velocities $\dot{x}_P, \dot{y}_P$, and accelerations $\ddot{x}_P, \ddot{y}_P$ of the joint P.

The velocity equations for the RRT dyad are obtained taking the derivative of the position equations

$$(x_M - x_P)(\dot{x}_M - \dot{x}_P) + (y_M - y_P)(\dot{y}_M - \dot{y}_P) = 0,$$

$$(\dot{x}_P - \dot{x}_N)\sin\theta + \dot{\theta}(x_P - x_N)\cos\theta - (\dot{y}_P - \dot{y}_N)\cos\theta + \dot{\theta}(y_P - y_N)\sin\theta = 0, \quad \text{(I.8.18)}$$

where the unknowns are the velocity components $\dot{x}_P$ and $\dot{y}_P$ of the joint P.

The acceleration equations for the RRT dyad are obtained taking the derivative of the velocity equations

$$(x_M - x_P)(\ddot{x}_M - \ddot{x}_P) + (\dot{x}_M - \dot{x}_P)^2 + (y_M - y_P)(\ddot{y}_M - \ddot{y}_P) + (\dot{y}_M - \dot{y}_P)^2 = 0,$$

$$(\ddot{x}_P - \ddot{x}_N)\sin\theta - (\ddot{y}_P - \ddot{y}_N)\cos\theta + [2(\dot{x}_P - \dot{x}_N)\cos\theta - \dot{\theta}(x_P - x_N)\sin\theta$$

$$+ 2(\dot{y}_P - \dot{y}_N)\sin\theta + \dot{\theta}(y_P - y_N)\cos\theta]\dot{\theta} + [(x_P - x_N)\cos\theta + (y_P - y_N)\sin\theta]\ddot{\theta} = 0,$$

$$\text{(I.8.19)}$$

where the unknowns are the acceleration components $\ddot{x}_P$ and $\ddot{y}_P$ of the joint P.

The *Mathematica*^{TM} function for the velocity and acceleration analysis is

```
VelAccRRT::usage = "VelAccRRT[xM,yM,xN,yN,xP,yP,
vMx,vMy,vNx,vNy,aMx,aMy,aNx,aNy,theta,omega,alpha]
computes the velocity and acceleration vectors for
RRT dyad"
   Begin["`Private`"]
   VelAccRRT[xM_,yM_,xN_,yN_,xP_,yP_,vMx_,vMy_,vNx_,vNy_,
aMx_,aMy_,aNx_,aNy_,theta_,omega_,alpha_]:=
   Block[
   { vPxSol, vPySol, aPxSol, aPySol, vPx, vPy, aPx, aPy, eqRRTv, eqRRTa,
eqRRT1v, eqRRT2v, solRRTv , eqRRR1a, eqRRR2a, solRRRa },
   (* Velocity *)
   eqRRT1v=(xM-xP) (vMx-vPxSol)+(yM-yP) (vMy-vPySol)==0;
   eqRRT2v = Sin[theta] (vPxSol-vNx) +
Cos[theta] omega (xP-xN) - Cos[theta] (vPySol-vNy) +
Sin[theta] omega (yP-yN) == 0 ;
   solRRTv=Solve[{eqRRT1v, eqRRT2v},{vPxSol, vPySol}] ;
   vPx = vPxSol/.solRRTv[[1]] ;
   vPy = vPySol/.solRRTv[[1]] ;
   (* Acceleration *)
   eqRRT1a = (xM-xP) (aMx-aPxSol) + (vMx-vPx)^2 +
(yM-yP) (aMy-aPySol) + (vMy-vPy)^2 == 0 ;
   eqRRT2a = Sin[theta] (aPxSol-aNx) -
Cos[theta] (aPySol-aNy) + ( 2 Cos[theta] (vPx-vNx) -
Sin[theta] dtheta (xP-xN) + 2 Sin[theta] (vPy-vNy) +
Cos[theta] dtheta (yP-yN)) dtheta +
(Cos[theta] (xP-xN) + Sin[theta] (yP-yN)) ddtheta == 0;
```

```
solRRTa=Solve[{eqRRT1a, eqRRT2a},{aPxSol, aPySol}];
aPx = aPxSol/.solRRTa[[1]] ;
aPy = aPySol/.solRRTa[[1]] ;
Return[ { vPx, vPy, aPx, aPy } ] ; ]
End[ ]
```

Angular Velocities and Accelerations

A *Mathematica*^{TM} function is used to compute the angular velocity and acceleration of a link. The input data are the coordinates x_M, y_M, x_N, y_N, the velocities $\dot{x}_M, \dot{y}_M, \dot{x}_N, \dot{y}_N$, the accelerations $\ddot{x}_M, \ddot{y}_M, \ddot{x}_N, \ddot{y}_N$ of two points M and N located on the link direction, and the angle θ between the link direction and the horizontal axis. The output data are the angular velocity $\omega = \dot{\theta}$ and the angular acceleration $\alpha = \ddot{\theta}$ of the link.

The slope of the line MN is

$$\tan \theta = \frac{y_M - y_N}{x_M - x_N}. \tag{I.8.20}$$

The derivative with respect to time of Eq. (I.8.20) is

$$[(y_M - y_N)\sin\theta + (x_M - x_N)\cos\theta]\omega = (v_{My} - v_{Ny})\cos\theta - (v_{Mx} - v_{Nx})\sin\theta. \tag{I.8.21}$$

The angular velocity ω is calculated from Eq. (I.8.21). The derivative with respect to time of Eq. (I.8.21) is

$$[(x_M - x_N)\cos\theta + (y_M - y_N)\sin\theta]\alpha = (a_{My} - a_{Ny})\cos\theta - (a_{Mx} - a_{Nx})\sin\theta$$

$$- [(y_M - y_N)\omega\cos\theta + 2(v_{My} - v_{Ny})\sin\theta - (x_M - x_N)\omega\sin\theta + 2(v_{Mx} - v_{Nx})\cos\theta]\omega. \tag{I.8.22}$$

Solving Eq. (I.8.22), the angular acceleration α is obtained.

The *Mathematica*^{TM} function for the angular velocity and acceleration analysis is

```
AngVelAcc::usage = "AngVelAcc[xM,yM,xN,yN,vMx,vMy,
vNx,vNy,aMx,aMy,aNx,aNy,theta] computes the angular velocity and
acceleration of a link."
Begin["`Private`"]
AngVelAcc[xM_,yM_,xN_,yN_,vMx_,vMy_,vNx_,vNy_,aMx_,aMy_,
aNx_,aNy_,theta_]:=
Block[
{ dtheta, ddtheta },
dtheta = ( Cos[theta] (vMy-vNy) - Sin[theta] (vMx-vNx))/
( Sin[theta] (yM-yN) + Cos[theta] (xM-xN) ) ;
ddtheta=( Cos[theta] (aMy-aNy) - Sin[theta] (aMx-aNx) -
(Cos[theta] dtheta (yM-yN) + 2 Sin[theta] (vMy-vNy) -
Sin[theta] dtheta (xM-xN) + 2 Cos[theta] (vMx-vNx) ) dtheta ) /
( Cos[theta] (xM-xN) + Sin[theta] (yM-yN) ) ;
Return[ { dtheta, ddtheta } ] ; ]
End[ ]
```

Absolute Velocities and Accelerations

A function is used to compute the velocity and acceleration of the point N, knowing the velocity and acceleration of the point M, both points N and M are located on a rigid link. The input data are the coordinates x_M, y_M, x_N, and y_N of the points M and N, the velocity and acceleration components $\dot{x}_M$, $\dot{y}_M$, $\ddot{x}_M$, $\ddot{y}_M$ of the point M, and the angular velocity and acceleration θ and α of the link. The output data are the velocity and acceleration components $\dot{x}_N$, $\dot{y}_N$, $\ddot{x}_N$, $\ddot{y}_N$ of the point N.

The following vectorial equation between the velocities $\mathbf{v}_N$ and $\mathbf{v}_M$ of the points N and M exists as

$$\mathbf{v}_N = \mathbf{v}_M + \boldsymbol{\omega} \times \mathbf{r}_{MN}, \tag{I.8.23}$$

where $\mathbf{v}_N = \dot{x}_N\mathbf{\imath} + \dot{y}_N\mathbf{\jmath}$, $\mathbf{v}_M = \dot{x}_M\mathbf{\imath} + \dot{y}_M\mathbf{\jmath}$, $\boldsymbol{\omega} = \omega\mathbf{k}$, and $\mathbf{r}_{MN} = (x_N - x_M)\mathbf{\imath} + (y_N - y_M)\mathbf{\jmath}$.

Equation (I.8.23) is projected on the $\mathbf{\imath}$ and $\mathbf{\jmath}$ directions to find the velocity components of the point N:

$$\dot{x}_N = \dot{x}_M - \omega(y_N - y_M),$$

$$\dot{y}_N = \dot{y}_M + \omega(x_N - x_M). \tag{I.8.24}$$

The following vectorial equation between the accelerations $\mathbf{a}_N$ and $\mathbf{a}_M$ of the points N and M can be written as

$$\mathbf{a}_N = \mathbf{a}_M + \boldsymbol{\alpha} \times \mathbf{r}_{MN} - \omega^2\mathbf{r}_{MN}, \tag{I.8.25}$$

where $\mathbf{a}_N = \ddot{x}_N\mathbf{\imath} + \ddot{y}_N\mathbf{\jmath}$, $\mathbf{a}_M = \ddot{x}_M\mathbf{\imath} + \ddot{y}_M\mathbf{\jmath}$, and $\boldsymbol{\alpha} = \alpha\mathbf{k}$.

The acceleration components of the point N are obtained from Eq. (I.8.25):

$$\ddot{x}_N = \ddot{x}_M - \alpha(y_N - y_M) - \omega^2(x_N - x_M),$$

$$\ddot{y}_N = \ddot{y}_M + \alpha(x_N - x_M) - \omega^2(y_N - y_M). \tag{I.8.26}$$

The *Mathematica*^{TM} function for the absolute velocity and acceleration analysis is

```
AbsVelAcc::usage = "AbsVelAcc[xM,yM,xN,yN,vMx,vMy,
aMx,aMy, dtheta,ddtheta] computes the absolute velocity and
acceleration vectors."
  Begin["`Private`"]
  AbsVelAcc[xM_,yM_,xN_,yN_,vMx_,vMy_,aMx_,aMy_,dtheta_,ddtheta_]:=
  Block[
  { vNx, vNy, aNx, aNy },
  vNx = vMx - dtheta (yN-yM) ;
  vNy = vMy + dtheta (xN-xM) ;
  aNx = aMx - ddtheta (yN-yM) - dtheta^2 (xN-xM) ;
  aNy = aMy + ddtheta (xN-xM) - dtheta^2 (yN-yM) ;
  Return[ { vNx, vNy, aNx, aNy } ] ; ]
  End[ ]
```

The functions **VelAccRRR**, **VelAccRRT**, **AngVelAcc**, and **AbsVelAcc** are included in the *Mathematica*^{TM} package **VelAcc**. The name of the source file for the package is **VelAcc.m** and is given in Program I.8.7.

R-RRR-RRT Mechanism

The position analysis of the planar R-RRR-RRT mechanism considered [see Fig I.8.4(a)] is presented in Subsection I.8.2. Given the angular velocity $\omega = \dot{\phi} = 3.14$ rad/s, calculate the velocities and the accelerations of the joints and the angular velocities and the accelerations of the links.

The *Mathematica*^{TM} packages **Driver**, **PosVec**, and **VelAcc** are loaded in the main program using the commands

```
<<Driver.m ;
<<PosVec.m ;
<<VelAcc.m ;
```

The angular velocity of the driver link is zero:

$$\alpha = \ddot{\phi} = 0.$$

Velocity and acceleration of the joint A
Since the joint A is the origin of the reference frame $xAyz$, the velocity and acceleration of the joint A are

$$\mathbf{v}_A = \mathbf{a}_A = \mathbf{0}.$$

Velocity and acceleration of the joint B
The velocity and acceleration components of the joint B are

$$v_{Bx} = -AB\omega \sin\phi = -1.110 \text{ m/s},$$
$$v_{By} = AB\omega \cos\phi = 1.110 \text{ m/s},$$
$$a_{Bx} = -AB\omega^2 \cos\phi - AB\alpha \sin\phi = -11.631 \text{ m/s}^2,$$
$$a_{By} = -AB\omega^2 \sin\phi + AB\alpha \cos\phi = -11.631 \text{ m/s}^2.$$

The numerical values for the velocity and acceleration components of the joint B are obtained using the *Mathematica*^{TM} function **Driver**:

```
vBx = Driver[xA,yA,AB,phi,omega,alpha][[3]] ;
vBy = Driver[xA,yA,AB,phi,omega,alpha][[4]] ;
aBx = Driver[xA,yA,AB,phi,omega,alpha][[5]] ;
aBy = Driver[xA,yA,AB,phi,omega,alpha][[6]] ;
```

Velocity and acceleration of the joint D

The velocity and acceleration of the joint D are

$$\mathbf{v}_D = \mathbf{a}_D = \mathbf{0}.$$

Velocity and acceleration of the joint C

To calculate the velocity components v_{Cx} and v_{Cy} of the joint C, the following equations are used:

$$(x_B - x_C)(v_{Bx} - v_{Cx}) + (y_B - y_C)(v_{By} - v_{Cy}) = 0,$$
$$(x_D - x_C)(v_{Dx} - v_{Cx}) + (y_D - y_C)(v_{Dy} - v_{Cy}) = 0. \tag{I.8.27}$$

The acceleration components a_{Cx} and a_{Cy} of the joint C are calculated from the equations

$$(x_B - x_C)(a_{Bx} - a_{Cx}) + (v_{Bx} - v_{Cx})^2 + (y_B - y_C)(a_{By} - a_{Cy}) + (v_{By} - v_{Cy})^2 = 0,$$
$$(x_D - x_C)(a_{Dx} - a_{Cx}) + (v_{Dx} - v_{Cx})^2 + (y_D - y_C)(a_{Dy} - a_{Cy}) + (v_{Dy} - v_{Cy})^2 = 0.$$
$$\tag{I.8.28}$$

The numerical solutions for the velocity and acceleration components of the joint C using the *Mathematica*^{TM} function **VelAccRRR** are

```
vCx=VelAccRRR[xB,yB,xD,yD,xC,yC,vBx,vBy,vDx,vDy,aBx,aBy,aDx,aDy][[1]];
vCy=VelAccRRR[xB,yB,xD,yD,xC,yC,vBx,vBy,vDx,vDy,aBx,aBy,aDx,aDy][[2]];
aCx=VelAccRRR[xB,yB,xD,yD,xC,yC,vBx,vBy,vDx,vDy,aBx,aBy,aDx,aDy][[3]];
aCy=VelAccRRR[xB,yB,xD,yD,xC,yC,vBx,vBy,vDx,vDy,aBx,aBy,aDx,aDy][[4]];
```

The input data are the coordinates of the joints B, D, and C, and the velocities and acceleration components of the joints B and D. The output data are the four elements of the vector returned by the function **VelAccRRR**, which are the velocity and acceleration components of the joint C.

Velocity and acceleration of the joint E

The numerical values for the angular velocity and acceleration of the link 3 using the *Mathematica*^{TM} function **AngVelAcc** are

```
omega3=AngVelAcc[xC,yC,xD,yD,vCx,vCy,vDx,vDy,aCx,aCy,
aDx,aDy,phi3][[1]];
    alpha3=AngVelAcc[xC,yC,xD,yD,vCx,vCy,vDx,vDy,aCx,aCy,
aDx,aDy,phi3][[2]];
```

The input data are the coordinates, velocities, and accelerations of the joints C and D, and the angle ϕ_3. The output data are the two components of the vector returned by the function **AngVelAcc**.

The velocity and the acceleration of the joint E are calculated with

$$\mathbf{v}_E = \omega_3 \times \mathbf{r}_{DE} \quad \text{and} \quad \mathbf{a}_E = \alpha_3 \times \mathbf{r}_{DE} - \omega^2 \mathbf{r}_{DE}.$$

The numerical solutions for the velocity and acceleration components of the joint E are obtained using the $Mathematica^{TM}$ function **AbsVelAcc**:

```
vEx=AbsVelAcc[xD,yD,xE,yE,vDx,vDy,aDx,aDy,omega3,alpha3][[1]];
vEy=AbsVelAcc[xD,yD,xE,yE,vDx,vDy,aDx,aDy,omega3,alpha3][[2]];
aEx=AbsVelAcc[xD,yD,xE,yE,vDx,vDy,aDx,aDy,omega3,alpha3][[3]];
aEy=AbsVelAcc[xD,yD,xE,yE,vDx,vDy,aDx,aDy,omega3,alpha3][[4]];
```

The input data are the coordinates of the joints D and E, and the velocity and acceleration components of the joint D. The output data are the four elements of the vector returned by the function **AbsVelAcc**, which are the velocity and acceleration components of the joint E.

Velocity and acceleration of the joint F
In this particular case, the angular velocity and acceleration of the link 5 are zero:

$$\omega_5 = \alpha_5 = \mathbf{0}.$$

The velocity and the acceleration of the point $P(-L_c, 0)$, on the sliding direction, are zero:

$$\mathbf{v}_P = \mathbf{a}_P = \mathbf{0}.$$

The velocity and acceleration components of the joint F are calculated using the $Mathematica^{TM}$ function **VelAccRRT**:

```
vFx=VelAccRRT[xE,yE,xP,yP,xF,yF,vEx,vEy,vPx,vPy,
aEx,aEy,aPx,aPy,phi5,omega5,alpha5][[1]];
vFy=VelAccRRT[xE,yE,xP,yP,xF,yF,vEx,vEy,vPx,vPy,
aEx,aEy,aPx,aPy,phi5,omega5,alpha5][[2]];
aFx=VelAccRRT[xE,yE,xP,yP,xF,yF,vEx,vEy,vPx,vPy,
aEx,aEy,aPx,aPy,phi5,omega5,alpha5][[3]];
aFy=VelAccRRT[xE,yE,xP,yP,xF,yF,vEx,vEy,vPx,vPy,
aEx,aEy,aPx,aPy, phi5,omega5,alpha5][[4]];
```

The input data are the coordinates of the joints E, P, and F, the velocities and acceleration components of the joints E and P, the angle ϕ_5, the angular velocity ω_5, and the angular acceleration α_5. The output data are the four elements of the vector returned by the function **VelAccRRT**, which are the velocity and acceleration components of the joint F.

The numerical values for the angular velocity ω_2 and the angular acceleration α_2 of the link 2 using the *Mathematica*^{TM} function **AngVelAcc** are

```
omega2=AngVelAcc[xB,yB,xC,yC,vBx,vBy,vCx,vCy,aBx,aBy,
aCx,aCy,phi2][[1]];
    alpha2=AngVelAcc[xB,yB,xC,yC,vBx,vBy,vCx,vCy,aBx,aBy,
aCx,aCy,phi2][[2]];
```

The input data are the coordinates, velocities, and accelerations of the joints B and C, and the angle ϕ_2. The numerical values for the angular velocity ω_4 and the angular acceleration α_4 of the link 4 are calculated with the *Mathematica*^{TM} function **AngVelAcc**:

```
omega4=AngVelAcc[xE,yE,xF,yF,vEx,vEy,vFx,vFy,aEx,aEy,
aFx,aFy,phi4][[1]];
    alpha4=AngVelAcc[xE,yE,xF,yF,vEx,vEy,vFx,vFy,aEx,aEy,
aFx,aFy,phi4][[2]];
```

The input data are the coordinates, velocities, and accelerations of the joints E and F, and the angle ϕ_4. The output data are the two components of the vector returned by the function **AngVelAcc**.

The *Mathematica*^{TM} program and the numerical results are shown in Program I.8.8.

R-RRT Mechanism

The position analysis of the planar R-RRT mechanism considered (Fig. I.8.5) has been presented in Subsection I.8.2. Given the angular velocity $\omega = 3.141$ rad/s, calculate the velocities and accelerations of the joints and the angular velocities and accelerations of the links.

The *Mathematica*^{TM} packages **Driver**, **Position**, and **VelAcc** are loaded in the main program using the commands

```
<<Driver.m ;
<<Position.m ;
<<VelAcc.m ;
```

The angular velocity of the driver link is zero:

$$\alpha = \ddot{\phi} = 0.$$

Velocity and acceleration of the joint A
Since the joint A is the origin of the reference frame $xAyz$, the velocity and acceleration of the joint A are

$$\mathbf{v}_A = \mathbf{a}_A = \mathbf{0}.$$

The velocity and the acceleration of the point C are zero:

$$\mathbf{v}_C = \mathbf{a}_C = \mathbf{0}.$$

Velocity and acceleration of the joint B

In order to calculate the velocity and acceleration of the point B, you need to calculate the velocity and acceleration of the point P, located on the driver link AB:

$$v_{Px} = -AP \sin \phi \omega = -1.110 \text{ m/s},$$

$$v_{Py} = AP \cos \phi \omega = 1.110 \text{ m/s},$$

$$a_{Px} = -AP\omega^2 \cos \phi - AP\alpha \sin \phi = -3.489 \text{ m/s}^2,$$

$$a_{Py} = -AP\omega^2 \sin \phi + AP\alpha \cos \phi = -3.489 \text{ m/s}^2.$$

The numerical values for the velocity and acceleration components of the point P using the *Mathematica*^{TM} function **Driver** are

```
vPx = Driver[xA,yA,AP,phi,omega,alpha][[3]] ;
vPy = Driver[xA,yA,AP,phi,omega,alpha][[4]] ;
aPx = Driver[xA,yA,AP,phi,omega,alpha][[5]] ;
aPy = Driver[xA,yA,AP,phi,omega,alpha][[6]] ;
```

The velocity components v_{Bx} and v_{By} of the point B are calculated using the equations

$$(x_C - x_B)(v_{Cx} - v_{Bx}) + (y_C - y_B)(v_{Cy} - v_{By}) = 0,$$

$$(v_{Bx} - v_{Px}) \sin \phi + \omega(x_B - x_P) \cos \phi - (v_{By} - v_{Py}) \cos \theta + \omega(y_B - y_P) \sin \phi = 0.$$
(I.8.29)

The acceleration components a_{Bx} and a_{By} of the point B are calculated using the equations

$$(x_C - x_B)(a_{Cx} - a_{Bx}) + (v_{Cx} - v_{Bx})^2 + (y_C - y_B)(a_{Cy} - a_{By}) + (v_{Cy} - v_{By})^2 = 0,$$

$$(a_{Bx} - a_{Px}) \sin \phi + \omega(v_{Bx} - v_{Px}) \cos \phi - (a_{By} - a_{Py}) \cos \phi + \omega(v_{By} - v_{Py}) \sin \phi$$

$$+ [(x_B - x_P) \cos \phi + (y_B - y_P) \sin \phi]\alpha + ((v_{Bx} - v_{Px}) \cos \phi - \omega(x_B - x_P) \sin \phi$$

$$+ (v_{By} - v_{Py}) \sin \phi + \omega(y_B - y_P) \cos \phi)\omega = 0.$$
(I.8.30)

The solutions for the velocity and acceleration components of the joint F are obtained using the *Mathematica*^{TM} function **VelAccRRT**:

```
vBx=VelAccRRT[xC,yC,xP,yP,xB,yB,vCx,vCy,vPx,vPy,
aCx,aCy,aPx,aPy,phi,omega,alpha][[1]];
   vBy=VelAccRRT[xC,yC,xP,yP,xB,yB,vCx,vCy,vPx,vPy,
aCx,aCy,aPx,aPy,phi,omega,alpha][[2]];
   aBx=VelAccRRT[xC,yC,xP,yP,xB,yB,vCx,vCy,vPx,vPy,
aCx,aCy,aPx,aPy,phi,omega,alpha][[3]];
   aBy=VelAccRRT[xC,yC,xP,yP,xB,yB,vCx,vCy,vPx,vPy,
aCx,aCy,aPx,aPy,phi,omega,alpha][[4]];
```

The input data are the coordinates of the points C, P, and B, the velocities and acceleration components of the points C and P, the angle ϕ, the angular velocity ω, and the angular acceleration α. The output data are the four elements of the vector returned by the function **VelAccRRT**, which are the velocity and acceleration components of the point B.

The *Mathematica^TM* program and the numerical results are shown in Program I.8.9.

R-RTR-RTR Mechanism

The planar R-RTR-RTR mechanism considered is shown in Figure I.8.6. Given the input data $AB = 0.14$ m, $AC = 0.06$ m, $AE = 0.25$ m, $CD = 0.15$ m, the angle of the driver link AB and the horizontal axis $\phi = 30°$, and the angular velocity $\omega = \dot{\phi} = 5.235$ rad/s, calculate the velocities and accelerations of the joints and the angular velocities and accelerations of the links. The Cartesian reference frame $xOyz$ is chosen, $A \equiv O$.

The *Mathematica^TM* packages **Driver** and **VelAcc** are loaded in the main program using the commands

```
<<Driver.m ;
<<VelAcc.m ;
```

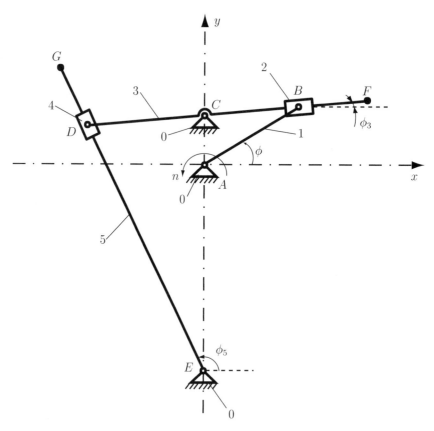

FIGURE I.8.6 *R-RTR-RTR mechanism.*

Position analysis

Position of the joint A. Since the joint A is the origin of the reference frame $xAyz$, the coordinates of the joint A are zero:

$$x_A = y_A = 0.$$

Position of the joint B. The coordinates of the joint B are

$$x_B = x_A + AB \cos \phi = 0.121 \text{ m},$$
$$y_B = y_A + AB \sin \phi = 0.070 \text{ m}. \qquad (\text{I}.8.31)$$

The coordinates of the joint B are obtained using the *Mathematica*^{TM} function **Driver**:

```
xB = Driver[xA,yA,AB,phi,omega,alpha][[1]] ;
yB = Driver[xA,xB,AB,phi,omega,alpha][[2]] ;
```

Position of the joint C. The coordinates of the joint C are

$$x_C = 0 \text{ m}, \ y_C = AC = 0.06 \text{ m}.$$

The angle ϕ_3 between the link 3 and the horizontal axis Ox is

$$\phi_3 = \arctan \frac{y_B - y_C}{x_B - x_C}.$$

Position of the joint D. The coordinates of the joint D are

$$x_D = x_C - CD \cos \phi_3 = -0.149 \text{ m},$$
$$y_D = y_C - CD \sin \phi_3 = 0.047 \text{ m}. \qquad (\text{I}.8.32)$$

Position of the joint E. The coordinates of the joint E are

$$x_E = 0 \text{ m}, \ y_E = -AE = -0.25 \text{ m}.$$

The angle ϕ_5 between the link 3 and the horizontal axis Ox is

$$\phi_5 = \pi + \arctan \frac{y_D - y_E}{x_D - x_E}.$$

Velocity and acceleration analysis

The angular velocity of the driver link is zero:

$$\alpha = \ddot{\phi} = 0.$$

Velocity and acceleration of the joint A. Since the joint A is the origin of the reference frame $xAyz$, the velocity and acceleration of the joint A are zero:

$$\mathbf{v}_A = \mathbf{a}_A = \mathbf{0}.$$

Velocity and acceleration of the joint B. The velocity and acceleration components of the joint B are

$$v_{Bx} = -AB\omega \sin \phi = -0.366 \text{ m/s},$$

$$v_{By} = AB\omega \cos \phi = 0.634 \text{ m/s},$$

$$a_{Bx} = -AB\omega^2 \cos \phi - AB\alpha \sin \phi = -3.323 \text{ m/s}^2,$$

$$a_{By} = -AB\omega^2 \sin \phi + AB\alpha \cos \phi = -1.919 \text{ m/s}^2.$$

The velocity and acceleration components of the joint B using the *Mathematica*^{TM} function **Driver** are

```
vBx = Driver[xA,yA,AB,phi,omega,alpha][[3]] ;
vBy = Driver[xA,yA,AB,phi,omega,alpha][[4]] ;
aBx = Driver[xA,yA,AB,phi,omega,alpha][[5]] ;
aBy = Driver[xA,yA,AB,phi,omega,alpha][[6]] ;
```

Velocity and acceleration of the joint C. The velocity and acceleration of the joint C are zero:

$$\mathbf{v}_C = \mathbf{a}_C = \mathbf{0}.$$

Velocity and acceleration of the joint D. The angular velocity ω_3 and angular acceleration α_3 of the link 3 are obtained using the *Mathematica*^{TM} function **AngVelAcc**:

```
omega3=AngVelAcc[xB,yB,xC,yC,vBx,vBy,vCx,vCy,aBx,aBy,
aCx,aCy,phi3][[1]];
    alpha3=AngVelAcc[xB,yB,xC,yC,vBx,vBy,vCx,vCy,aBx,aBy,
aCx,aCy,phi3][[2]];
```

The input data are the coordinates, velocities, and accelerations of the joints B and C, and the angle ϕ_3. The output data are the two components of the vector returned by the function **AngVelAcc**. To obtain the numerical values for the velocity and acceleration components of the joint D, the *Mathematica*^{TM} function **AbsVelAcc** is used:

```
vDx=AbsVelAcc[xC,yC,xD,yD,vCx,vCy,aCx,aCy,omega3,alpha3][[1]];
vDy=AbsVelAcc[xC,yC,xD,yD,vCx,vCy,aCx,aCy,omega3,alpha3][[2]];
aDx=AbsVelAcc[xC,yC,xD,yD,vCx,vCy,aCx,aCy,omega3,alpha3][[3]];
aDy=AbsVelAcc[xC,yC,xD,yD,vCx,vCy,aCx,aCy,omega3,alpha3][[4]];
```

The input data are the coordinates of the joints C and D, the velocity and acceleration components of the joint C, and the angular velocity and acceleration of the link 3. The output data are the four components of the vector returned by the function **AbsVelAcc**.

Velocity and acceleration of the joint E. The velocity and acceleration of the joint E are zero

$$\mathbf{v}_E = \mathbf{a}_E = \mathbf{0}.$$

The numerical values for the angular velocity ω_5 and angular acceleration α_5 of the link 5 using the *Mathematica*^{TM} function **AngVelAcc** are given by the commands

```
   omega5=AngVelAcc[xD,yD,xE,yE,vDx,vDy,vEx,vEy,
aDx,aDy,aEx,aEy,phi5][[1]];
   alpha5=AngVelAcc[xD,yD,xE,yE,vDx,vDy,vEx,vEy,
aDx,aDy,aEx,aEy,phi5][[2]];
```

The input data are the coordinates, velocities, and accelerations of the joints D and E, and the angle ϕ_5. The output data are the two components of the vector returned by the function **AngVelAcc**.

The *Mathematica*^{TM} program and the numerical results are shown in Program I.8.10.

I.8.4 Force Analysis

Force and Moment

A rigid link is shown in Figure I.8.7. The input data are the mass m, the acceleration vector of the center of mass $\mathbf{a}_{CM}$, the mass moment of inertia I_{CM}, and the angular acceleration $\boldsymbol{\alpha}$ of the link. The output data are the total force $\mathbf{F}$ and the moment of inertia $\mathbf{M}$ of the link.
   The total force $\mathbf{F}$ is

$$\mathbf{F} = \mathbf{F}_{in} + \mathbf{G},$$

where $\mathbf{F}_{in} = -m\,\mathbf{a}_{CM}$ is the inertia force, $\mathbf{G} = -m\,\mathbf{g}$ is the gravitational force, and $\mathbf{g} = -9.807\mathbf{k}$ m/s^2 is the gravitational acceleration.
   The moment of inertia M of the link is

$$\mathbf{M} = -I_{CM}\boldsymbol{\alpha},$$

where $\boldsymbol{\alpha} = \alpha\mathbf{k}$.
   The *Mathematica*^{TM} function **ForceMomentum** for the force analysis is

```
ForceMomentum::usage="ForceMomentum[m,aCM,ICM,ddtheta]
computes the total force and moment of a rigid link."
   Begin["`Private`"]
   ForceMomentum[m_,aCM_,ICM_,ddtheta_]:=
Block[
g, Fin, G, F, M ,
g = 9.807 ;
```

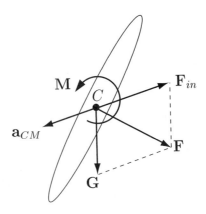

FIGURE I.8.7 *Rigid link.*

```
Fin = - m aCM ;
G = { 0, -m g, 0 } ;
F = Fin + G ;
M = - ICM { 0, 0, ddtheta } ;
Return[ { F, M } ] ; ]
End[ ]
```

Joint Force Computation

RRR dyad

Figure I.8.8 shows an RRR dyad with two links 2 and 3, and three pin joints M, N, and P. The input data are the total forces $\mathbf{F}_2$, $\mathbf{F}_3$ and the moments $\mathbf{M}_2$, $\mathbf{M}_3$ on the links 2 and 3, the position vectors $\mathbf{r}_M$, $\mathbf{r}_N$, $\mathbf{r}_P$ of the joints M, N, P, and position vectors $\mathbf{r}_{C2}, \mathbf{r}_{C3}$ of the centers of mass of the links 2 and 3. The output data are the joint reaction forces $\mathbf{F}_{12}$, $\mathbf{F}_{43}$, and $\mathbf{F}_{32}$.

The unknown joint reaction forces are

$$\mathbf{F}_{12} = F_{12x}\mathbf{i} + F_{12y}\mathbf{J},$$

$$\mathbf{F}_{43} = F_{43x}\mathbf{i} + F_{43y}\mathbf{J},$$

$$\mathbf{F}_{23} = -\mathbf{F}_{32} = F_{23x}\mathbf{i} + F_{23y}\mathbf{J}. \tag{I.8.33}$$

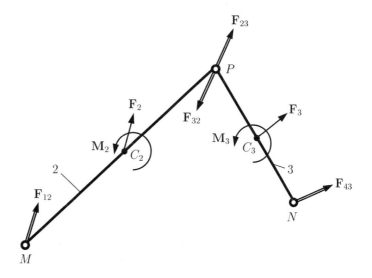

FIGURE I.8.8 *RRR dyad.*

To determine $\mathbf{F}_{12}$ and $\mathbf{F}_{43}$, the following equations are written:

- sum of all forces on links 2 and 3 is zero

$$\sum \mathbf{F}^{(2\&3)} = \mathbf{F}_{12} + \mathbf{F}_2 + \mathbf{F}_3 + \mathbf{F}_{43} = \mathbf{0},$$

or

$$\sum \mathbf{F}^{(2\&3)} \cdot \mathbf{i} = F_{12x} + F_{2x} + F_{3x} + F_{43x} = 0,$$
$$\sum \mathbf{F}^{(2\&3)} \cdot \mathbf{j} = F_{12y} + F_{2y} + F_{3y} + F_{43y} = 0. \qquad (\text{I.8.34})$$

- sum of moments of all forces and moments on link 2 about P is zero

$$\sum \mathbf{M}_P^{(2)} = (\mathbf{r}_M - \mathbf{r}_P) \times \mathbf{F}_{12} + (\mathbf{r}_{C2} - \mathbf{r}_P) \times \mathbf{F}_2 + \mathbf{M}_2 = \mathbf{0}. \qquad (\text{I.8.35})$$

- sum of moments of all forces and moments on link 3 about P is zero

$$\sum \mathbf{M}_P^{(3)} = (\mathbf{r}_N - \mathbf{r}_P) \times \mathbf{F}_{43} + (\mathbf{r}_{C3} - \mathbf{r}_P) \times \mathbf{F}_3 + \mathbf{M}_3 = \mathbf{0}. \qquad (\text{I.8.36})$$

The components F_{12x}, F_{12y}, F_{43x}, and F_{43y} are calculated from Eqs. (I.8.34), (I.8.35), and (I.8.36).

The reaction force $\mathbf{F}_{32} = -\mathbf{F}_{23}$ is computed from the sum of all forces on the link 2:

$$\sum \mathbf{F}^{(2)} = \mathbf{F}_{12} + \mathbf{F}_2 + \mathbf{F}_{32} = \mathbf{0},$$

or

$$\sum \mathbf{F}^{(2)} \cdot \mathbf{i} = F_{12x} + F_{2x} + F_{32x} = 0,$$

$$\sum \mathbf{F}^{(2)} \cdot \mathbf{j} = F_{12y} + F_{2y} + F_{32y} = 0. \qquad (I.8.37)$$

The *Mathematica*[TM] function **ForceRRR** for the RRR dyad joint force analysis is

```
ForceRRR::usage = "ForceRRR[F2,M2,F3,M3,rM,rN,rP,rC2,rC3]
computes the joint reactions for the RRR dyad."
  Begin["`Private`"]
  ForceRRR[F2_,M2_,F3_,M3_,rM_,rN_,rP_,rC2_,rC3_]:=
  Block[
  { F12, F12Sol, F12xSol, F12ySol, F43, F43Sol, F43xSol,
F43ySol, rPC2, rPC3, rPM, rPN, F32, eqRRR1, eqRRR2, eqRRR3,
eqRRR4, solRRR},
  F12Sol = { F12xSol, F12ySol, 0 } ;
  F43Sol = { F43xSol, F43ySol, 0 } ;
  rPC2 = rC2 - rP ;
  rPC3 = rC3 - rP ;
  rPM = rM - rP ;
  rPN = rN - rP ;
  eqRRR1=(F12Sol+F43Sol+F2+F3)[[1]]==0;
  eqRRR2=(F12Sol+F43Sol+F2+F3)[[2]]==0;
  eqRRR3=(Cross[rPC2,F2]+Cross[rPM,F12Sol]+M2)[[3]]==0;
  eqRRR4=(Cross[rPC3,F3]+Cross[rPN,F43Sol]+M3)[[3]]==0;
  solRRR = Solve[ {eqRRR1, eqRRR2, eqRRR3, eqRRR4},
{F12xSol,F12ySol,F43xSol,F43ySol} ] ;
  F12 = F12Sol/.solRRR[[1]] ;
  F43 = F43Sol/.solRRR[[1]] ;
  F32 = - F2 - F12 ;
  Return[ { F12, F43, F23 } ] ;
  ]
  End[ ]
```

RRT Dyad

Figure I.8.9 shows an RRT dyad with two links 2 and 3, two pin joints M and P, and one slider joint P. The input data are the total forces $\mathbf{F}_2$, $\mathbf{F}_3$ and moments $\mathbf{M}_2$, $\mathbf{M}_3$ on the links 2 and 3, the position vectors $\mathbf{r}_M$, $\mathbf{r}_N$, $\mathbf{r}_P$ of the joints M, N, P, and the position vector of the center of mass $\mathbf{r}_{C2}$ of the link 2. The output data are the joint reaction forces $\mathbf{F}_{12}$, $\mathbf{F}_{43}$, $\mathbf{F}_{23} = -\mathbf{F}_{32}$ and the position vector $\mathbf{r}_Q$ of the application point of the joint reaction force $\mathbf{F}_{43}$.

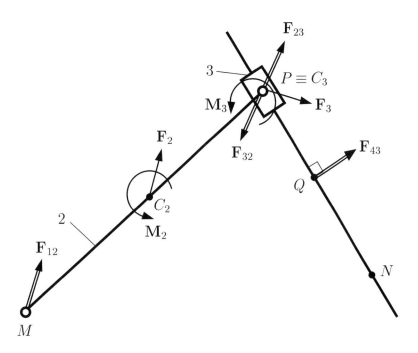

FIGURE I.8.9 *RRT dyad.*

The joint reaction $\mathbf{F}_{43}$ is perpendicular to the sliding direction $\mathbf{r}_{PN}$ or

$$\mathbf{F}_{43} \cdot \mathbf{r}_{PN} = (F_{43x}\mathbf{\imath} + F_{43y}\mathbf{J}) \cdot [(x_N - x_P)\mathbf{\imath} + (y_N - y_P)\mathbf{J}] = 0. \qquad (I.8.38)$$

In order to determine $\mathbf{F}_{12}$ and $\mathbf{F}_{43}$, the following equations are written:

• sum of all the forces on links 2 and 3 is zero

$$\sum \mathbf{F}^{(2\&3)} = \mathbf{F}_{12} + \mathbf{F}_2 + \mathbf{F}_3 + \mathbf{F}_{43} = \mathbf{0},$$

or

$$\sum \mathbf{F}^{(2\&3)} \cdot \mathbf{\imath} = F_{12x} + F_{2x} + F_{3x} + F_{43x} = 0,$$

$$\sum \mathbf{F}^{(2\&3)} \cdot \mathbf{J} = F_{12y} + F_{2y} + F_{3y} + F_{43y} = 0. \qquad (I.8.39)$$

- sum of moments of all forces and moments on link 2 about P is zero

$$\sum \mathbf{M}_P^{(2)} = (\mathbf{r}_M - \mathbf{r}_P) \times \mathbf{F}_{12} + (\mathbf{r}_{C2} - \mathbf{r}_P) \times \mathbf{F}_2 + \mathbf{M}_2 = \mathbf{0}. \qquad (I.8.40)$$

The components F_{12x}, F_{12y}, F_{43x}, and F_{43y} are calculated from Eqs. (I.8.38), (I.8.39), and (I.8.40).

The reaction force components F_{32x} and F_{32y} are computed from the sum of all the forces on the link 2:

$$\sum \mathbf{F}^{(2)} = \mathbf{F}_{12} + \mathbf{F}_2 + \mathbf{F}_{32} = \mathbf{0},$$

or

$$\sum \mathbf{F}^{(2)} \cdot \mathbf{1} = F_{12x} + F_{2x} + F_{32x} = 0,$$

$$\sum \mathbf{F}^{(2)} \cdot \mathbf{J} = F_{12y} + F_{2y} + F_{32y} = 0. \qquad (I.8.41)$$

To determine the application point $Q(x_Q, y_Q)$ of the reaction force $\mathbf{F}_{43}$, one can write sum of moments of all the forces and the moments on the link 3 about $C_3 \equiv P$ is zero:

$$\sum \mathbf{M}_P^{(3)} = (\mathbf{r}_Q - \mathbf{r}_P) \times \mathbf{F}_{43} + \mathbf{M}_3 = \mathbf{0}. \qquad (I.8.42)$$

If $\mathbf{M}_3 = \mathbf{0}$, then P is identical to Q ($P \equiv Q$).

If $\mathbf{M}_3 \neq \mathbf{0}$, an equation regarding the location of the point Q on the sliding direction $\mathbf{r}_{NP}$ is written as

$$\frac{y_Q - y_P}{x_Q - x_P} = \frac{y_N - y_P}{x_N - x_P}. \qquad (I.8.43)$$

From Eqs. (I.8.42) and (I.8.43) the coordinates x_Q and y_Q of the point Q are calculated.

The *Mathematica*^{TM} function **ForceRRT** for the RRT dyad joint force analysis is

```
ForceRRT::usage = "ForceRRT[F2,M2,F3,M3,rM,rN,rP,rC2]
computes the joint reactions for the RRR dyad."
  Begin["`Private`"]
  ForceRRT[F2_,M2_,F3_,M3_,rM_,rN_,rP_,rC2_]:=
  Block[
  { F12, F12Sol, F12xSol, F12ySol, F43, F43Sol, F43xSol,
F43ySol, F32, eqRRT1, eqRRT2, eqRRT3, eqRRT4, solRRT, rNC2,
rNM, rNP, rNQ, rQP, rQ, rQSol, xQSol, yQSol, eqRRTQ1, eqRRTQ2,
solRRTQ },
  F12Sol = { F12xSol, F12ySol, 0 } ;
  F43Sol = { F43xSol, F43ySol, 0 } ;
  rPC2 = rC2 - rP ;
  rPM = rM - rP ;
  rPN = rN - rP ;
  eqRRT1 = (F12Sol+F43Sol+F2+F3)[[1]] == 0 ;
  eqRRT2 = (F12Sol+F43Sol+F2+F3)[[2]] == 0 ;
```

```
eqRRT3 = (F43Sol.rPN) == 0 ;
eqRRT4 = (Cross[rNC2,F2]+Cross[rNM,F12Sol]+M2)[[3]]==0;
solRRT = Solve[ {eqRRT1, eqRRT2, eqRRT3, eqRRT4},
{F12xSol,F12ySol,F43xSol,F43ySol} ] ;
F12 = F12Sol/.solRRT[[1]] ;
F43 = F43Sol/.solRRT[[1]] ;
F32 = - F2 - F12 ;
F23 = - F32 ;
If[ M3[[3]]==0 , rQ = rP ,
rQSol = { xQSol, yQSol, 0 } ;
rQP = rP - rQSol ;
eqRRTQ1 = rPQ[[2]]/rPQ[[1]] - rPN[[2]]/rPN[[1]]==0;
eqRRTQ2 = Cross[rPQ,F43] + M3 == 0 ;
solRRTQ = Solve[ eqRRTQ1, eqRRTQ2 , xQSol,yQSol ] ;
rQ = rQSol/.solRRTQ[[1]] ; ] ;
Return[ { F12, F43, F23, rQ } ] ;
]
End[ ]
```

RTR Dyad

Figure I.8.10 shows an RTR dyad with two links 2 and 3, and one pin joint M, one slider joint P, and one pin joint P. The input data are the total forces $\mathbf{F}_2$, $\mathbf{F}_3$ and moments $\mathbf{M}_2$, $\mathbf{M}_3$ of the links 2 and 3, the position vectors $\mathbf{r}_M$, $\mathbf{r}_N$, $\mathbf{r}_P$ of the joints M, N, and the position vector of the center of mass $\mathbf{r}_{C2}$ of the link 2. The output data are the joint reaction forces $\mathbf{F}_{12}$, $\mathbf{F}_{43}$, $\mathbf{F}_{23} = -\mathbf{F}_{32}$ and the position vector $\mathbf{r}_Q$ of the application point of the joint reaction force $\mathbf{F}_{23}$.

The unknown joint reaction forces $\mathbf{F}_{12}$ and $\mathbf{F}_{43}$ are calculated from the relations:

- sum of all the forces on links 2 and 3 is zero

$$\sum \mathbf{F}^{(2\&3)} = \mathbf{F}_{12} + \mathbf{F}_2 + \mathbf{F}_3 + \mathbf{F}_{43} = \mathbf{0},$$

or

$$\sum \mathbf{F}^{(2\&3)} \cdot \mathbf{\imath} = F_{12x} + F_{2x} + F_{3x} + F_{43x} = 0,$$

$$\sum \mathbf{F}^{(2\&3)} \cdot \mathbf{\jmath} = F_{12y} + F_{2y} + F_{3y} + F_{43y} = 0. \tag{I.8.44}$$

- sum of the moments of all forces and moments on links 2 and 3 about M is zero

$$\sum \mathbf{M}_M^{(2\&3)} = (\mathbf{r}_P - \mathbf{r}_M) \times (\mathbf{F}_3 + \mathbf{F}_{43}) + (\mathbf{r}_{C2} - \mathbf{r}_M) \times \mathbf{F}_2 + \mathbf{M}_2 + \mathbf{M}_3 = \mathbf{0}. \tag{I.8.45}$$

- sum of all the forces on link 2 projected onto the sliding direction $\mathbf{r}_{MP}$ is zero

$$\mathbf{F}_2 \cdot \mathbf{r}_{MP} = (F_{43x}\mathbf{\imath} + F_{43y}\mathbf{\jmath}) \cdot [(x_P - x_M)\mathbf{\imath} + (y_P - y_M)\mathbf{\jmath}] = 0. \tag{I.8.46}$$

The components F_{12x}, F_{12y}, F_{43x}, and F_{43y} are calculated from Eqs. (I.8.44), (I.8.45), and (I.8.46).

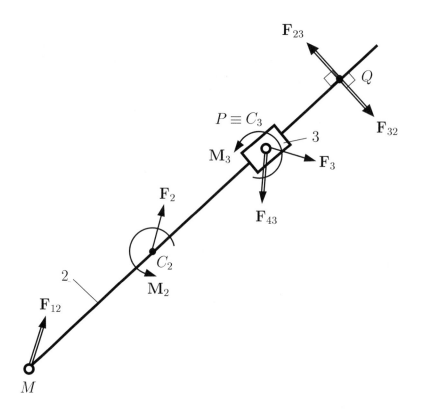

FIGURE I.8.10   *RTR dyad.*

The force components F_{32x} and F_{32y} are computed from the sum of all the forces on link 2:

$$\sum \mathbf{F}^{(2)} = \mathbf{F}_{12} + \mathbf{F}_2 + \mathbf{F}_{32} = \mathbf{0},$$

or

$$\sum \mathbf{F}^{(2)} \cdot \mathbf{i} = F_{12x} + F_{2x} + F_{32x} = 0,$$

$$\sum \mathbf{F}^{(2)} \cdot \mathbf{j} = F_{12y} + F_{2y} + F_{32y} = 0. \qquad (I.8.47)$$

To determine the application point $Q(x_Q, y_Q)$ of the reaction force $\mathbf{F}_{23}$, one can write sum of moments of all the forces and the moments on the link 3 about $C_3 \equiv P$ is zero:

$$\sum \mathbf{M}_P^{(3)} = (\mathbf{r}_Q - \mathbf{r}_P) \times \mathbf{F}_{23} + \mathbf{M}_3 = 0. \tag{I.8.48}$$

If $\mathbf{M}_3 = 0$, then P is identical to Q ($P \equiv Q$).

If $\mathbf{M}_3 \neq 0$, an equation regarding the location of the point Q on the sliding direction $\mathbf{r}_{NP}$ is written as

$$\frac{y_Q - y_M}{x_Q - x_M} = \frac{y_M - y_P}{x_M - x_P}. \tag{I.8.49}$$

From Eqs. (I.8.48) and (I.8.49) the coordinates x_Q and y_Q of the point Q are calculated. The *Mathematica*^{TM} function **ForceRTR** for the RTR dyad joint force analysis is

```
ForceRTR::usage = "ForceRTR[F2,M2,F3,M3,rM,rP,rC2]
computes the joint reactions for the RTR dyad."
  Begin["`Private`"]
  ForceRTR[F2_,M2_,F3_,M3_,rM_,rP_,rC2_]:=
  Block[
  { rC3, F12, F12Sol, F12xSol, F12ySol, F43, F43Sol, F43xSol,
F43ySol, F32, eqRTR1, eqRTR2, eqRTR3, eqRTR4, solRTR, rMC2,
rMP, rPQ, rQ, rQSol, xQSol, yQSol, eqRTRQ1, eqRTRQ2,
solRTRQ },
  F12Sol = { F12xSol, F12ySol, 0 } ;
  F43Sol = { F43xSol, F43ySol, 0 } ;
  rMC2 = rC2 - rM ;
  rMP = rP - rM ;
  eqRTR1 = (F12Sol+F43Sol+F2+F3) [[1]] == 0;
  eqRTR2 = (F12Sol+F43Sol+F2+F3) [[2]] == 0;
  eqRTR3 = (F12Sol+F2) .rMP == 0;
  eqRTR4 = (Cross[rMC2,F2]+Cross[rMP, (F3+F43Sol)]+M2+M3) [[3]] == 0;
  solRTR = Solve[{ eqRTR1, eqRTR2, eqRTR3, eqRTR4 } ,
{ F12xSol, F12ySol, F43xSol, F43ySol } ];
  F12 = F12Sol/.solRTR[[1]];
  F43 = F43Sol/.solRTR[[1]];
  F32 = - F2 - F12 ;
  F23 = - F32 ;
  If[ M3[[3]]==0 , rQ = rP ,
  rQSol = { xQSol, yQSol, 0 };
  rMQ = rQSol - rM ;
  rPQ = rQSol - rP ;
  eqRRTQ1 = rMQ[[2]]/rMQ[[1]] - rMP[[2]]/rMP[[1]] ==0;
  eqRRTQ2 = Cross[rPQ,F23] + M3 ==0;
  solRTRQ = Solve[{eqRTRQ1, eqRTRQ2}, {xQSol,yQSol}];
  rQ = rQSol/.solRTRQ[[1]];
  ];
  Return[ { F12, F43, F23, rQ } ];
  ]
  End[ ]
```

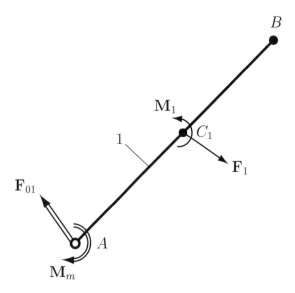

FIGURE I.8.11 *Driver link.*

Driver link

A driver link mechanism is shown in Figure I.8.11. The input data are the total force $\mathbf{F}_1$ and moment $\mathbf{M}_1$ on the driver link, the joint reaction force $\mathbf{F}_{21}$, the position vectors $\mathbf{r}_A$, $\mathbf{r}_B$ of the joints A, B, and the position vector of the center of mass $\mathbf{r}_{C1}$ of the driver link. The output data are the joint reaction force $\mathbf{F}_{01}$, and the moment of the motor $\mathbf{M}_m$ (equilibrium moment).

A force equation for the driver link is written to determine the joint reaction $\mathbf{F}_{01}$:

$$\sum \mathbf{F}^{(1)} = \mathbf{F}_{01} + \mathbf{F}_1 + \mathbf{F}_{21} = \mathbf{0},$$

or

$$\sum \mathbf{F}^{(1)} \cdot \mathbf{\imath} = F_{01x} + F_{1x} + F_{21x} = 0,$$

$$\sum \mathbf{F}^{(1)} \cdot \mathbf{\jmath} = F_{01y} + F_{1y} + F_{21y} = 0. \tag{I.8.50}$$

The sum of the moments about A_R for link 1 gives the equilibrium moment $\mathbf{M}_m$:

$$\sum \mathbf{M}_A^{(1)} = (\mathbf{r}_B - \mathbf{r}_A) \times \mathbf{F}_{21} + (\mathbf{r}_{C1} - \mathbf{r}_A) \times \mathbf{F}_1 + \mathbf{M}_m = \mathbf{0}. \tag{I.8.51}$$

The *Mathematica*[TM] function **FMDriver** for the driver link joint force analysis is

```
FMDriver::usage = "FMDriver[F1,M1,F21,rA,rB,rC1]
computes the joint reaction and torque of the motor."
   Begin["`Private`"]
   FMDriver[F1_,M1_,F21_,rA_,rB_,rC1_]:=
```

```
Block[
{ F01, rAC1, rAB, Mm },
F01 = - F1 - F21 ;
rAC1 = rC1 - rA ;
rAB = rB - rA ;
Mm = - Cross[rAC1,F1] - Cross[rAB,F21] - M1 ;
Return[ { F01, Mm } ] ;
]
End[ ]
```

The functions **ForceMomentum**, **ForceRRR**, **ForceRRT**, **ForceRTR**, and **FMDriver** are included in the *Mathematica*^{TM} package **Force**. The name of the source file for the package is **Force.m** and is given in Program I.8.11.

R-RRT Mechanism

The position, velocity, and acceleration analysis of the planar R-RRT mechanism considered are presented in Subsections I.8.2 and I.8.3. Given the external moment $\mathbf{M}_{ext} = -100 \, \text{sign}(\omega_3) \, \mathbf{k}$ N·m, applied on the link 3, calculate the motor moment $\mathbf{M}_m$ required for the dynamic equilibrium of the mechanism [Fig. I.8.12(a)]. All three links are rectangular prisms with the depth $d = 0.001$ m and the mass density $\rho = 8000$ Kg/m^3. The height of the links 1 and 3 is $h = 0.01$ m. The link 2 has the height $h_S = 0.02$ m, and the width $w_S = 0.05$ m. The center of mass location of the links $i = 1, 2, 3$ are designated by $C_i(x_{Ci}, y_{Ci}, 0)$.

The *Mathematica*^{TM} packages **Driver**, **Position**, **VelAcc**, and **Force** are loaded in the main program using the commands

```
<<Driver.m ;
<<Position.m ;
<<VelAcc.m ;
<<Force.m ;
```

Force and moment analysis
Link 1. The mass m_1, the acceleration of the center of mass $\mathbf{a}_{C1}$, and the mass moment of inertia I_{C1} of the link 1 are

$$m_1 = \rho \, AP \, h \, d,$$

$$\mathbf{a}_{C1} = (\mathbf{a}_A + \mathbf{a}_P)/2,$$

$$I_{C1} = m_1(AP^2 + h^2)/12.$$

The total force F_1 and moment M_1 of the link 1 are calculated using the *Mathematica*^{TM} function **ForceMomentum**:

```
F1=ForceMomentum[m1,aC1,IC1,alpha][[1]];
M1=ForceMomentum[m1,aC1,IC1,alpha][[2]];
```

The input data are the mass **m1**, the acceleration vector of the center of mass **aC1**, the mass moment of inertia **IC1**, and the angular acceleration **alpha** of the link 1. The output

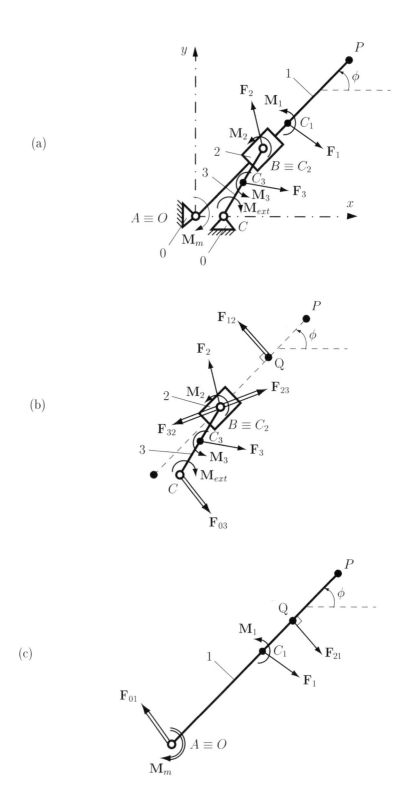

FIGURE I.8.12 *(a) R-RRT mechanism, (b) RRT (CBB) dyad, and (c) driver link.*

data are the two elements of the vector returned by the function **ForceMomentum**, which are the total force **F1** and moment of inertia **M1** of the link 1.

Link 2. The mass m_2, the acceleration of the center of mass $\mathbf{a}_{C2}$, and the mass moment of inertia I_{C2} of the link 2 are

$$m_2 = \rho \, w_S \, h_S \, d,$$

$$\mathbf{a}_{C2} = \mathbf{a}_B,$$

$$I_{C2} = m_2(w_S^2 + h_S^2)/12.$$

The total force F_2 and moment M_2 of the link 2 are computed using the *Mathematica*^{TM} function **ForceMomentum**:

```
F2=ForceMomentum[m2,aC2,IC2,alpha][[1]];
M1=ForceMomentum[m2,aC2,IC2,alpha][[2]];
```

The input data are the mass **m2**, the acceleration vector of the center of mass **aC2**, the mass moment of inertia **IC2**, and the angular acceleration **alpha2** of the link 2. The output data are the two elements of the vector returned by the function **ForceMomentum**, which are the total force **F2** and moment of inertia **M2** of the link 2.

Link 3. The mass m_3, the acceleration of the center of mass $\mathbf{a}_{C3}$, and the mass moment of inertia I_{C3} of the link 3 are

$$m_3 = \rho \, BC \, h \, d,$$

$$\mathbf{a}_{C3} = (\mathbf{a}_B + \mathbf{a}_C)/2,$$

$$I_{C3} = m_3(AB^2 + h^2)/12.$$

The total force F_3 and moment M_3 of the link 3 are calculated using the *Mathematica*^{TM} function **ForceMomentum**:

```
F3=ForceMoment[m3,aC3,IC3,alpha3][[1]];
M3=ForceMoment[m3,aC3,IC3,alpha3][[2]];
```

The input data are the mass **m3**, the acceleration vector of the center of mass **aC3**, the mass moment of inertia **IC3**, and the angular acceleration **alpha3** of the link 3. The output data are the two elements of the vector returned by the function **ForceMomentum**, which are the total force **F3** and moment of inertia **M3** of the link 3.

Joint reactions

The joint reactions for the dyad RRT (*CBB*) [Fig. I.8.12(b)] are computed using the *Mathematica*^{TM} function **ForceRRT**:

```
F03=ForceRRT[F3,M3+Mext,F2,M2,rC,rB,rA,rC3][[1]];
F12=ForceRRT[F3,M3+Mext,F2,M2,rC,rB,rA,rC3][[2]];
F23=ForceRRT[F3,M3+Mext,F2,M2,rC,rB,rA,rC3][[3]];
```

The input data are the total force **F3** and moment **M3+Mext** of the link 3, the total force **F2** and moment **M2** of the link 2, the position vectors **rC, rB, rA, rC3** of the joints C, B, A, and the center of mass C_3 of the link 3. The output data are the three elements of the vector returned by the function **ForceMomentum**, which are the joint reactions **F03, F12**, and **F23**.

The position vector of the application point Q of the joint reaction **F23** can be also computed using the *Mathematica*^{TM} function **ForceRRT**:

```
rQ=ForceRRT[F3,M3+Mext,F2,M2,rC,rB,rA,rC3][[4]];
```

The joint reaction and the moment of the motor [Fig. I.8.12(c)] are calculated using the *Mathematica*^{TM} function **FMDriver**:

```
F01=FMDriver[F1,M1,F21,rA,rB,rC1][[1]];
Mm=FMDriver[F1,M1,F21,rA,rB,rC1][[2]];
```

The input data are the total force **F1** and moment **M1** of the link 1, the joint reaction **F21=-F12**, the position vectors **rA, rB, rC1** of the joints A, B, and the center of mass C_1 of the link 1. The output data are the two elements of the vector returned by the function **FMDriver**, which are the joint reaction **F01** and moment **Mm** of the motor.

The *Mathematica*^{TM} program and the numerical results are shown in Program I.8.12.

R-RTR-RTR Mechanism

The position, velocity, and acceleration analysis of the planar R-RTR-RTR mechanism considered (see Figures I.8.6 and I.8.13) are presented in Subsection I.8.3. Given the external moment $\mathbf{M}_{ext} = -100\,\text{sign}(\omega_5)\,\mathbf{k}$ N·m, applied on the link 5, calculate the motor moment $\mathbf{M}_m$ required for the dynamic equilibrium of the mechanism. All five links are rectangular prisms with the depth $d = 0.001$ m and the mass density $\rho = 8000$ Kg/m^3. The heights of the links 1, 3, and 5 are $h = 0.01$ m. The link 2 has the height $h_S = 0.02$ m, and the width $w_S = 0.05$ m. The center of mass location of the links $i = 1, 2, 3, \ldots, 5$ are designated by $C_i(x_{Ci}, y_{Ci}, 0)$.

The *Mathematica*^{TM} packages **Driver, VelAcc**, and **Force** are loaded in the main program using the commands

```
<<Driver.m ;
<<VelAcc.m ;
<<Force.m ;
```

Force and moment analysis
Link 1. The mass m_1, the acceleration of the center of mass $\mathbf{a}_{C1}$, and the mass moment of inertia I_{C1} of the link 1 are

$$m_1 = \rho\, AB\, h\, d,$$

$$\mathbf{a}_{C1} = (\mathbf{a}_A + \mathbf{a}_B)/2,$$

$$I_{C1} = m_1(AB^2 + h^2)/12,$$

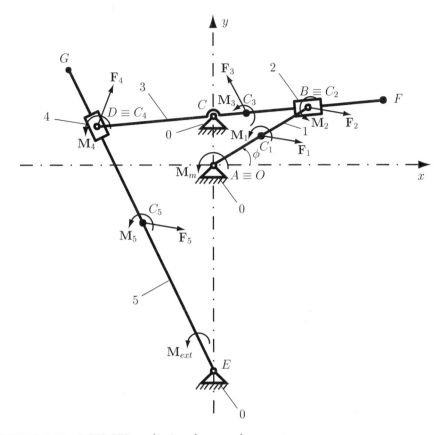

FIGURE I.8.13 *R-RTR-RTR mechanism: forces and moments.*

The total force F_1 and moment M_1 of the link 1 are computed using the *Mathematica*^{TM} function **ForceMomentum**:

```
F1=ForceMomentum[m1,aC1,IC1,alpha][[1]];
M1=ForceMomentum[m1,aC1,IC1,alpha][[2]];
```

The input data are the mass **m1**, the acceleration vector of the center of mass **aC1**, the mass moment of inertia **IC1**, and the angular acceleration **alpha** of the link 1. The output data are the two elements of the vector returned by the function **ForceMomentum**, which are the total force **F1** and moment of inertia **M1** of the link 1.

Link 2. The mass m_2, the acceleration of the center of mass $\mathbf{a}_{C2}$, and the mass moment of inertia I_{C2} of the link 2 are

$$m_2 = \rho\, w_S\, h_S\, d,$$

$$\mathbf{a}_{C2} = \mathbf{a}_B,$$

$$I_{C2} = m_2(w_S^2 + h_S^2)/12.$$

The total force F_2 and moment M_2 of the link 2 are calculated using the *Mathematica*[TM] function **ForceMomentum**:

```
F2=ForceMomentum[m2,aC2,IC2,alpha2][[1]];
M1=ForceMomentum[m2,aC2,IC2,alpha2][[2]];
```

The input data are the mass **m2**, the acceleration vector of the center of mass **aC2**, the mass moment of inertia **IC2**, and the angular acceleration **alpha2** of the link 2. The output data are the two elements of the vector returned by the function **ForceMomentum**, which are the total force **F2** and moment of inertia **M2** of the link 2.

Link 3. The coordinates of the center of mass C_3 of the link 3 are

$$x_{C3} = x_C + (DF/2 - CD)\cos\phi_3,$$

$$y_{C3} = y_C + (DF/2 - CD)\sin\phi_3.$$

The acceleration components of the center of mass of the link 3 are computed using the *Mathematica*[TM] function **AbsVelAcc**:

```
aC3x=AbsVelAcc[xC,yC,xC3,yC3,vCx,vCy,aCx,aCy,
omega3,alpha3][[3]];
aC3y=AbsVelAcc[xC,yC,xC3,yC3,vCx,vCy,aCx,aCy,
omega3,alpha3][[4]];
```

The mass m_3 and the mass moment of inertia I_{C3} of the link 3 are

$$m_3 = \rho\, DF\, h\, d,$$

$$I_{C3} = m_3(DF^2 + h^2)/12.$$

The total force F_3 and moment M_3 of the link 3 using the *Mathematica*[TM] function **ForceMomentum** are

```
F3=ForceMoment[m3,aC3,IC3,alpha3][[1]];
M3=ForceMoment[m3,aC3,IC3,alpha3][[2]];
```

The input data are the mass **m3**, the acceleration vector of the center of mass **aC3**, the mass moment of inertia **IC3**, and the angular acceleration **alpha3** of the link 3. The output data are the two elements of the vector returned by the function **ForceMomentum**, which are the total force **F3** and moment of inertia **M3** of the link 3.

Link 4. The mass m_4, the acceleration of the center of mass $\mathbf{a}_{C4}$, and the mass moment of inertia I_{C4} of the link 4 are

$$m_4 = \rho\, w_S\, h_S\, d,$$

$$\mathbf{a}_{C4} = \mathbf{a}_D,$$

$$I_{C4} = m_4(w_S^2 + h_S^2)/12.$$

The total force F_4 and moment M_4 of the link 4 using the *Mathematica*^{TM} function **ForceMomentum** are

```
F4=ForceMomentum[m4,aC4,IC4,alpha4][[1]];
M4=ForceMomentum[m4,aC4,IC4,alpha4][[2]];
```

The input data are the mass **m4**, the acceleration vector of the center of mass **aC4**, the mass moment of inertia **IC4**, and the angular acceleration **alpha4** of the link 4. The output data are the two elements of the vector returned by the function **ForceMomentum**, which are the total force **F4** and moment of inertia **M4** of the link 4.

Link 5. The coordinates of the center of mass C_5 of the link 5 are

$$x_{C5} = x_E + EG/2 \cos \phi_5,$$

$$y_{C5} = y_E + EG/2 \sin \phi_5.$$

The acceleration components of the center of mass of the link 3 using the *Mathematica*^{TM} function **AbsVelAcc** are

```
aC5x=AbsVelAcc[xE,yE,xC5,yC5,vEx,vEy,aEx,aEy,
omega5,alpha5][[3]];
aC5y=AbsVelAcc[xE,yE,xC5,yC5,vEx,vEy,aEx,aEy,
omega5,alpha5][[4]];
```

The mass m_5 and the mass moment of inertia I_{C5} of the link 5 are

$$m_5 = \rho \, EG \, h \, d,$$

$$I_{C5} = m_5(EG^2 + h^2)/12.$$

The total force F_5 and moment M_5 of the link 5 using the *Mathematica*^{TM} function **ForceMomentum** are

```
F5=ForceMomentum[m5,aC5,IC5,alpha5][[1]];
M5=ForceMomentum[m5,aC5,IC5,alpha5][[2]];
```

The input data are the mass **m5**, the acceleration vector of the center of mass **aC5**, the mass moment of inertia **IC5**, and the angular acceleration **alpha5** of the link 5. The output data are the two elements of the vector returned by the function **ForceMomentum**, which are the total force **F5** and moment of inertia **M5** of the link 5.

Joint reactions

The joint reactions for the dyad RTR (*EDD*) [Fig. I.8.14(a)] are computed using the *Mathematica*^{TM} function **ForceRTR**:

```
F05=ForceRTR[F5,M5+Mext,F4,M4,rE,rD,rC5][[1]];
F34=ForceRTR[F5,M5+Mext,F4,M4,rE,rD,rC5][[2]];
F54=ForceRTR[F5,M5+Mext,F4,M4,rE,rD,rC5][[3]];
```

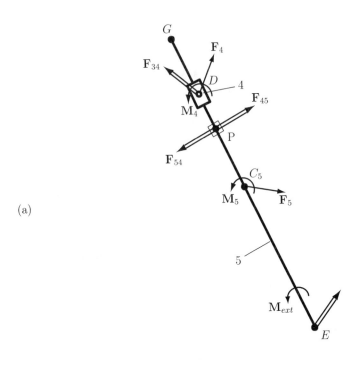

(a)

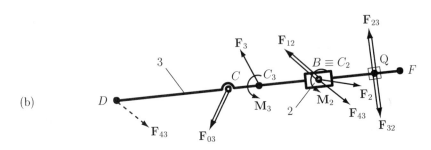

(b)

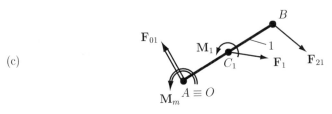

(c)

FIGURE I.8.14 *Joint reactions for: (a) RTR (EDD) dyad, (b) RTR (CBB) dyad, and (c) driver link.*

The input data are the total force **F5** and moment **M5+Mext** of the link 5, the total force **F4** and momentum **M4** of the link 4, the position vectors **rE**, **rD**, **rC5** of the joints E, D, and the center of mass C_5 of the link 5. The output data are the three elements of the vector returned by the function **ForceMomentum**, which are the joint reactions **F05**, **F34**, and **F45**.

The position vector of the application point P of the joint reaction **F45** can be also computed using the *Mathematica*^{TM} function **ForceRTR**:

```
rP=ForceRTR[F5,M5+Mext,F4,M4,rE,rD,rC5][[4]];
```

Next, consider the dyad RTR (*CBB*), shown in Figure I.8.14(b). The reaction force $\mathbf{F}_{43}$ acting at point D can be moved to a parallel position at point C_3 by adding the corresponding couple

$$\mathbf{M}_{43} = \mathbf{r}_{C_3 D} \times \mathbf{F}_{43}.$$

The joint reactions for the dyad RTR (*CBB*) using the *Mathematica*^{TM} function **ForceRTR** are

```
F03=ForceRTR[F3+F43,M3+M43,F2,M2,rC,rB,rC3][[1]];
F12=ForceRTR[F3+F43,M3+M43,F2,M2,rC,rB,rC3][[2]];
F32=ForceRTR[F3+F43,M3+M43,F2,M2,rC,rB,rC3][[3]];
```

The input data are the force **F3+F43** and moment **M3+M43** of the link 3, the total force **F2** and moment **M2** of the link 2, the position vectors **rC, rB, rC3** of the joints C, B, and the center of mass C_3 of the link 3. The output data are the three elements of the vector returned by the function **ForceMomentum**, which are the joint reactions **F03, F12,** and **F23**.

The position vector of the application point Q of the joint reaction **F12** can be also computed using the *Mathematica*^{TM} function **ForceRTR**:

```
rQ=ForceRTR[F3+F43,M3+M43,F2,M2,rC,rB,rC3][[4]];
```

The joint reaction and the moment of the motor [Fig. I.8.14(c)] are computed using the *Mathematica*^{TM} function **FMDriver**:

```
F01=FMDriver[F1,M1,F21,rA,rB,rC1][[1]];
Mm=FMDriver[F1,M1,F21,rA,rB,rC1][[2]];
```

The input data are the total force **F1** and moment **M1** of the link 1, the joint reaction **F21=-F12**, the position vectors **rA, rB, rC1** of the joints A, B, and center of mass C_1 of the link 1. The output data are the two elements of the vector returned by the function **FMDriver**, which are the joint reaction **F01** and moment **Mm** of the motor.

The *Mathematica*^{TM} program and the numerical results are shown in Program I.8.13.

Remark: All the packages must be placed in the *Mathematica*^{TM} folder for PC. For Macintosh OS X the packages must be placed in Home, Library, Mathematica, Applications.

I.8.5 Problems

I.8.1 Referring to Problem I.3.5 (Fig. I.3.20), write a *Mathematica*^{TM} program using packages for the position analysis of the mechanism.

I.8.2   Referring to Problem I.3.7 (Fig. I.3.22), write a *Mathematica^{TM}* program using packages for the position analysis of the mechanism.

I.8.3   Referring to Problem I.3.8 (Fig. I.3.23), write a *Mathematica^{TM}* program using packages for the position analysis of the mechanism.

I.8.4   Referring to Problem I.3.13 (Fig. I.3.28), write a *Mathematica^{TM}* program using packages for the position analysis of the mechanism.

I.8.5   Referring to Problem I.3.14 (Fig. I.3.29), write a *Mathematica^{TM}* program using packages for the position analysis of the mechanism.

I.8.6   Referring to Problem I.4.1 (Fig. I.3.16), write a *Mathematica^{TM}* program using packages for the velocity and acceleration analysis of the mechanism.

I.8.7   Referring to Problem I.4.3 (Fig. I.4.10), write a *Mathematica^{TM}* program using packages for the velocity and acceleration analysis of the mechanism.

I.8.8   Referring to Problem I.6.4 (Fig. I.3.19), write a *Mathematica^{TM}* program using packages for the equilibrium moment and the joint forces of the mechanism.

I.8.9   Referring to Problem I.6.16 (Fig. I.3.31), write a *Mathematica^{TM}* program using packages for the equilibrium moment and the joint forces of the mechanism.

I.8.10   Referring to Problem I.6.18 (Fig. I.3.33), write a *Mathematica^{TM}* program using packages for the equilibrium moment and the joint forces of the mechanism.

I.8.6   Programs

PROGRAM I.8.1

```
BeginPackage["Driver`"]

Driver::usage =
    "Driver[xA,yA,AB,phi,omega,alpha] computes the driver link
    position, velocity and acceleration

Begin["`Private`"]

Driver[xA_,yA_,AB_,phi_,omega_,alpha_]:=

Block[ { xB, yB ,vBx, vBy, aBx, aBy },

xB = xA + AB Cos[phi] ;
yB = yA + AB Sin[phi] ;

vBx = - AB omega Sin[phi] ;
vBy =   AB omega Cos[phi] ;

aBx = - AB omega^2 Cos[phi] - AB alpha Sin[phi] ;
aBy = - AB omega^2 Sin[phi] + AB alpha Cos[phi] ;
```

```
        Return[ { xB, yB, vBx, vBy, aBx, aBy } ] ;
        ]

        End[ ]

        EndPackage[ ]

        Driver`

        Driver[xA,yA,AB,phi,omega,alpha] computes
           the driver link position, velocity and acceleration vectors.

        Driver`Private`
        Driver`Private`
```

PROGRAM I.8.2

```
  (* Driver mechanism *)

  Apply[Clear,Names["Global`*"]] ;
  Off[General::spell];
  Off[General::spell1];

  <<Driver.m ;

  (* Input data *)
  AB = 0.20 ;        (* m *)
  phi = Pi/6 ;       (* rad *)
  omega = 5. ;       (* rad/s *)
  alpha = 0. ;       (* rad/s^2 *)

  (* Position of joint A *)
  xA = yA = 0 ;

  (* Position of joint B *)
  xB = Driver[xA,yA,AB,phi,omega,alpha][[1]] ;
  yB = Driver[xA,yA,AB,phi,omega,alpha][[2]] ;

  (* Velocity of joint B *)
  vBx = Driver[xA,yA,AB,phi,omega,alpha][[3]] ;
  vBy = Driver[xA,yA,AB,phi,omega,alpha][[4]] ;

  (* Acceleration of joint B *)
  aBx = Driver[xA,yA,AB,phi,omega,alpha][[5]] ;
  aBy = Driver[xA,yA,AB,phi,omega,alpha][[6]] ;

  Print["rB = ",{xB,yB,0}," [m]"];
  Print["vB = ",{vBx,vBy,0}," [m/s]"];
```

```
Print["aB = ",{aBx,aBy,0}," [m/s^2]"];

   rB = {0.173205, 0.1, 0} [m]

   vB = {-0.5, 0.866025, 0} [m/s]

   aB = {-4.33013, -2.5, 0} [m/s^2]
```

PROGRAM I.8.3

```
BeginPackage["Position`"]

PosRRR::usage =
    "PosRRR[xM,yM,xN,yN,MP,NP] computes the position vectors for RRR dyad"

Begin["`Private`"]

PosRRR[xM_,yM_,xN_,yN_,MP_,NP_]:=
Block[

{ xPSol, yPSol, xP1, yP1, xP2, yP2, eqRRR1, eqRRR2, solRRR },

eqRRR1 = (xM-xPSol)^2 + (yM-yPSol)^2 == MP^2 ;
eqRRR2 = (xN-xPSol)^2 + (yN-yPSol)^2 == NP^2 ;

solRRR = Solve[ { eqRRR1 , eqRRR2 }, { xPSol, yPSol } ] ;

xP1 = xPSol/.solRRR[[1]] ;
yP1 = yPSol/.solRRR[[1]] ;
xP2 = xPSol/.solRRR[[2]] ;
yP2 = yPSol/.solRRR[[2]] ;

Return[ { xP1, yP1, xP2, yP2 } ] ;
]

End[ ] (* PosRRR *)

PosRRT::usage =
    "PosRRT[xM,yM,xN,yN,MP,theta] computes the position vectors for RRT \
dyad"

Begin["`Private`"]

PosRRT[xM_,yM_,xN_,yN_,MP_,theta_]:=
Block[

{ xPSol, yPSol, xP1, yP1, xP2, yP2, eqRRT, solRRT, eqRRT1, eqRRT2 },

If[ (theta==Pi/2)||(theta==3*Pi/2),

    xP1 = xP2 = xN ;
```

```
        eqRRT = (xM-xN)^2 + (yM-yPSol)^2 == MP^2 ;
        solRRT = Solve[ eqRRT, yPSol ] ;
        yP1 = yPSol/.solRRT[[1]] ;
        yP2 = yPSol/.solRRT[[2]] ,

        eqRRT1 = (xM-xPSol)^2 + (yM-yPSol)^2 == MP^2 ;
        eqRRT2 = Tan[theta] == (yPSol-yN)/(xPSol-xN) ;
        solRRT = Solve[ { eqRRT1 , eqRRT2 }, { xPSol , yPSol } ] ;
        xP1 = xPSol/.solRRT[[1]] ;
        yP1 = yPSol/.solRRT[[1]] ;
        xP2 = xPSol/.solRRT[[2]] ;
        yP2 = yPSol/.solRRT[[2]] ;
  ] ;

  Return[ { xP1, yP1, xP2, yP2 } ] ;
  ]

  End[ ] (* PosRRT *)
  EndPackage[ ]

  Position`

  PosRRR[xM,yM,xN,yN,MP,NP] computes the position vectors for RRR dyad

  Position`Private`

  Position`Private`

  PosRRT[xM,yM,xN,yN,MP,theta] computes the position vectors for RRT dyad

  Position`Private`

  Position`Private`
```

PROGRAM I.8.4

```
  (* R-RTR-RRT mechanism *)

  Apply[Clear,Names["Global`*"]] ;
  Off[General::spell];
  Off[General::spell1];

  <<Driver.m ;
  <<Position.m ;

  (* Input data *)

  AB = 0.20 ;      (* m *)
```

```
AD = 0.40 ;        (* m *)
CD = 0.70 ;        (* m *)
CE = 0.30 ;        (* m *)
yE = 0.35 ;        (* m *)
phi = Pi/4 ;       (* rad *)
omega = 5. ;       (* rad/s *)
alpha = 0. ;       (* rad/s^2 *)

(* Position Vectors *)

xA = yA = 0 ;

xB = Driver[xA,yA,AB,phi,omega,alpha][[1]] ;
yB = Driver[xA,yA,AB,phi,omega,alpha][[2]] ;

xD = 0 ;
yD = -AD ;

phi3 = ArcTan[(yB-yD)/(xB-xD)] ;

xC = xD + CD Cos[phi3] ;
yC = yD + CD Sin[phi3] ;

phi5 = Pi ;

(* xP=0; yP=yE; *)

xE1 = PosRRT[xC,yC,0,yE,CE,phi5][[1]] ;
xE2 = PosRRT[xC,yC,0,yE,CE,phi5][[3]] ;

(* Choose the correct solution *)
If[ (xE1<xC), xE=xE1, xE=xE2 ] ;

Print["rB = ",{xB,yB,0}," [m]" ];
Print["phi3 = ",phi3*180/N[Pi]," [deg]"];
Print["rC = ",{xC,yC,0}," [m]" ];
Print["rE = ",{xE,yE,0}," [m]" ];

rB = {0.141421, 0.141421, 0} [m]

phi3 = 75.3612 [deg]

rC = {0.176907, 0.277277, 0} [m]

rE = {-0.114145, 0.35, 0} [m]
```

PROGRAM I.8.5
```
(* R-RRR-RRT mechanism *)

Apply [ Clear , Names [ "Global`*" ] ] ;
```

```
Off[General::spell];
Off[General::spell1];

<<Driver.m ;
<<Position.m ;

(* Input data *)

AB = 0.15 ;      (* m *)
BC = 0.40 ;      (* m *)
CD = 0.37 ;      (* m *)
CE = 0.23 ;      (* m *)
EF = CE ;        (* m *)
La = 0.30 ;      (* m *)
Lb = 0.45 ;      (* m *)
Lc = CD ;        (* m *)
phi = Pi/4 ;     (* rad *)
omega = N[Pi] ;  (* rad/s *)
alpha = 0. ;     (* rad/s^2 *)

(* Position Vectors *)

xA = yA = 0 ;

xB = Driver[xA,yA,AB,phi,omega,alpha][[1]] ;
yB = Driver[xA,yA,AB,phi,omega,alpha][[2]] ;

xD = La ;
yD = Lb ;

xC1 = PosRRR[xB,yB,xD,yD,BC,CD][[1]] ;
yC1 = PosRRR[xB,yB,xD,yD,BC,CD][[2]] ;
xC2 = PosRRR[xB,yB,xD,yD,BC,CD][[3]] ;
yC2 = PosRRR[xB,yB,xD,yD,BC,CD][[4]] ;

(* Choose the correct solution *)
If[ (yC1>yB), xC=xC1;yC=yC1, xC=xC2;yC=yC2 ] ;

phi3 = ArcTan[(yC-yD)/(xC-xD)] + Pi ;

xE = xC + CE Cos[phi3] ;
yE = yC + CE Sin[phi3] ;

xF = - Lc ;

phi5 = Pi/2 ;
```

```
yF1 = PosRRT[xE,yE,-Lc,0,EF,phi5][[2]] ;
yF2 = PosRRT[xE,yE,-Lc,0,EF,phi5][[4]] ;

(* Choose the correct solution *)
If[ (yF1<yE), yF=yF1, yF=yF2 ] ;

Print["Positions"];
Print["rB = ",{xB,yB,0}," [m]" ];
Print["rC = ",{xC,yC,0}," [m]" ];
Print["rE = ",{xE,yE,0}," [m]" ];
Print["rF = ",{xF,yF,0}," [m]" ];

phi2 = ArcTan[(yC-yB)/(xC-xB)] + Pi ;
phi4 = ArcTan[(yE-yF)/(xE-xF)] ;

Print["Angles"];
Print["phi = ",phi," [rad]"];
Print["phi2 = ",phi2," [rad]"];
Print["phi3 = ",phi3," [rad]"];
Print["phi4 = ",phi4," [rad]"];
```

Positions

rB = {0.106066, 0.106066, 0} [m]

rC = {-0.0696798, 0.46539, 0} [m]

rE = {-0.299481, 0.474956, 0} [m]

rF = {-0.37, 0.256034, 0} [m]

Angles

phi = $\dfrac{\pi}{4}$ [rad]

phi2 = 2.02569 [rad]

phi3 = 3.09999 [rad]

phi4 = 1.25917 [rad]

PROGRAM I.8.6
```
(* R-RRT mechanism *)

Apply [ Clear , Names [ "Global`*" ] ] ;
Off[General::spell];
Off[General::spell1];
```

```
        <<Driver.m ;
        <<Position.m ;

        (* Input data *)

        AC = 0.10 ;              (* m *)
        BC = 0.30 ;              (* m *)
        AP = 0.50 ;              (* m *)
        phi = Pi/4 ;             (* rad *)
        n = 30 ;                 (* rpm *)
        omega = N[Pi]*n/30 ;     (* rad/s *)
        alpha = 0. ;             (* rad/s^2 *)

        (* Position Vectors *)

        xA = yA = 0 ;

        xC = AC ;
        yC = 0 ;

        xP = Driver[xA,yA,AP,phi,omega,alpha][[1]] ;
        yP = Driver[xA,yA,AP,phi,omega,alpha][[2]] ;

        xB1 = PosRRT[xC,yC,xP,yP,BC,phi][[1]] ;
        yB1 = PosRRT[xC,yC,xP,yP,BC,phi][[2]] ;
        xB2 = PosRRT[xC,yC,xP,yP,BC,phi][[3]] ;
        yB2 = PosRRT[xC,yC,xP,yP,BC,phi][[4]] ;

        (* Choose the correct solution *)
        If[ (yB1>yC), xB=xB1;yB=yB1, xB=xB2;yB=yB2 ] ;

        Print["rB = ",{xB,yB,0}," [m]"];

        rB = {0.256155, 0.256155, 0} [m]
```

PROGRAM I.8.7

```
BeginPackage["VelAcc`"]

VelAccRRR::usage =
    "VelAccRRR[xM,yM,xN,yN,xP,yP,vMx,vMy,vNx,vNy,aMx,aMy,aNx,aNy]
computes the velocity

Begin["`Private`"]

VelAccRRR[xM_,yM_,xN_,yN_,xP_,yP_,vMx_,vMy_,vNx_,vNy_,aMx_,aMy_,
aNx_,aNy_]:=
Block[
```

```
{ vPxSol, vPySol, aPxSol, aPySol, vPx, vPy, aPx, aPy, eqRRR1v,
eqRRR2v, solRRRv, eqRRR1a, eqRRR2a, solRRRa },

(* Velocity *)

eqRRR1v = (xM-xP) (vMx-vPxSol) + (yM-yP) (vMy-vPySol) == 0 ;
eqRRR2v = (xN-xP) (vNx-vPxSol) + (yN-yP) (vNy-vPySol) == 0 ;

solRRRv = Solve[ { eqRRR1v , eqRRR2v }, { vPxSol, vPySol } ] ;

vPx = vPxSol/.solRRRv[[1]] ;
vPy = vPySol/.solRRRv[[1]] ;

(* Acceleration *)

eqRRR1a = (xM-xP) (aMx-aPxSol) + (vMx-vPx)^2 + (yM-yP) (aMy-aPySol)
+ (vMy-vPy)^2 == 0 ;
eqRRR2a = (xN-xP) (aNx-aPxSol) + (vNx-vPx)^2 + (yN-yP) (aNy-aPySol)
+ (vNy-vPy)^2 == 0 ;

solRRRa = Solve[ { eqRRR1a , eqRRR2a }, { aPxSol, aPySol } ] ;

aPx = aPxSol/.solRRRa[[1]] ;
aPy = aPySol/.solRRRa[[1]] ;

Return[ { vPx, vPy, aPx, aPy } ] ;
]

End[ ]  (* VelAccRRR *)


VelAccRRT::usage =
    "VelAccRRT[xM,yM,xN,yN,xP,yP,vMx,vMy,vNx,vNy,aMx,aMy,aNx,aNy,
theta,dtheta,ddtheta]

Begin["`Private`"]

VelAccRRT[xM_,yM_,xN_,yN_,xP_,yP_,vMx_,vMy_,vNx_,vNy_,aMx_,aMy_,aNx_,
aNy_,theta_,dtheta_,
Block[

{ vPxSol, vPySol, aPxSol, aPySol, vPx, vPy, aPx, aPy,
eqRRTv, eqRRTa, eqRRT1v, eqRRT2v, solRRTv , eqRRR1a, eqRRR2a, solRRRa },

(* Velocity *)

eqRRT1v = (xM-xP) (vMx-vPxSol) + (yM-yP) (vMy-vPySol) == 0 ;
```

```
eqRRT2v = Sin[theta] (vPxSol-vNx) + Cos[theta] dtheta (xP-xN)
- Cos[theta] (vPySol-vNy) + Sin[theta] dtheta (yP-yN) == 0 ;


solRRTv = Solve[ { eqRRT1v , eqRRT2v }, { vPxSol , vPySol } ] ;


vPx = vPxSol/.solRRTv[[1]] ;
vPy = vPySol/.solRRTv[[1]] ;


(* Acceleration *)


eqRRT1a = (xM-xP) (aMx-aPxSol) + (vMx-vPx)^2 + (yM-yP) (aMy-aPySol)
+ (vMy-vPy)^2 == 0 ;
eqRRT2a =
+ Sin[theta] (aPxSol-aNx) - Cos[theta] (aPySol-aNy)
+ ( 2 Cos[theta] (vPx-vNx) - Sin[theta] dtheta (xP-xN)
+ 2 Sin[theta] (vPy-vNy) + Cos[theta] dtheta (yP-yN) ) dtheta
+ ( Cos[theta] (xP-xN) + Sin[theta] (yP-yN) ) ddtheta == 0 ;


solRRTa = Solve[ { eqRRT1a , eqRRT2a }, { aPxSol , aPySol } ] ;


aPx = aPxSol/.solRRTa[[1]] ;
aPy = aPySol/.solRRTa[[1]] ;


Return[ { vPx, vPy, aPx, aPy } ] ;
]


End[ ] (* VelAccRRT *)



AngVelAcc::usage =
    "AngVelAcc[xM,yM,xN,yN,vMx,vMy,vNx,vNy,aMx,aMy,aNx,aNy,theta]
computes the angular velocity and


Begin["`Private`"]


AngVelAcc[xM_,yM_,xN_,yN_,vMx_,vMy_,vNx_,vNy_,aMx_,aMy_,aNx_,aNy_,
    theta_]:=
Block[

{ dtheta, ddtheta },

dtheta =
( Cos[theta] (vMy-vNy) - Sin[theta] (vMx-vNx) )
/ ( Sin[theta] (yM-yN) + Cos[theta] (xM-xN) ) ;
```

```
ddtheta =
( Cos[theta] (aMy-aNy) - Sin[theta] (aMx-aNx) -
( Cos[theta] dtheta (yM-yN) + 2 Sin[theta] (vMy-vNy)
- Sin[theta] dtheta (xM-xN) + 2 Cos[theta] (vMx-vNx) ) dtheta )
/ ( Cos[theta] (xM-xN) + Sin[theta] (yM-yN) ) ;

Return[ { dtheta, ddtheta } ] ;
]

End[ ] (* AngVelAcc *)


AbsVelAcc::usage =
     "AbsVelAcc[xM,yM,xN,yN,vMx,vMy,aMx,aMy,dtheta,ddtheta] computes
the absolute velocity and acceleration

Begin["`Private`"]

AbsVelAcc[xM_,yM_,xN_,yN_,vMx_,vMy_,aMx_,aMy_,dtheta_,ddtheta_]:=
Block[

{ vNx, vNy, aNx, aNy },

vNx = vMx - dtheta (yN-yM) ;
vNy = vMy + dtheta (xN-xM) ;

aNx = aMx - ddtheta (yN-yM) - dtheta^2 (xN-xM) ;
aNy = aMy + ddtheta (xN-xM) - dtheta^2 (yN-yM) ;

Return[ { vNx, vNy, aNx, aNy } ] ;
]

End[ ] (* AbsVelAcc *)


EndPackage[ ]


VelAcc`

VelAccRRR[xM,yM,xN,yN,xP,yP,vMx,vMy,vNx,vNy,aMx,aMy,aNx,
   aNy] computes the velocity and acceleration vectors for RRR dyad

VelAcc`Private`

VelAcc`Private`

VelAccRRT[xM,yM,xN,yN,xP,yP,vMx,vMy,vNx,vNy,aMx,aMy,aNx,aNy,theta,
   dtheta, ddtheta] computes the velocity and acceleration vectors for
   RRT dyad

VelAcc`Private`
```

VelAcc`Private`

AngVelAcc[xM,yM,xN,yN,vMx,vMy,vNx,vNy,aMx,aMy,aNx,aNy,
    theta] computes the angular velocity and acceleration of a link.

VelAcc`Private`

VelAcc`Private`

AbsVelAcc[xM,yM,xN,yN,vMx,vMy,aMx,aMy,dtheta,
    ddtheta] computes the absolute velocity and acceleration vectors.

VelAcc`Private`

VelAcc`Private`

PROGRAM I.8.8

```
(* R-RRR-RRT mechanism *)

Apply [ Clear , Names [ "Global`*" ] ] ;
Off[General::spell];
Off[General::spell1];

<<Driver.m ;
<<Position.m ;
<<VelAcc.m ;

(* Input data *)

AB = 0.15 ;              (* m *)
BC = 0.40 ;              (* m *)
CD = 0.37 ;              (* m *)
CE = 0.23 ;              (* m *)
EF = CE ;                (* m *)
La = 0.30 ;              (* m *)
Lb = 0.45 ;              (* m *)
Lc = CD ;                (* m *)
phi = Pi/4 ;             (* rad *)
omega = N[Pi]*100/30 ;   (* rad/s *)
alpha = 0. ;             (* rad/s^2 *)

(* Position Vectors *)

xA = yA = 0 ;

xB = Driver[xA,yA,AB,phi,omega,alpha][[1]] ;
yB = Driver[xA,yA,AB,phi,omega,alpha][[2]] ;

xD = La ;
yD = Lb ;
```

```
xC1 = PosRRR[xB,yB,xD,yD,BC,CD][[1]] ;
yC1 = PosRRR[xB,yB,xD,yD,BC,CD][[2]] ;
xC2 = PosRRR[xB,yB,xD,yD,BC,CD][[3]] ;
yC2 = PosRRR[xB,yB,xD,yD,BC,CD][[4]] ;

(* Choose the correct solution *)
If[ (yC1>yB), xC=xC1;yC=yC1, xC=xC2;yC=yC2 ] ;

phi3 = ArcTan[(yC-yD)/(xC-xD)] + Pi ;

xE = xC + CE Cos[phi3] ;
yE = yC + CE Sin[phi3] ;

xF = - Lc ;

phi5 = Pi/2 ;

xP = - Lc ;
yP = 0 ;

yF1 = PosRRT[xE,yE,xP,yP,EF,phi5][[2]] ;
yF2 = PosRRT[xE,yE,xP,yP,EF,phi5][[4]] ;

(* Choose the correct solution *)
If[ (yF1<yE), yF=yF1, yF=yF2 ] ;

phi2 = ArcTan[(yC-yB)/(xC-xB)] + Pi ;

phi4 = ArcTan[(yE-yF)/(xE-xF)] ;

Print["Positions"];
Print["rB = ",{xB,yB,0}," [m]"];
Print["rC = ",{xC,yC,0}," [m]"];
Print["rE = ",{xE,yE,0}," [m]"];
Print["rF = ",{xF,yF,0}," [m]"];

Print["Angles"];
Print["phi = ",phi," [rad]"];
Print["phi2 = ",phi2," [rad]"];
Print["phi3 = ",phi3," [rad]"];
Print["phi4 = ",phi4," [rad]"];

(* Velocity and acceleration vectors *)

vBx = Driver[xA,yA,AB,phi,omega,alpha][[3]] ;
vBy = Driver[xA,yA,AB,phi,omega,alpha][[4]] ;
aBx = Driver[xA,yA,AB,phi,omega,alpha][[5]] ;
aBy = Driver[xA,yA,AB,phi,omega,alpha][[6]] ;
```

```
vDx = 0 ; vDy = 0 ;
aDx = 0 ; aDy = 0 ;

vCx =
VelAccRRR[xB,yB,xD,yD,xC,yC,vBx,vBy,vDx,vDy,aBx,aBy,aDx,aDy][[1]] ;
vCy =
VelAccRRR[xB,yB,xD,yD,xC,yC,vBx,vBy,vDx,vDy,aBx,aBy,aDx,aDy][[2]] ;
aCx =
VelAccRRR[xB,yB,xD,yD,xC,yC,vBx,vBy,vDx,vDy,aBx,aBy,aDx,aDy][[3]] ;
aCy =
VelAccRRR[xB,yB,xD,yD,xC,yC,vBx,vBy,vDx,vDy,aBx,aBy,aDx,aDy][[4]] ;

omega3 =
AngVelAcc[xC,yC,xD,yD,vCx,vCy,vDx,vDy,aCx,aCy,aDx,aDy,phi3][[1]] ;

alpha3 =
AngVelAcc[xC,yC,xD,yD,vCx,vCy,vDx,vDy,aCx,aCy,aDx,aDy,phi3][[2]] ;

vEx = AbsVelAcc[xD,yD,xE,yE,vDx,vDy,aDx,aDy,omega3,alpha3][[1]] ;
vEy = AbsVelAcc[xD,yD,xE,yE,vDx,vDy,aDx,aDy,omega3,alpha3][[2]] ;

aEx = AbsVelAcc[xD,yD,xE,yE,vDx,vDy,aDx,aDy,omega3,alpha3][[3]] ;
aEy = AbsVelAcc[xD,yD,xE,yE,vDx,vDy,aDx,aDy,omega3,alpha3][[4]] ;

vPx = vPy = 0 ;
aPx = aPy = 0 ;

omega5 = alpha5 = 0 ;

vFx =
VelAccRRT[xE,yE,xP,yP,xF,yF,vEx,vEy,vPx,vPy,aEx,aEy,aPx,aPy,phi5,
omega5,alpha5][[1]] ;
vFy =
VelAccRRT[xE,yE,xP,yP,xF,yF,vEx,vEy,vPx,vPy,aEx,aEy,aPx,aPy,phi5,
omega5,alpha5][[2]] ;
aFx =
VelAccRRT[xE,yE,xP,yP,xF,yF,vEx,vEy,vPx,vPy,aEx,aEy,aPx,aPy,phi5,
omega5,alpha5][[3]] ;
aFy =
VelAccRRT[xE,yE,xP,yP,xF,yF,vEx,vEy,vPx,vPy,aEx,aEy,aPx,aPy,phi5,
omega5,alpha5][[4]] ;

Print["Velocities"];
Print["vB = ",{vBx,vBy,0}," [m/s]"];
Print["vC = ",{vCx,vCy,0}," [m/s]"];
Print["vE = ",{vEx,vEy,0}," [m/s]"];
Print["vF = ",{vFx,vFy,0}," [m/s]"];

Print["Accelerations"];
Print["aB = ",{aBx,aBy,0}," [m/s^2]"];
```

```
Print["aC = ",{aCx,aCy,0}," [m/s^2]"];
Print["aE = ",{aEx,aEy,0}," [m/s^2]"];
Print["aF = ",{aFx,aFy,0}," [m/s^2]"];

omega2 =
AngVelAcc[xB,yB,xC,yC,vBx,vBy,vCx,vCy,aBx,aBy,aCx,aCy,phi2][[1]] ;

alpha2 =
AngVelAcc[xB,yB,xC,yC,vBx,vBy,vCx,vCy,aBx,aBy,aCx,aCy,phi2][[2]] ;

omega4 =
AngVelAcc[xE,yE,xF,yF,vEx,vEy,vFx,vFy,aEx,aEy,aFx,aFy,phi4][[1]] ;

alpha4 =
AngVelAcc[xE,yE,xF,yF,vEx,vEy,vFx,vFy,aEx,aEy,aFx,aFy,phi4][[2]] ;

Print["Angular velocities"];
Print["omega = ",{0,0,omega}," [rad/s]"];
Print["omega2 = ",{0,0,omega2}," [rad/s]"];
Print["omega3 = ",{0,0,omega3}," [rad/s]"];
Print["omega4 = ",{0,0,omega4}," [rad/s]"];

Print["Angular accelerations"];
Print["alpha = ",{0,0,alpha}," [rad/s^2]"];
Print["alpha2 = ",{0,0,alpha2}," [rad/s^2]"];
Print["alpha3 = ",{0,0,alpha3}," [rad/s^2]"];
Print["alpha4 = ",{0,0,alpha4}," [rad/s^2]"];

Positions

rB = {0.106066, 0.106066, 0} [m]

rC = {-0.0696798, 0.46539, 0} [m]

rE = {-0.299481, 0.474956, 0} [m]

rF = {-0.37, 0.256034, 0} [m]

Angles

phi = $\frac{\pi}{4}$ [rad]

phi2 = 2.02569 [rad]

phi3 = 3.09999 [rad]

phi4 = 1.25917 [rad]
```

Velocities

vB = {-1.11072, 1.11072, 0} [m/s]

vC = {0.0702851, 1.68835, 0} [m/s]

vE = {0.113976, 2.73787, 0} [m/s]

vF = {0., 2.77458, 0} [m/s]

Accelerations

aB = {-11.6314, -11.6314, 0} [m/s^2]

aC = {7.42779, -7.11978, 0} [m/s^2]

aE = {12.0451, -11.5456, 0} [m/s^2]

aF = {0., -7.60013, 0} [m/s^2]

Angular velocities

omega = {0, 0, 10.472} [rad/s]

omega2 = {0, 0, -3.28675} [rad/s]

omega3 = {0, 0, -4.56707} [rad/s]

omega4 = {0, 0, -0.520622} [rad/s]

Angular accelerations

alpha = {0, 0, 0.} [rad/s^2]

alpha2 = {0, 0, -47.7584} [rad/s^2]

alpha3 = {0, 0, 18.391} [rad/s^2]

alpha4 = {0, 0, -55.1071} [rad/s^2]

PROGRAM I.8.9
(* R-RRT mechanism *)

```
Apply [ Clear , Names [ "Global`*" ] ] ;
Off[General::spell];
Off[General::spell1];

<<Driver.m ;
```

```
<<Position.m ;
<<VelAcc.m ;

(* Input data *)

AC = 0.10 ;            (* m *)
BC = 0.30 ;            (* m *)
AP = 0.50 ;            (* m *)
phi = Pi/4 ;           (* rad *)
n = 30 ;               (* rpm *)
omega = N[Pi]*n/30 ;   (* rad/s *)
alpha = 0. ;           (* rad/s^2 *)

(* Position Vectors *)

xA = yA = 0 ;

xC = AC ;
yC = 0 ;

xP = Driver[xA,yA,AP,phi,omega,alpha][[1]] ;
yP = Driver[xA,yA,AP,phi,omega,alpha][[2]] ;

xB1 = PosRRT[xC,yC,xP,yP,BC,phi][[1]] ;
yB1 = PosRRT[xC,yC,xP,yP,BC,phi][[2]] ;
xB2 = PosRRT[xC,yC,xP,yP,BC,phi][[3]] ;
yB2 = PosRRT[xC,yC,xP,yP,BC,phi][[4]] ;

(* Choose the correct solution *)
If[ (yB1>yC), xB=xB1;yB=yB1, xB=xB2;yB=yB2 ] ;

Print["Positions"];
Print["rB = ",{xB,yB,0}," [m]"];

(* Velocity and acceleration vectors *)

vPx = Driver[xA,yA,AP,phi,omega,alpha][[3]] ;
vPy = Driver[xA,yA,AP,phi,omega,alpha][[4]] ;
aPx = Driver[xA,yA,AP,phi,omega,alpha][[5]] ;
aPy = Driver[xA,yA,AP,phi,omega,alpha][[6]] ;

vCx = 0 ; vCy = 0 ;
aCx = 0 ; aCy = 0 ;

vBx =
VelAccRRT[xC,yC,xP,yP,xB,yB,vCx,vCy,vPx,vPy,aCx,aCy,aPx,aPy,phi,
omega,alpha][[1]] ;
vBy =
VelAccRRT[xC,yC,xP,yP,xB,yB,vCx,vCy,vPx,vPy,aCx,aCy,aPx,aPy,phi,
omega,alpha][[2]] ;
```

```
aBx =
VelAccRRT[xC,yC,xP,yP,xB,yB,vCx,vCy,vPx,vPy,aCx,aCy,aPx,aPy,phi,
omega,alpha][[3]] ;
aBy =
VelAccRRT[xC,yC,xP,yP,xB,yB,vCx,vCy,vPx,vPy,aCx,aCy,aPx,aPy,phi,
omega,alpha][[4]] ;

Print["Velocities"];
Print["vB = ",{vBx,vBy,0}," [m/s]"];

Print["Accelerations"];
Print["aB = ",{aBx,aBy,0}," [m/s^2]"];

phi3 = ArcTan[(yB-yC)/(xB-xC)] ;

omega3 = AngVelAcc[xB,yB,xC,yC, vBx,vBy,vCx,vCy,aBx,aBy,aCx,aCy][[1]] ;

alpha3 = AngVelAcc[xB,yB,xC,yC, vBx,vBy,vCx,vCy,aBx,aBy,aCx,aCy][[2]] ;

Print["Angular velocities"];
Print["omega3 = ",{0,0,omega3}," [rad/s]"];

Print["Angular accelerations"];
Print["alpha3 = ",{0,0,alpha3}," [rad/s^2]"];
```

Positions

rB = {0.256155, 0.256155, 0} [m]

Velocities

vB = {-0.999913, 0.609559, 0} [m/s]

Accelerations

aB = {-1.80234, -4.25501, 0} [m/s^2]

Angular velocities

omega3 = {0, 0, 3.90354} [rad/s]

Angular accelerations

alpha3 = {0, 0, -2.25292} [rad/s^2]

PROGRAM I.8.10

```
(* R-RTR-RTR mechanism *)

Apply [ Clear , Names [ "Global`*" ] ] ;
Off[General::spell];
Off[General::spell1];
```

```
<<Driver.m ;
<<VelAcc.m ;

(* Input data *)

AB = 0.14 ;           (* m *)
AC = 0.06 ;           (* m *)
AE = 0.25 ;           (* m *)
CD = 0.15 ;           (* m *)
phi = Pi/6 ;          (* rad *)
n = 50 ;              (* rpm *)
omega = n*N[Pi]/30 ;  (* rad/s *)
alpha = 0. ;          (* rad/s^2 *)

(* Position Vectors *)

xA = yA = 0 ;

xB = Driver[xA,yA,AB,phi,omega,alpha][[1]] ;
yB = Driver[xA,yA,AB,phi,omega,alpha][[2]] ;

xC = 0 ;
yC = AC ;

phi3 = ArcTan[(yB-yC)/(xB-xC)] ;

xD = xC - CD Cos[phi3] ;
yD = yC - CD Sin[phi3] ;

xE = 0 ;
yE = -AE ;

phi5 = Pi + ArcTan[(yD-yE)/(xD-xE)] ;

(* Velocity and acceleration vectors *)

vBx = Driver[xA,yA,AB,phi,omega,alpha][[3]] ;
vBy = Driver[xA,yA,AB,phi,omega,alpha][[4]] ;
aBx = Driver[xA,yA,AB,phi,omega,alpha][[5]] ;
aBy = Driver[xA,yA,AB,phi,omega,alpha][[6]] ;

vCx = vCy = 0 ;
aCx = aCy = 0 ;

omega3 =
AngVelAcc[xB,yB,xC,yC, vBx,vBy,vCx,vCy,aBx,aBy,aCx,aCy,phi3][[1]] ;
```

```
alpha3 =
AngVelAcc[xB,yB,xC,yC, vBx,vBy,vCx,vCy,aBx,aBy,aCx,aCy,phi3][[2]] ;

vDx = AbsVelAcc[xC,yC,xD,yD,vCx,vCy,aCx,aCy,omega3,alpha3][[1]] ;
vDy = AbsVelAcc[xC,yC,xD,yD,vCx,vCy,aCx,aCy,omega3,alpha3][[2]] ;
aDx = AbsVelAcc[xC,yC,xD,yD,vCx,vCy,aCx,aCy,omega3,alpha3][[3]] ;
aDy = AbsVelAcc[xC,yC,xD,yD,vCx,vCy,aCx,aCy,omega3,alpha3][[4]] ;

vEx = vEy = 0 ;
aEx = aEy = 0 ;

omega5 =
AngVelAcc[xD,yD,xE,yE, vDx,vDy,vEx,vEy,aDx,aDy,aEx,aEy,phi5][[1]] ;

alpha5 =
AngVelAcc[xD,yD,xE,yE, vDx,vDy,vEx,vEy,aDx,aDy,aEx,aEy,phi5][[2]] ;

Print["Positions"];
Print["rB = ",{xB,yB,0}," [m]"];
Print["rD = ",{xD,yD,0}," [m]"];

Print["Velocities"];
Print["vB = ",{vBx,vBy,0}," [m/s]"];
Print["vD = ",{vDx,vDy,0}," [m/s]"];

Print["Accelerations"];
Print["aB = ",{aBx,aBy,0}," [m/s^2]"];
Print["aD = ",{aDx,aDy,0}," [m/s^2]"];

Print["Angles"];
Print["phi = ",phi," [rad]"];
Print["phi3 = ",phi3," [rad]"];
Print["phi5 = ",phi5," [rad]"];

Print["Angular velocities"];
Print["omega = ",{0,0,omega}," [rad/s]"];
Print["omega3 = ",{0,0,omega3}," [rad/s]"];
Print["omega5 = ",{0,0,omega5}," [rad/s]"];

Print["Angular accelerations"];
Print["alpha = ",{0,0,alpha}," [rad/s^2]"];
Print["alpha3 = ",{0,0,alpha3}," [rad/s^2]"];
Print["alpha5 = ",{0,0,alpha5}," [rad/s^2]"];

Positions

rB = {0.121244, 0.07, 0} [m]

rD = {-0.149492, 0.0476701, 0} [m]
```

Velocities

vB = {-0.366519, 0.63483, 0} [m/s]

vD = {0.0671766, -0.814473, 0} [m/s]

Accelerations

aB = {-3.32396, -1.91909, 0} [m/s^2]

aD = {4.61708, -1.81183, 0} [m/s^2]

Angles

phi = $\frac{\pi}{6}$ [rad]

phi3 = 0.0822923 [rad]

phi5 = 2.03621 [rad]

Angular velocities

omega = {0, 0, 5.23599} [rad/s]

omega3 = {0, 0, 5.44826} [rad/s]

omega5 = {0, 0, 0.917134} [rad/s]

Angular accelerations

alpha = {0, 0, 0.} [rad/s^2]

alpha3 = {0, 0, 14.5681} [rad/s^2]

alpha5 = {0, 0, -5.77155} [rad/s^2]

PROGRAM I.8.11

```
BeginPackage["Force`"]

ForceMomentum::usage =
     "ForceMomentum[m,aCM,ICM,ddtheta] computes the total force and
moment of a rigid link."

Begin["`Private`"]

ForceMomentum[m_,aCM_,ICM_,ddtheta_]:=
Block[
```

```
          { g, Fin, G, F, M },

          g = 9.807 ;
          Fin = - m aCM ;
          G = { 0, -m g, 0 } ;
          F = Fin + G ;
          M = - ICM { 0, 0, ddtheta } ;

          Return[ { F, M } ] ;
          ]

          End[ ] (* ForceMomentum *)



          ForceRRR::usage =
              "ForceRRR[F2,M2,F3,M3,rM,rN,rP,rC2,rC3] computes the joint
          reactions for the RRR dyad."

          Begin["`Private`"]

          ForceRRR[F2_,M2_,F3_,M3_,rM_,rN_,rP_,rC2_,rC3_]:=
          Block[

          { F12, F12Sol, F12xSol, F12ySol, F43, F43Sol, F43xSol, F43ySol,
          rPC2, rPC3, rPM, rPN, F32, eqRRR1, eqRRR2, eqRRR3, eqRRR4, solRRR },

          F12Sol = { F12xSol, F12ySol, 0 } ;
          F43Sol = { F43xSol, F43ySol, 0 } ;

          rPC2 = rC2 - rP ;
          rPC3 = rC3 - rP ;
          rPM = rM - rP ;
          rPN = rN - rP ;

          eqRRR1 = (F12Sol+F43Sol+F2+F3)[[1]] == 0 ;
          eqRRR2 = (F12Sol+F43Sol+F2+F3)[[2]] == 0 ;
          eqRRR3 = (Cross[rPC2,F2]+Cross[rPM,F12Sol]+M2)[[3]] == 0 ;
          eqRRR4 = (Cross[rPC3,F3]+Cross[rPN,F43Sol]+M3)[[3]] == 0 ;

          solRRR = Solve[ {eqRRR1, eqRRR2, eqRRR3, eqRRR4},
          {F12xSol,F12ySol,F43xSol,F43ySol} ] ;

          F12 = F12Sol/.solRRR[[1]] ;
          F43 = F43Sol/.solRRR[[1]] ;

          F32 = - F2 - F12 ;
```

```
Return[ { F12, F43, F23 } ] ;
]

End[ ] (* ForceRRR *)

ForceRRT::usage =
    "ForceRRT[F2,M2,F3,M3,rM,rN,rP,rC2] computes the joint reactions
for the RRR dyad."

Begin["`Private`"]

ForceRRT[F2_,M2_,F3_,M3_,rM_,rN_,rP_,rC2_]:=
Block[

{ F12, F12Sol, F12xSol, F12ySol, F43, F43Sol, F43xSol,
F43ySol, F32, eqRRT1, eqRRT2, eqRRT3, eqRRT4, solRRT, rNC2, rNM,
rNP, rNQ, rQP, rQ, rQSol, xQSol, yQSol, eqRRTQ1, eqRRTQ2, solRRTQ },

F12Sol = { F12xSol, F12ySol, 0 } ;
F43Sol = { F43xSol, F43ySol, 0 } ;

rPC2 = rC2 - rP ;
rPM = rM - rP ;
rPN = rN - rP ;

eqRRT1 = (F12Sol+F43Sol+F2+F3)[[1]] == 0 ;
eqRRT2 = (F12Sol+F43Sol+F2+F3)[[2]] == 0 ;
eqRRT3 = (F43Sol.rPN) == 0 ;
eqRRT4 = (Cross[rPC2,F2]+Cross[rPM,F12Sol]+M2)[[3]] == 0 ;

solRRT = Solve[ {eqRRT1, eqRRT2, eqRRT3, eqRRT4},
{F12xSol,F12ySol,F43xSol,F43ySol} ] ;

F12 = F12Sol/.solRRT[[1]] ;
F43 = F43Sol/.solRRT[[1]] ;

F32 = - F2 - F12 ;
F23 = - F32 ;

If[ M3[[3]]==0 , rQ = rP ,

rQSol = { xQSol, yQSol, 0 } ;
rPQ = rQSol - rP ;

eqRRTQ1 = rPQ[[2]]/rPQ[[1]] - rPN[[2]]/rPN[[1]] == 0 ;
eqRRTQ2 = (Cross[rPQ,F43]+M3)[[3]] == 0 ;
```

```
solRRTQ = Solve[ {eqRRTQ1, eqRRTQ2} , {xQSol,yQSol} ] ;
rQ = rQSol/.solRRTQ[[1]] ;
] ;

Return[ { F12, F43, F23, rQ } ] ;
]

End[ ] (* ForceRRT *)


ForceRTR::usage =
    "ForceRTR[F2,M2,F3,M3,rM,rP,rC2] computes the joint reactions
for the RTR dyad."

Begin["`Private`"]

ForceRTR[F2_,M2_,F3_,M3_,rM_,rP_,rC2_]:=
Block[

{ rC3, F12, F12Sol, F12xSol, F12ySol, F43, F43Sol, F43xSol, F43ySol,
F32, eqRTR1, eqRTR2, eqRTR3, eqRTR4, solRTR, rMC2, rMP, rPQ, rQ,
rQSol, xQSol, yQSol, eqRTRQ1, eqRTRQ2, solRTRQ },

F12Sol = { F12xSol, F12ySol, 0 } ;
F43Sol = { F43xSol, F43ySol, 0 } ;

rMC2 = rC2 - rM ;
rMP = rP - rM ;

eqRTR1 = (F12Sol+F43Sol+F2+F3)[[1]] == 0 ;
eqRTR2 = (F12Sol+F43Sol+F2+F3)[[2]] == 0 ;
eqRTR3 = (F12Sol+F2).rMP == 0 ;
eqRTR4 = (Cross[rMC2,F2]+Cross[rMP,(F3+F43Sol)]+M2+M3)[[3]] == 0 ;

solRTR = Solve[ {eqRTR1, eqRTR2, eqRTR3, eqRTR4},
{F12xSol,F12ySol,F43xSol,F43ySol} ] ;

F12 = F12Sol/.solRTR[[1]] ;
F43 = F43Sol/.solRTR[[1]] ;

F32 = - F2 - F12 ;
F23 = - F32 ;

If[ M3[[3]]==0 , rQ = rP ,

rQSol = { xQSol, yQSol, 0 } ;
rMQ = rQSol - rM ;
rPQ = rQSol - rP ;
```

```
eqRTRQ1 = rMQ[[2]]/rMQ[[1]] - rMP[[2]]/rMP[[1]] == 0 ;
eqRTRQ2 = (Cross[rPQ,F23]+M3)[[3]] == 0 ;
solRTRQ = Solve[ {eqRTRQ1, eqRTRQ2} , {xQSol,yQSol} ] ;
rQ = rQSol/.solRTRQ[[1]] ;
] ;

Return[ { F12, F43, F23, rQ } ] ;
]

End[ ] (* ForceRTR *)



FMDriver::usage =
    "FMDriver[F1,M1,F21,rA,rB,rC1] computes the joint reaction and
torque of the motor."

Begin["`Private`"]

FMDriver[F1_,M1_,F21_,rA_,rB_,rC1_]:=
Block[

{ F01, rAC1, rAB, Mm },

F01 = - F1 - F21 ;

rAC1 = rC1 - rA ;
rAB = rB - rA ;
Mm = - Cross[rAC1,F1] - Cross[rAB,F21] - M1 ;

Return[ { F01, Mm } ] ;
]

End[ ] (* FMDriver *)


EndPackage[ ]
```

Force`

ForceMomentum[m,aCM,ICM,ddtheta] computes the total force and moment
of a rigid link.

Force`Private`

Force`Private`

ForceRRR[F2,M2,F3,M3,rM,rN,rP,rC2,rC3] computes the joint reactions
for the RRR dyad.

Force`Private`

Force`Private`

ForceRRT[F2,M2,F3,M3,rM,rN,rP,rC2] computes the joint reactions for the RRR dyad.

Force`Private`

Force`Private`

ForceRTR[F2,M2,F3,M3,rM,rP,rC2] computes the joint reactions for the RTR dyad.

Force`Private`

Force`Private`

FMDriver[F1,M1,F21,rA,rB,rC1] computes the joint reaction and torque of the motor.

Force`Private`

Force`Private`

PROGRAM I.8.12
```
(* R-RRT mechanism *)

Apply [ Clear , Names [ "Global`*" ] ] ;
Off[General::spell];
Off[General::spell1];

<<Driver.m ;
<<Position.m ;
<<VelAcc.m ;
<<Force.m ;

(* Input data *)

AC = 0.10 ;           (* m *)
BC = 0.30 ;           (* m *)
AP = 0.50 ;           (* m *)
wS = 0.05 ;           (* m *)
hS = 0.02 ;           (* m *)
phi = Pi/4 ;          (* rad *)
n = 30 ;              (* rpm *)
omega = N[Pi]*n/30 ;  (* rad/s *)
alpha = 0. ;          (* rad/s^2 *)
g = 9.807 ;           (* m/s^2 *)
```

```
rho = 8000 ;            (* Kg/m^3 *)
d = 0.001 ;             (* m *)
h = 0.01 ;              (* m *)
Mext = {0, 0, -100} ; (* Nm *)

(* Position Vectors *)

xA = yA = 0 ;

xC = AC ;
yC = 0 ;

xP = Driver[xA,yA,AP,phi,omega,alpha][[1]] ;
yP = Driver[xA,yA,AP,phi,omega,alpha][[2]] ;

xB1 = PosRRT[xC,yC,xP,yP,BC,phi][[1]] ;
yB1 = PosRRT[xC,yC,xP,yP,BC,phi][[2]] ;
xB2 = PosRRT[xC,yC,xP,yP,BC,phi][[3]] ;
yB2 = PosRRT[xC,yC,xP,yP,BC,phi][[4]] ;

(* Choose the correct solution *)
If[ (yB1>yC), xB=xB1;yB=yB1, xB=xB2;yB=yB2 ] ;

Print["Positions"];
Print["rB = ",{xB,yB,0}," [m]"];

(* Velocity and acceleration vectors *)

vPx = Driver[xA,yA,AP,phi,omega,alpha][[3]] ;
vPy = Driver[xA,yA,AP,phi,omega,alpha][[4]] ;
aPx = Driver[xA,yA,AP,phi,omega,alpha][[5]] ;
aPy = Driver[xA,yA,AP,phi,omega,alpha][[6]] ;

vCx = 0 ; vCy = 0 ;
aCx = 0 ; aCy = 0 ;

vBx = VelAccRRT[xC,yC,xP,yP,xB,yB,vCx,vCy,vPx,vPy,aCx,aCy,aPx,
aPy,phi,omega,alpha][[1]] ;
vBy = VelAccRRT[xC,yC,xP,yP,xB,yB,vCx,vCy,vPx,vPy,aCx,aCy,aPx,
aPy,phi,omega,alpha][[2]] ;
aBx = VelAccRRT[xC,yC,xP,yP,xB,yB,vCx,vCy,vPx,vPy,aCx,aCy,aPx,aPy,
phi,omega,alpha][[3]] ;
aBy = VelAccRRT[xC,yC,xP,yP,xB,yB,vCx,vCy,vPx,vPy,aCx,aCy,aPx,aPy,
phi,omega,alpha][[4]] ;

Print["Velocities"];
Print["vB = ",{vBx,vBy,0}," [m/s]"];
```

```
Print["Accelerations"];
Print["aB = ",{aBx,aBy,0}," [m/s^2]"];

phi3 = ArcTan[(yB-yC)/(xB-xC)] ;

omega3 = AngVelAcc[xB,yB,xC,yC, vBx,vBy,vCx,vCy,aBx,aBy,aCx,aCy][[1]] ;

alpha3 = AngVelAcc[xB,yB,xC,yC, vBx,vBy,vCx,vCy,aBx,aBy,aCx,aCy][[2]] ;

Print["Angular velocities"];
Print["omega3 = ",{0,0,omega3}," [rad/s]"];

Print["Angular accelerations"];
Print["alpha3 = ",{0,0,alpha3}," [rad/s^2]"];


aA = { 0, 0, 0 } ;
aB = {aBx, aBy, 0 } ;
aC = {0, 0, 0 } ;
aP = {aPx, aPy, 0} ;

(* Forces and moments *)

m1 = rho AP h d ;
aC1 = (aA+aP)/2 ;
IC1 = m1 (AP^2+h^2)/12 ;
F1 = ForceMomentum[m1,aC1,IC1,alpha][[1]] ;
M1 = ForceMomentum[m1,aC1,IC1,alpha][[2]] ;


m2 = rho wS hS d ;
aC2 = aB ;
IC2 = m2 (wS^2+hS^2)/12 ;
F2 = ForceMomentum[m2,aC2,IC2,alpha][[1]] ;
M2 = ForceMomentum[m2,aC2,IC2,alpha][[2]] ;


m3 = rho BC h d ;
aC3 = (aB+aC)/2 ;
IC3 = m1 (BC^2+h^2)/12 ;
F3 = ForceMomentum[m1,aC1,IC1,alpha3][[1]] ;
M3 = ForceMomentum[m1,aC1,IC1,alpha3][[2]] ;

Print["Forces and moments of links"];
Print["F1 = ",F1," [N]"];
Print["M1 = ",M1," [Nm]"];
Print["F2 = ",F2," [N]"];
Print["M2 = ",M2," [Nm]"];
Print["F3 = ",F3," [N]"];
Print["M3 = ",M3," [Nm]"];
```

```
rA = {xA, yA, 0} ;
rB = {xB, yB, 0} ;
rC = {xC, yC, 0} ;
rP = {xP, yP, 0} ;

rC3 = (rB+rC)/2 ;
F03 = ForceRRT[F3,M3+Mext,F2,M2,rC,rA,rB,rC3][[1]] ;
F12 = ForceRRT[F3,M3+Mext,F2,M2,rC,rA,rB,rC3][[2]] ;
F23 = ForceRRT[F3,M3+Mext,F2,M2,rC,rA,rB,rC3][[3]] ;

rQ = ForceRRT[F3,M3+Mext,F2,M2,rC,rA,rB,rC3][[4]] ;

F21 = - F12 ;
rC1 = (rA+rP)/2 ;
F01 = FMDriver[F1,M1,F21,rA,rB,rC1][[1]] ;
Mm = FMDriver[F1,M1,F21,rA,rB,rC1][[2]] ;

Print["Joint reactions"];
Print["F03 = ",F03," [N]"];
Print["F12 = ",F12," [N]"];
Print["F23 = ",F23," [N]"];
Print["rQ = ",rQ," [m]"];
Print["F01 = ",F01," [N]"];
Print["Mm = ",Mm," [N]"];
```

Positions

rB = {0.256155, 0.256155, 0} [m]

Velocities

vB = {-0.999913, 0.609559, 0} [m/s]

Accelerations

aB = {-1.80234, -4.25501, 0} [m/s^2]

Angular velocities

omega3 = {0, 0, 0.256155} [rad/s]

Angular accelerations

alpha3 = {0, 0, 0.256155} [rad/s^2]

Forces and moments of links

F1 = {0.0697886, -0.322491, 0} [N]

M1 = {0, 0, 0.} [Nm]

F2 = {0.0144187, -0.044416, 0} [N]

M2 = {0, 0, 0.} [Nm]

F3 = {0.0697886, -0.322491, 0} [N]

M3 = {0, 0, -0.000213548} [Nm]

Joint reactions

F03 = {242.56, -242.278, 0} [N]

F12 = {-242.645, 242.645, 0} [N]

F23 = {242.63, -242.6, 0} [N]

rQ = {0.256155, 0.256155, 0} [m]

F01 = {-242.714, 242.967, 0} [N]

Mm = {0., 0., 124.379} [N]

PROGRAM I.8.13

```
(* R-RTR-RTR mechanism *)

Apply [ Clear , Names [ "Global`*" ] ] ;
Off[General::spell];
Off[General::spell1];

<<Driver.m ;
<<VelAcc.m ;
<<Force.m ;

(* Input data *)

AB = 0.14 ;          (* m *)
DF = 0.40 ;          (* m *)
EG = 0.50 ;          (* m *)
CD = 0.15 ;          (* m *)
AC = 0.06 ;          (* m *)
AE = 0.25 ;          (* m *)
wS = 0.05 ;          (* m *)
hS = 0.02 ;          (* m *)
phi = Pi/6 ;         (* rad *)
n = 50 ;             (* rpm *)
```

```
omega = n*N[Pi]/30 ;    (* rad/s *)
alpha = 0. ;            (* rad/s^2 *)
g = 9.807 ;             (* m/s^2 *)
rho = 8000 ;            (* Kg/m^3 *)
d = 0.001 ;             (* m *)
h = 0.01 ;              (* m *)
Mext = {0, 0, -100} ;   (* Nm *)

(* Position Vectors *)

xA = yA = 0 ;

xB = Driver[xA,yA,AB,phi,omega,alpha][[1]] ;
yB = Driver[xA,yA,AB,phi,omega,alpha][[2]] ;

xC = 0 ;
yC = AC ;

phi2 = phi3 = ArcTan[(yB-yC)/(xB-xC)] ;

xD = xC - CD Cos[phi3] ;
yD = yC - CD Sin[phi3] ;

xE = 0 ;
yE = -AE ;

xF = xC+(DF-CD)Cos[phi3] ;
yF = yC+(DF-CD)Sin[phi3] ;

xG = xE + EG Cos[phi5] ;
yG = yE + EG Sin[phi5] ;

(* Velocity and acceleration vectors *)

vBx = Driver[xA,yA,AB,phi,omega,alpha][[3]] ;
vBy = Driver[xA,yA,AB,phi,omega,alpha][[4]] ;
aBx = Driver[xA,yA,AB,phi,omega,alpha][[5]] ;
aBy = Driver[xA,yA,AB,phi,omega,alpha][[6]] ;

vCx = vCy = 0 ;
aCx = aCy = 0 ;

omega2 = omega3 = AngVelAcc[xB,yB,xC,yC, vBx,vBy,vCx,vCy,aBx,aBy,aCx,
aCy,phi3][[1]] ;

alpha2 = alpha3 = AngVelAcc[xB,yB,xC,yC, vBx,vBy,vCx,vCy,aBx,aBy,aCx,
aCy,phi3][[2]] ;
```

```
vDx = AbsVelAcc[xC,yC,xD,yD,vCx,vCy,aCx,aCy,omega3,alpha3][[1]] ;
vDy = AbsVelAcc[xC,yC,xD,yD,vCx,vCy,aCx,aCy,omega3,alpha3][[2]] ;
aDx = AbsVelAcc[xC,yC,xD,yD,vCx,vCy,aCx,aCy,omega3,alpha3][[3]] ;
aDy = AbsVelAcc[xC,yC,xD,yD,vCx,vCy,aCx,aCy,omega3,alpha3][[4]] ;

vEx = vEy = 0 ;
aEx = aEy = 0 ;

phi4 = phi5 = Pi + ArcTan[(yD-yE)/(xD-xE)] ;

omega4 = omega5 = AngVelAcc[xD,yD,xE,yE, vDx,vDy,vEx,vEy,aDx,aDy,aEx,
aEy,phi5][[1]] ;

alpha4 = alpha5 = AngVelAcc[xD,yD,xE,yE, vDx,vDy,vEx,vEy,aDx,aDy,aEx,
aEy,phi5][[2]] ;

aA = { 0, 0, 0 } ;
aB = {aBx, aBy, 0 } ;
aC = {aCx, aCy, 0 } ;
aD = {aDx, aDy, 0 } ;
aE = {aEx, aEy, 0 } ;

(* Forces and moments *)

m1 = rho AB h d ;
aC1 = (aA+aB)/2 ;
IC1 = m1 (AB^2+h^2)/12 ;
F1 = ForceMomentum[m1,aC1,IC1,alpha][[1]] ;
M1 = ForceMomentum[m1,aC1,IC1,alpha][[2]] ;

m2 = rho wS hS d ;
aC2 = aB ;
IC2 = m2 (wS^2+hS^2)/12 ;
F2 = ForceMomentum[m2,aC2,IC2,alpha2][[1]] ;
M2 = ForceMomentum[m2,aC2,IC2,alpha2][[2]] ;

m3 = rho DF h d ;
xC3 = xC+(DF/2-CD)Cos[phi3] ;
yC3 = yC+(DF/2-CD)Sin[phi3] ;
aC3x = AbsVelAcc[xC,yC,xC3,yC3,vCx,vCy,aCx,aCy,omega3,alpha3][[3]] ;
aC3y = AbsVelAcc[xC,yC,xC3,yC3,vCx,vCy,aCx,aCy,omega3,alpha3][[4]] ;
aC3 = {aC3x, aC3y, 0} ;
IC3 = m3 (DF^2+h^2)/12 ;
F3 = ForceMomentum[m3,aC3,IC3,alpha3][[1]] ;
M3 = ForceMomentum[m3,aC3,IC3,alpha3][[2]] ;

m4 = rho wS hS d ;
```

```
aC4 = aD ;
IC4 = m4 (wS^2+hS^2)/12 ;
F4 = ForceMomentum[m4,aC4,IC4,alpha4][[1]] ;
M4 = ForceMomentum[m4,aC4,IC4,alpha4][[2]] ;

m5 = rho EG h d ;
xC5 = xE + EG/2 Cos[phi5] ;
yC5 = yE + EG/2 Sin[phi5] ;
aC5x = AbsVelAcc[xE,yE,xC5,yC5,vEx,vEy,aEx,aEy,omega5,alpha5][[3]] ;
aC5y = AbsVelAcc[xE,yE,xC5,yC5,vEx,vEy,aEx,aEy,omega5,alpha5][[4]] ;
aC5 = {aC5x, aC5y, 0} ;
IC5 = m5 (EG^2+h^2)/12 ;
F5 = ForceMomentum[m5,aC5,IC5,alpha5][[1]] ;
M5 = ForceMomentum[m5,aC5,IC5,alpha5][[2]] ;

Print["Forces and moments of links"];
Print["F1 = ",F1," [N]"];
Print["M1 = ",M1," [Nm]"];
Print["F2 = ",F2," [N]"];
Print["M2 = ",M2," [Nm]"];
Print["F3 = ",F3," [N]"];
Print["M3 = ",M3," [Nm]"];
Print["F4 = ",F4," [N]"];
Print["M4 = ",M4," [Nm]"];
Print["F5 = ",F5," [N]"];
Print["M5 = ",M5," [Nm]"];

(* Joint reactions *)

rA = {xA, yA, 0} ;
rB = {xB, yB, 0} ;
rC = {xC, yC, 0} ;
rD = {xD, yD, 0} ;
rE = {xE, yE, 0} ;
rF = {xF, yF, 0} ;
rG = {xG, yG, 0} ;

rC5 = {xC5,yC5,0} ;
F05 = ForceRTR[F5,M5+Mext,F4,M4,rE,rD,rC5][[1]] ;
F34 = ForceRTR[F5,M5+Mext,F4,M4,rE,rD,rC5][[2]] ;
F45 = ForceRTR[F5,M5+Mext,F4,M4,rE,rD,rC5][[3]] ;
rP = ForceRTR[F5,M5+Mext,F4,M4,rE,rD,rC5][[4]] ;

F43 = - F34 ;
rC3 = {xC3,yC3,0} ;
rC3D = rD-rC3 ;
M43 = Cross[rC3D,F43] ;
```

```
F03 = ForceRTR[F3+F43,M3+M43,F2,M2,rC,rB,rC3][[1]] ;
F12 = ForceRTR[F3+F43,M3+M43,F2,M2,rC,rB,rC3][[2]] ;
F23 = ForceRTR[F3+F43,M3+M43,F2,M2,rC,rB,rC3][[3]] ;
rQ = ForceRTR[F3+F43,M3+M43,F2,M2,rC,rB,rC3][[4]] ;

F21 = - F12 ;
rC1 = (rA+rB)/2 ;
F01 = FMDriver[F1,M1,F21,rA,rB,rC1][[1]] ;
Mm = FMDriver[F1,M1,F21,rA,rB,rC1][[2]] ;

Print["Joint reactions"];
Print["F05 = ",F05," [N]"];
Print["F34 = ",F34," [N]"];
Print["F45 = ",F45," [N]"];
Print["rP = ",rP," [m]"];
Print["F03 = ",F03," [N]"];
Print["F12 = ",F12," [N]"];
Print["F23 = ",F23," [N]"];
Print["rQ = ",rQ," [m]"];
Print["F01 = ",F01," [N]"];
Print["Mm = ",Mm," [Nm]"];
```

Forces and moments of links

F1 = {0.0186142, -0.0990915, 0} [N]

M1 = {0, 0, 0.} [Nm]

F2 = {0.0265917, -0.0631033, 0} [N]

M2 = {0, 0, -0.000028165} [Nm]

F3 = {0.0492489, -0.33315, 0} [N]

M3 = {0, 0, -0.00621962} [Nm]

F4 = {-0.0369367, -0.0639614, 0} [N]

M4 = {0, 0, 0.0000111583} [Nm]

F5 = {-0.0553516, -0.410666, 0} [N]

M5 = {0, 0, 0.00481155} [Nm]

Joint reactions

F05 = {268.165, 135.057, 0} [N]

F34 = {-268.072, -134.583, 0} [N]

F45 = {268.109, 134.647, 0} [N]

rP = {-0.149492, 0.0476701, 0} [m]

F03 = {-256.745, -272.179, 0} [N]

F12 = {-11.4028, 137.993, 0} [N]

F23 = {11.3762, -137.93, 0} [N]

rQ = {0.121243, 0.07, 0} [m]

F01 = {-11.4214, 138.092, 0} [N]

Mm = {0., 0., 17.5356} [Nm]

I.9 Simulation of Kinematic Chains with *Working Model*

This section serves as a tutorial to simulate the planar mechanism R-RTR shown in Figure I.9.1 by using *Working Model*. The mechanism has three links: the driver link (link 1), the slider (link 2), and the rocker (link 3).

Step 1: Opening *Working Model*

1. Click on the *Working Model* program icon to start the program.
2. Create a new *Working Model* document by selecting "New" from the "File" menu.

Toolbars create links, joints, and mechanism actuators.

3. Specify the units for the simulation.
4. Set up the workspace.

In the "View" menu: select "Workspace", check Coordinates and X,Y Axes from the Navigation box, and check all the objects from the Toolbars box except Simple; turn off Grid Snap and turn on Object Snap; select "Numbers and Units" and change the Unit System to SI (degrees); select "View Size" and choose the objects on the screen to be 1.0 times actual size.

Step 2: Creating the Links

This step creates the three moving links of the mechanism. The background serves as the fixed frame link (ground).

1. *Create the driver link.* Click on the rectangle tool in the toolbar to sketch out a rectangular body. Position the mouse at the first corner, click once, then move the mouse to the location of the opposite corner and click again. Four black boxes

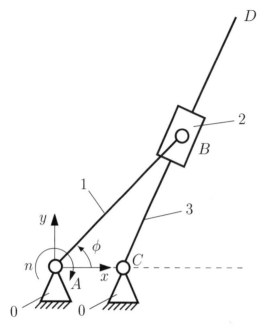

$\phi = \pi/6$
$AB = 0.1$ m
$AC = 0.05$ m
$CD = 0.15$ m
$n = 50$ rpm
$g = 9.807$ m/s^2

FIGURE I.9.1

appear around the link indicating that it has been selected. Modify its dimensions at the bottom of the screen accordingly to the height $h = 0.1$ m, and the width $w = 0.01$ m (Fig. I.9.2).

2. *Create the slider and the rocker.* Click on the "Rectangle" tool in the toolbar. The tool is now selected and it can be used multiple times. Sketch out two rectangular bodies: the rocker and the slider. Choose the widths $w = 0.15$ m for the rocker and $w = 0.040$ m for the slider. The height of the rocker is $h = 0.010$ m and the height of the slider is $h = 0.020$ m.

   The depth of all objects in *Working Model* is $d = 0.001$ m by default.

3. *Change the properties of the links.* Press the Shift key and click on the driver link, rocker, and slider, respectively. Select "Properties" in the "Window" menu and change the material to Steel, the coefficients of static and kinetic friction to 0.0 (no friction), the coefficient of restitution to 1.0 (perfect elastic), and the charge to 0.0 (no charge) as shown in Figure I.9.3.

Remark
The commands Zoom in and Zoom out can be used by clicking on the icons at the top of the screen in order to make the objects clearly visible.

Step 3: Connecting the Slider and the Rocker

1. Move the slider over the rocker.
2. Select the horizontal "Keyed Slot joint" icon at the left of the screen. The icon appears as a rectangle riding over a horizontal slot.

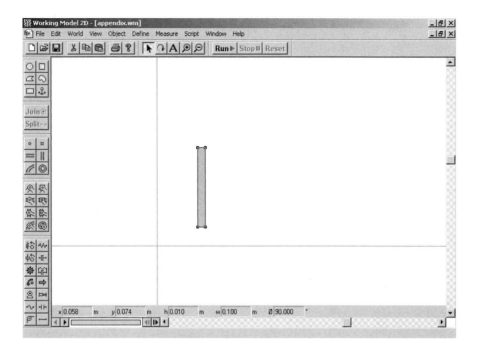

FIGURE I.9.2

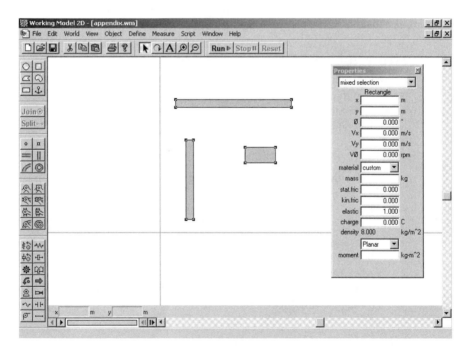

FIGURE I.9.3

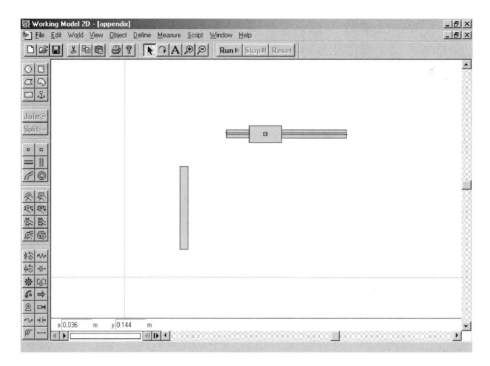

FIGURE I.9.4

3. Move the cursor over the snap point at the center of the slider and click the mouse button. The screen should look like Figure I.9.4.

Step 4: Adding a Motor to the Driver link

Similar to a pin joint, a motor has two attachment points. A motor automatically connects the top two bodies. If only one body were to lay beneath the motor, the motor would join the body to the background. The motor then applies a torque between the two bodies to which it is pinned.

1. Click on the "Motor" tool in the toolbox. This tool appears as a circle, sitting on a base with a point in its center. The cursor should now look like a small motor.
2. Place the cursor over the "snap point" on the center of axis and click the mouse button.
3. Click on the "Split" button in the toolbar. Click on the pin joint and drag it to the snap point at the bottom of the driver link.
4. Click on the "Join" button in the toolbar. Since the motor is fixed to the ground, the driver link moves in place.
5. Click on the driver link and change the value of the angle ϕ at the bottom of the screen to $-45°$ (Fig. I.9.5).

Step 5: Connecting the Driver Link and the Slider

1. Select the anchor tool.

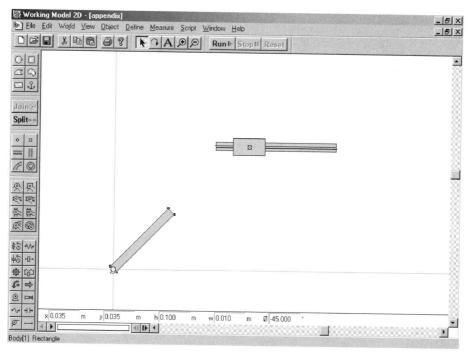

FIGURE I.9.5

2. Click on the driver link to anchor the link down. The anchor fixes the body to the ground during construction.
3. Click on the "Pin joint" tool.
4. Place the cursor over the upper end of the driver link. When an "X" appears around the pointer, click the mouse button.
5. Click on the "Split" button in the toolbar. *Working Model* creates two connected overlapping pin joints.
6. With the pointer tool selected, click on the pin joint and drag it to the snap point at the center of the slider (Fig. I.9.6).
7. Click on the "Join" button in the toolbar. *Working Model* merges the two pin joints into a single one, moving the unanchored link into place.
8. Click on the driver link. Select the "Move to front" option in the "Object" menu. This places the link in front of the rocker, making it visible, as shown in Figure I.9.7.

Step 6: Connecting the Rocker to the Ground

1. Click on the "Point element" in the toolbox. Place the cursor at any point on the ground and click the mouse button.
2. Modify the coordinates of the point accordingly to $x = 0.05$ m and $y = 0$.
3. Click on the "Pin joint" in the toolbox. Place the cursor on top of the point and click the mouse button. The pin joint is now fixed to the ground.
4. Using "Split" and "Join", connect the rocker to the pin joint.

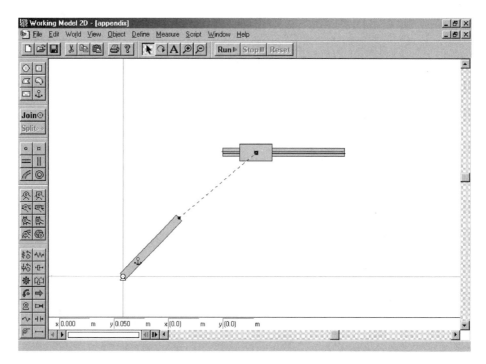

FIGURE I.9.6

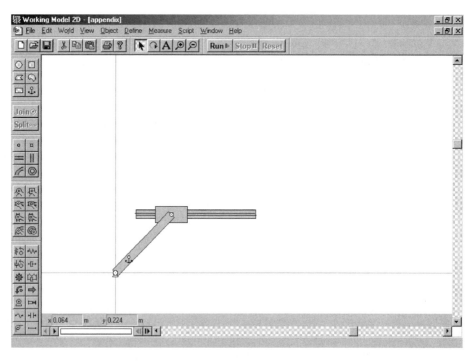

FIGURE I.9.7

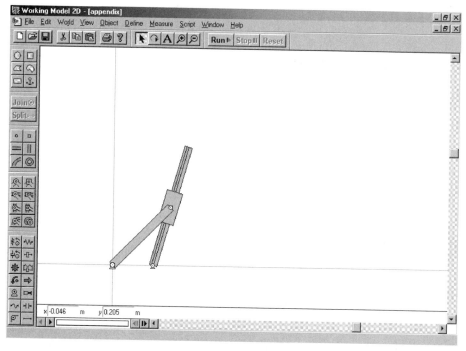

FIGURE I.9.8

5. Select the anchor, used to keep the driver link in position during building, and press the Delete key to remove it. The screen should look like that shown in Figure I.9.8.

Step 7: Adding an External Torque

1. Click on the "Torque" tool from the toolbox and then click on the rocker. This will apply an external torque to the rocker.
2. Select the torque and modify its value to $M_{ext} = 100$ N·m in the "Properties" menu.

Step 8: Measuring Positions, Velocities, Accelerations, Torques, and Forces

1. Select the driver link, then go to "Measure" menu and "Position" submenu. Apply the command "Rotation graph" to measure the rotation angle of the driver.
2. Click on the "Point element" from the toolbox and then click on the end point of the rocker. The point is now attached to the rocker. Go to "Measure" menu and apply the commands "Position", "Velocity", and "Acceleration" to measure the position, velocity, and acceleration of the point. Click on the arrow in the right upper corner of the measurement window to change it from graphic to numerical.
3. Select the motor, then go to "Measure" menu and apply the command "Torque Transmitted" to measure the torque of the motor.

4. Select the pin joint that connects the driver link to the ground. Go to "Measure" and apply the command "Force" to measure the reaction force between the ground and the driver.
5. Select the rigid keyed slot joint that connects the slider to the rocker. Go to "Measure" and apply the command "Force" to measure the reaction force between the slider and the rocker.
6. Select the pin joint that connects the rocker to the ground. Go to "Measure" and apply the command "Force" to measure the reaction force between the ground and the rocker.
7. Select the pin joint that connects the driver link to the slider. Go to "Measure" and apply the command "Force" to measure the reaction force between the slider and the driver.

Remark

When you select a joint to create a meter to measure the reaction force, the meter measures the forces exerted on the body located at the top when the joint was created. The components of the forces are given in terms of the local coordinate system. In order to measure the components of the pin joint forces in terms of the global coordinate systems, the angles of the two points that compose the pin joint are set to the value 0. The two points that compose a pin joint are seen by selecting the joint and opening the "Properties" window.

An example is shown in Figure I.9.9. The pin joint between the driver link and the slider (Fig. I.9.6), denoted by **Constraint[17]**, is composed of the two points **Point[15]** and **Point[16]**. Select **Point[15]** from the window "Properties" and change its angle to the

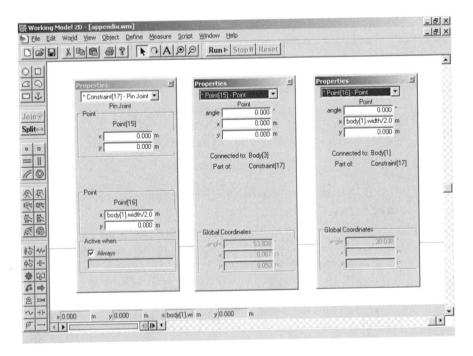

FIGURE I.9.9

value 0. Then select **Point[16]** and change its angle to 0. Now the components of the pin joint forces are measured in terms of the global coordinate system.

Step 9: Running the Simulation

1. With the pointer tool selected, select all the bodies. Select the "Do Not Collide" option in the "Object" menu.
2. Select "Numbers and Units" in the "View" menu. Select More Options and change the Rotational Velocity to Revs/min.
3. Double-click on the motor to open the "Properties" box. Modify the velocity of the motor to −50 rpm, as shown in Figure I.9.9.
4. Click on each graph and modify its label from "Window" menu and "Appearance" submenu.
5. Click on "Run" in the toolbar.

   Tape controls, which are used to run and view simulations, are located at the bottom of the screen.

6. Click on "Reset" in the toolbar. The simulation resets to the initial frame 0.

Remark
To increase the Simulation Accuracy, select "Accuracy" from the "World" menu and change the Animation Step to a larger value and the Integration Error to a smaller value. The screen should look like that shown in Figure I.9.10.

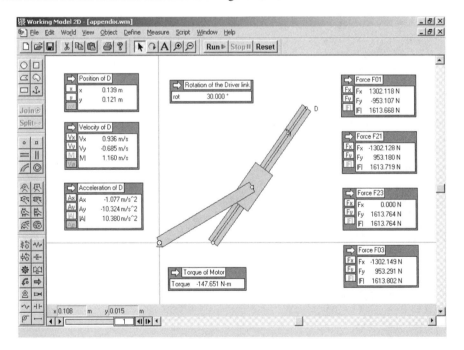

FIGURE I.9.10

Results

For the R-RTR mechanism (Fig. I.9.1), the following results are obtained: the motor torque is $M_{mot} = -147.651$ N·m; the position of the point D is $\mathbf{r}_D = 0.139\mathbf{i} + 0.121\mathbf{j}$ m, the velocity of the point D is $\mathbf{v}_D = 0.936\mathbf{i} - 0.685\mathbf{j}$ m/s; the acceleration of the point D is $\mathbf{a}_D = -1.077\mathbf{i} - 10.324\mathbf{j}$ m/s^2; the reaction force of the ground on the driver link is $\mathbf{F}_{01} = 1302\mathbf{i} - 953\mathbf{j}$ N; the reaction force of the slider on the driver link is $\mathbf{F}_{21} = -1302.128\mathbf{i} + 953.180\mathbf{j}$ N; the reaction force between the rocker and the slider is $F_{23} = 1613.764$ N; the reaction force of the ground on the rocker is $\mathbf{F}_{03} = -1302.15\mathbf{i} + 953.291\mathbf{j}$ N.

References

[1]   P. Antonescu, *Mechanisms*, Printech, Bucharest, 2003.

[2]   P. Appell, *Traité de Mécanique Rationnelle*, Gautier-Villars, Paris, 1941.

[3]   I. I. Artobolevski, *Mechanisms in Modern Engineering Design*, MIR, Moscow, 1977.

[4]   M. Atanasiu, *Mechanics [Mecanica]*, EDP, Bucharest, 1973.

[5]   H. Baruh, *Analytical Dynamics*, WCB/McGraw-Hill, Boston, 1999.

[6]   A. Bedford and W. Fowler, *Dynamics*, Addison Wesley, Menlo Park, CA, 1999.

[7]   A. Bedford and W. Fowler, *Statics*, Addison Wesley, Menlo Park, CA, 1999.

[8]   M. I. Buculei, *Mechanisms*, University of Craiova Press, Craiova, Romania, 1976.

[9]   M. I. Buculei, D. Bagnaru, G. Nanu, D. B. Marghitu, *Analysis of Mechanisms with Bars*, Scrisul Romanesc, Craiova, Romania, 1986.

[10]   T. Demian, et al., *Mechanisms — problems*, EDP, Bucharest, 1969.

[11]   A. G. Erdman, and G. N. Sandor, *Mechanisms Design*, Prentice-Hall, Upper Saddle River, NJ, 1984.

[12]   A. Ertas and J. C. Jones, *The Engineering Design Process*, John Wiley & Sons, New York, 1996.

[13]   F. Freudenstein, "An Application of Boolean Algebra to the Motion of Epicyclic Drives," Transaction of the ASME, *Journal of Engineering for Industry*, pp. 176–182, 1971.

[14]   J. H. Ginsberg, *Advanced Engineering Dynamics*, Cambridge University Press, Cambridge, 1995.

[15]   D. T. Greenwood, *Principles of Dynamics*, Prentice-Hall, Englewood Cliffs, NJ, 1998.

[16]   A. S. Hall, Jr., A. R. Holowenko, and H. G. Laughlin, *Theory and Problems of Machine Design*, McGraw-Hill, New York, 1961.

[17]   R. C. Hibbeler, *Engineering Mechanics — Statics and Dynamics*, Prentice-Hall, Upper Saddle River, NJ, 1995.

[18]   R. C. Juvinall and K. M. Marshek, *Fundamentals of Machine Component Design*, John Wiley & Sons, New York, 1983.

[19]   T. R. Kane, *Analytical Elements of Mechanics*, Vol. 1, Academic Press, New York, 1959.

[20] T. R. Kane, *Analytical Elements of Mechanics*, Vol. 2, Academic Press, New York, 1961.

[21] T. R. Kane and D. A. Levinson, "The Use of Kane's Dynamical Equations in Robotics", *MIT International Journal of Robotics Research*, No. 3, pp. 3–21, 1983.

[22] T. R. Kane, P. W. Likins, and D. A. Levinson, *Spacecraft Dynamics*, McGraw-Hill, New York, 1983.

[23] T. R. Kane and D. A. Levinson, *Dynamics*, McGraw-Hill, New York, 1985.

[24] J. T. Kimbrell, *Kinematics Analysis and Synthesis*, McGraw-Hill, New York, 1991.

[25] R. Maeder, *Programming in Mathematica*, Addison–Wesley Publishing Company, Redwood City, CA, 1990.

[26] N. H. Madsen, *Statics and Dynamics*, class notes, available at www.eng.auburn.edu/users/nmadsen/, 2004.

[27] N. I. Manolescu, F. Kovacs, and A. Oranescu, *The Theory of Mechanisms and Machines*, EDP, Bucharest, 1972.

[28] D. B. Marghitu, *Mechanical Engineer's Handbook*, Academic Press, San Diego, CA, 2001.

[29] D. B. Marghitu and M. J. Crocker, *Analytical Elements of Mechanisms*, Cambridge University Press, Cambridge, 2001.

[30] D. B. Marghitu and E. D. Stoenescu, *Kinematics and Dynamics of Machines and Machine Design*, class notes, available at www.eng.auburn.edu/users/marghitu/, 2004.

[31] J. L. Meriam and L. G. Kraige, *Engineering Mechanics: Dynamics*, John Wiley & Sons, New York, 1997.

[32] D. J. McGill and W. W. King, *Engineering Mechanics: Statics and an Introduction to Dynamics*, PWS Publishing Company, Boston, 1995.

[33] R. L. Mott, *Machine Elements in Mechanical Design*, Prentice Hall, Upper Saddle River, NJ, 1999.

[34] D. H. Myszka, *Machines and Mechanisms*, Prentice-Hall, Upper Saddle River, NJ, 1999.

[35] R. L. Norton, *Machine Design*, Prentice-Hall, Upper Saddle River, NJ, 1996.

[36] R. L. Norton, *Design of Machinery*, McGraw-Hill, New York, 1999.

[37] W. C. Orthwein, *Machine Component Design*, West Publishing Company, St. Paul, 1990.

[38] L. A. Pars, *A Treatise on Analytical Dynamics*, John Wiley & Sons, New York, 1965.

[39] R. M. Pehan, *Dynamics of Machinery*, McGraw-Hill, New York, 1967.

[40] I. Popescu, *Mechanisms*, University of Craiova Press, Craiova, Romania, 1990.

[41] I. Popescu and C. Ungureanu, *Structural Synthesis and Kinematics of Mechanisms with Bars*, Universitaria Press, Craiova, Romania, 2000.

[42] I. Popescu and D. B. Marghitu, "Dyad Classification for Mechanisms," *World Conference on Integrated Design and Process Technology*, Austin, TX, December 3–5, 2003.

[43]   I. Popescu, E. D. Stoenescu, and D. B. Marghitu, "Analysis of Spatial Kinematic Chains Using the System Groups," *8th International Congress on Sound and Vibration*, St. Petersburg, Russia, July 5–8, 2004.

[44]   M. Radoi and E. Deciu, *Mecanica*, EDP, Bucharest, 1981.

[45]   F. Reuleaux, *The Kinematics of Machinery*, Dover, New York, 1963.

[46]   C. A. Rubin, *The Student Edition of Working Model*, Addison–Wesley Publishing Company, Reading, MA, 1995.

[47]   I. H. Shames, *Engineering Mechanics — Statics and Dynamics*, Prentice-Hall, Upper Saddle River, NJ, 1997.

[48]   J. E. Shigley and C. R. Mischke, *Mechanical Engineering Design*, McGraw-Hill, New York, 1989.

[49]   J. E. Shigley and J. J. Uicker, *Theory of Machines and Mechanisms*, McGraw-Hill, New York, 1995.

[50]   R. W. Soutas-Little and D. J. Inman, *Engineering Mechanics: Statics and Dynamics*, Prentice-Hall, Upper Saddle River, NJ, 1999.

[51]   A. Stan and M. Grumarescu, *Mechanics Problems*, EDP, Bucharest, 1973.

[52]   A. Stoenescu, A. Ripianu, and M. Atanasiu, *Theoretical Mechanics Problems*, EDP, Bucharest, 1965.

[53]   A. Stoenescu and G. Silas, *Theoretical Mechanics*, ET, Bucharest, 1957.

[54]   E. D. Stoenescu, *Dynamics and Synthesis of Kinematic Chains with Impact and Clearance*, Ph.D. Dissertation, Mechanical Engineering, Auburn University, 2005.

[55]   L. W. Tsai, *Mechanism Design: Enumeration of Kinematic Structures According to Function*, CRC Press, Boca Raton, FL, 2001.

[56]   R. Voinea, D. Voiculescu, and V. Ceausu, *Mechanics [Mecanica]*, EDP, Bucharest, 1983.

[57]   K. J. Waldron and G. L. Kinzel, *Kinematics, Dynamics, and Design of Machinery*, John Wiley & Sons, New York, 1999.

[58]   C. E. Wilson and J. P. Sadler, *Kinematics and Dynamics of Machinery*, Harper Collins College Publishers, New York, 1991.

[59]   C. W. Wilson, *Computer Integrated Machine Design*, Prentice Hall, Upper Saddle River, NJ, 1997.

[60]   S. Wolfram, *Mathematica*, Wolfram Media/Cambridge University Press, Cambridge, 1999.

[61]   * * *, *The Theory of Mechanisms and Machines [Teoria mehanizmov i masin]*, Vassaia Scola, Minsku, Russia, 1970.

[62]   * * *, *Working Model 2D, Users Manual*, Knowledge Revolution, San Mateo, CA, 1996.

Part II   Machine Components

II.1   Stress and Deflection

II.1.1   Stress

In the design process, an important problem is to ensure that the strength of the mechanical element to be designed always exceeds the stress due to any load exerted on it.

Uniform distribution of stresses is an assumption that is frequently considered in the design process. Depending upon the way the force is applied to an element, the result is called *pure tension* (*compression*) or *pure shear*, respectively.

A tension load F is applied to the ends of a bar. If the bar is cut at a section remote from the ends and one piece is removed, the effect of the removed part can be replaced by applying a uniformly distributed force of magnitude σA to the cut end, where σ is the *normal stress* and A the cross section area of the bar. The stress σ is given by the following expression:

$$\sigma = \frac{F}{A}. \qquad (\text{II.1.1})$$

This uniform stress distribution requires that

- the bar be straight and made of a homogeneous material, and
- the line of action of the force contains the centroid of the section.

Equation (II.1.1) and the previous assumptions also hold for pure compression.

If a body is in shear (uniform stress distribution), the following equation can be used:

$$\tau = \frac{F}{A}, \qquad (\text{II.1.2})$$

where τ is the *shear stress*.

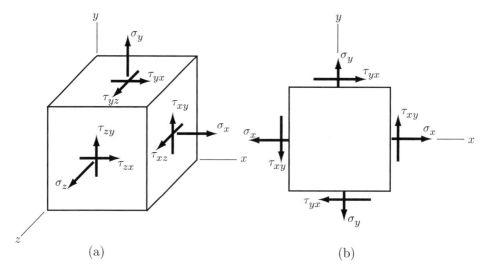

FIGURE II.1.1 *Stress element: (a) three-dimensional case; (b) planar case.*

Stress Components

A general three-dimensional *stress element* is illustrated in Figure II.1.1(a). Three normal positive stresses, σ_x, σ_y, and σ_z, and six positive shear stresses, $\tau_{xy}, \tau_{yx}, \tau_{yz}, \tau_{zy}, \tau_{zx}$, and τ_{xz}, are shown. Static equilibrium requires

$$\tau_{xy} = \tau_{yx}, \quad \tau_{yz} = \tau_{zy}, \quad \tau_{xz} = \tau_{zx}. \qquad \text{(II.1.3)}$$

The normal stresses σ_x, σ_y, and σ_z are called *tension* or *tensile stresses* and considered positive if they are oriented in the direction shown in the figure. *Shear stresses* on a positive face of an element are positive if they act in the positive direction of the reference axis. The first subscript of any shear stress component denotes the axis to which it is perpendicular. The second subscript denotes the axis to which the shear stress component is parallel.

A general two-dimensional stress element is shown in Figure II.1.1(b). The two normal stresses, σ_x and σ_y, respectively, are in the positive direction. Shear stresses are positive when they are in the clockwise (cw) direction and negative when they are in the counterclockwise (ccw) direction. Thus, τ_{yx} is positive (cw) and τ_{xy} is negative (ccw).

Mohr's Circle

The element in Figure II.1.1(b) is considered cut by an oblique plane at angle ϕ as shown in Figure II.1.2. The stresses σ and τ act on this oblique plane. The stresses σ and τ can be calculated with the formulas

$$\sigma = \frac{\sigma_x + \sigma_y}{2} + \frac{\sigma_x - \sigma_y}{2}\cos 2\phi + \tau_{xy}\sin 2\phi, \qquad \text{(II.1.4)}$$

$$\tau = -\frac{\sigma_x - \sigma_y}{2}\sin 2\phi + \tau_{xy}\cos 2\phi. \qquad \text{(II.1.5)}$$

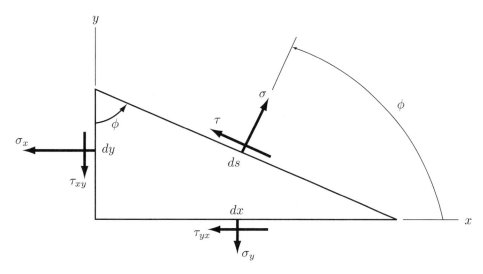

FIGURE II.1.2   *Normal and shear stresses on a planar surface.*

Differentiating Eq. (II.1.4) with respect to the angle ϕ and setting the result equal to zero yields

$$\tan 2\phi = \frac{2\tau_{xy}}{\sigma_x - \sigma_y}. \tag{II.1.6}$$

The solution of Eq. (II.1.6) gives two values for the angle 2ϕ defining the maximum normal stress σ_1 and the minimum normal stress σ_2. These minimum and maximum normal stresses are called the *principal stresses*. The corresponding directions are called the *principal directions*. The angle between the principal directions is 90°.

Similarly, differentiating Eq. (II.1.5) and setting the result to zero will result in the following relation:

$$\tan 2\phi = -\frac{\sigma_x - \sigma_y}{2\tau_{xy}}. \tag{II.1.7}$$

The solutions of Eq. (II.1.7) define the angles 2ϕ at which the shear stress τ reaches extreme values.

Equation (II.1.6) can be rewritten as

$$2\tau_{xy} \cos 2\phi = (\sigma_x - \sigma_y) \sin 2\phi,$$

or

$$\sin 2\phi = \frac{2\tau_{xy} \cos 2\phi}{\sigma_x - \sigma_y}. \tag{II.1.8}$$

Substituting Eq. (II.1.8) into Eq. (II.1.5) gives

$$\tau = -\frac{\sigma_x - \sigma_y}{2} \frac{2\tau_{xy}\cos 2\phi}{\sigma_x - \sigma_y} + \tau_{xy}\cos 2\phi = 0. \tag{II.1.9}$$

From Eq. (II.1.9) it results that the shear stress associated with both principal directions is zero.

Substituting $\sin 2\phi$ from Eq. (II.1.7) into Eq. (II.1.4) yields

$$\sigma = \frac{\sigma_x + \sigma_y}{2}. \tag{II.1.10}$$

The Eq. (II.1.10) states that the two normal stresses associated with the directions of the two maximum shear stresses are equal.

The analytical expressions for the two principal stresses can be obtained by manipulating Eqs. (II.1.6) and (II.1.4):

$$\sigma_1, \sigma_2 = \frac{\sigma_x + \sigma_y}{2} \pm \sqrt{\left(\frac{\sigma_x - \sigma_y}{2}\right)^2 + \tau_{xy}^2}. \tag{II.1.11}$$

Similarly, the maximum and minimum values of the shear stresses are obtained:

$$\tau_1, \tau_2 = \pm\sqrt{\left(\frac{\sigma_x - \sigma_y}{2}\right)^2 + \tau_{xy}^2}. \tag{II.1.12}$$

The Mohr's circle diagram (Fig. II.1.3) is a graphical method to visualize the stress state. The normal stresses are plotted along the abscissa axis of the coordinate system and the shear stresses along the ordinate axis. Tensile normal stresses are considered positive (σ_x and σ_y are positive in Fig. II.1.3) and compressive normal stresses are negative. Clockwise (cw) shear stresses are considered positive, while counterclockwise (ccw) shear stresses are negative.

The following notation is used: OA as σ_x, AB as τ_{xy}, OC as σ_y, and CD as τ_{yx}. The center of the Mohr's circle is at point E on the σ-axis. Point B has the stress coordinates σ_x, τ_{xy} on the x-faces and point D the stress coordinates σ_y, τ_{yx} on the y-faces. The angle 2ϕ between EB and ED is $180°$, hence the angle between x and y on the stress element is $\phi = 90°$. The maximum principal normal stress is σ_1 at point F, and the minimum principal normal stress is σ_2 at point G. The two extreme values of the shear stresses are plotted at points I and H, respectively. Thus, the Mohr's diagram is a circle of center E and diameter BD.

For three-dimensional stress elements it is considered a particular orientation when all shear stress components are zero. The principal directions are the normals to the faces for this particular orientation. Since the stress element is three-dimensional, there are three principal directions and three principal stresses, σ_1, σ_2 and σ_3, associated with the principal directions. In three dimensions, only six components of stress are required to specify the stress state, namely, σ_x, σ_y, σ_z, τ_{xy}, τ_{yz}, and τ_{zx}.

To plot Mohr's circles for triaxial stress, the principal normal stresses are ordered so that $\sigma_1 > \sigma_2 > \sigma_3$. The result is shown in Figure II.1.4(a). The three *principal shear*

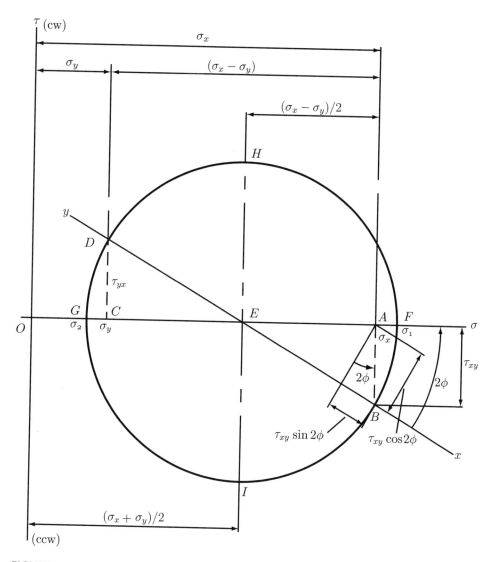

FIGURE II.1.3 *Mohr's circle.*

stresses, $\tau_{1/2}, \tau_{2/3}$ and $\tau_{1/3}$, are also shown in Figure II.1.4(a). Each of these shear stresses occurs on two planes, one of the planes being shown in Figure II.1.4(b). The principal shear stresses are

$$\tau_{1/2} = \frac{\sigma_1 - \sigma_2}{2}; \qquad \tau_{2/3} = \frac{\sigma_2 - \sigma_3}{2}; \qquad \tau_{1/3} = \frac{\sigma_1 - \sigma_3}{2}. \tag{II.1.13}$$

If $\tau_{max} = \tau_{1/3}$, then $\sigma_1 > \sigma_2 > \sigma_3$.

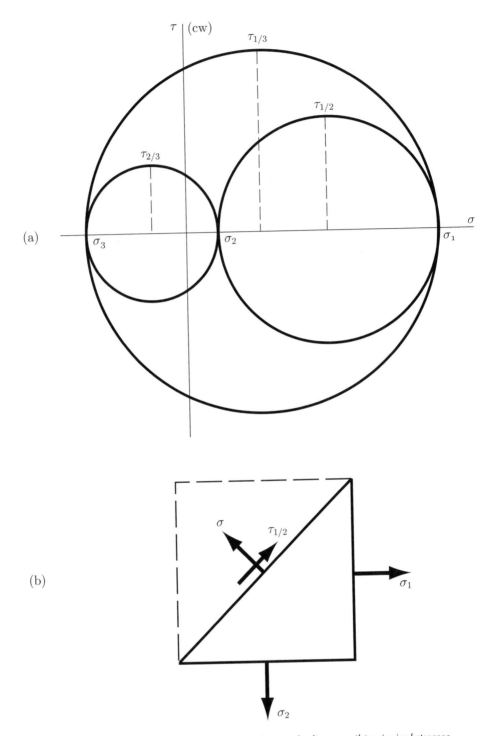

FIGURE II.1.4 *Mohr's circle application: (a) Mohr's circle diagram; (b) principal stresses.*

Elastic Strain

If a tensile load is applied to a straight bar, it becomes longer. The amount of elongation is called the *total strain*. The elongation per unit length of the bar ϵ is called *strain*. The strain is

$$\epsilon = \frac{\delta}{l},\tag{II.1.14}$$

where δ is the total elongation (total strain) of the bar of length l.

Shear strain γ is the change in a right angle of an element subjected to pure shear stresses.

Elasticity is a property of materials that allows them to regain the original geometry when the load is removed. The elasticity of a material can be expressed in terms of Hooke's law: the stress in a material is proportional to the strain that produced it, within certain limits

$$\sigma = E\,\epsilon, \quad \tau = G\gamma,\tag{II.1.15}$$

where E and G are constants of proportionality. The constant E is called the *modulus of elasticity* and the constant G is called the *shear modulus of elasticity* or the *modulus of rigidity*. A material that obeys Hooke's law is called *elastic*.

Substituting $\sigma = F/A$ and $\epsilon = \delta/l$ into Eq. (II.1.15), the expression for the total deformation δ of a bar loaded in axial tension or compression is

$$\delta = \frac{F\,l}{A\,E}.\tag{II.1.16}$$

When a tension load is applied to an elastic body, not only does an axial strain occur, but also a lateral strain occurs and the two strains are proportional to each other. This proportionality constant is called *Poisson's ratio* and is given by

$$\nu = \frac{\text{lateral strain}}{\text{axial strain}}.\tag{II.1.17}$$

The elastic constants are related by the following expression:

$$E = 2\,G\,(1 + \nu).\tag{II.1.18}$$

The *principal strains* are defined as the strains in the direction of the principal stresses. The shear strains are zero on the faces of an element aligned along the principal directions. Table II.1.1 lists the relationships for all types of stress. The values of Poisson's ratio ν for various materials are listed in Table II.1.2.

Shear and Moment

A beam supported by the reactions R_1 and R_2 and loaded by the transversal forces F_1, F_2 is shown in Figure II.1.5(a). The reactions R_1 and R_2 are considered positive since they act in the positive direction of the y-axis. The concentrated forces F_1 and F_2 are considered negative since they act in the negative y-direction. A cut is considered at a section located at x. Only the left-hand part of the beam with respect to the cut is taken as a free body.

TABLE II.1.1 Elastic stress-strain relations

Type of Stress	Principal strains	Principal stresses
Uniaxial	$\epsilon_1 = \dfrac{\sigma_1}{E}$ $\epsilon_2 = -\nu\epsilon_1$ $\epsilon_3 = -\nu\epsilon_1$	$\sigma_1 = E\epsilon_1$ $\sigma_2 = 0$ $\sigma_3 = 0$
Biaxial	$\epsilon_1 = -\dfrac{\sigma_1}{E} - \dfrac{\nu\sigma_2}{E}$ $\epsilon_2 = -\dfrac{\sigma_2}{E} - \dfrac{\nu\sigma_1}{E}$ $\epsilon_3 = -\dfrac{\nu\sigma_1}{E} - \dfrac{\nu\sigma_2}{E}$	$\sigma_1 = \dfrac{E(\epsilon_1 + \nu\epsilon_2)}{1 - \nu^2}$ $\sigma_2 = \dfrac{E(\epsilon_2 + \nu\epsilon_1)}{1 - \nu^2}$ $\sigma_3 = 0$
Triaxial	$\epsilon_1 = \dfrac{\sigma_1}{E} - \dfrac{\nu\sigma_2}{E} - \dfrac{\nu\sigma_3}{E}$ $\epsilon_2 = \dfrac{\sigma_2}{E} - \dfrac{\nu\sigma_1}{E} - \dfrac{\nu\sigma_3}{E}$ $\epsilon_3 = \dfrac{\sigma_3}{E} - \dfrac{\nu\sigma_1}{E} - \dfrac{\nu\sigma_2}{E}$	$\sigma_1 = \dfrac{E\epsilon_1(1 - \nu) + \nu E(\epsilon_2 + \epsilon_3)}{1 - \nu - 2\nu^2}$ $\sigma_2 = \dfrac{E\epsilon_2(1 - \nu) + \nu E(\epsilon_1 + \epsilon_3)}{1 - \nu - 2\nu^2}$ $\sigma_1 = \dfrac{E\epsilon_3(1 - \nu) + \nu E(\epsilon_1 + \epsilon_2)}{1 - \nu - 2\nu^2}$

Source: J. E. Shigley and C. R. Mischke, *Mechanical Engineering Design*, McGraw-Hill, New York, 1989. Reprinted with permission of McGraw-Hill.

TABLE II.1.2 Physical constants of materials

Material	Modulus of Elasticity E		Modulus of Rigidity G		Poisson's ratio	Unit weight w		
	Mpsi	GPa	Mpsi	GPa	ν	lb/in^3	lb/ft^3	kN/m^3
Aluminum (all alloy)	10.3	71.0	3.80	26.2	0.334	0.098	169	26.6
Beryllium copper	18.0	124.0	7.0	48.3	0.285	0.297	513	80.6
Brass	15.4	106.0	5.82	40.1	0.324	0.309	534	83.8
Carbon steel	30.0	207.0	11.5	79.3	0.292	0.282	487	76.5
Cast iron, gray	14.5	100.0	6.0	41.4	0.211	0.260	450	70.6
Copper	17.2	119.0	6.49	44.7	0.326	0.322	556	87.3
Douglas fir	1.6	11.0	0.6	4.1	0.33	0.016	28	4.3
Glass	6.7	46.2	2.7	18.6	0.245	0.094	162	25.4
Inconel	31.0	214.0	11.0	75.8	0.290	0.307	530	83.3
Lead	5.3	36.5	1.9	13.1	0.425	0.411	710	111.5
Magnesium	6.5	44.8	2.4	16.5	0.350	0.065	112	17.6
Molybdenum	48.0	331.0	17.0	117.0	0.307	0.368	636	100.0
Monel metal	26.0	179.0	9.5	65.5	0.320	0.319	551	86.6
Nickel silver	18.5	127.0	7.0	48.3	0.322	0.316	546	85.8
Nickel steel	30.0	207.0	11.5	79.3	0.291	0.280	484	76.0
Phosphor bronze	16.1	111.0	6.0	41.4	0.349	0.295	510	80.1
Stainless steel (18-8)	27.6	190.0	10.6	73.1	0.305	0.280	484	76.0

Source: J. E. Shigley and C. R. Mischke, *Mechanical Engineering Design*, McGraw-Hill, New York, 1989. Reprinted with permission of McGraw-Hill.

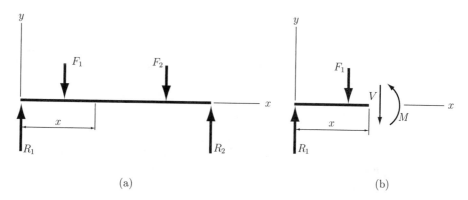

FIGURE II.1.5 *(a) Simply supported beam loaded by concentrated forces and (b) free-body diagram of the left-hand part of the beam.*

To ensure equilibrium, an internal shear force V and an internal bending moment M must act on the cut surface [Fig. II.1.5(b)]. The shear force is obtained by summing the forces to the left of the cut section. The bending moment is obtained by summing the moments of the forces to the left of the section with respect to an axis through the section. The shear force and the bending moment are related by the following expression:

$$V = \frac{dM}{dx}. \tag{II.1.19}$$

If bending is caused by a uniformly distributed load w (acting downward), then the relation between shear force and bending moment is

$$\frac{dV}{dx} = \frac{d^2M}{dx^2} = -w. \tag{II.1.20}$$

The units for w are units of force per unit of length. A general force distribution called *load intensity* can be expressed as

$$q = \lim_{\Delta x \to 0} \frac{\Delta F}{\Delta x}.$$

Integrating Eqs. (II.1.19) and (II.1.20) between two points on the beam of coordinates x_A and x_B yields

$$\int_{V_A}^{V_B} dV = \int_{x_A}^{x_B} q \, dx = V_B - V_A. \tag{II.1.21}$$

The above equation states that the changes in shear force from A to B are equal to the area of the loading diagram between x_A and x_B. Similarly,

$$\int_{M_A}^{M_B} dM = \int_{x_A}^{x_B} V \, dx = M_B - M_A, \tag{II.1.22}$$

which states that the changes in moment from A to B are equal to the area of the shear force diagram between x_A and x_B.

Table II.1.3 lists a set of five *singularity functions* that are useful in developing the general expressions for the shear force and the bending moment in a beam when it is loaded by concentrated forces or moments.

Normal Stresses in Pure Bending

The relationships for the normal stresses in beams are derived by considering that the beam is subjected to pure bending, the material is isotropic and homogeneous and obeys Hooke's law, the beam is initially straight with a constant cross section throughout the length, the beam axis of symmetry is in the plane of bending, and the beam cross sections remain plane during bending.

Figure II.1.6 shows a part of a beam on which a positive bending moment $\mathbf{M}_z = M\mathbf{k}$ ($\mathbf{k}$ being the unit vector associated with z-axis) is applied. A *neutral plane* is a plane that is coincident with the elements of the beam of zero strain. The xz plane is considered as the neutral plane. The x-axis is coincident with the *neutral axis* of the section and the y-axis is coincident with the axis of symmetry of the section.

Applying a positive moment on the beam, the upper surface will bend downward and, therefore, the neutral axis will also bend downward (Fig. II.1.6). Because of this fact, the section PQ initially parallel to RS will rotate through the angle $d\phi$ with respect to $P'Q'$. In Figure II.1.6, ρ is the radius of curvature of the neutral axis, ds is the length of a differential element of the neutral axis, and ϕ is the angle between the two adjacent sides, RS and $P'Q'$. The definition of the curvature is

$$\frac{1}{\rho} = \frac{d\phi}{ds}.$$ (II.1.23)

The deformation of the beam at distance y from the neutral axis is

$$dx = y d\phi,$$ (II.1.24)

and the strain is

$$\epsilon = -\frac{dx}{SD},$$ (II.1.25)

where the negative sign suggests that the beam is in compression. Equations (II.1.23), (II.1.24), and (II.1.25) give

$$\epsilon = -\frac{y}{\rho}.$$ (II.1.26)

Since $\sigma = E\epsilon$, the expression for stress is

$$\sigma = -\frac{Ey}{\rho}.$$ (II.1.27)

TABLE II.1.3 Singularity functions

Function	Graph of $f_n(x)$	Meaning
Concentrated moment (unit doublet)	$\langle x-a\rangle^{-2}$	$\langle x-a\rangle^{-2}=0 \quad x\neq a$ $\int_{-\infty}^{x}\langle x-a\rangle^{-2}dx=\langle x-a\rangle^{-1}$ $\langle x-a\rangle^{2}=\pm\infty \quad x=a$
Concentrated force (unit impulse)	$\langle x-a\rangle^{-1}$	$\langle x-a\rangle^{-1}=0 \quad x\neq a$ $\int_{-\infty}^{x}\langle x-a\rangle^{-1}dx=\langle x-a\rangle^{0}$ $\langle x-a\rangle^{-1}=+\infty \quad x=a$
Unit step	$\langle x-a\rangle^{0}$	$\langle x-a\rangle^{0}=\begin{cases}0 & x<a\\1 & x\geq a\end{cases}$ $\int_{-\infty}^{x}\langle x-a\rangle^{0}dx=\langle x-a\rangle^{1}$
Ramp	$\langle x-a\rangle^{1}$	$\langle x-a\rangle^{1}=\begin{cases}0 & x<a\\x-a & x\geq a\end{cases}$ $\int_{-\infty}^{x}\langle x-a\rangle^{1}dx=\dfrac{\langle x-a\rangle^{2}}{2}$
Parabolic	$\langle x-a\rangle^{2}$	$\langle x-a\rangle^{2}=\begin{cases}0 & x<a\\(x-a)^{2} & x\geq a\end{cases}$ $\int_{-\infty}^{x}\langle x-a\rangle^{2}dx=\dfrac{\langle x-a\rangle^{3}}{3}$

Source: J. E. Shigley and C. R. Mischke, *Mechanical Engineering Design*, McGraw-Hill, New York, 1989. Reprinted with permission of McGraw-Hill.

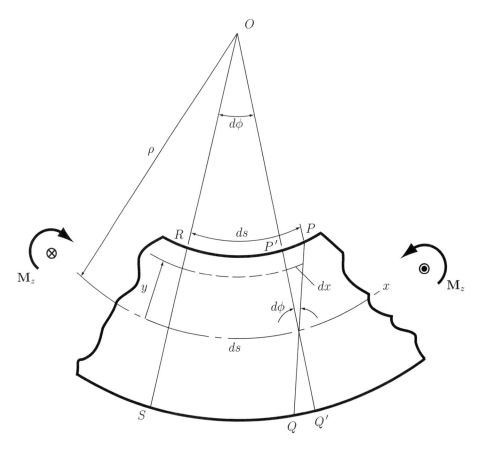

FIGURE II.1.6 *Normal stress in flexure.*

The force acting on an element of area dA is $\sigma\,dA$ and integrating this force

$$\int \sigma\,dA = -\frac{E}{\rho} \int y\,dA = 0. \tag{II.1.28}$$

Since the x-axis is the neutral axis, Eq. (II.1.28) states that the moment of the area about the neutral axis is zero. Thus, Eq. (II.1.28) defines the location of the neutral axis, that is, the neutral axis passes through the centroid of the cross-sectional area.

For equilibrium, the internal bending moment created by the stress σ must be the same as the external moment $\mathbf{M}_z = M\mathbf{k}$, namely

$$M = \int y\sigma\,dA = \frac{E}{\rho} \int y^2\,dA, \tag{II.1.29}$$

where the second integral is the second moment of area I about the z-axis,

$$I = \int y^2\,dA. \tag{II.1.30}$$

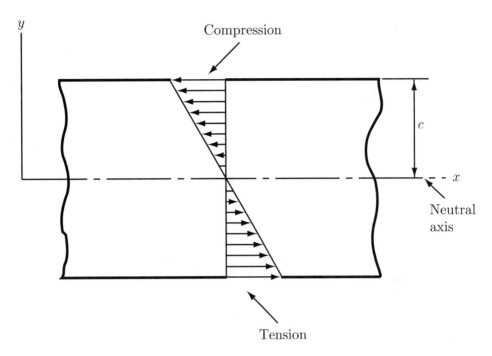

FIGURE II.1.7 *Bending stress in flexure.*

From Eqs. (II.1.29) and (II.1.30)

$$\frac{1}{\rho} = \frac{M}{EI},$$ (II.1.31)

is obtained. Eliminating ρ from Eqs. (II.1.27) and (II.1.31) yields

$$\sigma = -\frac{My}{I}.$$ (II.1.32)

Equation (II.1.32) states that the stress σ is directly proportional to the bending moment M and the distance y from the neutral axis (Fig. II.1.7). The maximum stress is

$$\sigma = \frac{Mc}{I},$$ (II.1.33)

where $c = y_{max}$. Equation (II.1.33) can also be written in the following two forms

$$\sigma = \frac{M}{I/c}, \qquad \sigma = \frac{M}{Z},$$ (II.1.34)

where $Z = I/c$ is called the *section modulus*.

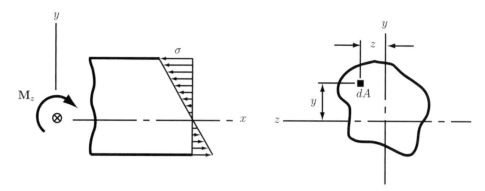

FIGURE II.1.8 *Normal stress of asymmetrical section beam.*

Normal Stresses in Beams with Asymmetrical Sections

If the plane of bending coincides with one of the two principal axes of the section, the results of the previous section can be applied to beams with asymmetrical sections.

From Eq. (II.1.28), the stress at a distance y from the neutral axis is

$$\sigma = -\frac{Ey}{\rho}. \tag{II.1.35}$$

The force on the element of area dA shown in Figure II.1.8 is

$$dF = \sigma \ dA = -\frac{Ey}{\rho} dA.$$

The moment of this force about the y-axis gives

$$M_y = \int z \ dF = \int \sigma z \ dA = -\frac{E}{\rho} \int yz \ dA, \tag{II.1.36}$$

where the integral is across the section. The last integral in Eq. (II.1.36) is the product of inertia I_{yz}. If the bending moment on the beam is in the plane of one of the principal axes then

$$I_{yz} = \int yz \ dA = 0. \tag{II.1.37}$$

Hence, the relations developed in the preceding section can be applied to beams having asymmetrical sections only if $I_{yz} = 0$.

Shear Stresses in Beams

In the general case, beams have both shear forces and bending moments acting upon them. A beam of constant cross section subjected to a shear force $\mathbf{V} = V\mathbf{j}$ and a bending moment

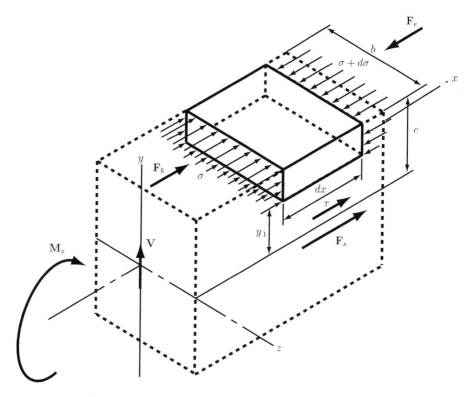

FIGURE II.1.9 *Shear stresses.*

$\mathbf{M}_z = M\mathbf{k}$ ($\mathbf{j}$ and $\mathbf{k}$ are the unit vectors corresponding to the y- and z-axes) is considered in Figure II.1.9. The relationship between V and M is

$$V = \frac{dM}{dx}. \tag{II.1.38}$$

An element of length dx located at a distance y_1 above the neutral axis is considered. Because of the shear force, the bending moment is not constant along the x-axis. The bending moment M on the beginning side of the section produces the normal stress σ and the bending moment $M + dM$ on the end side of the section produces the normal stress $\sigma + d\sigma$. The normal stress σ generates the force $\mathbf{F}_b = F_b\,\mathbf{1}$ and the the normal stress $\sigma + d\sigma$ generates the force $\mathbf{F}_e = -F_e\,\mathbf{1}$. Since $F_e > F_b$, the resultant of these forces would cause the element to slide in the $-x$ direction. To ensure equilibrium, the resultant must be balanced by a shear force acting in the $+x$ direction on the bottom of the section. A shear stress τ generates the shear force $\mathbf{F}_s = F_s\,\mathbf{1}$.

The force on the beginning side is

$$F_b = \int_{y_1}^{c} \sigma \, dA, \tag{II.1.39}$$

where the integration is from the bottom of the element $y = y_1$ to the top $y = c$ and dA is a small element of area on the face. Using the expression $\sigma = \dfrac{My}{I}$, Eq. (II.1.39) yields

$$F_b = \frac{M}{I} \int_{y_1}^{c} y \, dA. \tag{II.1.40}$$

The force acting on the end face is calculated in a similar way:

$$F_e = \int_{y_1}^{c} (\sigma + d\sigma) \, dA = \frac{M + dM}{I} \int_{y_1}^{c} y \, dA. \tag{II.1.41}$$

The force on the bottom face is

$$F_s = \tau b \, dx, \tag{II.1.42}$$

where b is the width of the element and $b \, dx$ is the area of the bottom face.

The sum of all the forces on x direction gives

$$\sum F_x = F_b - F_e + F_s = 0, \tag{II.1.43}$$

or

$$F_s = F_e - F_b = \frac{M + dM}{I} \int_{y_1}^{c} y \, dA - \frac{M}{I} \int_{y_1}^{c} y \, dA = \frac{dM}{I} \int_{y_1}^{c} y \, dA. \tag{II.1.44}$$

Substituting Eq. (II.1.42) for F_e and solving for shear stress gives

$$\tau = \frac{dM}{dx} \frac{1}{Ib} \int_{y_1}^{c} y \, dA. \tag{II.1.45}$$

Using Eq. (II.1.38), the shear stress formula becomes

$$\tau = \frac{V}{Ib} \int_{y_1}^{c} y \, dA. \tag{II.1.46}$$

The integral

$$Q = \int_{y_1}^{c} y \, dA \tag{II.1.47}$$

is the first moment of area of the vertical face about the neutral axis. Therefore, Eq. (II.1.46) can be rewritten as

$$\tau = \frac{VQ}{Ib}, \tag{II.1.48}$$

where I is the second moment of area of the section about the neutral axis.

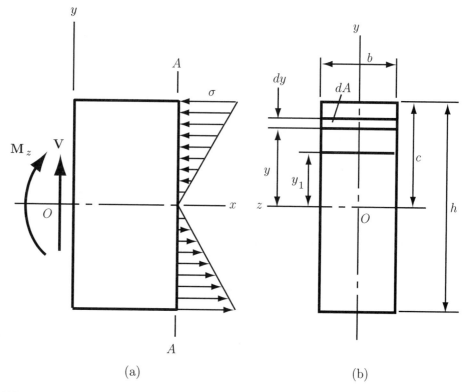

FIGURE II.1.10 *Stresses in beam with rectangular cross section: (a) side view; (b) cross-section view.*

Figure II.1.10 shows a part of a beam with a rectangular cross section. A shear force $\mathbf{V} = V\,\mathbf{J}$ and a bending moment $\mathbf{M}_z = M\mathbf{k}$ act on the beam. Due to the bending moment, a normal stress σ is produced on a cross section of the beam, such as $A - A$. The beam is in compression above the neutral axis and in tension below. An element of area dA located at a distance y above the neutral axis is considered. With $dA = b\,dy$, Eq. (II.1.47) becomes

$$Q = \int_{y_1}^{c} y\,dA = b \int_{y_1}^{c} y\,dy = \left. \frac{by^2}{2} \right|_{y_1}^{c} = \frac{b}{2}(c^2 - y_1^2). \qquad \text{(II.1.49)}$$

Substituting Eq. (II.1.49) into Eq. (II.1.48) gives

$$\tau = \frac{V}{2I}(c^2 - y_1^2). \qquad \text{(II.1.50)}$$

Equation (II.1.50) represents the general equation for shear stress in a beam of rectangular cross section. The expression for the second moment of area I for a rectangular section is

$$I = \frac{bh^3}{12},$$

TABLE II.1.4 Variation of shear stress $\tau = C\dfrac{V}{A}$

Distance y_1	0	0.2c	0.4c	0.6c	0.8c	c
Factor C	1.50	1.44	1.26	0.96	0.54	0

Source: J. E. Shigley and C. R. Mischke, *Mechanical Engineering Design*, McGraw-Hill, New York, 1989. Reprinted with permission of McGraw-Hill.

and, substituting $h = 2c$ and $A = bh = 2bc$, the expression for I becomes

$$I = \frac{Ac^2}{3}.\tag{II.1.51}$$

Substituting Eq. (II.1.51) into Eq. (II.1.50) yields

$$\tau = \frac{3V}{2A}\left(1 - \frac{y_1^2}{c^2}\right) = C\frac{V}{A}.\tag{II.1.52}$$

The values C versus y_1 are listed in Table II.1.4 [20]. The maximum shear stress is obtained for $y_1 = 0$, that is

$$\tau_{max} = \frac{3V}{2A},\tag{II.1.53}$$

and the zero shear stress is obtained at the outer surface where $y_1 = c$. Formulas for the maximum flexural shear stress for the most commonly used shapes are listed in Table II.1.5.

Torsion

A *torque vector* is a moment vector collinear with an axis of a mechanical element, causing the element to twist about that axis. A torque $\mathbf{T}_x = T\,\mathbf{\imath}$ applied to a solid round bar is shown in Figure II.1.11. The angle of twist is given by the following relation:

$$\theta = \frac{Tl}{GJ},\tag{II.1.54}$$

where T is the torque, l the length, G the modulus of rigidity, and J the polar second moment of area. Since the shear stress is zero at the center and maximum at the surface for a solid round bar, the shear stress is proportional to the radius ρ, namely

$$\tau = \frac{T\rho}{J}.\tag{II.1.55}$$

If r is the radius to the outer surface, then

$$\tau_{max} = \frac{Tr}{J}.\tag{II.1.56}$$

TABLE II.1.5 Formulas for maximum shear stress due to bending

Beam shape	Formula
Rectangular	$\tau_{max} = \dfrac{3V}{2A}$
Circular	$\tau_{max} = \dfrac{4V}{3A}$
Hollow round	$\tau_{max} = \dfrac{2V}{A}$
Structural (Web)	$\tau_{max} = \dfrac{V}{A_{web}}$

Source: J. E. Shigley and C. R. Mischke, *Mechanical Engineering Design*, McGraw-Hill, New York, 1989. Reprinted with permission of McGraw-Hill.

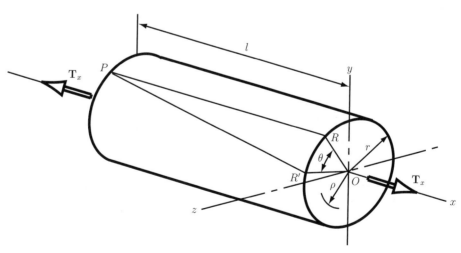

FIGURE II.1.11 *Bar in torsion.*

For a solid round section with the diameter d, the polar second moment of area is

$$J = \frac{\pi \, d^4}{32},$$

and for a hollow round section with the outside diameter d_o and inside diameter d_i, it is

$$J = \frac{\pi \, (d_o^4 - d_i^4)}{32}.$$

For a rotating shaft, the torque T can be expressed in terms of power and speed:

$$H = \frac{2\pi \, Tn}{33000(12)} = \frac{FV}{33000} = \frac{Tn}{63000}, \qquad \text{(II.1.57)}$$

where H is the power in hp, T is the torque in lb·in, n is the shaft speed in rpm, F is the force in lb, and V is the velocity in ft/min. For SI units the power is

$$H = T \, \omega, \qquad \text{(II.1.58)}$$

where H is the power in W, T is the torque in N·m, and ω is the angular velocity in rad/s. The torque T can be approximated by

$$T = 9.55 \frac{H}{n}, \qquad \text{(II.1.59)}$$

where H is in W and n is in rpm.

For rectangular sections, the following approximate formula applies [20]:

$$\tau_{max} = \frac{T}{w \, t^2} \left(3 + 1.8 \frac{t}{w} \right), \qquad \text{(II.1.60)}$$

where w and t are the width and the thickness of the bar, respectively ($t < w$).

II.1.2 Deflection

A *rigid* element does not bend, deflect, or twist when an external force or moment is exerted on it. Conversely, a *flexible* element changes its geometry when an external force, moment, or torque is applied. Therefore, *rigidity* and *flexibility* are terms that apply to particular situations.

The property of a material that enables it to regain its original geometry after having been deformed is called *elasticity*. A straight beam of length l, which is simply supported at the ends and loaded by the transversal force F, is considered in Figure II.1.12(a). If the elastic limit of the material is not exceeded (as indicated by the graph), the deflection y of the beam is linearly related to the force and, therefore, the beam can be described as a *linear spring*.

The case of a straight beam supported by two cylinders is illustrated in Figure II.1.12(b). As the force F is applied to the beam, the length between the supports decreases and,

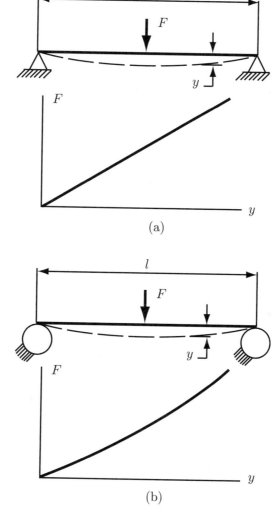

Source: J. E. Shigley, and C. R. Mischke, *Mechanical Engineering Design*, McGraw-Hill, Inc., 1989.

FIGURE II.1.12 *Beam described as a (a) linear spring; (b) nonlinear stiffening spring. Reprinted with permission of McGraw-Hill.*

therefore, a larger force is needed to deflect a short beam than that required for a long one. Hence, the more this beam is deflected, the stiffer it becomes. The force is not linearly related to the deflection, and, therefore, the beam can be described as a *nonlinear stiffening spring*.

The *spring rate* is defined as

$$k(y) = \lim_{\Delta y \to 0} \frac{\Delta F}{\Delta y} = \frac{dF}{dy}, \qquad \text{(II.1.61)}$$

where y is measured at the point of application of F in the direction of F $[F = F(y)]$. For a linear spring, k is a constant called the *spring constant*, and Eq. (II.1.61) becomes

$$k = \frac{F}{y}.$$

(II.1.62)

The total extension or deformation of a uniform bar is

$$\delta = \frac{Fl}{AE},$$

(II.1.63)

where F is the force applied on the bar, l the length of the bar, A the cross-section area, and E the modulus of elasticity. From Eqs. (II.1.62) and (II.1.63), the spring constant of an axially loaded bar is obtained:

$$k = \frac{AE}{l}.$$

(II.1.64)

If a uniform round bar is subjected to a torque T, the angular deflection is

$$\theta = \frac{Tl}{GJ},$$

(II.1.65)

where T is the torque, l the length of the bar, G the modulus of rigidity, and J the polar moment of area. Multiplying Eq. (II.1.65) by $180/\pi$ and substituting $J = \pi d^4/32$ (for a solid round bar), the expression for θ becomes

$$\theta = \frac{583.6Tl}{Gd^4},$$

(II.1.66)

where θ is in degrees and d is the diameter of the round cross section. The torsional spring rate is defined as

$$k = \frac{T}{\theta} = \frac{GJ}{l}.$$

(II.1.67)

If a beam is subjected to a positive bending moment M, the beam will deflect downward. The relationship between the curvature of the beam and the external moment M is

$$\frac{1}{\rho} = \frac{M}{EI},$$

(II.1.68)

where ρ is the radius of curvature, E the modulus of elasticity, and I the second moment of area. The curvature of a plane curve is

$$\frac{1}{\rho} = \frac{d^2y/dx^2}{[1 + (dy/dx)^2]^{3/2}},$$

(II.1.69)

where y is the deflection of the beam at any point of coordinate x along its length. The slope of the beam at point x is

$$\theta = \frac{dy}{dx}. \tag{II.1.70}$$

If the slope is very small, that is, $\theta \approx 0$, then the denominator of Eq. (II.1.69) is expressed as

$$\left[1 + \left(\frac{dy}{dx}\right)^2 \right]^{3/2} = [1 + \theta^2]^{3/2} \approx 1.$$

Hence, Eq. (II.1.68) yields

$$\frac{M}{EI} = \frac{d^2y}{dx^2}. \tag{II.1.71}$$

Differentiating successively Eq. (II.1.71) two times gives

$$\frac{V}{EI} = \frac{d^3y}{dx^3}, \tag{II.1.72}$$

$$\frac{q}{EI} = \frac{d^4y}{dx^4}, \tag{II.1.73}$$

where q is the load intensity and V the shear force:

$$V = \frac{dM}{dx} \quad \text{and} \quad \frac{dV}{dx} = \frac{d^2M}{dx^2} = q.$$

The above relations can be arranged as follows:

$$\frac{q}{EI} = \frac{d^4y}{dx^4}, \tag{II.1.74}$$

$$\frac{V}{EI} = \frac{d^3y}{dx^3}, \tag{II.1.75}$$

$$\frac{M}{EI} = \frac{d^2y}{dx^2}, \tag{II.1.76}$$

$$\theta = \frac{dy}{dx}, \tag{II.1.77}$$

$$y = f(x). \tag{II.1.78}$$

Figure II.1.13 shows a beam of length $l = 10$ in. loaded by the uniform load $w = 10$ lb/in. All quantities are positive if upward and negative if downward. Figure II.1.13 also shows the shear force, bending moment, slope and deflection diagrams. The values of these quantities at the ends of the beam, that is at $x = 0$ and $x = l$, are called *boundary values*. For example, the bending moment and the deflection are zero at each end because the beam is simply supported.

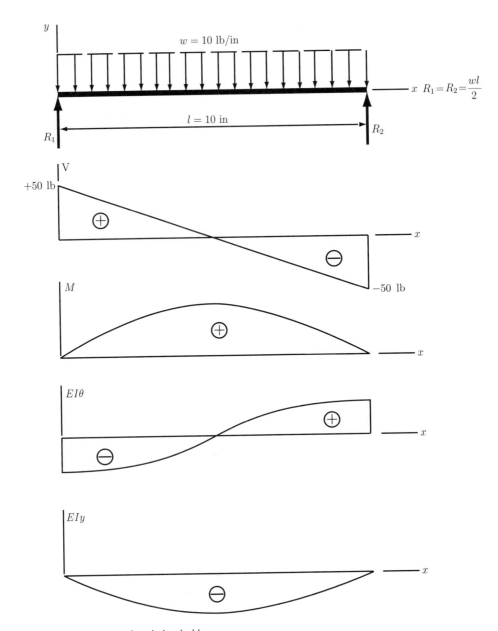

FIGURE II.1.13   *Uniformly loaded beam.*

Deflections Analysis Using Singularity Functions

A simply supported beam acted upon by a concentrated load at the distance a from the origin of the xy coordinate system is shown in Figure II.1.14. The analytical expression for the deflection of the beam will be calculated using the singularity functions. The deflection

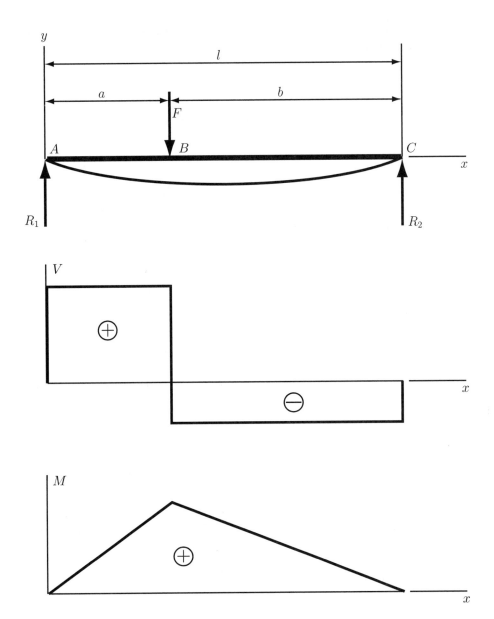

FIGURE II.1.14 *Simply supported beam loaded by a concentrated force.*

of the beam in between the supports $(0 < x < l)$ will be determined. Thus, Eq. (II.1.74) yields

$$EI\frac{d^4y}{dx^4} = q = -F\langle x - a\rangle^{-1}. \tag{II.1.79}$$

Due to the range chosen for x, the reactions R_1 and R_2 do not appear in the above equation. Integrating from 0 to x Eq. (II.1.79) and using Eq. (II.1.75) gives

$$EI\frac{d^3y}{dx^3} = V = -F\langle x - a \rangle^0 + C_1, \qquad \text{(II.1.80)}$$

where C_1 is an integration constant. Using Eq. (II.1.76) and integrating again we obtain

$$EI\frac{d^2y}{dx^2} = M = -F\langle x - a \rangle^1 + C_1 x + C_2, \qquad \text{(II.1.81)}$$

where C_2 is also an integration constant. We can determine the constants C_1 and C_2 by considering two boundary conditions. The boundary condition can be $M = 0$ at $x = 0$ applied to Eq. (II.1.81) which gives $C_2 = 0$ and $M = 0$ at $x = l$ also applied to Eq. (II.1.81) which gives

$$C_1 = \frac{F(l - a)}{l} = \frac{Fb}{l}.$$

Substituting C_1 and C_2 in Eq. (II.1.81) gives

$$EI\frac{d^2y}{dx^2} = M = \frac{Fbx}{l} - F\langle x - a \rangle^1. \qquad \text{(II.1.82)}$$

Integrating Eq. (II.1.82) twice accordingly to Eqs. (II.1.77) and (II.1.78) yields

$$EI\frac{dy}{dx} = EI\theta = \frac{Fbx^2}{2l} - \frac{F\langle x - a \rangle^2}{2} + C_3, \qquad \text{(II.1.83)}$$

$$EIy = \frac{Fbx^3}{6l} - \frac{F\langle x - a \rangle^3}{6} + C_3 x + C_4. \qquad \text{(II.1.84)}$$

The integration constants C_3 and C_4 in the above equations can be evaluated by considering the boundary conditions $y = 0$ at $x = 0$ and $y = 0$ at $x = l$. Substituting the first boundary condition in Eq. (II.1.84) yields $C_4 = 0$. The second condition substituted in Eq. (II.1.84) yields

$$0 = \frac{Fbl^2}{6} - \frac{Fb^3}{6} + C_3 l,$$

or

$$C_3 = -\frac{Fb}{6l}(l^2 - b^2).$$

Substituting C_3 and C_4 in Eq. (II.1.84), the analytical expression for the deflection y is obtained:

$$y = \frac{F}{6EIl}[bx(x^2 + b^2 - l^2) - l\langle x - a \rangle^3]. \qquad \text{(II.1.85)}$$

The shear force and bending moment diagrams are shown in Figure II.1.14.

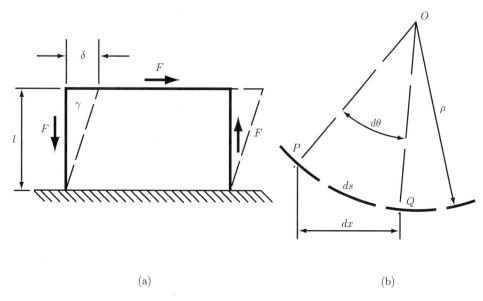

FIGURE II.1.15 *Strain energy due to: (a) direct shear; (b) bending.*

Strain energy

The work done by the external forces on a deforming elastic member is transformed into *strain*, or *potential energy*. If y is the distance a member is deformed, then the strain energy is

$$U = \frac{F}{2}y = \frac{F^2}{2k}. \tag{II.1.86}$$

where $y = \frac{F}{k}$. In the above equation, F can be a force, moment or torque.

For tension (compression) and torsion, the potential energy is, respectively,

$$U = \frac{F^2 l}{2AE}, \tag{II.1.87}$$

and

$$U = \frac{T^2 l}{2GJ}. \tag{II.1.88}$$

Figure II.1.15(a) shows an element with one side fixed. The force F places the element in pure shear and the work done is $U = F\delta/2$. The shear strain is $\gamma = \delta/l = \tau/G = F/AG$. Therefore, the strain energy due to shear is

$$U = \frac{F^2 l}{2AG}. \tag{II.1.89}$$

The expression for the strain energy due to bending can be developed by considering a section of a beam as shown in Figure II.1.15(b). The section PQ of the elastic curve has the length ds and the radius of curvature ρ. The strain energy is $dU = (M/2)d\theta$. Since $\rho d\theta = ds$, the strain energy becomes

$$dU = \frac{Mds}{2\rho}.$$
(II.1.90)

Considering Eq. (II.1.68), ρ can be eliminated in Eq. (II.1.90) and

$$dU = \frac{M^2 ds}{2EI}.$$
(II.1.91)

The strain energy due to bending for the whole beam can be obtained by integrating Eq. (II.1.91) and considering that $ds \approx dx$ for small deflections of the beam, that is

$$U = \int \frac{M^2 dx}{2EI}.$$
(II.1.92)

The strain energy stored in a unit volume u can be obtained by dividing Eqs. (II.1.87), (II.1.88), and (II.1.89) by the total volume lA

$$u = \frac{\sigma^2}{2E} \quad \text{tension and compression,}$$

$$u = \frac{\tau^2}{2G} \quad \text{direct shear,}$$

$$u = \frac{\tau_{max}^2}{4G} \quad \text{torsion.}$$

Even if shear is present and the beam is not very short, Eq. (II.1.92) still gives good results. The expression of the strain energy due to shear loading of a beam can be approximated by considering Eq. (II.1.89) multiplied by a correction factor C. The values of C depend upon the shape of the cross section of the beam. Thus, the strain energy due to shear in bending is

$$U = \int \frac{CV^2 dx}{2AG},$$
(II.1.93)

where V is the shear force. Table II.1.6 lists the values of the correction factor C for various cross sections.

Castigliano's theorem provides an approach to deflection analysis.

Castigliano's theorem: When forces act on a systems subject to small elastic displacements, the displacement corresponding to any force, collinear with the force, is equal to the partial derivative of the total strain energy with respect to that force.

Castigliano's theorem can be written as

$$\delta_i = \frac{\partial U}{\partial F_i},$$
(II.1.94)

TABLE II.1.6   Strain energy correction factors for shear

Beam cross-sectional shape	Factor C
Rectangular	1.50
Circular	1.33
Tubular, round	2.00
Box sections	1.00
Structural sections	1.00

Source: A. P. Boresi, O. M. Sidebottom, F. B. Seely, and J. O. Smith, *Advanced Mechanics of Materials*, 3rd ed., John Wiley & Sons, New York, 1978, p. 173. Reprinted with permission of John Wiley & Sons, Inc.

where δ_i is the displacement of the point of application of the force F_i in the direction of F_i and U is the strain energy. For example, applying Castigliano's theorem for the cases of axial and torsional deflections and considering the expressions for the strain energy given by Eqs. (II.1.87) and (II.1.88), the following relations are obtained:

$$\delta = \frac{\partial}{\partial F}\left(\frac{F^2 l}{2AE}\right) = \frac{Fl}{AE}, \tag{II.1.95}$$

$$\theta = \frac{\partial}{\partial T}\left(\frac{T^2 l}{2GJ}\right) = \frac{Tl}{GJ}. \tag{II.1.96}$$

Even though no force or moment act at a point, Castigliano's theorem can be used to determine the deflection:

- Consider a fictitious force or moment P_i at the point of interest and calculate the expression of the strain energy including the energy due to that dummy force or moment.
- Find the expression for the deflection using Eq. (II.1.96) where the differentiation will be performed with respect to the fictitious force or moment P_i, that is,

$$\delta_i = \frac{\partial U}{\partial P_i}. \tag{II.1.97}$$

- Solve Eq. (II.1.97) and set $P_i = 0$, since P_i is a fictitious force or moment.

Compression

The analysis and design of *compression members* depend upon whether these members are loaded in tension or in torsion. The term *column* is applied to those members for which failure is not produced because of pure compression. Columns are classified according to their length and to whether the loading is central or eccentric. The problem of compression members is to find the *critical load* that produces the failure of the member. Next, the approach presented by Shigley and Mischke will be presented [20].

1. **Long columns with central loading**
   Figure II.1.16 shows long columns of length l having applied an axial load P and various *end conditions*. The load P is applied along the vertical symmetry axis of

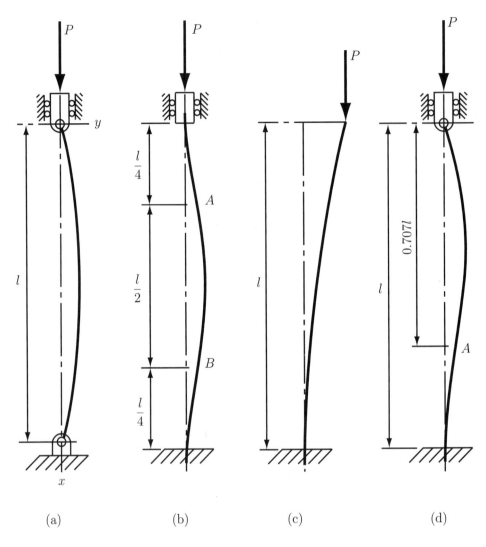

(a)   (b)   (c)   (d)

Source : J. E. Shigley, and C. R. Mischke, *Mechanical Engineering Design*, McGraw-Hill, Inc., 1989.

FIGURE II.1.16   *Long columns: (a) both round ends column; (b) both fixed ends column; (c) one free end and one fixed end column; (d) one rounded end and one fixed end column. Reprinted with permission of McGraw-Hill.*

the column. The end conditions shown in Figure II.1.16 are:

- rounded (or pivoted) — rounded ends [Fig. II.1.16(a)]
- fixed — fixed ends [Fig. II.1.16(b)]
- free — fixed ends [Fig. II.1.16(c)]
- rounded — fixed ends [Fig. II.1.16)(d)].

To develop the relationship between the critical load P_{cr} and the column material and geometry, the situation shown in Figure II.1.16(a) is considered. The figure shows that the bar is bent in the positive y direction and, thus, a negative moment is required:

$$M = -Py. \tag{II.1.98}$$

Equations (II.1.76) and (II.1.98) give

$$\frac{d^2y}{dx^2} = -\frac{P}{EI}y, \tag{II.1.99}$$

or

$$\frac{d^2y}{dx^2} + \frac{P}{EI}y = 0, \tag{II.1.100}$$

with the solution

$$y = A \sin \sqrt{\frac{P}{EI}}x + B \cos \sqrt{\frac{P}{EI}}x, \tag{II.1.101}$$

where A and B are constants of integration which can be determined by considering the boundary conditions $y = 0$ at $x = 0$ and $y = 0$ at $x = l$. Substituting the two boundary conditions in Eq. (II.1.101), results in $B = 0$ and

$$0 = A \sin \sqrt{\frac{P}{EI}}l. \tag{II.1.102}$$

If $A = 0$ is considered into Eq. (II.1.102), the trivial solution of no *buckling* is obtained.

If $A \neq 0$, then

$$\sin \sqrt{\frac{P}{EI}}l = 0, \tag{II.1.103}$$

which is satisfied if $\left(\sqrt{P/EI}\right) l = n\pi$, where $n = 1, 2, 3, \ldots$.

The critical load associated with $n = 1$ is called the *first critical load* and is given by the following expression:

$$P_{cr} = \frac{\pi^2 EI}{l^2}. \tag{II.1.104}$$

Equation (II.1.104) is called *Euler column formula* and applied only for columns with rounded ends. Substituting Eq. (II.1.104) into Eq. (II.1.101), the equation of the deflection curve is obtained:

$$y = A \sin \frac{\pi x}{l}. \tag{II.1.105}$$

The minimum critical load occurs for $n = 1$.

Consider the relation $I = Ak^2$ for the second moment of area I, where A is the cross-section area and k the radius of gyration. Equation (II.1.104) can be rewritten as

$$\frac{P_{cr}}{A} = \frac{\pi^2 E}{(l/k)^2},$$ (II.1.106)

where the ratio l/k is called the *slenderness ratio* and P_{cr}/A is the *critical unit load*. The critical unit load is the load per unit area which can place the column in *unstable equilibrium*. Equation (II.1.106) shows that the critical unit load depends only upon the modulus of elasticity and the slenderness ratio.

Figure II.1.16(b) depicts a column with both ends fixed. The inflection points are at A and B located at a distance $l/4$ from the ends. The distance AB is the same curve as for a column with rounded ends. Substituting the length l by $l/2$ in Eq. (II.1.104), the expression for the first critical load is

$$P_{cr} = \frac{\pi^2 EI}{(l/2)^2} = \frac{4\pi^2 EI}{l^2}.$$ (II.1.107)

Figure II.1.16(c) shows a column with one end free and the other one fixed. The curve of the free-fixed ends column is equivalent to half of the curve for columns with rounded ends. Therefore, if $2l$ is substituted in Eq. (II.1.104) for l, then the critical load for this case is obtained:

$$P_{cr} = \frac{\pi^2 EI}{(2l)^2} = \frac{\pi^2 EI}{4l^2}.$$ (II.1.108)

Figure II.1.16(d) shows a column with one end fixed and the other one rounded. The inflection point is the point A located at a distance of $0.707l$ from the rounded end. Therefore,

$$P_{cr} = \frac{\pi^2 EI}{(0.707l)^2} = \frac{2\pi^2 EI}{l^2}.$$ (II.1.109)

The above situations can be summarized by writing the Euler equation in the following forms:

$$P_{cr} = \frac{C\pi^2 EI}{l^2} \qquad \frac{P_{cr}}{A} = \frac{C\pi^2 E}{(l/k)^2},$$ (II.1.110)

where C is called the *end-condition constant*. It can have one of the values listed in Table II.1.7.

Figure II.1.17 shows the unit load P_{cr}/A as a function of the slenderness ratio l/k. The curve PQR is obtained. The quantity S_y corresponds to point Q and represents the yield strength of the material. From the graph it results that any compression member having the l/k value less than $(l/k)_Q$ should be treated as a pure compression member, while all others can be treated as Euler columns. In practice, this fact is not true. Several tests showed the failure of columns with the slenderness

---

TABLE II.1.7 End condition constants for Euler columns

Column end conditions	End-condition constant C		
	Theoretical value	*Conservative value*	*Recommended value* [*]
Fixed-free	1/4	1/4	1/4
Rounded-rounded	1	1	1
Fixed-rounded	2	1	1.2
Fixed-fixed	4	1	1.2

[*] To be used only with liberal factors of safety when the column load is accurately known.
Source: J. E. Shigley and C. R. Mischke, *Mechanical Engineering Design*, 5th ed., New York, McGraw-Hill, 1989, p. 123. Reprinted with permission of McGraw-Hill.

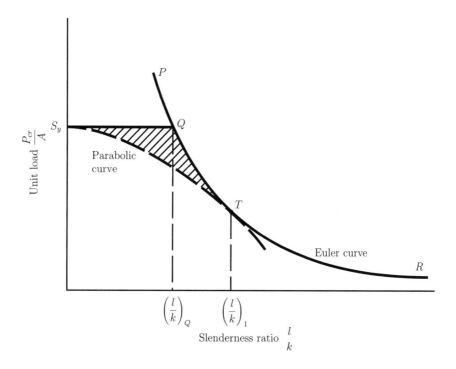

Source: J. E. Shigley, and C. R. Mischke, *Mechanical Engineering Design*, McGraw-Hill, Inc., 1989.

FIGURE II.1.17 *Euler's curve. Reprinted with permission of McGraw-Hill.*

ratio below or very close to point Q. For this reason, neither simple compression methods nor the Euler column equation should be used when the slenderness ratio is near $(l/k)_Q$. The solution in this case is to consider a point T on the Euler curve of Figure II.1.17 such that, if the slenderness ratio corresponding to T is $(l/k)_1$, the

Euler equation should be used only when the actual slenderness ratio of the column is greater than $(l/k)_1$. Point T can be selected such that $P_{cr}/A = S_y/2$.
From Eq. (II.1.110),
the slenderness ratio $(l/k)_1$ is obtained:

$$\left(\frac{l}{k}\right)_1 = \left(\frac{2\pi^2 CE}{S_y}\right)^{1/2}. \tag{II.1.111}$$

2. **Intermediate-length columns with central loading**
When the actual slenderness ratio l/k is less than $(l/k)_1$ (the region in Fig. II.1.17 where the Euler formula is not suitable), the *parabolic* or *J. B. Johnson formula* can be used:

$$\frac{P_{cr}}{A} = a - b\left(\frac{l}{k}\right)^2, \tag{II.1.112}$$

where a and b are constants that can be obtained by fitting a parabola to the Euler curve in Figure II.1.17 (the dashed line ending at T). The constants are

$$a = S_y, \tag{II.1.113}$$

and

$$b = \left(\frac{S_y}{2\pi}\right)^2 \frac{1}{CE}. \tag{II.1.114}$$

Substituting Eqs. (II.1.113) and (II.1.114) into Eq. (II.1.112) yields

$$\frac{P_{cr}}{A} = S_y - \left(\frac{S_y}{2\pi}\frac{l}{k}\right)^2 \frac{1}{CE}, \tag{II.1.115}$$

which can be applied if $\frac{l}{k} \leq \left(\frac{l}{k}\right)_1$.

3. **Columns with eccentric loading**
Figure II.1.18(a) shows a column acted upon by a force P that is applied at a distance e,
also called eccentricity, from the centroidal axis of the column. The free-body diagram is shown in Figure II.1.18(b). Equating the sum of moments about the origin O to zero gives

$$\sum M_O = M + Pe + Py = 0. \tag{II.1.116}$$

Substituting M from Eq. (II.1.116) into Eq. (II.1.76), a nonhomogeneous second order differential equation is obtained:

$$\frac{d^2 y}{dx^2} + \frac{P}{EI}y = -\frac{Pe}{EI}. \tag{II.1.117}$$

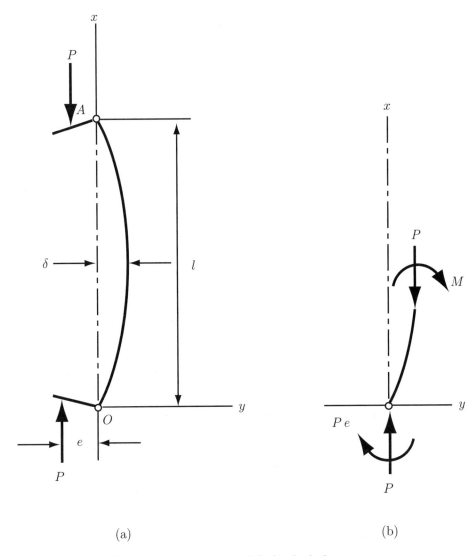

(a)   (b)

FIGURE II.1.18 *(a) Eccentric loaded column and (b) free-body diagram.*

Considering the following boundary conditions

$$x = 0, \quad y = 0,$$

$$x = \frac{l}{2}, \quad \frac{dy}{dx} = 0,$$

and substituting $x = l/2$ in the resulting solution, the maximum deflection δ and the maximum bending moment M_{max} are obtained:

$$\delta = e \left[\sec\left(\frac{1}{2}\sqrt{\frac{P}{EI}} \right) - 1 \right], \tag{II.1.118}$$

$$M_{max} = -P(e + \delta) = -Pe \sec\left(\frac{1}{2}\sqrt{\frac{P}{EI}}\right). \qquad \text{(II.1.119)}$$

At $x = l/2$, the compressive stress σ_c is maximum and can be calculated by adding the axial component produced by the load P and the bending component produced by the bending moment M_{max}, that is

$$\sigma_c = \frac{P}{A} - \frac{Mc}{I} = \frac{P}{A} - \frac{Mc}{Ak^2}. \qquad \text{(II.1.120)}$$

Substituting Eq. (II.1.119) into Eq. (II.1.120) yields

$$\sigma_c = \frac{P}{A}\left[1 + \frac{ec}{k^2}\sec\left(\frac{1}{2k}\sqrt{\frac{P}{EA}}\right)\right]. \qquad \text{(II.1.121)}$$

Considering the yield strength S_y of the column material as σ_c and manipulating Eq. (II.1.121) gives

$$\frac{P}{A} = \frac{S_{yc}}{1 + (ec/k^2)\sec[(l/2k)\sqrt{P/AE}]}. \qquad \text{(II.1.122)}$$

This equation is called the *secant column formula* and the term ec/k^2 is the *eccentricity ratio*. Since Eq. (II.1.122) cannot be solved explicitly for the load P, root-finding techniques using numerical methods can be applied.

4. **Short compression member**

A short compression member is illustrated in Figure II.1.19. At point D of coordinate y, the compressive stress in the x-direction has two components, one due to the axial load P which is equal to P/A and the other due to the bending moment which is equal to My/I. The compressive stress is

$$\sigma_c = \frac{P}{A} + \frac{My}{I} = \frac{P}{A} + \frac{PeyA}{IA} = \frac{P}{A}\left(1 + \frac{ey}{k^2}\right), \qquad \text{(II.1.123)}$$

where $k = (I/A)^{1/2}$ is the radius of gyration, y the coordinate of point D, and e the eccentricity of loading. Setting $\sigma_c = 0$ and solving, the y-coordinate of a line parallel to the x-axis along which the normal stress is zero is obtained:

$$y = -\frac{k^2}{e}. \qquad \text{(II.1.124)}$$

If $y = c$ at point B in Figure II.1.19, the largest compressive stress is obtained. Hence, substituting $y = c$ in Eq. (II.1.123) gives

$$\sigma_c = \frac{P}{A}\left(1 + \frac{ec}{k^2}\right). \qquad \text{(II.1.125)}$$

For design or analysis, Eq. (II.1.125) can be used only if the range of lengths for which the equation is valid is known. For a strut, it is desired that the effect of

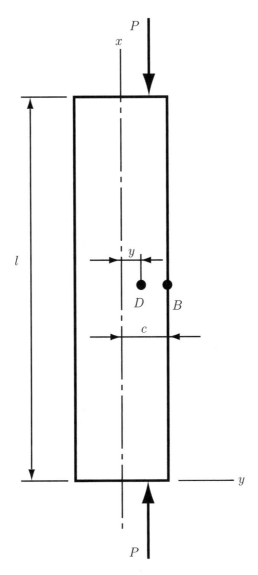

FIGURE II.1.19 *Short compression member.*

bending deflection be within a certain small percentage of eccentricity. If the limiting percentage is 1% of e, then the slenderness ratio is bounded by

$$\left(\frac{1}{k}\right)_2 = 0.282 \left(\frac{AE}{P_{cr}}\right)^{1/2}.$$
(II.1.126)

Therefore, the limiting slenderness ratio for using Eq. (II.1.125) is given by Eq. (II.1.126).

II.1.3 Examples

EXAMPLE II.1.1:

For a stress element having $\sigma_x = 100$ MPa and $\tau_{xy} = 60$ MPa (cw), find the principal stresses and directions on a stress element with respect to the xy system. Plot the maximum and minimum shear stresses τ_1 and τ_2, and find the corresponding normal stresses on another stress element. The stress components that are not given are taken as zero.

Solution First, the Mohr's circle diagram corresponding to the given data will be constructed. Then, using the diagram the stress components will be calculated. Finally, the stress components will be drawn.

The first step to construct Mohr's diagram is to draw the σ- and τ-axes [Fig. II.1.20(a)] and locate the points A of $\sigma_x = 100$ MPa and C of $\sigma_y = 0$ MPa on the σ-axis.

Then, $\tau_{xy} = 60$ MPa is represented in the cw direction and $\tau_{yx} = 60$ MPa in the ccw direction. Hence, point B has the coordinates $\sigma_x = 100$ MPa, $\tau_{xy} = 60$ MPa and point D has the coordinates $\sigma_x = 0$ MPa, $\tau_{yx} = -60$ MPa. The line BD is the diameter and point $E(0, 50)$ the center of the Mohr's circle. The intersections of the circle with the σ-axis give the principal stresses σ_1 and σ_2 at points F and G, respectively.

The x-axis of the stress elements is line EB and the y-axis line ED. The segments BA and AE have the length of 60 and 50 MPa, respectively. The length of segment BE is

$$BE = HE = \tau_1 = \sqrt{(60)^2 + (50)^2} = 78.1 \text{ MPa}.$$

Since the intersection E is 50 MPa from the origin, the principal stresses are

$$\sigma_1 = 50 + 78.1 = 128.1 \text{ MPa}, \qquad \sigma_2 = 50 - 78.1 = -28.1 \text{ MPa}.$$

The angle 2ϕ with respect to the x-axis cw to σ_1 is

$$2\phi = \tan^{-1} \frac{60}{50} = 50.2°.$$

For the first stress element, the x- and y-axes are parallel to the original axes as shown in Figure II.1.20(b). The angle ϕ is in the same direction as the angle 2ϕ in the Mohr's circle diagram. Thus, measuring 25.1° (half of 50.2°) clockwise from x-axis, σ_1-axis is located. The σ_2-axis will be at 90° with respect with the σ_1-axis, as shown in Figure II.1.20(b).

For the second stress element, the two extreme shear stresses occur at the points H and I in Figure II.1.20(a). The two normal stresses corresponding to these shear

---

EXAMPLE II.1.1: *Cont'd*

stresses are equal each to 50 MPa. Point H is 39.8° ccw from point B in the Mohr's circle diagram. Therefore, the stress element is oriented 19.9° (half of 39.8°) ccw from x as shown in Figure II.1.20(c).

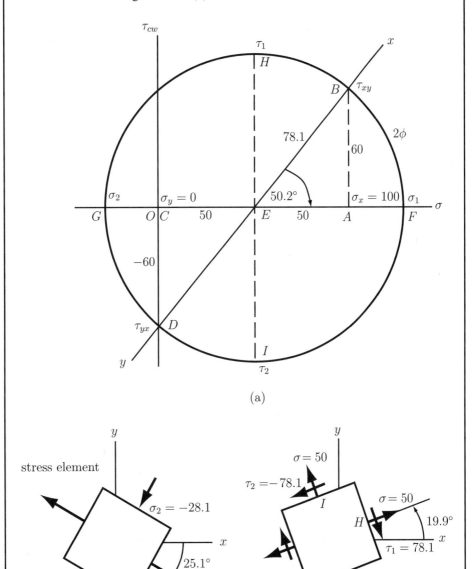

(a)

(b)

(c)

FIGURE II.1.20 *Mohr's circle application for Example II.1.1.*

EXAMPLE II.1.2:

Develop the expressions for the load, the shear force, and the bending moment for the beam in Figure II.1.21.

Solution   The beam shown in Figure II.1.21 is loaded with the concentrated forces F_1 and F_2. The reactions R_1 and R_2 are also concentrated loads. Thus, using Table II.1.3, the load intensity has the following expression:

$$q(x) = R_1 \langle x \rangle^{-1} - F_1 \langle x - l_1 \rangle^{-1} - F_2 \langle x - l_2 \rangle^{-1} + R_2 \langle x - l \rangle^{-1}.$$

The shear force is $V = 0$ at $x = -\infty$. Hence,

$$V(x) = \int_{-\infty}^{x} q(x)\, dx = R_1 \langle x \rangle^0 - F_1 \langle x - l_1 \rangle^0 - F_2 \langle x - l_2 \rangle^0 + R_2 \langle x - l \rangle^0.$$

A second integration yields

$$M(x) = \int_{-\infty}^{x} V(x)\, dx = R_1 \langle x \rangle^1 - F_1 \langle x - l_1 \rangle^1 - F_2 \langle x - l_2 \rangle^1 + R_2 \langle x - l \rangle^1.$$

To calculate the reactions R_1 and R_2 the functions $V(x)$ and $M(x)$ are evaluated at x slightly larger than l. At that point, both shear force and bending moment must be zero. Therefore, $V(x) = 0$ at x slightly larger than l, that is,

$$V = R_1 - F_1 - F_2 + R_2 = 0.$$

Similarly, the moment equation yields

$$M = R_1 l - F_1(l - l_1) - F_2(l - l_2) = 0.$$

The preceding two equations can be solved to obtain the reaction forces R_1 and R_2.

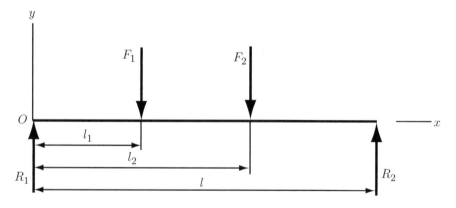

FIGURE II.1.21   *Free-body diagram for simply supported beam loaded by concentrated forces in Example II.1.2.*

EXAMPLE II.1.3:

A cantilever beam with a uniformly distributed load w is shown in Figure II.1.22. The load w acts on the portion $a \leq x \leq l$. Determine the expressions of the shear force and the bending moment.

Solution The moment M_1 and the force R_1 are the support reactions. Using Table II.1.2, the load intensity function is

$$q(x) = -M_1 \langle x \rangle^{-2} + R_1 \langle x \rangle^{-1} - w \langle x - a \rangle^0.$$

Integrating successively two times gives

$$V(x) = \int_{-\infty}^{x} q(x) \, dx = -M_1 \langle x \rangle^{-1} + R_1 \langle x \rangle^0 - w \langle x - a \rangle^1,$$

$$M(x) = \int_{-\infty}^{x} V(x) \, dx = -M_1 \langle x \rangle^0 + R_1 \langle x \rangle^1 - \frac{w}{2} \langle x - a \rangle^2.$$

The reactions can be calculated by evaluating $V(x)$ and $M(x)$ at x slightly larger than l and observing that both V and M are zero in this region. Shear force equation yields

$$-M_1 \cdot 0 + R_1 - w(l - a) = 0,$$

which can be solved to obtain the reaction R_1. The moment equations give

$$-M_1 + R_1 l - \frac{w}{2}(l - a) = 0,$$

which can be solved to obtain the moment M_1.

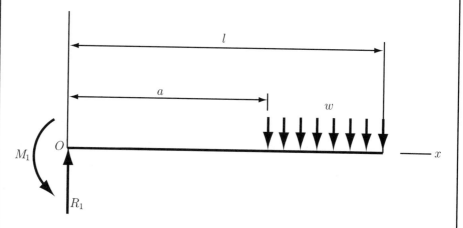

FIGURE II.1.22 *Free-body diagram for cantilever beam with a uniformly distributed load in Example II.1.3.*

EXAMPLE II.1.4:

Determine the diameter of a solid round shaft OC, shown in Figure II.1.23, such that the bending stress does not exceed 10 kpsi. The transversal loads are $F_A = 800$ lb and $F_B = 300$ lb. The length of the shaft is $l = 36$ in., $a = 12$ in., and $b = 16$ in.

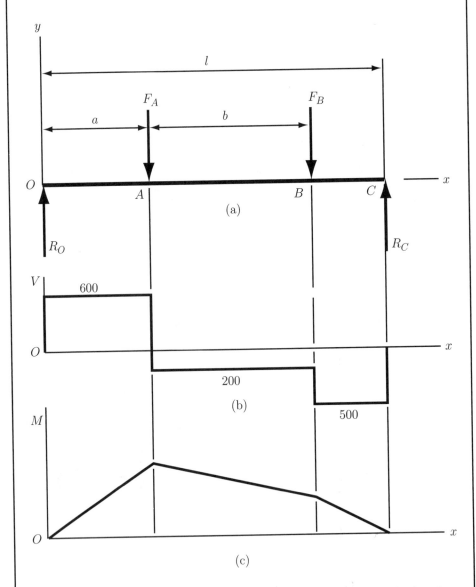

FIGURE II.1.23   Loading diagram for Example II.1.4: (a) free-body diagram; (b) shear force diagram; (c) bending moment diagram.

EXAMPLE II.1.4: *Cont'd*

Solution   The moment equation for the shaft about C yields

$$\sum M_C = -lR_O + (l-a)F_A + (l-a-b)F_B = -36R_O + 24(800) + 8(300) = 0.$$

This equation gives $R_O = 600$ lb.

The force equation for the shaft with respect to the y-axis is

$$\sum F_y = R_O - F_A - F_B + R_C = R_O - 800 - 300 + R_C,$$

yielding $R_C = 500$ lb.

The shear force and the bending moment diagrams are shown in Figure II.1.23(b) and (c).

The maximum bending moment is

$$M = 600(12) = 7200 \ \text{lb} \cdot \text{in}.$$

The section modulus is

$$\frac{I}{c} = \frac{\pi d^3}{32} = 0.0982 d^3.$$

Then, the bending stress is

$$\sigma = \frac{M}{I/c} = \frac{7200}{0.0982 d^3}.$$

Considering $\sigma = 10\,000$ psi and solving for d, it results:

$$d = \sqrt[3]{\frac{7200}{0.0982(10000)}} = 1.94 \ \text{in}.$$

EXAMPLE II.1.5:

Figure II.1.24 [20] shows the link 1 with the length $l = 4$ in., the width $w = 1.25$ in., and the thickness $t = 0.25$ in. The link is loaded by the force $F = 1000$ lb. at the distance $a = 1$ in. This force causes the twisting and bending of a shaft 2 with the diameter $D = 0.75$ in. and the length $L = 5$ in. Find: (a) the force, the moment, and the torque at the origin A and (b) the maximum torsional stress and the bending stress in the arm BC.

Solution   The free-body diagrams of links 1 and 2 are shown in Figure II.1.25.

The force and torque at point C are

$$\mathbf{F} = -1000\mathbf{j} \ \text{lb}, \qquad \mathbf{T} = -1000\mathbf{k} \ \text{lb} \cdot \text{in}.$$

Continued

EXAMPLE II.1.5: *Cont'd*

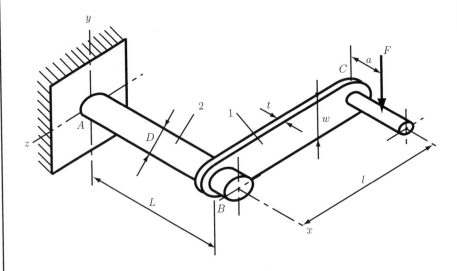

Source : J. E. Shigley, and C. R. Mischke, *Mechanical Engineering Design*, McGraw-Hill, Inc.,1989.

FIGURE II.1.24 *Crank mechanism for Example II.1.5. Reprinted with permission of McGraw-Hill.*

The force, moment, and torque at the end B of the arm BC are

$$\mathbf{F} = 1000\mathbf{j} \text{ lb}, \qquad \mathbf{M} = 4000\mathbf{\imath} \text{ lb} \cdot \text{in.}, \qquad \mathbf{T} = 1000\mathbf{k} \text{ lb} \cdot \text{in.}$$

The force, moment, and torque at the end B of the shaft AB are

$$\mathbf{F} = -1000\mathbf{j} \text{ lb}, \qquad \mathbf{T} = -4000\mathbf{\imath} \text{ lb} \cdot \text{in}, \qquad \mathbf{M} = -1000\mathbf{k} \text{ lb} \cdot \text{in.}$$

The force, moment, and torque at the end A of the shaft AB are

$$\mathbf{F} = 1000\mathbf{j} \text{ lb}, \qquad \mathbf{M} = 6000\mathbf{k} \text{ lb} \cdot \text{in.}, \qquad \mathbf{T} = 4000\mathbf{\imath} \text{ lb} \cdot \text{in.}$$

For the arm BC, the bending stress will reach a maximum near the shaft at B. The bending stress for the rectangular cross section of the arm is

$$\sigma = \frac{M}{I/c} = \frac{6M}{bh^2} = \frac{6(4000)}{0.25(1.25)^2} = 61440 \text{ psi.}$$

The torsional stress is

$$\tau_{max} = \frac{T}{wt^2}\left(3 + 1.8\frac{t}{w}\right) = \frac{1000}{1.25(0.25)^2}\left(3 + 1.8\frac{0.25}{1.25}\right) = 43008 \text{ psi.}$$

EXAMPLE II.1.5: *Cont'd*

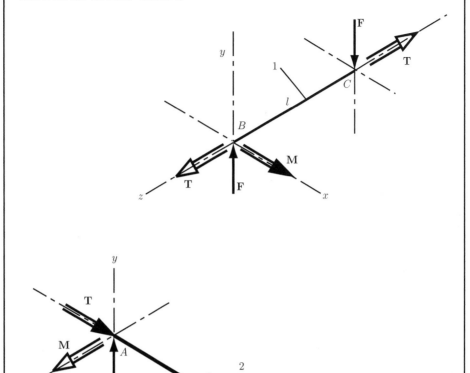

Source : J. E. Shigley, and C. R. Mischke, *Mechanical Engineering Design*, McGraw-Hill, Inc., 1989.

FIGURE II.1.25 *Free-body diagrams of the crank mechanism for Example II.1.5. Reprinted with permission of McGraw-Hill.*

EXAMPLE II.1.6:

Figure II.1.26 shows a beam AC loaded by the uniform distributed force w between B and C. Find the analytical expression for the deflection y as a function of x.

Continued

EXAMPLE II.1.6: *Cont'd*

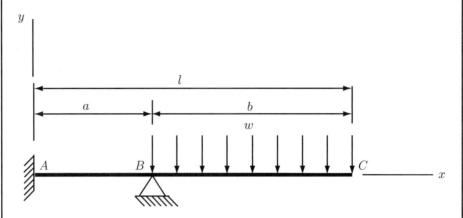

FIGURE II.1.26 *Cantilever beam loaded by a uniformly distributed force at the free end for Example II.1.6.*

Solution The loading equation for x in the range $0 < x < l$ is

$$q = R_B \langle x - a \rangle^{-1} - w \langle x - a \rangle^0, \tag{II.1.127}$$

where R_B is the reaction at B. Integrating this equation four times accordingly to Eqs. (II.1.74)–(II.1.78) yields

$$V = R_B \langle x - a \rangle^0 - w \langle x - a \rangle^1 + C_1, \tag{II.1.128}$$

$$M = R_B \langle x - a \rangle^1 - \frac{w}{2} \langle x - a \rangle^2 + C_1 x + C_2, \tag{II.1.129}$$

$$EI\theta = \frac{R_B}{2} \langle x - a \rangle^2 - \frac{w}{6} \langle x - a \rangle^3 + \frac{C_1}{2} x^2 + C_2 x + C_3, \tag{II.1.130}$$

$$EIy = \frac{R_B}{6} \langle x - a \rangle^3 - \frac{w}{24} \langle x - a \rangle^4 + \frac{C_1}{6} x^3 + \frac{C_2}{2} x^2 + C_3 x + C_4. \tag{II.1.131}$$

The integration constants C_1 to C_4 are found using the boundary conditions.

At $x = 0$ both $EI\theta = 0$ and $EIy = 0$. This gives $C_3 = 0$ and $C_4 = 0$.

At $x = 0$ the shear force is equal to the reaction at A. Therefore, Eq. (II.1.128) gives $V(0) = R_A = C_1$.

The deflection must be zero at $x = a$. Thus, Eq. (II.1.131) yields

$$\frac{C_1}{6} a^3 + \frac{C_2}{2} a^2 = 0 \quad \text{or} \quad C_1 \frac{a}{3} + C_2 = 0. \tag{II.1.132}$$

At the free end, at $x = l$, the moment must be zero. For this boundary condition Eq. (II.1.129) gives

$$R_B(l - a) - \frac{w}{2}(l - a)^2 + C_1 l + C_2 = 0,$$

EXAMPLE II.1.6: *Cont'd*

and using the notation $l - a = b$, the equation resulted from the sum of the forces in the y direction, namely $R_B = R_A + wb = -C_1 + wb$,

$$C_1 a + C_2 = -\frac{wb^2}{2}. \qquad (\text{II.1.133})$$

Solving Eqs. (II.1.132) and (II.1.133) simultaneously for C_1 and C_2 gives

$$C_1 = R_A = \frac{3wb^2}{4a}, \qquad C_2 = \frac{wb^2}{4}.$$

Therefore, the reaction R_B is obtained:

$$R_B = -R_A + wb = \frac{wb}{4a}(4a + 3b).$$

Equation (II.1.129) for $x = 0$ gives

$$M(0) = M_A = C_2 = \frac{wb^2}{4}.$$

The analytical expression for the deflection curve is obtained by substituting the expressions for R_B and the constants C_1, C_2, C_3, and C_4 in Eq. (II.1.131), that is

$$EIy = \frac{wb}{24a}(4a + 3b)\langle x - a \rangle^3 - \frac{w}{24}\langle x - a \rangle^4 - \frac{wb^2 x^3}{8a} + \frac{wb^2 x^2}{8}.$$

EXAMPLE II.1.7:

Consider a simply supported beam of length l and rectangular cross section as shown in Figure II.1.27. A uniformly distributed load w is applied to the beam. Find the strain energy due to shear.

Solution The shear force at an arbitrary distance x from the origin is

$$V = R_1 - wx = \frac{wl}{2} - wx.$$

The strain energy given by Eq. (II.1.93) with $C = 1.5$ (see Table II.1.1) is

$$U = \frac{1.5}{2AG} \int_0^l \left(\frac{wl}{2} - wx \right)^2 dx = \frac{3w^2 l^3}{48AG}.$$

Continued

EXAMPLE II.1.7: *Cont'd*

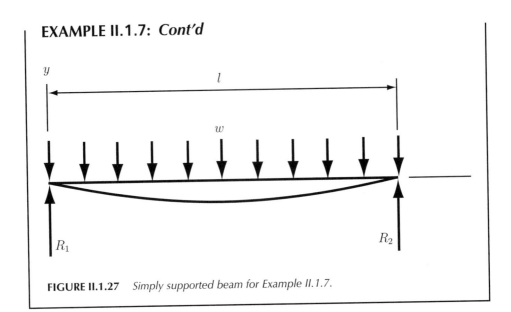

FIGURE II.1.27   *Simply supported beam for Example II.1.7.*

EXAMPLE II.1.8:

A concentrated load F is applied to the end of a cantilever beam (Fig. II.1.28). Find the strain energy by neglecting the shear.

Solution   The bending moment at any point x along the beam has the expression $M = -Fx$. Substituting M into Eq. (II.1.92), the strain energy is

$$U = \int_0^l \frac{F^2 x^2}{2EI}\, dx = \frac{F^2 l^3}{6EI}.$$

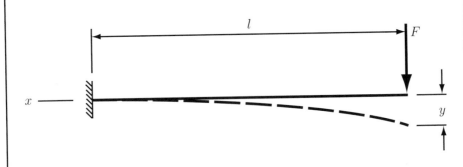

FIGURE II.1.28   *Cantilever beam for Example II.1.8.*

EXAMPLE II.1.9:

A cantilever of length l is loaded by a transversal force F at a distance a as shown in Figure II.1.29. Find the maximum deflection of the cantilever if shear is neglected.

Solution   The maximum deflection of the cantilever will be at its free end. To apply Castigliano's theorem, a fictitious force Q is considered at that point. The bending moments corresponding to the segments OA and AB are, respectively,

$$M_{OA} = F(x - a) + Q(x - l)$$

$$M_{AB} = Q(x - l).$$

The total strain energy is obtained:

$$U = \int_0^a \frac{M_{OA}^2}{2EI}\, dx + \int_a^l \frac{M_{AB}^2}{2EI}\, dx.$$

Applying Castigliano's theorem, the deflection is

$$y = \frac{\partial U}{\partial Q} = \frac{1}{2EI}\left[\int_0^a 2M_{OA}\frac{\partial M_{OA}}{\partial Q}\, dx + \int_a^l 2M_{AB}\frac{\partial M_{AB}}{\partial Q}\, dx\right].$$

Since

$$\frac{\partial M_{OA}}{\partial Q} = \frac{\partial M_{AB}}{\partial Q} = x - l,$$

the expression for the deflection becomes

$$y = \frac{F}{EI}\left\{\int_0^a [F(x - a) + Q(x - l)](x - l)\, dx + \int_a^l [Q(x - l)](x - l)\, dx\right\}.$$

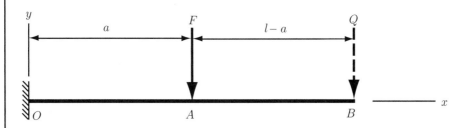

FIGURE II.1.29   *Cantilever beam for Example II.1.9.*

Continued

EXAMPLE II.1.9: *Cont'd*

Since Q is a dummy force, setting $Q = 0$ in the previous equation gives

$$y = \frac{F}{EI} \int_0^a (x - a)(x - l)\, dx = \frac{a^2(3l - a)}{6EI}.$$

II.1.4 Problems

II.1.1 Find the total elongation of a straight bar of length L if a tensile load F is applied at the ends of the bar. The cross section of the bar is A and the modulus of elasticity is E.

II.1.2 The bar in Figure II.1.30 has a constant cross section and is held rigidly between the walls. An axial load F is applied to the bar at a distance a from the left end. The length of the bar is l. Find the reactions of the walls upon the bar.

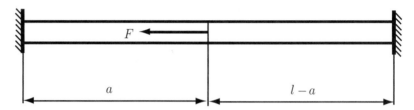

FIGURE II.1.30 *Bar for Problem II.1.2.*

II.1.3 Consider a straight bar of uniform cross-section A loaded with the axial load F. Find (a) the normal and shearing stress intensities on a plane inclined at an angle ϕ to the axis of the bar and (b) the magnitude and direction of the maximum shearing stress.

II.1.4 A straight bar with the uniform cross section of 1.2 in.^{2} is acted upon by an axial force of 14 000 lb at each end. Determine (a) the normal and shearing stress intensities on a plane inclined at an angle $45°$ to the axis of the bar and (b) the maximum shearing stress.

II.1.5 A plane element in a body is subjected to a normal stress in the x-direction of $\sigma_x = 12\,000$ lb/in.^{2}, as well as shearing stress (cw) of $\tau_{xy} = 4000$ lb/in.^{2} Determine (a) the normal and shearing stress intensities on a plane inclined at an angle $30°$ to the normal stress and (b) the maximum shearing stress on the inclined plane.

II.1.6 A plane element in a body is subjected to a normal compressive stress in the x-direction of $\sigma_x = -12\,000$ lb/in^2, as well as shearing stress (ccw) of $\tau_{xy} = -4000$ lb/in.^{2} Determine (a) the normal and shearing stress intensities on a plane inclined at an angle $30°$ to the normal stress and (b) the maximum shearing stress on the inclined plane.

II.1.7 A plane element in a body is subjected to a normal stress in the x-direction of $\sigma_x = 12\,000$ lb/in.^{2}, a normal stress in the y-direction of $\sigma_x = 15\,000$ lb/in.^{2}, as well as shearing stress (cw) of $\tau_{xy} = 8000$ lb/in.^{2} Determine (a) the principal stresses and their directions and (b) the maximum shearing stresses and the directions of the planes on which they occur.

II.1.8 A plane element in a body is subjected to a normal compressive stress in the x-direction of $\sigma_x = -12\,000$ lb/in^2, a normal stress in the y-direction of $\sigma_x = 15\,000$ lb/in.^{2}, as well as shearing stress (ccw) of $\tau_{xy} = -8000$ lb/in.^{2} Determine (a) the principal stresses and their directions and (b) the maximum shearing stresses and the directions of the planes on which they occur.

II.1.9 A bolted joint is shown in Figure II.1.31. The diameter of the bolt is 0.75 in. and the force is $F = 8000$ lb. Determine the average shearing stress across either of the planes $a - a$ or $b - b$.

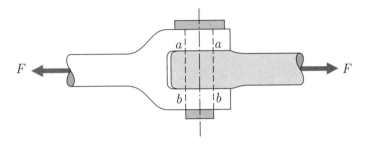

FIGURE II.1.31 *Bolted join for Problem II.1.9.*

II.1.10 Two plates are joined by a single rivet of 1-in. diameter as shown in Figure II.1.32. The load is $F = 9000$ lb and the rivet holes are 1/16 in. larger in diameter than the rivet. The rivet fills the hole completely. Find the average shearing stress developed in the rivet.

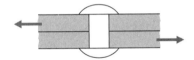

FIGURE II.1.32 *Plates joined by a rivet for Problem II.1.10.*

II.1.11 The fillet weld is a common type of weld used for joining two plates as shown in Figure II.1.33. The dimensions in Figure II.1.33 are $a = 8$ in., $b = 7$ in., and

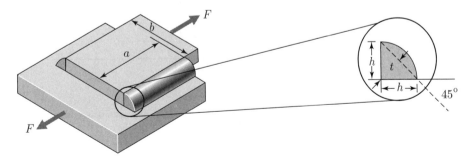

FIGURE II.1.33 *Fillet weld for Problem II.1.11.*

$h = 0.5$ in. The throat of the weld is t. The allowable working stress for shear loading is 11 000 lb/in.^{2} Determine the allowable tensile force F that is applied midway between the two welds. Only shearing stresses are considered in the weld.

II.1.12 The shafts and gears are usually fastened together by means of a key, as shown in Figure II.1.34. Consider the gear with the radius $R = 10$ in. subject to a force F of 1000 lb. The shaft has a radius $r = 1$ in. The dimensions of the key are $t = b = 1/2$ in. and $L = 3$ in. Determine the shear stress on a horizontal plane through the key.

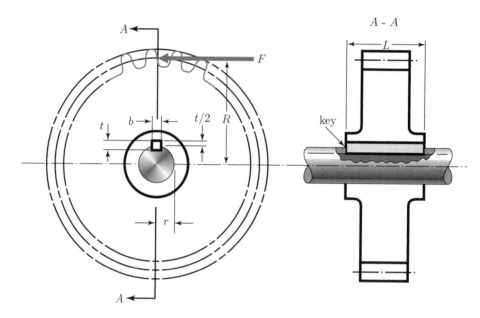

FIGURE II.1.34 *Gear with key for Problem II.1.12.*

II.1.13 Consider the simply supported beam, shown in Figure II.1.35, subjected to a concentrated moment M. Find the equation of the deflection curve.

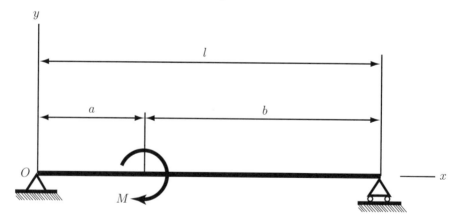

FIGURE II.1.35 *Simply supported beam for Problem II.1.13.*

II.1.14 Consider a simply supported beam subjected to a uniform load distributed, w, over a portion of its length as indicated in Figure II.1.36. Determine the equation of the deflection curve.

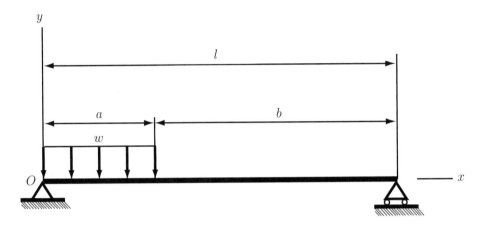

FIGURE II.1.36 *Simply supported beam for Problem II.1.14.*

II.1.15 Consider the cantilever beam of Figure II.1.37, subjected to a uniform load distributed, w, over a portion of its length. Find the equation of the deflection curve.

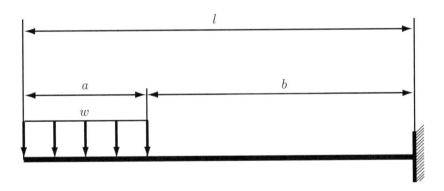

FIGURE II.1.37 *Cantilever beam for Problem II.1.15.*

II.1.16 A sphere of weight W is falling freely through a height h above a cantilever beam as shown in Figure II.1.38. The beam is struck at its tip by the sphere. Determine the total deflection of the tip. Neglect the weight of the beam.

II.1.17 Consider the simply supported beam, shown in Figure II.1.39, loaded by a concentrated moment M at the left end. Find the equation of the deflection curve and the slope at the left end using Castigliano's theorem.

II.1.18 The overhanging beam of Figure II.1.40 is loaded by two equal forces F. Find the deflection at the left end using Castigliano's theorem.

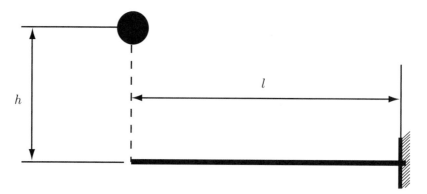

FIGURE II.1.38 *Cantilever beam with weight for Problem II.1.16.*

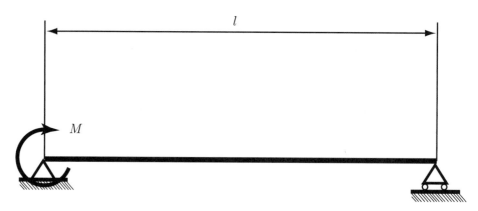

FIGURE II.1.39 *Simply supported beam for Problem II.1.17.*

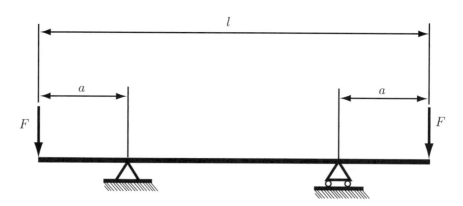

FIGURE II.1.40 *Overhanging beam for Problem II.1.18.*

II.1.19 Determine the slenderness ratio for a wood column 10×10 in. in cross section and 30 ft. long.

II.1.20 A steel bar with the rectangular cross section of 2×2 in. and pined at each end is subjected to axial compression. The critical unit load of the material is $33\,000$ lb/in.^{2} and $E = 30 \times 10^6$ lb/in.^{2} Find (a) the minimum length for which Euler's equation may be used to determine the buckling load and (b) the critical load if the bar is 75 in. long.

II.2   Fatigue

A periodic stress oscillating between some limits applied to a machine member is called *repeated, alternating,* or *fluctuating* stress. The failure of the machine members under the action of these stresses is called *fatigue failure*. A small crack is enough to initiate the fatigue failure. The crack progresses rapidly since the stress concentration effect becomes greater around it. If the stressed area decreases in size, the stress increases in magnitude and if the remaining area is small, the member can fail. A member failed because of fatigue shows two distinct regions. The first one is due to the progressive development of the crack, while the other one is due to the sudden fracture.

II.2.1   Endurance Limit

The strength of materials acted upon by fatigue loads can be determined by performing a fatigue test provided by R. R. Moore's high-speed rotating beam machine (Fig. II.2.1). During the test, the specimen is subjected to pure bending by using weights and rotated with constant velocity. For a particular magnitude of the weights, one records the number of revolutions at which the specimen fails. Then, a second test is performed for a specimen identical with the first one, but the magnitude of the weight is reduced. Again, the number of revolutions at which the fatigue failure occurs is recorded. The process is repeated several times. The intensity of the reversed stress causing failure after a given number of cycles is the *fatigue strength* corresponding to that number of loading cycles. Finally, the fatigue strengths considered for each test are plotted against the corresponding number of revolutions. The resulting chart is called the S–N *diagram*. The S–N curves are plotted on log–log coordinates.

Numerous tests have established that the ferrous materials have an *endurance limit* defined as the highest level of alternating stress that can be withstood indefinitely by a test specimen without failure. The symbol for endurance limit is S'_e. The endurance limit can be related to the tensile strength through some relationships. For example, for steel, Mischke

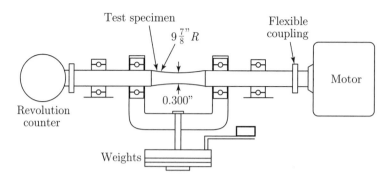

Source: R. C. Juvinall and K. M. Marshek, "Fundamentals of Machine Component Design", John Wiley & Sons, Inc., 1991.

FIGURE II.2.1   *Rotating beam fatigue testing machine. Reprinted with permission of John Wiley & Sons, Inc.*

predicted the following relationships [11]:

$$S'_e = \begin{cases} 0.50\,S_{ut}, & S_{ut} \leq 200 \text{ kpsi (1400 MPa)} \\ 100 \text{ kpsi}, & S_{ut} > 200 \text{ kpsi} \\ 700 \text{ MPa}, & S_{ut} > 1400 \text{ MPa} \end{cases} \quad \text{(II.2.1)}$$

where S_{ut} (or S_u) is the ultimate strength in tension. Table II.2.1 lists the values of the endurance limit for various classes of cast iron that is polished or machined. The symbol S'_e refers to the endurance limit of the test specimen that can be significantly different from the endurance limit S_e of any machine element subjected to any kind of loads. The endurance limit S_e can be affected by several factors called *modifying factors*. Some of these factors are the surface factor k_S, the gradient (size) factor k_G, or the load factor k_L. Thus, the endurance limit of a member can be related to the endurance limit of the test specimen by the following relationship:

$$S_e = k_S\, k_G\, k_L\, S'_e. \quad \text{(II.2.2)}$$

TABLE II.2.1   Endurance Limit for Various Classes of Cast Iron (Polished or Machined)

ASTM* number	Tensile strength S_{ut}, kpsi	Compressive strength S_{uc}, kpsi	Shear modulus of Rupture S_{su}, kpsi	Modulus of elasticity, Mpsi		Endurance limit S_e, kpsi	Brinell hardness H_B	Fatigue stress concentration factor K_f
				Tension	Torsion			
20	22	83	26	9.6–14	3.9–5.6	10	156	1.00
25	26	97	32	11.5–14.8	4.6–6.0	11.5	174	1.05
30	31	109	40	13–16.4	5.2–6.6	14	201	1.10
35	36.5	124	48.5	14.5–17.2	5.8–6.9	16	212	1.15
40	42.5	140	57	16–20	6.4–7.8	18.5	235	1.25
50	52.5	164	73	18.8–22.8	7.2–8.0	21.5	262	1.35
60	62.5	187.5	88.5	20.4–23.5	7.8–8.5	24.5	302	1.50

*ASTM, American Society for Testing Materials.
Source: J. E. Shigley and C. R. Mischke, *Mechanical Engineering Design*, 5th ed., New York, McGraw-Hill, 1989. Reprinted with permission of McGraw-Hill.

TABLE II.2.2 Surface Finish Factors

Surface	Factor a		Exponent
finish	kpsi	MPa	b
Ground	1.34	1.58	−0.085
Machined or cold-drawn	2.70	4.51	−0.256
Hot-rolled	14.4	57.7	−0.718
As forged	39.9	272.0	−0.995

Source: J. E. Shigley and C. R. Mischke, *Mechanical Engineering Design*, New York, McGraw-Hill, 1989. Reprinted with permission of McGraw-Hill.

Surface factor k_S.

The influence of the surface of the specimen is described by the modification factor k_S which depends upon the quality of the finishing. The following formula describes the surface factor [20]:

$$k_S = a S_{ut}^b, \tag{II.2.3}$$

where S_{ut} is the ultimate tensile strength. Some values for a and b are listed in Table II.2.2.

Gradient (size) factor k_G.

The results of the tests performed to evaluate the size factor in the case of bending and torsion are as follows [20]:

$$k_G = \begin{cases} \left(\dfrac{d}{0.3} \right)^{-0.1133} \text{ in.} & 0.11 \leq d \leq 2 \text{ in.} \\[3mm] \left(\dfrac{d}{7.62} \right)^{-0.1133} \text{ mm} & 2.79 \leq d \leq 51 \text{ mm,} \end{cases} \tag{II.2.4}$$

where d is the diameter of the test bar. To apply Eq. (II.2.4) for a nonrotating round bar in bending or for a noncircular cross section, an effective dimension, d_e, is introduced [20]. This dimension is obtained by considering the volume of material stressed at and above 95% of the maximum stress and a similar volume in the rotating beam specimen. When these two volumes are equated, the lengths cancel each other out and only the areas have to be considered.

Some recommended values for k_G are given in reference [27]:

• for bending and torsion:

$$\begin{aligned} d \leq 8 \text{ mm} \qquad & k_G = 1, \\ 8 \text{ mm} < d \leq 250 \text{ mm} \quad & k_G = 1.189 \, d_e^{-0.097}, \\ d > 250 \text{ mm} \qquad & 0.6 \leq k_G \leq 0.75, \end{aligned} \tag{II.2.5}$$

• for axial loading: $\qquad\qquad\qquad\qquad k_G = 1.$

TABLE II.2.3 Summary of Modifying Factors for Bending, Axial Loading, and Torsion of Ductile Materials

10^3-cycle strength
Bending loads: $S = 0.9\,S_u$
Axial loads: $S = 0.75\,S_u$
Torsional loads: $S = 0.9\,S_{us}$, $S_{us} \approx 0.8 S_u$ for steel; $S_{us} \approx 0.7 S_u$ for other ductile materials

10^6-cycle strength (endurance limit)
$S_e = k_S k_G k_L S'_e$, where S'_e is the specimen endurance limit
$\qquad\qquad S'_e = 0.5 S_u$ for steel, lacking better data

Surface factor, k_S	(see Fig. II.2.2)		
	bending	axial	torsion
Load factor, k_L	1	1	0.58
Gradient factor, k_G			
diameter < (0.4 in. or 10 mm)	1	0.7 – 0.9	1
(0.4 in. or 10 mm) < diameter < (2 in. or 50 mm)	0.9	0.7 – 0.9	0.9

for (2 in. or 50 mm) < diameter < (4 in. or 100 mm) reduce the factors by about 0.1
for (4 in. or 100 mm) < diameter < (6 in. or 150 mm) reduce the factors by about 0.2

Source: R. C. Juvinall and K. M. Marshek, *Fundamentals of Machine Component Design*, New York, John Wiley & Sons, 1991. Reprinted with permission of John Wiley & Sons, Inc.

Load factor k_L.
Tests revealed that the load factor has the following values [20]:

$$k_L = \begin{cases} 0.923, & \text{axial loading,} \quad S_{ut} \leq 220 \text{ kpsi (1520 MPa)}, \\ 1, & \text{axial loading,} \quad S_{ut} > 220 \text{ kpsi (1520 MPa)}, \\ 1, & \text{bending,} \\ 0.577 & \text{torsion and shear.} \end{cases} \qquad (\text{II.2.6})$$

Juvinall and Marshek present a summary of all modifying factors for bending, axial loading, and torsion used for fatigue of ductile materials and listed in Table II.2.3 [7].

II.2.2 Fluctuating Stresses

In design problems the stress frequently fluctuates without passing through zero. The components of the stresses are depicted in Figure II.2.3(a), where σ_{min} is minimum stress, σ_{max} the maximum stress, σ_a the stress amplitude or the alternating stress, σ_m the midrange or the mean stress, σ_r the stress range. The steady stress or static stress, σ_s, can have any value between σ_{min} and σ_{max} and exists because of a fixed load. It is usually independent

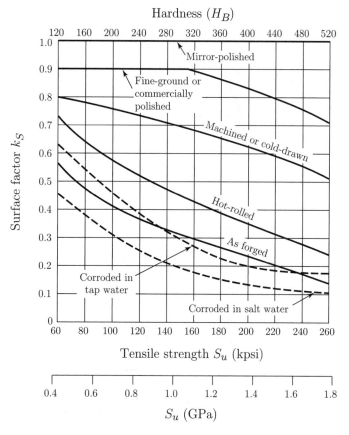

Hardness (H_B)

Surface factor k_S

Mirror-polished

Fine-ground or commercially polished

Machined or cold-drawn

Hot-rolled

As forged

Corroded in tap water

Corroded in salt water

Tensile strength S_u (kpsi)

S_u (GPa)

Source: R. C. Juvinall and K. M. Marshek, "Fundamentals of Machine Component Design", John Wiley & Sons, Inc., 1991.

FIGURE II.2.2 *Surface factor k_S. Reprinted with permission of John Wiley & Sons, Inc.*

of the varying portion of the load. The following relations between the stress components exist as

$$\sigma_m = \frac{\sigma_{max} + \sigma_{min}}{2}, \tag{II.2.7}$$

$$\sigma_a = \frac{\sigma_{max} - \sigma_{min}}{2}. \tag{II.2.8}$$

The fluctuating stresses are described by the stress ratios

$$R = \frac{\sigma_{min}}{\sigma_{max}}, \qquad A = \frac{\sigma_a}{\sigma_m}. \tag{II.2.9}$$

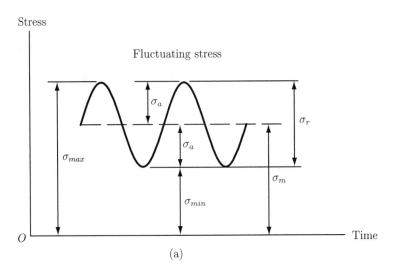

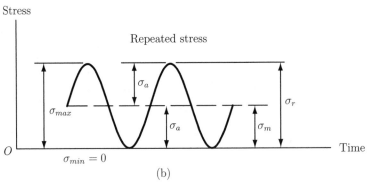

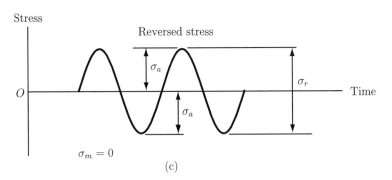

FIGURE II.2.3 *Time varying stresses: (a) sinusoidal fluctuating stress; (b) repeated stress; (c) reversed sinusoidal stress.*

A fluctuating stress is a combination of static plus completely reversed stress. Figure II.2.3(a) shows a sinusoidal fluctuating stress, Figure II.2.3(b) represents a repeated stress, and Figure II.2.3(c) is a completely reversed sinusoidal stress.

II.2.3 Constant Life Fatigue Diagram

Figure II.2.4 illustrates the graphical representation of various combinations of mean and alternating stress in relation to yielding and various fatigue life [7]. This diagram is called the *constant life fatigue diagram* because it has lines corresponding to a constant 10^6 cycle or "infinite" life, constant 10^5 cycle, constant 10^4 cycle, and so forth. The horizontal axis ($\sigma_a = 0$) corresponds to static loading. The point $A(\sigma_m = S_y, \sigma_a = 0)$ represents the yield strength. For ductile materials the point $A'(-S_y, 0)$ represents the compressive yield strength. The point $B(S_u, 0)$ represents the ultimate tensile strength. At the point $A''(\sigma_m = 0, \sigma_a = S_y)$ the stress fluctuates between $+S_y$ and $-S_y$. The points on line AA'' correspond to fluctuations having a tensile peak of S_y. The points on line $A'A''$ correspond to fluctuations having a compressive peak of S_y. Within the triangle $AA'A''$ there are all the combinations with no yielding. The points $C, D, E,$ and F correspond to $\sigma_m = 0$ for various values of fatigue life and are obtained from the S–N diagram. The lines $CB, DB, EB,$ and FB are the estimated lines of constant life. These lines are called the *Goodman lines*.

The area $A'HCGA$ corresponds to a life of at least 10^6 cycles and no yielding. For a life of at least 10^6 cycles and yielding, in addition to the area $A'HCGA$, the area AGB and the

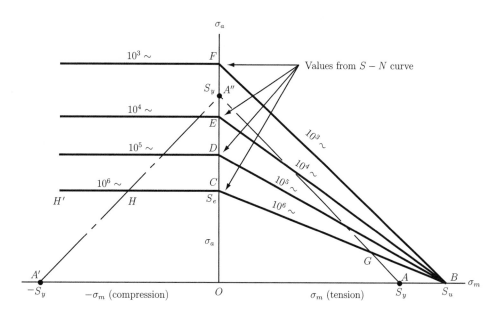

Source: R. C. Juvinall and K. M. Marshek, "Fundamentals of Machine Component Design", John Wiley & Sons, Inc., 1991.

FIGURE II.2.4 *Constant life fatigue diagram. Reprinted with permission of John Wiley & Sons, Inc.*

area to the left of the line $A'H$ may be used. The area $HCGA''H$ corresponds to less than 10^6 cycles of life and no yielding.

II.2.4 Fatigue Life for Randomly Varying Loads

For most mechanical parts acted upon by randomly varying stresses, the prediction of fatigue life is not an easy task. Instead of a single reversed stress σ for n cycles, a part is subjected to σ_1 for n_1 cycles, σ_2 for n_2 cycles, and so forth, and the problem is to estimate the fatigue life of the part. The procedure for dealing with this situation is the *linear cumulative damage rule* (or Miner's rule) and can be expressed by the following equation [7, 20]:

$$\frac{n_1}{N_1} + \frac{n_2}{N_2} + \cdots + \frac{n_k}{N_k} = 1 \quad \text{or} \quad \sum_{j=1}^{j=k} \frac{n_j}{N_j} = 1 \tag{II.2.10}$$

where $n_1, n_2, \cdots, n_k$ represent the number of cycles at specific overstress levels $\sigma_1, \sigma_2, \cdots, \sigma_k$ and $N_1, N_2, \cdots, N_k$ represent the life (in cycles) at these overstress levels, as taken from the appropriate S–N curve. Fatigue failure is predicted when Eq. (II.2.10) holds.

II.2.5 Variable Loading Failure Theories

There are various techniques for plotting the results of the fatigue failure test of a part subjected to fluctuating stress. One of them is called the *modified Goodman diagram* and is shown in Figure II.2.5 [20]. For this diagram the mean stress σ_m is plotted on the abscissa and the other stress components (S_e, S_y, S_u) on the ordinate (tension is the positive direction). The mean stress line makes a 45° angle with the abscissa from the origin to the tensile strength. Lines are constructed to S_e (above the origin) and to $-S_e$ (below the origin) as shown in Figure II.2.5. Yielding can be considered as a criterion of failure if $\sigma_{max} > S_y$.

Another way to display the results is shown in Figure II.2.6 using the strengths [20]. The fatigue limit S_e (or the finite life strength S_f) is plotted on the ordinate. The tensile yield strength S_{yt} is plotted on both coordinate axes. The ultimate tensile strength S_{ut} is plotted on the abscissa. The alternating strength is S_a as a limiting value of σ_a and is plotted on the ordinate. The mean strength is S_m as a limiting value of σ_m and is plotted on the abscissa.

Four criteria of failure are shown in the diagram in Figure II.2.6, that is, Soderberg, the modified Goodman, Gerber, and yielding. Only the Soderberg criterion guards against yielding as shown in Figure II.2.6.

The Soderberg, Goodman, and yield criterion are described by the equation of a straight line in intercept form:

$$\frac{S_m}{a} + \frac{S_a}{b} = 1, \tag{II.2.11}$$

where a and b are the S_m and S_a intercepts, respectively. The equation for the Soderberg line is

$$\frac{S_m}{S_{yt}} + \frac{S_a}{S_e} = 1. \tag{II.2.12}$$

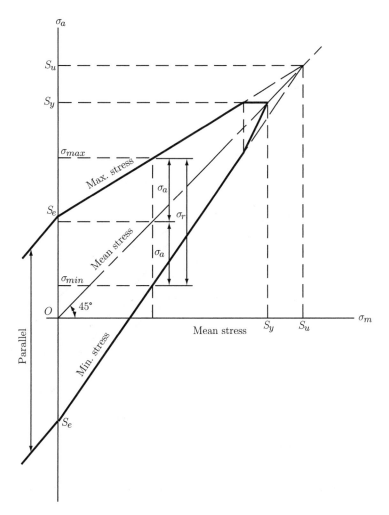

Source: J. E. Shigley, and C. R. Mischke, "Mechanical Engineering Design", McGraw-Hill, Inc., 1989.

FIGURE II.2.5 *Modified Goodman diagram. Reprinted with permission of McGraw-Hill.*

Similarly, the equation for the modified Goodman line is

$$\frac{S_m}{S_{ut}} + \frac{S_a}{S_e} = 1.$$ (II.2.13)

The yielding line is described by the equation

$$\frac{S_m}{S_{yt}} + \frac{S_a}{S_y} = 1.$$ (II.2.14)

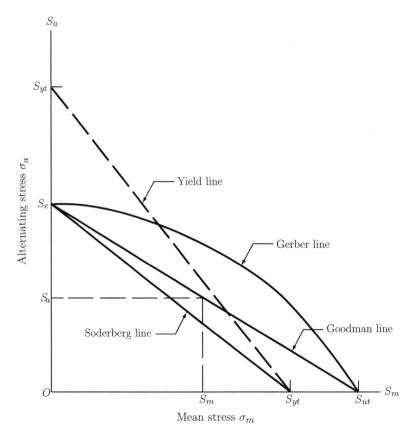

Source: J. E. Shigley, and C. R. Mischke, "Mechanical Engineering Design", McGraw-Hill, Inc., 1989.

FIGURE II.2.6 *Various failure theories. Reprinted with permission of McGraw-Hill.*

The Gerber criterion is also called the *Gerber parabolic relation* because the curve can be modeled by a parabolic equation of the form

$$\frac{S_a}{S_e} + \left(\frac{S_m}{S_{yt}}\right)^2 = 1. \tag{II.2.15}$$

The curve representing the Gerber theory is a better predictor since it passes through the central region of the failure points.

If each strength in Eqs. (II.2.12) to (II.2.15) is divided by a safety factor SF, the stresses σ_a and σ_m can replace S_a and S_m. Therefore, the Soderberg equation becomes

$$\frac{\sigma_a}{S_e} + \frac{\sigma_m}{S_y} = \frac{1}{SF}, \tag{II.2.16}$$

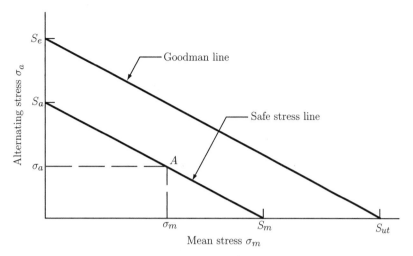

Source: J. E. Shigley, and C. R. Mischke, "Mechanical Engineering Design",
McGraw-Hill, Inc., 1989.

FIGURE II.2.7   *Safe stress line. Reprinted with permission of McGraw-Hill.*

the modified Goodman equation becomes

$$\frac{\sigma_a}{S_e} + \frac{\sigma_m}{S_{ut}} = \frac{1}{SF},$$
(II.2.17)

and the Gerber equation becomes

$$\frac{SF\,\sigma_a}{S_e} + \left(\frac{SF\,\sigma_m}{S_{ut}}\right)^2 = 1.$$
(II.2.18)

Figure II.2.7 shows the explanation of Eq. (II.2.17) [20]. A safe stress line through point A of coordinates σ_m, σ_m is drawn parallel to the modified Goodman line. The safe stress line is the locus of all points of coordinates σ_m, σ_a for which the same safety factor SF is considered, that is, $S_m = SF\,\sigma_m$ and $S_a = SF\,\sigma_a$.

Table 2.4 lists the values of the tensile strength and Table 2.5 gives the yield strength for various materials.

II.2.6   Examples

EXAMPLE II.2.1:

Estimate the S–N curve for a precision steel part for torsional loading. The part has the cross-section diameter under 2 in. and has a fine ground surface. The material has the ultimate tensile strength $S_u = 110$ kpsi and the yield strength $S_y = 77$ kpsi. Use the empirical relationships given in Table II.2.3.

Continued

TABLE II.2.4 Tensile Strength

Material name	Tensile strength	
	ultimate MPa	yield MPa
AISI 1006 Steel, cold drawn	330	285
AISI 1006 Steel, hot rolled bar, 19–32 mm round	295	165
AISI 1006 Steel, cold drawn bar, 19–32 mm round	330	285
AISI 1008 Steel, hot rolled bar, 19–32 mm round	305	170
AISI 1008 Steel, cold drawn bar, 19–32 mm round	340	285
AISI 1010 Steel, cold drawn	365	305
AISI 1010 Steel, hot rolled bar, 19–32 mm round or thickness	325	180
AISI 1010 Steel, cold drawn bar, 19–32 mm round or thickness	365	305
AISI 1012 Steel, cold drawn	370	310
AISI 1012 Steel, hot rolled bar, 19–32 mm round or thickness	330	185
AISI 1012 Steel, cold drawn bar, 19–32 mm round or thickness	370	310
AISI 1015 Steel, cold drawn	385	325
AISI 1015 Steel, cold drawn, 19–32 mm round	385	325
AISI 1015 Steel, hot rolled, 19–32 mm round	345	190
AISI 1015 Steel, as rolled	420	315
AISI 1015 Steel, normalized at 925°C (1700°F)	425	325
AISI 1015 Steel, annealed at 870°C (1600°F)	385	285
AISI 1016 Steel, cold drawn, 19–32 mm round	420	350
AISI 1016 Steel, hot rolled, 19–32 mm round	380	205
AISI 1017 Steel, cold drawn	405	340
AISI 1017 Steel, hot rolled, 19–32 mm round	365	200
AISI 1018 Steel, cold drawn	440	370
AISI 1018 Steel, hot rolled, quenched, and tempered	475	275
AISI 1018 Steel, hot rolled, 19–32 mm round	400	220
AISI 1018 Steel, as cold drawn, 16–22 mm round	485	415
AISI 1018 Steel, as cold drawn, 22–32 mm round	450	380
AISI 1018 Steel, as cold drawn, 32–50 mm round	415	345
AISI 1018 Steel, as cold drawn, 50–76 mm round	380	310
AISI 1019 Steel, cold drawn	455	379
AISI 1019 Steel, hot rolled, 19–32 mm round	407	224
AISI 1020 Steel, cold rolled	420	350
AISI 1020 Steel, hot rolled, 19–32 mm round	380	205
AISI 1020 Steel, as rolled	450	330
AISI 1020 Steel, normalized at 870°C (1600°F)	440	345
AISI 1020 Steel, as rolled, 25 mm round	472	384
AISI 1021 Steel, cold drawn	470	395
AISI 1022 Steel, cold drawn round (19 – 32 mm)	475	400
AISI 1022 Steel, as rolled	505	360

AISI, American Iron and Steel Institute.
Source: MatWeb — Material Property Data available at: http://www.matweb.com/.

TABLE II.2.5 Yield Strength

Material	Condition	Yield strength (MPa)
Cyclic properties		
Steel		
1015	**Normalized**	228
4340	**tempered**	1172
1045	Q&T 80°F	—
1045	Q&T 360°F	1720
1045	Q&T 500°F	1275
1045	Q&T 600°F	965
4142	Q&T 80°F	2070
4142	Q&T 400°F	1720
4142	Q&T 600°F	1340
4142	Q&T 700°F	1070
4142	Q&T 840°F	900
Aluminum		
1100	**Annealed**	97
2014	T6	462
2024	T351	379
5456	H311	234
7075	T6	469

Q&T, Quenched and tempered.

EXAMPLE II.2.1: *Cont'd*

Solution   According to Table II.2.3, the 10^3-cycle peak alternating strength for torsional loaded material is $S = 0.9\, S_{us}$, and for steel is $S_{us} = 0.8\, S_u$. It results:

$$S = 0.9\, S_{us} = 0.9\,(0.8)\, S_u = 0.9\,(0.8)\,(110) = 79.2 \text{ kpsi} \quad \text{for} \quad N = 10^3 \text{ cycles.}$$

The 10^6-cycle peak alternating strength (endurance limit) for torsional loaded ductile material is $S_e = k_S k_G k_L S'_e$. The endurance limit of the test specimen, for $S_u = 110$ kpsi < 200 kpsi, is given by Eq. (II.2.1):

$$S'_e = 0.5\, S_u = 0.5\,(110) = 55 \text{ kpsi.}$$

The surface factor is found from Figure II.2.2, for fine ground surface, $k_S = 0.9$. The other modifying factors for the endurance limit are given in Table II.2.3. The gradient (size) factor is $k_G = 0.9$ for $d < 2$ in. The load factor for torsional load is $k_L = 0.58$. The endurance limit is

$$S_e = k_S\, k_G\, k_L\, S'_e = 0.9\,(0.9)\,(0.58)\,(55) = 25.839 \text{ kpsi} \quad \text{for} \quad N = 10^6 \text{ cycles.}$$

Continued

EXAMPLE II.2.1: *Cont'd*

The S–N diagram is plotted on log–log coordinates. For log $10^3 = 3$ it results:

$$\log S = \log 79.2 = 1.898.$$

For log $10^6 = 6$ it results:

$$\log S_e = \log 25.839 = 1.412.$$

The estimated S–N curve is plotted in Figure II.2.8, and the *Mathematica*^{TM} program is given in Program II.2.1.

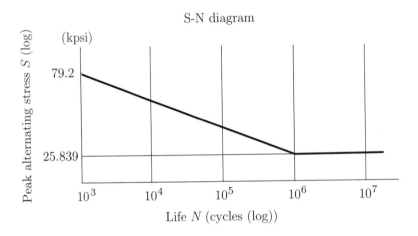

FIGURE II.2.8 *S–N diagram for Example II.2.1.*

EXAMPLE II.2.2:

A precision steel part is subjected to fluctuating axial loading. The part has the cross-section diameter under 8 mm and has a fine ground surface. The material has the ultimate tensile strength $S_u = 1100$ MPa and the yield strength $S_y = 715$ MPa. Find the 10^6 cycle strength (endurance limit).

Solution   The endurance limit of the test specimen for $S_u = 1100$ MPa < 1400 MPa, is given by Eq. (II.2.1):

$$S_e' = 0.5\, S_u = 0.5\,(1100) = 550 \text{ MPa}.$$

Equation (II.2.3) gives the surface factor

$$k_S = a S_{ut}^b = 1.58(1100)^{-0.085} = 0.871,$$

EXAMPLE II.2.2: *Cont'd*

where $a = 1.58$ and $b = -0.085$ are obtained from Table II.2.2. The gradient (size) factor is $k_G = 1$ for axial loading from Eq. (II.2.5). Equation (II.2.6) gives the load factor $k_L = 0.923$ for axial loading and $S_u = 1100$ MPa < 1520 MPa. The endurance limit is

$$S_e = k_S\, k_G\, k_L\, S'_e = 0.871\, (1)\, (0.923)\, (550) = 442.286 \text{ MPa} \quad \text{for} \quad N = 10^6 \text{ cycles.}$$

The *Mathematica*^{TM} program is given in Program II.2.2.

EXAMPLE II.2.3:

A 15 mm diameter steel bar has a fine ground surface with the ultimate strength $S_u = 1100$ MPa and the yield strength $S_y = 715$ MPa.

(a) Using Table II.2.3, estimate the S–N curve and the family of constant life fatigue curves for bending load. Estimate the bending fatigue life for 5×10^4 cycles.

(b) Determine the fatigue strength corresponding to 10^6 cycles and to 5×10^4 cycles for the case of zero-to-maximum (rather than completely reversed) load fluctuations for bending and no yielding.

Solution

(a) According to Table II.2.3, the 10^3-cycle peak alternating strength for bending load is

$$S = 0.9\, S_u = 0.9\, (1100) = 990 \text{ MPa.}$$

The 10^6-cycle peak alternating strength (endurance limit) is $S_e = k_S k_G k_L S'_e$. The endurance limit of the test specimen for $S_u = 1100$ MPa < 1400 MPa is given by Eq. (II.2.1):

$$S'_e = 0.5\, S_u = 0.5\, (1100) = 550 \text{ MPa.}$$

The surface factor is found from Figure II.2.2, for fine ground surface and $S_u = 1100$ MPa, $k_S = 0.89$. The gradient (size) factor is $k_G = 0.9$ for 10 mm $< d = 15$ mm < 50 mm from Table II.2.3. The load factor for bending load is $k_L = 1$ from Table II.2.3. The endurance limit is

$$S_e = k_S\, k_G\, k_L\, S'_e = 0.89\, (0.9)\, (1)\, (550) = 440.55 \text{ MPa.}$$

The S–N diagram is plotted on log–log coordinates. For $\log 10^3 = 3$ it results:

$$\log S = \log 990 = 2.995.$$

Continued

EXAMPLE II.2.3: *Cont'd*

For $\log 10^6 = 6$ it results:

$$\log S_e = \log 440.55 = 2.644.$$

The estimated S–N line is plotted in Figure II.2.6. The equation of the S–N line on log–log coordinates is

$$y = mx + b = -0.117x + 3.347,$$

where the slope m is

$$m = \frac{\log S_e - \log S}{6 - 3} = \frac{2.644 - 2.995}{6 - 3} = -0.117,$$

and the y-intercept b is

$$b = \log S - 3m = 2.995 - 3(-0.117) = 3.347.$$

The log of 10^4-cycle peak alternating strength is

$$\log S_4 = 4m + b = 4(-0.117) + 3.347 = 2.878,$$

and the 10^4-cycle peak alternating strength is

$$S_4 = 10^{\log S_4} = 10^{2.878} = 755.826 \text{ MPa}.$$

The log of 10^5-cycle peak alternating strength is

$$\log S_5 = 5m + b = 5(-0.117) + 3.347 = 2.761,$$

and the 10^5-cycle peak alternating strength is

$$S_5 = 10^{\log S_5} = 10^{2.761} = 577.043 \text{ MPa}.$$

The log of fatigue life for $N = 5 \times 10^4$ cycles is

$$\log S_N = m \, \log(5 \times 10^4) + b = (-0.117)(4.698) + 3.347 = 2.796,$$

and the fatigue life for $N = 5 \times 10^4$ cycles is

$$S_N = 10^{\log S_N} = 10^{2.796} = 625.883 \text{ MPa}.$$

The estimated S–N curve and the $\sigma_m - \sigma_a$ curves for $10^3, 10^4, 5 \times 10^4, 10^5$, and 10^6 cycles of life are plotted on Figure II.2.9.

EXAMPLE II.2.3: *Cont'd*

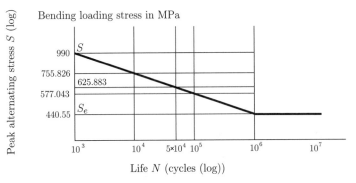

S-N diagram

Bending loading stress in MPa

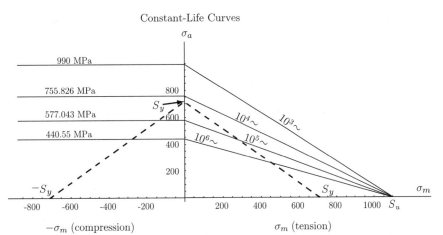

Constant-Life Curves

FIGURE II.2.9 *S–N diagram and constant life fatigue diagram for Example II.2.3(a).*

(b) The case of zero-to-maximum load fluctuations corresponds to $\sigma_m = \sigma_a$.
This is represented by the line OA on Figure II.2.10. The $\sigma_m - \sigma_a$ curves for
10^6 and 5×10^4 cycles of life are plotted on Figure II.2.10. The line equation
corresponding to 10^6 cycles of life is

$$\sigma_a = S_e \left(1 - \frac{\sigma_m}{S_u} \right) = 440.55 \left(1 - \frac{\sigma_m}{1100} \right).$$

The intersection of the line OA with the 10^6 cycles line is the point B of
coordinates $\sigma_m = \sigma_a = 314.566$ MPa. For infinite life (10^6 cycles of life)
$\sigma_{max} = \sigma_a + \sigma_m = 629.132$ MPa.

Continued

EXAMPLE II.2.3: *Cont'd*

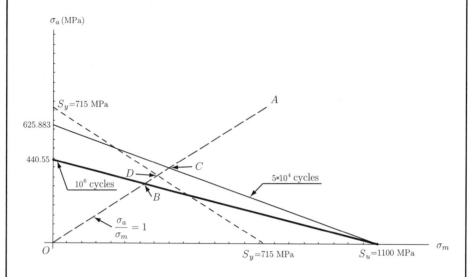

FIGURE II.2.10 $\sigma_m - \sigma_a$ curves for Example II.2.3(b).

The line equation corresponding to 5×10^4 cycles of life is

$$\sigma_a = S_e \left(1 - \frac{\sigma_m}{S_u} \right) = 625.883 \left(1 - \frac{\sigma_m}{1100} \right).$$

The intersection of the line OA with the 5×10^4 cycles line is the point C of coordinates $\sigma_m = \sigma_a = 398.91$ MPa. For this case
$\sigma_{max} = \sigma_a + \sigma_m = 797.819$ MPa greater than the yield strength $S_y = 715$ MPa and this is not permitted.
The line equation corresponding to the $S_y - S_y$ line is

$$\sigma_a = S_y \left(1 - \frac{\sigma_m}{S_y} \right) = 715 \left(1 - \frac{\sigma_m}{715} \right).$$

The intersection of the line OA with the $S_y - S_y$ line is the point D of coordinates $\sigma_m = \sigma_a = 357.5$ MPa. If no yielding is permitted the point D is selected and $\sigma_{max} = \sigma_a + \sigma_m = 715$ MPa $\leq S_y = 715$ MPa. The *Mathematica*[TM] program is given in Program II.2.3.

EXAMPLE II.2.4:

A round steel part with the ultimate strength $S_u = 110$ kpsi and the yield strength $S_y = 77$ kpsi has average machined surfaces. The diameter of the part is less than 1 in. The part is subjected to an axial load fluctuating between 1000 and 6000 lb.

EXAMPLE II.2.4: *Cont'd*

A safety factor of 2 is applied to the loads. Determine the required diameter for infinite life (10^6 cycles of life) and for 10^3 cycles of life. No yielding is permitted.

Solution   According to Table II.2.3, the 10^3-cycle peak alternating strength is

$$S = 0.75 \, S_u = 0.75 \, (110) = 82.5 \text{ kpsi.}$$

The surface factor is $k_S = 0.74$ (from Fig. II.2.2 for machined surfaces and $S_u = 110$ kpsi). The gradient (size) factor is $k_G = 0.8$ (between 0.7 and 0.9 from Table II.2.3). The load factor for axial load is $k_L = 1$ from Table II.2.3. The endurance limit of the test specimen for $S_u = 110$ kpsi < 200 kpsi, is

$$S'_e = 0.5 \, S_u = 0.5 \, (110) = 55 \text{ kpsi.}$$

The endurance limit (10^6-cycle peak alternating strength) is

$$S_e = k_S \, k_G \, k_L \, S'_e = 0.74 \, (0.8) \, (1) \, (55) = 32.55 \text{ kpsi.}$$

The $\sigma_m - \sigma_a$ curves for 10^3 and 10^6 cycles of life are plotted on Figure II.2.8.
    The following relations between the stress components exist:

$$\sigma_m = SF \frac{F_m}{A}, \qquad \sigma_a = SF \frac{F_a}{A},$$

where A is the unknown area, $SF = 2$ is the safety factor, and

$$F_m = \frac{F_{max} + F_{min}}{2}, \qquad F_a = \frac{F_{max} - F_{min}}{2},$$

with $F_{max} = 6$ kip and $F_{min} = 1$ kip.
    The ratio between the alternating stress and the mean stress is $\sigma_a/\sigma_m = 0.714$, and the equation of the line OA on Figure II.2.11 is

$$\sigma_a = 0.714 \, \sigma_m.$$

At the point O, $\sigma_a = \sigma_m = 0$.
    Moving out along line OA, the area A of the part is decreasing.
    The intersection of the line OA with the 10^6 cycles line (infinite life) is the point B of coordinates $\sigma_a = 23.020$ kpsi, $\sigma_m = 32.228$ kpsi. At this point $\sigma_a = SF \, F_a/A$ and the area A is determined as

$$A = \frac{\pi \, d^2}{4} = SF \, \frac{F_a}{\sigma_a} = \frac{2(2.5)}{23.020}.$$

The required diameter for infinite life is $d = 0.525$ in. This diameter is within the range for the gradient factor $k_G = 0.8$.

Continued

EXAMPLE II.2.4: *Cont'd*

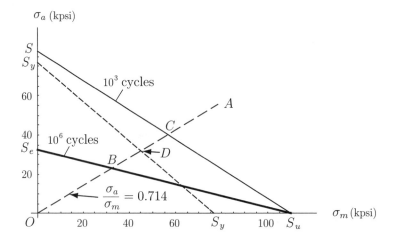

FIGURE II.2.11 $\sigma_m - \sigma_a$ *curves for Example II.2.4.*

The intersection of the line OA with the 10^3 cycles line is the point C of coordinates $\sigma_a = 40.243$ kpsi, $\sigma_m = 56.341$ kpsi, and $\sigma_{max} = 96.585$ kpsi $> S_y = 77$ kpsi. This is not permitted because yielding is unacceptable. The line equation corresponding to the $S_y - S_y$ line is

$$\sigma_a = S_y \left(1 - \frac{\sigma_m}{S_y}\right) = 77 \left(1 - \frac{\sigma_m}{77}\right).$$

The intersection of the line OA with the $S_y - S_y$ line is the point D of coordinates $\sigma_a = 32.083$ kpsi, $\sigma_m = 44.916$ kpsi. If no yielding is permitted the point D is selected and the diameter is selected based on this point. At point D the area A is

$$A = \frac{\pi d^2}{4} = SF \frac{F_a}{\sigma_a} = \frac{2(2.5)}{32.083},$$

and the required diameter is $d = 0.445$ in. A diameter smaller than 0.445 in. would cause yielding. The *Mathematica*™ program is given in Program II.2.4.

EXAMPLE II.2.5:

Consider a 30 mm diameter steel bar having the ultimate tensile strength $S_u = 950$ MPa and the yield strength $S_y = 600$ MPa. The part has a hot rolled surface finish and is subjected to axial loading. The fluctuating stress of the bar for a typical $t = 5$ seconds of operation includes, in order, two cycles with minimum stress $\sigma_{minI} = -100$ MPa and maximum stress $\sigma_{maxI} = 300$ MPa, and three cycles with minimum stress $\sigma_{minII} = -100$ MPa and maximum stress $\sigma_{maxII} = 400$ MPa, (Fig. II.2.12).

EXAMPLE II.2.5: *Cont'd*

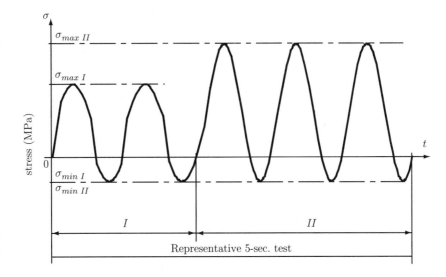

FIGURE II.2.12 *Stress time plot for Example II.2.5.*

For the axial loading of the part the 10^3 cycle peak strength is $S = 712.5$ MPa and the 10^6 cycle peak strength (endurance limit) is $S_e = 180.5$ MPa. Estimate the life of the part when operating continuously.

Solution For the first two cycles of fluctuation ($n_I = 2$), with minimum stress $\sigma_{minI} = -100$ MPa and maximum stress $\sigma_{maxI} = 300$ MPa, the mean stress and the alternating stress are

$$\sigma_{mI} = \frac{\sigma_{maxI} + \sigma_{minI}}{2} = \frac{300 + (-100)}{2} = 100 \text{ MPa},$$

$$\sigma_{aI} = \frac{\sigma_{maxI} - \sigma_{minI}}{2} = \frac{300 - (-100)}{2} = 200 \text{ MPa}.$$

The point A of coordinates $\sigma_{mI} = 100$ MPa, $\sigma_{aI} = 200$ MPa on the $\sigma_m - \sigma_a$ plot in Figure II.2.13(a) is connected by a straight line to the point $\sigma_m = S_u = 950$ MPa on the horizontal axis. The slope of this line equation (line AS_u) is

$$m_I = \frac{\sigma_{aI}}{\sigma_{mI} - S_u} = \frac{200}{100 - 950} = -0.235.$$

The intersection of the line AS_u with the vertical axis ($\sigma_m = 0$) is the y-intercept

$$S_I = -m_I S_u = (-0.235)(950) = 223.529 \text{ MPa}.$$

Continued

EXAMPLE II.2.5: *Cont'd*

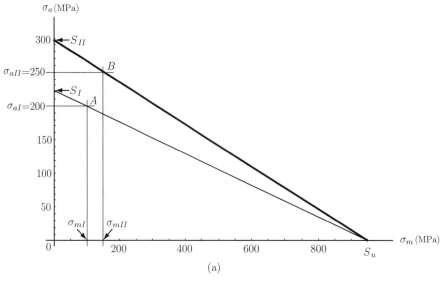

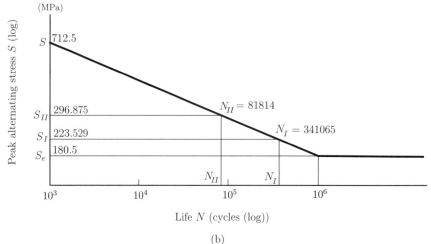

FIGURE II.2.13 (a) $\sigma_m - \sigma_a$ curves and (b) S–N diagram for Example II.2.5.

For the second three cycles of fluctuation ($n_{II} = 3$), with minimum stress $\sigma_{minII} = -100$ MPa and maximum stress $\sigma_{maxII} = 400$ MPa, the mean stress and the alternating stress are

$$\sigma_{mII} = \frac{\sigma_{maxII} + \sigma_{minII}}{2} = \frac{400 + (-100)}{2} = 150 \text{ MPa},$$

$$\sigma_{aII} = \frac{\sigma_{maxII} - \sigma_{minII}}{2} = \frac{400 - (-100)}{2} = 250 \text{ MPa}.$$

EXAMPLE II.2.5: *Cont'd*

The point B of coordinates $\sigma_{mII} = 150$ MPa, $\sigma_{all} = 250$ MPa on the $\sigma_m - \sigma_a$ plot in Figure II.2.13(a) is connected by a straight line to the point $\sigma_m = S_u = 950$ MPa on the horizontal axis. The slope of this line equation (line BS_u) is

$$m_{II} = \frac{\sigma_{all}}{\sigma_{mII} - S_u} = \frac{250}{150 - 950} = -0.312.$$

The intersection of the line BS_u with the vertical axis ($\sigma_m = 0$) is the y-intercept

$$S_{II} = -m_{II}\, S_u = (-0.312)(950) = 296.875 \text{ MPa}.$$

The lines $S_I A S_u$ and $S_{II} B S_u$ are Goodman lines (constant life) and the points A and B correspond to the same fatigue lives as the points S_I and S_{II}. These fatigue lives are determined from the S–N diagram (log–log coordinates) in Figure II.2.13(b). For $\log 10^3 = 3$ it results:

$$\log S = \log 712.5 = 2.852.$$

For $\log 10^6 = 6$ it results

$$\log S_e = \log 180.5 = 2.256.$$

The estimated S–N line is plotted in Figure II.2.13.(b). The equation of the S–N line on log–log coordinates is

$$y = mx + b = -0.198\, x + 3.449.$$

For $\log S_I = \log 223.529 = 2.349$ it results:

$$\log N_I = \frac{\log S_I - b}{m} = \frac{2.349 - 3.449}{-0.198} = 5.532,$$

and the number of cycles is

$$N_I = 10^{\log N_I} = 10^{5.532} = 341065 \text{ cycles}.$$

For $\log S_{II} = \log 296.875 = 2.472$ it results:

$$\log N_{II} = \frac{\log S_{II} - b}{m} = \frac{2.472 - 3.449}{-0.198} = 4.912,$$

and the number of cycles is

$$N_{II} = 10^{\log N_{II}} = 10^{4.912} = 81814 \text{ cycles}.$$

Continued

EXAMPLE II.2.5: *Cont'd*

Adding the portions of life consumed by cycles *I* and *II* gives

$$C = \frac{n_I}{N_I} + \frac{n_{II}}{N_{II}} = \frac{2}{341065} + \frac{3}{81814} = 0.0000425325.$$

This means that the estimated life corresponds to $1/C$ of $t = 5$-second duration. The life of the part is

$$l = t/C = 117557 \text{ s} = 1959.28 \text{ min} = 32.6547 \text{ h}$$

The *Mathematica*™ program for this example is given in Program II.2.5.

EXAMPLE II.2.6:

A 2 in. tension bar is machined from a material with the ultimate tensile strength S_u = 97 kpsi and the yield strength S_y = 68 kpsi. This part is to withstand a fluctuating tensile load varying from 3 to 60 kip. The endurance limit is S_e = 29.488 kpsi. Using the modified Goodman theory, determine the safety factor under the assumption that (a) σ_m remains fixed; (b) σ_a remains fixed; and (c) the ratio σ_a/σ_m is constant.

Solution   The equation for the modified Goodman line is

$$\frac{S_m}{S_u} + \frac{S_a}{S_e} = 1, \quad \text{or} \quad \frac{S_m}{97} + \frac{S_a}{29.488} = 1.$$

This equation is plotted in Fig. II.2.14. The alternating and mean loads are

$$F_a = \frac{F_{max} - F_{min}}{2} = \frac{60 - 3}{2} = 28.5 \text{ kip,}$$

$$F_m = \frac{F_{max} + F_{min}}{2} = \frac{60 + 3}{2} = 31.5 \text{ kip.}$$

The alternating and mean stresses are found to be

$$\sigma_a = \frac{4 F_a}{\pi d^2} = \frac{4(28.5)}{3.141(2)^2} = 9.071 \text{ kpsi,}$$

$$\sigma_m = \frac{4 F_m}{\pi d^2} = \frac{4(31.5)}{3.141(2)^2} = 10.026 \text{ kpsi.}$$

These data are shown in Fig. II.2.14.

(a) Using $S_m = \sigma_m = 10.026$ kpsi, Eq. (II.2.13) yields

$$S_a = S_e \left(1 - \frac{S_m}{S_u} \right) = 29.488 \left(1 - \frac{10.026}{97} \right) = 26.439 \text{ kpsi.}$$

EXAMPLE II.2.6: *Cont'd*

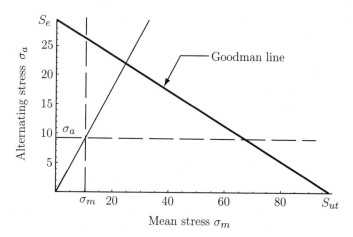

FIGURE II.2.14 $\sigma_m - \sigma_a$ *curves for Example II.2.6.*

The safety factor is

$$SF = \frac{S_a}{\sigma_a} = \frac{26.439}{9.071} = 2.914.$$

(b) Using $S_a = \sigma_a = 9.071$ kpsi, Eq. (II.2.13) yields

$$S_m = S_u \left(1 - \frac{S_a}{S_e}\right) = 97 \left(1 - \frac{19.071}{29.488}\right) = 67.158 \text{ kpsi.}$$

The safety factor is

$$SF = \frac{S_m}{\sigma_m} = \frac{67.158}{10.026} = 6.697.$$

(c) From Eq. (II.2.9):

$$A = \frac{S_a}{S_m} = \frac{\sigma_a}{\sigma_m} = \frac{9.071}{10.026} = 0.904.$$

Equation (II.2.13) yields

$$S_m = \frac{S_e \, S_u}{S_e + A \, S_u} = \frac{29.488 \, (97)}{29.488 + 0.904 \, (97)} = 24.395 \text{ kpsi,}$$

and the safety factor is

$$SF = \frac{S_m}{\sigma_m} = \frac{24.395}{10.026} = 2.433.$$

These data are plotted on Figure II.2.14. The *Mathematica*™ program for this example is given in Program 2.6.

II.2.7   Problems

II.2.1   Using the empirical relationships given in Table II.2.3, estimate the S–N curve for a precision steel part for axial and bending loading. The part has the cross-section diameter under 2 in. and has a machined surface. The material has the ultimate tensile strength S_u = 110 kpsi and the yield strength S_y = 77 kpsi.

II.2.2   A precision steel part with the diameter under 8 mm is subjected to fluctuating bending and torsional loading. The part has a hot rolled surface. The material has the ultimate tensile strength S_u = 1100 MPa and the yield strength S_y = 715 MPa. Find the 10^6 cycle strength (endurance limit).

II.2.3   Plot the S–N curves for bending, axial, and torsional loading of a steel shaft with the diameter d = 1.5 in. The shaft was machined from steel having tensile properties S_u = 90 kpsi and S_y = 75 kpsi. Find the fatigue strength for 6×10^4 cycles.

II.2.4   A steel bar having the ultimate strength S_u = 950 MPa and the yield strength S_y = 600 MPa has a hot rolled surface finish. The surface factor is k_S = 0.475, the gradient factor is k_G = 0.8, and the load factor is k_L = 1. Determine the fatigue strength at 5×10^4 cycles for reversed axial loading.

II.2.5   A steel round link has the diameter d = 25 mm is subjected to an axial load fluctuating between 1000 and 6000 lb. The link has a hot rolled surface finish. The ultimate tensile strength of the material is S_u = 950 MPa and the yield strength is S_y = 600 MPa. Find the fatigue strength corresponding to 10^6 cycles.

II.2.6   A 2 in. diameter shaft is machined from AISI 4320 steel having S_u = 140 kpsi and S_y = 90 kpsi. Estimate the $\sigma_m - \sigma_a$ curves for bending load.

II.2.7   A 15 mm diameter steel bar has a forged surface with the ultimate strength S_u = 1100 MPa and the yield strength S_y = 715 MPa. (a) Using Table II.2.3 estimate the S–N curve and the family of constant life fatigue curves for axial load. Estimate the fatigue life for 4×10^5 cycles. (b) Determine the fatigue strength corresponding to 10^6 cycles and to 4×10^4 cycles for the case of zero-to-maximum (rather than completely reversed) load fluctuations for bending and no yielding.

II.2.8   A round steel link with the ultimate strength S_u = 110 kpsi and the yield strength S_y = 77 kpsi has ground surfaces. The diameter of the link is less than 2 in. The link is subjected to a bending load fluctuating between 1000 and 6000 lb. A safety factor of 1.5 is applied to the loads. Determine the required diameter for infinite life (10^6 cycles of life) and for 10^3 cycles of life.

II.2.9   Consider a 30 mm diameter steel bar having the ultimate tensile strength S_u = 950 MPa and the yield strength S_y = 600 MPa. The part has a hot rolled surface finish and is subjected to axial loading. The fluctuating stress of the bar for a typical t = 6 seconds of operation includes, in order, two cycles with minimum stress of -100 MPa and maximum stress of 300 MPa, three cycles with minimum stress of -150 MPa and maximum stress of 400 MPa, and four cycles with minimum stress of -200 MPa and maximum stress of 600 MPa. Estimate the life of the part when operating continuously.

II.2.10   A steel part with the diameter 20 mm has the ultimate tensile strength S_u = 1100 MPa and the yield strength S_y = 715 MPa. The part has a hot rolled surface finish and is subjected to bending loading. The fluctuating stress of the bar for 10 seconds of operation includes, in order, two cycles with zero minimum stress and maximum stress equal to 100 MPa, two cycles with minimum stress equal to −50 MPa and maximum stress equal to 300 MPa, and two cycles with minimum stress equal to 100 MPa and maximum stress equal to 500 MPa. Determine the fatigue life of the part.

II.2.11 A tension bar is machined from AISI 1050 (S_u = 100 kpsi and S_y = 84 kpsi). This bar has the diameter $d = 1.5$ in. and is subjected to a fluctuating tensile load varying from 0 to 50 kip. The endurance limit is S_e = 25 kpsi. Using the modified Goodman theory, determine the safety factor under the assumption that σ_m and σ_a remain constant.

II.2.12 Repeat the previous problem considering that the ratio σ_a/σ_m is constant.

II.2.8   Programs

PROGRAM II.2.1

```
Apply[Clear,Names["Global`*"]];
Off[General::spell];
Off[General::spell1];

"steel bar, d<2in., fine ground surface"

Su=110.; (* ultimate strength [kpsi] *)
Sy=77.;  (* yield strength [kpsi] *)

Print["ultimate strength [kpsi] Su=",Su];
Print["yield strength [kpsi] Sy=",Sy];

"10^3 cycle strength S"
"torsional load: S=0.9 Sue; Sue=0.8 Su (for steel); Sue=0.7 Su (other)"
S=(0.9)(0.8) Su;
Print["torsional: S=",S];

"10^6 cycle strength (endurance limits) Se=kS kG kL Se'"

"endurance limit of test speciment Se'"
"Se'=0.5 Su (for Su<200 kpsi); Se'=100 kpsi (for Su>200 kpsi)"
"bending, axial, torsion: Se'=0.5 Su "
Sep=0.5 Su;
Print[" Se'=",Sep];

"modifying factors for endurance limit"

"surface factor kS (bending, axial, torsion)"
"fine ground surface"
kS=0.9;
Print["surface factor kS=",kS];
```

```
"size (gradient) factor kG"
"bending and torsion: kG=1 (for d<0.4 in.); kG=0.9 (for 0.4<d<2.0 in)"
kG=0.9;
Print["bending and torsion: size factor kG=",kG];

"load factor kL"
"bending and axial: kL=1; torsion: kL=0.58"
kL=0.58;

"endurance limit Se=kS kG kL Se'"
Se=kS kG kL Sep;
Print["torsion: Se=",Se];


"S-N diagram"
LS=Log[10,S];
LSe=Log[10,Se];
Lm=(LSe-LS)/(6-3);
Lb=LS-3 Lm;

Print["N=10^3 cycles, S = ",S," [kpsi]; Log[N]=3, Log[S] = ",LS];
Print["N=10^6 cycles, Se = ",Se," [kpsi]; Log[N]=6, Log[Se] = ",LSe];

LSN=Lm x+Lb;

SNG=Plot[LSN,{x,3,6},AxesOrigin→{3,LSe},
    AxesLabel→{"Log[N]","Log[S]"}];

Show[SNG,Ticks→{{3,4,5,6},{LSe,LS}},
    GridLines→{{3,4,5,6},{LSe,LS}},
    PlotRange→{{3,6},{LSe-0.15,LS+0.15}},AxesOrigin→{3,LSe-0.15},
    AxesLabel→{"Log[N]","Log[S]"}];
```

steel bar, d<2in., fine ground surface

ultimate strength [kpsi] Su=110.

yield strength [kpsi] Sy=77.

10^3 cycle strength S

torsional load: S=0.9 Sue; Sue=0.8 Su (for steel); Sue=0.7 Su (other)

torsional: S=79.2

10^6 cycle strength (endurance limits) Se=kS kG kL Se'

endurance limit of test speciment Se'

Se'=0.5 Su (for Su<200 kpsi); Se'=100 kpsi (for Su>200 kpsi)

bending, axial, torsion: Se'=0.5 Su

Se'=55.

modifying factors for endurance limit

surface factor kS (bending, axial, torsion)

fine ground surface

surface factor kS=0.9

size (gradient) factor kG

bending and torsion: kG=1 (for d<0.4 in.); kG=0.9 (for 0.4<d<2.0 in)

bending and torsion: size factor kG=0.9

load factor kL

bending and axial: kL=1; torsion: kL=0.58

endurance limit Se=kS kG kL Se'

torsion: Se=25.839

S-N diagram

N=10^3 cycles, S = 79.2 [kpsi]; Log[N]=3, Log[S] = 1.89873

N=10^6 cycles, Se = 25.839 [kpsi]; Log[N]=6, Log[Se] = 1.41228

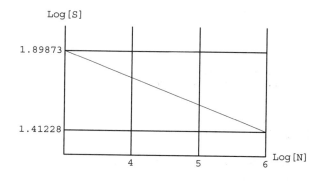

PROGRAM II.2.2

```
Apply[Clear,Names["Global`*"]];
Off[General::spell];
Off[General::spell1];

"steel bar, d<=8 mm., fine ground surface, axial loading"
d=8;
Print["diameter d<= ",d," mm "];
Su=1100.; (* ultimate strength [MPa] *)
Sy=715.;  (* yield strength [MPa] *)

Print["ultimate strength [MPa] Su=",Su];
Print["yield strength [MPa] Sy=",Sy];
```

```
"10^6 cycle strength (endurance limits) Se=kS kG kL Se'"

"endurance limit of test speciment Se'"
"Se'=0.5 Su (for Su<200 kpsi); Se'=100 kpsi (for Su>200 kpsi)"
"Se'=0.5 Su (for Su<1400 MPa); Se'=700 MPa (for Su>1400 MPa)"
"Se'=0.5 Su "
Sep=0.5 Su;
Print["Se'=",Sep," MPa"];

"modifying factors for endurance limit"

"surface factor kS: kS= a (Su)^b "
"surface finish: fine ground => a=1.58; b=-0.085"
a=1.58; b=-0.085;
kS=a (Su)^b;
Print["surface factor kS=",kS];

"size (gradient) factor kG"
"bending and torsion:"
"d<=8 mm ;    kG=1 "
"8 mm <= d <= 250 mm ; kG=1.189 (def)^(-0.097)"
"d > 250 mm ; 0.6 <= kG <= 0.75"
"axial: kG=1"
kG=1;

"load factor kL"
"axial: Su <= 1520 MPa (220 kpsi); kL=0.923"
"axial: Su > 1520 MPa (220 kpsi); kL=1"
"bending: kL=1"
"torsion and shear: kL=0.577"
kL=0.923;

"endurance limit Se=kS kG kL Se'"
Se=kS kG kL Sep;
Print["axial: Se=",Se," MPa"];
```

steel bar, d<=8 mm., fine ground surface, axial loading

diameter d<= 8 mm

ultimate strength [MPa] Su=1100.

yield strength [MPa] Sy=715.

10^6 cycle strength (endurance limits) Se=kS kG kL Se'

endurance limit of test speciment Se'

Se'=0.5 Su (for Su<200 kpsi); Se'=100 kpsi (for Su>200 kpsi)

Se'=0.5 Su (for Su<1400 MPa); Se'=700 MPa (for Su>1400 MPa)

Se'=0.5 Su

Se'=550. MPa

modifying factors for endurance limit

surface factor kS: kS= a (Su)^b

surface finish: fine ground => a=1.58; b=-0.085

surface factor kS=0.871242

size (gradient) factor kG

bending and torsion:

d<=8 mm ; kG=1

8 mm <= d <= 250 mm ; kG=1.189 (def)^(-0.097)

d > 250 mm ; 0.6 <= kG <= 0.75

axial: kG=1

load factor kL

axial: Su <= 1520 MPa (220 kpsi); kL=0.923

axial: Su > 1520 MPa (220 kpsi); kL=1

bending: kL=1

torsion and shear: kL=0.577

endurance limit Se=kS kG kL Se'

axial: Se=442.286 MPa

PROGRAM II.2.3

```
Apply[Clear,Names["Global`*"]];
Off[General::spell];
Off[General::spell1];

(*Input data: steel bar, fine ground surface*)

Su=1100; (* ultimate strength [MPa] *)
Sy=715;  (* yield strength [MPa] *)

Print["ultimate strength [MPa] Su=",Su];
Print["yield strength [MPa] Sy=",Sy];

"10^3 cycle strength S"
"bending loads: S=0.9 Su "
S=0.9 Su;
"Print["bending: S=",S];
```

```
"10^6 cycle strength (endurance limit) Se=kS kG kL Se'"

"endurance limit of test speciment Se'"
"Se'=0.5 Su (for Su<1400 MPa); Se'=700 MPa (for Su>1400 MPa)"
"bending, axial, torsion: Se'=0.5 Su "
Sep=0.5 Su;
Print[" Se'=",Sep];

"modifying factors for endurance limit"

"surface factor (kS) (bending, axial, torsion)"
"print["ultimate strength [kpsi] Su=",Su/6.895];
"fine ground surface"
kS=0.89;
Print["surface factor kS=",kS];

"size (gradient) factor (kG)"
"bending and torsion: kG=1 (for d<10 mm); kG=0.9 (for <10<d<50 mm)"
"axial: kG=0.7-0.9"
kG=0.9;
"Print["bending: size factor KG=",kG];

"load factor (kL)"
"bending and axial: kL=1; torsion: kL=0.58"
kL=1;

"endurance limit Se=kS kG kL Se' "
Se=kS kG kL Sep;
Print["bending: Se=", Se];

"bending: S-N diagram:
LS=Log[10,S];
LSe=Log[10,Se];
Lm=(LSe-LS)/(6-3);
Lb=LS-3 Lm;
Print["S-N line slope: m=",Lm];
Print["S-N line y-intercept: b=",Lb];

LS4=4 Lm+Lb;
S4=10^LS4;
LS5=5 Lm+Lb;
S5=10^LS5;

NN=5 10^4;
LN=N[Log[10,NN]];
LSN=Lm Log[10,NN]+Lb;
SN=10^LSN;

Print["N=10^3 cycles, S  = ",S," [MPa]; Log[N]=3, Log[S] = ",LS];
```

```
Print["N=10^4 cycles, S4 = ",S4," [MPa]; Log[N]=4, Log[S4] = ",LS4];
Print["N=5 10^4 cycles, SN = ",SN," [MPa]; Log[NN]=",LN,",
    Log[SN] = ",LSN];
Print["N=10^5 cycles, S5 = ",S5," [MPa]; Log[N]=5, Log[S5] = ",LS5];
Print["N=10^6 cycles, Se = ",Se," [MPa]; Log[N]=6, Log[Se] = ",Le];

lineSN=Lm x+Lb;
SNG=Plot[lineSN, {x,3,6},AxesOrigin]→{3,LSe},
    AxesLabel→{"Log[N]", "Log[S]"}];
Show[SNG,Ticks→{{3,4,LN,5,6},{LSe,LS5,LSN,LS4,LS}},
    GridLines→{{3,4,LN,5,6},{LSe,LS5,LSN,LS4,LS}},
    PlotRange→{{3,6},{LSe-0.15,LS+0.15}},AxesOrigin→{3,LSe-0.15},
    AxesLabel→{"Log[N]","Log[S]"}];


"constant=life fatigue diagram"
eqS=Sy (1-x/Sy);
GS=Plot[{eqS},{x,0,Sy},PlotStyle->{Dashing[{0.01,0.01}]}];
eq3=S (1-x/Su);
eq4=S4 (1-x/Su);
eq5=S5 (1-x/Su);
eq6=S6 (1-x/Su);
eqN=SN (1-x/Su);
CF=Plot[{eq3,eq4,eq5,eq6,eqN},{x,0,Su}, Axes->True];

Show[{GS,CF},AxesLabel→{"σm","σa"}];
C2=Plot[{eq6,eqN},{x,0,Su}, Axes->True];
Print["eq. for Sy - Sy:",eqS,"=0"]
Print["eq. for N=10^6 cycles:",eq6,"=0"]
Print["eq. for N=5 x 10^4 cycles:",eqN,"=0"]

ms=1.;
eqSS=ms x;
GSS=Plot[{eqSS},{x,0,Su},PlotStyle->{Dashing[{0.05,0.05}]}];
Show[{GS,C2,GSS},AxesLabel→{"σm","σa"}];

"intersection of σa=σm with 10^6 cycle line:"
sol6=Solve[{y==ms x,y==eq6},{x,y}];
siga6=y/.sol6[[1]];
sigm6=x/.sol6[[1]];
Print["σa=",siga6,", σm=",sigm6, " [MPa]" ];
Print["σmax=",2 siga6," [MPa]" ];

"intersection of σa=σm with 5 x 10^6 cycle line:"
sol6b=Solve[{y==ms x,y==eqN},{x,y}];
siga6b=y/.sol6b[[1]];
sigm6b=x/.sol6b[[1]];
Print["σa=",siga6b,", σm=",sigm6b, " [MPa]" ];
Print["σmax=",2 siga6b," [MPa]" ];
```

```
"intersection of σa=σm with SySy line"
solS=Solve[{y==ms x,y==eqS},{x,y}];
sigaS=y/.solS[[1]];
sigmS=x/.solS[[1]];
Print["σa=",sigaS,", σm=",sigmS, " [MPa]" ];
Print["σmax=",2 sigaS," [MPa]" ];

"intersection of 5 10^6 cycle line with SySy line"
sol=Solve[{y==eqN,y==eqS},{x,y}];
siga=y/.sol[[1]];
sigm=x/.sol[[1]];

Print["σa=",siga,", σm=",sigm, "[MPa]" ];
Print["σmax=", 2 siga, " [MPa]" ];
```

ultimate strength [MPa] Su=1100

yield strength [MPa] Sy=715

10^3 cycle strength S

bending loads: S=0.9 Su

bending: S=990.

10^6 cycle strength (endurance limit) Se=kS kG kL Se'

endurance limit of test speciment Se'

Se'=0.5 Su (for Su<1400 MPa); Se'=700 MPa (for Su>1400 MPa)

bending, axial, torsion: Se'=0.5 Su

 Se'=550.

modifying factors for endurance limit

surface factor (kS) (bending, axial, torsion)

ultimate strength [kpsi] Su=159.536

fine ground surface

surface factor kS=0.89

size (gradient) factor (kG)

bending and torsion: kG=1 (for d<10 mm); kG=0.9 (for <10<d<50 mm)

axial: kG=0.7-0.9

bending: size factor kG=0.9

load factor (kL)

bending and axial: kL=1; torsion: kL=0.58

endurance limit Se=kS kG kL Se'

bending: Se=440.55

bending: S-N diagram

S-N line slope: m=-0.117213

S-N line y-intercept: b=3.34728

N=10^3 cycles, S = 990. [MPa]; Log[N]=3, Log[S] = 2.99564

N=10^4 cycles, S4 = 755.826 [MPa]; Log[N]=4, Log[S4] = 2.87842

N=5 10^4 cycles, SN = 625.883 [MPa]; Log[NN]=4.69897, Log[SN] = 2.79649

N=10^5 cycles, S5 = 577.043 [MPa]; Log[N]=5, Log[S5] = 2.76121

N=10^6 cycles, Se = 440.55 [MPa]; Log[N]=6, Log[Se] = 2.644

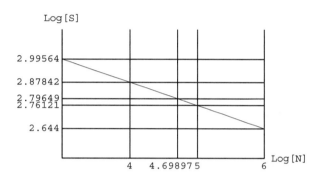

constant-life fatigue diagram

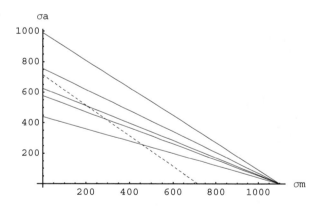

eq. for Sy - Sy:715 $(1-\dfrac{x}{715})$=0

eq. for N=10^6 cycles:440.55 $(1-\dfrac{x}{1100})$=0

eq. for N=5 x 10^4 cycles:625.883$(1-\dfrac{x}{1100})$=0

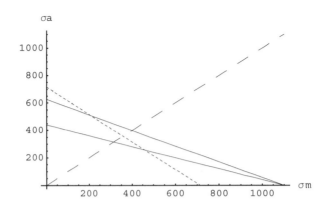

intersection of σa=σm with 10^6 cycle line:

σa=314.566, σm=314.566 [MPa]

σmax=629.132 [MPa]

intersection of σa=σm with 5 x 10^6 cycle line:

σa=398.91, σm=398.91 [MPa]

σmax=797.819 [MPa]

intersection of σa=σm with SySy line

σa=357.5, σm=357.5 [MPa]

σmax=715. [MPa]

intersection of 5 10^6 cycle line with SySy line

σa=508.24, σm=206.76 [MPa]

σmax=1016.48 [MPa]

PROGRAM II.2.4

```
Apply[Clear,Names["Global`*"]];
Off[General::spell];
Off[General::spell1];
(*d=1 in. machined surfaces*)
(* Input data *)
Su=110; Sy=77;(*kpsi*)
Print["ultimate strength [kpsi] Su=",Su];
Print["yield strength [kpsi] Sy=",Sy];

"10^6 cycle strength (endurance limit) Se=kS kG kL Se'"
"kS=0.74; machined surfaces Fig.2"
kS=0.74;
"kG=0.8; between 0.7 and 0.9; Table 3"
kG=0.8;
```

```
"kL=1; Table 3"
kL=1;
"endurance limit of test speciment Se'"
"Se'=0.5 Su (for Su<200 kpsi)"
Sep=0.5 Su;
Print["Se'=",Sep," [kpsi]"];
Se=kS kG kL Sep;
Print["10^6 cycle strength Se=",Se," [kpsi]"];

"10^3 cycle strength S"
"axial loads: S=0.75 Su "
S=0.75 Su;
Print["10^3 cycle strength S=",S," [kpsi]"];

"constant-life fatigue diagram"
eqS=Sy (1-x/Sy);
GS=Plot[{eqS},{x,0,Sy},PlotStyle->{Dashing[{0.01,0.01}]}];
eq3=S (1-x/Su);
eq6=Se (1-x/Su);
CF=Plot[{eq3,eq6},{x,0,Su}, Axes->True];

Show[{GS,CF},AxesLabel→{"σm","σa"}];
Print["eq. for Sy - Sy:",eqS,"=0"];
Print["eq. for N=10^3 cycles:",eq3,"=0"];
Print["eq. for N=10^6 cycles:",eq6,"=0"];

Fmax=6 (*kip*);
Fmin=1 (*kip*);
SF=2;
Fa=(Fmax-Fmin)/2;
Fm=(Fmax+Fmin)/2;
Sa=SF Fa/A; Sm=SF Fm/A;
ms=N[Sa/Sm];

Print["OA: σa=",ms," σm"];

eqSS=ms x;
GSS=Plot[{eqSS},{x,0,Su},PlotStyle->{Dashing[{0.05,0.05}]}];

Show[{GS,CF,GSS},AxesLabel→{"σm","σa"}];

"intersection of OA with 10^6 cycle line:"
sol6=Solve[{y==ms x,y==eq6},{x,y}];
siga6=y/.sol6[[1]];
sigm6=x/.sol6[[1]];

Print["σa=",siga6,", σm",sigm6, " [kpsi]" ];

Print["σmax=",siga6+sigm6," [kpsi]" ];
A6=SF Fa/siga6;
d6=Sqrt[4 A6/N[Pi]];
```

```
Print["d=",d6," [in]" ];

"intersection of OA with 10^3 cycle line:"
sol3=Solve[{y==ms x,y==eq3},{x,y}];
siga3=y/.sol3[[1]];
sigm3=x/.sol3[[1]];
Print["σa=",siga3,", σm=",sigm3, " [kpsi]" ];
Print["σmax=",siga3+sigm3," [kpsi]" ];

"intersection of OA with Sy-Sy line"
solS=Solve[{y==ms x,y==eqS},{x,y}];
sigaS=y/.solS[[1]];
sigmS=x/.solS[[1]];

Print["σa=",sigaS,", σm=",sigmS, " [kpsi]" ];

Print["σmax=",sigaS+sigmS," [kpsi]" ];

AS=SF Fa/sigaS;
dS=Sqrt[4 AS/N[Pi]];
Print["d=",dS," [in]" ];
```

ultimate strength [kpsi] Su=110

yield strength [kpsi] Sy=77

10^6 cycle strength (endurance limit) Se=kS kG kL Se'

kS=0.74; machined surfaces Fig.2

kG=0.8; between 0.7 and 0.9; Table 3

kL=1; Table 3

endurance limit of test speciment Se'

Se'=0.5 Su (for Su<200 kpsi)

Se'=55. [kpsi]

10^6 cycle strength Se=32.56 [kpsi]

10^3 cycle strength S

axial loads: S=0.75 Su

10^3 cycle strength S=82.5 [kpsi]

constant-life fatigue diagram

eq. for Sy - Sy:77 $(1 - \frac{x}{77})$=0

eq. for N=10^3 cycles:82.5 $(1 - \frac{x}{110})$=0

eq. for N=10^6 cycles:32.56 $(1 - \frac{x}{110})$=0

OA: σa=0.714286 σm

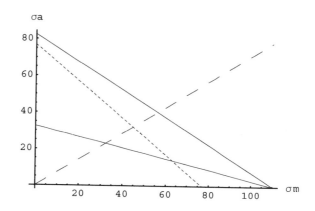

intersection of OA with 10^6 cycle line:

σa=23.0204, σm=32.2285 [kpsi]

σmax=55.2489 [kpsi]

d=0.525877 [in]

intersection of OA with 10^3 cycle line:

σa=40.2439, σm=56.3415 [kpsi]

σmax=96.5854 [kpsi]

intersection of OA with Sy-Sy line

σa=32.0833, σm=44.9167 [kpsi]

σmax=77. [kpsi]

d=0.445451 [in]

PROGRAM II.2.5

```
Apply[Clear,Names["Global`*"]];
Off[General::spell];
Off[General::spell1];

(*Input data: steel bar, fine ground surface*)

Su=950; (* ultimate strength [MPa] *)
Sy=600;  (* yield strength [MPa] *)

Print["ultimate strength [MPa] Su=",Su];
Print["yield strength [MPa] Sy=",Sy];

σminI=-100.;

σmaxI=300.;

σaI=(σmaxI-σminI)/2;
```

```
σmI=(σmaxI+σminI)/2;

mI=σaI/(σmI-Su);
SI=-mI Su;

Print["σmI =",σmI,"; σaI=",σaI];
Print["mI=",mI];
Print["SI=",SI];

σminII=-100.;

σmaxII=400;

σaII=(σmaxII-σminII)/2;

σmII=(σmaxII+σminII)/2;

mII=σaII/(σmII-Su);
SII=-mII Su;

Print["σmII=",σmII,"; σaII=",σaII];
Print["mII=",mII];
Print["SII=",SII];

yI=mI x+SI;
yII=mII x+SII;

graph=Plot[{yI,yII},{x,0,Su},AxesLabel→{"σm","σa"}, GridLines→
{{smI,smII},{saI,saII}}];

"10^3 cycle strength S"
"axial loads: S=0.75 Su"
S=0.75 Su;
Print["axial: S=",S];

"10^6 cycle strength (endurance limits) Se=kL kG kL Se'"

"endurance limit of test speciment Se'"
"Se'=0.5 Su (for Su<1400 MPa); Se'=700 MPa (for Su>1400 MPa)"
"bending, axial, torsion: Se'=0.5 Su "
Sep=0.5 Su;
Print[" Se'=",Sep];

"modifying factors for endurance limit"

"surface factor kS (bending, axial, torsion)"
Print["ultimate strength [kpsi] Su=",Su/6.895];
"fine ground surface"
kS=0.475;
Print["surface factor kS=",kS];

"size (gradient) factor kG"
```

```
"bending and torsion: kG=1 (for d<10 mm); kG=0.9 (for <10<d<50 mm)"
"axial: kG=0.7-0.9"
kG=0.8;
Print["axial: size factor kG=",kG];

"load factor kL"
"bending and axial: kL=1; torsion: kc=0.58"
kL=1;

"endurance limit Se=kS kG kL Se'"
Se=kS kG kL Sep;
Print["axial: Se=",Se];

"S-N diagram"
LS=Log[10,S];
LSe=Log[10,Se];
Lm=(LSe-LS)/(6-3);
Lb=LS-3 Lm;
Print["m=",Lm," b=",Lb];

LSI=Log[10,SI];
LNI=(LSI-Lb)/Lm;
NI=10^LNI;

LSII=Log[10,SII];
LNII=(LSII-Lb)/Lm;
NII=10^LNII;

Print["N=10^3 cycles, S = ",S," [MPa]; Log[N]=3, Log[S] = ",LS];
Print["N=10^6 cycles, Se = ",Se," [MPa]; Log[N]=6, Log[Se] = ",LSe];

Print["SI = ",SI," [MPa]; Log[SI] = ",LSI," =>"];
Print["Log[NI]= ",LNI," => NI = ",NI," cycles"];

Print["SII = ",SII," [MPa]; Log[SII] = ",LSII," =>"];
Print["Log[NII]= ",LNII," => NII = ",NII," cycles"];

LSN=Lm x+Lb;
SNG=Plot[LSN,{x,3,6},AxesOrigin→{3,LSe},
     AxesLabel→{"Log[N]","Log[S]"}];
Show[SNG,Ticks→{{3,4,5,6},{LSe,LSI,LSII,LS}},
    GridLines→{{3,4,5,6},{LSe,LSI,LSII,LS}},
    PlotRange→{{3,6},{LSe-0.15,LS+0.15}},AxesOrigin→{3,LSe-0.15},
    AxesLabel→{"Log[N]","Log[S]"}];


nI=2;nII=3;
p=nI/NI+nII/NII;
t=5;
Li=t/p;
```

```
Print["nI/NI+nII/NII= ",p];
Print["life of the part = ",Li," s = ",Li/60," min = ",Li/3600," h "];
```

ultimate strength [MPa] Su=950

yield strength [MPa] Sy=600

σmI=100.; σaI=200.

mI=-0.235294

SI=223.529

σmII=150.; σaII=250.

mII=-0.3125

SII=296.875

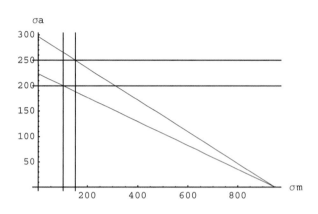

10^3 cycle strength S

axial loads: S=0.75 Su

axial: S=712.5

10^6 cycle strength (endurance limits) Se=kL kG kL Se'

endurance limit of test speciment Se'

Se'=0.5 Su (for Su<1400 MPa); Se'=700 MPa (for Su>1400 MPa)

bending, axial, torsion: Se'=0.5 Su

Se'=475.

modifying factors for endurance limit

surface factor kS (bending, axial, torsion)

ultimate strength [kpsi] Su=137.781

fine ground surface

surface factor kS=0.475

size (gradient) factor kG

bending and torsion: kG=1 (for d<10 mm); kG=0.9 (for <10<d<50 mm)

axial: kG=0.7-0.9

axial: size factor kG=0.8

load factor kL

bending and axial: kL=1; torsion: kc=0.58

endurance limit Se=kS kG kL Se'

axial: Se=180.5

S-N diagram

m=-0.198769 b=3.44909

N=10^3 cycles, S = 712.5 [MPa]; Log[N]=3, Log[S] = 2.85278

N=10^6 cycles, Se = 180.5 [MPa]; Log[N]=6, Log[Se] = 2.25648

SI = 223.529 [MPa]; Log[SI] = 2.34933 =>

Log[NI]= 5.53284 => NI = 341065. cycles

SII = 296.875 [MPa]; Log[SII] = 2.47257 =>

Log[NII]= 4.91283 => NII = 81814. cycles

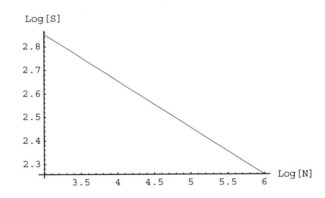

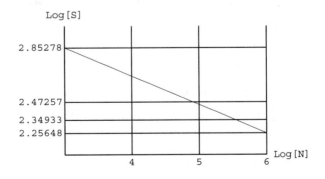

nI/NI+nII/NII= 0.0000425325

life of the part = 117557. s = 1959.28 min = 32.6547 h

PROGRAM II.2.6

```
Apply[Clear,Names["Global`*"]];
Off[General::spell];
Off[General::spell1];

d=2.;
Print["d=",d," in."];

"fluctuating tensile load"
Fmin=3;
Fmax=60.;

Print["Fmin=",Fmin," kip"];
Print["Fmax=",Fmax," kip"];

Su=97;
Sy=68;

Print["ultimate strength Su=",Su," kpsi"];
Print["yield strength Sy=",Su," kpsi"];

"endurance limit of test speciment Se'=0.5 Su"
Sep=0.5 Su;
Print["Se'=",Sep," kpsi"];
"modifying factors for endurance limit"
kS=0.76;
Print["surface factor kS=",kS];
kG=0.8;
Print["size factor kG=",kG];
kL=1;
Print["load factor kL=",kL];
Se=kS kG kL Sep;
Print["endurance limit Se=kS kG kL Se'=",Se," kpsi"];

Fa=(Fmax-Fmin)/2;
Fm=(Fmax+Fmin)/2;
Print["Fa=Fm=Fmax/2=",Fa," kpsi"];

σa=4 Fa/(N[Pi] d^2);

Print["σa=4 Fa/(π d^2)=",σa," kpsi"];

σm=4 Fm/(N[Pi] d^2);

Print["σm=4 Fm/(π d^2)=",σm," kpsi"];
```

```
"overload on σa"
"modified Goodman: Sa/Se+Sm/Su=1 => Sa=Se(1-Sm/Su)"

Sam=Se(1-Sm/Su)/.{Sm->σm};

"for Sm=σm =>"

Print["Sa=Se(1-σm/Su)=",Sam," kpsi"];

SFa=Sam/σa;

Print["safety factor SFa=Sa/σa=",SFa];

"overload on σm"
"modified Goodman: Sa/Se+Sm/Su=1 =>Sm=Su(1-Sa/Se)"

Sma=Su(1-Sa/Se)/.{Sa->σa};

"for Sa=σa =>"

Print["Sm=Su(1-Sa/Se)=",Sma," kpsi"];

SFm=Sma/σm;

Print["safety factor SFm=Sm/σm=",SFm];

"σa/σm = constant"

r=σa/σm;

Print["σa/σm = Sa/Sm = r=",r];
"modified Goodman: Sa/Se+Sm/Su=1 => Sm=Se Su(Se+r Su)"
Smr=Se Su/(Se+r Su);
Print["Sm=Se Su/(Se+r Su)=",Smr," kpsi"];

SFr=Smr/σm;

Print["safety factor SFr=Sm/σm=",SFr];

"σm-σa diagram"
eq=Se (1-x/Su);

eqm=σa;

eqa=r x;

CF=Plot[{eq,eqm,eqa},{x,0,Su},GridLines→{{σm},{σa}},
PlotRange→{{0,Su},{0,Se}}, Axes->True];

d=2. in.

fluctuating tensile load

Fmin=3 kip

Fmax=60. kip

ultimate strength Su=97 kpsi
```

yield strength Sy=97 kpsi

endurance limit of test speciment Se'=0.5 Su

Se'=48.5 kpsi

modifying factors for endurance limit

surface factor kS=0.76

size factor kG=0.8

load factor kL=1

endurance limit Se=kS kG kL Se'=29.488 kpsi

Fa=Fm=Fmax/2=28.5 kpsi

$\sigma a=4$ Fa$/(\pi$ d^2$)=9.07183$ kpsi

$\sigma m=4$ Fm$/(\pi$ d^2$)=10.0268$ kpsi

overload on σa

modified Goodman: Sa/Se+Sm/Su=1 => Sa=Se(1-Sm/Su)

for Sm=σm =>

Sa=Se(1-σm/Su)=26.4399 kpsi

safety factor SFa=Sa/σa=2.9145

overload on σm
modified Goodman: Sa/Se+Sm/Su=1 => Sm=Su(1-Sa/Se)

for Sa=σa =>

Sm=Su(1-Sa/Se)=67.1584 kpsi

safety factor SFm=Sm/σm=6.69792

$\sigma a/\sigma m$ = constant

$\sigma a/\sigma m$ = Sa/Sm = r=0.904762

modified Goodman: Sa/Se+Sm/Su=1 => Sm=Se Su(Se+r Su)

Sm=Se Su/(Se+r Su)=24.3952 kpsi

safety factor SFr=Sm/σm=2.43301

$\sigma m-\sigma a$ diagram

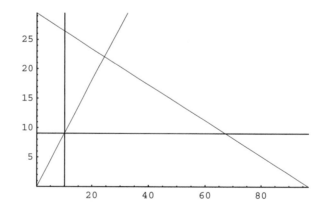

II.3   Screws

Fasteners may be classified as threaded, fixed, and locking. Threaded fasteners include bolts, studs, and various forms of screws. Fixed fasteners include welds, solders, brazing, adhesives, and rivets. Locking fasteners may be used separately or in conjunction with other fasteners and include washers, keys, splines, springs, and pins.

Threaded fasteners such as screws, nuts, and bolts are important components of mechanical structures and machines. Screws may be used as removable fasteners or as devices for moving loads.

II.3.1   Screw Thread

A screw thread is a uniform wedge-shaped section in the form of a helix on the external or internal surface of a cylinder (straight thread) or a cone (taper thread).

The basic arrangement of a helical thread wound around a cylinder is illustrated in Figure II.3.1. The terminology of an external screw threads is:

- *Pitch* — denoted by p is the distance, parallel to the screw axis, between corresponding points on adjacent thread forms having uniform spacing.
- *Major diameter* — denoted by d is the largest (outside) diameter of a screw thread.
- *Minor diameter* — denoted by d_r or d_1, is the smallest diameter of a screw thread.
- *Pitch diameter* — denoted by d_m or d_2 is the imaginary diameter for which the width of the threads and the grooves are equal.

The standard geometry of a basic profile of an external thread is shown in Figure II.3.2, and it is basically the same for both Unified (inch series) and ISO (International Standards Organization, metric) threads.

The *lead* denoted by l is the distance the nut moves parallel to the screw axis when the nut is given one turn (distance a threaded section moves axially in one revolution). A screw with two or more threads cut beside each other is called a *multiple-threaded* screw. The lead is equal to twice the pitch for a double-threaded screw, and up to three times the

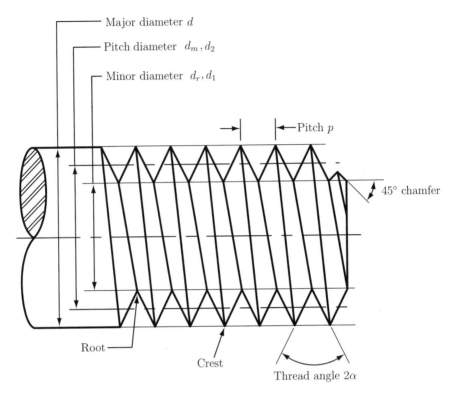

FIGURE II.3.1 *Terminology of an external screw thread.*

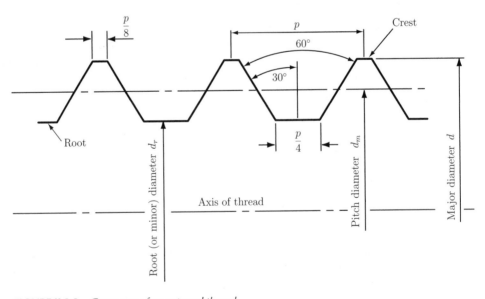

FIGURE II.3.2 *Geometry of an external thread.*

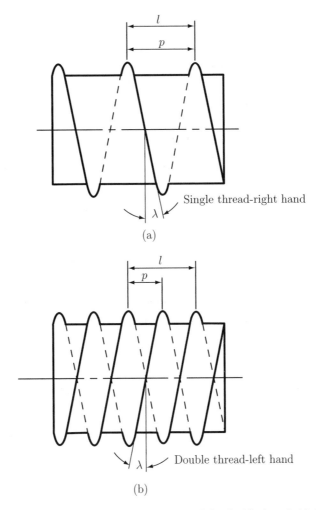

Single thread-right hand

(a)

Double thread-left hand

(b)

FIGURE II.3.3 *(a) Single threaded right-hand screw and (b) double-threaded left-hand screw.*

pitch for a triple-threaded screw. The pitch p, lead l, and lead angle λ are represented in Figure II.3.3. Figure II.3.3(a) shows a single thread right-hand screw and Figure II.3.3(b) shows a double-threaded left-hand screw. If a thread traverses a path in a clockwise and receding direction when viewed axially, it is a *right-hand thread*. All threads are assumed to be right-hand, unless otherwise specified.

A standard geometry of an ISO profile, M (metric) profile, with 60° symmetric threads is shown in Figure II.3.4. In Figure II.3.4 $D\,(d)$ is the basic major diameter of the internal (external) thread, $D_1\,(d_1)$ is the basic minor diameter of the internal (external) thread, $D_2\,(d_2)$ is the basic pitch diameter, and $H = 0.5\,\sqrt{3}\,p$.

Metric threads are specified by the letter M preceding the nominal major diameter in millimeters and the pitch in millimeters per thread. For example:

$$M\,14 \times 2$$

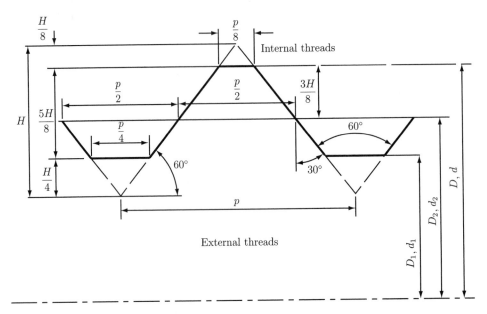

FIGURE II.3.4 *Standard geometry of an ISO profile.*

M is the SI thread designation, 14 mm is the outside (major) diameter, and the pitch is 2 mm per thread.

Screw size in the Unified system is designated by the size number for major diameter (in.), the number of treads per in., and the thread form and series, like this:

$$\frac{5"}{8} - 18 \text{ UNF}$$

$\frac{5"}{8}$ is the outside (major) diameter where the double tick marks mean inches, and 18 threads per in. Some Unified thread series are:

UNC, Unified National Coarse
UNEF, Unified National Extra Fine
UNF, Unified National Fine
UNS, Unified National Special
UNR, Unified National Round (round root)

The UNR series threads have improved fatigue strengths.

Figure II.3.5(a) shows different types of thread delineation on a drawing: detailed, schematic, and simplified thread representation. The schematic representation is realistic and is used frequently in assembly drawings. The simplified thread representation is used widely because its ease of drawing. Typical screw heads are illustrated in Figure II.3.5(b).

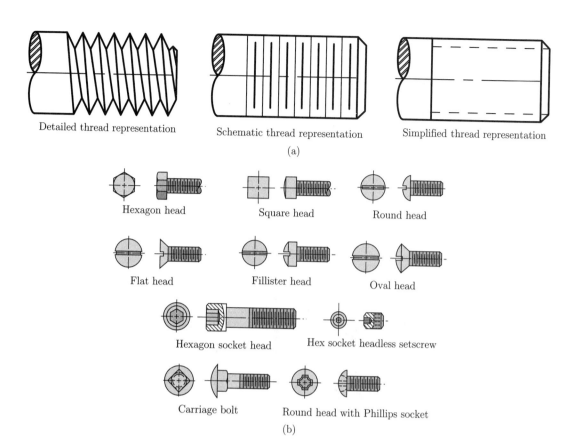

Detailed thread representation   Schematic thread representation   Simplified thread representation

(a)

Hexagon head   Square head   Round head

Flat head   Fillister head   Oval head

Hexagon socket head   Hex socket headless setscrew

Carriage bolt   Round head with Phillips socket

(b)

FIGURE II.3.5   *(a) Thread representation and (b) typical screw heads.*

II.3.2   Power Screws

For applications that require power transmission, the Acme (Fig. II.3.6) and square threads (Fig. II.3.7) are used.

Power screws are used to convert rotary motion to linear motion of the meeting member along the screw axis. These screws are used to lift weights (screw-type jacks) or exert large forces (presses, tensile testing machines). The power screws can also be used to obtain precise positioning of the axial movement.

A square-threaded power screw with a single thread having the pitch diameter d_m, the pitch p, and the helix angle λ is considered in Figure II.3.8. Consider that a single thread of the screw is unrolled for exactly one turn. The edge of the thread is the hypotenuse of a right triangle and the height is the lead. The base of the right triangle is the circumference of the pitch diameter circle (Fig. II.3.9). The lead angle λ is the helix angle of the thread.

The screw is loaded by an axial compressive force F (Figs. II.3.8 and II.3.9).

The force diagram for lifting the load is shown in Figure II.3.9(a), (the force P_r is positive). The force diagram for lowering the load is shown in Figure II.3.9(b), (the force

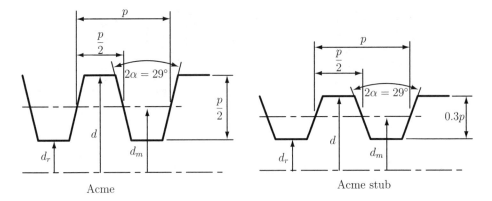

FIGURE II.3.6 *Acme threads.*

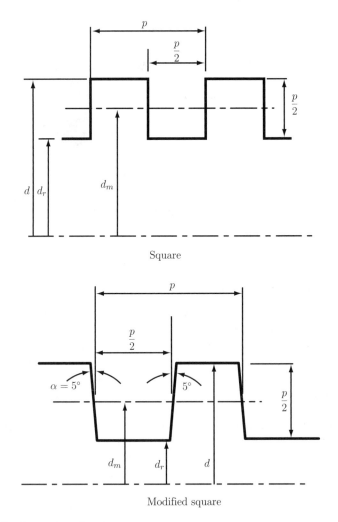

FIGURE II.3.7 *Square threads.*

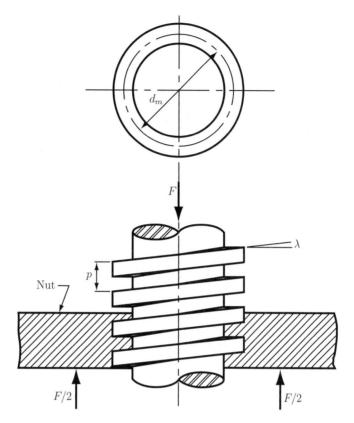

FIGURE II.3.8 *Power screw.*

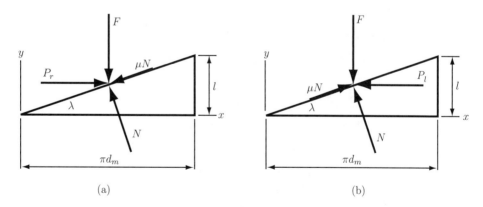

FIGURE II.3.9 *Force diagrams for (a) lifting the load and (b) lowering the load.*

P_l is negative). The friction force is

$$F_f = \mu N,$$

where μ is the coefficient of dry friction and N is the normal force. The friction force is acting opposite to the motion.

The equilibrium of forces for raising the load gives

$$\sum F_x = P_r - N \sin \lambda - \mu N \cos \lambda = 0, \tag{II.3.1}$$

$$\sum F_y = F + \mu N \sin \lambda - N \cos \lambda = 0. \tag{II.3.2}$$

Similarly, for lowering the load one may write the equations

$$\sum F_x = -P_l - N \sin \lambda + \mu N \cos \lambda = 0, \tag{II.3.3}$$

$$\sum F_y = F - \mu N \sin \lambda - N \cos \lambda = 0. \tag{II.3.4}$$

Eliminating N and solving for P_r

$$P_r = \frac{F (\sin \lambda + \mu \cos \lambda)}{\cos \lambda - \mu \sin \lambda}, \tag{II.3.5}$$

and for lowering the load

$$P_l = \frac{F (\mu \cos \lambda - \sin \lambda)}{\cos \lambda + \mu \sin \lambda}. \tag{II.3.6}$$

Using the relation

$$\tan \lambda = l/(\pi d_m),$$

and dividing the equations by $\cos \lambda$ one may obtain

$$P_r = \frac{F[(l\pi d_m) + \mu]}{1 - (\mu l \pi d_m)}, \tag{II.3.7}$$

$$P_l = \frac{F[\mu - (l\pi d_m)]}{1 + (\mu l \pi d_m)}. \tag{II.3.8}$$

The moment required to overcome the thread friction and to raise the load is

$$M_r = P_r \frac{d_m}{2} = \frac{F d_m}{2} \left(\frac{l + \pi \mu d_m}{\pi d_m - \mu l} \right). \tag{II.3.9}$$

The moment required to lower the load (and to overcome a part of the friction) is

$$M_l = \frac{Fd_m}{2} \left(\frac{\pi \mu d_m - l}{\pi d_m + \mu l} \right). \tag{II.3.10}$$

When the lead, l, is large or the friction, μ, is low the load will lower itself. In this case the screw will spin without any external effort, and the moment M_l in Eq. (II.3.10) will be negative or zero. When the moment is positive, $M_l > 0$, the screw is said to be *self-locking*.

The condition for self-locking is

$$\pi \mu d_m > l.$$

Dividing both sides of this inequality by $\pi \, d_m$, and using $l/(\pi \, d_m) = \tan \lambda$, yields

$$\mu > \tan \lambda. \tag{II.3.11}$$

The self-locking is obtained whenever the coefficient of friction is equal to or greater than the tangent of the thread lead angle.

The moment, M_0, required only to raise the load when the friction is zero, $\mu = 0$, is obtained from Eq. (II.3.9):

$$M_0 = \frac{Fl}{2\pi}. \tag{II.3.12}$$

The screw efficiency e can be defined as

$$e = \frac{M_0}{M_r} = \frac{Fl}{2\pi M_r}. \tag{II.3.13}$$

For square threads the normal thread load, F, is parallel to the axis of the screw (Figs. II.3.7 and II.3.8). The preceding equations can be applied for square threads.

For Acme threads (Fig. II.3.6) or other threads, the normal thread load is inclined to the axis due to the thread angle 2α and the lead angle λ.

The screw threads in normal and axial planes are shown in Figure II.3.10(a). The angle α_n is the thread angle measured in normal plane. The relation between the thread angle measured in axial plane, α in Figure II.3.6, and the thread angle measured in normal plane, α_n in Figure II.3.10(a), is

$$\tan \alpha_n = \frac{s}{H} = \frac{s}{H \cos \lambda} \cos \lambda = \tan \alpha \cos \lambda,$$

or

$$\tan \alpha_n = \tan \alpha \cos \lambda. \tag{II.3.14}$$

The screw thread forces in normal plane are represented in Figure II.3.10(b). The force diagram for lifting the load is shown in Figure II.3.10(b). The equilibrium of forces for

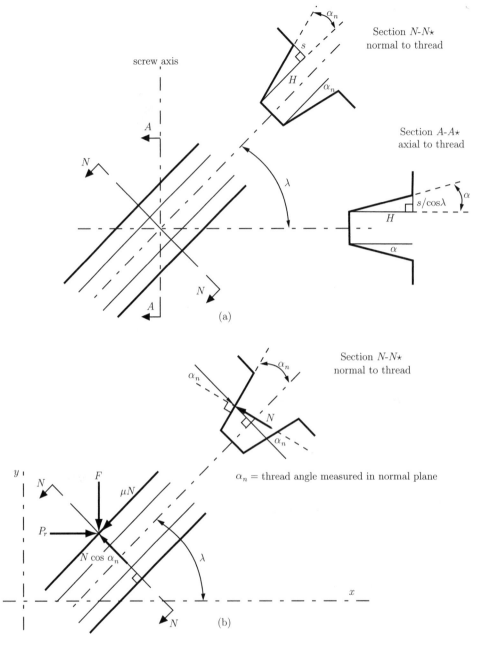

FIGURE II.3.10 *(a) Acme screw threads in normal and axial planes and (b) force diagram.*

raising the load gives

$$\sum F_x = P_r - N \cos \alpha_n \sin \lambda - \mu N \cos \lambda = 0, \qquad (II.3.15)$$

$$\sum F_y = -F + N \cos \alpha_n \cos \lambda - \mu N \sin \lambda = 0. \qquad (II.3.16)$$

Eliminating N and solving for P_r,

$$P_r = F \frac{\mu \cos \lambda + \sin \lambda \cos \alpha_n}{\cos \lambda \cos \alpha_n - \mu \sin \lambda} = F \frac{\mu + \tan \lambda \cos \alpha_n}{\cos \alpha_n - \mu \tan \lambda}. \qquad \text{(II.3.17)}$$

The moment required to overcome the thread friction and to raise the load is

$$M_r = P_r \frac{d_m}{2} = \frac{F \, d_m}{2} \left(\frac{\mu + \tan \lambda \cos \alpha_n}{\cos \alpha_n - \mu \tan \lambda} \right). \qquad \text{(II.3.18)}$$

Using the relation

$$\tan \lambda = l/(\pi d_m),$$

the following expression is obtained:

$$M_r = \frac{F d_m}{2} \left(\frac{\mu \pi d_m + l \cos \alpha_n}{\pi d_m \cos \alpha_n - \mu l} \right). \qquad \text{(II.3.19)}$$

Similarly, the moment required to lower the load and to overcome a part of the friction is

$$M_l = \frac{F d_m}{2} \left(\frac{\mu \pi d_m - l \cos \alpha_n}{\pi d_m \cos \alpha_n + \mu l} \right). \qquad \text{(II.3.20)}$$

For power screws the square thread ($\alpha_n = 0$) is more efficient than the Acme thread. The Acme thread adds an additional friction due to the wedging action. It is easier to machine an Acme thread than a square thread.

In general, when the screw is loaded axially, a thrust bearing or thrust collar may be used between the rotating and stationary links to carry the axial component (Fig. II.3.11). The load is concentrated at the mean collar diameter d_c. The moment required is

$$M_c = \frac{F \mu_c d_c}{2}, \qquad \text{(II.3.21)}$$

where μ_c is the coefficient of collar friction.

II.3.3 Force Analysis for a Square-Threaded Screw

Consider a square-threaded jack under the action of an axial load F and a moment M about the axis of the screw [Fig. II.3.12(a)]. The screw has the mean radius r_m and the lead l. The force exerted by the frame thread on the screw thread is R. The angle θ made by R with the normal to the thread is the angle of friction [Fig. II.3.12(b)]:

$$\tan \theta = \mu = \frac{F_f}{N}.$$

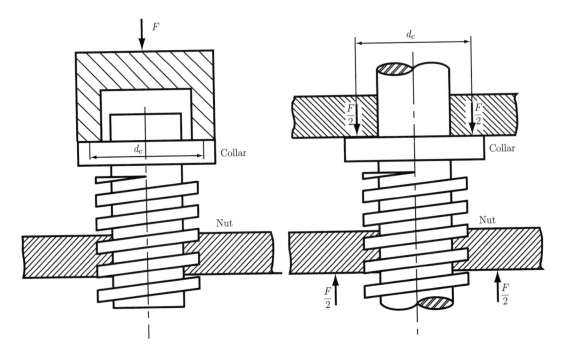

FIGURE II.3.11 *Thrust collar.*

The unwrapped thread of the screw shown in Figure II.3.12(b) is for lifting the load. The force equilibrium equation in the axial direction is

$$F = R\cos(\lambda + \theta),$$

where λ is the helix angle, $\tan \lambda = l/(2\pi r_m)$. The moment of R about the vertical axis of the screw is $Rr_m \sin(\lambda + \theta)$. The moment equilibrium equation for the screw becomes

$$M = Rr_m \sin(\lambda + \theta).$$

Combining the expression for F and M gives

$$M = M_r = Fr_m \tan(\lambda + \theta). \qquad (\text{II.3.22})$$

The force required to push the thread up is $P = M/r_m$.

The moment required to lower the load by unwinding the screw is obtained in a similar manner:

$$M = M_l = Fr_m \tan(\theta - \lambda). \qquad (\text{II.3.23})$$

If $\theta < \lambda$ the screw will unwind by itself.

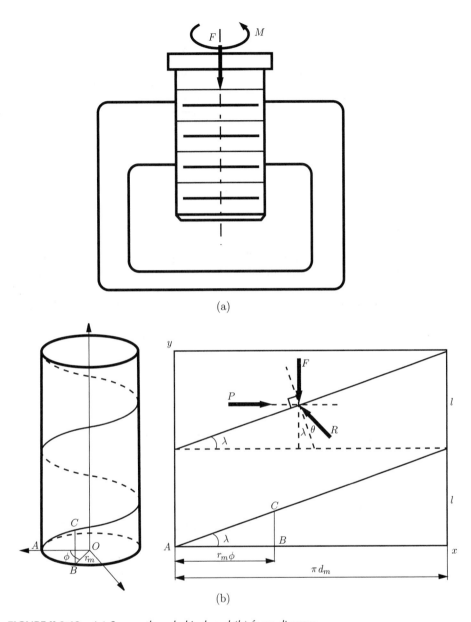

FIGURE II.3.12 *(a) Square-threaded jack and (b) force diagram.*

II.3.4 Threaded Fasteners

In general bolts are used to hold parts together. External forces tend to pull, or slide, the parts apart. Figure II.3.13(a) shows two parts connected with a bolt. An external force, F_e, acts on the joint and tends to separate the two parts. The free-body diagram of a portion of this joint without the external load is shown in Figure II.3.13(b). In this figure the nut has been initially tightened to a preload force F_i. The initial bolt axial load F_{b0} and the clamping

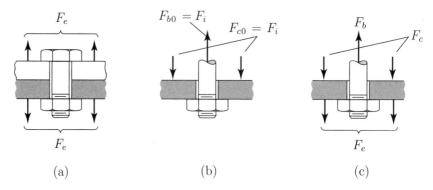

F_e
$F_{b0} = F_i$
$F_{c0} = F_i$
F_b
F_c
F_e
F_e

(a)   (b)   (c)

FIGURE II.3.13 *(a) Force diagrams for two parts connected with a bolt; (b) free-body diagram of a portion without the external load; and (c) free-body diagram of a portion with the external load.*

force between the two plates, F_{c0}, are both equal to the preload force F_i: $F_{b0} = F_{c0} = F_i$. The free-body diagram of the portion with the external load F_e is shown in Figure II.3.13(c). Equilibrium requires an increase in F_b and a decrease in F_c. The separating force F_e must be equal to the sum of the increased bolt force ΔF_b plus the decreased clamping force ΔF_c:

$$F_e = \Delta F_b + \Delta F_c. \tag{II.3.24}$$

The bolt and the clamped members elongate the same amount:

$$\delta = \frac{\Delta F_b}{k_b} = \frac{\Delta F_c}{k_c}, \tag{II.3.25}$$

where k_b and k_c are the spring constants for the bolt and clamped parts, respectively. From Eqs. (II.3.24) and (II.3.25) the elongation is

$$\delta = \frac{F_e}{k_b + k_c}. \tag{II.3.26}$$

The bolt axial load F_b and the clamping force F_c are

$$F_b = F_i + \Delta F_b = F_i + \frac{k_b}{k_b + k_c} F_e, \tag{II.3.27}$$

$$F_c = F_i - \Delta F_c = F_i - \frac{k_c}{k_b + k_c} F_e. \tag{II.3.28}$$

The *joint constant* is defined as a dimensionless stiffness parameter given by

$$C = \frac{k_b}{k_b + k_c}. \tag{II.3.29}$$

Equations (II.3.27) and (II.3.28) will become

$$F_b = F_i + C F_e, \tag{II.3.30}$$

$$F_c = F_i - (1 - C) F_e. \tag{II.3.31}$$

The general axial deflection equation

$$\delta = \frac{F\,l}{A\,E},$$

gives the spring constant (stiffness)

$$k = \frac{F}{\delta} = \frac{A\,E}{l},$$

where A is the cross-sectional area, E is the modulus of elasticity, and l is the length.

Bolt Stiffness

A bolt with thread is considered as a shaft with a variable section. The minor diameter (root diameter), d_r, is used for the threaded section of the bolt, and the major diameter (crest diameter), d, is used for the unthreaded section of the bolt (shank). The stiffness of the bolt is

$$\frac{1}{k_b} = \frac{1}{k_{\text{thread}}} + \frac{1}{k_{\text{shank}}}.$$

For a bolt with a shank having a constant major diameter the spring constant is [6]:

$$\frac{1}{k_b} = \frac{4}{\pi\,E}\left(\frac{l_{se}}{d^2} + \frac{l_{te}}{d_r^2}\right) = \frac{4}{\pi\,E}\left(\frac{l_s + 0.4d}{d^2} + \frac{l_t + 0.4d_r}{d_r^2}\right), \qquad \text{(II.3.32)}$$

where l_s is the length of the unthreaded section and l_t is the length of the threaded section (Fig. II.3.14). The effective lengths of the unthreaded and threaded sections are l_{se} and l_{te}, respectively. The modulus of elasticity of the bolt is E.

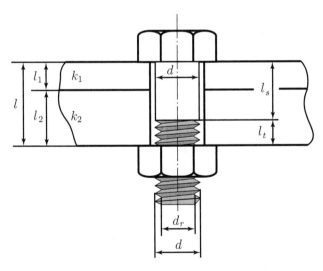

FIGURE II.3.14   *Geometry of a bolt and nut assembly.*

Shigley and Mischke [20] proposed the following expressions for the stiffness of the unthreaded section of the bolt, k_s:

$$k_s = \frac{A_s E}{l_s} = \frac{\pi d^2 E}{4 l_s},$$ (II.3.33)

and for the stiffness of the threaded section of the bolt, k_t:

$$k_t = \frac{A_t E}{l_t}.$$ (II.3.34)

The *tensile stress area*, A_t is defined as [6]:

$$A_t = 0.7854 \, (d - 0.9743/n)^2 \quad \text{in.}^2$$ (II.3.35)

for UN thread profiles where d is in inches and n is the number of threads per inch:

$$A_t = 0.7854(d - 0.9382p)^2 \quad \text{mm}^2,$$ (II.3.36)

for M thread profiles with the major diameter d and the pitch p in millimeters. The tensile stress area is also given in Tables II.3.1 and II.3.2 [6].

TABLE II.3.1 Tensile Stress Areas for Metric Threads

Major diameter (d mm)	Coarse threads (MC)		Fine threads (MF)	
	Pitch (p mm)	Tensile stress area (A_t mm^2)	Pitch (p mm)	Tensile stress area (A_t mm^2)
1	0.25	0.460	—	—
1.6	0.35	1.27	0.20	1.57
2	0.4	2.07	.25	2.45
2.5	0.45	3.39	.35	3.70
3	0.5	5.03	.35	5.61
4	0.7	8.78	.5	9.79
5	0.8	14.2	.5	16.1
6	1	20.1	.75	22
8	1.25	36.6	1	39.2
10	1.5	58.0	1.25	61.2
12	1.75	84.3	1.25	92.1
16	2	157	1.5	167
20	2.5	245	1.5	272
24	3	353	2	384
30	3.5	561	2	621
36	4	817	3	865
42	4.5	1121	—	—
48	5	1473	—	—

Source: B. G. Hamrock, B. Jacobson, and S. R. Schmid, *Fundamentals of Machine Elements*, New York, McGraw-Hill, 1999. Reprinted with permission of McGraw-Hill.

TABLE II.3.2 Tensile Stress Areas for UN Threads

Major diameter (d in.)	Coarse threads (UNC)		Fine threads (UNF)	
	Number of threads per in. (n)	Tensile stress area (A_t in.^{2})	Number of threads per in. (n)	Tensile stress area (A_t in.^{2})
0.0600	—	—	80	0.00180
0.0730	64	0.00263	72	0.00278
0.0860	56	0.00370	64	0.00394
0.0990	48	0.00487	56	0.00523
0.1120	40	0.00604	48	0.00661
0.1250	40	0.00796	44	0.00830
0.1380	32	0.00909	40	0.01015
0.1640	32	0.0140	36	0.01474
0.1900	24	0.0175	32	0.0200
0.2160	24	0.0242	28	0.0258
0.3500	20	0.0318	28	0.0364
0.3125	18	0.0524	24	0.0580
0.3750	16	0.0775	24	0.0878
0.4735	14	0.1063	20	0.1187
0.5000	13	0.1419	20	0.1599
0.5625	12	0.182	18	0.203
0.6250	11	0.226	18	0.256
0.7500	10	0.334	16	0.373
0.8750	9	0.462	14	0.509
1.000	8	0.606	12	0.663
1.125	7	0.763	12	0.856
1.250	7	0.969	12	1.073
1.375	6	1.155	12	1.315
1.500	6	1.405	12	1.581
1.750	5	1.90	—	—
2.000	4.5	2.50	—	—

Source: B. G. Hamrock, B. Jacobson, and S. R. Schmid, *Fundamentals of Machine Elements*, New York, McGraw-Hill, 1999. Reprinted with permission of McGraw-Hill.

Thus, the bolt stiffness is

$$k_b = \frac{A_s A_t E}{A_s l_t + A_t l_s}.$$

(II.3.37)

Stiffness of the clamped parts

Difficulties commonly arise in estimating the stiffness of the clamped parts or the joint stiffness. The clamped parts may consist of a combination of different materials. The parts may represent "springs" in series, as shown in Figure II.3.14. The stiffness of the clamped parts is

$$\frac{1}{k_c} = \frac{1}{k_1} + \frac{1}{k_2} + \cdots + \frac{1}{k_i} + \cdots$$

Shigley and Mischke [20] proposed the following expression for the joint stiffness:

$$k_i = \frac{0.577\,\pi\,E_i\,d}{2\ln\left(5\dfrac{0.577\,l_i + 0.5\,d}{0.577\,l_i + 2.5\,d}\right)} \qquad \text{(II.3.38)}$$

Wileman et al. [25] obtained an exponential expression using finite element analysis:

$$k_i = E_i\,d\,A\,e^{Bd/l_i}, \qquad \text{(II.3.39)}$$

the numerical constants are:

$A = 0.78715,\ B = 0.62873$ for steel;
$A = 0.79670,\ B = 0.63816$ for aluminum;
$A = 0.79568,\ B = 0.63553$ for copper, and
$A = 0.77871,\ B = 0.61616$ for gray cast iron.

Bolt Preload

The initial tensile force F_i is defined as [7]:

$$F_i = K\,A_t\,S_p, \qquad \text{(II.3.40)}$$

where A_t is the tensile stress area of the thread and S_p is the *proof strength* of the material [7, 20]. The proof strength of steel bolts is given in Tables II.3.3 and II.3.4 for various sizes [6]. The International Organization for Standardization (ISO) defines a metric grade number as a range of 4.6 to 12.9 (Table II.3.3) and the Society of Automotive Engineers (SAE) specifies grade number from 1 to 8. The higher grade numbers represent greater strength. The bolt grades are numbered according to the tensile strength. The constant K is 0.75 for reused connections and 0.90 for permanent connections. The *proof load* is defined as $F_p = A_t\,S_p$ and is the maximum load that a bolt can withstand without acquiring a permanent set.

TABLE II.3.3 Proof Strength of Steel Bolts (ISO)

Metric grade	Major diameter (d, mm)	Ultimate tensile strength (S_u, MPa)	Yield strength (S_y, MPa)	Proof strength (S_p, MPa)
4.6	M5–M36	400	240	225
4.8	M1.6–M16	420	340*	310
5.8	M5–M24	520	415*	380
8.8	M17–M36	830	660	600
9.8	M1.6–M16	900	720*	650
10.9	M6–M36	1040	940	830
12.9	M1.6–M36	1220	1100	970

* Yield strength approximate and not included in standard.

Source: B. G. Hamrock, B. Jacobson, and S. R. Schmid, *Fundamentals of Machine Elements*, New York, McGraw-Hill, 1999. Reprinted with permission of McGraw-Hill.

TABLE II.3.4 Proof Strength of Steel Bolts (SAE)

SAE grade	Range of major diameter (in.)	Ultimate tensile strength (S_u, kpsi)	Yield strength (S_y, kpsi)	Proof strength (S_p, kpsi)
1	$\frac{1}{4}-1\frac{1}{2}$	60	36	33
2	$\frac{1}{4}-\frac{3}{4}$	74	57	55
	$\frac{3}{4}-1\frac{1}{2}$	60	36	33
4	$\frac{1}{4}-1\frac{1}{2}$	115	100	65
5	$\frac{1}{4}-1$	120	92	85
	$1-1\frac{1}{2}$	105	81	74
7	$\frac{1}{4}-1\frac{1}{2}$	133	115	105
8	$\frac{1}{4}-1\frac{1}{2}$	150	130	120

Source: B. G. Hamrock, B. Jacobson, and S. R. Schmid, *Fundamentals of Machine Elements*, New York, McGraw-Hill, 1999. Reprinted with permission of McGraw-Hill.

Static Loading of the Joint

The bolt stress can be calculated from Eq. (II.3.30):

$$\sigma_b = \frac{F_b}{A_t} = \frac{F_i}{A_t} + C\frac{F_e}{A_t}, \tag{II.3.41}$$

where A_t is the tensile stress area. The limiting value for the bolt stress, σ_b, represents the proof strength, S_p. A safety factor n_b is introduced for the bolt stress and Eq. (II.3.41) becomes

$$S_p = \frac{F_i}{A_t} + C\frac{F_{max}\,n_b}{A_t}. \tag{II.3.42}$$

The safety factor is not applied to the preload stress F_i/A_t. The bolt failure safety factor is

$$n_{bf} = \frac{S_p A_t - F_i}{C F_{max,b}}, \tag{II.3.43}$$

where $F_{max,b}$ is the maximum external load applied to the bolt.

Separation occurs when the clamping force is zero, $F_c = 0$. The safety factor against separation of the parts of the joint is obtained from Eq. (II.3.31) with $F_c = 0$ and has the expression

$$n_s = \frac{F_i}{F_{max}(1 - C)}, \tag{II.3.44}$$

where F_{max} is the maximum external load applied to joint.

II.3.5 Examples

EXAMPLE II.3.1: Double square-thread power screw

A double square-thread power screw (Fig. II.3.15) has the major diameter $d = 40$ mm and the pitch $p = 6$ mm. The coefficient of friction of the thread is $\mu = 0.08$ and the coefficient of collar friction is $\mu_c = 0.1$. The mean collar diameter is $d_c = 45$ mm. The external load on the screw is $F = 8$ kN.

Find:

- (a) the lead, the pitch (mean) diameter, and the minor diameter;
- (b) the moment required to raise the load;
- (c) the moment required to lower the load;
- (d) the efficiency of the device.

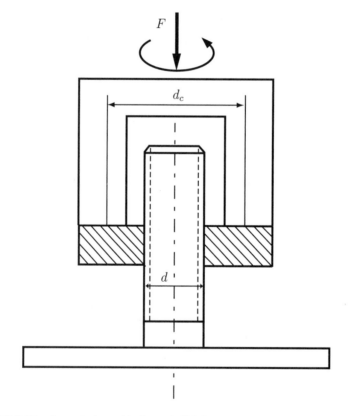

FIGURE II.3.15 *Screw jack used in Example II.3.1.*

EXAMPLE II.3.1: Double square-thread power screw — *Cont'd*

Solution

(a) From Figure II.3.7: the minor diameter is

$$d_r = d - p = 40 - 6 = 34 \text{ mm,}$$

the pitch (mean) diameter is

$$d_m = d - p/2 = 40 - 3 = 37 \text{ mm.}$$

The lead is

$$l = 2p = 2(6) = 12 \text{ mm.}$$

(b) The moment required to raise the load is [Eqs. (II.3.9) and (II.3.21)]:

$$M_r = \frac{Fd_m}{2}\left(\frac{l + \pi\mu d_m}{\pi d_m - \mu l}\right) + \frac{F\mu_c d_c}{2}$$

$$= \frac{8(10^3)(37)(10^{-3})}{2}\left[\frac{12 + 0.08(37)\pi}{37\pi - 0.08(12)}\right] + \frac{8(10^3)(0.1)(45)(10^{-3})}{2}$$

$$= 45.344 \text{ N m.}$$

(c) The moment required to lower the load is [Eqs. (II.3.10) and (II.3.21)]:

$$M_l = \frac{Fd_m}{2}\left(\frac{\pi\mu d_m - l}{\pi d_m + \mu l}\right) + \frac{F\mu_c d_c}{2}$$

$$= \frac{8(10^3)(37)(10^{-3})}{2}\left[\frac{0.08(37)\pi - 12}{37\pi + 0.08(12)}\right] + \frac{8(10^3)(0.1)(45)(10^{-3})}{2}$$

$$= 14.589 \text{ N m.}$$

The screw is not self-locking:

$$\pi\mu d_m - l = 0.08(37)\pi - 12 = -2.700 < 0.$$

(d) The overall efficiency is [Eq. (II.3.13)]:

$$e = \frac{Fl}{2\pi M_r} = \frac{8(10^3)(12)(10^{-3})}{2(45.344)\pi} = 0.336.$$

The *Mathematica*™ program for this example is given in Program II.3.1.

EXAMPLE II.3.2: Acme-thread power screw

A double-thread Acme screw is used in a jack to raise a load of 2000 lb (Fig. II.3.16). The major diameter of the screw is $d = 2$ in. A plain thrust collar is used. The mean diameter of the collar is $d_c = 3$ in. The coefficient of friction of the thread is $\mu = 0.12$ and the coefficient of collar friction is $\mu_c = 0.09$.

Determine:

(a) the screw pitch, lead, thread depth, mean pitch diameter, and helix angle;
(b) the starting moment for raising and for lowering the load;
(c) the efficiency of the jack.

Solution

(a) The preferred pitches for Acme threads are [20]:

d [in.]	$\frac{1}{4}$	$\frac{5}{16}$	$\frac{3}{8}$	$\frac{1}{2}$	$\frac{5}{8}$	$\frac{3}{4}$	$\frac{7}{8}$	1	$1\frac{1}{4}$	$1\frac{1}{2}$	$1\frac{3}{4}$	2	$2\frac{1}{2}$	3
p [in.]	$\frac{1}{16}$	$\frac{1}{14}$	$\frac{1}{12}$	$\frac{1}{10}$	$\frac{1}{8}$	$\frac{1}{6}$	$\frac{1}{6}$	$\frac{1}{5}$	$\frac{1}{5}$	$\frac{1}{4}$	$\frac{1}{4}$	$\frac{1}{4}$	$\frac{1}{3}$	$\frac{1}{2}$

For the major diameter, $d = 2$ in., the preferred screw pitch is $p = 0.25$ in.

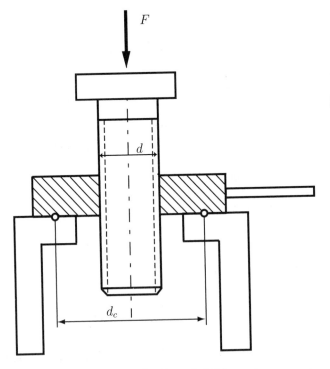

FIGURE II.3.16   *Acme screw jack used in Example II.3.2.*

EXAMPLE II.3.2: Acme-thread power screw — *Cont'd*

Because of the double thread, the lead is

$$l = 2p = 2(0.25) = 0.5 \text{ in.}$$

The pitch (mean) diameter is (see Fig. II.3.6):

$$d_m = d - p/2 = 2 - 0.25/2 = 1.875 \text{ in.}$$

The helix angle is

$$\lambda = \tan^{-1} \frac{l}{\pi d_m} = \tan^{-1} \frac{0.5}{1.875\pi} = 4.851°.$$

(b) The starting friction is about one-third higher than running friction and the coefficients of starting friction are [7]:

$$\mu_s = \frac{4}{3}\mu = \frac{4}{3}(0.12) = 0.16 \quad \text{and} \quad \mu_{sc} = \frac{4}{3}\mu_c = \frac{4}{3}(0.09) = 0.12.$$

The angle α_n is calculated with the formula

$$\alpha_n = \tan^{-1}(\tan\alpha\cos\lambda) = \tan^{-1}(\tan 14.5°\cos 4.851°) = 14.450°,$$

where $\alpha = 14.5°$ (see Fig. II.3.6). The moment for lifting the load is

$$M_{rs} = \frac{Fd_m}{2}\left(\frac{\mu_s\pi d_m + l\cos\alpha_n}{\pi d_m\cos\alpha_n - \mu_s l}\right) + \frac{F\mu_{cs}d_c}{2}$$

$$= \frac{2000(1.875)}{2}\left(\frac{0.16\pi(1.875) + 0.5\cos 14.450°}{\pi(1.875)\cos 14.450° - (0.16)(0.5)}\right) + \frac{2000(0.12)(3)}{2}$$

$$= 835.626 \text{ lb in.}$$

Similarly, the moment required to lower the load and to overcome a part of the friction is

$$M_{ls} = \frac{Fd_m}{2}\left(\frac{\mu_s\pi d_m - l\cos\alpha_n}{\pi d_m\cos\alpha_n + \mu_s l}\right) + \frac{F\mu_{cs}d_c}{2}$$

$$= \frac{2000(1.875)}{2}\left(\frac{0.16\pi(1.875) - 0.5\cos 14.450°}{\pi(1.875)\cos 14.450° + (0.16)(0.5)}\right) + \frac{2000(0.12)(3)}{2}$$

$$= 508.562 \text{ lb in.}$$

Continued

EXAMPLE II.3.2: Acme-thread power screw — *Cont'd*

(c) Changing the coefficient of friction to the running values of μ and μ_c, the moment for lifting the load is

$$M_r = \frac{F d_m}{2} \left(\frac{\mu \pi d_m + l \cos \alpha_n}{\pi d_m \cos \alpha_n - \mu l} \right) + \frac{F \mu_c d_c}{2}$$

$$= \frac{2000(1.875)}{2} \left(\frac{0.12\pi(1.875) + 0.5 \cos 14.450°}{\pi(1.875) \cos 14.450° - (0.12)(0.5)} \right) + \frac{2000(0.09)(3)}{2}$$

$$= 665.667 \text{ lb in.}$$

With both friction coefficients zero, the moment to raise the load is

$$M_0 = \frac{F d_m}{2} \left(\frac{l \cos \alpha_n}{\pi d_m \cos \alpha_n} \right) = \frac{F l}{2\pi} = \frac{2000(0.5)}{2\pi} = 159.155 \text{ lb in.}$$

(d) The efficiency is the ratio of friction-free moment to actual moment, or

$$e = \frac{F l}{2\pi M_r} = \frac{M_0}{M_r} = \frac{159.155}{665.667} = 23.909 \, \%.$$

If the collar friction is neglected ($\mu_c = 0$), the efficiency

$$e = \frac{\cos \alpha_n - \mu \tan \lambda}{\cos \alpha_n + \mu \cos \lambda},$$

function of λ is plotted in Figure II.3.17 where $\alpha_n = \tan^{-1}(\tan 14.5° \cos \lambda)$ and $\mu = \{0.05;\ 0.12;\ 0.15\}$.

The *Mathematica*^{TM} program for this example is given in Program II.3.2.

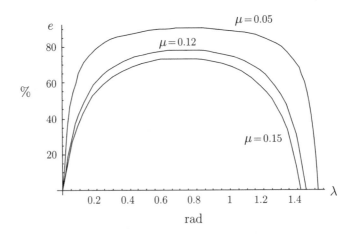

FIGURE II.3.17 *Efficiency of Acme screw thread (the collar friction is neglected).*

EXAMPLE II.3.3:

A hexagonal bolt made of steel with the modulus of elasticity $E_b = 206.8$ GPa is used to join two parts made of cast iron. The modulus of elasticity for gray cast iron is $E_c = 100.0$ GPa. The thread major diameter is $d = 14$ mm, the root (minor) diameter is $d_r = 12$ mm, and the pitch is $p = 2$ mm. The bolt has the axial length $l = 50$ mm and the threaded portion of the bolt is $l/2$. The cast iron parts have the same axial length, $l/2$. Determine:

(a) the stiffness of the bolt using Eq. (II.3.32);
(b) the stiffness of the clamped parts using Eq. (II.3.38);
(c) the stiffness of the clamped parts using Eq. (II.3.39).

Solution

(a) The lengths of the threaded section and unthreaded section of the bolt are $l_t = l_s = l/2 = 25$ mm. The effective length of the threaded bolt is

$$l_{te} = l_t + 0.4d_r = 25 + 0.4(12) = 29.8 \text{ mm} = 0.0298 \text{ m}.$$

The effective length of the unthreaded bolt is

$$l_{se} = l_s + 0.4d = 25 + 0.4(14) = 30.6 \text{ mm} = 0.0306 \text{ m}.$$

The minor diameter area is

$$A_r = \pi d_r^2/4 = \pi(0.012)^2/4 = 113.097 \times 10^{-6} \text{ m}^2.$$

The major diameter area is

$$A_s = \pi d^2/4 = \pi(0.014)^2/4 = 153.938 \times 10^6 \text{ m}^2.$$

The stiffness of the threaded portion is

$$k_t = \frac{A_r E_b}{l_{te}} = \frac{(113.097 \times 10^{-6})(206.8 \times 10^9)}{0.0298} = 7.8485 \times 10^8 \text{ N/m}.$$

The stiffness of the unthreaded portion is

$$k_s = \frac{A_s E_b}{l_{se}} = \frac{(153.938 \times 10^{-6})(206.8 \times 10^9)}{0.0306} = 1.04034 \times 10^9 \text{ N/m}.$$

The bolt stiffness is

$$k_b = \frac{k_t k_s}{k_t + k_s} = \frac{(7.8485 \times 10^8)(1.04034 \times 10^9)}{7.8485 \times 10^8 + 1.04034 \times 10^9} = 4.47357 \times 10^8 \text{ N/m}.$$

Continued

EXAMPLE II.3.3: *Cont'd*

(b) Shigley and Mischke [20] proposed the following expression for the stiffness of the clamped parts [see Eq. (II.3.38)]:

$$k_c = \frac{0.577\pi E_c d}{2\ln\left(5\dfrac{0.577l + 0.5d}{0.577l + 2.5d}\right)} = \frac{0.577\pi(100.0 \times 10^9)(0.014)}{2\ln\left(5\dfrac{0.577(0.05) + 0.5(0.014)}{0.577(0.05) + 2.5(0.014)}\right)}$$

$$= 1.22925 \times 10^9 \text{ N/m}.$$

(c) Wileman et al. [25] proposed the expression for the stiffness of the clamped parts [Eq. (II.3.39)], with the numerical constants $A = 0.77871$ and $B = 0.61616$ for gray cast iron:

$$k_c = E_c \, d \, A \, e^{Bd/l} = (100.0 \times 10^9)(0.014)(0.77871)e^{0.61616(0.014)/0.05}$$

$$= 1.29548 \times 10^9 \text{ N/m}.$$

The *Mathematica*^TM program for this example is given in Program II.3.3.

EXAMPLE II.3.4:

A hexagonal bolt and nut assembly is used to join two parts. The bolt and nut are made of steel (modulus of elasticity $E_b = 30$ Mpsi for steel). One part is made of steel and the other part is made of cast iron (modulus of elasticity $E_c = 12$ Mpsi for cast iron). The thread major diameter is $d = 5/8$ in., the root (minor) diameter is $d_r = 0.5135$ in. There are 11 threads per inch, hence the pitch is $p = 1/11$ in. The assembly and the bolt have the axial length $l = 1.5$ in. The length of the threaded portion of the bolt is $l/2$. The axial length of the cast iron part is $l/2$.
   Determine:

(a) the stiffness of the bolt using Eq. (II.3.37);
(b) the stiffness of the clamped parts using Eq. (II.3.38);
(c) the stiffness of the clamped parts using Eq. (II.3.39).

Solution

(a) The lengths of the threaded section and unthreaded section of the bolt are $l_t = l_s = l/2 = 0.75$ in. The major diameter area is

$$A_s = \pi d^2/4 = \pi(5/8)^2/4 = 0.306 \text{ in.}^2$$

EXAMPLE II.3.4: *Cont'd*

The tensile stress area is given by Eq. (II.3.36):

$$A_t = 0.7854(d - 0.9743\,p)^2 = 0.7854 \left(\frac{5}{8} - 0.9743\frac{1}{11} \right)^2 = 0.226 \text{ in.}^2$$

The stiffness of the unthreaded portion is

$$k_s = A_s E_b / l_s = 0.306(30)/0.75 = 12.271 \text{ Mlb/in.}$$

The stiffness of the threaded portion is

$$k_t = A_t E_b / l_t = 0.226\,(30)/0.75 = 9.040 \text{ Mlb/in.}$$

The bolt stiffness is

$$k_b = \frac{k_s k_t}{k_s + k_t} = \frac{2.271(9.040)}{2.271 + 9.040} = 5.205 \text{ Mlb/in.}$$

(b) The axial length of the cast iron clamped part and steel clamped part are $l_1 = l_2 = l/2 = 0.75$ in. Using Eq. (II.3.38), the stiffness of the cast iron clamped part is

$$k_1 = \frac{0.577\pi E_c d}{2\ln\left(5\dfrac{0.577l_1 + 0.5d}{0.577l_1 + 2.5d}\right)} = \frac{0.577\pi(12)(5/8)}{2\ln\left[5\dfrac{0.577(0.75) + 0.5(5/8)}{0.577(0.75) + 2.5(5/8)}\right]}$$

$$= 10.882 \text{ Mlb/in.}$$

and the stiffness of the steel clamped part is

$$k_2 = \frac{0.577\pi E_b d}{2\ln\left(5\dfrac{0.577l_2 + 0.5d}{0.577l_2 + 2.5d}\right)} = \frac{0.577\pi(30)(5/8)}{2\ln\left[5\dfrac{0.577(0.75) + 0.5(5/8)}{0.577(0.75) + 2.5(5/8)}\right]}$$

$$= 27.206 \text{ Mlb/in.}$$

The resulting stiffness of the clamped parts is

$$k_c = \frac{k_1 k_2}{k_1 + k_2} = \frac{10.882(27.206)}{10.882 + 27.206} = 7.773 \text{ Mlb/in.}$$

(c) Using Eq. (II.3.39), the stiffness of the cast iron clamped part is ($A = 0.77871$ and $B = 0.61616$ for gray cast iron):

$$k_1 = E_c dA e^{Bd/l_1} = 12(5/8)(0.77871)e^{0.61616(5/8)/0.75}$$

$$= 9.759 \text{ Mlb/in.}$$

Continued

EXAMPLE II.3.4: *Cont'd*

and the stiffness of the steel clamped part is ($A = 0.78715$ and $B = 0.62873$ for steel):

$$k_2 = E_b dA e^{Bd/l_2} = 30(5/8)(0.78715)e^{0.62873(5/8)/0.75}$$

$$= 24.923 \text{ Mlb/in.}$$

The resulting stiffness of the clamped parts is

$$k_c = \frac{k_1 k_2}{k_1 + k_2} = \frac{9.759(24.923)}{9.759 + 24.923} = 7.013 \text{ Mlb/in.}$$

The *Mathematica*™ program for this example is given in Program II.3.4.

EXAMPLE II.3.5:

A bolt made from cold-drawn steel with the stiffness k_b is used to clamp two steel plates with the stiffness k_c. The elasticities are such that $k_c = 5 k_b$. The plates and the bolt have the same length. The external joint separating force fluctuates continuously between 0 and 6000 lb.
   Determine:

(a) the minimum required value of initial preload to prevent loss of compression of the plates;
(b) if the preload is 6500 lb, find the minimum force in the plates for fluctuating load.

Solution

(a) Compression of the plates is lost when $F_c = 0$ when maximum load is applied. Equation (II.3.28) becomes

$$F_i = F_c + F_e \frac{k_c}{k_b + k_c} = 0 + 6000 \frac{5}{1+5} = 5000 \text{ lb.}$$

(b) Minimum force in plates occurs when fluctuating load is maximum. From Eq. (II.3.28) with $F_i = 6500$ lb, it results:

$$F_c = F_i - F_e \frac{k_c}{k_b + k_c} = 6500 - 6000 \frac{5}{1+5} = 1500 \text{ lb.}$$

EXAMPLE II.3.6:

A bolt and nut assembly is used to join two parts made of cast iron. The bolt has the thread major diameter $d = 10$ mm, the pitch $p = 1.5$ mm, and a 4.8 grade. The clamped plates have a stiffness k_c six times the bolt stiffness k_b. The assembly and the bolt have the axial length and the cast iron parts have the same axial length. Determine the maximum load for bolt and joint failure assuming a reused connection and a static safety factor of $n_f = 2$.

Solution   The joint constant is

$$C = \frac{k_b}{k_b + k_c} = \frac{k_b}{k_b + 6k_b} = \frac{1}{7}.$$

The tensile stress area is

$$A_t = 0.7854(d - 0.9382\,p)^2 = 0.7854[10 - 0.9382(1.5)]^2 = 57.989 \text{ mm}^2.$$

From Table II.3.3, for a 4.8 grade the proof strength is $S_p = 310$ MPa. The proof load is

$$F_p = A_t S_p = (57.989 \times 10^{-6})(310 \times 10^6) = 17\,976.8 \text{ N}.$$

The preload for reused connections is

$$F_i = 0.75 F_p = 0.75(17976.8) = 13\,482.6 \text{ N}.$$

The maximum external load applied to the bolt is

$$F_{max,b} = \frac{S_p A_t - F_i}{n_f\,C} = \frac{(310 \times 10^6)(57.989 \times 10^{-6}) - 13\,482.6}{2\,(1/7)} = 15\,729.7 \text{ N}.$$

Equation (II.3.44) gives the maximum external load applied to the joint before separation as

$$F_{max} = \frac{F_i}{n_f\,(1 - C)} = \frac{13\,482.6}{2[1 - (1/7)]} = 7864.84 \text{ N}.$$

EXAMPLE II.3.7:

A number of N identical bolts, 1" - 8 UNC grade 5, are used to join two members. The joint constant is $C = 0.5$ and the separating force is 60 kip. Assume that the bolts may be reused when the joint is taken apart. Find the number of bolts (N) for a design safety factor of 2.

Continued

EXAMPLE II.3.7: *Cont'd*

Solution From Table II.3.2 for $d = 1$ in. and $n = 8$, the tensile stress area is $A_t = 0.606$ in.^{2} From Table II.3.4 for grade 5, the proof strength is $S_p = 85$ kpsi. The recommended preload for reused connections is

$$F_i = 0.75\, A_t\, S_p = 0.75\, (0.606)\, (85) = 38.632 \text{ kip}.$$

For N bolts Eq. (II.3.43) can be written as

$$n_f = \frac{S_p A_t - F_i}{C\,(F_{max,b}/N)}, \tag{II.3.45}$$

or

$$N = \frac{C\, n_f\, F_{max,b}}{S_p A_t - F_i} = \frac{0.5(2)(60)}{(85)(0.606) - 38.632} = 4.659.$$

Five bolts are selected. Using Eq. (II.3.45) with $N = 5$, the safety factor is

$$n_{fc} = \frac{S_p A_t - F_i}{C\,(F_{max,b}/N)} = \frac{(85)(0.606) - 38.632}{0.5(60/5)} = 2.146,$$

which is greater than the required safety factor of 2; therefore five bolts will be used for the recommended preload in tightening.

EXAMPLE II.3.8:

The support block of a machine is attached to the ground with two screws. The machine applies a tensile static load of 10 kN to the block.

(a) Select appropriate metric screws of class 5.8 for the block attachment;
(b) Find the appropriate tightening moment. Use a safety factor of 4 based on proof strength.

Solution

(a) The load of 10 kN is applied equally by each screw and the bolt load is axial tension. The nominal load for each of the two bolts is $F = 5$ kN. With a safety factor of $n_b = 4$, the design overload for each bolt is $n_b F = 4\,(5) = 20$ kN. For static loading of a ductile material the stress equation is $\sigma = P/A$. When P is equal to the design overload, σ is equal to the proof strength and

$$S_p = \frac{n_b F}{A_t}. \tag{II.3.46}$$

EXAMPLE II.3.8: *Cont'd*

For class 5.8 a proof strength of $S_p = 380$ MPa is selected from Table II.3.3. Equation (II.3.46) gives the tensile stress area:

$$A_t = \frac{n_b F}{S_p} = \frac{20\,000}{380 \times 10^6} = 52.631 \times 10^{-6} \text{ m}^2 = 52.631 \text{ mm}^2.$$

From Table II.3.1 an appropriate standard size is M 10×1.5 ($A_t = 58.0$ mm^2).

(b) The initial tightening tension F_i is defined as

$$F_i = KA_t S_p = 0.9(58.01 \times 10^{-6})(380 \times 10^6) = 19\,836 \text{ N},$$

where the constant K is 0.90 for permanent connections. Juvinall and Marshek give an estimated tightening moment for standard screw threads [7] as

$$T = 0.2\, F_i d = 0.2(19836)(10 \times 10^{-3}) = 39.672 \text{ Nm}. \qquad \text{(II.3.47)}$$

II.3.6 Problems

II.3.1 The double square-threaded screw has the major diameter $d = 1$ in. and the pitch $p = 0.2$ in. The coefficient of friction in the threads is 0.15. A moment $M = 60$ lb-in. is applied about the axis of the screw (Fig. II.3.18). Find the axial force required to advance the screw: (a) to the right, and (b) to the left.

II.3.2 A double square-thread power screw has a pitch (mean) diameter of 30 mm and a pitch of 4 mm (Fig. II.3.15). The coefficient of friction of the thread is 0.08 and the coefficient of collar friction is also 0.08. The mean collar diameter is 40 mm. The external load on the screw is 6.4 kN. Determine the moment required to lower the load and the overall efficiency.

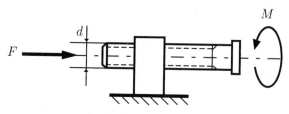

FIGURE II.3.18 *Double square-threaded screw.*

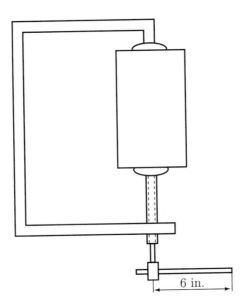

FIGURE II.3.19 *C-clamp.*

II.3.3 A power screw has a double square thread with a mean diameter of 40 mm and a pitch of 12 mm. The coefficient of friction in the thread is 0.15. Determine if the screw is self-locking.

II.3.4 The single-threaded screw of a vise has a mean diameter of 1 in. and has 5 square threads per in. The coefficient of static friction in the thread is 0.20. Determine the helix angle and the fiction angle for the thread.

II.3.5 A triple-thread Acme screw is used in a jack (as shown in Fig. II.3.16) to raise a load of 4000 lb. The major diameter of the screw is 3 in. A plain thrust collar is used. The mean diameter of the collar is 4 in. The coefficient of friction of the thread is 0.08 and the coefficient of collar friction is 0.1. Determine: (a) the screw pitch, lead, thread depth, mean pitch diameter, and helix angle; (b) the starting moment for raising and for lowering the load; (c) the efficiency of the jack.

II.3.6 A C-clamp develops a 250-lb clamping force (Fig. II.3.19). The clamp uses a 1/2-in. Acme single thread. The collar of the clamp has a mean diameter of 5/8 in. The coefficients of running friction are estimated as 0.1 for both the collar and the screw. Estimate the force required at the end of a 6-in. handle.

II.3.7 A bolt made of steel is used to join two parts made of cast iron. The thread major diameter is $d = 14$ mm, the root (minor) diameter is $d_r = 12$ mm, and the pitch is $p = 2$ mm. The bolt has the axial length $l = 60$ mm and the threaded portion of the bolt is $l/3$. The cast iron parts have the same axial length $l/2$. Determine: (a) the stiffness of the bolt using Eq. (II.3.32); (b) the stiffness of the clamped parts using Eq. (II.3.38); (c) the stiffness of the clamped parts using Eq. (II.3.39).

II.3.8 A hexagonal bolt and nut assembly is used to join two parts, as illustrated in Figure II.3.20. The bolt and nut are made of steel. One part is made of steel and

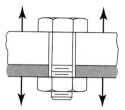

FIGURE II.3.20 *Bolt and nut assembly.*

the other part is made of cast iron. The thread major diameter is $d = 5/8$ in., the root (minor) diameter is $d_r = 0.5135$ in., and there are $n = 11$ threads per in. The assembly and the bolt have the axial length $l = 1.8$ in. The length of the threaded portion of the bolt is $l/3$. The axial length of the cast iron part is $l/3$. Determine: (a) the stiffness of the bolt using Eq. (II.3.37); (b) the stiffness of the clamped parts using Eq. (II.3.38); (c) the stiffness of the clamped parts using Eq. (II.3.39).

II.3.9 A bolt M 12 × 1.25 ISO grade 5.8 made of steel is used to join two parts made of cast iron. The assembly and the bolt have the axial length $l = 80$ mm and the threaded portion of the bolt is $l/4$. The cast iron parts have the same axial length $l/2$. The external joint separating force fluctuates continuously between 0 and 20 kN. Determine the minimum required value of initial preload to prevent loss of compression of the parts.

II.3.10 A bolt M 10 × 1 ISO grade 4.8 made of steel is used to join two plates made of cast iron and steel. The bolt and the assembly have the axial length $l = 60$ mm and the threaded portion of the bolt is $l/2$. The cast iron plate has the axial length $l/4$. The external joint separating force fluctuates continuously between 0 and 20 kN. The bolt is tightened to an initial tension of 5 kN. Determine the minimum force in the plates.

II.3.11 A bolt made from steel has the stiffness k_b. Two steel plates are held together by the bolt and have a stiffness k_c. The elasticities are such that $k_c = 7 k_b$. The plates and the bolt have the same length. The external joint separating force fluctuates continuously between 0 and 2500 lb. Determine: (a) the minimum required value of initial preload to prevent loss of compression of the plates, and (b) if the preload is 3500 lb, find the minimum force in the plates for fluctuating load.

II.3.12 Repeat the previous problem, but consider that the external joint separating force varies between 0 and 8500 lb.

II.3.13 A bolt and nut assembly is used to join two parts made of cast iron. The bolt has the thread (3/4)"–16 UNF, SAE grade 5. The clamped plates have a stiffness k_c six times the bolt stiffness k_b. The assembly and the bolt have the axial length and the cast iron parts have the same axial length. Determine the maximum load for bolt and joint failure assuming a reused connection and a static safety factor of $n_f = 2$.

II.3.14 A number of N identical bolts, M 10 × 1.5, ISO grade 4.6, are used to join two members. The joint constant is $C = 0.45$ and the separating force is 6 kN. Assume

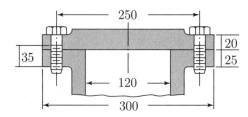

FIGURE II.3.21 *Bolted pressure vessel.*

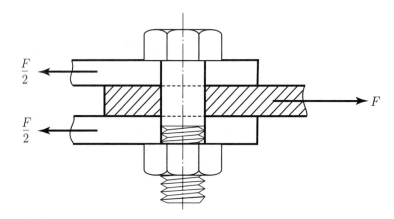

FIGURE II.3.22 *Bolt with two shear planes.*

that the bolts may be reused when the joint is taken apart. Find the number of bolts (*N*) for a design safety factor of 2.

II.3.15 Figure II.3.21 shows the connection of a cylinder head to a pressure vessel using 10 identical steel bolts, M 16 × 2, ISO grade 8.8. The parts are made of steel. All the dimensions are illustrated in Figure II.3.21 and are all in mm. The static pressure inside the pressure vessel is 5.5 MPa. Determine the load safety factor.

II.3.16 Repeat the previous problem for the pressure vessel made of cast iron and the cover plate made of aluminum (for aluminum, $E = 70$ GPa).

II.3.17 A rotating shaft applies a load of 20 kN on a block. Select the appropriate size for the two screws of class 4.8 for the block attachment and find the tightening moment.

II.3.18 Figure II.3.22 shows an M 8 × 1.25, ISO grade 4.6, steel bolt tightened to its full proof load. The bolt is loaded in double shear (the bolt has two shear planes). The clamped plates are made of steel with the coefficient of friction approximately 0.3. Determine the force *F* the joint is capable to withstand.

II.3.7 Programs

PROGRAM II.3.1

```
Apply[Clear, Names["Global`*"]];
Off[General::spell];
Off[General::spell1];
" input data "
" load F [N] "
" major diameter d [mm] "
" screw pitch p [mm] "
" coefficient of friction for thread µ "
" collar diameter dc [mm] "
" coefficient of friction for collar µc "
data = {F → 8000, d → 40, p → 6, µ -> 0.08, dc → 45, µc -> 0.1}
"solution"
l = 2 p;
Print["lead for double thread l = ", l, " = ", l /. data, " [mm]"]
dr = d - p;
Print["minor diameter dr = ", dr, " = ", dr /. data, " [mm]"]
dm = d - p/2;
Print["mean (pitch) diameter dm = ", dm, " = ", dm /. data, " [mm]"]
Mr = 0.5 F dm (1 + π µ dm)/(π dm - µ l) + 0.5 F dc µc;
Print["moment to raise load
Mr = 0.5 F dm ( 1 + π µ dm )/( π dm - µ l ) + 0.5 F dc µc "]
Print["Mr = ", Mr /. data, " [N mm] = ", 10^(-3)Mr /. data, " [N m]"]
Ml = 0.5 F dm (π µ dm - 1)/(π dm + µ l) + 0.5 F dc µc;
Print[
  "moment to lower load Ml = 0.5 F dm ( π µ dm - 1 )/( π dm + µ l)
                         + 0.5 F dc µc "]
Print["Ml = ", Ml /. data, " [N mm] = ", 10^(-3)Ml /. data, " [N m]"]
sf = (π µ dm - 1);
"sef-locking condition: ( π µ dm - 1 ) > 0 "
Print["( π µ dm - 1 ) = ", sf /. data]
If[(sf /. data) > 0, Print["the screw is sef-locking"],
 Print["the screw is not sef-locking"]]
e = Fl/(2 π Mr);
Print["efficiency e = F l/( 2 π Mr ) = ", e /. data]
```

input data

load F [N]

major diameter d [mm]

screw pitch p [mm]

coefficient of friction for thread μ

collar diameter dc [mm]

coefficient of friction for collar μc

$\{F \rightarrow 8000, d \rightarrow 40, p \rightarrow 6, \mu \rightarrow 0.08, dc \rightarrow 45, \mu c \rightarrow 0.1\}$

solution

lead for double thread l = 2 p = 12 [mm]

minor diameter dr = d - p = 34 [mm]

mean (pitch) diameter dm = d - $\frac{p}{2}$ = 37 [mm]

moment to raise load Mr = 0.5 F dm (l + π μ dm)/(π dm - μ l)
    + 0.5 F dc μc

Mr = 45344.7 [N mm] = 45.3447 [N m]

moment to lower load Ml = 0.5 F dm (π μ dm - l)/(π dm + μ l)
    + 0.5 F dc μc

Ml = 14589.3 [N mm] = 14.5893 [N m]

sef-locking condition: (π μ dm - l) > 0

(π μ dm - l) = -2.70089

the screw is not sef-locking

efficiency e = F l/(2 π Mr) = 0.336949

PROGRAM II.3.2

```
(* Acme double thread power screw *)
Apply[Clear, Names["Global`*"]];
Off[General::spell];
Off[General::spell1];

(*Given data*)
d = 2.; (* in *) (* major diameter *)
F = 2000.; (* lb *) (* weight of the load *)
dc = 3.; (* in *) (* mean diameter of plain thrust collar *)
μ = 0.12; μc = 0.09; (* coefficients of friction *)

(*Assumption: coefficient of starting friction is about one-
    third higher than the coefficient of friction (running friction)
    μs=(4/3) μ ; μcs=(4/3) μc
*)

(* a *)
p = 1/4.; (* in *) (* screw pitch; Standard sizes of power screw threads)
    for d=2 in. there are 4 (four) threads per inch, p=(1 in.)/4=0.25 in. *)
l = 2 p; (* lead, because of the double thread *)
dm = d - p/2; (* mean diameter of thread contact from figure *)
λ = ArcTan[l/(π dm)]; (* rad *) (* lead angle *)
```

```
(* b *)
(*For starting, increase the coefficients of running friction by
    one third*)
μs = (4/3) μ;
μcs = (4/3) μc ;
(* coefficients of static friction *)
α = 14.5 * π/180; (* rad *)
(* from Figure , 2 α=29 deg yields α=14.5 deg (Acme)*)
αn = ArcTan[Tan[α]*Cos[λ]]; (* thread angle in the normal plane *)
(* the torque for raising the load: *)
(* Moment using coefficients of static friction *)
Mrs = 0.5 F dm (π μs dm + 1 Cos[αn])/(π dm Cos[αn] - μs 1 ) +
        0.5 F dc μcs ;
(* Moment using coefficients of running friction *)
Mr = 0.5 F dm ( π μ dm + 1 Cos[αn])/(π dm Cos[αn] - μ 1 ) +
        0.5 F dc μc;
(* for lowering the load*)
Mls = 0.5 F dm ( π μs dm - 1 Cos[αn])/(π dm Cos[αn]+ μs 1 )+
        0.5 F dc μcs ;
Ml = 0.5 F dm ( π μ dm - 1 Cos[αn])/(π dm Cos[αn] + μ 1 ) + 0.5 F dc μc;

(* c *)
(* Efficiency (the ratio of friction-free moment to actual moment) *)
Mr0 = F 1 /(2 π);
e = F 1 /(2 π Mr);

Print["Screw pitch p = (1 in.)/4 = ", p, " in"];
Print["Lead 1 = 2 p = ", 1, " in"];
Print["Mean pitch diameter dm = d-p/2 = ", dm, " in"];
Print["Lead angle λ=ArcTan[1/(π dm)] = ", N[λ]*180 /Pi, " deg"];
Print["Thread angle αn=ArcTan[Tan[α]*Cos[λ]] = ", αn * 180/Pi, " deg"];
Print["Moment for raising the load with starting friction = ", Mrs,
        " lb in"];
Print["Moment for lowering the load with starting friction = ", Mls,
        " lb in"];
Print["Moment for raising the load with running friction = ", Mr,
        " lb in"];
Print["Moment for raising the load with no friction = ", N[Mr0],
        " lb in"];
Print["Efficiency (the ratio of friction-free moment to actual moment)
        e = ", e, " %"];
deg = 180/N[Pi];

μ = 0.05;
graph1 = Plot[(Cos[αn] - μ * Tan[λ])/(Cos[αn] + μ * Cos[λ]) * 100,
    {λ, 0.0, 1.5192087045738232 }, PlotLabel → "Efficiency e(λ) (%)
        for μ =0.05"];
```

```
μ = 0.12;
graph2 = Plot[(Cos[αn] - μ * Tan[λ])/(Cos[αn] + μ * Cos[λ]) * 100,
    {λ, 0.0, 1.4475046197126569}, PlotLabel → "Efficiency e(λ) (%)
       for μ =0.12"];

μ = 0.15;
graph3 =
  Plot[(Cos[αn] - μ * Tan[λ])/(Cos[αn] + μ * Cos[λ]) * 100,
      {λ, 0.0, 1.4171173868835931}, PlotLabel → "Efficiency e(λ)
      (%) for μ =0.15"];

Show[graph1, graph2, graph3];
```

Screw pitch p = (1 in.)/4 = 0.25 in

Lead l = 2 p = 0.5 in

Mean pitch diameter dm = d-p/2 = 1.875 in

Lead angle λ=ArcTan[l/(π dm)] = 4.85179 deg

Thread angle αn=ArcTan[Tan[α]*Cos[λ]] = 14.4502 deg

Moment for raising the load with starting friction = 835.626 lb in

Moment for lowering the load with starting friction = 508.562 lb in

Moment for raising the load with running friction = 665.667 lb in

Moment for raising the load with no friction = 159.155 lb in

Efficiency (the ratio of friction-free moment to actual moment)
  e = 0.239091 %

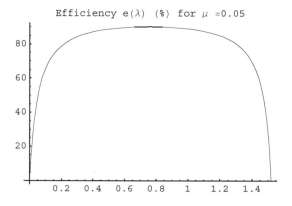

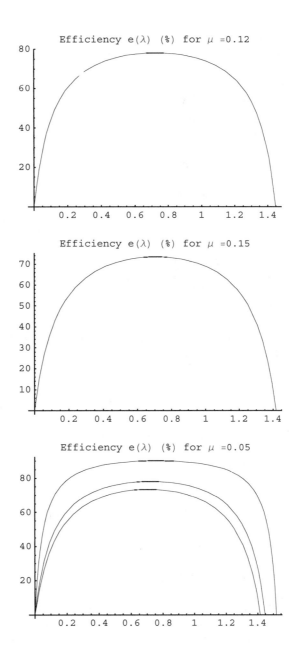

PROGRAM II.3.3

```
Apply[Clear,Names["Global`*"]];
Off[General::spell];
Off[General::spell1];

d=14. 10^(-3); (* m *) (* major diameter *)
dr=12. 10^(-3); (* m *) (* minor diameter *)
```

```
p=2. 10^(-3); (* m *) (* pitch *)
1=50. 10^(-3); (* m *) (* joint axial length *)

Print["major thread diameter d=",d," m"];
Print["minor thread diameter dr=",dr," m"];
Print["pitch p=",p," m"];
Print["joint axial length l=",l," m"];

"modulus of elasticity [GPa] E=206.8 for steel"
Eb=206.8 10^9;
Print["modulus of elasticity of the bolt Eb=",Eb," Pa"];

"a. Bolt stiffness kb"
lt=l/2;
lte=lt+0.4 dr;
Print["effective length of threaded portion of the bolt lte=lt+0.4
    dr=", lte," m"];
ls=l/2;
lse=ls+0.4 d;
Print["effective length of the unthreaded portion of the bolt
    lse=ls+0.4 d=",lse," m"];

Ar=N[Pi] dr^2/4.;
Print["minor diameter area Ar=Pi dr^2/4=",Ar," m^2"];
As=N[Pi] d^2/4.;
Print["major diameter area As=Pi d^2/4=",As," m^2"];

kt=Ar Eb/lte;
Print["stiffness of the threaded portion kt=Ar Eb/lte=",kt," N/m"];
ks=As Eb/lse;
Print["stiffness of the unthreaded portion ks=As Eb/lse=",ks," N/m"];
kb=kt ks/(kt+ks);
Print["bolt stiffness kb=kt ks/(kt+ks)=",kb," N/m"];

Print["joint axial length l=",l," m"];
l1=l2=l/2;
"axial length of the parts l1=l2=l/2"
"modulus of elasticity [GPa] E=100.0 for gray cast iron"
Ec=100 10^9;
Print["modulus of elasticity of the part Ec=",Ec," Pa"];
"Joint Stiffness kc"

"b. Stiffness of clamped parts (Shigley)"
Print["J.E.Shigley & C.R.Mischke, Mechanical Engineering Design,
        McGraw-Hill 1989"];
kc=0.577 N[Pi] Ec d/(2 Log[5 (0.577 l + 0.5 d)/(0.577 l + 2.5 d)]);
Print["kc=0.577 Pi Ec d/(2 Log[5 (0.577 l + 0.5 d)/(0.577 l + 2.5 d)])"]
```

Print["stiffness of clamped parts (Shigley) kc=",kc," N/m"];

"c. Stiffness of clamped parts (Wileman)"
Print["J.Wileman et al.,1990"];
"constants A=0.78715; B=0.62873 for steel"
"constants A=0.77871; B=0.61616 for cast iron"
"joint stiffness: kc = Ec d A e^(B d/l)"
"parts of the joint: cast iron "
A=0.77871;
B=0.61616;
kcw=Ec d A E^(B d/l);
Print["joint stiffness (Wileman) kc=",kcw," N/m"];

major thread diameter d=0.014 m

minor thread diameter dr=0.012 m

pitch p=0.002 m

joint axial length l=0.05 m

modulus of elasticity [GPa] E=206.8 for steel

modulus of elasticity of the bolt Eb=2.068 $\times$ 10^{11} Pa

a. Bolt stiffness kb

effective length of threaded portion of the bolt lte=lt+0.4
  dr=0.0298 m

effective length of the unthreaded portion of the bolt lse=ls+0.4
  d=0.0306 m

minor diameter area Ar=Pi dr^2/4=0.000113097 m^2

major diameter area As=Pi d^2/4=0.000153938 m^2

stiffness of the threaded portion kt=Ar Eb/lte=7.8485 $\times$ 10^8 N/m

stiffness of the unthreaded portion ks=As Eb/lse=1.04034 $\times$ 10^9 N/m

bolt stiffness kb=kt ks/(kt+ks)=4.47357 $\times$ 10^8 N/m

joint axial length l=0.05 m

axial length of the parts l1=l2=l/2

modulus of elasticity [GPa] E=100.0 for gray cast iron

modulus of elasticity of the part Ec=100000000000 Pa

Joint Stiffness kc

b. Stiffness of clamped parts (Shigley)

J.E.Shigley & C.R.Mischke, Mechanical Engineering Design,
  McGraw-Hill 1989

kc=0.577 Pi Ec d/(2 Log[5 (0.577 1 + 0.5 d)/(0.577 1 + 2.5 d)])

stiffness of clamped parts (Shigley) kc=1.22925 × 10^9 N/m

c. Stiffness of clamped parts (Wileman)

J.Wileman et al.,1990

constants A=0.78715; B=0.62873 for steel

constants A=0.77871; B=0.61616 for cast iron

joint stiffness: kc = Ec d A e^(B d/l)

parts of the joint: cast iron

joint stiffness (Wileman) kc=1.29548 × 10^9 N/m

PROGRAM II.3.4

```
Apply[Clear,Names["Global`*"]];
Off[General::spell];
Off[General::spell1];

"coarse thread"
d=5/8.; (* in *) (* major diameter *)
dr=0.5135; (* in *) (* minor diameter *)
n=11; (* threads per inch *)
p=1./n;
(* At=226 in^2 *)
l=1.5 ; (* in *) (* joint axial length *)

Print["major thread diameter d=",d," in."];
Print["minor thread diameter dr=",dr," in."];
Print["pitch p=1/n=",p," in."];
Print["joint axial length l=",l," in."];

l1=l2=1/2;
"axial length of the parts l1=l2=1/2"

"modulus of elasticity [Mpsi] E=30 for steel"
"modulus of elasticity [Mpsi] E=12 for cast iron"

Eb=30;
Ec=12;
Print["modulus of elasticity of the bolt Eb=",Eb," Mpsi"];
Print["modulus of elasticity of the part Ec=",Ec," Mpsi"];

Print["-----------------------"];
"a. Bolt stiffness kb"
Print["-----------------------"];
```

```
Print["J.E.Shigley & C.R.Mischke, Mechanical Engineering Design,
    McGraw-Hill 1989"];
lt=1/2;
Print["length of threaded portion of the bolt lt=",lt];
ls=1/2;
Print["length of unthreaded portion of the bolt ls=",ls];
As=N[Pi] d^2/4.;
Print["major diameter area As=Pi d^2/4=",As];
At=0.7854(d-0.9743 p)^2;
Print["tensile stress area At=0.7854(d-0.9743 p)^2=",At];
ks=As Eb/ls;
Print["stiffness of the unthreaded portion ks=As Eb/ls=",ks," Mlb/in"];
kt=At Eb/lt;
Print["stiffness of the threaded portion kt=At Eb/lt=",kt," Mlb/in"];
kb=ks kt/(ks+kt);
Print["bolt stiffness kb=kt ks/(kt+ks)=",kb," Mlb/in"];


Print["-----------------------"];
Print["Stiffness of the parts kc"];
Print["-----------------------"];


Print["b. J.E.Shigley & C.R.Mischke, Mechanical Engineering Design,
    McGraw-Hill 1989"];

k1=0.577 N[Pi] Ec d/(2 Log[5 (0.577 l1+0.5 d)/(0.577 l1+2.5 d)]);
Print["k1=0.577 Pi Ec d/(2 Log[5 (0.577 l1 +0.5 d)/(0.577 l1 +2.5 d)])"]
Print["stiffness of the cast iron part k1=",k1," Mlb/in"];

k2=0.577 N[Pi] Eb d/(2 Log[5 (0.577 l2+0.5 d)/(0.577 l2+2.5 d)]);
Print["k2=0.577 Pi Eb d/(2 Log[5 (0.577 l2 +0.5 d)/(0.577 l2 +2.5 d)])"]
Print["stiffness of the steel part k2=",k2," Mlb/in"];
kc=k1 k2/(k1+k2);
Print["resulting stiffness of clamped parts: kc=k1 k2/(k1+k2)=",kc," Mlb/in"];
Print["c. J.Wileman et al,1990"];
"constants A=0.78715; B=0.62873 for steel"
"constants A=0.77871; B=0.61616 for cast iron"
"joint stiffness: kc = Ec d A e^(B d/L)"

"parts of the joint: cast iron "
A1=0.77871;
B1=0.61616;
kw1=Ec d A1 E^(B1 d/l1);
Print["stiffness k1=Ec d1 A1 E^(B1 d1/l1)=",kw1," Mlb/in"];

"parts of the joint: steel "
A2=0.78715;
B2=0.62873;
```

```
kw2=Eb d A2 E^(B2 d/l2);
Print["joint stiffness: k2=Eb d A2 E^(B2 d/l2)=",kw2," Mlb/in"];

kwc=kw1 kw2/(kw1+kw2);
Print["resulting stiffness of clamped parts: kc=k1 k2/(k1+k2)=",kwc,"
  Mlb/in"];
```

coarse thread

major thread diameter d=0.625 in.

minor thread diameter dr=0.5135 in.

pitch p=1/n=0.0909091 in.

joint axial length l=1.5 in.

axial length of the parts l1=l2=1/2

modulus of elasticity [Mpsi] E=30 for steel

modulus of elasticity [Mpsi] E=12 for cast iron

modulus of elasticity of the bolt Eb=30 Mpsi

modulus of elasticity of the part Ec=12 Mpsi

----

a. Bolt stiffness kb

J.E.Shigley & C.R.Mischke, Mechanical Engineering Design,
  McGraw-Hill 1989

length of threaded portion of the bolt lt=0.75

length of unthreaded portion of the bolt ls=0.75

major diameter area As=Pi d^2/4=0.306796

tensile stress area At=0.7854(d-0.9743 p)^2=0.226002

stiffness of the unthreaded portion ks=As Eb/ls=12.2718 Mlb/in

stiffness of the threaded portion kt=At Eb/lt=9.04009 Mlb/in

bolt stiffness kb=kt ks/(kt+ks)=5.20547 Mlb/in

Stiffness of the parts kc

----

b. J.E.Shigley & C.R.Mischke, Mechanical Engineering Design,
  McGraw-Hill 1989

---

k1=0.577 Pi Ec d/(2 Log[5 (0.577 l1 +0.5 d)/(0.577 l1 +2.5 d)])

stiffness of the cast iron part k1=10.8826 Mlb/in

k2=0.577 Pi Eb d/(2 Log[5 (0.577 l2 +0.5 d)/(0.577 l2 +2.5 d)])

stiffness of the steel part k2=27.2065 Mlb/in

resulting stiffness of clamped parts: kc=k1 k2/(k1+k2)=7.77327 Mlb/in

c. J.Wileman et al,1990

constants A=0.78715; B=0.62873 for steel

constants A=0.77871; B=0.61616 for cast iron

joint stiffness: kc = Ec d A e^(B d/L)

parts of the joint: cast iron

stiffness k1=Ec d1 A1 E^(B1 d1/l1)=9.75962 Mlb/in

parts of the joint: steel

joint stiffness: k2=Eb d A2 E^(B2 d/l2)=24.9232 Mlb/in

resulting stiffness of clamped parts: kc=k1 k2/(k1+k2)=7.0133 Mlb/in

II.4 Rolling Bearings

II.4.1 Generalities

A bearing is a connector that permits the connected parts to rotate or to move relative to one another. Often one of the parts is fixed, and the bearing acts as a support for the moving part. Most bearings support rotating shafts against either transverse (radial) or thrust (axial) forces. To minimize friction, the contacting surfaces in a bearing may be partially or completely separated by a film of liquid (usually oil) or gas. These are *sliding bearings*, and the part of the shaft that turns in the bearing is the journal. Under certain combinations of force, speed, fluid viscosity, and bearing geometry, a fluid film forms and separates the contacting surfaces in a sliding bearing, and this is known as a *hydrodynamic* film. The *hydrostatic film* is the oil film that can be developed with a separate pumping unit that supplies pressurized oil.

The surfaces in a bearing can also be separated by balls, rollers, or needles; these are known as *rolling bearings*. Because shaft speed is required for the development of a hydrodynamic film, the starting friction in hydrodynamic bearings is higher than in rolling bearings. To minimize friction for metal-to-metal contact, in hydrodynamic bearings, materials with low coefficient of friction have been developed (bronze alloys and babbitt metal).

The principal advantage of the rolling bearings is the ability to operate at friction levels considerably lower at start-up, the friction coefficient having the values $\mu = 0.001 - 0.003$. Other advantages over bearings with sliding contact are: accurate shaft alignment for long periods of time, easy lubrication, little attention, easy replacement in case of failure, and heavy momentary overloads without failure.

The rolling bearings have the following disadvantages: design and processing of the shaft and house are more complicated, higher cost, more noise for higher speeds, and lower resistance to impact forces.

II.4.2 Classification

The important parts of rolling bearings are illustrated in Figure II.4.1: these include the outer ring, inner ring, rolling element, and separator (retainer). The role of the separator is to maintain an equal distance between the rolling elements. The races are the outer ring or the inner ring of a bearing. The raceway is the path of the rolling element on either ring or the bearing.

Rolling bearings can be classified using the following criteria (Fig. II.4.2):

- the rolling element shape: ball bearings [Fig. II.4.2(a)–(f)], roller bearings [cylinder, Fig. II.4.2(g) and (h), cone, Fig. II.4.2(i), barrel, Fig. II.4.2(j)], and needle bearings [Fig. II.4.2(k)];
- the direction of the principal force: radial bearings [Fig. II.4.2(a)(b)(g)(h)], thrust bearings [Fig. II.4.2(d)(e)], radial-thrust bearings [Fig. II.4.2(c)(i)], or thrust-radial bearings [Fig. II.4.2(f)];
- the number of rolling bearing rows: rolling bearings with one row [Fig. II.4.2(a)(d)(g)(k)], with two rows [Fig. II.4.2(b)(e)(h)].

The radial bearing is primarily designed to support a force perpendicular to the shaft axis. The thrust bearing is primarily designed to support a force parallel to the shaft axis.

Single row rolling bearings are manufactured to take radial forces and some thrust forces. The angular contact bearings provide a greater thrust capacity. Double row bearings are made to carry heavier radial and thrust forces. The single row bearings will withstand a small misalignment or deflection of the shaft. The self-aligning bearings [Fig. II.4.2(f)] are used for severe misalignments and deflections of the shaft.

Cylinder roller bearings provide a greater force than ball bearings of the same size because of the greater contact area. This type of bearing will not take thrust forces. Tapered (cone) roller bearings combine the advantages of ball and cylinder roller bearings, because they can take either radial or thrust forces, and they have high force capacity.

Needle bearings are used where the radial space is limited, and when the separators are used they have high force capacity. In many practical cases they are used without the separators.

II.4.3 Geometry

Figure II.4.3 shows a ball bearing with the *pitch diameter* given by

$$d_m \approx \frac{d_0 + d}{2},$$

(II.4.1)

where d_0 is the outer diameter of the ball bearing and d is the bore.

Exactly, the *pitch diameter* can be calculated as

$$d_m = \frac{D_i + D_e}{2},$$

(II.4.2)

where D_i is the race diameter of the inner ring and D_e is the race diameter of the outer ring.

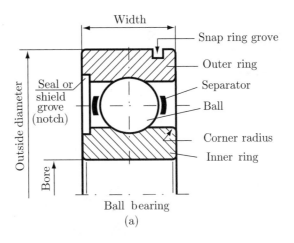

Ball bearing
(a)

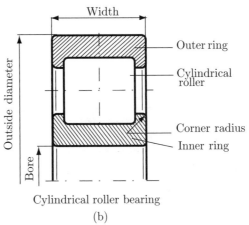

Cylindrical roller bearing
(b)

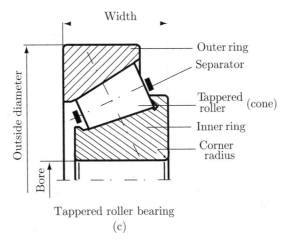

Tappered roller bearing
(c)

FIGURE II.4.1 *Rolling bearing nomenclature.*

BALL BEARINGS

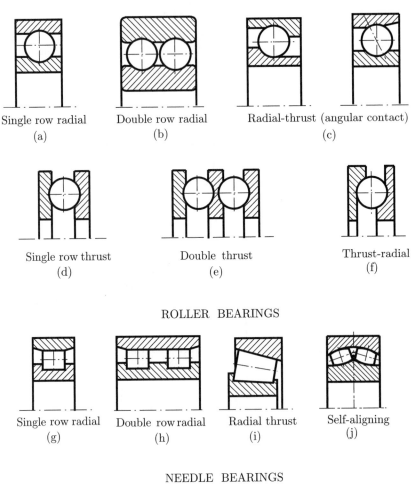

Single row radial
(a)

Double row radial
(b)

Radial-thrust (angular contact)
(c)

Single row thrust
(d)

Double thrust
(e)

Thrust-radial
(f)

ROLLER BEARINGS

Single row radial
(g)

Double row radial
(h)

Radial thrust
(i)

Self-aligning
(j)

NEEDLE BEARINGS

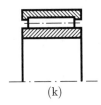

(k)

FIGURE II.4.2 *Rolling bearing classification.*

In general the ball bearings are manufactured with a clearance between the balls and the raceways. The clearance measured in the radial plane is the *diametral clearance*, s_d, and is computed with the relation (Fig. II.4.3):

$$s_d = D_e - D_i - 2D, \qquad (\text{II.4.3})$$

where D is the ball diameter.

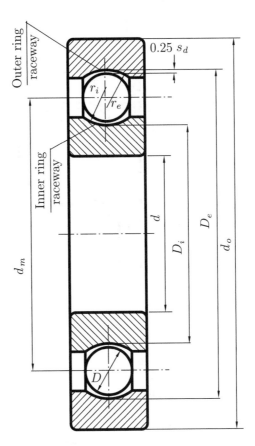

FIGURE II.4.3 *Ball-bearing geometry.*

Because a radial ball bearing has a diametral clearance in the no-load state, the bearing also has an axial clearance. Removal of this axial clearance causes the ball raceway contact to assume an oblique angle with the radial plane. Angular contact ball bearings are designed to operate under thrust force and the clearance built into the unloaded bearing along with the raceway groove curvatures determines the bearing-free contact angle. Due to the diametral clearance for a radial ball bearing there is what is known as *free endplay*, s_a, (Fig. II.4.4). In Figure II.4.4 the center of the outer ring raceway circle is O_e, the center of the inner ring raceway circle is O_i.

The distance between the centers O_e and O_i is

$$A = r_e + r_i - D, \tag{II.4.4}$$

where r_e is the radius of the outer ring raceway and r_i is the radius of the inner ring raceway.

If the raceway groove curvature radius is $r = f D$, where f is a dimensionless coefficient, then

$$A = (f_e + f_i - 1) D = B D, \tag{II.4.5}$$

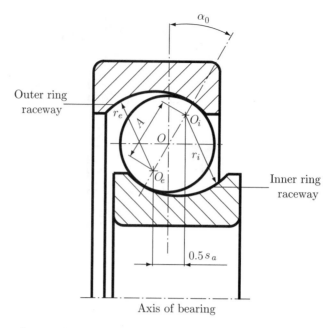

FIGURE II.4.4 *Clearance for radial ball bearing.*

where $B = f_e + f_i - 1$ is defined as the *total curvature of the bearing*. In the above formula $r_e = f_e D$ and $r_i = f_i D$, where f_e and f_i are dimensionless coefficients.

The *free contact angle*, α_0, is the angle made by the line passing through the points of contact of the ball and both raceways and a plane perpendicular to the bearing axis of rotation (Fig. II.4.4). The magnitude of the free contact angle can be written as

$$\sin \alpha_0 = 0.5 \, s_a/A. \tag{II.4.6}$$

The diametral clearance can allow the ball bearing to misalign slightly under no load. The *free angle of misalignment*, θ, is defined as the maximum angle through which the axis of the inner ring can be rotated with respect to the axis of the outer ring before stressing bearing components:

$$\theta = \theta_i + \theta_e, \tag{II.4.7}$$

where θ_i [Fig. II.4.5(a)] is the misalignment angle for the inner ring

$$\cos \theta_i = 1 - \frac{s_d \left[(2f_i - 1) D - s_d/4 \right]}{2d_m \left[d_m + (2f_i - 1) D + s_d/2 \right]}, \tag{II.4.8}$$

and θ_e, [Fig. II.4.5(b)] is the misalignment angle for the outer ring

$$\cos \theta_e = 1 - \frac{s_d \left[(2f_e - 1) D - s_d/4 \right]}{2d_m \left[d_m - (2f_e - 1) D + s_d/2 \right]}. \tag{II.4.9}$$

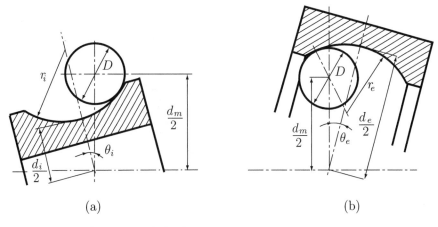

FIGURE II.4.5 *Misalignment angle for the rings.*

With the following trigonometric identity:

$$\cos\theta_i + \cos\theta_e = 2\cos\left[(\theta_i + \theta_e)/2\right]\cos\left[(\theta_i - \theta_e)/2\right], \tag{II.4.10}$$

and with the approximation $\theta_i - \theta_e \approx 0$, the free angle of misalignment becomes

$$\theta = 2\arccos\left[(\cos\theta_i + \cos\theta_e)/2\right], \tag{II.4.11}$$

or

$$\theta = 2\arccos\left\{1 - \frac{s_d}{4d_m}\left[\frac{(2f_i - 1)D - s_d/4}{d_m + (2f_i - 1)D + s_d/2} + \frac{(2f_e - 1)D - s_d/4}{d_m - (2f_e - 1)D + s_d/2}\right]\right\}. \tag{II.4.12}$$

II.4.4 Static Loading

In Figure II.4.6(a) a single row radial thrust (angular contact) ball bearing is shown. The *contact angle*, α, is the angle of the axis of contact between balls and races. For a single row radial ball bearing the angle α is zero. If F_r is the radial force applied to the ball, then the normal force to be supported by the ball is

$$F = \frac{F_r}{\cos\alpha}, \tag{II.4.13}$$

and the axial force, F_a (or F_t), is

$$F_a = F\sin\alpha. \tag{II.4.14}$$

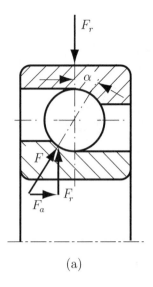

(a)

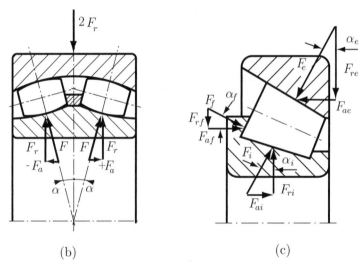

(b)   (c)

FIGURE II.4.6 *Static loading for rolling bearing.*

For self-aligning roller bearings [Fig. II.4.6(b)] the above relations are valid for each roller, and the total axial force is zero.

For taper roller bearings [Fig. II.4.6(c)] there are three contact angles: α_i the contact angle for the inner ring, α_e the contact angle for the outer ring, and α_f the contact angle for the frontal face.

The normal and axial forces for the inner ring are

$$F_i = \frac{F_{ri}}{\cos \alpha_i} \quad \text{and} \quad F_{ai} = F_{ri} \tan \alpha_i, \tag{II.4.15}$$

where F_{ri} is the radial force acting on the inner ring. The normal and axial forces for the outer ring are

$$F_e = \frac{F_{re}}{\cos \alpha_e} \quad \text{and} \quad F_{ae} = F_{re} \tan \alpha_e, \qquad \text{(II.4.16)}$$

where F_{re} is the radial force acting on the outer ring. The normal and axial forces for the frontal face are

$$F_f = \frac{F_{rf}}{\cos \alpha_f} \quad \text{and} \quad F_{af} = F_{rf} \tan \alpha_f, \qquad \text{(II.4.17)}$$

where F_{rf} is the radial force acting on the frontal face.

The equilibrium equations for radial and axial directions are

$$F_{ri} - F_{rf} - F_{re} = 0 \quad \text{or} \quad F_{ri} - F_f \cos \alpha_f - F_e \cos \alpha_e = 0, \qquad \text{(II.4.18)}$$
$$F_{ai} + F_{af} - F_{ae} = 0 \quad \text{or} \quad F_{ri} \tan \alpha_i + F_f \sin \alpha_f - F_e \sin \alpha_e = 0. \qquad \text{(II.4.19)}$$

From Eqs. (II.4.18) and (II.4.19) the forces F_e and F_f are obtained:

$$F_e = \frac{F_{ri} \left(\sin \alpha_f + \tan \alpha_i \cos \alpha_f \right)}{\sin \left(\alpha_e + \alpha_f \right)}, \qquad \text{(II.4.20)}$$

$$F_f = \frac{F_{ri} \left(\sin \alpha_e - \tan \alpha_i \cos \alpha_e \right)}{\sin \left(\alpha_e + \alpha_f \right)}. \qquad \text{(II.4.21)}$$

II.4.5 Standard Dimensions

The Annular Bearing Engineers Committee (ABEC) of the Anti-Friction Bearing Manufacturers Association (AFBMA) has established four primary grades of precision, designated ABEC 1, 5, 7, and 9 for ball bearings. The standard grade is ABEC 1 and is adequate for most normal applications. The other grades have progressively finer tolerances. The AFBMA Roller Bearing Engineers Committee has established RBEC standards 1 and 5 for cylindrical roller bearings.

The bearing manufacturers have established standard dimensions (Fig. II.4.7 and Table II.4.1) for ball and straight roller bearings in the metric sizes, which define the bearing bore d, the outside diameter d_0, the width w, the fillet sizes on the shaft and housing shoulders r, the shaft diameter d_S, and the housing diameter d_H.

For a given bore, there is an assortment of widths and outside diameters. Furthermore, the outside diameters selected are such that, for a particular outside diameter, one can usually find a variety of bearings having different bores. That is why the bearings are made in various proportions for different series (Fig. II.4.8): extra-extra-light series (LL00), extra-light series (L00), light series (200), and medium series (300).

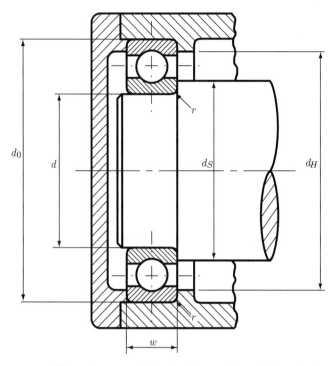

r - maximum fillet radius on a shaft and in housing that will clear the bearing corner radius

Shaft and housing shoulder dimensions

FIGURE II.4.7 *Standard dimensions for rolling bearing, shaft, and housing shoulder.*

TABLE II.4.1 Bearing Dimensions

Bearing basic number	d [mm]	Ball bearings						Roller bearings				
		d_0 [mm]	w [mm]	r [mm]	d_S [mm]	d_H [mm]		d_0 [mm]	w [mm]	r [mm]	d_S [mm]	d_H [mm]
L00	10	26	8	0.30	12.7	23.4		—	—	—	—	—
200	10	30	9	0.64	13.8	26.7		—	—	—	—	—
300	10	35	11	0.64	14.8	31.2		—	—	—	—	—
L01	12	28	8	0.30	14.5	25.4		—	—	—	—	—
201	12	32	10	0.64	16.2	28.4		—	—	—	—	—
301	12	37	12	1.02	17.7	32.0		—	—	—	—	—
L02	15	32	9	0.30	17.5	29.2		—	—	—	—	—
202	15	35	11	0.64	19.0	31.2		—	—	—	—	—
302	15	42	13	1.02	21.2	36.6		—	—	—	—	—
L03	17	35	10	0.30	19.8	32.3		35	10	0.64	20.8	32.0
203	17	40	12	0.64	22.4	34.8		40	12	0.64	20.8	36.3
303	17	47	14	1.02	23.6	41.1		47	14	1.02	22.9	41.4
L04	20	42	12	0.64	23.9	38.1		42	12	0.64	24.4	36.8
204	20	47	14	1.02	25.9	41.7		47	14	1.02	25.9	42.7
304	20	52	15	1.02	27.7	45.2		52	15	1.02	25.9	46.2

TABLE II.4.1 Bearing Dimensions (*continued*)

Bearing basic number	d [mm]	Ball bearings					Roller bearings				
		d_0 [mm]	w [mm]	r [mm]	d_S [mm]	d_H [mm]	d_0 [mm]	w [mm]	r [mm]	d_S [mm]	d_H [mm]
L05	25	47	12	0.64	29.0	42.9	47	12	0.64	29.2	43.4
205	25	52	15	1.02	30.5	46.7	52	15	1.02	30.5	47.0
305	25	62	17	1.02	33.0	54.9	62	17	1.02	31.5	55.9
L06	30	55	13	1.02	34.8	49.3	47	9	0.38	33.3	43.9
206	30	62	16	1.02	36.8	55.4	62	16	1.02	36.1	56.4
306	30	72	19	1.02	38.4	64.8	72	19	1.52	37.8	64.0
L07	35	62	14	1.02	40.1	56.1	55	10	0.64	39.4	50.8
207	35	72	17	1.02	42.4	65.0	72	17	1.02	41.7	65.3
307	35	80	21	1.52	45.2	70.4	80	21	1.52	43.7	71.4
L08	40	68	15	1.02	45.2	62.0	68	15	1.02	45.7	62.7
208	40	80	18	1.02	48.0	72.4	80	18	1.52	47.2	72.9
308	40	90	23	1.52	50.8	80.0	90	23	1.52	49.0	81.3
L09	45	75	16	1.02	50.8	68.6	75	16	1.02	50.8	69.3
209	45	85	19	1.02	52.8	77.5	85	19	1.52	52.8	78.2
309	45	100	25	1.52	57.2	88.9	100	25	2.03	55.9	90.4
L10	50	80	16	1.02	55.6	73.7	72	12	0.64	54.1	68.1
210	50	90	20	1.02	57.7	82.3	90	20	1.52	57.7	82.8
310	50	110	27	2.03	64.3	96.5	110	27	2.03	61.0	99.1
L11	55	90	18	1.02	61.7	83.1	90	18	1.52	62.0	83.6
211	55	100	21	1.52	65.0	90.2	100	21	2.03	64.0	91.4
311	55	120	29	2.03	69.8	106.2	120	29	2.03	66.5	108.7
L12	60	95	18	1.02	66.8	87.9	95	18	1.52	67.1	88.6
212	60	110	22	1.52	70.6	99.3	110	22	2.03	69.3	101.3
312	60	130	31	2.03	75.4	115.6	130	31	2.54	72.9	117.9
L13	65	100	18	1.02	71.9	92.7	100	18	1.52	72.1	93.7
213	65	120	23	1.52	76.5	108.7	120	23	2.54	77.0	110.0
313	65	140	33	2.03	81.3	125.0	140	33	2.54	78.7	127.0
L14	70	110	20	1.02	77.7	102.1	110	20	—	—	—
214	70	125	24	1.52	81.0	114.0	125	24	2.54	81.8	115.6
314	70	150	35	2.03	86.9	134.4	150	35	3.18	84.3	135.6
L15	75	115	20	1.02	82.3	107.2	115	20	—	—	—
215	75	130	25	1.52	86.1	118.9	130	25	2.54	85.6	120.1
315	75	160	37	2.03	92.7	143.8	160	37	3.18	90.4	145.8
L16	80	125	22	1.02	88.1	116.3	125	22	2.03	88.4	117.6
216	80	140	26	2.03	93.2	126.7	140	26	2.54	91.2	129.3
316	80	170	39	2.03	98.6	152.9	170	39	3.18	96.0	154.4
L17	85	130	22	1.02	93.2	121.4	130	22	2.03	93.5	122.7
217	85	150	28	2.03	99.1	135.6	150	28	3.18	98.0	139.2
317	85	180	41	2.54	105.7	160.8	180	41	3.96	102.9	164.3

TABLE II.4.1 Bearing Dimensions (*continued*)

Bearing basic number	d [mm]	Ball bearings					Roller bearings				
		d_0 [mm]	w [mm]	r [mm]	d_S [mm]	d_H [mm]	d_0 [mm]	w [mm]	r [mm]	d_S [mm]	d_H [mm]
L18	85	140	24	1.02	99.6	129.0	140	24	—	—	—
218	85	160	30	2.03	104.4	145.5	160	30	3.18	103.1	147.6
318	85	190	43	2.54	111.3	170.2	190	43	3.96	108.2	172.7
L19	90	145	24	1.52	104.4	134.1	145	24	—	—	—
219	90	170	32	2.03	110.2	154.9	170	32	3.18	109.0	157.0
319	90	200	45	2.54	117.3	179.3	200	45	3.96	115.1	181.9
L20	95	150	24	1.52	109.5	139.2	150	24	2.54	109.5	141.7
220	95	180	34	2.03	116.1	164.1	180	34	3.96	116.1	167.1
320	95	215	47	2.54	122.9	194.1	215	47	4.75	122.4	194.6
L21	100	160	26	2.03	116.1	146.8	160	26	—	—	—
221	100	190	36	2.03	121.9	173.5	190	36	3.96	121.4	175.3
321	100	225	49	2.54	128.8	203.5	225	49	4.75	128.0	203.5
L22	105	170	28	2.03	122.7	156.5	170	28	2.54	121.9	159.3
222	105	200	38	2.03	127.8	182.6	200	38	3.96	127.3	183.9
322	105	240	50	2.54	134.4	218.2	240	50	4.75	135.9	217.2
L24	120	180	28	2.03	132.6	166.6	180	28	—	—	—
224	120	215	40	2.03	138.2	197.1	215	40	4.75	139.2	198.9
324	120	—	—	—	—	—	260	55	6.35	147.8	235.2
L26	130	200	33	2.03	143.8	185.4	200	33	3.18	143.0	188.2
226	130	230	40	2.54	149.9	210.1	230	40	4.75	149.1	213.9
326	130	280	58	3.05	160.0	253.0	280	58	6.35	160.3	254.5
L28	140	210	33	2.03	153.7	195.3	210	33	—	—	—
228	140	250	42	2.54	161.5	228.6	250	42	4.75	161.5	232.4
328	140	—	—	—	—	—	300	62	7.92	172.0	271.3
L30	150	225	35	2.03	164.3	209.8	225	35	3.96	164.3	212.3
230	150	270	45	2.54	173.0	247.6	270	45	6.35	174.2	251.0
L32	160	240	38	2.03	175.8	223.0	240	38	—	—	—
232	160	—	—	—	—	—	290	48	6.35	185.7	269.5
L36	180	280	46	2.03	196.8	261.6	280	46	4.75	199.6	262.9
236	180	—	—	—	—	—	320	52	6.35	207.5	298.2
L40	200	—	—	—	—	—	310	51	—	—	—
240	200	—	—	—	—	—	360	58	7.92	232.4	334.5
L44	220	—	—	—	—	—	340	56	—	—	—
244	220	—	—	—	—	—	400	65	9.52	256.0	372.1
L48	240	—	—	—	—	—	360	56	—	—	—
248	240	—	—	—	—	—	440	72	9.52	279.4	408.4

Source: R. C. Juvinall and K. M. Marshek, *Fundamentals of Machine Component Design*, New York, John Wiley & Sons, 1991. Reprinted with permission of John Wiley & Sons, Inc.

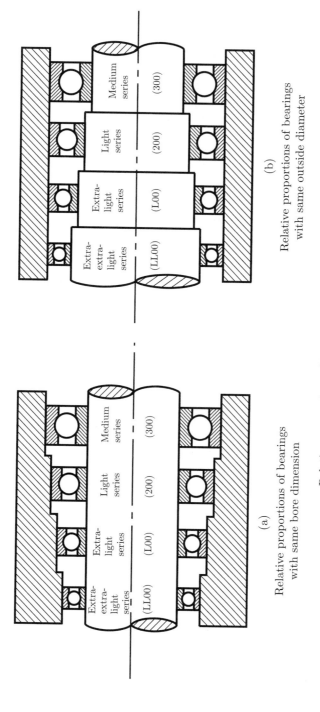

(a)

Relative proportions of bearings
with same bore dimension

(b)

Relative proportions of bearings
with same outside diameter

Relative proportions of bearings of different series

Source: R. C. Juvinall, K. M. Marshek, Fundamentals of machine component design,
John Wiley & Sons, 1991, New York

FIGURE II.4.8 *Different series for rolling bearing. Reprinted with permission of John Wiley & Sons, Inc.*

II.4.6 Bearing Selection

Bearing manufacturers' catalogues identify bearings by number, give complete dimensional information, list rated load capacities, and furnish details concerning mounting, lubrication, and operation. Lubrication (fine oil mist or spray) is important in high-speed bearing applications. For ball bearings nonmetallic separators permit highest speeds.

The size of bearing selected for an application is usually influenced by the size of shaft required (for strength and rigidity considerations) and by the available space. In addition, the bearing must have a high load rating to provide a good combination of life and reliability.

Juvinall and Marshek [7] proposed the following expressions for bearing selection.

Life Requirement

The *life* of an individual ball or roller bearing is the number of revolutions (or hours at some constant speed) that the bearings run before the first evidence of fatigue develops in the material of either the rings or of any of the rolling elements. Bearing applications usually require lives different from that used for the catalogue rating. Palmgren determined that ball bearing life varies inversely with approximately the third power of the force. Later studies have indicated that this exponent ranges between 3 and 4 for various rolling-element bearings. Many manufacturers retain Palmgren's exponent of 3 for ball bearings and use 10/3 for roller bearings. Following the recommendation of other manufacturers, the exponent 10/3 will be used for both bearing types. Thus, the *life required* by the application is

$$L = L_R \, (C/F_r)^{10/3} \,, \tag{II.4.22}$$

where C is the *rated capacity*, from Table II.4.2, L_R is the life corresponding to rated capacity (i.e., $L_R = 9 \times 10^7$ revolutions), and F_r is radial force involved in the application.

The values of the rated capacity in Table II.4.2 correspond to a constant radial load that 90 of a group of identical bearings can endure a rating life of $L_R = 9 \times 10^7$ revolutions without surface fatigue failure.

The required value of the rated capacity for the application is

$$C_{req} = F_r \, (L/L_R)^{3/10} \,. \tag{II.4.23}$$

For a group of apparently identical bearings the *rating life*, L_R, is the life in revolutions (at a given constant speed and force) that 90% of the group tested bearings will exceed before the first evidence of fatigue develops. For example, the Timken Company rates the bearings for 3000 h at a speed of 500 rpm operation. The corresponding life is

$$L = (3000 \, \text{h}) \left(\frac{60 \, \text{min}}{\text{h}} \right) \left(\frac{500 \, \text{rev}}{\text{min}} \right) = 9 \left(10^7 \right) \, \text{rev}.$$

Different manufacturers' catalogues use different values of L_R (some use 10^6 revolutions).

Reliability Requirement

Tests show that the *median life* of rolling-element bearings is about 5 times the standard 10% failure fatigue life. The *standard life* is commonly designated as the L_{10} life (sometimes

TABLE II.4.2 Bearing Rated Capacities, C, for 90×10^6 Revolution Life (L_R) with 90% Reliability

d [mm]	Radial ball, $\alpha = 0°$			Angular ball, $\alpha = 25°$			Roller		
	L00 xlt [kN]	200 lt [kN]	300 med [kN]	L00 xlt [kN]	200 lt [kN]	300 med [kN]	1000 xlt [kN]	1200 lt [kN]	1300 med [kN]
10	1.02	1.42	1.90	1.02	1.10	1.88	—	—	—
12	1.12	1.42	2.46	1.10	1.54	2.05	—	—	—
15	1.22	1.56	3.05	1.28	1.66	2.85	—	—	—
17	1.32	2.70	3.75	1.36	2.20	3.55	2.12	3.80	4.90
20	2.25	3.35	5.30	2.20	3.05	5.80	3.30	4.40	6.20
25	2.45	3.65	5.90	2.65	3.25	7.20	3.70	5.50	8.50
30	3.35	5.40	8.80	3.60	6.00	8.80	—	8.30	10.0
35	4.20	8.50	10.6	4.75	8.20	11.0	—	9.30	13.1
40	4.50	9.40	12.6	4.95	9.90	13.2	7.20	11.1	16.5
45	5.80	9.10	14.8	6.30	10.4	16.4	7.40	12.2	20.9
50	6.10	9.70	15.8	6.60	11.0	19.2	—	12.5	24.5
55	8.20	12.0	18.0	9.00	13.6	21.5	11.3	14.9	27.1
60	8.70	13.6	20.0	9.70	16.4	24.0	12.0	18.9	32.5
65	9.10	16.0	22.0	10.2	19.2	26.5	12.2	21.1	38.3
70	11.6	17.0	24.5	13.4	19.2	29.5	—	23.6	44.0
75	12.2	17.0	25.5	13.8	20.0	32.5	—	23.6	45.4
80	14.2	18.4	28.0	16.6	22.5	35.5	17.3	26.2	51.6
85	15.0	22.5	30.0	17.2	26.5	38.5	18.0	30.7	55.2
90	17.2	25.0	32.5	20.0	28.0	41.5	—	37.4	65.8
95	18.0	27.5	38.0	21.0	31.0	45.5	—	44.0	65.8
100	18.0	30.5	40.5	21.5	34.5	—	20.9	48.0	72.9
105	21.0	32.0	43.5	24.5	37.5	—	—	49.8	84.5
110	23.5	35.0	46.0	27.5	41.0	55.0	29.4	54.3	85.4
120	24.5	37.5	—	28.5	44.5	—	—	61.4	100.1
130	29.5	41.0	—	33.5	48.0	71.0	48.9	69.4	120.1
140	30.5	47.5	—	35.0	56.0	—	—	77.4	131.2
150	34.5	—	—	39.0	62.0	—	58.7	83.6	—
160	—	—	—	—	—	—	—	113.4	—
180	47.0	—	—	54.0	—	—	97.9	140.1	—
200	—	—	—	—	—	—	—	162.4	—
220	—	—	—	—	—	—	—	211.3	—
240	—	—	—	—	—	—	—	258.0	—

Source: New Departure-Hyatt Bearing Division, General Motors Corporation.

as the B_{10} life) and this life corresponds to 10% failures. It means that this is the life for which 90% have not failed, and corresponds to 90% reliability ($r = 90\%$). Using the general Weibull equation together with extensive experimental data, a life adjustment *reliability factor*, K_r, is recommended. The life adjustment reliability factor K_r is plotted in Figure II.4.9. This factor is applicable to both ball and roller bearings. The rated bearing life for any given reliability (greater than 90%) is thus the product $K_r L_R$. Incorporating this

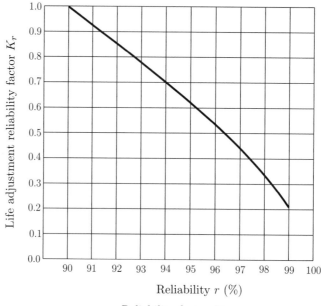

Reliability factor K_r

Source: R. C. Juvinall, K. M. Marshek, Fundamentals of machine
component design, John Wiley & Sons, 1991, New York

FIGURE II.4.9 *Life adjustment reliability factor. Reprinted with permission of John Wiley & Sons, Inc.*

factor into Eqs. (II.4.22) and (II.4.23) gives

$$L = K_r L_R (C/F_r)^{3.33},$$

$$C_{req} = F_r \left(\frac{L}{K_r L_R} \right)^{0.3}. \qquad (II.4.24)$$

Influence of Axial Force

For ball bearings (load angle $\alpha = 0°$) any combination of radial force (F_r) and thrust force (F_a) results in approximately the same life as does a pure *radial equivalent force*, F_e, calculated as

- for $0.00 < F_a/Fr < 0.35 \implies F_e = F_r$;
- for $0.35 < F_a/F_r < 10.0 \implies F_e = F_r [1 + 1.115(F_a/F_r - 0.35)]$;
- for $F_a/Fr > 10.0 \implies F_e = 1.176 F_a$. Standard values of load angle α for angular ball bearings are $15°$, $25°$, and $35°$. Only the $25°$ angular ball bearings will be discussed here. The radial equivalent force, F_e, for angular ball bearings with $\alpha = 25°$ is
- for $0.00 < F_a/Fr < 0.68 \implies F_e = F_r$;
- for $0.68 < F_a/F_r < 10.0 \implies F_e = F_r [1 + 0.87(F_a/F_r - 0.68)]$;
- for $F_a/Fr > 10.0 \implies F_e = 0.911 F_a$.

TABLE II.4.3 Application Factors K_a

Type of application	Ball bearing	Roller bearing
Uniform force, no impact	1.0	1.0
Gearing	1.0–1.3	1.0
Light impact	1.2–1.5	1.0–1.1
Moderate impact	1.5–2.0	1.1–1.5
Heavy impact	2.0–3.0	1.5–2.0

Source: R. C. Juvinall and K. M. Marshek, *Fundamentals of Machine Component Design*, New York, John Wiley & Sons, 1991. Reprinted with permission of John Wiley & Sons, Inc.

Shock Force

The standard bearing rated capacity is for the condition of uniform force without shock, which is a desirable condition. In many applications there are various degrees of shock loading. This has the effect of increasing the nominal force by an *application factor, K_a*. In Table II.4.3 some representative sample values of K_a are given. The force application factor in Table II.4.3 serves the same purpose as factors of safety.

Substituting F_e for F_r and adding K_a, Eq. (II.4.24) gives

$$L = K_r L_R \left(\frac{C}{K_a F_e} \right)^{3.33},$$

$$C_{req} = K_a F_e \left(\frac{L}{K_r L_R} \right)^{0.3}. \tag{II.4.25}$$

When more specific information is not available, Table II.4.4 can be used as a guide for the life of a bearing in industrial applications. Table II.4.4 contains recommendations

TABLE II.4.4 Representative Bearing Design Lives

Type of application	Design life $k\,h$ (thousands of hours)
Instruments and apparatus for infrequent use	0.1–0.5
Aircraft engines	0.5–2.0
Machines used intermittently, where service interruption is of minor importance	4–8
Machines used intermittently, where reliability is of great importance	8–14
Machines for 8-hour service, but not every day	14–20
Machines for 8-hour service, every working day	20–30
Machines for continuous 24-hour service	50–60
Machines for continuous 24-hour service where reliability is of extreme importance	100–200

Source: R. C. Juvinall and K. M. Marshek, *Fundamentals of Machine Component Design*, New York, John Wiley & Sons, 1991. Reprinted with permission of John Wiley & Sons, Inc.

on bearing life for some classes of machinery. The information has been accumulated by experience.

Shigley and Mischke [20] proposed the following expressions for bearing selection.

- The minimum basic load rating (load for which 90% of the bearings from a given group will survive 1 million revolutions) is defined as

$$C_s = P L^{1/a},$$

where P is the design load, L is the design life in millions of revolutions, and $a = 3$ for ball bearings and $a = 10/3$ for roller bearings. The equivalent radial load is

$$P_e = X V F_r + Y F_a,$$

where F_r is the radial force and F_a is the thrust force. For a rotating inner ring $V = 1$, and for a rotating outer ring $V = 1.2$. The AFBMA recommendations are based on the ratio of the thrust force F_a to the basic static load rating C_0, and a variable reference value,

$$e = 0.513 \left(\frac{F_a}{C_0} \right)^{0.236}.$$

- The static load rating C_0 is tabulated in bearing catalogues. The X and Y factors have the values

  - for $F_a/(V F_r) > e \implies X{=}0.56$ and $Y = 0.840 \left(\dfrac{F_a}{C_0} \right)^{-0.247}$,

  - for $F_a/(V F_r) \le e \implies X{=}1$ and $Y = 0$.

II.4.7 Examples

EXAMPLE II.4.1:

Select a light series (200) radial ball bearing for a machine for continuous 24-hour service. The machine rotates at the angular speed of 1000 rpm. The radial force is $F_r{=}1.5$ kN, and the thrust force is $F_a{=}1.8$ kN, with light impact (Fig. II.4.10).

Solution For $F_a/F_r = 1.8/1.5{=}1.2$, the equivalent radial force (for radial ball bearings with $0.35 < F_a/F_r < 10.0$) is

$$F_e = F_r \left[1 + 1.115(F_a/F_r - 0.35)\right] = 2.921 \text{ kN}.$$

From Table II.4.3 choose (conservatively) $K_a = 1.5$ for light impact. From Table II.4.4 choose (conservatively) 60 000 hour life. The life in revolutions is

$$L = 1000 \text{ rpm} \times 60\,000 \text{ h} \times 60 \text{ min/h} = 3600 \times 10^6 \text{ rev}.$$

EXAMPLE II.4.1: *Cont'd*

For standard 90% reliability ($K_r=1$, Fig. II.4.9), and for $L_R = 90 \times 10^6$ rev (for use with Table. II.4.2), Eq. (II.4.25) gives

$$C_{req} = K_a F_e \left(\frac{L}{K_r L_R} \right)^{0.3}$$

$$= (1.5)(2.921)(3\,600/90)^{0.3} = 13.253\,\text{kN}$$

From Table II.4.2 with 13.253 kN for 200 series $\Longrightarrow C = 13.6$ kN and $d = 60$ mm bore. From Table II.4.1 with 60 mm bore and 200 series the bearing number is 212.

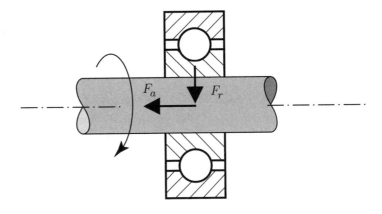

FIGURE II.4.10   *Radial ball bearing for Example II.4.1.*

EXAMPLE II.4.2:

A no. 305 radial contact ball bearing carries a radial load of 4 kN, 5 kN, and 6 kN for, respectively, 50%, 40%, and 10% of the time. The loads are uniform, so that $K_a = 1$. The bearing supports a shaft that rotates with 2000 rpm. Determine the B_{10} life and the median life of the bearing.

Solution   For no. 305 radial contact bearing from Table II.4.2, the rated capacity is $C = 5.9$ kN with $L_R = 90 \times 10^6$ and standard 90% reliability ($K_r = 1$). Equation (II.4.25) gives

$$L = K_r L_R \left(\frac{C}{K_a F_e} \right)^{3.33},$$

where $K_a = 1$ and $F_e = F_r$.

Continued

EXAMPLE II.4.2: *Cont'd*

The corresponding life is calculated from the previous relation:

$$N_1 = (1)(90 \times 10^6)\left[\frac{5.9}{(1)(4)}\right]^{3.33} = 3.283 \times 10^8 \text{ rev, for } F_r = 4 \text{ kN};$$

$$N_2 = (1)(90 \times 10^6)\left[\frac{5.9}{(1)(5)}\right]^{3.33} = 1.561 \times 10^8 \text{ rev, for } F_r = 5 \text{ kN};$$

$$N_3 = (1)(90 \times 10^6)\left[\frac{5.9}{(1)(6)}\right]^{3.33} = 8.510 \times 10^7 \text{ rev, for } F_r = 6 \text{ kN}.$$

The Miner rule is

$$\frac{n_1}{N_1} + \frac{n_2}{N_2} + \frac{n_3}{N_3} = 1, \tag{II.4.26}$$

where

$n_1 = (50\%) n X = (0.5)(2000 \text{ rpm}) X = 1000 X \text{ rev, for 50\% of the time;}$

$n_2 = (40\%) n X = (0.4)(2000 \text{ rpm}) X = 800 X \text{ rev, for 40\% of the time;}$

$n_3 = (10\%) n X = (0.1)(2000 \text{ rpm}) X = 200 X \text{ rev, for 10\% of the time.}$

The minutes of operation are $X = B_{10}$ and the shaft rotates with $n = 2000$ rpm. Equation (II.4.26) gives

$$\frac{1000 X}{3.283 \times 10^8} + \frac{800 X}{1.561 \times 10^8} + \frac{200 X}{8.510 \times 10^7} = 1,$$

or $X = B_{10} = 95072.6$ min ($= 1584.54$ hs).

The median life is approximately 5 times the B_{10} life or 475,363 min (7922.72 hs).

II.4.8 Problems

II.4.1 A no. 208 radial ball bearing has a 4000-hour B_{10} life at 1200 rpm. Find the bearing radial capacity.

II.4.2 A radial ball bearing has a given radial load F and a given life L. Find the radial load F_{new} if the life of the bearing is tripled, $L_{new} = 3L$.

II.4.3 A no. 207 angular ball bearing is selected to carry a radial load of 250 lb and a thrust load of 150 lb at 1000 rpm. Determine the bearing life B_{10} for steady loading.

II.4.4 A ball bearing can withstand a radial load of 4 kN and a thrust load of 6 kN at a speed of 600 rpm. The bearing is intended for an aircraft engine with heavy impacts. Select an angular ball bearing for this application.

II.4.5 The life of a bearing for 90% reliability is 12000 hours. Determine the lives of the bearing for, respectively, 60% and 98% reliability.

II.4.6 A ball bearing carries a radial load of 2.5 kN and a thrust load of 1.5 kN at 900 rpm. The application is considered to be light to moderate with respect to shock loading. The required life is 6000 hours with only 4% probability of failure. Select a suitable ball bearing: (a) radial ball bearing and (b) angular ball bearing.

II.4.7 Repeat the previous problem for 3 kN radial load, 2 kN thrust load at 1200 rpm, and 2% probability of failure.

II.4.8 A no. 211 radial ball bearing is intended for a continuous one-shift (8 h/d) operation at 1000 rpm. The radial load varies in such a way that 60% of the time the load is 5 kN and 40% of the time the load is 10 kN. The application factor is $K_a = 1.5$ (light to moderate impact). Estimate the B_{10} life and the median life of the radial ball bearing.

II.4.9 A no. 207 radial-contact ball bearing supports a shaft that rotates 1500 rpm. A radial load varies in such a way that 30%, 30%, and 40% of the time the load is 5, 2, and 10 kN. The loads are uniform, so that $K_a = 1$. Estimate the B_{10} life and the median life of the bearing.

II.4.10 The shaft shown in Figure II.4.11 rotates at 900 rpm and is supported by radial ball bearings at points A and B. The length dimension of the shaft is $l = 300$ mm. The radial force acting on the shaft at R is on the yz plane and has the magnitude $F_R = 900$ N. The angle of the force F_R with the z-axis is $\alpha = 45°$. The thrust load (along x-axis) on the bearing at A is $F_{aA} = 500$ N and at B is $F_{aB} = 500$ N. The bearings are subjected to steady loading with 98% reliability and 30 000 hours of life. Select radial ball bearings for A and B.

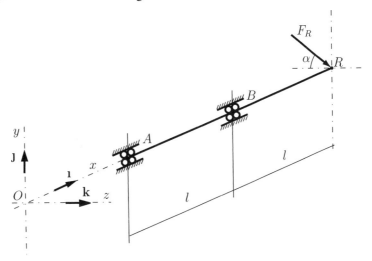

FIGURE II.4.11 *Sketch of the shaft for Problem II.4.10.*

II.4.11 Figure II.4.12 shows two bearings at A and B supporting a shaft that rotates at 1000 rpm. The loads acting on the shaft are at point P, $\mathbf{F}_P = F_{Py}\mathbf{J} + F_{Py}\mathbf{k} = -500\,\mathbf{J} + 600\,\mathbf{k}$ N, and at point R, $\mathbf{F}_R = F_{Ry}\mathbf{J} = -1000\,\mathbf{J}$ N. The loading is light to moderate impact. The length dimensions are $s = 100$ mm and $l = 200$ mm. The required life is 5000 hours with only 2% probability of failure. Select identical ball bearings for A and B.

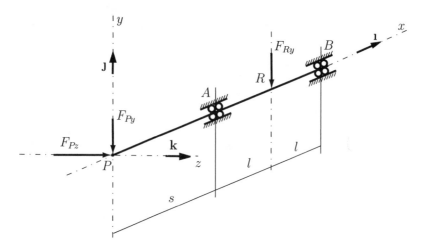

FIGURE II.4.12   *Sketch of the shaft for Problem II.4.11.*

II.4.12   Figure II.4.13 shows a countershaft a with two rigidly connected gears 1 and 2. The angular speed of the countershaft is 300 rpm. The force on the countershaft gear 1

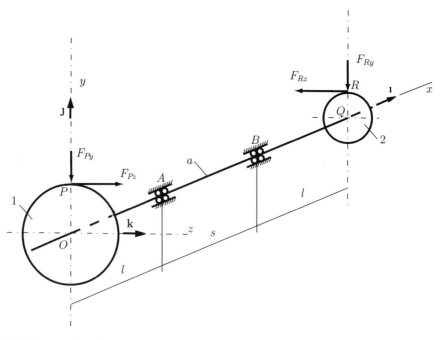

FIGURE II.4.13   *Sketch of the shaft for Problem II.4.12.*

at P is $\mathbf{F}_P = F_{Py}\mathbf{J} + F_{Pz}\mathbf{k} = -200\mathbf{J} + 500\mathbf{k}$ N, and the force on the gear 2 at R is $\mathbf{F}_R = F_{Ry}\mathbf{J} + F_{Rz}\mathbf{k} = -600\mathbf{J} - 1500\mathbf{k}$ N. The radius of the gear 1 is $OP = 0.15$ m and the radius of the gear 2 is $QR = 0.05$ m. The distance between the bearings is $s = 100$ mm and the other distance is $l = 25$ mm. The gear reducer is a part of an industrial machine intended for continuous one-shift (8h/d). Select identical extra-light series (L00) ball bearings for A and B.

II.5   Lubrication and Sliding Bearings

Lubrication reduces the friction, wear, and heating of machine parts in relative motion. A lubricant is a substance that is inserted between the moving parts.

II.5.1   Viscosity

Newton'S Law of Viscous Flow

A surface of area A is moving with the linear velocity V on a film of lubricant as shown in Figure II.5.1(a). The thickness of the lubricant is s and the deforming force acting on the film is F. The layers of the fluid in contact with the moving surface have the velocity $v = V$ and the layers of the fluid in contact with the fixed surface have the velocity $v = 0$.

Newton's law of viscous flow states that the shear stress τ in a fluid is proportional to the rate of change of the velocity v with respect to the distance y from the fixed surface,

$$\tau = \frac{F}{A} = \mu \frac{\partial v}{\partial y},$$

(II.5.1)

where μ is a constant, the *absolute viscosity*, or the *dynamic viscosity*. The derivate $\frac{\partial v}{\partial y}$ is the rate of change of velocity with distance and represents the rate of shear, or the velocity gradient. Thicker oils have a higher viscosity value causing relatively higher shear stesses at the same shear rate. For a constant velocity gradient Eq. (II.5.1) can be written as [Fig. II.5.1(b)]:

$$\tau = \mu \frac{V}{s}.$$

(II.5.2)

The unit of viscosity μ, for U.S. Customary units, is pound–second per square inch, (lb–s/in^2), or reyn (from Osborne Reynolds).

In SI units the viscosity is expressed as newton-seconds per square meter, (N–s/m^2), or pascal–second (Pa·s).

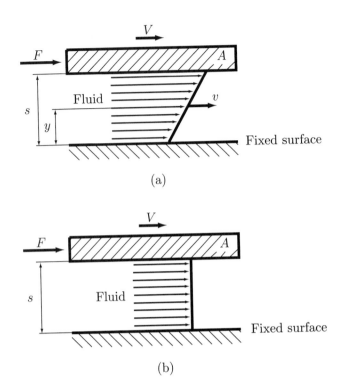

(a)

(b)

FIGURE II.5.1 *Slider bearing.*

The conversion factor between the two is the same as for stress:

$$1 \, \text{reyn} = 1 \, \text{lb} \cdot \text{s/in}^2 = 6890 \, \text{N} \cdot \text{s/m}^2 = 6890 \, \text{Pa} \cdot \text{s}.$$

The reyn and pascal–second are such large units that microreyn (μreyn) and millipascal second (mPa·s) are more commonly used. The former standard metric unit of viscosity was the poise (shortening of Jean Louis Marie Poiseuille, French physician and physiologist). One centipoise, cp, is equal to one millipascal-second

$$1 \, \text{cp} = 1 \, \text{mPa} \cdot \text{s}.$$

Dynamic viscosities are usually measured under high shear conditions, for example, the cylinder viscometer in which the viscous shear torque is measured between two cylinders. The *kinematic viscosity* is defined as absolute viscosity, μ, divided by mass density, ρ

$$\nu = \frac{\mu}{\rho}. \tag{II.5.3}$$

The units for kinematic viscosity are length^2/time, as cm^2/s, which is the Stoke (St). Using SI units: $1 \, \text{m}^2/\text{s} = 10^4 \, \text{St}$ and $1 \, \text{cSt}$ (centistoke) $= 1 \, \text{mm}^2/\text{s}$. The physical principle of measurement is based on the rate at which a fluid flows vertically downward under gravity

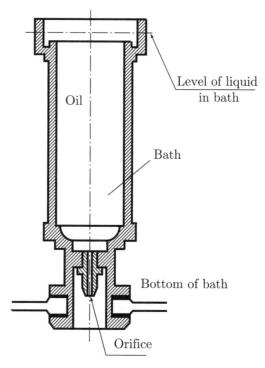

FIGURE II.5.2 *Saybolt Universal Viscometer.*

through a small-diameter tube. Liquid viscosities are determined by measuring the time required for a given quantity of the liquid to flow by gravity through a precision opening. For lubricating oils, the Saybolt Universal Viscometer, shown in Figure II.5.2, is an instrument used to measure the viscosity. The viscosity measurements are *Saybolt seconds*, or *SUS* (Saybolt Universal Seconds), *SSU* (Saybolt Seconds Universal), and *SUV* (Saybolt Universal Viscosity).

With a Saybolt Universal Viscometer one can measure the kinematic viscosity, v. Absolute viscosities can be obtained from Saybolt viscometer measurements by the equations

$$\mu \text{ (mPa·s or cp)} = \left(0.22\,t - \frac{180}{t}\right)\rho, \qquad (\text{II.5.4})$$

and

$$\mu \text{ (}\mu\text{reyn)} = 0.145\left(0.22\,t - \frac{180}{t}\right)\rho, \qquad (\text{II.5.5})$$

where ρ is the mass density in grams per cubic centimeter, g/cm^3 (which is also called specific gravity) and t is the time in seconds. For petroleum oils the mass density at different temperatures is

$$\rho = 0.89 - 0.00063 \left(^0C - 15.6\right) \ \text{g/cm}^3,$$ (II.5.6)

or

$$\rho = 0.89 - 0.00035 \left(^0F - 60\right) \ \text{g/cm}^3.$$ (II.5.7)

The Society of Automotive Engineers (SAE) classifies oils according to viscosity. Any viscosity grade should be proceeded by the initials SAE. It should be noted that SAE is not a performance category, it only refers to the viscosity of the oil.

Two series of SAE viscosity are defined (Table II.5.1): monograde oils or single viscosity oils and multigrade oils (those with the suffix W), which are essentially for winter conditions.

For the monograde oils (grades without the W), the viscosity is measured in cSt at 100°C (212°F). A monograde oil only has one part, such as SAE 30, or SAE 40. The number after

TABLE II.5.1 SAE Viscosity Grades for Engine Oils[a, b]

SAE viscosity grade	Low temperature (°C), cranking viscosity[c], cp max	Low temperature (°C) pumping viscosity[d], cp max with no yield stress[d]	Low-shear-rate kinematic viscosity[e] (cSt) at 100°C min	Low-shear-rate kinematic viscosity[e] (cSt)at 100°C max	High-shear-rate viscosity[f] (cp) at 150°C min
0W	6200 at −35	60,000 at −40	3.8	—	—
5W	6600 at −30	60,000 at −35	3.8	—	—
10W	7000 at −25	60,000 at −30	4.1	—	—
15W	7000 at −20	60,000 at −25	5.6	—	—
20W	9500 at −15	60,000 at −20	5.6	—	—
25W	13,000 at −10	60,000 at −15	9.3	—	—
20	—	—	5.6	< 9.3	2.6
30	—	—	9.3	< 12.5	2.9
40	—	—	12.5	< 16.3	2.9 (0W-40, 5W-40, 10W-40 grades)
40	—	—	12.5	< 16.3	3.7 (15W-40, 20W-40, 25W-40, 40 grades)
50	—	—	16.3	< 21.9	3.7
60	—	—	21.9	< 26.1	3.7

[a] 1 cp = 1mpa·s;   1 cSt = 1 mm^2/s
[b] All values are critical specifications as defined by ASTM D3244.
[c] ASTM D5293
[d] ASTM D4684: Note that the presence of any yield stress detectable by this method constitutes a failure regardless of viscosity.
[e] ASTM D445
[f] ASTM D4683, CEC L-36-A-90 (ASTM D4741) or D5481
Source: SAE J300 ©1999, Society of Automotive Engineers, Inc.

SAE gives a measure of the viscosity of the oil at high temperature (100°C). The lower the number the thinner the oil is at high temperature. Thus an SAE 30 is a thinner, or less viscous, oil than an SAE 40.

A multigrade oil (multiviscosity graded motor oil) must meet the viscosity standard at the W temperature and the SAE viscosity requirement at 100°C. For the W grades (W from winter), viscosity is measured by two test methods, one in the cold cranking simulator and the other in a pumping test that evaluates borderline pumping temperature. The cold cranking simulator reports dynamic viscosity in cp at temperatures that depend upon the grade. Additionally, the viscosity of motor oils with a W suffix is measured in cSt at 100°C. A multigrade oil is an oil that has two parts, such as SAE 15W-40 or 20W-50. For the multigrade oil SAE-10W20, the first number (10W) refers to the viscosity grade at low temperatures (W), whereas the second number (20) refers to the viscosity grade at high temperatures. The lower the W number the lower the viscosity of the oil. Therefore an SAE 5W oil is a lower viscosity oil than an SAE 10W oil.

The multigrade oils SAE 10W-30 and SAE 15W-30 have a similar high temperature viscosity as indicated by the 30. The 10W-30 oil is a thinner oil than the SAE 15W-30 at cold temperatures as indicated by the W number (10 < 15). Therefore, in cold temperatures, the SAE 10W-30 oil is better than the SAE 15W-30 oil. In winter it is beneficial to move from an SAE 15W-30 oil to an SAE 10W-30 oil.

During summer the ambient temperatures are high and the oil tends to be thinner, so a more viscous oil should be used. The SAE 10W-20 is a thinner oil than the SAE 10W-30 or SAE 15W-30 at high temperatures, as indicated by the second number (20 < 30). Therefore, in warm temperatures, a thicker oil (either an SAE 10W-30 or SAE 15W-30), could offer better engine protection than SAE 10W-20. In summer SAE 10W type oil is not required, which is why the SAE 15W-40 is favored.

There is much discussion about mineral oils versus synthetic oils and the relative performance of each type. The synthetic oils offer certain advantages over mineral oils in terms of low temperature performance and high temperature oxidation stability. Synthetic oils are very expensive, and properly formulated mineral oils are more than suitable for most engine applications. A synthetic oil can be considered for very cold temperatures, or for applications that may need a high level of oxidation protection. The manufacturer's recommendations should be followed.

Petroleum products can be graded according to the ISO Viscosity Classification System, approved by the International Standards Oreganization (ISO). Each ISO viscosity grade number corresponds to the mid-point of a viscosity range expressed in centistokes (cSt) at 40°C (the viscosity of the ISO grades, however, is measured at 40°C instead of 100°F = 37.8°C, which results in a slightly more viscous lubricant for each corresponding grade). For the ISO 3448 viscosity classification system the ISO VG 22 lubricant refers to a viscosity grade of 22 cSt ± 10% at 40°C. The kinematic viscosity limits are 19.8 cSt (min.) and 24.2 cSt (max.), and the mid-point viscosity is 22 cSt.

In Figures II.5.3, II.5.4, and II.5.5 the absolute viscosity function of temperature for typical SAE numbered oils is shown. Grease is a non-Newtonian material that does not begin to flow until a shear stress exceeding a yield point is applied.

The *viscosity index* (VI) measures the variation in viscosity with temperature. The viscosity index, on the Dean and Davis scale, of Pennsylvania oils is VI = 100. The viscosity index, on the same scale, of Gulf Coast oils is VI = 0. Other oils are rated intermediately. Nonpetroleum-base lubricants have widely varying viscosity indices. Silicone oils

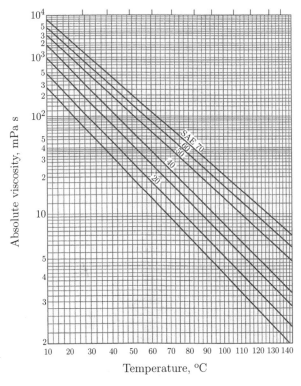

Source: R. C. Juvinall, K. M. Marshek, Fundamentals of machine component design, John Wiley & Sons, 1991

FIGURE II.5.3 *Absolute viscosity (mPa·s) function of temperature (°C). Reprinted with permission of John Wiley & Sons, Inc.*

have relatively little variation of viscosity with temperature. The viscosity index improvers (additives) can increase viscosity index of petroleum oils.

II.5.2 Petroff's Equation

Hydrodynamic lubrication is defined when the surfaces of the bearing are separated by a film of lubricant and does not depend upon the introduction of the lubricant under pressure. The pressure is created by the motion of the moving surface. Hydrostatic lubrication is defined when the lubricant is introduced at a pressure sufficiently high to separate the surfaces of the bearing.

A hydrodynamic bearing (hydrodynamic lubrication) is considered in Figure II.5.6. There is no lubricant flow in the axial direction and the bearing carries a very small load. The radius of the shaft is R, the radial clearance is c, and the length of the bearing is L (Fig. II.5.6). The shaft rotates with the angular speed n rev/s and its surface velocity is $V = 2\pi R n$.

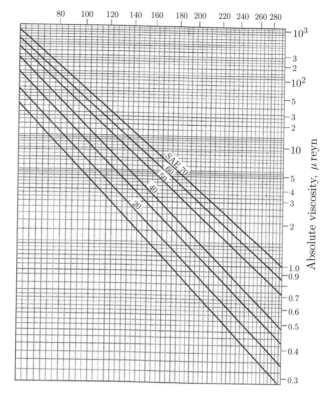

Temperature, °F

Source: R. C. Juvinall, K. M. Marshek, Fundamentals of machine component design, John Wiley & Sons, 2000

FIGURE II.5.4 *Absolute viscosity (µreyn) function of temperature (°F). Reprinted with permission of John Wiley & Sons, Inc.*

From Eq. (II.5.2) the shearing stress is

$$\tau = \mu \frac{V}{s} = \frac{2\pi R \mu n}{c}. \qquad (\text{II.5.8})$$

The force required to shear the film is the stress times the area,

$$F = \tau A,$$

where $A = 2\pi RL$.

The friction torque is the force times the lever arm:

$$T_f = FR = (\tau A) R = \left(\frac{2\pi R \mu n}{c} 2\pi RL \right) R = \frac{4\pi^2 \mu n L R^3}{c}. \qquad (\text{II.5.9})$$

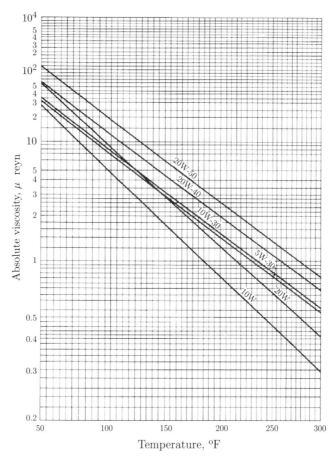

Source: J. E. Shigley, C. R. Mitchell, Mechanical engineering design, McGraw-Hill, 1983, New York

FIGURE II.5.5 *Absolute viscosity function of temperature for multigrade oils. Reprinted with permission of McGraw-Hill.*

If a small radial load W is applied on the bearing, the pressure P (the radial load per unit of projected bearing area) is

$$P = \frac{W}{2RL}.$$

The friction force is fW, where f is the coefficient of friction, and the friction torque is

$$T_f = f\,W\,R = f\,(2RLP)\,R = 2R^2 fLP. \tag{II.5.10}$$

Equations (II.5.9) and (II.5.10) can be equated and the coefficient of friction is

$$f = 2\pi^2 \left(\frac{\mu\,n}{P}\right)\left(\frac{R}{c}\right). \tag{II.5.11}$$

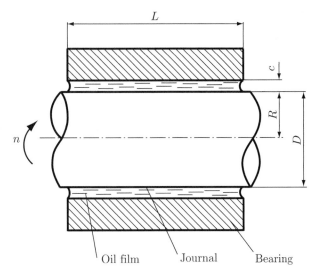

FIGURE II.5.6 *Hydrodynamic bearing.*

This is called Petroff's law or Petroff's equation. In Petroff's equation there are two important bearing parameters: the dimensionless variable, $\left(\dfrac{\mu n}{P}\right)$ and the clearance ratio $\left(\dfrac{R}{c}\right)$ with the order between 500 to 1000.

The bearing *characteristic number*, or Sommerfeld number S, is given by

$$S = \frac{\mu n}{P}\left(\frac{R}{c}\right)^2, \tag{II.5.12}$$

where R is the journal radius (in.), c is the radial clearance (in.), μ is the absolute viscosity (reyn), n is the speed (rev/s), and P is the pressure (psi).

The power loss in SI units is calculated with the relation

$$H = 2\pi T_f n, \quad \text{W} \tag{II.5.13}$$

where H = power (W), n = shaft speed (rev/s), T_f = torque (N·m), or

$$H = \frac{T_f n}{9549}, \quad \text{kW} \tag{II.5.14}$$

where H = power (kW), n = shaft speed (rpm), and T_f = torque (N·m).

The power loss in British units is

$$H = \frac{T_f n}{5252}, \quad \text{hp} \tag{II.5.15}$$

where H = power (hp), n = shaft speed (rpm), and T_f = torque (lb·ft).

II.5.3   Hydrodynamic Lubrication Theory

In Figure II.5.7 a small element of lubricant film of dimensions dx, dy, and dz is shown. The normal forces, due to the pressure, act upon right and left sides of the element. The shear forces, due to the viscosity and to the velocity, act upon the top and bottom sides of the element. The equilibrium of forces give

$$p \, dx \, dz + \tau \, dx \, dz - \left(p + \frac{dp}{dx}dx \right) dy \, dz - \left(\tau + \frac{\partial \tau}{\partial y}dy \right) dx \, dz = 0, \qquad \text{(II.5.16)}$$

which reduces to

$$\frac{dp}{dx} = \frac{\partial \tau}{\partial y}. \qquad \text{(II.5.17)}$$

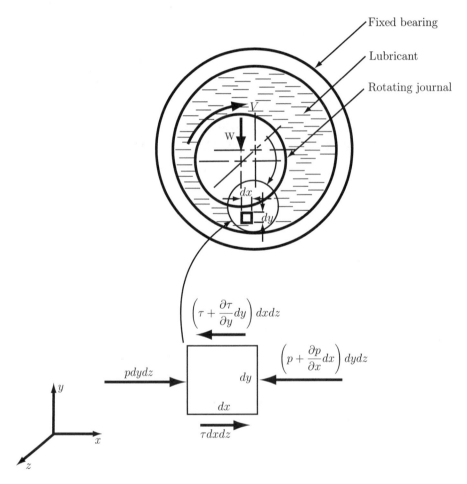

FIGURE II.5.7   *Pressure and forces on an element of lubricant film.*

In Eq. (II.5.17) the pressure of the film p is constant in y direction and depends only on the coordinate x, $p = p(x)$. The shear stress τ is calculated from Eq. (II.5.1):

$$\tau = \mu \frac{\partial v(x, y)}{\partial y}.$$ (II.5.18)

The velocity v of any particle of lubricant depends on both coordinates x and y, $v = v(x, y)$. From Eqs. (II.5.17) and (II.5.18), it results

$$\frac{dp}{dx} = \mu \frac{\partial^2 v}{\partial^2 y},$$ (II.5.19)

or

$$\frac{\partial^2 v}{\partial^2 y} = \frac{1}{\mu} \frac{dp}{dx}.$$ (II.5.20)

Holding x constant and integrating twice with respect to y gives

$$\frac{\partial v}{\partial y} = \frac{1}{\mu} \left(\frac{dp}{dx} y + C_1 \right),$$ (II.5.21)

and

$$v = \frac{1}{\mu} \left(\frac{dp}{dx} \frac{y^2}{2} + C_1 x + C_2 \right).$$ (II.5.22)

The constants C_1 and C_2 are calculated using the boundary conditions: for $y = 0 \implies v = 0$, and for $y = s \implies v = V$.

With C_1 and C_2 values computed, Eq. (II.5.18) gives the equation for the velocity distribution of the lubricant film across any yz plane:

$$v = \frac{1}{2\mu} \frac{dp}{dx} \left(y^2 - s y \right) + \frac{V}{s} y.$$ (II.5.23)

Equation (II.5.23) gives the velocity distribution of the lubricant in the film as a function of the coordinate y and the pressure gradient dp/dx. The velocity distribution is a superposition of a parabolic distribution, the first term, onto a linear distribution, the second term (and shown as a dashed line in Fig. II.5.8).

The volume of lubricant Q flowing across the section for width of unity in the z direction is

$$Q = \int_0^s v(x, y) dy = \frac{Vs}{2} - \frac{s^3}{12\mu} \frac{dp}{dx}.$$ (II.5.24)

For incompressible lubricant the flow is the same for any section:

$$\frac{dQ}{dx} = 0.$$

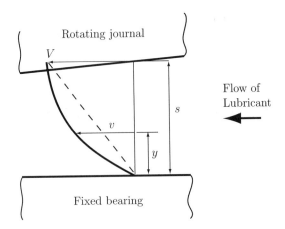

Rotating journal

V

Flow of
Lubricant

s

v

y

Fixed bearing

FIGURE II.5.8 *Velocity distribution.*

By differentiating Eq. (II.5.24), one can write

$$\frac{dQ}{dx} = \frac{V}{2}\frac{ds}{dx} - \frac{d}{dx}\left(\frac{s^3}{12\mu}\frac{dp}{dx}\right),$$

or

$$\frac{d}{dx}\left(\frac{s^3}{\mu}\frac{dp}{dx}\right) = 6V\frac{ds}{dx}, \qquad\qquad (II.5.25)$$

which is the classical Reynolds equation for one-dimensional flow.

The following assumptions were made:

- The fluid is Newtonian, incompressible, of constant viscosity, and experiences no inertial or gravitational forces.
- The fluid has a laminar flow, with no slip at the boundary surfaces.
- The fluid experiences negligible pressure variation over its thickness.
- The journal radius can be considered infinite.

The Reynolds equation for two-dimensional flow is (the z direction is included):

$$\frac{\partial}{\partial x}\left(\frac{s^3}{\mu}\frac{\partial p}{\partial x}\right) + \frac{\partial}{\partial z}\left(\frac{s^3}{\mu}\frac{\partial p}{\partial z}\right) = 6V\frac{\partial s}{\partial x}. \qquad\qquad (II.5.26)$$

For short bearings, one can neglect the x term in the Reynolds equation

$$\frac{\partial}{\partial z}\left(\frac{s^3}{\mu}\frac{\partial p}{\partial z}\right) = 6V\frac{\partial h}{\partial x}. \qquad\qquad (II.5.27)$$

Equation (II.5.27) can be used for analysis and design.

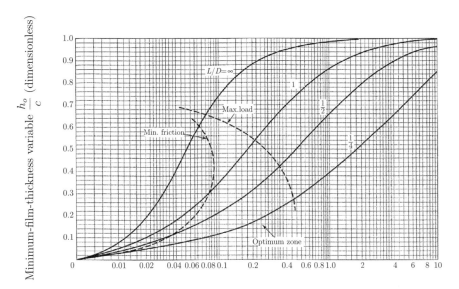

$$\text{Bearing characteristic number, } S = \left(\frac{R}{c}\right)^2 \frac{\mu n}{P}$$

Source: A. A. Raimondi, J. Boyd, A solution for the finite journal bearing and its application to analysis and design, Part I, II, and III, Trans. ASLE, Vol. 1, No. 1, pp. 159 - 209, in Lubrication Science and Technology. Pergamon Press, New York, 1958

FIGURE II.5.9 *Minimum film thickness variable.*

II.5.4 Design Charts

Raimondi and Boyd have transformed the solutions of the Reynolds Eq. (II.5.27) to chart form. The charts provide accurate solutions for bearings of all proportions. Some charts are shown in Figures II.5.9 to II.5.15. The quantities given in the charts are shown in Figure II.5.16. The Raimondi and Boyd charts give plots of dimensionless bearing parameters as functions of the bearing characteristic number, or Sommerfeld variable, S. The S scale on the charts is logarithmic except for a linear portion between 0 and 0.01.

II.5.5 Examples

EXAMPLE II.5.1:

The Saybolt kinematic viscosity of an oil corresponds to 60 seconds at 90°C (Fig. II.5.2). What is the corresponding absolute viscosity in millipascal-seconds (or centipoises) and in microreyns?

Continued

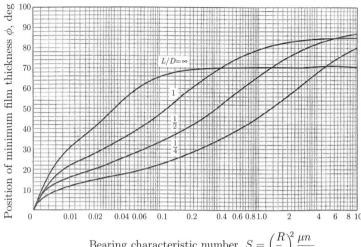

$$\text{Bearing characteristic number, } S = \left(\frac{R}{c}\right)^2 \frac{\mu n}{P}$$

Source: A. A. Raimondi, J. Boyd, A solution for the finite journal bearing and its application to analysis and design, Part I, II, and III, Trans. ASLE, Vol. 1, No. 1, pp. 159 - 209, in Lubrication Science and Technology. Pergamon Press, New York, 1958.

FIGURE II.5.10 *Position angle of minimum film thickness.*

EXAMPLE II.5.1: *Cont'd*

Solution   From Eq. (II.5.6), the mass density of the oil is

$$\rho = 0.89 - 0.00063(^\circ C - 15.6) = 0.89 - 0.00063(90 - 15.6)$$

$$= 0.843 \text{ g/cm}^3.$$

From Eq. (II.5.4), the absolute viscosity in centipoise is

$$\mu = \left(0.22\,t - \frac{180}{t}\right)\rho = \left[(0.22)(60) - \frac{180}{60}\right]0.843$$

$$= 8.598 \text{ cp (or 8.598 mPa} \cdot \text{s)}.$$

From Eq. (II.5.5), the absolute viscosity in microreyns, μreyn, is

$$\mu = 0.145\left(0.22t - \frac{180}{t}\right)\rho = 0.145\left[(0.22)(60) - \frac{180}{60}\right]0.843$$

$$= 1.246 \ \mu\text{reyn}.$$

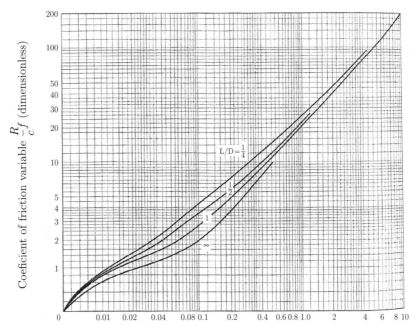

Source: A. A. Raimondi, J. Boyd, A solution for the finite journal bearing and its application to analysis and design, Part I, II, and III, Trans. ASLE, Vol. 1, No. 1, pp. 159 - 209, in Lubrication Science and Technology. Pergamon Press, New York, 1958.

FIGURE II.5.11   *Friction variable.*

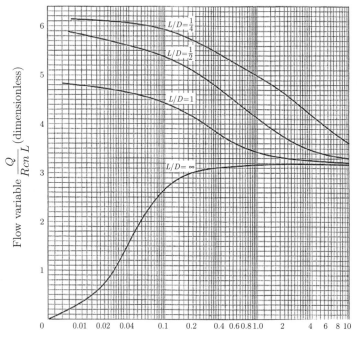

Bearing characteristic number, $S = \left(\dfrac{R}{c}\right)^2 \dfrac{\mu n}{P}$

Source: A. A. Raimondi, J. Boyd, A solution for the finite journal bearing and its application to analysis and design, Part I, II, and III, Trans. ASLE, Vol. 1, No. 1, pp. 159 - 209, in Lubrication Science and Technology. Pergamon Press, New York, 1958.

FIGURE II.5.12 *Flow variable.*

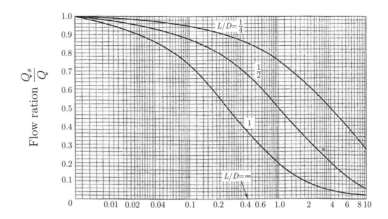

Source: A. A. Raimondi, J. Boyd, A solution for the finite journal bearing and its application to analysis and design, Part I, II, and III, Trans. ASLE, Vol. 1, No. 1, pp. 159 - 209, in Lubrication Science and Technology. Pergamon Press, New York, 1958.

FIGURE II.5.13 *Flow ratio (side leakage flow/total flow).*

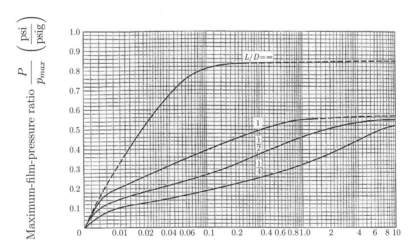

Source: A. A. Raimondi, J. Boyd, A solution for the finite journal bearing and its application to analysis and design, Part I, II, and III, Trans. ASLE, Vol. 1, No. 1, pp. 159 - 209, in Lubrication Science and Technology. Pergamon Press, New York, 1958.

FIGURE II.5.14 *Pressure ratio.*

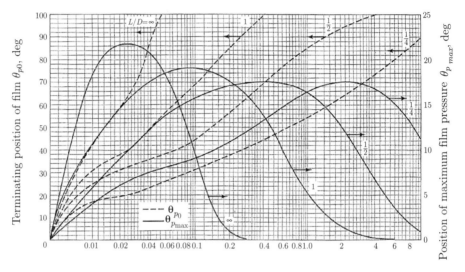

Bearing characteristic number, $S = \left(\dfrac{R}{c}\right)^2 \dfrac{\mu n}{P}$

Source: A. A. Raimondi, J. Boyd, A solution for the finite journal bearing and its application to analysis and design, Part I, II, and III, Trans. ASLE, Vol. 1, No. 1, pp. 159 - 209, in Lubrication Science and Technology. Pergamon Press, New York, 1958.

FIGURE II.5.15 *Terminating position of the oil film and position of the maximum pressure.*

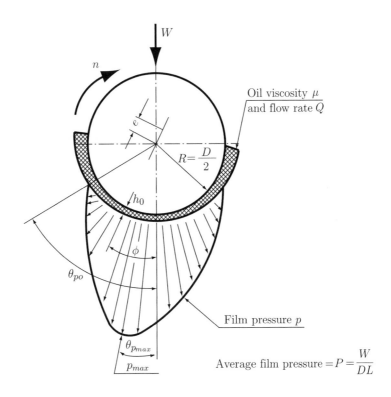

W

n

Oil viscosity μ
and flow rate Q

e

$R = \dfrac{D}{2}$

h_0

ϕ

θ_{po}

Film pressure p

$\theta_{p_{max}}$

p_{max}

Average film pressure $= P = \dfrac{W}{DL}$

h_0 the minimum oil film thickness
$\theta_{p_{max}}$ the angular position of the point of maximum pressure
θ_{po} the terminating position of the oil film
p_{max} the maximum pressure
ϕ the position angle of minimum film thickness

$e = c \text{ - } h_0$ the eccentricity

Source: J. E. Shigley, C. R. Mischke, Mechanical engineering design,
McGraw-Hill Inc., 1989, New York.

FIGURE II.5.16 *Notation for Raimondi and Boyd charts. Reprinted with permission of McGraw-Hill.*

EXAMPLE II.5.2:

A shaft with a 120-mm diameter (Fig. II.5.17), is supported by a bearing of 100-mm length with a diametral clearance of 0.2 mm and is lubricated by oil having a viscosity of 60 mPa·s. The shaft rotates at 720 rpm. The radial load is 6000 N. Find the bearing coefficient of friction and the power loss.

Solution The pressure is calculated with the relation

$$P = \frac{W}{2RL} = \frac{6000}{2(0.06)(0.1)} = 500\,000 \text{ N/m}^2 = 500\,000 \text{ Pa},$$

where $W = 6000$ N, $R = 0.06$ m, and $L = 0.1$ m.

From Eq. (II.5.11), the coefficient of friction is

$$f = 2\pi^2 \left(\frac{\mu\, n}{P}\right)\left(\frac{R}{c}\right) = 2\pi^2 \left[\frac{(0.06)(12)}{500\,000}\right]\left(\frac{60}{0.1}\right) = 0.017,$$

where $\mu = 60$ mPa·s $= 0.06$ Pa·s, $n = 720$ rev/min $= 12$ rev/s, $R = 60$ mm, and $c = 0.1$ mm.

The friction torque is calculated with

$$T_f = f\,W\,R = (0.017)(6000)(0.06) = 6.139 \text{ N} \cdot \text{m}$$

The power loss is

$$H = 2\pi\,T_f\,n = 2\pi\,(6.139)(12) = 462.921 \text{ N} \cdot \text{m/s} = 462.921 \text{ W}.$$

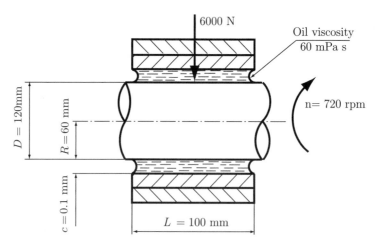

FIGURE II.5.17 *Journal bearing for Example II.5.2.*

EXAMPLE II.5.3:

A journal bearing has the diameter $D = 2.5$ in., the length $L = 0.625$ in., and the radial clearance $c = 0.002$ in., as shown in Figure II.5.18. The shaft rotates at 3600 rpm. The journal bearing supports a constant load, $W = 1500$ lb. The lubricant film is SAE 40 oil at atmospheric pressure. The average temperature of the oil film is $T_{avg} = 140°F$.

Find the minimum oil film thickness, h_0, the bearing coefficient of friction, f, the maximum pressure, p_{max}, the position angle of minimum film thickness, ϕ, the angular position of the point of maximum pressure $\theta_{p_{max}}$, the terminating position of the oil film θ_{po}, the total oil flow rate, Q, and the flow ratio (side flow/total flow) Q_s/Q.

Solution The pressure is

$$P = \frac{W}{LD} = \frac{1500}{(0.625)(2.5)} = 960 \text{ psi.}$$

The dynamic viscosity is $\mu = 5 \times 10^{-6}$ reyn (SAE 40, $T_{avg} = 140°F$), from Figure II.5.4.

The Sommerfeld number is

$$S = \left(\frac{R}{c}\right)^2 \frac{\mu n}{P} = \left(\frac{1.25}{0.002}\right)^2 \left[\frac{(5 \times 10^{-6})(60)}{960}\right] = 0.12.$$

For all charts $S = 0.12$ and $L/D = 0.25$ are used.

- From Figure II.5.9, the minimum film thickness variable is $h_o/c = 0.125$ and the minimum film thickness is $h_0 = 0.125\,c = 0.00025$ in.

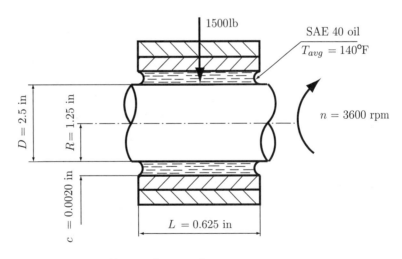

FIGURE II.5.18 *Journal bearing for Example II.5.3.*

Continued

EXAMPLE II.5.3: *Cont'd*

- From Figure II.5.11, the friction variable is $(R/c)f = 5$ and the coefficient of friction is $f = 5\,c/R = 0.00832$.
- From Figure II.5.14, the pressure ratio is $P/p_{max} = 0.2$ and the maximum film pressure is $p_{max} = P/0.2 = 4800$ psi.
- From Figure II.5.10, the position angle of minimum film thickness is $\phi = 24°$ (see Fig. II.5.16).
- From Figure II.5.15, the terminating position of the oil film is $\theta_{po} = 33°$ and the angular position of the point of maximum pressure is $\theta_{p_{max}} = 9.5°$ (see Fig. II.5.16).
- From Figure II.5.12, the flow variable is $\dfrac{Q}{R\,c\,n\,L} = 5.9$ and the total flow is $Q = 0.553$ in^3/s.
- From Figure II.5.13, the flow ratio (side leakage flow/total flow) is $Q_s/Q = 0.94$. Of the volume of oil Q pumped by the rotating journal, an amont Q_s flows out the ends. The side leakage that must be made up by the oil represents 94% of the flow. The remaining 6% of the flow is recirculated.

EXAMPLE II.5.4:

A journal shaft of a gear train has a rotational speed of 2200 rpm and a radial load of 2200 lb. The shaft is lubricated with an SAE 30 oil and the average film temperature is 180°F. Determine the values of the clearance c for the two edges of the optimum zone for the bearing characteristic number for $L/D = 1$ (Fig. II.5.9).

Solution   From Figure II.5.9, the optimum values for the bearing characteristic number are [7]:

L/D	S for min. friction	S for max. load
1	0.082	0.21

where D is the diameter and L is the length of the bearing.

The absolute viscosity of an SAE 30 oil at 180°F is [7] (see Fig. II.5.4):

$$\mu = 1.87\ \mu\text{reyn} = 1.87 \times 10^{-6}\text{reyn}.$$

From Table II.5.2, the representative unit sleeve bearing load, for gear reducers, arbitrarily is selected $P = 250$ psi.

With $L = D$ the bearing length is

$$L = D = \sqrt{\frac{W}{P}} = \sqrt{\frac{2200}{250}} = 2.966 \text{ in.}$$

A diameter of $D = 3$ in. is selected.

TABLE II.5.2 Representative Unit Sleeve Bearing Loads in Current Practice

Application	Units load $P = \dfrac{W_{max}}{LD}$	
Relatively steady load		
Electric motors	0.8–1.5 MPa	120–250 psi
Steam turbines	1.0–2.0 MPa	150–300 psi
Gear reducers	0.8–1.5 MPa	120–250 psi
Centrifuge pumps	0.6–1.2 MPa	100–180 psi
Rapidly fluctuating loads		
Diesel engines		
Main bearings	6–12 MPa	900–1700 psi
Connecting rod bearings	8–15 MPa	1150–2300 psi
Automotive gasoline engines		
Main bearings	4–5 MPa	600–750 psi
Connecting rod bearings	10–15 MPa	1700–2300 psi

Source: R. C. Juvinall and K. M. Marshek, *Fundamentals of Machine Component Design*, New York, John Wiley & Sons, 1991. Reprinted with permission of John Wiley & Sons, Inc.

EXAMPLE II.5.4: *Cont'd*

The clearance is calculated from the relation

$$S = \left(\frac{\mu\,n}{P} \right) \left(\frac{R}{c} \right)^2 ,$$

where $n = 2200/60 = 36.666$ rps and $R = 3/2 = 1.5$ in.
   For minimum friction $S_{min} = 0.082$ the clearance is

$$c = R\sqrt{\frac{\mu\,n}{P\,S_{min}}} = (1.5)\sqrt{\frac{(1.87 \times 10^{-6})\,(36.666)}{(250)\,(0.082)}} = 0.002743 \text{ in.}$$

and for maximum load $S_{max} = 0.21$ the clearance is

$$c = R\sqrt{\frac{\mu\,n}{P\,S_{max}}} = (1.5)\sqrt{\frac{(1.87 \times 10^{-6})\,(36.666)}{(250)\,(0.21)}} = 0.001714 \text{ in.}$$

EXAMPLE II.5.5:

The oil lubricated bearing of a steam turbine has the diameter $D = 160$ mm, (Fig. II.5.19). The angular velocity of the rotor shaft is $n = 2400$ rpm. The radial load is $W = 18$ kN. The lubricant is SAE 20, controlled to an average temperature of 78°C.

Continued

EXAMPLE II.5.5: *Cont'd*

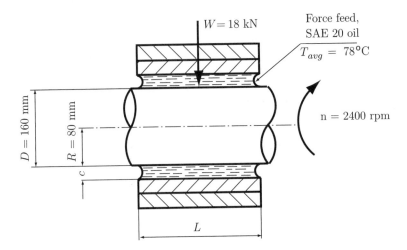

FIGURE II.5.19 *Journal bearing for Example II.5.5.*

Find the bearing length, the radial clearance, the corresponding values of the minimum oil film thickness, the coefficient of friction, and the friction power loss.

Solution From Table II.5.2, for steam turbine (1 to 2 MPa range) the unit load $P = 1.5$ MPa is arbitrarily selected. The bearing length is

$$L = \frac{W}{PD} = \frac{18\,000}{(1.5)(160)} = 75 \text{ mm}.$$

Arbitrarily round this up to $L = 80$ mm to give $L/D = 1/2$ for convenient use of the Raimondi and Boyd charts.

With $L = 80$ mm, P is given by the relation

$$P = \frac{W}{LD} = \frac{18\,000}{(80)\,(160)} = 1.406 \text{ MPa}.$$

From Figure II.5.3, the viscosity of SAE 20 oil at 78°C is $\mu = 9.75$ mPa·s. From Figure II.5.9 the optimum values for the bearing characteristic number are [7]:

L/D	S for min. friction	S for max. load
1/2	0.037	0.35

For minimum friction $S_{min} = 0.037$ the clearance is

$$c = R\sqrt{\frac{\mu\,n}{P\,S_{min}}} = (0.08)\sqrt{\frac{(9.75 \times 10^{-3})\,(40)}{(1.406 \times 10^{6})\,(0.037)}} = 0.219 \times 10^{-3} \text{ m}.$$

EXAMPLE II.5.5: *Cont'd*

and for maximum load $S_{max} = 0.35$ the clearance is

$$c = R\sqrt{\frac{\mu n}{P S_{max}}} = (0.08)\sqrt{\frac{(9.75 \times 10^{-3})(40)}{(1.406 \times 10^6)(0.35)}} = 0.071 \times 10^{-3} \text{ m}.$$

The minimum oil film thickness, h_o, is calculated from the ratio, h_o/c obtained from Figure II.5.9, and the coefficient of friction, f, is calculated from the ratio Rf/c obtained from Figure II.5.11.

The values of S, h_o, and f function of c, ($0.048 \text{ mm} \leq c \leq 0.243 \text{ mm}$), with c extending to either side of the optimum range are listed below.

| c | S | h_0 | f |
mm		mm	
0.0482629	0.762	0.0284751	0.00965258
0.0712126	0.350	0.0302654	0.00774437
0.1125970	0.140	0.0292752	0.00619284
0.1445050	0.085	0.0281784	0.00559955
0.2080650	0.041	0.0249678	0.00494154
0.2190230	0.037	0.0240926	0.00479113
0.2432370	0.030	0.0243237	0.00486475

Figure II.5.20 shows h_o and f function of c, and indicates a good operation.

For the minimum acceptable oil film thickness, h_o, the following empirical relations are given (Trumpler empirical equation) [7]:

$$h_o \geq h_{omin} = 0.0002 + 0.00004D \quad (h_o \text{ and } D \text{ in in.}),$$

$$h_o \geq h_{omin} = 0.005 + 0.00004D \quad (h_o \text{ and } D \text{ in m.}). \quad \text{(II.5.28)}$$

For $D = 160$ mm, the minimum acceptable oil film thickness is

$$h_{omin} = 0.005 + 0.00004(160) = 0.0114 \text{ mm}.$$

The minimum film thickness using a safety factor of $C_s = 2$ applied to the load, and assuming an "extreme case" of $c = 0.243$ mm, is calculated as follows:

- the Sommerfeld number is

$$S = \left(\frac{\mu n}{C_s P}\right)\left(\frac{R}{c}\right)^2 = \frac{(9.75 \times 10^{-3})(40)}{(2)(1.406 \times 10^6)}\left(\frac{80}{0.243}\right)^2 = 0.015.$$

- from Figure II.5.9 using $S = 0.015$ $\dfrac{h_o}{c} = 0.06$ is obtained, and the minimum film thickness is $h_o = 0.0145$ mm.

This value satisfies the condition $h_o = 0.0145 \geq h_{omin} = 0.0114$.

Continued

EXAMPLE II.5.5: *Cont'd*

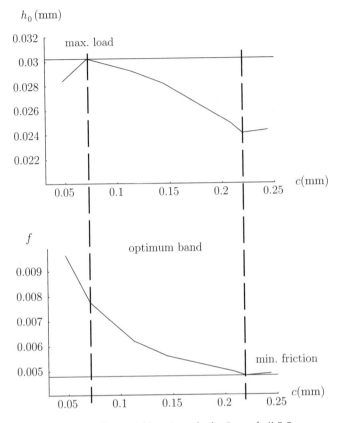

FIGURE II.5.20 *Variation of h_o and f function of c for Example II.5.5.*

For the tightest bearing fit, where $c = 0.048$ mm and $f = 0.009$, the friction torque is

$$T_f = \frac{W f D}{2} = \frac{(18\,000)(0.009)(0.16)}{2} = 13.899 \text{ N} \cdot \text{m},$$

and the friction power is

$$\text{friction power} = \frac{n T_f}{9549} = \frac{(2400 \text{ rpm})(13.899 \text{ N} \cdot \text{m})}{9549} = 3.493 \text{ kW}.$$

The *Mathematica*[TM] program for this example is given in Program II.5.1.

II.5.6 Problems

II.5.1 Determine the density of a SAE 40 oil at 160°F.

II.5.2 From a Saybolt viscometer the kinematic viscosity of an oil corresponds to 60 seconds at 120°C. Find the corresponding absolute viscosity in millipascal-seconds and in microreyns.

II.5.3 An engine oil has a kinematic viscosity at 93°C corresponding to 50 seconds, as determined from a Saybolt viscometer. Find the corresponding SAE number.

II.5.4 A 100 mm diameter shaft is supported by a bearing of 100-mm length with a diametral clearance of 0.075 mm. It is lubricated by SAE 20 oil at the operating temperature of 70°C. The shaft rotates 3000 rpm and carries a radial load of 5000 N. Estimate the bearing coefficient of friction and power loss using the Petroff approach.

II.5.5 A journal bearing 4 in. in diameter and 6 in. long is lubricated with an SAE 10 oil with the average temperature of 130°F. The diametral clearance of the bearing is 0.0015 in. The shaft rotates at 2000 rpm. Find the friction torque and the power loss.

II.5.6 A shaft with the diameter D, rotational speed n, and radial load W is supported by an oil-lubricated bearing of length L and radial clearance c. There is no eccentricity between the bearing and the journal, and no lubricant flow in the axial direction. Determine the bearing coefficient of friction and the power loss. Numerical data is as follows: (a) $D = 0.2$ m, $L = 0.15$ m, $c = 0.075$ mm, $n = 1200$ rpm, $W = 6.5$ kN, and $\mu = 32$ mPa·s (for SAE 10 oil at 40°C); (b) $D = 1.5$ in., $L = 1.5$ in., $c = 0.0015$ in., $n = 30$ rps, $W = 500$ lb, and $\mu = 4$ μreyn.

II.5.7 A journal bearing of 200-mm diameter, 100-mm length, and 0.1-mm radial clearance carries a load of 20 kN. The shaft rotates at 1000 rpm. The bearing is lubricated by SAE 20 oil and the average temperature of the oil film is estimated at 70°C. Determine the minimum oil film thickness bearing coefficient of friction, maximum pressure within the oil film, angles ϕ, $\theta_{p\,max}$, θ_{po}, total oil flow rate through the bearing, and side leakage.

II.5.8 A journal bearing of 2 in. diameter, 2 in. length, and 0.001 in. radial clearance supports a load of 1500 lb when the shaft rotates 1000 rpm. The lubrication oil is SAE 30 supplied at atmospheric pressure. The average temperature of the oil film is 140°F. Using the Raimondi and Boyd charts, determine the minimum oil film thickness bearing coefficient of friction, maximum pressure within the oil film, angles ϕ, $\theta_{p\,max}$, θ_{po}, total oil flow rate, and fraction of the flow rate that is recirculated oil flow.

II.5.9 A full journal bearing has the diameter of 60 mm and an L/D ratio of unity and runs at a speed of 200 rpm. The radial clearance is 0.04 mm and the oil supply is SAE 30 at the temperature of 60°C. The radial load is 3000 N. Determine the minimum oil film thickness bearing coefficient of friction and its angular location, the maximum pressure within the oil film and its angular location, and the side flow.

II.5.10 A 5 kN load is applied to a 100 mm diameter shaft rotating at 2000 rpm. A journal bearing is used to carry the load. The journal bearing with a diameter to length ratio of 0.25 is lubricated with an SAE 40 oil with an inlet temperature of 40°C. Determine the minimum oil film thickness bearing coefficient of friction and its angular location, the maximum pressure within the oil film and its angular location, and the side flow.

II.5.11 A shaft rotates at 1000 rpm and is lubricated with an SAE 30 oil at 80°C. The radial load is 30 kN. Determine the values of the clearance c for the two edges of the optimum zone for the bearing characteristic number for the ratio $L/D = 1$.

II.5.12 A journal bearing for a gear train rotates at 1000 rpm and applies a force of 10 kN to the bearing. An SAE 20 oil is used and the average temperature is expected to be 60°C. A proportion of $L/D = 1/2$ is desired. Find: (a) the value of D, and (b) the values of c corresponding for the two edges of the optimum zone for the bearing characteristic number.

II.5.13 The oil-lubricated journal bearing of a gear reducer has the diameter $D = 100$ mm. The angular speed of the shaft is 1200 rpm and the radial load is 20 kN. The lubricant is SAE 30, controlled to an average temperature of 65°C. Find the bearing length, the radial clearance, and the corresponding values of the minimum oil film thickness, the coefficient of friction, the friction power loss, and the oil flows.

II.5.7 Programs

Program II.5.1

```
Apply[Clear,Names["Global`*"]];
Off[General::spell];
Off[General::spell1];
d=160. 10^-3 ; (* m *)
R=d/2 ; (* m *)
n0=2400; (* rpm *)
n=n0/60.; (* rps *)
W=18000; (* N *)
"mu=9.75 mPa s"
mu=9.75*10^(-3); (* Pa s *)
"select P=1.5 MPa unit load for steam turbine"
P=1.5 10^6; (*Pa*)
L=W/(P d); (* m *)
Print["L=W/(P D)= ",L];
L=0.080;
"Select L=80 mm "
P=W/(d L);
Print["P= ",P," Pa"];
Smin=0.037;
Smax=0.35;

(* S=(R/c)^2 (mu n/P) *)
cf=R/(Smin P/(mu n))^0.5;
cl=R/(Smax P/(mu n))^0.5;
Print[" min. friction => c=",cf," m"];
Print[" max load => c=",cl," m"];

S1=0.762;  H1=0.59;  F1=16.;  q1=4.3;  qs1=0.56;
Smax=0.35; H1=0.425; F1=8.7;  q1=4.8;  qs1=0.72;
S2=0.140;  H2=0.26;  F2=4.4;  q2=5.25; qs2=0.84;
S3=0.085;  H3=0.195; F3=3.1;  q3=5.45; qs3=0.88;
S4=0.041;  H4=0.12;  F4=1.9;  q4=5.6 ; qs4=0.92 ;
Smin=0.037;Hf=0.11;  Ff=1.75; qf=5.65; qsf=0.93;
S5=0.030;  H5=0.1;   F5=1.6;  q5=5.7 ; qs5=0.94 ;
```

```
c1=R/(S1 P/(mu n))^0.5 ;
(* c1=R/(Smax P/(mu n))^0.5;*)
c2=R/(S2 P/(mu n))^0.5 ;
c3=R/(S3 P/(mu n))^0.5 ;
c4=R/(S4 P/(mu n))^0.5 ;

(* cf=R/(Smin P/(mu n))^0.5;*)
c5=R/(S5 P/(mu n))^0.5 ;

h1=c1 H1 ;
hl=cl Hl ;
h2=c2 H2 ;
h3=c3 H3 ;
h4=c4 H4 ;
hf=cf Hf ;
h5=c5 H5 ;

ff=Ff cf/R;
fl=Fl cl/R;
f1=F1 c1/R;
f2=F2 c2/R;
f3=F3 c3/R;
f4=F4 c4/R;
f5=F5 c5/R;

c1m=c1 10^3;
clm=cl 10^3;
c2m=c2 10^3;
c3m=c3 10^3;
c4m=c4 10^3;
cfm=cf 10^3;
c5m=c5 10^3;

h1m=h1 10^3;
hlm=hl 10^3;
h2m=h2 10^3;
h3m=h3 10^3;
h4m=h4 10^3;
hfm=hf 10^3;
h5m=h5 10^3;

Print["c=",c1 ,  " S=",S1," h0=",h1," f=",f1];
Print["c=cl=",cl," S=Smax=",Smax," h0=",hl," f=",fl];
Print["c=",c2 ,  " S=",S2," h0=",h2," f=",f2];
Print["c=",c3 ,  " S=",S3," h0=",h3," f=",f3];
Print["c=",c4 ,  " S=",S4," h0=",h4," f=",f4];
Print["c=cf=",cf," S=Smin=",Smin," h0=",hf," f=",ff];
Print["c=",c5 ,  " S=",S5," h0=",h5," f=",f5];
```

```
Print["c[mm]        S        h[mm]        f"];
Print[c1m,"      ",S1,"      ",h1m,"      ",f1];
Print[c1m,"      ",Smax,"        ",h1m,"      ",f1];
Print[c2m,"      ",S2,"      ",h2m,"      ",f2];
Print[c3m,"      ",S3,"      ",h3m,"      ",f3];
Print[c4m,"      ",S4,"      ",h4m,"      ",f4];
Print[cfm,"      ",Smin,"        ",hfm,"      ",ff];
Print[c5m,"      ",S5,"      ",h5m,"      ",f5];

g1=ListPlot[10^3{{c1,h1},{c1,h1},{c2,h2},{c3,h3},{c4,h4},{cf,hf},{c5,h5}},
PlotJoined->True,
Axes->True,AxesOrigin→{0.03,0.02},PlotRange→{{0.03,0.25},{0.02,0.032}},
AxesLabel→{"c","ho"}, GridLines→{{c1m},{h1m}}];
g2=ListPlot[{{c1m,f1},{c1m,f1},{c2m,f2},{c3m,f3},{c4m,f4},{cfm,ff},{c5m,f5}},
PlotJoined->True,
Axes->True,AxesOrigin→{0.03,0.004},PlotRange→{{0.03,0.25},{0.004,0.0099}},
AxesLabel→{"c","f"}, GridLines→{{cfm},{ff}}];

homin=0.005+0.00004 d 10^3;
Print[" homin = 0.005+0.00004 D = ",homin," mm"];
"extreme case c=0.243237 mm =>"
Cs=2;
S=(R/c5)^2 (mu n/(Cs P));
Print["S=",S];
" from figure => h0/c=0.06"
hos=0.06 c5;
Print[" ho = 0.06 c = ",hos 10^3," mm > homin "];
"friction torque for tightest bearing fit"
"c=0.0482629 f=0.00965258"
Tf=W f1 d/2;
Print["Tf = W f d/2 = ",Tf," N m"];
FP=n0 Tf/9549;
Print["friction power = n[rpm] Tf[N m]/9549 = ",FP," kW"];

mu=9.75 mPa s

select P=1.5 MPa unit load for steam turbine

L=W/(P D)= 0.075

Select L=80 mm

P= 1.40625 × 10^6 Pa

 min. friction => c=0.000219023 m

 max load => c=0.0000712126 m

c=0.0000482629 S=0.762 h0=0.0000284751 f=0.00965258

c=c1=0.0000712126 S=Smax=0.35 h0=0.0000302654 f=0.00774437
```

```
c=0.000112597 S=0.14 h0=0.0000292752 f=0.00619284
c=0.000144505 S=0.085 h0=0.0000281784 f=0.00559955
c=0.000208065 S=0.041 h0=0.0000249678 f=0.00494154
c=cf=0.000219023 S=Smin=0.037 h0=0.0000240926 f=0.00479113
c=0.000243237 S=0.03 h0=0.0000243237 f=0.00486475
c[mm]   S   h[mm]   f
0.0482629 0.762 0.0284751 0.00965258
0.0712126 0.35  0.0302654 0.00774437
0.112597  0.14  0.0292752 0.00619284
0.144505  0.085 0.0281784 0.00559955
0.208065  0.041 0.0249678 0.00494154
0.219023  0.037 0.0240926 0.00479113
0.243237  0.03  0.0243237 0.00486475
```

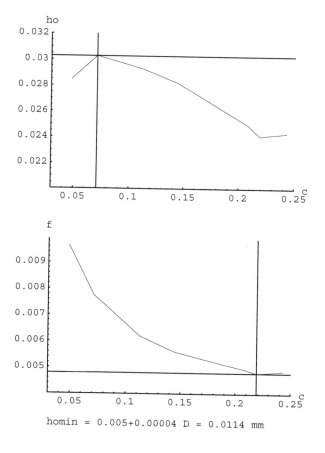

homin = 0.005+0.00004 D = 0.0114 mm

extreme case c=0.243237 mm =>

S=0.015

 from figure => h0/c=0.06

 ho = 0.06 c = 0.0145942 mm > homin

friction torque for tightest bearing fit

c=0.0482629 f=0.00965258

Tf = W f d/2 = 13.8997 N m

friction power = n[rpm] Tf[N m]/9549 = 3.49349 kW

II.6   Gears

II.6.1   Introduction

Gears are toothed elements that transmit rotary motion from one shaft to another. Gears are generally rugged and durable and their power transmission efficiency is as high as 98%. Gears are usually more costly than chains and belts. The American Gear Manufacturers Association (AGMA) has established standard tolerances for various degrees of gear manufacturing precision. *Spurs gears* are the simplest and most common type of gears. They are used to transfer motion between parallel shafts, and they have teeth that are parallel to the shaft axes.

II.6.2   Geometry and Nomenclature

The basic requirement of gear-tooth geometry is the condition of angular velocity ratios that are exactly constant, i.e., the angular velocity ratio between a 30-tooth and a 90-tooth gear must be precisely 3 in every position. The action of a pair of gear teeth satisfying this criterion is called *conjugate gear-tooth action.*

> *Law of conjugate gear-tooth action:* The common normal to the surfaces at the point of contact of two gears in rotation must always intersect the line of centers at the same point P, called the *pitch point.*

The law of conjugate gear-tooth action can be satisfied by various tooth shapes, but the one of current importance is the involute of the circle. An *involute* (of the circle) is the curve generated by any point on a taut thread as it unwinds from a circle, called the *base circle* [Fig. II.6.1(a)]. The involute can also be defined as the locus of a point on a taut string that is unwrapped from a cylinder. The circle that represents the cylinder is the base circle. Figure II.6.1(b) represents an involute generated from a base circle of radius r_b starting at the point A. The radius of curvature of the involute at any point I is given by

$$\rho = \sqrt{r^2 - r_b^2},\tag{II.6.1}$$

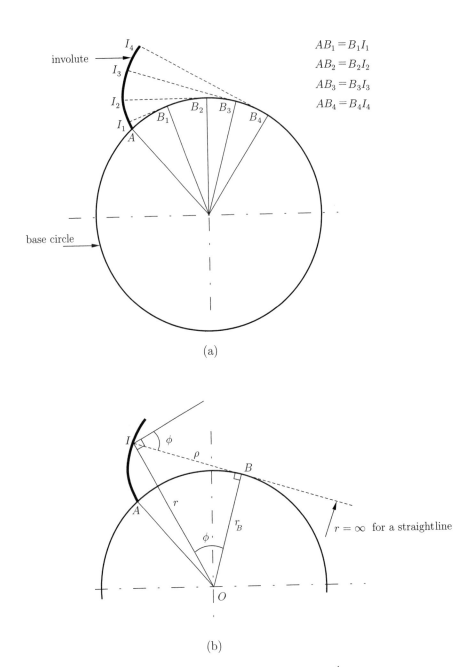

$AB_1 = B_1 I_1$
$AB_2 = B_2 I_2$
$AB_3 = B_3 I_3$
$AB_4 = B_4 I_4$

involute

base circle

(a)

$r = \infty$ for a straight line

(b)

FIGURE II.6.1 *(a) Development of involute curve; (b) pressure angle.*

where $r = OI$. The *involute pressure angle* at I is defined as the angle between the normal to the involute IB and the normal to OI, $\phi = \angle IOB$.

In any pair of meshing gears, the smaller of the two is called the pinion and the larger one the gear. The term "gear" is used in a general sense to indicate either of the members and also in a specific sense to indicate the larger of the two. The angular velocity ratio between

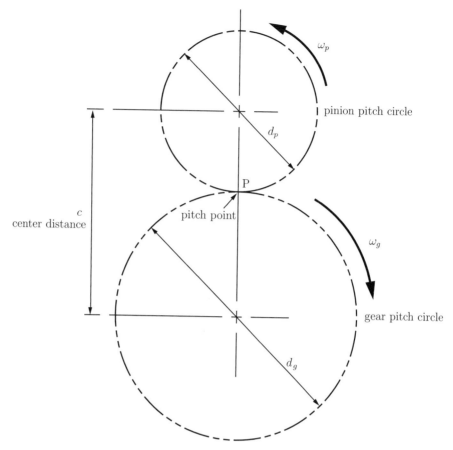

FIGURE II.6.2 *Gears of pitch diameter d rotating at angular velocity ω.*

a pinion and a gear is (Fig. II.6.2):

$$i = \omega_p/\omega_g = -d_g/d_p, \tag{II.6.2}$$

where ω is the angular velocity and d is the pitch diameter, and the minus sign indicates that the two gears rotate in opposite directions. The *pitch circles* are the two circles, one for each gear, that remain tangent throughout the engagement cycle. The point of tangency is the pitch point. The diameter of the pitch circle is the pitch diameter. If the angular speed is expressed in rpm then the symbol n is preferred instead of ω. The diameter (without a qualifying adjective) of a gear always refers to its pitch diameter. If other diameters (base, root, outside, etc.) are intended, they are always specified. Similarly, d, without subscripts, refers to pitch diameter. The pitch diameters of a pinion and gear are distinguished by subscripts p and g (d_p and d_g, are their symbols, see Fig. II.6.2). The *center distance* is

$$c = (d_p + d_g)/2 = r_p + r_g, \tag{II.6.3}$$

where $r = d/2$ is the *pitch circle radius*.

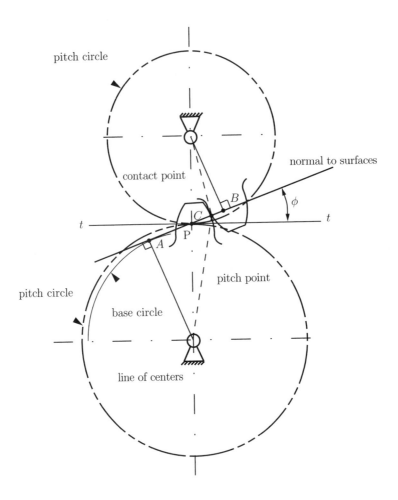

FIGURE II.6.3 *Pressure angle φ.*

In Figure II.6.3, line *tt* is the common tangent to the pitch circles at the pitch point and *AB* is the common normal to the surfaces at *C*, the point of contact of two gears. The angle of *AB* with the line *tt* is called the *pressure angle*, ϕ. The most common pressure angle used, with both English and SI units, is 20°. In the United States 25° is also standard, and 14.5° was formerly an alternative standard value. The pressure angle affects the force that tends to separate the two meshing gears.

The involute profiles are augmented outward beyond the pitch circle by a distance called the *addendum*, *a*, [Fig. II.6.4]. The outer circle is usually termed the *addendum circle*, $r_a = r + a$. Similarly, the tooth profiles are extended inward from the pitch circle, a distance called the *dedendum*, *b*. The involute portion can extend inward only to the base circle. A fillet at the base of the tooth merges the profile into the dedendum circle. The fillet decreases the bending stress concentration. The *clearance* is the amount by which the dedendum in a given gear exceeds the addendum of its meshing gear.

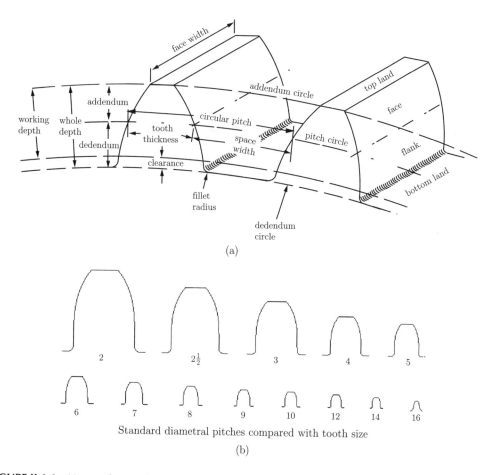

FIGURE II.6.4 *Nomenclature of gear teeth.*

The *circular pitch* is designated as p, and measured in inches (English units) or millimeters (SI units). If N is the number of teeth in the gear (or pinion), then

$$p = \pi d/N, \quad p = \pi d_p/N_p, \quad p = \pi d_g/N_g. \tag{II.6.4}$$

More commonly used indices of gear-tooth size are *diametral pitch*, P_d (used only with English units), and *module*, m (used only with SI). Diametral pitch is defined as the number of teeth per inch of pitch diameter [see Fig. II.6.4]:

$$P_d = N/d, \quad P_d = N_p/d_p, \quad P_d = N_g/d_g. \tag{II.6.5}$$

Module m, which is essentially the complement of P_d, is defined as the pitch diameter in millimeters divided by the number of teeth (number of millimeters of pitch diameter per tooth):

$$m = d/N, \quad m = d_p/N_p, \quad m = d_g/N_g. \tag{II.6.6}$$

One can easily verify that

$$p P_d = \pi \quad (p \text{ in inches}; \; P_d \text{ in teeth per inch})$$

$$p/m = \pi \quad (p \text{ in millimeters}; m \text{ in millimeters per tooth})$$

$$m = 25.4/P_d.$$

With English units the word "pitch", without a qualifying adjective, denotes diametral pitch (a "12-pitch gear" refers to a gear with $P_d = 12$ teeth per inch of pitch diameter). With SI units "pitch" means circular pitch (a "gear of pitch = 3.14 mm" refers to a gear having a circular pitch of $p = 3.14$ mm).

Standard diametral pitches, P_d (English units), in common use are

1 to 2 by increments of 0.25
2 to 4 by increments of 0.5
4 to 10 by increments of 1
10 to 20 by increments of 2
20 to 40 by increments of 4.

With SI units, commonly used standard values of module m are

0.2 to 1.0 by increments of 0.1
1.0 to 4.0 by increments of 0.25
4.0 to 5.0 by increments of 0.5.

Addendum, minimum dedendum, whole depth, and clearance for gears with English units in common use are [5]:

	$14\frac{1}{2}°$ Full-depth involute or composite	$20°$ Full-depth involute	$20°$ Stub involute
addendum a	$1/P_d$	$1/P_d$	$0.8/P_d$
minimum dedendum b	$1.157/P_d$	$1.157/P_d$	$1/P_d$
whole depth	$2.157/P_d$	$2.157/P_d$	$1.8/P_d$
clearance	$0.157/P_d$	$0.157/P_d$	$0.2/P_d$

For SI units the standard values for full-depth involute teeth with a pressure angle of $20°$ are addendum $a = m$ and minimum dedendum $b = 1.25\,m$.

II.6.3 Interference and Contact Ratio

The contact of segments of tooth profiles which are not conjugate is called *interference*. The involute tooth form is only defined outside the base circle. In some cases, the dedendum will extend below the base circle, then the portion of tooth below the base circle will not be an involute and will interfere with the tip of the tooth on the meshing gear, which is an involute. Interference will occur, preventing rotation of the meshing gears, if either of

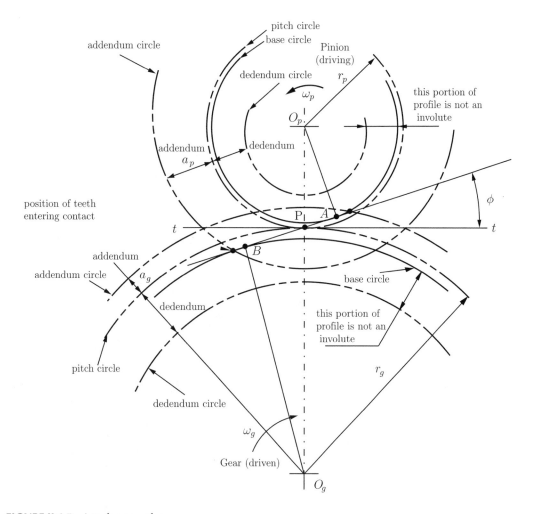

FIGURE II.6.5 *Interference of spur gears.*

the addendum circles extends beyond tangent points A and B (Fig. II.6.5), which are called interference points. In Figure II.6.5 both addendum circles extend beyond the interference points.

The maximum possible addendum circle radius, of pinion or gear, without interference is

$$r_{a(max)} = \sqrt{r_b^2 + c^2 \sin^2 \phi},$$
(II.6.7)

where $r_b = r \cos \phi$ is the base circle radius of pinion or gear. The base circle diameter is

$$d_b = d \cos \phi.$$
(II.6.8)

The average number of teeth in contact as the gears rotate together is the contact ratio CR, which is calculated from the following equation (for external gears):

$$CR = \frac{\sqrt{r_{ap}^2 - r_{bp}^2} + \sqrt{r_{ag}^2 - r_{bg}^2} - c\sin\phi}{p_b}, \qquad (II.6.9)$$

where r_{ap}, r_{ag} are addendum radii of the pinion and gear, and r_{bp}, r_{bg} are base circle radii of the pinion and gear. The base pitch, p_b, is computed with

$$p_b = \pi d_b/N = p\cos\phi. \qquad (II.6.10)$$

The base pitch is like the circular pitch except that it represents an arc of the base circle rather than an arc of the pitch circle.

For internal gears the contact ratio is

$$CR = \frac{\sqrt{r_{ap}^2 - r_{bp}^2} - \sqrt{r_{ag}^2 - r_{bg}^2} + c\sin\phi}{p_b}, \qquad (II.6.11)$$

The greater the contact ratio, the smoother and quieter the operation of the gears. If the contact ratio is 2 then two pairs of teeth are in contact at all the times. The acceptable values for contact ratio are $CR > 1.2$.

Gears are commonly specified according to AGMA Class Number, a code that denotes important quality characteristics. Quality numbers denote tooth-elements tolerances. The higher the number, the tighter the tolerance. Gears are heat treated by case hardening, nitriding, precipitation hardening, or through hardening. In general, harder gears are stronger and last longer than soft ones.

II.6.4   Ordinary Gear Trains

A gear train is any collection of two or more meshing gears. Figure II.6.6(a) shows a simple gear train with three gears in series. The train ratio is computed with the relation

$$i_{13} = \frac{\omega_1}{\omega_3} = \frac{\omega_1}{\omega_2}\frac{\omega_2}{\omega_3} = \left(-\frac{N_2}{N_1}\right)\left(-\frac{N_3}{N_2}\right) = \frac{N_3}{N_1}. \qquad (II.6.12)$$

Only the sign of the overall ratio is affected by the intermediate gear 2 which is called an *idler*.

Figure II.6.6(b) shows a compound gear train, without idler gears, with the train ratio

$$i_{14} = \frac{\omega_1}{\omega_2}\frac{\omega_{2'}}{\omega_3}\frac{\omega_{3'}}{\omega_4} = \left(-\frac{N_2}{N_1}\right)\left(-\frac{N_3}{N_{2'}}\right)\left(-\frac{N_4}{N_{3'}}\right) = -\frac{N_2 N_3 N_4}{N_1 N_{2'} N_{3'}}. \qquad (II.6.13)$$

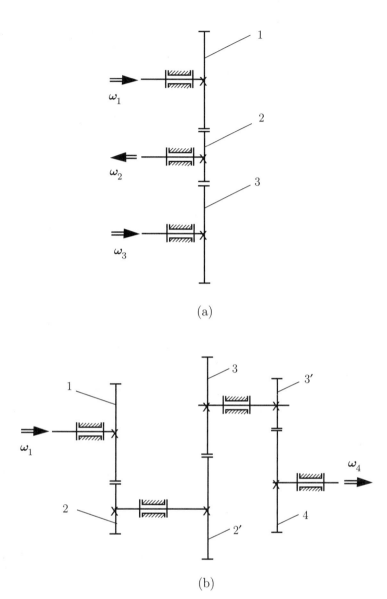

(a)

(b)

FIGURE II.6.6 *Gear train; (a) simple gear trains, and (b) compound gear trains.*

II.6.5 Epicyclic Gear Trains

When at least one of the gear axes rotates relative to the frame in addition to the gear's own rotation about its own axes, the train is called a *planetary gear train* or *epicyclic gear train*. The term "epicyclic" comes from the fact that points on gears with moving axes of rotation describe epicyclic paths. When a generating circle (planet gear) rolls on the outside of another circle, called a *directing circle* (sun gear), each point on the generating circle describes an epicycloid, as shown in Figure II.6.7.

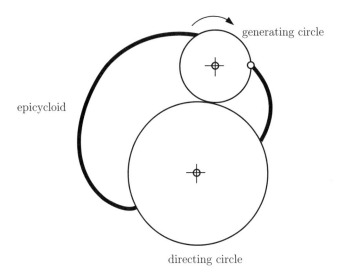

generating circle

epicycloid

directing circle

FIGURE II.6.7 *Epicycloid curve.*

Generally, the more planet gears there are the greater the torque capacity of the system. For better load balancing, new designs have two sun gears and up to 12 planetary assemblies in one casing.

In the case of simple and compound gears it is not difficult to visualize the motion of the gears and the determination of the speed ratio is relatively easy. In the case of epicyclic gear trains it is often difficult to visualize the motion of the gears. A systematic procedure, using the contour method, is presented below. The contour method is applied to determine the distribution of velocities for epicyclic gear trains.

The velocity equations for a simple closed kinematic chain are [1, 24]:

$$\sum_{(i)} \boldsymbol{\omega}_{i,i-1} = \mathbf{0} \quad \text{and} \quad \sum_{(i)} \mathbf{r}_{A_i} \times \boldsymbol{\omega}_{i,i-1} = \mathbf{0},$$

where $\boldsymbol{\omega}_{i,i-1}$ is the relative angular velocity of the rigid body (i) with respect to the rigid body $(i-1)$ and $\mathbf{r}_{A_i}$ is the position vector of the joint between the rigid body (i) and the rigid body $(i-1)$ with respect to a "fixed" reference frame.

The epicyclic (planetary) gear train shown in Figure II.6.8 consists of a central gear 2 (sun gear) and another gear 3 (planet gear) in mesh with 2 at B. Gear 3 is carried by the arm 1 hinged at A. The ring gear 4 meshes with the planet gear 3 and pivots at A, so it can be easily tapped as an output member. The sun gear and the ring gear are concentric. The sun gear, the ring gear, and the arm can be accessed to tap the angular velocity and torque either as an input or an output. There are four moving bodies 1, 2, 3, and 4, $(n = 4)$ connected by:

- Four full joints ($c_5 = 4$): one hinge between the arm 1 and the planet gear 3 at C, one hinge between the frame 0 and the shaft of the sun gear 2 at A, one hinge between the frame 0 and the ring gear 4 at A, and one hinge between the frame 0 and the arm 1 at A.
- Two half joints ($c_4 = 2$): one between the sun gear 2 and the planet gear 3, and one between the planet gear 3 and the ring gear 4.

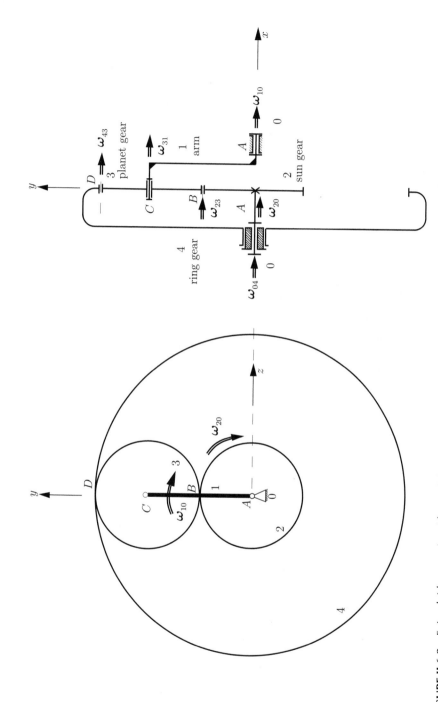

FIGURE II.6.8 *Epicycloid gear train with two DOF.*

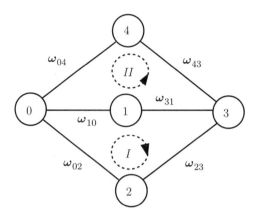

FIGURE II.6.9 *Contour diagram of the epicycloid gear train with two DOF.*

The system has two degrees of freedom (DOF): $M = 3\,n - 2\,c_5 - c_4 = 3 \cdot 4 - 2 \cdot 4 - 2 = 2$. The sun gear has N_2-tooth external gear, the planet gear has N_3-tooth external gear, and the ring gear has N_4-tooth internal gear.

If the arm and the sun gear rotate with input angular speeds ω_1 and ω_2, find the absolute output angular velocity of the ring gear.

The velocity analysis is carried out using the contour method. The system shown in Figure II.6.8 has a total of five elements ($p = 5$): the frame 0 and four moving links 1, 2, 3, and 4. There are six joints ($l = 6$): four full joints and two half joints. The number of independent contours is given by

$$n_c = l - p + 1 = 6 - 5 + 1 = 2.$$

This gear system has two independent contours. The graph of the kinematic chain is represented in Figure II.6.9.

First Contour
The first contour is formed by the elements 0, 1, 3, 2, and 0 (clockwise path). For the velocity analysis, the following vectorial equations can be written

$$\boldsymbol{\omega}_{10} + \boldsymbol{\omega}_{31} + \boldsymbol{\omega}_{23} + \boldsymbol{\omega}_{02} = \mathbf{0},$$
$$\mathbf{r}_{AC} \times \boldsymbol{\omega}_{31} + \mathbf{r}_{AB} \times \boldsymbol{\omega}_{23} = \mathbf{0}, \tag{II.6.14}$$

where the input angular velocities are

$$\boldsymbol{\omega}_{10} = \omega_1\,\mathbf{l} \quad \text{and} \quad \boldsymbol{\omega}_{02} = -\omega_2\,\mathbf{l},$$

and the unknown angular velocities are

$$\boldsymbol{\omega}_{31} = \omega_{31}\,\mathbf{l} \quad \text{and} \quad \boldsymbol{\omega}_{23} = \omega_{23}\,\mathbf{l}.$$

The sign of the relative angular velocities is selected positive and then the numerical computation gives the true orientation of the vectors.

The vectors $\mathbf{r}_{AB}$, $\mathbf{r}_{AC}$, and $\mathbf{r}_{AD}$ are defined as

$$\mathbf{r}_{AB} = x_B \, \mathbf{\imath} + y_B \, \mathbf{\jmath}, \quad \mathbf{r}_{AC} = x_C \, \mathbf{\imath} + y_C \, \mathbf{\jmath}, \quad \mathbf{r}_{AD} = x_D \, \mathbf{\imath} + y_D \, \mathbf{\jmath}, \tag{II.6.15}$$

where

$$y_B = r_2 = m \, N_2/2,$$
$$y_C = r_2 + r_3 = m \, (N_2 + N_3)/2,$$
$$y_D = r_2 + 2 \, r_3 = m \, N_2/2 + m \, N_3.$$

The module of the gears is m. Equation (II.6.14) becomes

$$\omega_1 \, \mathbf{\imath} + \omega_{31} \, \mathbf{\imath} + \omega_{23} \, \mathbf{\imath} - \omega_2 \, \mathbf{\imath} = \mathbf{0},$$

$$\begin{vmatrix} \mathbf{\imath} & \mathbf{\jmath} & \mathbf{k} \\ x_C & y_C & 0 \\ \omega_{31} & 0 & 0 \end{vmatrix} + \begin{vmatrix} \mathbf{\imath} & \mathbf{\jmath} & \mathbf{k} \\ x_B & y_B & 0 \\ \omega_{23} & 0 & 0 \end{vmatrix} = \mathbf{0}. \tag{II.6.16}$$

Equation (II.6.16) can be projected on a "fixed" reference frame $xOyz$ as

$$\omega_1 + \omega_{31} + \omega_{23} - \omega_2 = 0,$$
$$y_C \, \omega_{31} + y_B \, \omega_{23} = 0. \tag{II.6.17}$$

Equation (II.6.17) represents a system of two equations with two unknowns, ω_{31} and ω_{23}. Solving the algebraic equations, the following values are obtained:

$$\omega_{31} = N_2 \, (\omega_1 - \omega_2)/N_3,$$
$$\omega_{23} = -\omega_1 + \omega_2 - N_2 \, (\omega_1 - \omega_2)/N_3.$$

Second Contour

The second closed contour contains the elements 0, 1, 3, 4, and 0 (Fig. II.6.9). The contour velocity equations can be written as (counterclockwise path)

$$\boldsymbol{\omega}_{10} + \boldsymbol{\omega}_{31} + \boldsymbol{\omega}_{43} + \boldsymbol{\omega}_{04} = \mathbf{0},$$
$$\mathbf{r}_{AC} \times \boldsymbol{\omega}_{31} + \mathbf{r}_{AD} \times \boldsymbol{\omega}_{43} = \mathbf{0}, \tag{II.6.18}$$

where the known angular velocities are $\boldsymbol{\omega}_{10}$, $\boldsymbol{\omega}_{31}$, and the unknown angular velocities are

$$\boldsymbol{\omega}_{43} = \omega_{43} \, \mathbf{\imath} \quad \text{and} \quad \boldsymbol{\omega}_{04} = \omega_{04} \, \mathbf{\imath}.$$

Equation (II.6.18) can be written as

$$\omega_1\, \mathbf{1} + \omega_{31}\, \mathbf{1} + \omega_{43}\, \mathbf{1} + \omega_{04}\, \mathbf{1} = \mathbf{0},$$

$$\begin{vmatrix} \mathbf{1} & \mathbf{J} & \mathbf{k} \\ x_C & y_C & 0 \\ \omega_{31} & 0 & 0 \end{vmatrix} + \begin{vmatrix} \mathbf{1} & \mathbf{J} & \mathbf{k} \\ x_D & y_D & 0 \\ \omega_{43} & 0 & 0 \end{vmatrix} = \mathbf{0}. \qquad (\text{II.6.19})$$

From Eq. (II.6.19), the absolute angular velocity of the ring gear is

$$\omega_{40} = -\omega_{04} = \frac{2N_2\omega_1 + 2N_3\omega_1 - N_2\omega_2}{N_2 + 2N_3}.$$

II.6.6   Differential

Figure II.6.10(a) is a schematic drawing of the ordinary bevel-gear automotive differential. The drive shaft pinion 1 and the ring gear 2 are normally hypoid gears. The ring gear 2 acts as the planet carrier for the planet gear 3, and its speed can be calculated as for a simple gear train when the speed of the drive shaft is given. Sun gears 4 and 5 are connected, respectively, to each rear wheel.

When the car is traveling in a straight line, the two sun gears rotate in the same direction with exactly the same speed. Thus for straight-line motion of the car, there is no relative motion between the planet gear 3 and ring 2. The planet gear 3, in effect, serves only as a key to transmit motion from the planet carrier to both wheels.

When the vehicle is making a turn, the wheel on the inside of the turn makes fewer revolutions than the wheel with a larger turning radius. Unless this difference in speed is accommodated in some manner, one or both of the tires would have to slide in order to make the turn. The differential permits the two wheels to rotate at different velocities while at the same time delivering power to both. During a turn, the planet gear 3 rotates about their own axes, thus permitting gears 4 and 5 to revolve at different velocities. The purpose of a differential is to differentiate between the speeds of the two wheels. In the usual passenger-car differential, the torque is divided equally whether the car is traveling in a straight line or on a curve. Sometimes the road conditions are such that the tractive effort developed by the two wheels is unequal. In this case the total tractive effort available will be only twice that at the wheel having the least traction, because the differential divides the torque equally. If one wheel should happen to be resting on snow or ice, the total effort available is very small and only a small torque will be required to cause the wheel to spin. Thus, the car sits with one wheel spinning and the other at rest with no tractive effort. And, if the car is in motion and encounters slippery surfaces, then all traction as well as control is lost.

It is possible to overcome the disadvantages of the simple bevel-gear differential by adding a coupling unit which is sensitive to wheel speeds. The object of such a unit is to cause most of the torque to be directed to the slow-moving wheel. Such a combination is then called a *limited-slip differential.*

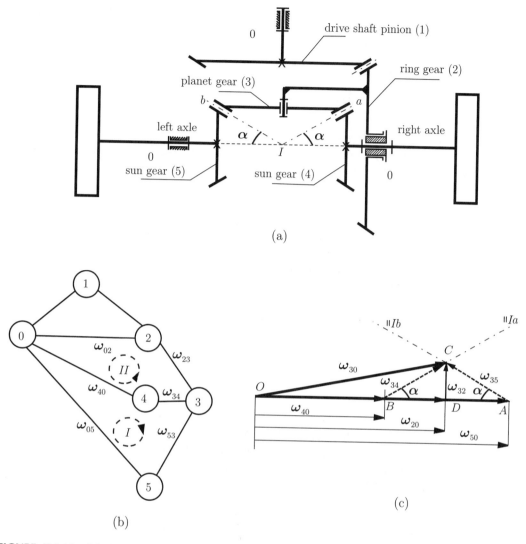

FIGURE II.6.10 *(a) Automotive differential planetary gear train; (b) graph attached to the differential mechanism; (c) angular velocities diagram.*

Angular Velocities Diagram

The velocity analysis is carried out using the contour equation method and the graphical angular velocities diagram.

There are five moving elements (1, 2, 3, 4, and 5) connected by

- Five full joints ($c_5 = 5$): one between the frame 0 and the drive shaft pinion gear 1, one between the frame 0 and the ring gear 2, one between the planet carrier arm 2 and the planet gear 3, one between the frame 0 and the sun gear 4, and one between the frame 0 and the sun gear 5.

- Three half joints ($c_4 = 3$): one between the drive shaft pinion gear 1 and the ring gear 2, one between the planet gear 3 and the sun gear 4, and one between the planet gear 3 and the sun gear 5.

The system possesses two DOF:

$$M = 3n - 2c_5 - c_4 = 3 \cdot 5 - 2 \cdot 5 - 3 = 2.$$

The input data are the absolute angular velocities of the two wheels, ω_{40} and ω_{50}.

The system shown in Figure II.6.10(a) has six elements (0, 1, 2, 3, 4, and 5) and eight joints ($c_4 + c_5$). The number of independent contours is given by

$$n_c = 8 - p + 1 = 8 - 6 + 1 = 3.$$

This gear system has three independent contours. The graph of the kinematic chain and the independent contours are represented in Figure II.6.10(b).

The first closed contour contains the elements 0, 4, 3, 5, and 0 (clockwise path). For the velocity analysis, the following vectorial equations can be written

$$\boldsymbol{\omega}_{40} + \boldsymbol{\omega}_{34} + \boldsymbol{\omega}_{53} + \boldsymbol{\omega}_{05} = \mathbf{0},$$

or

$$\boldsymbol{\omega}_{40} + \boldsymbol{\omega}_{34} = \boldsymbol{\omega}_{50} + \boldsymbol{\omega}_{35}. \qquad (II.6.20)$$

The unknown angular velocities are $\boldsymbol{\omega}_{34}$ and $\boldsymbol{\omega}_{35}$. The relative angular velocity of the planet gear 3 with respect to the sun gear 4 is parallel to the Ia line and the relative angular velocity of the planet gear 3 with respect to the sun gear 5 is parallel to Ib. Equation (II.6.20) can be solved graphically as shown in Figure II.6.10(c). The vectors OA and OB represent the velocities ω_{50} and ω_{40}. At A and B two parallels at Ib and Ia are drawn. The intersection between the two lines is the point C. The vector BC represents the relative angular velocity of the planet gear 3 with respect to the sun gear 4, and the vector AC represents the relative angular velocity of the planet gear 3 with respect to the sun gear 5.

The absolute angular velocity of planet gear 3 is

$$\boldsymbol{\omega}_{30} = \boldsymbol{\omega}_{40} + \boldsymbol{\omega}_{34}.$$

The vector OC represents the absolute angular velocity of planet gear.

The second closed contour contains the elements 0, 4, 3, 2, and 0 (counterclockwise path). For the velocity analysis, the following vectorial equations can be written

$$\boldsymbol{\omega}_{40} + \boldsymbol{\omega}_{34} + \boldsymbol{\omega}_{23} + \boldsymbol{\omega}_{02} = \mathbf{0}. \qquad (II.6.21)$$

Using the velocities diagram [Fig. II.6.10(c)] the vector DC represents the relative angular velocity of the planet gear 3 with respect to the ring gear 2, $\boldsymbol{\omega}_{23}$, and the OD represents the absolute angular velocity of the ring gear 2, $\boldsymbol{\omega}_{20}$.

Figure II.6.10(c) gives

$$\omega_{20} = |OD| = \frac{1}{2}(\omega_{40} + \omega_{50}),$$

$$\omega_{32} = |DC| = \frac{1}{2}(\omega_{50} - \omega_{40})\tan\alpha. \tag{II.6.22}$$

When the car is traveling in a straight line, the two sun gears rotate in the same direction with exactly the same speed, $\omega_{50} = \omega_{40}$, and there is no relative motion between the planet gear and the ring gear, $\omega_{32} = 0$. When the wheels are jacked up, $\omega_{50} = -\omega_{40}$, the absolute angular velocity of the ring gear 2 is zero.

II.6.7 Gear Force Analysis

The force between meshing teeth (neglecting the sliding friction) can be resolved at the pitch point (P in Fig. II.6.11) into two components:

- tangential component F_t, which accounts for the power transmitted;
- radial component F_r, which does no work but tends to push the gears apart.

The relationship between these components is

$$F_r = F_t \tan\phi, \tag{II.6.23}$$

where ϕ is the pressure angle.

The pitch line velocity in feet per minute is equal to

$$V = \pi\,d\,n/12 \quad \text{(ft/min)}, \tag{II.6.24}$$

where d is the pitch diameter in inches of the gear rotating n rpm.

In SI units,

$$V = \pi\,d\,n/60\,000 \quad \text{(m/s)}, \tag{II.6.25}$$

where d is the pitch diameter in millimeters of the gear rotating n rpm.

The transmitted power in horsepower is

$$H = F_t\,V/33\,000 \quad \text{(hp)}, \tag{II.6.26}$$

where F_t is in pounds and V is in feet per minute.

In SI units the transmitted power in watts is

$$H = F_t\,V \quad \text{(W)}, \tag{II.6.27}$$

where F_t is in newtons and V is in meters per second.

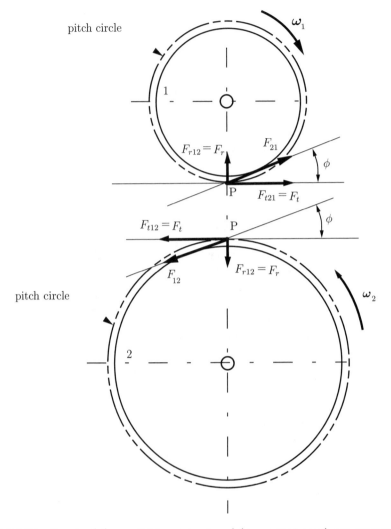

FIGURE II.6.11 *Gear tooth forces. Driving pinion 1 and driven gear 2 are shown separately.*

The transmitted torque can be expressed as

$$M_t = 63\,000\,H/n \quad (\text{lb·in}), \tag{II.6.28}$$

where H is in horsepower and n is in rpm.
In SI units,

$$M_t = 9549\,H/n \quad (\text{N·m}), \tag{II.6.29}$$

where the power H is in kW and n is in rpm.

For the gear force analysis the following assumptions will be made:

- all the gears mesh along their pitch circles
- friction losses are negligible
- all the tooth loads are transferred at the pitch point
- centripetal forces will not be considered.

II.6.8 Strength of Gear Teeth

Hall et al. present an analysis of the strength of gear teeth [5]. The flank of the driver tooth makes contact with the tip of the driven tooth at the beginning of action between a pair of gear teeth. The total load F is carried by one tooth, and is normal to the tooth profile (see Fig. II.6.12). The bending stress at the base of the tooth is produced by the tangential load component F_t, which is perpendicular to the centerline of the tooth. The friction and the radial component F_r are neglected. The parabola shown in Figure II.6.12 outlines a beam

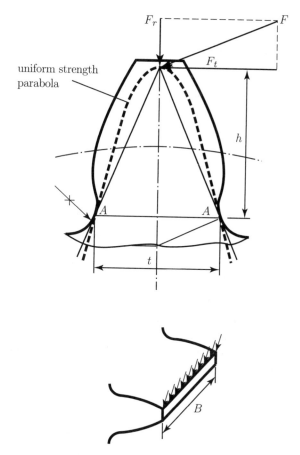

FIGURE II.6.12 *Load carried by the gear tooth.*

of uniform strength. The weakest section of the gear tooth is at section $A - A$ where the parabola is tangent to the tooth outline.

The bending stress σ is

$$\sigma = \frac{Mc}{I} = \frac{M(t/2)}{Bt^3/12} = \frac{6M}{Bt^2} = \frac{6F_th}{Bt^2},$$

and

$$F_t = \sigma B(t^2/6h) = \sigma B(t^2/6hp)p, \qquad (\text{II}.6.30)$$

where $M = F_th$ is the bending moment, h is the distance between the section $A - A$ and the point where the load is applied, and t is the tooth thickness. In Eq. (II.6.30) B is the face width and is limited to a maximum of 4 times the circular pitch, i.e., $B = kp$, where $k \leq 4$.

The form factor

$$y = \frac{t^2}{6hp}, \qquad (\text{II}.6.31)$$

is a dimensionless quantity tabulated in Table II.6.1.

Substituting y in the Eq. (II.6.30) gives

$$F_t = \sigma Bpy, \qquad (\text{II}.6.32)$$

or

$$F_t = \sigma p^2 ky = \sigma \pi^2 ky/P_d^2, \qquad (\text{II}.6.33)$$

which is the Lewis equation.

In the design problem the diameters are either known or unknown.

- If the diameters are unknown the stress is

$$\sigma = \frac{2M_t P_d^3}{k\pi^2 yN}, \qquad (\text{II}.6.34)$$

where M_t is the torque on the weaker gear, $k = 4$ (upper limit), and N is the number of teeth on the weaker gear. The minimum number of teeth, N, is usually limited to 15. The stress σ should be less than or equal to allowable stress.
- If the diameters are known then the allowable value for the ratio P_d^2/y which controls the design is

$$P_d^2/y = \sigma k\pi^2/F_t, \qquad (\text{II}.6.35)$$

where σ is the allowable stress, $k = 4$ (upper limit), $F_t = 2M_t/d$ is the transmitted force, and M_t is the torque on the weaker gear.

TABLE II.6.1 Form factors γ — for use in Lewis strength equation

Number of Teeth	$14\frac{1}{2}°$ Full-depth involute or composite	20° Full-depth involute	20° stub involute
12	0.067	0.078	0.099
13	0.071	0.083	0.103
14	0.075	0.088	0.108
15	0.078	0.092	0.111
16	0.081	0.094	0.115
17	0.084	0.096	0.117
18	0.086	0.098	0.120
19	0.088	0.100	0.123
20	0.090	0.102	0.125
21	0.092	0.104	0.127
23	0.094	0.106	0.130
25	0.097	0.108	0.133
27	0.099	0.111	0.136
30	0.101	0.114	0.139
34	0.104	0.118	0.142
38	0.106	0.122	0.145
43	0.108	0.126	0.147
50	0.110	0.130	0.151
60	0.113	0.134	0.154
75	0.115	0.138	0.158
100	0.117	0.142	0.161
150	0.119	0.146	0.165
300	0.122	0.150	0.170
Rack	0.124	0.154	0.175

Source: A. S. Hall, A. R. Holowenko, and H. G. Laughlin, *Theory and Problems of Machine Design*, Schaum's Outline Series, New York, McGraw-Hill, 1961. Reprinted with permission of McGraw-Hill.

If the diameters are known, design for the largest number of teeth; if the diameters are unknown, design for the smallest pitch diameters possible. The most economical design is given by the largest diametral pitch.

From Eq. (II.6.33) the force that can be transmitted to a gear tooth is a function of the product $\sigma_0 \gamma$. For two gears in contact the weaker gear will have the smaller $\sigma_0 \gamma$ value. For gears made of the same material, the smaller gear will be the weaker and control design.

Allowable Tooth Stresses

The allowable stress for gear tooth design is

$$\text{Allowable } \sigma = \sigma_0 \left(\frac{600}{600 + V} \right) \quad \text{for } V < 2000 \text{ ft/min}$$

$$= \sigma_0 \left(\frac{1200}{1200 + V} \right) \quad \text{for } 2000 < V < 4000 \text{ ft/min}$$

$$= \sigma_0 \left(\frac{78}{78 + \sqrt{V}} \right) \quad \text{for } V > 4000 \text{ ft/min}, \qquad (\text{II.6.36})$$

where σ_0 is the endurance strength in psi, and V is the pitch line velocity in ft/min. The endurance strength is:

$\sigma_0 = 8000$ psi for cast iron,
$\sigma_0 = 12\,000$ psi for bronze, and
$\sigma_0 = [10\,000, \ldots, 50\,000]$ psi for carbon steels.

In general, $\sigma_0 \approx (1/3)$ ultimate strength of the material.

Dynamic Tooth Loads

The dynamic forces on the teeth are produced by velocity changes due to inaccuracies of the tooth profiles, and misalignments in mounting, spacing, tooth deflection, and so forth. The dynamic load F_d proposed by Buckingham is

$$F_d = \frac{0.05\,V(B\,C + F_t)}{0.05\,V + \sqrt{B\,C + F_t}} + F_t, \qquad \text{(II.6.37)}$$

where F_d is the dynamic load (lb), $F_t = M_t/r$, and C is a constant that depends on the tooth material, form, and the accuracy of the tooth cutting process (tooth error, e). The constant C is tabulated in Table II.6.2. Figure II.6.13(a) shows the relation of permissible errors in tooth profiles function of pitch line velocity, V, and Fig. II.6.13(b) represents the connection between the errors e and diametral pitch, P_d. The dynamic force F_d must be less than the allowable endurance load, $F_0 = \sigma_0\,B\,\gamma\,p$.

Wear Tooth Loads

The wear load F_w is

$$F_w = d_p\,B\,K\,Q, \qquad \text{(II.6.38)}$$

TABLE II.6.2 Values of deformation Factor C—for dynamic load check

Materials			Tooth error inches			
Pinion	Gear	Involute tooth form	0.0005	0.001	0.002	0.003
cast iron	cast iron	14½°	400	800	1600	2400
steel	cast iron	14½°	550	1100	2200	3300
steel	steel	14½°	800	1600	3200	4800
cast iron	cast iron	20° full depth	415	830	1660	2490
steel	cast iron	20° full depth	570	1140	2280	3420
steel	steel	20° full depth	830	1660	3320	4980
cast iron	cast iron	20° stub	430	860	1720	2580
steel	cast iron	20° stub	590	1180	2360	3540
steel	steel	20° stub	860	1720	3440	5160

Source: A. S. Hall, A. R. Holowenko, and H. G. Laughlin, *Theory and Problems of Machine Design*, Schaum's Outline Series, New York, McGraw-Hill, 1961. Reprinted with permission of McGraw-Hill.

permissible error, e (in.)

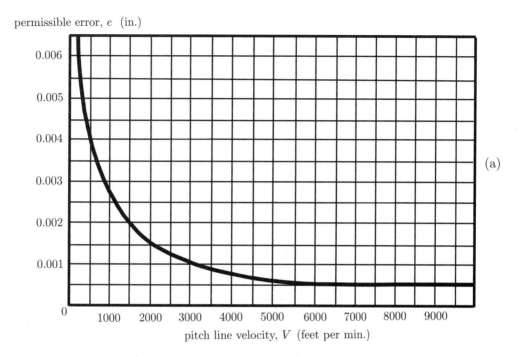

pitch line velocity, V (feet per min.)

error, e (in.)

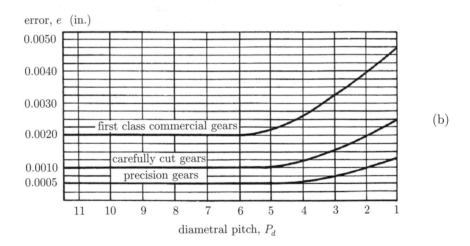

diametral pitch, P_d

Source: A. S. Hall, A. R. Holowenko, and H. G. Laughlin, Theory and Problems of Machine Design, Schaum's Outline Series McGraw-Hill, 1961.

FIGURE II.6.13 *(a) Errors in tooth profiles versus pitch line velocity and (b) errors in tooth profiles versus diametral pitch. Reprinted with permission of McGraw-Hill.*

where d_p is the pitch diameter of the smaller gear (pinion), K is the stress factor for fatigue, $Q = 2N_g/(N_p + N_g)$, N_g is the number of teeth on gear, and N_p is the number of teeth on pinion.

The stress factor for fatigue has the following expression:

$$K = \frac{\sigma_{es}^2(\sin \phi)(1/E_p + 1/E_g)}{1.4}, \tag{II.6.39}$$

where σ_{es} is the surface endurance limit of a gear pair (psi), E_p is the modulus of elasticity of the pinion material (psi), E_g is the modulus of elasticity of the gear material (psi), and ϕ is the pressure angle. Values for the modulus of elasticity are [7]:

material	E (psi)	E (GPa)
steel	30×10^6	207
cast iron	19×10^6	131
aluminum bronze	17.5×10^6	121
tin bronze	16×10^6	110

An estimated value for surface endurance is

$$\sigma_{es} = (400)(\text{BHN}) - 10\,000 \text{ psi}, \tag{II.6.40}$$

where BHN may be approximated by the average Brinell Hardness Number of the gear and pinion. The wear load F_w is an allowable load and must be greater than the dynamic load F_d. Table II.6.3 presents several tentative values of K for various materials and tooth forms.

TABLE II.6.3   Values for surface endurance limit σ_{es} and stress fatigue factor K

Average Brinell Hardness Number of steel pinion and steel gear		Surface Endurance limit σ_{es}	Stress Fatigue Factor K $14\frac{1}{2}°$	$20°$
150		50,000	30	41
200		70,000	58	79
250		90,000	96	131
300		110,000	144	196
400		150,000	268	366
Brinell Hardness Number				
Steel pinion	*Gear*			
150	C.I.	50,000	44	60
200	C.I.	70,000	87	119
250	C.I.	90,000	144	196
150	Phosphor Bronze	50,000	46	62
200	Phosphor Bronze	65,000	73	100
C.I. Pinion	C.I. Gear	80,000	152	208
C.I. Pinion	C.I. Gear	90,000	193	284

Source: A. S. Hall, A. R. Holowenko, and H. G. Laughlin, *Theory and Problems of Machine Design*, Schaum's Outline Series, New York, McGraw-Hill, 1961. Reprinted with permission of McGraw-Hill.

II.6.9 Examples

EXAMPLE II.6.1:

Two involute spur gears of module 5, with 19 and 28 teeth operate at a pressure angle of 20°. Determine whether there will be interference when standard full-depth teeth are used. Find the contact ratio.

Solution A standard full-depth tooth has the addendum of $a = m = 5$ mm. The gears will mesh at their pitch circles, and the pitch circle radii of pinion and gear are

$$r_p = m N_p/2 = 5\,(19)/2 = 47.5\,\text{mm},$$

and

$$r_g = m N_g/2 = 5\,(28)/2 = 70\,\text{mm}.$$

The theoretical center distance is

$$c = (d_p + d_g)/2 = r_p + r_g = 47.5 + 70 = 117.5\,\text{mm}.$$

The base circle radii of pinion and gear are

$$r_{bp} = r_p \cos\phi = 47.5\,\cos 20° = 44.635\,\text{mm},$$

and

$$r_{bg} = r_g \cos\phi = 70\,\cos 20° = 65.778\,\text{mm}.$$

The addendum circle radii of pinion and gear are

$$r_{ap} = r_p + a = m(N_p + 2)/2 = 52.5\,\text{mm},$$

and

$$r_{ag} = r_g + a = m(N_g + 2)/2 = 75\,\text{mm}.$$

The maximum possible addendum circle radii of pinion and gear, without interference, are

$$r_{a(max)p} = \sqrt{r_{bp}^2 + c^2 \sin^2\phi} = 60.061\,\text{mm} > r_{ap} = 52.5\,\text{mm},$$

and

$$r_{a(max)g} = \sqrt{r_{bg}^2 + c^2 \sin^2\phi} = 77.083\,\text{mm} > r_{ag} = 75\,\text{mm}.$$

Continued

EXAMPLE II.6.1: *Cont'd*

Clearly, the use of standard teeth would not cause interference.
The contact ratio is

$$CR = \frac{\sqrt{r_{ap}^2 - r_{bp}^2} + \sqrt{r_{ag}^2 - r_{bg}^2} - c \sin \phi}{\pi m \cos \phi} = 1.590,$$

which should be a suitable value ($CR > 1.2$).

EXAMPLE II.6.2:

A planetary gear train is shown in Figure II.6.14(a). The system consists of an input sun gear 1 and a planet gear 2 in mesh with 1 at B. Gear $2'$ is fixed on the shaft of gear 2. The system of gears 2 and $2'$ is carried by the arm 3. The gear $2'$ meshes with the fixed frame 0 at E.

There are three moving gears (1, 2, and 3) connected by:

- Three full joints ($c_5 = 3$): one at A, between the frame 0 and the sun gear 1; one at C, between the arm 3 and the planet gear system 2; and another at D, between the frame 0 and the arm 3.
- Two half joints ($c_4 = 2$): one at B, between the sun gear 1 and the planet gear 2, and another at E, between the planet gear $2'$ and the frame 0. The system has one degree of freedom. The sun gear has a radius of the pitch circle equal to r_1, the planet gear 2 has a radius of the pitch circle equal to r_2, the arm 3 has a length equal to r_3, and the planet gear $2'$ has a radius of the pitch circle equal to r_4.

The sun gear rotates with the input angular velocity ω_1. Find the speed ratio i_{13} between the sun gear 1 and the arm 3.

Solution The system shown in Figure II.6.14(a) has four elements (0, 1, 2, 3) and five joints. The number of independent loops is given by

$$n_c = l - p + 1 = 5 - 4 + 1 = 2.$$

This gear system has two independent contours. The diagram representing the kinematic chain and the independent contours is shown in Figure II.6.14(b).

The position vectors $\mathbf{r}_{AB}$, $\mathbf{r}_{AC}$, $\mathbf{r}_{AD}$, and $\mathbf{r}_{AE}$ are defined as follows:

$$\mathbf{r}_{AB} = x_B \mathbf{\imath} + y_B \mathbf{\jmath} = x_B \mathbf{\imath} + r_1 \mathbf{\jmath},$$

$$\mathbf{r}_{AC} = x_C \mathbf{\imath} + y_C \mathbf{\jmath} = x_C \mathbf{\imath} + (r_1 + r_2) \mathbf{\jmath},$$

$$\mathbf{r}_{AD} = x_D \mathbf{\imath} + y_D \mathbf{\jmath} = x_D \mathbf{\imath} + (r_1 + r_2 - r_3) \mathbf{\jmath},$$

$$\mathbf{r}_{AE} = x_E \mathbf{\imath} + y_E \mathbf{\jmath} = x_E \mathbf{\imath} + (r_1 + r_2 - r_4) \mathbf{\jmath}.$$

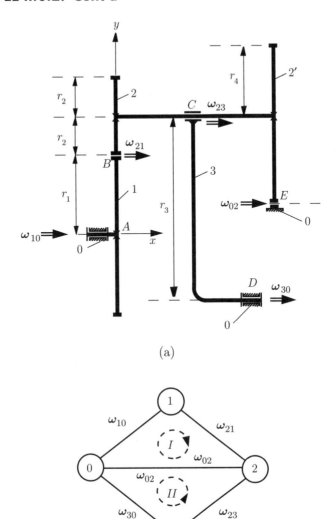

(a)

(b)

FIGURE II.6.14 *(a) Planetary gear train for Example II.6.2; (b) contour diagram.*

FIRST CONTOUR. The first closed contour contains the elements 0, 1, 2, and 0 (following the clockwise path). For the velocity analysis, the following vectorial equations can be written:

$$\boldsymbol{\omega}_{10} + \boldsymbol{\omega}_{21} + \boldsymbol{\omega}_{02} = \mathbf{0},$$

$$\mathbf{r}_{AB} \times \boldsymbol{\omega}_{21} + \mathbf{r}_{AE} \times \boldsymbol{\omega}_{02} = \mathbf{0}, \qquad (\text{II.6.41})$$

Continued

EXAMPLE II.6.2: *Cont'd*

where the input angular velocity is

$$\boldsymbol{\omega}_{10} = \omega_{10}\,\mathbf{I},$$

and the unknown angular velocities are

$$\boldsymbol{\omega}_{21} = \omega_{21}\,\mathbf{I},$$

$$\boldsymbol{\omega}_{02} = \omega_{02}\,\mathbf{I}.$$

Equation (II.6.41) becomes

$$\omega_1\,\mathbf{I} + \omega_{21}\,\mathbf{I} + \omega_{02}\,\mathbf{I} = \mathbf{0},$$

$$\begin{vmatrix} \mathbf{I} & \mathbf{J} & \mathbf{k} \\ x_B & y_B & 0 \\ \omega_{21} & 0 & 0 \end{vmatrix} + \begin{vmatrix} \mathbf{I} & \mathbf{J} & \mathbf{k} \\ x_E & y_E & 0 \\ \omega_{02} & 0 & 0 \end{vmatrix} = \mathbf{0}. \qquad \text{(II.6.42)}$$

Equation (II.6.42) projected onto a "fixed" reference frame $xOyz$ is

$$\omega_1 + \omega_{21} + \omega_{02} = 0,$$

$$y_B\,\omega_{21} + y_E\,\omega_{02} = 0. \qquad \text{(II.6.43)}$$

Equation (II.6.43) represents a system of two equations with two unknowns, ω_{21} and ω_{02}. Solving the algebraic equations, the following value is obtained for the absolute angular velocity of planet gear 2:

$$\omega_{20} = -\omega_{02} = -\frac{r_1\,\omega_1}{r_2 - r_4}. \qquad \text{(II.6.44)}$$

SECOND CONTOUR. The second closed contour contains the elements 0, 3, 2, and 0 (counterclockwise path). For the velocity analysis, the following vectorial equations can be written:

$$\boldsymbol{\omega}_{30} + \boldsymbol{\omega}_{23} + \boldsymbol{\omega}_{02} = \mathbf{0},$$

$$\mathbf{r}_{AD} \times \boldsymbol{\omega}_{30} + \mathbf{r}_{AC} \times \boldsymbol{\omega}_{23} + \mathbf{r}_{AE} \times \boldsymbol{\omega}_{02} = \mathbf{0}, \qquad \text{(II.6.45)}$$

The unknown angular velocities are

$$\boldsymbol{\omega}_{30} = \omega_{21}\,\mathbf{I},$$

$$\boldsymbol{\omega}_{23} = \omega_{23}\,\mathbf{I}.$$

EXAMPLE II.6.2: *Cont'd*

Solving Eq. (II.6.45), the following value is obtained for the absolute angular velocity of the arm 3:

$$\omega_{30} = \frac{r_1 \, r_4 \, \omega_1}{r_3 \, (-r_2 + r_4)}.$$ (II.6.46)

The speed ratio is

$$i_{13} = \frac{\omega_{10}}{\omega_{30}} = \frac{\omega_1}{\omega_{30}} = \frac{r_3 \, (-r_2 + r_4)}{r_1 \, r_4}.$$ (II.6.47)

EXAMPLE II.6.3:

Figure II.6.15 shows a planetary gear train. The schematic representation of the planetary gear train is depicted in Figure II.6.16(a). The system consists of an input sun gear 1 and a planet gear 2 in mesh with 1 at B. Gear 2 is carried by the arm S fixed on the shaft of gear 3, as shown. Gear 3 meshes with the output gear 4 at F. The fixed ring gear 4 meshes with the planet gear 2 at D.

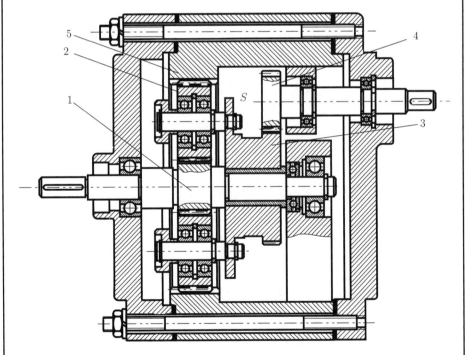

FIGURE II.6.15   *Drawing for the planetary gear train for Example II.6.3.*

Continued

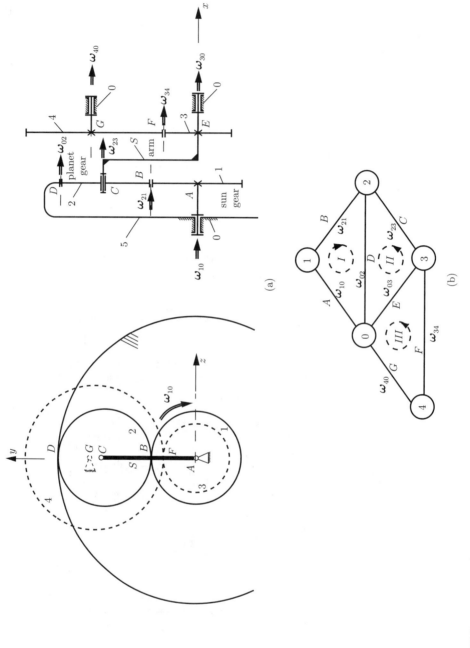

FIGURE II.6.16   (a) Schematic representation of the planetary gear train for Example II.6.3; (b) contour diagram.

EXAMPLE II.6.3: *Cont'd*

There are four moving gears (1, 2, 3, and 4) connected by:

- Four full joints ($c_5 = 4$): one at A, between the frame 0 and the sun gear 1; one at C, between the arm S and the planet gear 2; one at E, between the frame 0 and the gear 3, and another at G, between the frame 0 and the gear 3.
- Three half joints ($c_4 = 3$): one at B, between the sun gear 1 and the planet gear 2; one at D, between the planet gear 2 and the ring gear; and another at F, between the gear 3 and the output gear 4. The module of the gears is $m = 5$ mm. The system has one degree of freedom. The sun gear has $N_1 = 19$ external gear teeth, the planet gear has $N_2 = 28$ external gear teeth, and the fixed ring gear has $N_5 = 75$ internal gear teeth. Gear 3 has $N_3 = 18$ external gear teeth, and the output gear has $N_4 = 36$ external gear teeth. The sun gear rotates with an input angular speed $n_1 = 2970$ rpm ($\omega_1 = \omega_{10} = \pi n_1/30 = 311.018$ rad/s). Find the absolute output angular velocity of gear 4, the velocities of the pitch points B and F, and the velocity of joint C.

Solution   The velocity analysis is carried out using the contour equation method. The system shown in Figure II.6.16(a) has five elements (0, 1, 2, 3, 4) and seven joints. The number of independent loops is given by

$$n_c = l - p + 1 = 7 - 5 + 1 = 3.$$

This gear system has three independent contours. The diagram representing the kinematic chain and the independent contours is shown in Figure II.6.16(b).

The position vectors $\mathbf{r}_{AB}$, $\mathbf{r}_{AC}$, $\mathbf{r}_{AD}$, $\mathbf{r}_{AF}$, and $\mathbf{r}_{AG}$ are defined as follows:

$$\mathbf{r}_{AB} = x_B \imath + y_B \jmath = x_B \imath + r_1 \jmath = x_B \imath + \frac{mN_1}{2} \jmath,$$

$$\mathbf{r}_{AC} = x_C \imath + y_C \jmath = x_C \imath + (r_1 + r_2) \jmath = x_C \imath + \frac{m(N_1 + N_2)}{2} \jmath.$$

$$\mathbf{r}_{AD} = x_D \imath + y_D \jmath = x_D \imath (r_1 + 2r_2) \jmath = x_D \imath + \frac{m(N_1 + 2N_2)}{2} \jmath,$$

$$\mathbf{r}_{AF} = x_F \imath + y_F \jmath = x_F \imath + r_3 \jmath = x_F \imath + \frac{mN_3}{2} \jmath,$$

$$\mathbf{r}_{AG} = x_G \imath + y_G \jmath = x_G \imath + (r_3 + r_4) \jmath = x_G \imath + \frac{m(N_3 + N_4)}{2} \jmath.$$

FIRST CONTOUR.   The first closed contour contains the elements 0, 1, 2, and 0 (following the clockwise path). For the velocity analysis, the following vectorial equations can be written:

$$\sum_{(i)} \boldsymbol{\omega}_{i,i-1} = \mathbf{0} \quad \Longrightarrow \quad \boldsymbol{\omega}_{10} + \boldsymbol{\omega}_{21} + \boldsymbol{\omega}_{02} = \mathbf{0},$$

$$\sum_{(i)} \mathbf{r}_{A_i} \times \boldsymbol{\omega}_{i,i-1} = \mathbf{0} \quad \Longrightarrow \quad \mathbf{r}_{AB} \times \boldsymbol{\omega}_{21} + \mathbf{r}_{AD} \times \boldsymbol{\omega}_{02} = \mathbf{0}, \qquad \text{(II.6.48)}$$

Continued

EXAMPLE II.6.3: *Cont'd*

where the input angular velocity is

$$\boldsymbol{\omega}_{10} = \omega_{10}\,\mathbf{I},$$

and the unknown angular velocities are

$$\boldsymbol{\omega}_{21} = \omega_{21}\,\mathbf{I}, \quad \boldsymbol{\omega}_{02} = \omega_{02}\,\mathbf{I}.$$

The sign of the relative angular velocities is selected to be positive, and then the numerical results give the real orientation of the vectors.

Equation (II.6.48) becomes

$$\omega_1\,\mathbf{I} + \omega_{21}\,\mathbf{I} + \omega_{02}\,\mathbf{I} = \mathbf{0},$$

$$\begin{vmatrix} \mathbf{I} & \mathbf{J} & \mathbf{k} \\ x_B & y_B & 0 \\ \omega_{21} & 0 & 0 \end{vmatrix} + \begin{vmatrix} \mathbf{I} & \mathbf{J} & \mathbf{k} \\ x_D & y_D & 0 \\ \omega_{02} & 0 & 0 \end{vmatrix} = \mathbf{0}. \tag{II.6.49}$$

Equation (II.6.49) projected on a "fixed" reference frame $xOyz$ is

$$\omega_1 + \omega_{21} + \omega_{02} = 0,$$

$$y_B\,\omega_{21} + y_D\,\omega_{02} = 0. \tag{II.6.50}$$

Equation (II.6.50) represents a system of two equations with two unknowns, ω_{21} and ω_{02}. Solving the algebraic equations, the following value is obtained for the absolute angular velocity of the planet gear 2:

$$\omega_{20} = -\omega_{02} = -\frac{N_1\,\omega_1}{2\,N_2} = -\frac{19\,(311.018)}{2\,(28)} = -105.524\,\text{rad/s}. \tag{II.6.51}$$

SECOND CONTOUR. The second closed contour contains the elements 0, 3, 2, and 0 (following the counterclockwise path). For the velocity analysis, the following vectorial equations can be written:

$$\boldsymbol{\omega}_{30} + \boldsymbol{\omega}_{23} + \boldsymbol{\omega}_{02} = \mathbf{0},$$

$$\mathbf{r}_{AE} \times \boldsymbol{\omega}_{30} + \mathbf{r}_{AC} \times \boldsymbol{\omega}_{23} + \mathbf{r}_{AD} \times \boldsymbol{\omega}_{02} = \mathbf{0}. \tag{II.6.52}$$

The unknown angular velocities are

$$\boldsymbol{\omega}_{30} = \omega_{21}\,\mathbf{I}, \quad \boldsymbol{\omega}_{23} = \omega_{23}\,\mathbf{I}.$$

Solving Eq. (II.6.52) the following value is obtained for the absolute angular velocity of gear 3 and arm S:

$$\omega_{30} = \frac{N_1\,\omega_1}{2\,(N_1 + N_2)} = \frac{19\,(311.018)}{2\,(19 + 28)} = 62.865\,\text{rad/s}.$$

EXAMPLE II.6.3: *Cont'd*

The absolute angular velocity of the output gear 4 is

$$\omega_{40} = -\frac{\omega_{30} N_3}{N_4} = -\frac{N_1 N_3 \omega_1}{2(N_1 + N_2)N_4} = -\frac{19(18)(311.018)}{2(19+28)(36)}$$

$$= -31.432 \text{ rad/s}.$$

LINEAR VELOCITIES OF PITCH POINTS. The velocity of the pitch point B is

$$v_B = \omega_{10} r_1 = 311.018(0.005)(19)/2 = 14.773 \text{ m/s},$$

and the velocity of the pitch point F is

$$v_F = \omega_{40} r_4 = 31.432(0.005)(36)/2 = 2.828 \text{ m/s}.$$

The velocity of the joint C is

$$v_C = \omega_{30}(r_1 + r_2) = 62.865(0.005)(19+28)/2 = 7.386 \text{ m/s}.$$

GEAR GEOMETRICAL DIMENSIONS. For standard external gear teeth the addendum is $a = m$.

Gear 1

pitch circle diameter $d_1 = mN_1 = 95.0$ mm;
addendum circle diameter $d_{a1} = m(N_1 + 2) = 105.0$ mm;
dedendum circle diameter $d_{d1} = m(N_1 - 2.5) = 82.5$ mm.

Gear 2

pitch circle diameter $d_2 = mN_2 = 140.0$ mm;
addendum circle diameter $d_{a2} = m(N_2 + 2) = 150.0$ mm;
dedendum circle diameter $d_{d2} = m(N_2 - 2.5) = 127.5$ mm.

Gear 3

pitch circle diameter $d_3 = mN_3 = 90.0$ mm;
addendum circle diameter $d_{a3} = m(N_3 + 2) = 100.0$ mm;
dedendum circle diameter $d_{d3} = m(N_3 - 2.5) = 77.5$ mm.

Gear 4

pitch circle diameter $d_4 = mN_4 = 180.0$ mm;
addendum circle diameter $d_{a4} = m(N_4 + 2) = 190.0$ mm;
dedendum circle diameter $d_{d4} = m(N_4 - 2.5) = 167.5$ mm.

Gear 5 (internal gear)

pitch circle diameter $d_5 = mN_5 = 375.0$ mm;
addendum circle diameter $d_{a5} = m(N_5 - 2) = 365.0$ mm;
dedendum circle diameter $d_{d5} = m(N_5 + 2.5) = 387.5$ mm.

Continued

EXAMPLE II.6.3: *Cont'd*

NUMBER OF PLANET GEARS. The number of necessary planet gears, k, is given by the assembly condition

$$(N_1 + N_5)/k = INTEGER,$$

and for the planetary gear train $k = 2$ planet gears. The vicinity condition between the sun gear and the planet gear,

$$m (N_1 + N_2) \sin(\pi/k) > d_{a2}$$

is thus verified.

EXAMPLE II.6.4:

Figure II.6.17 shows a two-stage gear reducer with identical pairs of gears. An electric motor with the power $H = 2$ kW and $n_1 = 900$ rpm is coupled to the shaft a. On this shaft the input driver gear 1 is rigidly connected with the number of teeth, $N_1 = N_p = 17$. The speed reducer uses a countershaft b with two rigidly connected gears, 2 and $2'$, having $N_2 = N_g = 51$ teeth and $N_{2'} = N_p = 17$ teeth. The output gear 3 has $N_3 = N_g = 51$ teeth and is rigidly fixed to the shaft c coupled to

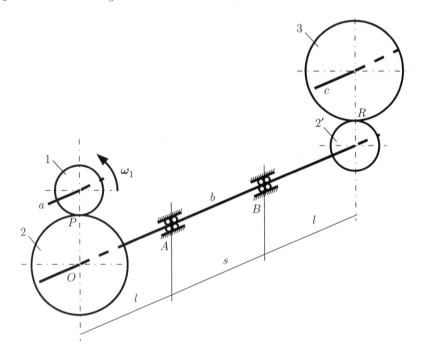

FIGURE II.6.17 *Two-stage gear reducer for Example II.6.4.*

EXAMPLE II.6.4: *Cont'd*

the driven machine. The input shaft a and output shaft c are collinear. The countershaft b turns freely in bearings A and B. The gears mesh along the pitch diameter and the shafts are parallel. The diametral pitch for each stage is $P_d = 5$, and the pressure angle is $\phi = 20°$. The distance between the bearings is $s = 100$ mm, and the distance $l = 25$ mm (Fig. II.6.17). The gear reducer is a part of an industrial machine intended for continuous one-shift (8 h pd). Select identical extra-light series (L00) ball bearings for A and B.

Solution

GEOMETRY. The pitch diameters of pinions 1 and $2'$ are $d_1 = d_{2'} = d_p = N_p/P_d = 17/5 = 3.4$ in. The pitch diameters of gears 2 and 3 are $d_2 = d_3 = d_g = N_g/P_d = 51/5 = 10.2$ in. The circular pitch is $p = \pi/P_d = 3.14/5 = 0.63$ in.

ANGULAR SPEEDS. The following relation exists for the first stage:

$$\frac{n_1}{n_2} = \frac{N_2}{N_1} \quad \Rightarrow \quad n_2 = n_1 \frac{N_1}{N_2} = 900 \frac{17}{51} = 300 \text{ rpm},$$

and for the second stage:

$$\frac{n_2}{n_3} = \frac{N_3}{N_{2'}} \quad \Rightarrow \quad n_3 = n_2 \frac{N_{2'}}{N_3} = 300 \frac{17}{51} = 100 \text{ rpm}.$$

The angular speed of the coutershaft b is $n_b = n_2 = 300$ rpm, and the angular speed of the driven shaft c is $n_c = n_3 = 100$ rpm.

TORQUE CARRIED BY EACH OF THE SHAFTS. The relation between the power H_a of the motor and the torque M_a in shaft a is

$$H_a = \frac{M_a \, n_a}{9549},$$

and the torque M_a in shaft a is

$$M_a = \frac{9549 \, H_a}{n_a} = \frac{9549 \, (2 \text{ kW})}{900 \text{ rpm}} = 21.22 \text{ N} \cdot \text{m}.$$

The torque in shaft b is

$$M_b = \frac{9549 \, H_a}{n_b} = M_a \frac{N_2}{N_1} = 21.22 \frac{51}{17} = 63.66 \text{ N} \cdot \text{m},$$

and the torque in shaft c is

$$M_c = \frac{9549 \, H_a}{n_c} = M_b \frac{N_3}{N_{2'}} = 63.66 \frac{51}{17} = 190.98 \text{ N} \cdot \text{m}.$$

Continued

EXAMPLE II.6.4: *Cont'd*

BEARING REACTIONS. All the gear radial and tangential force is transferred at the pitch point P. The tangential force on the motor pinion is

$$F_t = \frac{M_a}{r_p} = \frac{21.22}{0.0431} = 492.34 \text{ N},$$

where $r_p = d_p/2 = 1.7$ in $= 0.0431$ m. The radial force on the motor pinion is

$$F_r = F_t \tan \phi = 492.34 \tan 20° = 179.2 \text{ N}.$$

The force on the motor pinion 1 at P (Fig. II.6.17) is

$$\mathbf{F}_{21} = F_{r21}\mathbf{J} + F_{t21}\mathbf{k} = 179.2\mathbf{J} - 492.34\mathbf{k} \text{ N}.$$

The force on the countershaft gear 2 at P is

$$\mathbf{F}_{12} = -\mathbf{F}_{21} = F_{r12}\mathbf{J} + F_{t12}\mathbf{k} = -179.2\mathbf{J} + 492.34\mathbf{k} \text{ N}.$$

The forces on the countershaft pinion $2'$ at R are three times as large, i.e.,

$$F_{t'} = \frac{M_b}{r_p} = \frac{63.66}{0.0431} = 1477 \text{ N},$$

$$F_{r'} = F_{t'} \tan \phi = 1477 \tan 20° = 537.6 \text{ N},$$

and

$$\mathbf{F}_{32'} = F_{r32'}\mathbf{J} + F_{t32'}\mathbf{k} = -537.6\mathbf{J} - 1477\mathbf{k} \text{ N}.$$

The unknown forces applied to bearings A and B can be written as

$$\mathbf{F}_A = F_{Ay}\mathbf{J} + F_{Az}\mathbf{k},$$

$$\mathbf{F}_B = F_{By}\mathbf{J} + F_{Bz}\mathbf{k}.$$

The sketch of the countershaft as a free body in equilibrium is shown in Figure II.6.18. To determine these forces two vectorial equations are used. Sum of moments of all forces that act on the countershaft with respect to A are zero:

$$\sum \mathbf{M}_A = \mathbf{r}_{AP} \times \mathbf{F}_{12} + \mathbf{r}_{AR} \times \mathbf{F}_{32'} + \mathbf{r}_{AB} \times \mathbf{F}_B$$

$$= \begin{vmatrix} \mathbf{I} & \mathbf{J} & \mathbf{k} \\ -l & r_2 & 0 \\ 0 & F_{r12} & F_{t12} \end{vmatrix} + \begin{vmatrix} \mathbf{I} & \mathbf{J} & \mathbf{k} \\ s+l & r_{2'} & 0 \\ 0 & F_{r32'} & F_{t32'} \end{vmatrix}$$

$$+ \begin{vmatrix} \mathbf{I} & \mathbf{J} & \mathbf{k} \\ s & 0 & 0 \\ 0 & F_{By} & F_{Bz} \end{vmatrix} = \mathbf{0},$$

EXAMPLE II.6.4: *Cont'd*

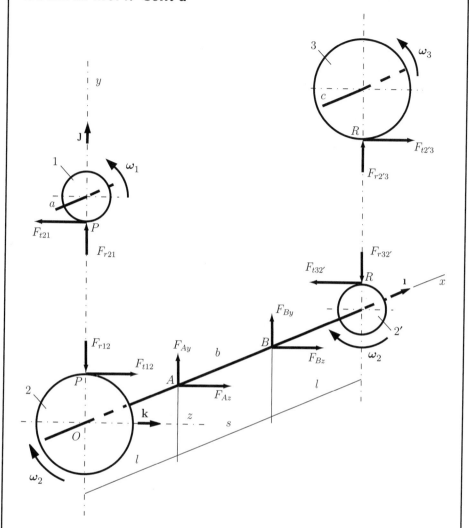

FIGURE II.6.18 *Free-body diagrams for Example II.6.4.*

or

$$\sum \mathbf{M}_A \cdot \mathbf{J} = lF_{t12} - (s+l)F_{t32'} - sF_{Bz} = 0,$$

$$\sum \mathbf{M}_A \cdot \mathbf{k} = -lF_{r12} + (s+l)F_{r32'} + sF_{By} = 0.$$

From the above equations $F_{By} = 627.2$ N, and $F_{Bz} = 1969.33$ N. The radial force at B is

$$F_B = \sqrt{F_{By}^2 + F_{Bz}^2} = 2066.8 \quad \text{N}.$$

Continued

EXAMPLE II.6.4: *Cont'd*

Sum of all forces that act on the countershaft are zero:

$$\sum \mathbf{F} = \mathbf{F}_{12} + \mathbf{F}_A + \mathbf{F}_B + \mathbf{F}_{32'} = \mathbf{0},$$

or

$$-F_{r12} + F_{Ay} + F_{By} - F_{r32'} = 0,$$
$$F_{t12} + F_{Az} + F_{Bz} - F_{t32'} = 0. \quad \text{(II.6.53)}$$

From Eq. (II.6.53) $F_{Ay} = 89.6$ N, and $F_{Az} = -984.67$ N. The radial force at A is

$$F_A = \sqrt{F_{Ay}^2 + F_{Az}^2} = 988.73 \text{ N}.$$

BALL BEARING SELECTION. Since the radial force at B is greater than the radial force at A, $F_B > F_A$, the bearing selection will be based on bearing B. The equivalent radial force for radial ball bearings is $F_e = F_B = 2066.8$ N. The ball bearings operate 8 hours per day, 5 days per week.

From Table II.3.3 choose $K_a = 1.1$ for gearing. From Table II.3.4 choose (conservatively) 30 000-hour life.

The life in revolutions is

$$L = 300 \text{ rpm} \times 30\,000 \text{ h} \times 60 \text{ min/h} = 540 \times 10^6 \text{ rev}.$$

For standard 90% reliability ($K_r = 1$, Fig. II.3.9), and for $L_R = 90 \times 10^6$ rev (for use with Table II.3.2), the rated capacity is

$$C_{req} = K_a F_e \left(\frac{L}{K_r L_R} \right)^{0.3}$$

$$= (1.1)(2066.8) \left[\frac{540 \times 10^6}{(1)\,90 \times 10^6} \right]^{0.3} = 3891.67 \text{ N}$$

$$\approx 3.9 \text{ kN}.$$

From Table II.3.2 with 3.9 kN for L00 series $\Longrightarrow C = 4.2$ kN and $d = 35$ mm bore.

From Table II.3.1 with 35 mm bore and L00 series the bearing number is L07. The shaft size requirement may necessitate use of a larger bore bearing.

EXAMPLE II.6.5:

Figure II.6.19(a) shows a gear set. Gear 1 is the driving or input gear; it rotates with the angular speed ω_{10}, ($\omega_{10} > 0$), and transmits an unknown motor torque M_{mot}. The output (driven) gear 2 is attached to a shaft that drives a machine. The external

EXAMPLE II.6.5: *Cont'd*

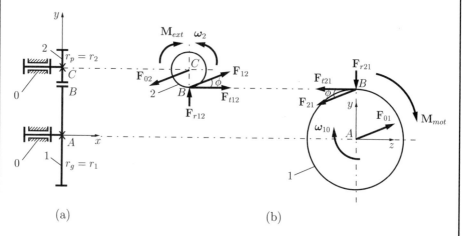

FIGURE II.6.19 *(a) Two gears in contact for Example II.6.5; (b) free-body diagrams of the gears.*

torque exerted by the machine on the gear 2 is opposite to the absolute angular velocity of the output gear, ω_{20}, and is given by

$$\mathbf{M}_{ext} = -|\mathbf{M}_{ext}|\frac{\omega_{20}}{|\omega_{20}|}.$$

The radii of the pitch circles of the two gears in contact are r_1 and r_2, and the pressure angle is ϕ. Find the motor torque (equilibrium moment) M_{mot} and the bearing reactions in terms of r_1, r_2, ω_{10}, and $|\mathbf{M}_{ext}|$. Use the following numerical application: $r_1 = 1$ m, $r_2 = 0.5$ m, $\phi = 20°$, $\omega_{10} = \dfrac{\pi}{3}$ rad/s, and $|\mathbf{M}_{ext}| = 400$ N·m.

Solution The angular speed ratio between the gears is

$$\frac{\omega_{10}}{\omega_{20}} = -\frac{r_1}{r_2}.$$

Thus, the angular speed ω_{20} of the output gear is

$$\omega_{20} = -\frac{r_2\,\omega_{10}}{r_1} = -\frac{1(\pi/3)}{0.5} = -\frac{2\pi}{3}\ \text{rad/s}.$$

The angular velocity vector of the output gear is

$$\omega_{20} = \omega_{20}\,\mathbf{1} = -\frac{2\pi}{3}\ \mathbf{1}\ \text{rad/s}.$$

The free-body diagrams of the gears are shown in Figure II.6.19(b). The external torque exerted by the machine on the gear 2 is

$$\mathbf{M}_{ext} = M_{ext}\,\mathbf{1} = -|\mathbf{M}_{ext}|\frac{\omega_{20}}{|\omega_{20}|} = -400\frac{-(2\pi/3)\,\mathbf{1}}{2\pi/3} = 400\,\mathbf{1}\ \text{N·m}.$$

Continued

EXAMPLE II.6.5: *Cont'd*

Gear 2

The moment equation for gear 2 with respect to its center C gives

$$\sum \mathbf{M}_C^{(2)} = \mathbf{r}_{CB} \times \mathbf{F}_{12} + \mathbf{M}_{ext} = \mathbf{0}, \qquad \text{(II.6.54)}$$

where $\mathbf{F}_{12} = F_{r12}\mathbf{J} + F_{t12}\mathbf{k}$ is the reaction force of gear 1 on gear 2, and $\mathbf{r}_{CB} = -r_2\mathbf{J}$. Equation (II.6.54) becomes

$$\begin{vmatrix} \mathbf{I} & \mathbf{J} & \mathbf{k} \\ 0 & -r_2 & 0 \\ 0 & F_{r12} & F_{t12} \end{vmatrix} + M_{ext}\mathbf{I} = \mathbf{0}. \qquad \text{(II.6.55)}$$

From Eq. (II.6.55) the tangential force is

$$F_{t12} = \frac{M_{ext}}{r_2} = \frac{400}{0.5} = 800 \ \text{N}.$$

The radial reaction force F_{r12} is

$$F_{r12} = F_{t12} \tan \phi = \frac{M_{ext}}{r_2} \tan \phi = 800 \tan 20° = 291.176 \ \text{N}.$$

The force equation for gear 2 gives

$$\sum \mathbf{F}^{(2)} = \mathbf{F}_{12} + \mathbf{F}_{02} = \mathbf{0},$$

and the reaction force of the ground on gear 2 is

$$\mathbf{F}_{02} = -\mathbf{F}_{12} = -291.176\mathbf{J} - 800\mathbf{k} \ \text{N}.$$

Gear 1 (driver)

The moment equation for gear 1 with respect its center A gives

$$\sum \mathbf{M}_A^{(1)} = \mathbf{r}_{AB} \times \mathbf{F}_{21} + \mathbf{M}_{mot} = \mathbf{0}, \qquad \text{(II.6.56)}$$

where $\mathbf{F}_{21} = -\mathbf{F}_{12}$, and $\mathbf{r}_{AB} = r_1\mathbf{J}$. Equation (II.6.56) becomes

$$\begin{vmatrix} \mathbf{I} & \mathbf{J} & \mathbf{k} \\ 0 & r_1 & 0 \\ 0 & -F_{r12} & -F_{t12} \end{vmatrix} + M_{mot}\mathbf{I} = \mathbf{0},$$

and the motor torque M_{mot} is

$$M_{mot} = F_{t12} \, r_1 = \frac{r_1}{r_2} M_{ext} = \frac{1}{0.5} 400 = 800 \ \text{N} \cdot \text{m}.$$

EXAMPLE II.6.5: *Cont'd*

The force equation for gear 1 is

$$\sum \mathbf{F}^{(1)} = \mathbf{F}_{21} + \mathbf{F}_{01} = \mathbf{0},$$

and the reaction force of the ground on gear 1 is

$$\mathbf{F}_{01} = -\mathbf{F}_{21} = \mathbf{F}_{12} = 291.176\mathbf{j} + 800\mathbf{k} \text{ N}.$$

EXAMPLE II.6.6:

A planetary gear train is shown in Figure II.6.20(a). The planet gear 2 rotates around the sun gear 1. The arm 3 is connected to the planet gear at the point C (pin joint) and to the ground 0 at the point D (pin joint). The sun gear is connected to the

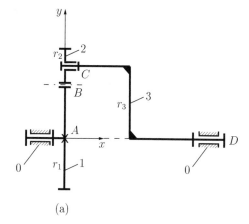

(a)

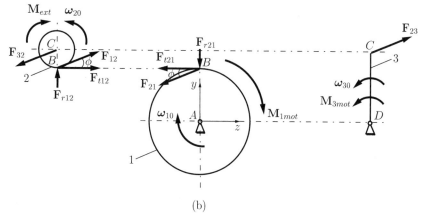

(b)

FIGURE II.6.20   (a) Gear train for Example II.6.6; (b) free-body diagrams of the gears.

Continued

EXAMPLE II.6.6: *Cont'd*

ground at the point A (pin joint). A motor drives the sun gear with the angular speed $\omega_{10} = 2\pi/3$ rad/s ($n_1 = 20$ rpm). A second motor is connected to the arm and has the angular speed $\omega_{30} = -\pi/3$ rad/s ($n_3 = -10$ rpm). The radii of the pitch circles of the sun gear 1 and planet gear 2 are $r_1 = 1$ m and $r_2 = 0.5$ m, respectively.

An external moment $\mathbf{M}_{ext} = -|\mathbf{M}_{ext}|\dfrac{\boldsymbol{\omega}_{20}}{|\boldsymbol{\omega}_{20}|}$ acts on the planet gear 2, where $|\mathbf{M}_{ext}| = 400$ N·m. The pressure angle of the gears is $\phi = 20°$. Find the equilibrium moments (motor moments) M_{1mot} and M_{3mot} that act on gear 1 and arm 3, and the reaction forces.

Solution First, the angular velocity $\boldsymbol{\omega}_{20}$ of the planet gear will be calculated. The gear train has one contour: $0 - A - 1 - B - 2 - C - 3 - D - 0$.
For this contour the relations between the relative angular velocities of the links are

$$\boldsymbol{\omega}_{10} + \boldsymbol{\omega}_{21} + \boldsymbol{\omega}_{32} + \boldsymbol{\omega}_{03} = \mathbf{0},$$

$$\mathbf{r}_{AB} \times \boldsymbol{\omega}_{21} + \mathbf{r}_{AC} \times \boldsymbol{\omega}_{32} = \mathbf{0}, \qquad \text{(II.6.57)}$$

where $\boldsymbol{\omega}_{10} = \omega_{10}\,\mathbf{i}$, $\boldsymbol{\omega}_{03} = -\boldsymbol{\omega}_{30} = \omega_{30}\,\mathbf{i}$, $\mathbf{r}_{AB}$, and $\mathbf{r}_{AC}$ are known.
Equation (II.6.57) can be solved simultaneously with respect to the two unknowns, $\boldsymbol{\omega}_{21}$ and $\boldsymbol{\omega}_{32}$. The solutions are

$$\boldsymbol{\omega}_{21} = -3\pi\,\mathbf{i}\ \text{rad/s} \qquad \text{and} \qquad \boldsymbol{\omega}_{32} = 2\pi\,\mathbf{i}\ \text{rad/s}.$$

The angular speed $\boldsymbol{\omega}_{20}$ of the planet gear 2 is

$$\boldsymbol{\omega}_{20} = \boldsymbol{\omega}_{10} + \boldsymbol{\omega}_{21} = -\frac{7\pi}{3}\mathbf{i}\ \text{rad/s}\ \ (n_2 = -70\,\text{rpm}).$$

Figure II.6.20(b) shows the free-body diagrams of gear 2, gear 1, and arm 3. The external torque $\mathbf{M}_{ext}$ on the driven gear 2 is

$$\mathbf{M}_{ext} = M_{ext}\,\mathbf{i} = -|\mathbf{M}_{ext}|\frac{\boldsymbol{\omega}_{20}}{|\boldsymbol{\omega}_{20}|} = -400\,\frac{-7\pi/3\,\mathbf{i}}{7\pi/3} = 400\,\mathbf{i}\ \text{N}\cdot\text{m},$$

and the force analysis starts with the driven planet gear.

Gear 2
The sum of the moments with respect to the center C for the planet gear 2 gives

$$\sum \mathbf{M}_C^{(2)} = \mathbf{r}_{CB} \times \mathbf{F}_{12} + \mathbf{M}_{ext} = \mathbf{0}, \ \ \text{or} \ \ \begin{vmatrix} \mathbf{i} & \mathbf{j} & \mathbf{k} \\ 0 & -r_2 & 0 \\ 0 & F_{r12} & F_{t12} \end{vmatrix} + M_{ext}\mathbf{i} = \mathbf{0}. \ \ \text{(II.6.58)}$$

where $\mathbf{F}_{12} = F_{r12}\,\mathbf{j} + F_{t12}\,\mathbf{k}$ and $\mathbf{r}_{CB} = -r_2\,\mathbf{j}$. From Eq. (II.6.58), it results:

$$F_{t12} = \frac{M_{ext}}{r_2} = \frac{400}{0.5} = 800\ \text{N}.$$

EXAMPLE II.6.6: *Cont'd*

The radial reaction force F_{r12} is

$$F_{r12} = F_{t12} \tan \phi = \frac{M_{ext}}{r_2} \tan \phi = 800 \, \tan 20° = 291.176 \, \text{N}.$$

The sum of the forces for the planet gear 2 is

$$\sum \mathbf{F}^{(2)} = \mathbf{F}_{12} + \mathbf{F}_{32} = \mathbf{0},$$

and the reaction force of arm 3 on gear 2 is

$$\mathbf{F}_{32} = -\mathbf{F}_{12} = F_{32y}\,\mathbf{J} + F_{32z}\,\mathbf{k} = -291.176\,\mathbf{J} - 800\,\mathbf{k} \ \text{N}.$$

Gear 1
The sum of moments for gear 1 with respect to its center A is

$$\sum \mathbf{M}_A^{(1)} = \mathbf{r}_{AB} \times \mathbf{F}_{21} + \mathbf{M}_{1mot} = \mathbf{0},$$

where $\mathbf{F}_{21} = -\mathbf{F}_{12}$ and $\mathbf{r}_{AB} = r_1\,\mathbf{J}$. The motor torque (equilibrium moment) M_{1mot} is

$$M_{1mot} = -F_{t21}\,r_1 = F_{t12}\,r_1 = \frac{M_{ext}}{r_2}\,r_1 = 800 \, \text{N} \cdot \text{m}.$$

The sum of the forces for the gear 1 is

$$\sum \mathbf{F}^{(1)} = \mathbf{F}_{21} + \mathbf{F}_{01} = \mathbf{0},$$

and the reaction force of the ground on gear 1 is

$$\mathbf{F}_{01} = -\mathbf{F}_{21} = 291.176\,\mathbf{J} + 800\,\mathbf{k} \ \text{N}.$$

Arm 3
The sum of the moments for arm 3 with respect to the point D is

$$\sum \mathbf{M}_D^{(3)} = \mathbf{r}_{DC} \times \mathbf{F}_{23} + \mathbf{M}_{3mot} = \mathbf{0},$$

where $\mathbf{F}_{23} = -\mathbf{F}_{32}$ and $\mathbf{r}_{DC} = (r_1 + r_2)\mathbf{J}$. The motor torque (equilibrium moment) M_3 can be computed as

$$M_{3mot} = -F_{23z}\,(r_1 + r_2) = -F_{t12}\,(r_1 + r_2) = -\frac{M_{ext}}{r_2}\,(r_1 + r_2) = -1200 \, \text{N} \cdot \text{m}.$$

Continued

EXAMPLE II.6.6: *Cont'd*

For arm 3 the sum of forces equation is

$$\sum \mathbf{F}^{(3)} = \mathbf{F}_{23} + \mathbf{F}_{03} = \mathbf{0},$$

and the reaction force of the ground on arm 3 is

$$\mathbf{F}_{03} = -\mathbf{F}_{23} = -291.176\,\mathbf{J} - 800\,\mathbf{k}\ \text{N}.$$

EXAMPLE II.6.7:

A planetary gear train with one degree of freedom is shown in Figure II.6.21(a). The sun gear 1 is connected to the ground with a pin joint at point A. The arm 3 is connected with pin joints to the planet gear 2 at point C and to the ground at point D. The planet gear 2 is also in contact to gear 4 (an internal gear) which is fixed to the ground ($4 = 0$). The angular speed of the motor that drives the sun gear is $\omega_{10} = 2\pi/3$ rad/s ($n_1 = 20$ rpm). The radii of the pitch circles of the sun gear 1 and planet gear 2 are $r_1 = 1$ m and $r_2 = 0.5$ m. An external moment $\mathbf{M}_{ext} = -|\mathbf{M}_{ext}|\dfrac{\boldsymbol{\omega}_{30}}{|\boldsymbol{\omega}_{30}|}$ acts on the driven arm 3, where $|\mathbf{M}_{ext}| = 400$ N·m. The pressure angle of the gears is $\phi = 20°$. Find the equilibrium moment (motor moment) M_{mot} that acts on the sun gear and the reaction forces.

Solution

Contour $0 - A - 1 - B - 2 - E - 0$

For the relative angular velocities of the gears the following relations can be written:

$$\boldsymbol{\omega}_{10} + \boldsymbol{\omega}_{21} + \boldsymbol{\omega}_{02} = \mathbf{0},$$
$$\mathbf{r}_{AB} \times \boldsymbol{\omega}_{21} + \mathbf{r}_{AE} \times \boldsymbol{\omega}_{02} = \mathbf{0},$$

where $\boldsymbol{\omega}_{10} = 2\pi/3\,\mathbf{1}$ rad/s. The solutions of the system are $\boldsymbol{\omega}_{21}$ and $\boldsymbol{\omega}_{02}$:

$$\boldsymbol{\omega}_{21} = -4\pi/3\,\mathbf{1}\ \text{rad/s}, \quad \boldsymbol{\omega}_{02} = 2\pi/3\,\mathbf{1}\ \text{rad/s}.$$

The angular speed $\boldsymbol{\omega}_{20}$ of the planet gear 2 can be computed as

$$\boldsymbol{\omega}_{20} = -\boldsymbol{\omega}_{02} = -2\pi/3\,\mathbf{1}\ \text{rad/s}.$$

Contour $0 - E - 2 - C - 3 - D - 0$

EXAMPLE II.6.7: *Cont'd*

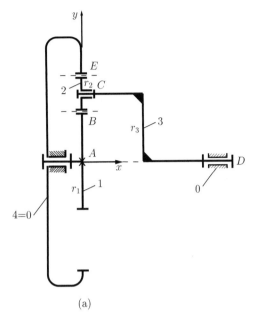

(a)

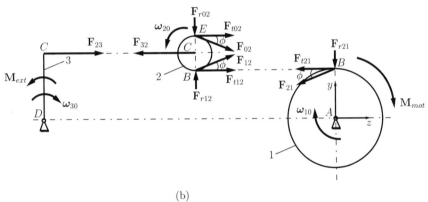

(b)

FIGURE II.6.21 *(a) Planetary gear train for Example II.6.7; (b) free-body diagrams.*

For this contour the relative angular velocity equations are

$$\boldsymbol{\omega}_{20} + \boldsymbol{\omega}_{32} + \boldsymbol{\omega}_{03} = \mathbf{0},$$

$$\mathbf{r}_{AE} \times \boldsymbol{\omega}_{20} + \mathbf{r}_{AC} \times \boldsymbol{\omega}_{32} = \mathbf{0}.$$

The relative angular speed $\boldsymbol{\omega}_{32}$ is

$$\boldsymbol{\omega}_{32} = \frac{26.6\,\pi}{30}\,\mathbf{\imath}\ \ \text{rad/s}.$$

Continued

EXAMPLE II.6.7: *Cont'd*

The angular speed ω_{30} of the arm 3 is

$$\omega_{30} = -\omega_{03} = \omega_{20} + \omega_{32} = \frac{6.6\,\pi}{30}\,\imath \ \text{rad/s}.$$

Figure II.6.21(b) shows the free-body diagrams of arm 3, gear 2, and gear 1. The external torque $\mathbf{M}_{ext}$ on the driven arm 3 is

$$\mathbf{M}_{ext} = M_{ext}\,\imath = -|\mathbf{M}_{ext}|\,\frac{\omega_{30}}{|\omega_{30}|} = -400\,\imath \ \text{N}\cdot\text{m},$$

and the force analysis starts with the driven planet gear.

Arm 3 (driven)
The moment equation with respect to point D for arm 3 is

$$\sum \mathbf{M}_D^{(3)} = \mathbf{r}_{DC} \times \mathbf{F}_{23} + \mathbf{M}_{ext} = \mathbf{0}, \tag{II.6.59}$$

where $\mathbf{F}_{23} = F_{23y}\,\mathbf{J} + F_{23z}\,\mathbf{k}$. Equation (II.6.59) gives

$$\begin{vmatrix} \imath & \mathbf{J} & \mathbf{k} \\ x_C - x_D & r_1 + r_2 & 0 \\ 0 & F_{32y} & F_{32z} \end{vmatrix} + M_{ext}\imath = \mathbf{0}. \tag{II.6.60}$$

From Eq. (II.6.60), it results:

$$F_{23z} = -\frac{M_{ext}}{r_1 + r_2} = -\frac{-400}{1 + 0.5} = 266.666 \ \text{N}.$$

Gear 2
The sum of the moments for gear 2 with respect to its center C is

$$\sum \mathbf{M}_C^{(2)} = \mathbf{r}_{CE} \times \mathbf{F}_{02} + \mathbf{r}_{CB} \times \mathbf{F}_{12} = \mathbf{0},$$

or

$$\begin{vmatrix} \imath & \mathbf{J} & \mathbf{k} \\ 0 & r_2 & 0 \\ 0 & F_{r02} & F_{t02} \end{vmatrix} + \begin{vmatrix} \imath & \mathbf{J} & \mathbf{k} \\ 0 & -r_2 & 0 \\ 0 & F_{r12} & F_{t12} \end{vmatrix} = \mathbf{0},$$

or

$$r_2\,F_{t02} - r_2\,F_{t12} = 0 \implies F_{t02} = F_{t12}.$$

The radial component F_r tends to push the gears apart:

$$F_{r12} = F_{t12}\tan\phi \quad \text{and} \quad F_{r02} = -F_{t12}\tan\phi.$$

EXAMPLE II.6.7: *Cont'd*

For the planet gear 2 the force equation is

$$\sum \mathbf{F}^{(2)} = \mathbf{F}_{12} + \mathbf{F}_{02} + \mathbf{F}_{32} = \mathbf{0}, \qquad (\text{II}.6.61)$$

where $\mathbf{F}_{12} = F_{r12}\mathbf{J} + F_{t12}\mathbf{k}$, $\mathbf{F}_{02} = -F_{r12}\mathbf{J} + F_{t12}\mathbf{k}$, and $\mathbf{F}_{32} = -\mathbf{F}_{23}$. Equation (II.6.61) gives

$$F_{r12} - F_{r12} - F_{23y} = 0 \implies F_{23y} = 0,$$

$$F_{t12} + F_{t12} - F_{23z} = 0 \implies F_{t12} = \frac{F_{23z}}{2} = \frac{266.666}{2} = 133.333 \text{ N}.$$

The radial reaction forces F_{r12} and F_{r02} are

$$F_{r12} = -F_{r02} = F_{t12} \tan \phi = 133.333 \tan 20° = 48.529 \text{ N}.$$

For arm 3 the force equation is

$$\sum \mathbf{F}^{(3)} = \mathbf{F}_{23} + \mathbf{F}_{03} = \mathbf{0},$$

and the reaction force $\mathbf{F}_{03}$ is

$$\mathbf{F}_{03} = -\mathbf{F}_{23} = -266.666\,\mathbf{k} \text{ N}.$$

Gear 1 (driver)

For gear 1 the sum of the moments with respect to its center A is

$$\sum \mathbf{M}_A^{(1)} = \mathbf{r}_{AB} \times \mathbf{F}_{21} + \mathbf{M}_{mot} = \mathbf{0},$$

where $\mathbf{F}_{21} = -\mathbf{F}_{12}$. The motor torque M_{mot} is

$$M_{mot} = -F_{t21}\, r_1 = F_{t12}\, r_1 = 133.333 \text{ N} \cdot \text{m}.$$

The sum of the forces for the gear 1 is

$$\sum \mathbf{F}^{(1)} = \mathbf{F}_{21} + \mathbf{F}_{01} = \mathbf{0},$$

and the reaction force of the ground on gear 1 is

$$\mathbf{F}_{01} = -\mathbf{F}_{21} = \mathbf{F}_{12} = 48.529\,\mathbf{J} + 133.333\,\mathbf{k} \text{ N}.$$

The *Mathematica*^{TM} program for this example is given in Program II.6.1.

EXAMPLE II.6.8:

A driver pinion made of steel with the endurance strength $\sigma_0 = 15\,000$ psi rotates at $n_p = 1200$ rpm. The pinion is surface hardened to BHN 250. The gear is made of cast iron with the endurance strength $\sigma_0 = 8000$ psi and rotates at $n_g = 300$ rpm. The teeth have standard $20°$ stub involute profiles. The maximum power to be transmitted is 33 horse power (hp). Determine the proper diametral pitch, numbers of teeth, and face width for the gears from the standpoint of strength, dynamic load, and wear.

Solution The diameters of the gears are unknown. In order to determine the smallest diameter gears that can be used, the minimum number of teeth for the pinion will be selected, $N_p = 16$. Then the number of teeth for the gear is

$$N_g = -N_p \, i = N_p \frac{n_p}{n_g} = 16 \, \frac{1200}{300} = 64,$$

where $i = -n_p/n_g = 4$ is the speed ratio. Next it will be determined which is weaker, the gear or the pinion. The load-carrying capacity of the tooth is a function of the $\sigma_0 \gamma$ product.

From Table II.6.1 for a $20°$ stub involute gear with 16 teeth, the form factor is $\gamma_p = 0.115$. For the pinion, the load-carrying capacity is

$$F_p = \sigma_{0p} \gamma_p = 15\,000 \,(0.115) = 1725.$$

From Table II.6.1 for a $20°$ stub involute gear with 64 teeth, the calculated form factor is $\gamma_g = 0.155$. For the gear, the load-carrying capacity is

$$F_g = \sigma_{0g} \gamma_g = 8000 \,(0.155) = 1240.53.$$

Since $F_g < F_p$ the gear is weaker and the gear will be analyzed.

The moment transmitted by the gear is

$$M_t = \frac{63\,000\,H}{n_g} = \frac{63\,000\,(33)}{300} = 6930 \;\text{lb} \cdot \text{in.}$$

The diameters are unknown and the induced stress is

$$\sigma = \frac{2\,M_t\,P_d^3}{k\,\pi^2\,\gamma_g\,N_g} = \frac{2(6930)P_d^3}{4\pi^2(0.155)(64)} = 35.375\,P_d^3, \qquad \text{(II.6.62)}$$

where a maximum value of $k = 4$ was considered.

To determine an approximate P_d, assume the allowable stress

$$\sigma \approx \frac{\sigma_0}{2} = \frac{8000}{2} = 4000 \;\text{psi.}$$

EXAMPLE II.6.8: *Cont'd*

Equation (II.6.62) yields

$$4000 = 35.375 \, P_d^3 \quad \Longrightarrow \quad P_d \approx 4.835.$$

Try a diametral pitch $P_d = 5$ teeth per inch. Then the pitch diameter is

$$d_g = \frac{N_g}{P_d} = \frac{64}{5} = 12.8 \text{ in.}$$

The pitch line velocity is

$$V = d_g \, \pi \, n_g / 12 = 12.8 \, \pi \, (300)/12 = 1005.31 \text{ ft/min.}$$

The allowable stress for $V < 2000$ ft/min is given by Eq. (II.6.36):

$$\sigma = 8000 \left(\frac{600}{600 + 1005.31} \right) = 2990.08 \text{ psi.}$$

Using Eq. (II.6.62) the induced stress is

$$\sigma = \frac{2 \, M_t \, P_d^3}{k \, \pi^2 \, \gamma_g \, N_g} = \frac{2(6930)(5^3)}{4\pi^2(0.155)(64)} = 4421.96 \text{ psi.}$$

The gear is weak because the induced stress is larger than the allowable stress. Try a stronger tooth and select $P_d = 4$ teeth per inch. The pitch diameter is

$$d_g = \frac{N_g}{P_d} = \frac{64}{4} = 16 \text{ in.}$$

The pitch line velocity is

$$V = d_g \, \pi n_g / 12 = 16 \, \pi \, (300)/12 = 1256.64 \text{ ft/min.}$$

Because the pitch line velocity is less than 2000 ft/min, the allowable stress is

$$\sigma = 8000 \left(\frac{600}{600 + 1256.64} \right) = 2585.32 \text{ psi.}$$

Using Eq. (II.6.62) the induced stress is

$$\sigma = \frac{2(6930)(4^3)}{4\pi^2(0.155)(64)} = 2264.04 \text{ psi.}$$

The gear is strong because the induced stress is smaller than the allowable stress.

Continued

EXAMPLE II.6.8: *Cont'd*

The parameter k can be reduced from the maximum value of $k = 4$. Equation (II.6.62) with the allowable stress, $\sigma = 2585.32$ psi, gives

$$2585.32 = \frac{2(6930)(4^3)}{k\,\pi^2(0.155)(64)} \implies k = 3.502.$$

The face width is

$$B = k\,p = k\,(\pi/P_d) = 3.502\,(\pi/4) = 2.751 \text{ in.},$$

where the circular pitch is

$$p = \pi/P_d = \pi/4 = 0.785 \text{ in.}$$

Then

$$P_d = 4, B = 2\frac{3}{4} \text{ in.}, r_g = d_g/2 = 16/2 = 8 \text{ in.}$$

and

$$r_p = -r_g/i = 8/4 = 2 \text{ in.}$$

The center distance is

$$c = r_p + r_g = 2 + 8 = 10 \text{ in.}$$

The addendum of the gears is

$$a = 0.8/P_d = 0.8/4 = 0.2 \text{ in.},$$

while the minimum dedendum for $20°$ full-depth involute gears is

$$b = 1/P_d = 1/4 = 0.25 \text{ in.}$$

The radii of the base circle for the pinion and the gear are

$$r_{bp} = r_p \cos \phi = 2 \cos 20° = 1.879 \text{ in.}$$

and

$$r_{bg} = r_g \cos \phi = 8 \cos 20° = 7.517 \text{ in.},$$

respectively.

EXAMPLE II.6.8: *Cont'd*

The maximum possible addendum circle radius of pinion or gear without interference is

$$r_{a(max)} = \sqrt{r_b^2 + c^2 \sin^2 \phi}.$$

Hence, for the pinion and for the gear

$$r_{ap(max)} = \sqrt{1.879^2 + 10^2 \sin^2 20°} = 3.902 \text{ in.},$$

$$r_{ag(max)} = \sqrt{7.517^2 + 10^2 \sin^2 20°} = 8.259 \text{ in.}$$

The addendum radii of the meshing pinion and gear are

$$r_{ap} = r_p + a = 2 + 0.2 = 2.2 \text{ in.}$$

and

$$r_{ag} = r_g + a = 8 + 0.2 = 8.2 \text{ in.}$$

Since $r_{a(max)} > r_a$, there is no interference.

The contact ratio is calculated from the equation:

$$CR = \frac{\sqrt{r_{ap}^2 - r_{bp}^2} + \sqrt{r_{ag}^2 - r_{bg}^2} - c \sin \phi}{p_b},$$

where the base pitch is $p_b = \pi d_b/N = p \cos \phi = 0.785 \cos 20° = 0.738$ in. The contact ratio is $CR = 1.353$, which is a suitable value (> 1.2).

Next, the tentative design will be checked from the standpoint of dynamic load and wear effects.

The allowable endurance load is

$$F_0 = \sigma_0 B \gamma_g p = 8000(2.75)(0.155)(\pi/4) = 2679.36 \text{ lb.}$$

Equation (II.6.38) gives the allowable wear load

$$F_w = d_p B K Q = 4(2.75)(170.11)(1.6) = 2993.94 \text{ lb,}$$

where

$$Q = \frac{2 N_g}{N_p + N_g} = \frac{2 \cdot 64}{16 + 64} = 1.6,$$

Continued

EXAMPLE II.6.8: *Cont'd*

the surface endurance limit of a gear pair is

$$\sigma_{es} = (400)(\text{BHN}) - 10\,000 = (400)(250) - 10\,000 = 9000 \text{ psi},$$

and the stress factor for fatigue is

$$K = \frac{\sigma_{es}^2 (\sin\phi)(1/E_p + 1/E_g)}{1.4} = \frac{9000^2(\sin 20°)\left(\dfrac{1}{30 \times 10^6} + \dfrac{1}{19 \times 10^6}\right)}{1.4}$$

$$= 170.11 \text{ lb.}$$

For $V = 1256.64$ ft/min from Figure II.6.13(a), the permissible error is 0.00225 in. From Figure II.6.13(b), for carefully cut gears with $P_d = 4$, the tooth error is $e = 0.0012$ in. From Table II.6.2 the deformation factor for dynamic load check is $C = 1416$.
The dynamic load F_d proposed by Buckingham is

$$F_d = \frac{0.05\, V(B\,C + F_t)}{0.05\, V + \sqrt{B\,C + F_t}} + F_t$$

$$= \frac{0.05\,(1256.64)[(2.75)\,(1416) + 866.25]}{0.05\,(1256.64) + \sqrt{(2.75)\,(1416) + 866.25}} + 866.25$$

$$= 3135.7 \text{ lb.}$$

where $F_t = M_t/r_g = 6930/8 = 866.25$ lb.
    The design is unsatisfactory because the dynamic force F_d must be less than the allowable endurance load F_0 and less than the wear load F_w.
    From Figure II.6.13(b) select a precision gear with an error of action $e = 0.00051$ for $P_d = 4$. From Table II.6.2 the deformation factor for dynamic load check is $C = 601.8$. Recalculating the dynamic load F_d for $C = 601.8$ gives

$$F_d = \frac{0.05\,(1256.64)[(2.75)\,(601.8) + 866.25]}{0.05\,(1256.64) + \sqrt{(2.75)\,(601.8) + 866.25}} + 866.25$$

$$= 2267.89 \text{ lb.}$$

The design is satisfactory because the dynamic force F_d is less than the allowable endurance load F_0 and less than the wear load F_w:

$$F_d < F_0 \text{ and } F_d < F_w.$$

The *Mathematica*™ program for this example is given in Program II.6.2.

II.6.10 Problems

II.6.1 A planetary gear train is shown in Figure II.6.22. Gear 1 has $N_1 = 36$ external gear teeth, gear 2 has $N_2 = 40$ external gear teeth, and the gear 2' has $N_{2'} = 21$ external gear teeth. Gears 2 and 2' are fixed on the same shaft, CC'. Gear 3 has $N_3 = 30$ external gear teeth, and gear 3' has $N_{3'} = 24$ external gear teeth. Gears 3 and 3' are fixed on the same shaft, EE'. The planet gear 4 has $N_4 = 18$ external gear teeth, and the planet gear 5 has $N_5 = 14$ external gear teeth. Gear 1 rotates with a constant input angular speed $n_1 = 320$ rpm. The module of the gears is $m = 30$ mm. The pressure angle of the gears is $20°$. (a) Determine whether there will be interference when standard full-depth teeth are used and find the contact ratios of the meshing gears. (b) Find the angular velocity of the output planet arm 6, ω_6. (c) Find the equilibrium moment on gear 1 if an external moment $\mathbf{M}_{ext} = -|\mathbf{M}_{ext}| \dfrac{\omega_6}{|\omega_6|}$ acts on the arm 6, where $|\mathbf{M}_{ext}| = 600$ N·m.

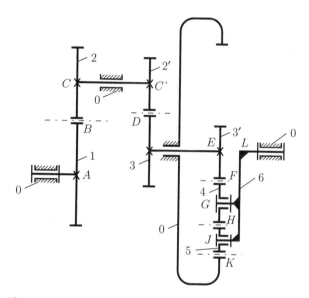

FIGURE II.6.22 *Planetary gear train for Problem II.6.1.*

II.6.2 The planetary gear train considered in Figure II.6.23 has gears with the same module $m = 24$ mm. The sun gear 1 has $N_1 = 22$ external gear teeth, the planet gear 2 has $N_2 = 18$ external gear teeth, gear 3 has $N_3 = 20$ external gear teeth, and gear 4 has $N_4 = 54$ external gear teeth. Gears 3 and 3' are fixed on the same shaft. The sun gear 1 rotates with an input angular speed $n_1 = 290$ rpm, and arm 5 rotates with $n_5 = 110$ rpm. The pressure angle of the gears is $20°$. (a) Find the number of DOF for the planetary gear train. (b) Determine whether there will be interference when standard full-depth teeth are used and find the contact ratios of the meshing gears. (c) Find the angular velocity of the output gear 4, ω_4. (d) Find the

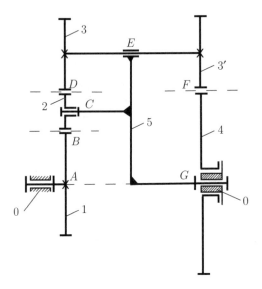

FIGURE II.6.23 *Planetary gear train for Problem II.6.2.*

equilibrium moments on gear 1 and arm 5 if an external moment
$\mathbf{M}_{ext} = -|\mathbf{M}_{ext}| \dfrac{\omega_4}{|\omega_4|}$ acts on gear 4, where $|\mathbf{M}_{ext}| = 800$ N·m.

II.6.3 A planetary gear train is depicted in Figure II.6.24. Gear 1 has $N_1 = 15$ external gear teeth, gear 2 has $N_2 = 27$ external gear teeth, gear $2'$ has $N_{2'} = 18$ external

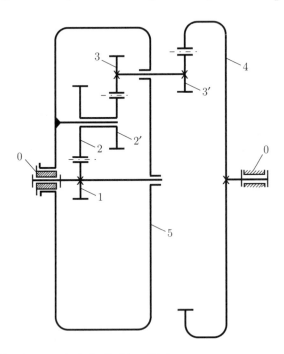

FIGURE II.6.24 *Planetary gear train for Problem II.6.3.*

gear teeth, gear 3 has $N_3 = 24$ external gear teeth, and gear $3'$ has $N_{3'} = 16$ external gear teeth. The planet gears 2 and $2'$ are fixed on the same link and the planet gears 3 and $3'$ are fixed on the same shaft. Gear 1 rotates with the input angular speed $n_1 = 440$ rpm, and the arm 5 rotates at $n_5 = 80$ rpm. The module of the gears is $m = 24$ and the pressure angle of the gears is $20°$. (a) Determine whether there will be interference when standard full-depth teeth are used and find the contact ratios of the meshing gears. (b) Find the angular velocity of the output gear 4, $\boldsymbol{\omega}_4$. (c) Find the equilibrium moments on gear 1 and arm 5 if an external moment $\mathbf{M}_{ext} = -|\mathbf{M}_{ext}| \dfrac{\boldsymbol{\omega}_4}{|\boldsymbol{\omega}_4|}$ acts on the gear 4, where $|\mathbf{M}_{ext}| = 1000$ N·m.

II.6.4 A planetary gear train is shown in Figure II.6.25. Gear 1 has $N_1 = 11$ external gear teeth, the planet gear 2 has $N_2 = 22$ external gear teeth, gear $2'$ has $N_{2'} = 17$ external gear teeth, gear 3 has $N_3 = 51$ internal gear teeth, the sun gear $3'$ has $N_{3'} = 12$ external gear teeth, and the planet gear 4 has $N_4 = 32$ external gear teeth. Gears 2 and $2'$ are fixed on the same shaft and gears 3 and $3'$ are fixed on the same shaft. The sun gear 1 rotates with an input angular speed of $n_1 = 550$ rpm. The module of the gears is $m = 26$ mm and the pressure angle of the gears is $20°$. (a) Determine whether there will be interference when standard full-depth teeth are used and find the contact ratios of the meshing gears. (b) Find the angular velocity of the arm 5, $\boldsymbol{\omega}_5$. (c) Find the joint forces if an electric motor with the power $H = 4$ kW is coupled to gear 1.

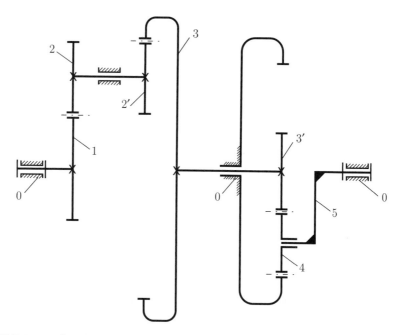

FIGURE II.6.25 *Planetary gear train for Problem II.6.4.*

II.6.5 A planetary gear train is shown in Figure II.6.26. The ring gear 1 has $N_1 = 60$ internal gear teeth, the planet gear 2 has $N_2 = 25$ external gear teeth, and the planet

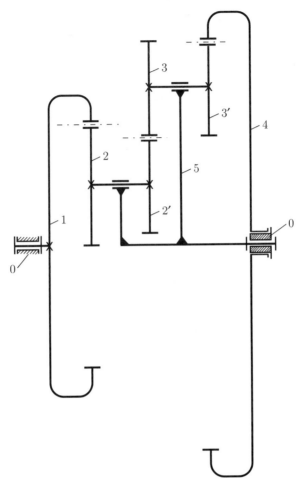

FIGURE II.6.26 *Planetary gear train for Problem II.6.5.*

gear $2'$ has $N_{2'} = 15$ external gear teeth. The gears 2 and $2'$ are fixed on the same shaft. The planet gear 3 has $N_3 = 20$ teeth and the gears 3 and $3'$ are fixed on the same shaft. The ring gear 4 has $N_4 = 90$ internal gear teeth. Gear 1 rotates with the input angular speed $n_1 = 100$ rpm, and arm 5 rotates at $n_5 = -150$ rpm (n_1 is opposite to n_5). The module of the gears is $m = 28$ and the pressure angle of the gears is $20°$. (a) Determine whether there will be interference when standard full-depth teeth are used and find the contact ratios of the meshing gears. (b) Find the angular velocity of the output ring gear 4, ω_4. (c) Find the joint forces and the equilibrium moments on gear 1 and arm 5 if an external moment

$$\mathbf{M}_{ext} = -|\mathbf{M}_{ext}|\frac{\omega_4}{|\omega_4|} \text{ acts on 4, where } |\mathbf{M}_{ext}| = 1000 \text{ N·m.}$$

II.6.6 A planetary gear train is shown in Figure II.6.27. The sun gear 1 has $N_1 = 11$ teeth, the planet gear 2 has $N_2 = 19$ teeth, gear 3 has $N_3 = 40$ internal gear teeth, gear 4 has $N_4 = 29$ external gear teeth, and gear 5 has $N_5 = 24$ external gear teeth. Gear 1

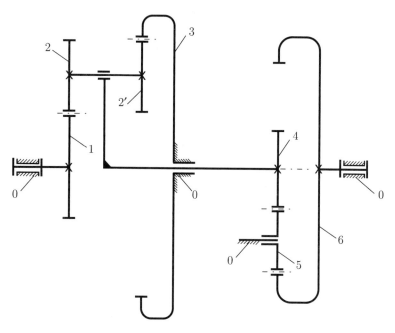

FIGURE II.6.27 *Planetary gear train for Problem II.6.6.*

rotates with a constant input angular speed $n_1 = 200$ rpm. The module of the gears is $m = 22$ and the pressure angle of the gears is $20°$. (a) Determine whether there will be interference when standard full-depth teeth are used and find the contact ratios of the meshing gears. (b) Find the angular velocity of the output ring gear 6, ω_6. (c) An external moment $\mathbf{M}_{ext} = -|\mathbf{M}_{ext}|\dfrac{\omega_6}{|\omega_6|}$ acts on gear 6, where $|\mathbf{M}_{ext}| = 900$ N·m. Find the joint forces and the equilibrium moment on gear 1.

II.6.7 A planetary gear train is shown in Figure II.6.28. The sun ring gear 1 has $N_1 = 28$ teeth, the planet gear 2 has $N_2 = 21$ teeth, and the planet gear $2'$ has $N_{2'} = 16$ teeth. Gears 2 and $2'$ are fixed on the same shaft. Gear 1 rotates with the input angular speed $n_1 = 370$ rpm. The module of the gears is $m = 20$ and the pressure angle of the gears is $20°$. (a) Determine whether there will be interference when standard full-depth teeth are used and find the contact ratios of the meshing gears. (b) Find the angular velocity of the arm gear 4, ω_4. (c) An external moment $\mathbf{M}_{ext} = -|\mathbf{M}_{ext}|\dfrac{\omega_4}{|\omega_4|}$ acts on the arm gear 4, where $|\mathbf{M}_{ext}| = 800$ N·m. Find the joint forces and the equilibrium moment on gear 1.

II.6.8 A planetary gear train is shown in Figure II.6.29. The ring gear 1 has $N_1 = 75$ internal gear teeth, the planet gear 2 has $N_2 = 35$ teeth, the planet gear $2'$ has $N_{2'} = 20$ teeth, gear 3 has $N_3 = 11$ teeth, gear 4 has $N_4 = 13$ external gear teeth, and the ring gear 5 has $N_5 = 50$ internal gear teeth. Gears 2 and $2'$ are fixed on the same shaft. The module of the gears is $m = 42$ and the pressure angle of the gears is $20°$. (a) Determine whether there will be interference when standard full-depth teeth are used and find the contact ratios of the meshing gears. (b) Find the angular

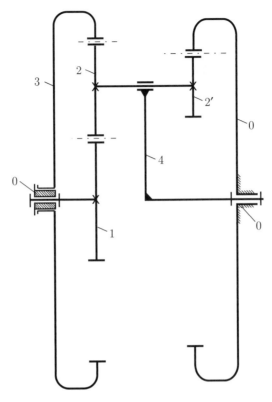

FIGURE II.6.28 *Planetary gear train for Problem II.6.7.*

velocity of the ring gear 5, ω_5. (c) An external moment $\mathbf{M}_{ext} = -|\mathbf{M}_{ext}|\dfrac{\omega_5}{|\omega_5|}$ acts on the output ring gear 5, where $|\mathbf{M}_{ext}| = 500$ N·m. Find the joint forces and the equilibrium moment.

II.6.9 A planetary gear train is shown in Figure II.6.30. An electric motor with the power $H = 2$ kW and 319 rpm is coupled to sun gear 1. Gear 1 has $N_1 = 12$ teeth, the planet gear 2 has $N_2 = 17$ teeth, gear 3 has $N_3 = 20$ teeth, gear 4 has $N_4 = 11$ teeth, gear $4'$ has $N_{4'} = 17$ external gear teeth, and the ring gear 5 has $N_5 = 51$ internal gear teeth. Gears 4 and $4'$ are fixed on the same shaft. The module of the gears is $m = 33$ mm and the pressure angle of the gears is 20°. (a) Find the angular velocity of the output ring gear 5. (b) Find the joint forces.

II.6.10 The planetary gear train considered is shown in Figure II.6.31. The sun gear 1 has $N_1 = 22$ teeth, the planet gear 2 has $N_2 = 20$ teeth, the planet gear $2'$ has $N_{2'} = 35$ teeth, the sun gear 4 has $N_4 = 15$ teeth, and the planet gear 5 has $N_5 = 16$ teeth. Gears 2 and $2'$ are fixed on the same shaft. The ring gear 3 rotates with the input angular speed $n_3 = 200$ rpm and the ring gear 6 rotates at the input angular speed $n_6 = 150$ rpm. The module of the gears is $m = 24$ mm. Find the absolute angular velocity of gear 1.

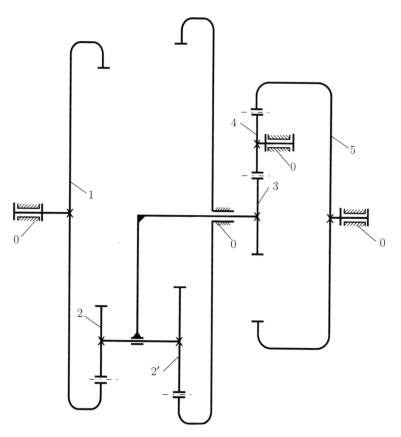

FIGURE II.6.29 *Planetary gear train for Problem II.6.8.*

II.6.11 The planetary train with two planet gears is shown in Figure II.6.32. The sun gear 1 has an angular speed of 600 rpm and is driven by a motor with the moment 20 N·m. The planet gears are 2 and 2″, each having $N_2 = N_{2'} = 20$ teeth. The ring gear 4 has $N_4 = 70$ teeth. A brake holds the ring gear 4 fixed. The module of the gears is $m = 2$ mm and the pressure angle of the gears is 20°. The arm 3 drives a machine. Determine: (a) the circular pitch of the gears; (b) the angular speed of the arm; (c) the pitch line velocity of each gear; (d) the joint forces; (e) the output moment; (f) the moment to be applied to the ring to keep it fixed.

II.6.12 A driver spur pinion of cast steel with the endurance strength $\sigma_0 = 20\,000$ psi rotates at $n = 1500$ rpm and transmits 35 hp. The driven gear is made of cast iron with the endurance strength $\sigma_0 = 8000$ psi. The transmission ratio is 3.5 to 1 (external gearing). Both gears have 14.5° pressure angles, and full-depth involute gear teeth. Design for strength and determine the smallest diameter gears and the face width.

II.6.13 A cast steel spur pinion ($\sigma_0 = 15\,000$ psi) rotating at 900 rpm is to drive a bronze spur gear ($\sigma_0 = 12\,000$ psi) at 300 rpm. The power to be transmitted is 10 hp.

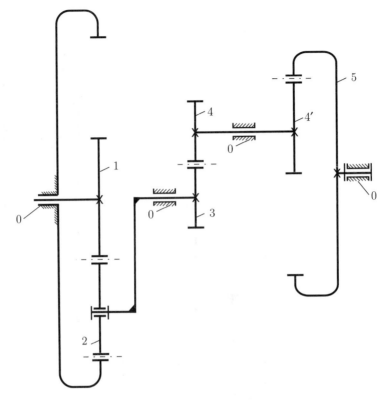

FIGURE II.6.30 *Planetary gear train for Problem II.6.9.*

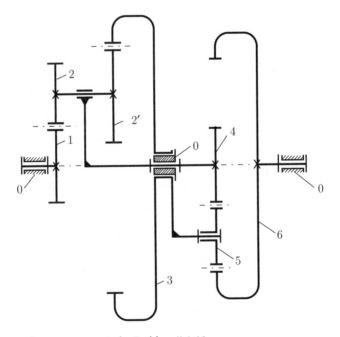

FIGURE II.6.31 *Planetary gear train for Problem II.6.10.*

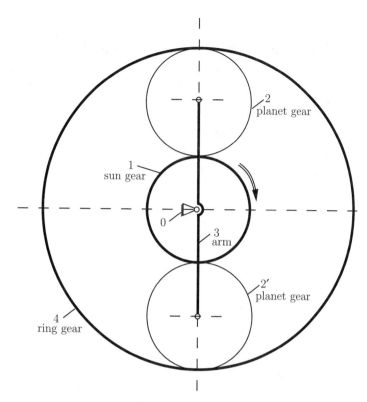

FIGURE II.6.32 *Planetary gear train for Problem II.6.11.*

The teeth have standard 20° stub involute profiles. Determine the smallest diameter gear that can be used and the necessary face width.

II.6.14 Two spur gears are to be designed with a minimum size. The following requirements are given: speed of the pinion 600 rpm, power to be transmitted 15 hp, velocity ratio 3 to 1 external gearing, endurance strength for pinion $\sigma_0 = 30\,000$ psi, endurance strength for gear $\sigma_0 = 20\,000$ psi, tooth profile 20° stub involute. For strength design determine the necessary face width and diametral pitch.

II.6.15 A driver made of mild steel pinion with the endurance strength $\sigma_0 = 15\,000$ psi rotates at $n_1 = 1750$ rpm and transmits 6 hp. The transmission ratio is $i = -3.5$. The gear is made of bronze and has the endurance strength $\sigma_0 = 12\,000$ psi. The gears have 20° pressure angles, and full-depth involute gear teeth. Design a gear with the smallest diameter that can be used. No less than 15 teeth are to be used on either gear.

II.6.16 A pinion made of cast iron with the endurance strength $\sigma_0 = 20\,000$ psi rotates at $n_p = 900$ rpm and transmits 30 hp. The transmission ratio is $i = -7/3$. The gear is made of cast iron and has the endurance strength $\sigma_0 = 8000$ psi. The gears have 20° pressure angles, and full-depth involute gear teeth. The diameter of the pinion is $d_p = 4$ in. Design for the greatest number of teeth.

II.6.11   Working Model Simulation for Gear Trains

Working Model Simulation for Example II.6.5

The gear set shown in Figure II.6.19(a) will be simulated using Working Model software. Gear 1 ($r_1 = r_g = 1$ m) rotates with the angular speed $n_1 = 10$ rpm. The external torque exerted on the pinion 2 ($r_2 = r_p = 0.5$ m) is $M_{ext} = 400$ N·m.

Step 1: Opening Working Model

1. Click on the Working Model program icon to start the program.
2. Create a new Working Model document by selecting "New" from the "File" menu.
3. Set up the workspace.

In the "View" menu: select "Workspace", check Coordinates and X,Y Axes from the Navigation box, and check all the objects from the Toolbars box except Simple; turn off Grid Snap and turn on Object Snap; select "Numbers and Units" and change the Unit System to SI (degrees).

Step 2: Creating the gear
This step creates the two gears for the system.

1. *Create the gear 1.* Click on the "Circle" tool in the toolbar to sketch out a disk. Click on the disk and modify its radius at the bottom of the screen to $r_1 = 1$ m.
2. *Create the pinion 2.* Click on the "Circle" tool in the toolbar to sketch out a disk. Click on the disk and modify its radius at the bottom of the screen to $r_2 = 0.5$ m.
3. *Change the properties of the gears.* Press the Shift key and click on the main gear and the pinion, respectively. Select "Properties" in the "Window" menu and change the material to Steel, the coefficients of static and kinetic friction to 0.0 (no friction), the coefficient of restitution to 1.0 (perfect elastic), and the charge to 0.0 (no charge), as shown in Figure II.6.33.

Remark In order to make the objects clearly visible the commands "Zoom in" and "Zoom out" can be used by clicking on the icons at the top of the screen.

Step 3: Connecting the gears to the ground
This step connects a motor to gear 1 and the pinion 2 to the ground using a pin joint.

1. Select the gear 1 and modify its center coordinates at the bottom of the screen to $x = 0$ and $y = 0$ (the center of axis).
2. Click on the "Motor" tool, place the cursor over the "snap point" on the center of the main gear and then click again. This connects the motor to the ground and the gear.
3. Select "Numbers and Units" in the "View" menu and change the "Rot. Velocity" to Revs/min. Select the "Properties" box in the "Window" menu and change the "value" to $n_1 = 10$ rpm.
   The screen should look like that shown in Figure II.6.34.
4. Select the pinion and modify its center coordinates at the bottom of the screen to $x = 0$ and $y = 1.5$ m.

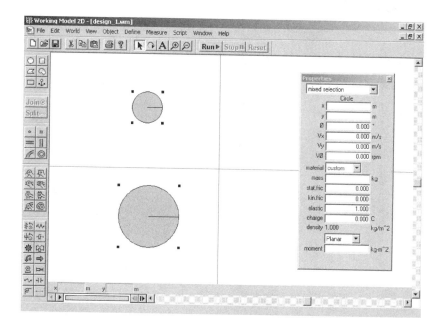

FIGURE II.6.33

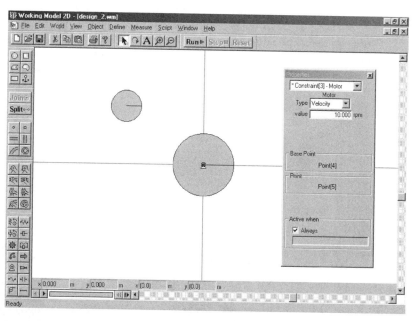

FIGURE II.6.34

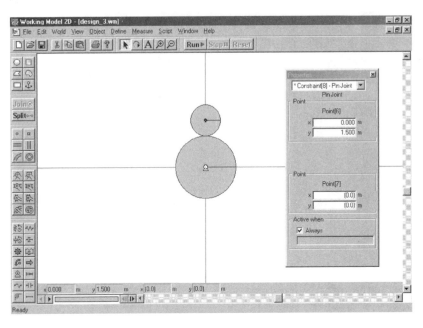

FIGURE II.6.35

5. Click on the "Pin joint" tool and then click again on the center of the pinion. This connects the pinion to the ground with the pin joint.

The screen should look like that shown in Figure II.6.35.

Step 4: Connecting the gears
This step connects the gear and the pinion using the "Gear" tool.

1. Click on the "Gear" tool from the toolbox and then click on the centers of the gear and the pinion, respectively. This connects the two gears with a rigid rod.

By default, each pair of gears has a rigid rod constraint between the two mass centers. The rod maintains a constant distance between the two objects.

The screen should look like that shown in Figure II.6.36.

Step 5: Running the simulation

1. Click on "Run" in the toolbar to start the simulation.
2. Click on "Reset" in the toolbar. The simulation resets to the initial frame 0.
3. Select the pinion, then go to "Measure" menu and "Velocity" submenu. Apply the "Rotational graph" command to measure the rotational velocity of the pinion. Click on the arrow in the right upper corner of the measurement window to change it from graphic to numerical. Select the gear and apply the same command to measure the rotational velocity of the gear.

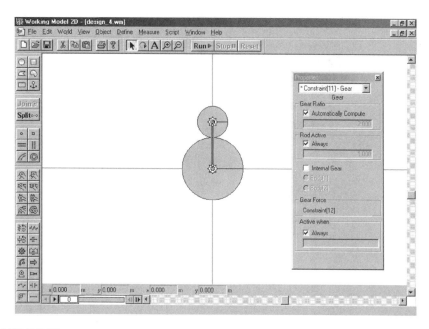

FIGURE II.6.36

Step 6: Adding an external torque

1. Click on the "Torque" tool from the toolbox and then click on the pinion (anywhere). This will apply an external torque to the pinion.
2. Select the torque and modify its value to $M_{ext} = 400$ N·m in the "Properties" menu. Apply the command "Torque" from the "Measure" menu to measure the torque applied.
3. Select the motor and apply the command "Torque Transmitted" from the "Measure" menu to measure the torque of the motor.

The screen should look like that shown in Figure II.6.37.

Results

The angular speed of the gear $n_1 = 10$ rpm and the external torque $M_{ext} = 400$ N·m are given. It results in the angular speed of the pinion $n_2 = -20$ rpm and the motor torque $M_{mot} = 800$ N·m.

Working Model Simulation for Example II.6.6

The planetary gear train with two DOF shown in Figure II.6.20(a) will be simulated using Working Model software. The sun gear 1 ($r_1 = 1$ m) has the angular speed $n_1 = 20$ rpm. The planet gear 2 has the pitch radius $r_2 = 0.5$ m. The arm has the angular speed $n_3 = -10$ rpm. The external torque exerted on the gear 2 is $M_{ext} = 400$ N·m.

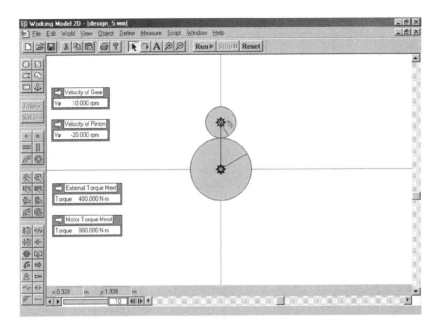

FIGURE II.6.37

Step 1: Creating the gears and the arm

1. *Create the sun gear 1.* Click on the "Circle" tool in the toolbar and sketch out a disk. Click on the disk and modify its radius at the bottom of the screen as $r_1 = 1$ m.
2. *Create the planet gear 2.* Click on the "Circle" tool in the toolbar and sketch out a disk. Click on the disk and modify its radius at the bottom of the screen as $r_2 = 0.5$ m.
3. *Create arm 3.* Click on the "Rectangle" tool in the toolbar and sketch out a rectangle. Click on the rectangle and modify its the dimensions as $h = 1.5$ m and $w = 0.1$ m. The rod created with a set of gears cannot have torques applied to it or have an anchor or a motor placed on it. That is why the rectangle 3 will be used instead of the rod to model the arm connected to the planet gear 2 and the ground.
4. Select the arm 3 and modify the coordinates of its center as $x = 0$ and $y = 0.75$ m at the bottom of the screen.
5. Select the planet gear 2 and modify the coordinates of its center as $x = 0$ and $y = 1.5$ m at the bottom of the screen.

Step 2: Connecting the planet gear and the arm

1. Click on the "Pin joint" tool on the toolbox and connect the planet gear and the rectangle by clicking again on the center of the circle.
2. Click on the "Motor" tool on the toolbox and then click again on the center of the axis. This connects the motor to the ground and the arm. Click on the motor and change the value of the velocity to 10 rpm in the "Properties" window ($n_3 = -10$ rpm).

The screen should look like that shown in Figure II.6.38.

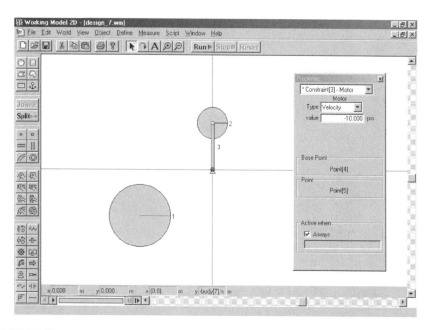

FIGURE II.6.38

Step 3: Connecting the sun gear to the ground

1. Click on the "Motor" tool on the toolbox and then click again on the center of the sun gear 1.
2. Select the motor and open the "Properties" window. Change the value of the velocity to 20 rpm ($n_1 = 20$ rpm), as shown in Figure II.6.39.
3. In the "Properties" window, select the base point of the motor and change its coordinates to $x = 0$ and $y = 0$. This moves the motor along with the sun gear to the center of axis (the motor is still connected to the ground).

Step 4: Connecting the gears

1. Click on the sun gear 1 and select "Move to front" from the "Object" menu. Do the same command for the planet gear 2.
2. Click on the "Gear" tool from the toolbox. With the gear selected, click on the center of the sun gear 1 and then again on the center of the planet gear 2. The two circles are now connected with a gear.
3. Click on the rectangle 3 and select the command "Bring to front" from the "Object" menu.

The screen should look like that shown in Figure II.6.40.

Step 5: Running the simulation

1. Select all the bodies and choose the command "Do not collide" from the "Object" menu.
2. Click on "Run" in the toolbar to start the simulation.

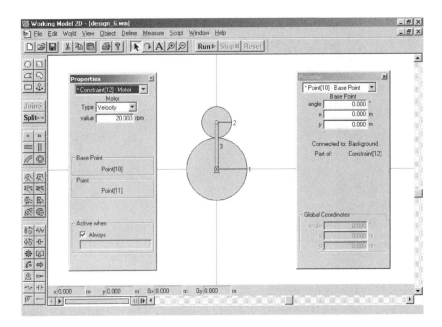

FIGURE II.6.39

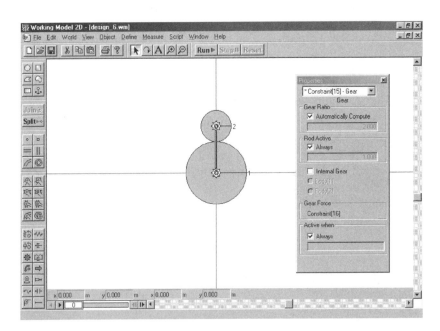

FIGURE II.6.40

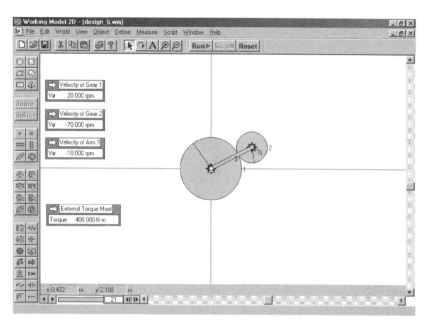

FIGURE II.6.41

3. Click on "Reset" in the toolbar. The simulation resets to the initial frame 0.
4. Click on the planet gear 2 and select the "Measure" menu and the "Velocity" and "Rotational graph" submenus to measure the rotational velocity n_2. Click on the arrow in the right upper corner of the measurement window to change it from graphic to numerical. Apply the same command to visualize the rotational velocity n_1 of the sun gear 1.

Step 6: Adding an external torque

1. Click on the "Torque" tool from the toolbox and then click on gear 2 (anywhere on the disk). This will apply an external torque to gear 2.
2. Select the torque and modify its value to $M_{ext} = 400$ N·m in the "Properties" menu.
3. Select the submenu "Torque" from the "Measure" menu to measure the torque applied.

The screen should look like that shown in Figure II.6.41.

Results
The angular speed of gear 1, $n_1 = 20$ rpm, the angular speed of arm 3, $n_3 = -10$ rpm, and the external torque, $M_{ext} = 400$ N·m, are given. It results in the angular speed of gear 2, $n_2 = -70$ rpm, the motor torque on gear 1, $M_{1mot} = 800$ N·m, and the motor torque on arm 3, $M_{3mot} = -1200$ N·m.

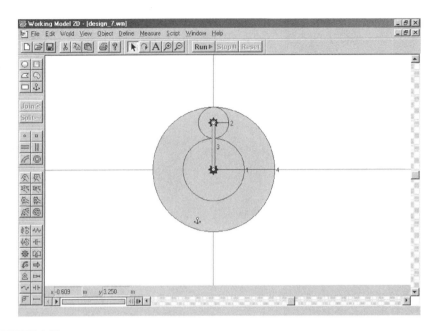

FIGURE II.6.42

Working Model Simulation for Example II.6.7

The planetary gear train with one DOF shown in Figure II.6.21(a) will be simulated using Working Model software. The sun gear 1 ($r_1 = 1$ m) has the angular speed $n_1 = 20$ rpm. The planet gear 2 ($r_2 = 0.5$ m) is connected to the sun gear 1 and the fixed ring gear 4. The external torque exerted on arm 3 is $M_{ext} = -400$ N·m.

Step 1: Creating the gears and the arm

1. Open a new file and make a drawing as shown in Figure II.6.41, following the steps from the previous example of planetary gears.
2. Select the motor connected to the rectangle 3 and erase it using the "Delete" command from the "Edit" menu.
3. Click on the "Pin joint" tool from the toolbox and then click again on the end of the rectangle 3. This connects arm 3 to the ground with a pin joint.
4. Click on the "Circle" tool from the toolbox and draw a disk with the radius $r_4 = 2$ m. Modify the coordinates of its center as $x = y = 0$ at the bottom of the screen. Select the command "Send to back" from the "Object" menu.
5. Click on the "Anchor" tool from the toolbox and then click again on gear 4. This fixes gear 4 to the ground.

The screen should look like that shown in Figure II.6.42.

Step 2: Connecting gear 2 and gear 4

1. Click on gear 2 and select the command "Bring to front" from the "Object" menu. Apply the same command to gear 4.

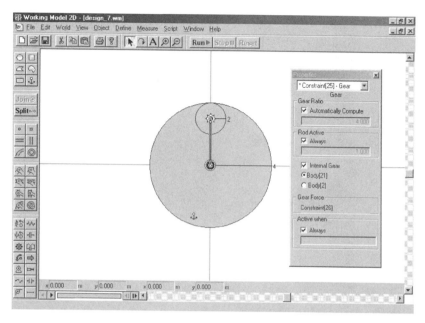

FIGURE II.6.43

2. Click on the "Gear" tool, then click on the center of gear 4 and the center of gear 2, respectively. Double-click on the gear and check the box "Internal gear" on the "Properties" window. Choose gear 4 as internal gear.

The screen should look like that shown in Figure II.6.43.

Step 3: Running the simulation

1. Select gear 1 and choose "Bring to front" command. Apply the same command to arm 3.
2. Select all the bodies and choose the command "Do not collide" from the "Object" menu.
3. Click on "Run" in the toolbar to start the simulation.
4. Click on "Reset" in the toolbar. The simulation resets to the initial frame 0.
5. Click on the planet gear 2 and select the "Measure" menu and the "Velocity" and "Rotational graph" submenus to measure the rotational velocity n_2. Click on the arrow in the right upper corner of the measurement window to change it from graphical to numerical. Apply the same command to visualize the rotational velocity n_1 of the sun gear 1 and the rotational velocity n_3 of the arm 3.

Step 4: Adding an external torque

1. Click on the "Torque" tool from the toolbox and then click on arm 3 (anywhere on the rectangle). This will apply an external torque to arm 3.
2. Select the torque and modify its value to $M_{ext} = -400$ N·m in the "Properties" menu.

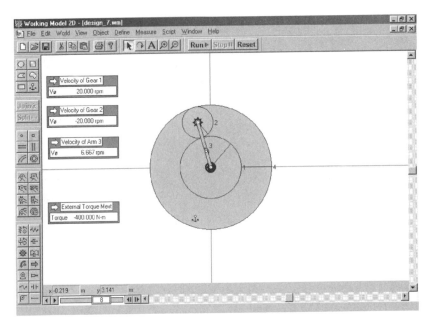

FIGURE II.6.44

3. Select the submenu "Torque" from the "Measure" menu to measure the torque applied.

The screen should look like that shown in Figure II.6.44.

Results
The angular speed of gear 1, $n_1 = 20$ rpm, and the external torque, $M_{ext} = -400$ N·m, are given. It results in the angular speed of gear 2, $n_2 = -20$ rpm, the angular speed of arm 3, $n_3 = 6.667$ rpm, and the motor torque on gear 1, $M_{mot} = 133.333$ N·m.

II.6.12   Programs

PROGRAM II.6.1

```
Apply[Clear, Names["Global`*"]];
Off[General::spell];
Off[General::spell1];

(*Input data*)

n10 = 20;
fi = 20 N[Pi]/180;
r1 = 1.;
r2 = 0.5;
mext = 400;
```

```
Print["sun gear speed n1 = ", n10, " rpm"];
Print["r1 = ", r1, " m"];
Print["r2 = ", r2, " m"];
Print["pressure angle fi = ", fi, " rad"];
Print["|Mext| = ", mext, " Nm"];

d1 = 2 r1;
d2 = 2 r2;

(* Position of joint B *)
xB = 0;
yB = r1;
rB = {xB, yB, 0};

(* Position of joint C *)
xC = 0;
yC = r1 + r2;
rC = {xC, yC, 0};

(* Position of joint E *)
xE = 0;
yE = r1 + 2 r2;
rE = {xE, yE, 0};

(* Position of joint D *)
rD = {xD, 0, 0};

" Contour 0-1-2-0 "

(*Relative velocities*)

n10v = {n10, 0, 0};
n21vSol = {n21Sol, 0, 0};
n02vSol = {n02Sol, 0, 0};

"n10 + n21 + n02 = 0"
"rB x n21 + rE x n02 = 0"
eqIk = (n10v + n21vSol + n02vSol)[[1]] == 0;
eqIi = (Cross[rB, n21vSol] + Cross[rE, n02vSol])[[3]] == 0;
eqIk = (n10v + n21vSol + n02vSol)[[1]] == 0
eqIi = (Cross[rB, n21vSol] + Cross[rE, n02vSol])[[3]] == 0;
solI = Solve[{eqIk, eqIi}, {n21Sol, n02Sol}];
n21v = n21vSol /. solI[[1]];
n02v = n02vSol /. solI[[1]];

Print["n21 = ", n21v, " rpm"];
Print["n02 = ", n02v, " rpm"];
```

```
(*Absolute velocities*)

n20v = -n02v;

Print["n20 = ", n20v, "rpm"];

" Contour 0-2-3-0 "

(*Relative velocities*)

n32vSol = {n32Sol, 0, 0};
n03vSol = {n03Sol, 0, 0};

"n20 + n32 + n03 = 0 "
"rE x n20 +  rC x n32 = 0"
eqIk = (n20v + n32vSol + n03vSol)[[1]] == 0;
eqIi = (Cross[rE, n20v] + Cross[rC, n32vSol])[[3]] == 0;

solI = Solve[{eqIk, eqIi}, {n32Sol, n03Sol}];
n32v = n32vSol /. solI[[1]];
n03v = n03vSol /. solI[[1]];

Print["n32 = ", n32v, " rpm"];
Print["n03 = ", n03v, " rpm"];

(*Absolute velocities*)
n30v = -n03v;
Print["n30 = ", n30v, " rpm"];
Print["arm speed n3 = ", n30v[[1]], " rpm"];

Mext = -Sign[n30v[[1]]] {mext, 0, 0};
Print["Mext = - Sign[n3] |Mext|=", Mext, " rpm"];

"arm 3"
"F23s={0,F23y,F23z}"
F23s = {0, F23ys, F23zs};
"arm 3: sumM_D = rDC x F23 + Mext = 0 =>"
MD = (Cross[rC - rD, F23s] + Mext)[[1]];
s3 = Solve[MD == 0, F23zs];
F23z = F23zs /. s3[[1]];
Print["F23z =", F23z, " N"];

"planet gear 2"
"F12={0,Fr12,Ft12}"
"F02={0,Fr02,Ft02}"
"gear 2: sumM_C = rCE x F02 + rCB x F12 = 0"
"=> r2 Ft02 - r2 Ft12 = 0 => Ft02=Ft12"
"=> F02={0,-Fr12,Ft12}"
"gear 2: sumF = F12 - F23 + F02 = 0"
```

```
"y-axis: F23y = 0"
"z-axis: Ft12 = F23z/2"
"Fr12=Ft12 Tan[fi]"
F23 = {0, 0, F23z};
Ft12 = F23z/2;
Fr12 = Ft12 Tan[fi];
F12 = {0, Fr12, Ft12};
F02 = {0, -Fr12, Ft12};
Print["F23 =", F23, " N"];
Print["F12 =", F12, " N"];
Print["F02 =", F02, " N"];
"arm 3: F03 = -F23 =>"
Print["F03 =", -F23, " N"];

"sun gear 1"
"gear 1: sumM_A = rAB x F21 + Mmot = 0 =>"
MA = (Cross[rB, -F12] + {Mmot, 0, 0})[[1]];
s1 = Solve[MA == 0, Mmot];
Mms = Mmot /. s1[[1]];
Mmos = {Mms, 0, 0};
Print["Mmot =", Mmos, " Nm"];
"gear 1: sumF = F21 + F01 = 0 =>"
Print["F01 = -F21 = F12= ", F12, " N"];
```

sun gear speed n1 = 20 rpm

r1 = 1. m

r2 = 0.5 m

pressure angle fi = 0.349066 rad

|Mext| = 400 Nm

 Contour 0-1-2-0

n10 + n21 + n02 = 0

rB x n21 + rE x n02 = 0

20 + n02Sol + n21Sol == 0

n21 = {-40., 0, 0} rpm

n02 = {20., 0, 0} rpm

n20 = {-20., 0, 0}rpm

 Contour 0-2-3-0

n20 + n32 + n03 = 0

rE x n20 + rC x n32 = 0

n32 = {26.6667, 0, 0} rpm

n03 = {-6.66667, 0, 0} rpm

n30 = {6.66667, 0, 0} rpm

arm speed n3 = 6.66667 rpm

Mext = - Sign[n3] |Mext|={-400, 0, 0} rpm

arm 3

F23s={0,F23y,F23z}

arm 3: sumM_D = rDC x F23 + Mext = 0 =>

F23z =266.667 N

planet gear 2

F12={0,Fr12,Ft12}

F02={0,Fr02,Ft02}

gear 2: sumM_C = rCE x F02 + rCB x F12 = 0

=> r2 Ft02 - r2 Ft12 = 0 => Ft02=Ft12

=> F02={0,-Fr12,Ft12}

gear 2: sumF = F12 - F23 + F02 = 0

y-axis: F23y = 0

z-axis: Ft12 = F23z/2

Fr12=Ft12 Tan[fi]

F23 ={0, 0, 266.667} N

F12 ={0, 48.5294, 133.333} N

F02 ={0, -48.5294, 133.333} N

arm 3: F03 = -F23 =>

F03 ={0, 0, -266.667} N

sun gear 1

gear 1: sumM_A = rAB x F21 + Mmot = 0 =>

Mmot ={133.333, 0, 0} Nm

gear 1: sumF = F21 + F01 = 0 =>

F01 = -F21 = F12= {0, 48.5294, 133.333} N

PROGRAM II.6.2

```
Apply[Clear, Names["Global`*"]];
Off[General::spell];
Off[General::spell1];
```

```
(*Input data*)
"stub involute profile"
(*pressure angle φ=20 deg stub involute*)
σ0p = 15000; (*endurance strength steel psi*)
σ0g = 8000; (*endurance strength cast iron psi*)
φ = 20.π/180;
np = 1200.; (*rpm*)
ng = 300.; (*rpm*)
i = -np/ng;(*transmission ratio*)
H = 33.; (*hp*);
BHN = 250;

Print["endurance strength for pinion - σ0p = ", σ0p, " psi"];
Print["endurance strength for gear - σ0g = ", σ0g, " psi"];
Print["transmission ratio i = ", i];
Print["pressure angle φ = ", φ, " rad = ", φ180/π, " deg"];
Print["speed of the pinion np = ", np, " rpm"];
Print["speed of the gear ng = ", ng, " rpm"];
Print["power trasmitted H = ", H, " hp"];
Print["Brinell Hardness Number BHN = ", BHN];

Np = 16;
Print["select Np = ", Np, " teeth" ];
Ng = -i Np;
Print["Ng = -i Np = ", Ng, " teeth" ];
γp = 0.115;
Print["table 1, Np = ", Np, " => form factor γp = ", γp ];
x1 = 60.; y1 = 0.154;
x2 = 75; y2 = 0.158;
m1 = (y2 - y1)/(x2 - x1);
b1 = y2 - m1x2;
γg = m1Ng + b1;
Print["table 1, Ng = ", Ng, " => form factor γg = ", γg ];
Fp = σ0p γp;
Print["load carrying capacity pinion: Fp = σ0p γp = ", Fp, ""];
Fg = σ0g γg;
Print["load carrying capacity gear: Fg = σ0g γg = ", Fg, ""];
If[Fp < Fg,
  Print["Fp<Fg => pinion is weaker => design will be based on the pinion"];
  Nw = Np; γ = γp; σ0 = σ0p; n = np ,
  Print["Fp>Fg => gear is weaker => design will be based on the gear"];
  Nw = Ng; γ = γg; σ0 = σ0g; n = ng ];
Mt = 63000 H/n;
Print["moment trasmitted Mt=63000 H/n = ", Mt, " lb-in"]
k = 4;
Print["face width factor k = ", k]
σ = σ0 / 2.;
```

```
Print["assume allowable stress σ = σ0/2 = ", σ, " psi"];
"σ=(2 Mt Pd^3 )/(k π^2 γ N) =>"
eq1 = σ == (2 Mt Pd^3 ) / (k π^2 γ Nw)
Pds = Solve[eq1, Pd];
Print["Pd = ", Pd /. Pds[[3]]];
P = Input["Select diametral pitch Pd"];
Print["try Pd = ", P, " teeth per inch"];
dw = N[Nw /P];
Print["pitch diameter dg = Ng/Pd = ", dw, " in"];
V = dw π n /12.;
Print["pitch line velocity V = d π n/12 = ", V, " ft/min" ];
If[V < 2000, σ = 600 σ0 / (600 + V);
   Print["allowable stress for V<2000 ft/min is σ = 600 σ0/(600+V) = ",
     σ, " psi"]
   ];
If[2000 < V < 4000, σ = 1200 σ0 /(1200 + V);
   Print[
     "allowable stress for 2000<V<4000 ft/min is σ = 1200 σ0/(1200+V) = ",
     σ, " psi"]
   ];
If[V > 4000, σ = 78 σ0 /(78 + V^0.5);
   Print["allowable stress for V>4000 ft/min is σ = 78 σ0/(78+V^0.5) = ",
     σ, " psi"]
   ];

σi = (2 Mt P^3 )/(k π^2 γ Nw);
Print["induced stress σi=(2 Mt Pd^3 )/(k π^2 γ N) = ", σi, " psi"];
While[σi > σ,
   Print[
     "σi>σ => weak because the induced stress is greater than the allowable
        stress"];
   P = Input["Select a stronger diametral pitch Pd"];
   Print["try a stronger tooth, Pd = ", P, " teeth per inch"];

   dw = N[Nw/P];
   Print["pitch diameter: d = N/Pd = ", dw , " in"];
   V = dw π n/12.;
   Print["pitch line velocity V = d π n/12 = ", V, " ft/min" ];
   If[V < 2000, σ = 600 σ0/(600 + V);
    Print["allowable stress for V<2000 ft/min is σ = 600 σ0/(600+V) = ",
      σ, " psi"]
   ];
   If[2000 < V < 4000, σ = 1200 σ0 /(1200 + V);
    Print[
      "allowable stress for 2000<V<4000 ft/min is σ = 1200 σ0/(1200+V) = ",
      σ, " psi"]
    ];
```

```
   If[V > 4000, σ = 78 σ0 /(78 + V^0.5);
     Print["allowable stress for V>4000 ft/min is σ = 78 σ0/(78+V^0.5) = ",
       σ, " psi"]
   ];

   σi = (2 Mt P^3 )/(k π^2 γ Nw);
     Print["induced stress σi=(2 Mt Pd^3 )/(k π^2 γ N) = ", σi, " psi"]
   ];
Print["σi<σ => strong because the induced stress is less than the
   allowable stress"]

"k is reduced from the maximum value k=4"
"σ=(2 Mt Pd^3 )/(k π^2 γ N) => k"
eq2 = σ == (2 Mt P^3 )/(kp π^2 γ Nw);
kps = Solve[eq2, kp];
k = kp /. kps[[1]];
Print["reduce k to k = ", k];
P = N[π/P];
Print["circular pitch p = π/Pd = ", p, " in."];
B = k p;
Print["face width B = k p = ", B, " in."];
B = 2.75;
Print["select face width B = ", B, " in."];
a = 0.8 / P;
Print["addendum a = 0.8/Pd = ",a, " in."];
b = 1. /P;
Print["min. dedendum b = 1/Pd = ", b, " in."];
rp = Np/(2. P);
Print["radius of pitch diameter for pinion: rp = Np/(2 Pd) = ", rp, " in."];
rg = Ng/(2. P);
Print["radius of pitch diameter for gear: rg = Ng/(2 Pd) = ", rg, " in."];
rbp = rp Cos[ϕ];
rbg = rg Cos[ϕ];
Print["radius of base diameter for pinion: rbp = rp Cos[ϕ] = ", rbp, " in."];
Print["radius of base diameter for gear: rbg = rg Cos[ϕ] = ", rbg, " in."];
c = rp + rg;
Print["center distance c=rp+rg= ", c, " in"];
"radius of maximum possible addendum circle ra(max)=
   (rb^2+(c Sin[ϕ])^2)^.5"
ramp = (rbp^2 + (c Sin[ϕ])^2)^.5;
ramg = (rbg^2 + (c Sin[ϕ])^2)^.5;
Print["ra(max)p = (rbp^2+(c Sin[ϕ])^2)^.5 = ", ramp, " in."];
Print["ra(max)g = (rbg^2+(c Sin[ϕ])^2)^.5 = ", ramg, " in."];
"radius of addendum circle ra = r + a"
rap = rp + a;
rag = rg + a;
Print["rap = rp + a = ", rap, " in."];
Print["rag = rg + a = ", rag, " in."];
```

```
If[ramp > rap && ramg > rag,
  Print["ra(max)>ra => no interference "], Print["interference "]];

pb = p Cos[ϕ];
Print["base pitch pb = p Cos[ϕ] = ", pb, " in."];
CR = ((rap^2 - rbp^2)^.5 + (rag^2 - rbg^2)^.5 - c Sin[ϕ])/pb;
Print[
  "contact ratio CR = ((rap^2-rbp^2)^.5+(rag^2-rbg^2)^.5-c Sin[ϕ])/
    pb = ", CR ];
"contact ratio CR > 1.2 "
F0 = σ0 B γ p;
Print["endurance load F0 = σ0 B γ p = ", F0, " lb"];

σes = 400 BHN - 10000;
Print["endurance limit of the gear pair σes = 400 BHN - 10000 = ", σes,
  " psi"];
Q = 2 Ng/(Np + Ng);
Print["Q = 2 Ng/(Np+Ng) = ", Q];
Ep = 30. 10^6 (*psi*);
Eg = 19. 10^6 (*psi*);
K = σes^2 Sin[ϕ](1/Ep + 1/Eg)/1.4;
Print["stress fatigue factor K = σes^2 Sin[ϕ] (1/Ep+1/Eg)/1.4 = ", K];
dp = 2 rp;
Fw = dp B K Q;
Print["wear load Fw = dp B K Q = ", Fw, " lb"];
F = Mt/rg;
Print["F = Mt/rg = ", F, " lb"];
Print["for V = ", V, " ftm => from fig 13(a) permissible error =
  0.00225 in. "];
Print["from fig 13(a) for first class comercial
    gears with Pd=4 => error e = 0.0025 in. > permissible error "];
Print["from fig 13(b) for carefully cut gears with Pd=4 => error
  e = 0.0012 in. "];
(*deformation factor C*)
x1 = 0.001; y1 = 1180; (*y1=1140;*)
x2 = 0.002; y1 = 2360; (*y2=2280;*)
m1 = (y2 - y1) / (x2 - x1)
b1 = y2 - m1 x2;
e = 0.0012;
Print["select tooth error e = ", e, " in."];
Cf = m1 e + b1;
Print[
  "from table 2 calculate the deformation factor for dynamic load
    check C = ", Cf];
Fd = 0.05 V (B Cf + F)/(0.05 V + Sqrt[B Cf + F])+ F;
Print[
  "dynamic load Fd = 0.05 V (B C+F)/(0.05 V + Sqrt[ B C+F]) + F = ", Fd,
    " lb"];
```

```
If[Fd > F0, Print["F0 = ", F0, " approx Fd = ", Fd, " within ",
  (1 - F0 /Fd)100, " %"],
  Print["Fd<F0"]];
If[Fd > Fw, Print["Fw = ", Fw, " approx Fd = ", Fd, " within ",
  (1 - Fw /Fd)100, " %"],
  Print["Fd<Fw"]];

Print["from fig 13(b) for precision gears with Pd=4 => error
  e = 0.00051 in. "];
e = 0.00051;
Cf = m1 e + b1;
Print[
  "from table 2 calculate the deformation factor for dynamic load
    check C = ", Cf];
Fd = 0.05 V (B Cf + F)/(0.05 V + Sqrt[B Cf + F])+ F;
Print[
  "dynamic load Fd = 0.05 V (B C+F)/(0.05 V + Sqrt[ B C+F]) + F = ",
    Fd, " lb"];
If[Fd > F0, Print["F0 = ", F0, " approx Fd = ", Fd, " within ",
  (1 - F0 /Fd)100, " %"],
  Print["Fd<F0"]];
If[Fd > Fw, Print["Fw = ", Fw, " approx Fd = ", Fd, " within ",
  (1 - Fw /Fd)100, " %"],
  Print["Fd<Fw"]];
```

stub involute profile

endurance strength for pinion - $\sigma 0p$ = 15000 psi

endurance strength for gear - $\sigma 0g$ = 8000 psi

transmission ratio i = -4.

pressure angle ϕ = 0.349066 rad = 20. deg

speed of the pinion np = 1200. rpm

speed of the gear ng = 300. rpm

power trasmitted H = 33. hp

Brinell Hardness Number BHN = 250

select Np = 16 teeth

Ng = -i Np = 64. teeth

table 1, Np = 16 => form factor γp = 0.115

table 1, Ng = 64. => form factor γg = 0.155067

load carrying capacity pinion: Fp = $\sigma 0p$ γp = 1725.

load carrying capacity gear: Fg = $\sigma 0g$ γg = 1240.53

Fp>Fg => gear is weaker => design will be based on the gear

moment trasmitted Mt=63000 H/n = 6930. lb-in

face width factor k = 4

assume allowable stress $\sigma = \sigma 0/2$ = 4000. psi

σ=(2 Mt Pd^3)/(k π^2 γ N) =>

4000. == 35.3757 Pd^3

Pd = 4.83561

try Pd = 5 teeth per inch

pitch diameter dg = Ng/Pd = 12.8 in

pitch line velocity V = d π n/12 = 1005.31 ft/min

allowable stress for V<2000 ft/min is σ = 600 σ0/(600+V) = 2990.08 psi

induced stress σi=(2 Mt Pd^3)/(k π^2 γ N) = 4421.96 psi

σi>σ => weak because the induced stress is greater than the allowable stress

try a stronger tooth, Pd = 4 teeth per inch

pitch diameter: d = N/Pd = 16. in

pitch line velocity V = d π n/12 = 1256.64 ft/min

allowable stress for V<2000 ft/min is σ = 600 σ0/(600+V) = 2585.32 psi

induced stress σi=(2 Mt Pd^3)/(k π^2 γ N) = 2264.04 psi

σi<σ => strong because the induced stress is less than the allowable stress

k is reduced from the maximum value k=4

σ=(2 Mt Pd^3)/(k π^2 γ N) => k

reduce k to k = 3.50292

circular pitch p = π/Pd = 0.785398 in.

face width B = k p = 2.75119 in.

select face width B = 2.75 in.

addendum a = 0.8/Pd = 0.2 in.

min. dedendum b = 1/Pd = 0.25 in.

radius of pitch diameter for pinion: rp = Np/(2 Pd) = 2. in.

radius of pitch diameter for gear: rg = Ng/(2 Pd) = 8. in.

radius of base diameter for pinion: rbp = rp Cos[ϕ] = 1.87939 in.

radius of base diameter for gear: rbg = rg Cos[ϕ] = 7.51754 in.

center distance c=rp+rg= 10. in

radius of maximum possible addendum circle ra(max)=

  (rb^2+(c Sin[ϕ]))^2^.5

ra(max)p = (rbp^2+(c Sin[ϕ])^2)^.5 = 3.90255 in.

ra(max)g = (rbg^2+(c Sin[ϕ])^2)^.5 = 8.25901 in.

radius of addendum circle ra = r + a

rap = rp + a = 2.2 in.

rag = rg + a = 8.2 in.

ra(max)>ra => no interference

base pitch pb = p Cos[ϕ] = 0.738033 in.

contact ratio CR = ((rap^2-rbp^2)^.5+(rag^2-rbg^2)^.5-c Sin[ϕ])/pb =
  1.35303

contact ratio CR > 1.2

endurance load F0 = σ0 B γ p = 2679.36 lb

endurance limit of the gear pair σes = 400 BHN - 10000 = 90000 psi

Q = 2 Ng/(Np+Ng) = 1.6

stress fatigue factor K = σes^2 Sin[ϕ] (1/Ep+1/Eg)/1.4 = 170.11

wear load Fw = dp B K Q = 2993.94 lb

F = Mt/rg = 866.25 lb

for V = 1256.64 ftm => from fig 13(a) permissible error = 0.00225 in.

from fig 13(a) for first class comercial

  gears with Pd=4 => error e = 0.0025 in. > permissible error

from fig 13(b) for carefully cut gears with Pd=4 => error e = 0.0012 in.

select tooth error e = 0.0012 in.

from table 2 calculate the deformation factor for dynamic load check
  C = 1416.

dynamic load Fd = 0.05 V (B C+F)/(0.05 V + Sqrt[B C+F]) + F = 3135.11 lb

F0 = 2679.36 approx Fd = 3135.11 within 14.5369 %

Fw = 2993.94 approx Fd = 3135.11 within 4.50291 %

from fig 13(b) for precision gears with Pd=4 => error e = 0.00051 in.

from table 2 calculate the deformation factor for dynamic load check
  C = 601.8

dynamic load Fd = 0.05 V (B C+F)/(0.05 V + Sqrt[B C+F]) + F =
  2267.58 lb

Fd<F0

Fd<Fw

II.7   Mechanical Springs

Springs are mechanical elements that exert forces or torques and absorb energy. The absorbed energy is usually stored and later released. Springs are made of metal. For light loads the metal can be replaced by plastics. Some applications which require minimum spring mass use structural composite materials. Blocks of rubber can be used as springs in bumpers and vibration isolation mountings of electric or combustion motors.

II.7.1   Material for Springs [20, 21]

The hot and cold working processes are used for springs manufacturing. Plain carbon steels, alloy steels, corrosion-resisting steels, or nonferrous materials can be used for spring manufacturing. Spring materials are compared by an examination of their tensile strengths which require the material, its processing, and the wire size. The tensile strength S_{ut} is a linear function of the wire diameter d, which is estimated by

$$S_{ut} = \frac{A}{d^m},$$ (II.7.1)

where the constant A and the exponent m are presented in Table II.7.1.

The torsional yield strength can be obtained by assuming that the tensile yield strength is between 60 and 90% of the tensile strength. Using the distortions–energy theory, the torsional yield strength is

$$S_{sy} = 0.5777\, S_y,$$ (II.7.2)

and for steels it is

$$0.35 S_{ut} \leq S_{sy} \leq 0.52 S_{ut}.$$ (II.7.3)

TABLE II.7.1 Constants of Tensile Strengths Expression

		A	
Material	m	kpsi	MPa
Music wire	0.163	186	2060
Oil-tempered wire	0.193	146	1610
Hard-drawn wire	0.201	137	1510
Chrome vanadium	0.155	173	1790
Chrome silicom	0.091	218	1960

Source: Barnes Group, Inc., Design Handbook, Barnes Group, Inc., Bristol, Conn., 1981. Reprinted with permission.

For static application, the *maximum allowable torsional stress* τ_{all} may be used instead of S_{sy}

$$S_{sy} = \tau_{all} = \begin{cases} 0.45 S_{ut} & \text{cold-drawn carbon steel;} \\ 0.50 S_{ut} & \text{hardened and tempered carbon} \\ & \text{and low-alloy steel;} \\ 0.35 S_{ut} & \text{austenitic stainless steel} \\ & \text{and nonferrous alloys.} \end{cases} \tag{II.7.4}$$

Figure II.7.1 shows the minimum tensile strength of commonly used spring wire materials.

II.7.2 Helical Extension Springs

Extension springs [Fig. II.7.2(a)] are used for maintaining the torsional stress in the wire. The initial tension is the external force, F, applied to the spring. Spring manufacturers recommended that the initial tension be

$$\tau_{initial} = (0.4 - 0.8) \frac{S_{ut}}{C}, \tag{II.7.5}$$

where S_{ut} is the tensile strength in psi. The constant C is the spring index, defined by $C = \dfrac{D}{d}$, where D is the mean diameter of the coil and d is the diameter of the wire [Fig. II.7.2(a)]. The bending stress, which occurs in section $A - A$, is

$$\sigma = \frac{16FD}{\pi d^3} \left(\frac{r_1}{r_3} \right), \tag{II.7.6}$$

and torsional stress, which occurs in section $B - B$, is

$$\tau = \frac{FD}{\pi d^3} \left(\frac{r_4}{r_2} \right). \tag{II.7.7}$$

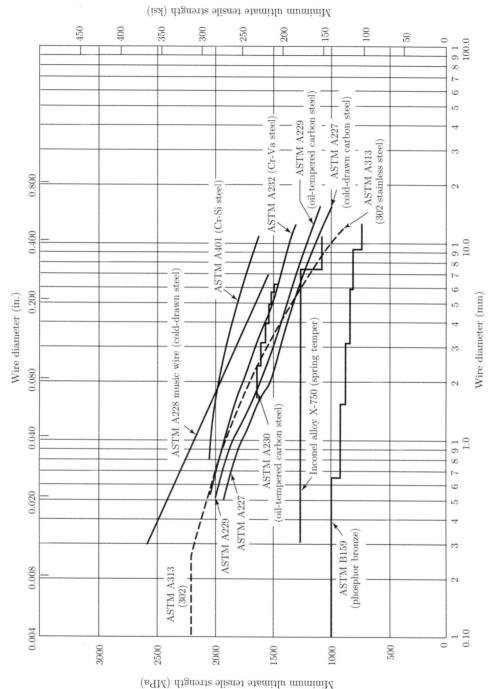

FIGURE II.7.1 *Tensile strength of spring wire materials.*

Source: Barnes Group, Inc., Design Handbook, Barnes Group, Inc., Bristol, Conn., 1981

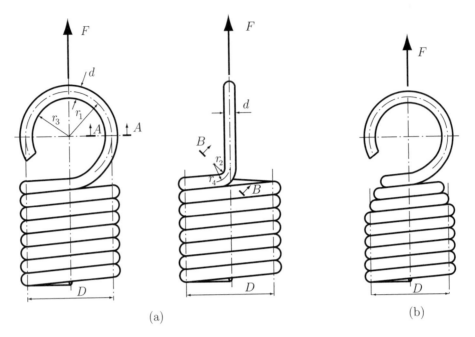

(a)   (b)

FIGURE II.7.2 *Extension springs.*

In practical application the radius r_4 is greater than twice the wire diameter. Hook stresses can be further reduced by winding the last few coils with a decreasing diameter D [Fig. II.7.2(b)]. This lowers the nominal stress by reducing the bending and torsional moment arms.

II.7.3 Helical Compression Springs

The helical springs are usually made of circular cross-section wire or rod (Fig. II.7.3). These springs are subjected to a torsional component and to a shear component. There is also an additional stress effect due to the curvature of the helix.

Shear Stress, τ

The total shear stress, τ (psi), induced in a helical spring is

$$\tau = \frac{Tr}{J} + \frac{F}{A} = \frac{16\,T}{\pi d^3} + \frac{4\,F}{\pi d^2} = \frac{8\,F\,D}{\pi d^3} + \frac{4F}{\pi d^2}, \qquad (\text{II.7.8})$$

where

$T = FD/2$, is the torque, lb in,
$r = d/2$ is the wire radius, in,
F is the axial load, lb,
$A = \pi d^2/4$ is cross-section area, in.^{2}, and
$J = \pi d^4/32$ is the polar second moment of inertia, in in.^{4}.

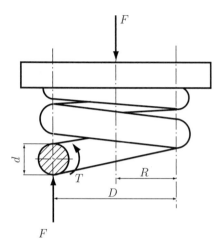

FIGURE II.7.3 *Helical compression spring.*

The shear stress expressed in Eq. (II.7.8) can be rewritten as

$$\tau = K_s \frac{8FD}{\pi d^3},$$ (II.7.9)

where K_s is the shear stress multiplication factor

$$K_s = \frac{2C+1}{2C}.$$ (II.7.10)

The spring index $C = \dfrac{D}{d}$ is in the range 6 to 12.

Curvature Effect

The curvature of the wire increases the stress on the inside of the spring and decreases it on the outside. The stress equation is a function of the factor K_s which can be replaced by a correction factor K_B or K_w

$$\tau = K_B \frac{8FD}{\pi d^3}, \quad \text{or} \quad \tau = K_w \frac{8FD}{\pi d^3},$$ (II.7.11)

where K_B is called the Bergstrasser factor (preferred factor),

$$K_B = \frac{4C+2}{4C-3},$$ (II.7.12)

and K_w is the Wahl factor and is given by

$$K_w = \frac{4C-1}{4C-4} + \frac{0.615}{C}.$$ (II.7.13)

The factor K_B or K_w corrects both curvature and direct shear effects. The effect of the curvature alone is defined by the curvature correction factor K_c, which can be obtained as

$$K_c = \frac{K_B}{K_s}. \tag{II.7.14}$$

Deflection, δ

The deflection-force relations are obtained using Castigliano's theorem. The total strain energy for a helical spring is

$$U = U_t + U_s = \frac{T^2 l}{2GJ} + \frac{F^2 l}{2AG}, \tag{II.7.15}$$

where

$$U_t = \frac{T^2 l}{2GJ}, \tag{II.7.16}$$

is the torsional component of the energy, and

$$U_s = \frac{F^2 l}{2AG}, \tag{II.7.17}$$

is the shear component of the energy. The spring load is F, the torsion torque is T, the length of the wire is l, the second moment of inertia is J, the cross-section area of the wire is A, and the modulus of rigidity is G.

Substituting $T = FD/2$, $l = \pi DN$, $J = \pi d^4/32$, and $A = \pi d^2/4$ in Eq. (II.7.15), one may obtain the total strain energy as

$$U = \frac{4F^2 D^3 N}{d^4 G} + \frac{2F^2 DN}{d^2 G}, \tag{II.7.18}$$

where $N = N_a$ is the number of active coils.

Applying Castigliano's theorem, the deflection of the helical spring is

$$\delta = \frac{\partial U}{\partial F} = \frac{8FD^3 N}{d^4 G} + \frac{4FDN}{d^2 G}. \tag{II.7.19}$$

Using the spring index $C = D/d$, the deflection becomes

$$\delta = \frac{8FD^3 N}{d^4 G}\left(1 + \frac{1}{2C^2}\right) \approx \frac{8FD^3 N}{d^4 G}. \tag{II.7.20}$$

Spring Rate

The general relationship between force and deflection can be written as

$$F = F(\delta). \tag{II.7.21}$$

The *spring rate* is then defined as

$$k(\delta) = \lim_{\Delta\delta \to 0} \frac{\Delta F}{\Delta\delta} = \frac{dF}{d\delta}, \tag{II.7.22}$$

where δ must be measured in the direction of the load F and at the point of application of F. Because most of the force-deflection equations that treat the springs are linear, k is constant and is named the *spring constant*. For this reason Eq. (II.7.22) may be written as

$$k = \frac{F}{\delta}. \tag{II.7.23}$$

From Eq. (II.7.20), with the substitution $C = D/d$, the spring rate for a helical spring under an axial load is

$$k = \frac{G\,d^4}{8\,D^3 N} = \frac{G\,d}{8\,C^3 N}. \tag{II.7.24}$$

For springs in parallel having individual spring rates, k_i [Fig. II.7.4(a)], the spring rate k is

$$k = k_1 + k_2 + k_3. \tag{II.7.25}$$

For springs in series, with individual spring rates, k_i [Fig. II.7.4(b)], the spring rate k is

$$k = \frac{1}{\dfrac{1}{k_1} + \dfrac{1}{k_2} + \dfrac{1}{k_3}}. \tag{II.7.26}$$

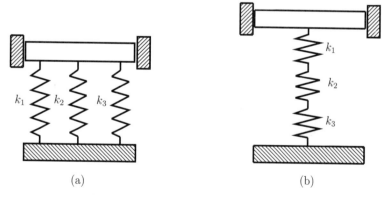

(a)   (b)

FIGURE II.7.4   *(a) Springs in parallel; (b) springs in series.*

Spring Ends

For helical springs the ends can be specified as shown in Figure II.7.5: (a) plain ends; (b) plain and ground ends; (c) squared ends; (d) squared and ground ends. A spring with plain ends [Fig. II.7.5(a)], has a noninterrupted helicoid and the ends are the same as if a long spring had been cut into sections. A spring with plain and ground ends [Fig. II.7.5(b)], or squared ends [Fig. II.7.5(c)], is obtained by deforming the ends to a zero-degree helix angle. Springs should always be both squared and ground [Fig. II.7.5(d)], because a better transfer of the load is obtained. Table II.7.2 presents the type of ends and how that affects the number of coils and the spring length. In Table II.7.2, N_a is the number of active coils, and d is the wire diameter.

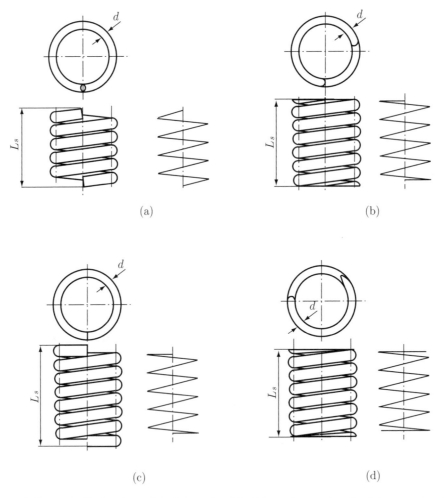

(a)   (b)

(c)   (d)

FIGURE II.7.5 *Helical springs: (a) plain ends; (b) plain and ground ends; (c) squared ends; (d) squared and ground ends.*

TABLE II.7.2 Type of Spring Ends

Term	End coils, N_e	Total coil, N_t	Free length, L_0	Solid length, L_s	Pitch, p
Plain	0	N_a	$pN_a + d$	$d(N_t + 1)$	$(L_0 - d)/N_a$
Plain and ground	1	$N_a + 1$	$p(N_a + 1)$	dN_t	$L_0/(N_a + 1)$
Squared or closed	2	$N_a + 2$	$pN_a + 3d$	$d(N_t + 1)$	$(L_0 - 3d)/N_a$
Squared and ground	2	$N_a + 2$	$pN_a + 2d$	dN_t	$(L_0 - 2d)/N_a$

Source: Barnes Group, Inc., Design Handbook, Barnes Group, Inc., Bristol, Conn., 1981. Reprinted with permission.

Stability

The springs in compression will buckle when the deflection is too large. Figure II.7.6 gives the stability zones for two end conditions.

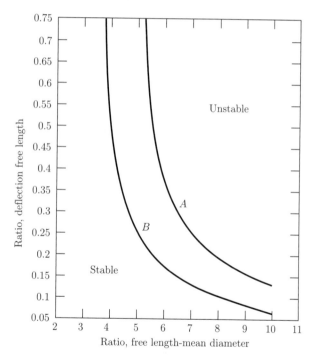

A- end plates are constrained parallel

B- one end plate is free to tip

Source: R. C. Juvinall and K. M. Marshek, Fundamentals of Machine Component Design, 3rd ed., John Wiley & Sons, New York, 2000.

FIGURE II.7.6 *Stability zone for springs in compression. Reprinted with permission of John Wiley & Sons, Inc.*

II.7.4 Torsion Springs

The helical torsion springs (Fig. II.7.7) are used in door hinges, automobile starters, and for any application where torque is required. Torsion springs are of two general types: helical [Fig. II.7.8(a)] and spiral [Fig. II.7.8(b)]. The primary stress in torsion springs is bending. The bending moment Fa is applied to each end of the wire. The highest stress acting inside of the wire is

$$\sigma_i = \frac{K_i Mc}{I},$$ (II.7.27)

where the factor for inner surface stress concentration K_i is given in Figure II.7.9, and I is the moment of inertia. The distance from the neutral axis to the extreme fiber for a round solid bar is $c = d/2$, and for a rectangular bar it is $c = h/2$.

For a solid round bar section $I = \pi d^4/64$, and for a rectangular bar $I = bh^3/12$. Substituting the product Fa for bending moment and the equations for section properties of round and rectangular wire one may write,

- *Round wire:*

$$\frac{I}{c} = \frac{\pi d^3}{32}, \quad \sigma_i = \frac{32Fa}{\pi\, d^3} K_{i,round}.$$ (II.7.28)

- *Rectangular wire:*

$$\frac{I}{c} = \frac{bh^2}{6}, \quad \sigma_i = \frac{6\,Fa}{bh^2} K_{i,rectangular}.$$ (II.7.29)

The angular deflection of a beam subjected to bending is

$$\theta = \frac{ML}{EI},$$ (II.7.30)

where M is the bending moment, L the beam length, E the modulus of elasticity, and I the momentum of inertia.

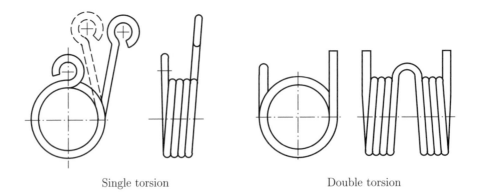

Single torsion   Double torsion

FIGURE II.7.7   *Helical torsion springs.*

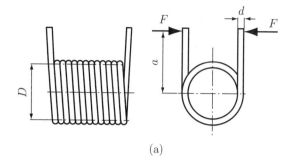

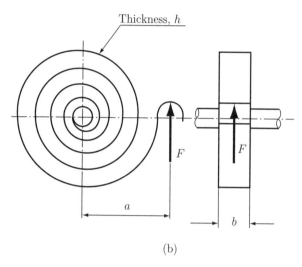

FIGURE II.7.8 *Torsion springs: (a) helical and (b) spiral.*

Equation (II.7.30) can be used for helical and spiral torsion springs. Helical torsion springs and spiral springs can be made from thin rectangular wire. Round wire is often used in noncritical applications.

II.7.5 Torsion Bar Spring

The torsion bar spring, shown in Figure II.7.10, is used in automotive suspension. The stress, angular deflection, and spring rate equation are

$$\tau = \frac{Tr}{J}, \tag{II.7.31}$$

$$\theta = \frac{Tl}{JG}, \tag{II.7.32}$$

$$k = \frac{JG}{l}, \tag{II.7.33}$$

where T is the torque, $r = d/2$ is the bar radius, l is the length of the spring, G is the modulus of rigidity, and J is the second polar moment of area.

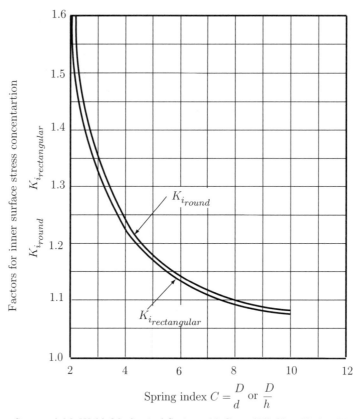

Source: A.M. Wahl, Mechanical Springs, McGraw-Hill, New York, 1963.

FIGURE II.7.9 *Stress concentration factor. Reprinted with permission of McGraw-Hill.*

For a solid round section, J is

$$J = \frac{\pi\, d^4}{32}.$$

(II.7.34)

For a solid rectangular section

$$J = \frac{bh^3}{12}.$$

(II.7.35)

For solid round rod of diameter d, Eqs. (II.7.31), (II.7.32), and (II.7.33), become

$$\tau = \frac{16T}{\pi d^3},$$

(II.7.36)

$$\theta = \frac{32Tl}{\pi d^4 G},$$

(II.7.37)

$$k = \frac{\pi d^4 G}{32l}.$$

(II.7.38)

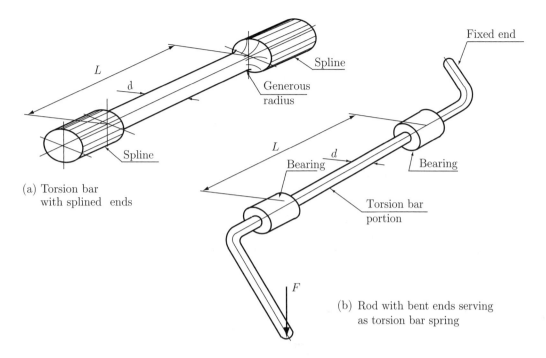

FIGURE II.7.10 *Torsion bar spring.*

II.7.6 Multi-Leaf Spring

The multi-leaf spring can be a simple cantilever [Fig. II.7.11(a)], or the semi-elliptic leaf [Fig. II.7.11(b)]. The design of the multi-leaf springs is based on force, F, length, L, deflection, and stress relationships. The multi-leaf spring may be considered as a triangular plate [Fig. II.7.12(a)], cut into n strips of width b, or stacked in a graduated manner [Fig. II.7.12(b)].

To support transverse shear N_e more extra full-length leaves are added on the graduated stack, as shown in Figure II.7.13. The number N_e is always one less than the total number of full-length leaves, N.

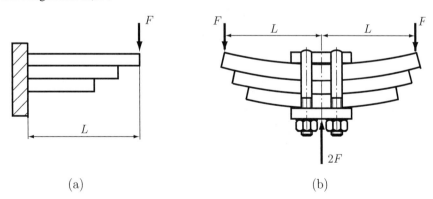

(a)   (b)

FIGURE II.7.11 *Multi-leaf spring: (a) simple cantilever and (b) semi-elliptic.*

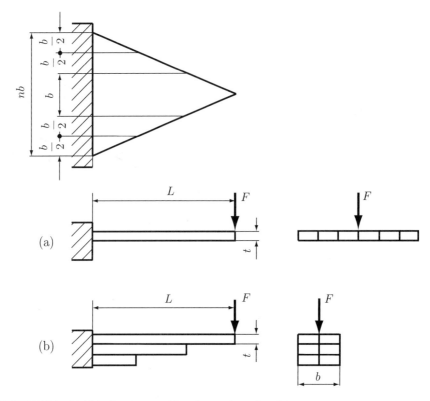

FIGURE II.7.12 *Multi-leaf spring considered as a triangular plate.*

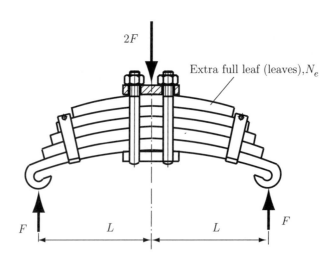

FIGURE II.7.13 *Multi-leaf spring: extra full-length leaves.*

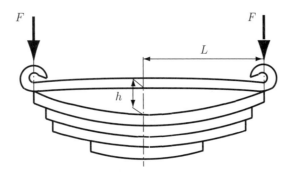

FIGURE II.7.14 *Multi-leaf spring: gap between the extra full-length leaves.*

The pre-stressed leaves have a different radius of curvature than the graduated leaves. This will leave a gap h between the extra full-length leaves and the graduated leaves before assembly (Fig. II.7.14).

Bending Stress, σ_e

The bending stress in the extra full-length leaves installed without initial pre-stress is

$$\sigma_e = \frac{18FL}{bt^2(3N_e + 2N_g)}, \tag{II.7.39}$$

where F is the total applied load at the end of the spring (lb), L is the length of the cantilever or half the length of the semi-elliptic spring (in.), b is the width of each spring leaf (in.), t is the thickness of each spring leaf (in.), N_e is the number of extra full length leaves, and N_g is the number of graduated leaves.

Bending Stress, σ_g

For graduated leaves assembled with extra full-length leaves without initial pre-stress, the bending stress is

$$\sigma_g = \frac{12FL}{bt^2(3N_e + 2N_g)} = \frac{2\sigma_e}{3}. \tag{II.7.40}$$

Deflection of a Multi-leaf Spring, δ

The deflection of a multi-leaf spring with graduated and extra full-length leaves is

$$\delta = \frac{12Fl^3}{bt^2E(3N_e + 2N_g)}, \tag{II.7.41}$$

where E is the modulus of elasticity (psi).

Bending Stress, σ

The bending stress of multi-leaf springs without extra leaves or with extra full-length pre-stressed leaves which have the same stress after the full load has been applied is

$$\sigma = \frac{6Fl}{Nbt^2}. \tag{II.7.42}$$

where N is the total number of leaves.

Gap

The gap between pre-assembled graduated leaves and extra full-length leaves (Fig. II.7.14) is

$$h = \frac{2FL^3}{Nbt^3E}. \tag{II.7.43}$$

II.7.7 Belleville Springs

The Belleville springs are made from tapered washers [Fig. II.7.15(a)] stacked in series, parallel, or a combination of parallel-series, as shown in Figure II.7.15(b). The load-deflection and stress-deflection are

$$F = \frac{E\,\delta}{\left(1 - \mu^2\right)(d_o/2)^2 M}[(h - \delta/2)(h - \delta)t + t^3], \tag{II.7.44}$$

$$\sigma = \frac{E\,\delta}{\left(1 - \mu^2\right)(d_o/2)^2 M}[C_1(h - \delta/2) + C_2 t], \tag{II.7.45}$$

where F is the axial load (lb), δ is the deflection (in.), t is the thickness of the washer (in.), h, is the free height minus thickness (in.), E is the modulus of elasticity (psi), σ is the stress at inside circumference (psi), d_o is the outside diameter of washer (in.), d_i is the inside diameter of washer (in.) and μ is the Poisson's ratio. The constants M, C_1, and C_2 are given by the equations

$$M = \frac{6}{\pi\,log_e(d_o/d_i)}\left(\frac{d_o/d_i - 1}{d_o/d_i}\right)^2,$$

$$C_1 = \frac{6}{\pi\,log_e(d_o/d_i)}\left[\frac{d_o/d_i - 1}{log_e(d_o/d_i)} - 1\right],$$

$$C_2 = \frac{6}{\pi\,log_e(d_o/d_i)}\left[\frac{d_o/d_i - 1}{2}\right].$$

II.7.8 Elastic Potential Energy and Virtual Work

A particle in static equilibrium position is considered. The static equilibrium position of the particle is determined by the forces that act on it. The *virtual displacement*, $\delta\mathbf{r}$, is any

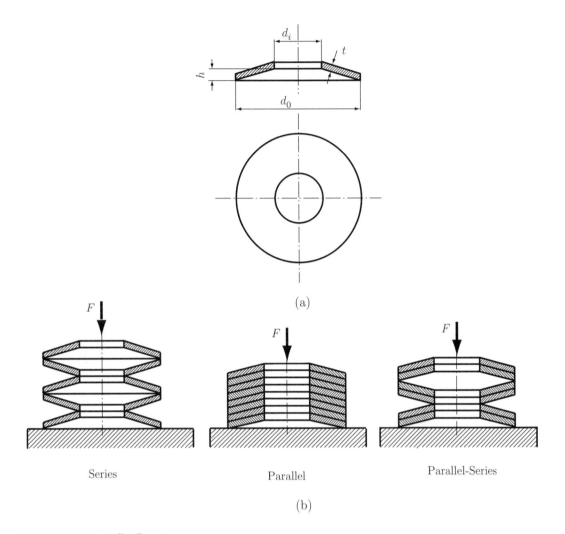

FIGURE II.7.15 *Belleville springs.*

arbitrary small displacement away from this natural position and consistent with the system constraints. The term *virtual* is used to indicate that the displacement does not really exist but only is assumed to exist. The *virtual work* is the work done by any force $\mathbf{F}$ acting on the particle during the virtual displacement $\delta\mathbf{r}$:

$$\delta U = \mathbf{F} \cdot \delta\mathbf{r} = F\,\delta r\,\cos\alpha,$$

where α is the angle between $\mathbf{F}$ and $\delta\mathbf{r}$ ($|\delta\mathbf{r}| = \delta r$). The actual infinitesimal change in position $d\mathbf{r}$ can be integrated and the infinitesimal virtual or assumed movement $\delta\mathbf{r}$ cannot be integrated. Mathematically, both quantities are first-order differentials. A virtual displacement may also be a rotation $\delta\theta$ of a body. The virtual work done by a couple M during a virtual angular displacement $\delta\theta$ is $\delta U = M\,\delta\theta$. The force F or couple M remain constant during any infinitesimal virtual displacement.

Consider a particle in equilibrium position as a result of the forces $\mathbf{F}_1, \mathbf{F}_2, \ldots, \mathbf{F}_n$. For an assumed virtual displacement $\delta\mathbf{r}$ of the particle away from its equilibrium position, the total virtual work done on the particle is

$$\delta U = \Sigma \mathbf{F} \cdot \delta \mathbf{r} = \Sigma F_x\, \delta x + \Sigma F_y\, \delta y + \Sigma F_z\, \delta z = 0.$$

The sum is zero, since $\Sigma \mathbf{F} = \mathbf{0}$. The equation $\delta U = 0$ is therefore an alternative statement of the equilibrium conditions for a particle. This condition of zero virtual work for equilibrium is both necessary and sufficient.

The principle of virtual work for a single particle can be extended to a rigid body treated as a system of small elements or particles rigidly attached to one another. Because the virtual work done on each particle of the body in equilibrium is zero, it results that the virtual work done on the entire rigid body is zero.

All the internal forces appear in pairs of equal, opposite, and collinear forces, and the net work done by these forces during any movement is zero. Only the virtual work done by external forces are taken into account in the evaluation of $\delta U = 0$ for the entire body.

The principle of virtual work will be extended to the equilibrium of an interconnected ideal system of rigid bodies. The *ideal systems* are systems composed of two or more rigid bodies linked together by mechanical connections which are incapable of absorbing energy through elongation or compression, and in which friction is small enough to be neglected. There are two types of forces that act in such an interconnected system:

- Active forces are external forces capable of doing virtual work during possible virtual displacements.
- Joint forces are forces in the connections between members. During any possible movement of the system or its parts, the net work done by the joint forces at the connections is zero, because the joint forces always exist in pairs of equal and opposite forces.

Principle of Virtual Work: The work done by external active forces on an ideal mechanical system in equilibrium is zero for any and all virtual displacements consistent with the constraints.

Mathematically, the principle can be expressed as

$$\delta U = 0. \tag{II.7.46}$$

The advantage of the method of virtual work is that relations between the active forces can be determined directly without reference to the joint forces. The method is useful in determining the position of equilibrium of a system under known forces. The method of virtual work cannot be applied for the system where the internal friction in a mechanical system is appreciable (the work done by internal friction should be included).

Elastic Potential Energy

The work done on an elastic body is stored in the body in the form of elastic potential energy, V_e. The potential energy is available to do work on some other body during the compression or extension. A spring can store and release potential energy. Consider a spring that is being compressed by a force F. The spring is elastic and linear, and the force F is directly proportional to the deflection x:

$$F = kx,$$

where k is the spring constant or stiffness of the spring. The work done on the spring by F during dx is

$$dU = F\, dx.$$

The elastic potential energy of the spring for a compression x is the total work done on the spring:

$$V_e = \int_0^x F\, dx = \int_0^x kx\, dx = \frac{1}{2}kx^2.$$

For an increase in the compression of the spring from x_1 to x_2, the work done on the spring equals its change in elastic potential energy:

$$\Delta V_e = \int_{x_1}^{x_2} kx\, dx = \frac{1}{2}k(x_2^2 - x_1^2).$$

During a virtual displacement δx of the spring, the virtual work done on the spring is the virtual change in elastic potential energy:

$$\delta V_e = F\, \delta x = kx\, \delta x.$$

When the spring is in tension rather than compression, the work and energy relations are the same as those for compression, where x represents the elongation of the spring rather than its compression.

A torsional spring resists the rotation of a shaft or another body and the resisting moment is

$$M = K\theta,$$

where K is the torsional stiffness. The potential energy becomes

$$V_e = \int_0^\theta K\theta\, d\theta = \frac{1}{2}K\theta^2,$$

which is analogous to the expression for the linear extension spring.

The units of elastic potential energy are the same as those of work and are expressed in joules (J) in SI units and in foot-pounds (ft.-lb) in U.S. customary units.

Gravitational Potential Energy

For an upward displacement δh of a body, the weight $W = mg$ does negative work, $\delta U = -mg\,\delta h$. If the body has a downward displacement δh the weight does positive work, $\delta U = +mg\,\delta h$. The gravitational potential energy V_g of a body is defined as the work done on the body by a force equal and opposite to the weight in bringing the body to the position under consideration from some arbitrary datum plane where the potential energy is defined to be zero. The potential energy is the negative of the work done by the weight. If $V_g = 0$ at $h = 0$ (datum plane), then at a height h above the datum plane, the gravitational potential energy of the body is $V_g = mgh$. If the body is a distance h below the datum plane, its gravitational potential energy is $-mgh$.

Remarks

(1) The datum plane for zero potential energy is arbitrary because only the change in potential energy matters, and this change is the same no matter where the datum plane is located.

(2) The gravitational potential energy is independent of the path followed in arriving at a particular level h.

The virtual change in gravitational potential energy is

$$\delta V_g = mg\,\delta h,$$

where δh is the upward virtual displacement of the mass center of the body. The units of gravitational potential energy are the same as those for work and elastic potential energy, joules (J) in SI units and foot-pounds (ft.-lb) in U.S. customary units.

Consider a linear spring attached to a body of mass m. The work done by the linear spring on the body is the negative of the change in the elastic potential energy of the spring. The work done by the gravitational force or weight mg is the negative of the change in gravitational potential energy.

The total virtual work δU is the sum of the work $\delta U'$ done by all active forces (other than spring forces and weight forces) and the work $-(\delta V_e + \delta V_g)$ done by the spring and weight forces. The virtual work equation $\delta U = 0$ becomes

$$\delta U' - (\delta V_e + \delta V_g) = 0, \quad \text{or} \quad \delta U' = \delta V, \tag{II.7.47}$$

where $V = V_e + V_g$ is the total potential energy of the system.

Thus, for a mechanical system with elastic members and bodies which undergo changes in position the principle of virtual work is:

For a mechanical system in equilibrium the virtual work done by all external forces (other than the gravitational and spring forces) equals the change in the elastic and potential energy of the system for any virtual displacements consistent with the constraints.

Stability of Equilibrium

Consider a mechanical system where no work is done on the system by nonpotential forces. With $\delta U' = 0$, the virtual work relation becomes

$$\delta(V_e + V_g) = 0, \quad \text{or} \quad \delta V = 0. \tag{II.7.48}$$

Equation (II.7.48) expresses that the equilibrium configuration of a mechanical system is one for which the total potential energy V of the system has a stationary value.

For a system of one degree of freedom where the potential energy and its derivatives are continuous functions of the single variable, x describes the configuration the equilibrium condition $\delta V = 0$ is equivalent mathematically to

$$\frac{dV}{dx} = 0. \tag{II.7.49}$$

Equation (II.7.49) states that a mechanical system is in equilibrium when the derivative of its total potential energy is zero. For systems with several degrees of freedom, the partial derivative of V with respect to each coordinate in turn must be zero for equilibrium.

There are three conditions under which Eq. (II.7.49) applies:

- the total potential energy is a minimum (stable equilibrium)
- the total potential energy is a maximum (unstable equilibrium)
- the total potential energy is constant (neutral equilibrium).

When a function and its derivatives are continuous, the second derivative is positive at a point of minimum value of the function and negative at a point of maximum value of the function. Thus, the mathematical conditions for equilibrium and stability of a system with a single degree of freedom x are:

$$\text{equilibrium: } \frac{dV}{dx} = 0; \quad \text{stable: } \frac{d^2V}{dx^2} > 0; \quad \text{unstable: } \frac{d^2V}{dx^2} < 0.$$

II.7.9   Examples

EXAMPLE II.7.1:

A hardened and oil-tempered steel wire is used for a helical compression spring. The wire diameter is $d = 0.105$ in., and the outside diameter of the spring is $D_0 = 1.225$ in. The ends are plain and the number of total turns is $N_t = 8$. Find: (a) the torsional yield strength; (b) the static load corresponding to the yield strength; (c) the rate of the spring; (d) the deflection that would be caused by the static load found in part (b); (e) the solid length of the spring; (f) the length of the spring so that no permanent change of the free length occurs when the spring is compressed solid and then released; (g) the pitch of the spring for the free length, and (h) the stability of the spring.

Continued

EXAMPLE II.7.1: *Cont'd*

Solution

(a) From Eq. (II.7.4), the torsional yield strength for hardened and tempered carbon and low-alloy steel is

$$S_{sy} = 0.50 S_{ut}.$$

The minimum tensile strength given from Eq. (II.7.1) is

$$S_{ut} = \frac{A}{d^m},$$

where, from Table II.7.1, the constant $A = 146$ kpsi and the exponent $m = 0.193$.

The minimum tensile strength is

$$S_{ut} = \frac{A}{d^m} = \frac{146}{(0.105)^{0.193}} = 225.561 \text{ kpsi.}$$

The torsional yield strength is

$$S_{sy} = 0.50 \, S_{ut} = 0.50 \, (225.561) = 112.78 \text{ kpsi.}$$

(b) To calculate the static load F corresponding to the yield strength it is necessary to find the spring index, C, and the shear stress correction factor, K_s. The mean diameter D is the difference between the outside diameter and the wire diameter d:

$$D = D_0 - d = 1.225 - 0.105 = 1.12 \text{ in.}$$

The spring index is

$$C = \frac{D}{d} = \frac{1.12}{0.105} = 10.666$$

From Eq. (II.7.10), the shear stress correction factor is

$$K_s = \frac{2C + 1}{2C} = \frac{2\,(10.666) + 1}{2\,(10.666)} = 1.046.$$

Using the torsional yield strength instead of shear stress, Eq. (II.7.8) gives the static load:

$$F = \frac{\pi d^3 S_{sy}}{8 \, K_s D} = \frac{\pi \,(0.105^3) \,(112.78)(10^3)}{8\,(1.046)\,(1.12)} = 43.726 \text{ lb.}$$

EXAMPLE II.7.1: *Cont'd*

(c) From Table II.7.2, the number of active coils is $N_a = N_t = 8$. For $N = N_a$, the spring rate is calculated using Eq. (II.7.24):

$$k = \frac{Gd}{8C^3 N_a} = \frac{(11.5)(10^6)(0.105)}{8(10.666^3)(8)} = 15.546 \text{ lb/in},$$

where $G = 11.5$ Mpsi.

(d) The deflection of the spring is

$$\delta = \frac{F}{k} = \frac{43.726}{15.546} = 2.812 \text{ in.}$$

(e) The solid length, L_s, is calculated using Table II.7.2:

$$L_s = d(N_t + 1) = 0.105(8 + 1) = 0.945 \text{ in.}$$

(f) The free length of the spring is the solid length plus the deflection caused by the load,

$$L_0 = \delta + L_s = 2.812 + 0.945 = 3.757 \text{ in.}$$

(g) From Table II.7.2 the pitch, p, is calculated with the relation

$$p = \frac{L_0 - d}{N_a} = \frac{3.757 - 0.105}{8} = 0.456 \text{ in.}$$

(h) Buckling is checked for the worst case of deflection:

$$\frac{\delta}{L_0} = \frac{2.812}{3.757} = 0.748 \quad \text{and} \quad \frac{\delta}{D} = \frac{2.812}{1.12} = 2.511.$$

Reference to Figure II.7.6 indicates that this spring is outside the buckling region, even if one end plate is free to tip.

EXAMPLE II.7.2:

In a vertical plane two uniform links, each of mass m and length l, are connected and constrained as shown in Figure II.7.16(a). The spring is not stretched when the links are horizontal ($\theta = 0$). The angle θ increases with the application of the known horizontal force F. Determine the spring stiffness k which will produce equilibrium at a given angle θ.

Continued

EXAMPLE II.7.2: *Cont'd*

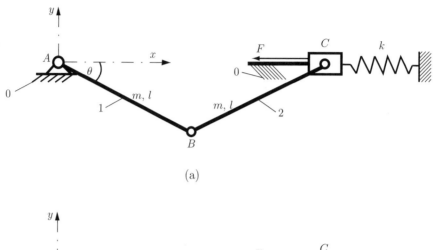

(a)

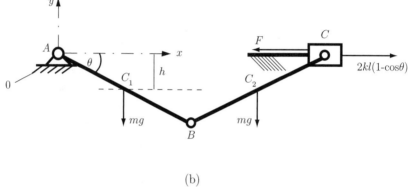

(b)

FIGURE II.7.16   *(a) Mechanism and (b) force diagram for Example II.7.2.*

Solution   The ideal mechanical system has one degree of freedom. The displacement of every link can be expressed in term of the angle θ.
The spring deflection is

$$x = 2l - 2l \cos\theta = 2l(1 - \cos\theta).$$

The force diagram is shown in Figure II.7.16(b). The joint forces are not included in the diagram. The elastic potential energy of the spring is

$$V_e = \frac{1}{2}kx^2 = 2kl^2(1 - \cos\theta)^2.$$

The virtual change in elastic potential energy is

$$\delta V_e = \delta\left[2kl^2(1 - \cos\theta)^2\right] = 2kl^2\,\delta(1 - \cos\theta)^2 = 4kl^2(1 - \cos\theta)\sin\theta\,\delta\theta.$$

EXAMPLE II.7.2: *Cont'd*

The gravitational potential energy is

$$V_g = -2\,m\,g\,h = -2\,m\,g\,\left(\frac{l}{2}\sin\theta\right) = -m\,g\,l\,\sin\theta.$$

The datum for zero gravitational potential energy was taken through the support at A. The virtual change in gravitational potential energy is

$$\delta V_g = \delta(-m\,g\,l\,\sin\theta) = -m\,g\,l\,\cos\theta\,\delta\theta.$$

The virtual work done by the active external force F is

$$\delta U' = F\,\delta = F\,\delta\,[2\,l\,(1-\cos\theta)] = 2\,F\,l\,\delta(1-\cos\theta) = 2\,F\,l\,\sin\theta\,\delta\theta.$$

The virtual work equation $\delta U' = \delta V_e + \delta V_g$ gives

$$2\,F\,l\,\sin\theta\,\delta\theta = 4\,k\,l^2\,(1-\cos\theta)\,\sin\theta\,\delta\theta - m\,g\,l\,\cos\theta\,\delta\theta.$$

The stiffness of the spring is

$$k = \frac{F\,\sin\theta + m\,g\,\cos\theta}{2\,k\,(1-\cos\theta)\,\sin\theta}.$$

EXAMPLE II.7.3:

Figure II.7.17 shows a uniform bar of mass m and length l that slides freely in the vertical and horizontal directions. The spring has the stiffness k and is not compressed when the bar is vertical. Find the equilibrium positions and examine the stability. There are no external active forces.

Solution   The ideal mechanical system has one degree of freedom and the displacement of the bar can be expressed in terms of the angle θ. The spring is undeformed when $\theta = 0$. The datum for zero gravitational potential energy is the horizontal x-axis. The spring deflection is

$$y = l - l\,\cos\theta = l\,(1-\cos\theta).$$

The elastic potential energy of the spring is

$$V_e = \frac{1}{2}k\,y^2 = \frac{1}{2}k\,l^2\,(1-\cos\theta)^2.$$

Continued

EXAMPLE II.7.3: *Cont'd*

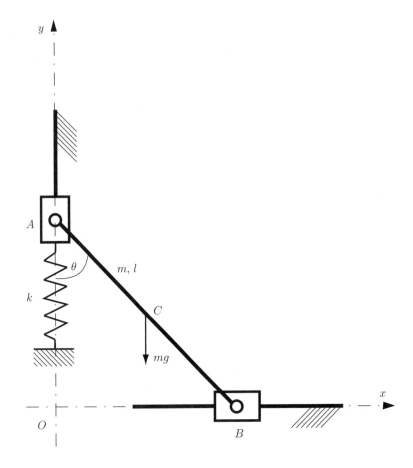

FIGURE II.7.17   *Mechanism for Example II.7.3.*

The gravitational potential energy is

$$V_g = m\,g\,h = m\,g\left(\frac{l}{2}\cos\theta\right) = \frac{1}{2}m\,g\,l\,\cos\theta.$$

The total potential energy is

$$V = V_e + V_g = \frac{1}{2}k\,l^2\,(1 - \cos\theta)^2 + \frac{1}{2}m\,g\,l\,\cos\theta.$$

The equilibrium position is obtained by differentiating the total potential energy and setting it to zero

$$\frac{dV}{d\theta} = kl^2(1 - \cos\theta)\sin\theta - \frac{mgl\sin\theta}{2} = l\sin\theta\left[kl(1 - \cos\theta) - \frac{mg}{2}\right] = 0.$$

EXAMPLE II.7.3: *Cont'd*

The two solutions to this equation are the equilibrium positions:

$$\sin\theta = 0 \quad \text{and} \quad \cos\theta = 1 - \frac{mg}{2kl}.$$

The sign of the second derivative of the potential energy for each of the two equilibrium positions will determine the stability of the system. The second derivative of the total potential energy is

$$\frac{d^2 V}{d\theta^2} = kl^2 \sin^2\theta + kl^2(1 - \cos\theta)\cos\theta - \frac{mgl\cos\theta}{2}.$$

Solution 1: $\sin\theta = 0, \theta = 0 \implies$

$$\frac{d^2 V}{d\theta^2} = 0 + kl^2(1 - 1)(1) - \frac{mgl}{2} = -\frac{mgl}{2} < 0.$$

Equilibrium for $\theta = 0$ is never stable.

Solution 2: $\cos\theta = 1 - \dfrac{mg}{2kl} \implies$

$$\frac{d^2 V}{d\theta^2} = mg\left(k - \frac{mg}{4k}\right).$$

For $k > mg/(4k) \implies d^2V/d\theta^2 > 0$ the equilibrium position is stable.
For $k < mg/(4k) \implies d^2V/d\theta^2 < 0$ the equilibrium position is unstable.

II.7.10 Problems

II.7.1 A helical compression spring is made of hard-drawn spring steel. The wire diameter is 2 mm and the outside diameter is 24 mm. There are 9 total coils. The ends are plain and ground. Determine: (a) the free length when the spring is compressed solid and the stress is not greater than the yield strength; (b) the force needed to compress the spring to its solid length; (c) the rate of the spring; (d) the stability of the spring.

II.7.2 A helical compression spring with squared and ground ends has an outside diameter of 1 in. and a wire diameter of 0.074 in. The solid length is 0.9 in. and the free length is 3 in. Find the spring rate and the approximate load at the solid length.

II.7.3 A helical compression spring has squared and ground ends. A minimum force of 50 lb is applied to compress the spring and the length cannot exceed 3 in. As the force is increased to 120 lb the length is 0.75 in. shorter. The outside diameter of the spring is $D_0 = 1.15$ in. and the wire diameter is $d = 0.157$ in. The number of total turns is 6.38. The spring material is oil-tempered ASTM 229 wire. Find: (a) the torsional yield strength; (b) the static load corresponding to the yield strength; (c) the rate of the spring; (d) the deflection that would be caused by the static load found in part (b); (e) the solid length of the spring; (f) the length of the spring so that no permanent change of the free length occurs when the spring is compressed solid and then released; (g) the pitch of the spring for the free length; and (h) the stability of the spring.

II.7.4 A helical compression spring made of steel with closed ends has an outside diameter of 56 mm and a wire diameter of 3 mm. The number of total coils is 13 and the free length is 100 mm. Find: (a) the spring rate; (b) the force required to close the spring to its solid length and the stress due to this force.

II.7.5 Two bars, 1 and 2, each of mass m and length l are connected and constrained as shown in Figure II.7.18. The angle θ is between the link 1 and the vertical axes. The spring of stiffness k is not stretched in the position where $\theta = 0$. Find the force F which will produce equilibrium at the angle θ.

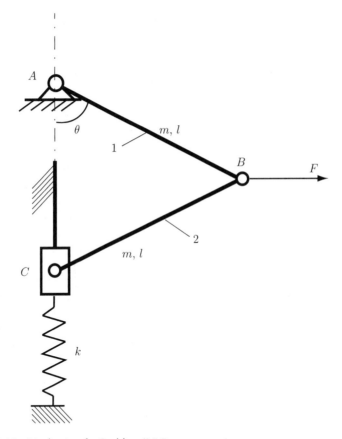

FIGURE II.7.18 *Mechanism for Problem II.7.5.*

II.7.6 Figure II.7.19 shows a mechanism with two links, 1 and 2. Link 1 has the mass $m_1 = m$ and the length $l_1 = l$. Link 2 has the mass $m_2 = 2\,m$ and the length $l_2 = 2\,l$. The spring is unstretched in the position $\theta = 0$. A known vertical force F is applied on link 2 at D. Determine the spring stiffness k which will establish an equilibrium at a given angle θ.

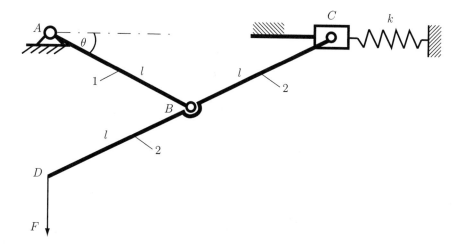

FIGURE II.7.19   *Mechanism for Problem II.7.6.*

II.7.7 For the mechanism shown in Figure II.7.20, link 1 has the mass $m_1 = 2\,m$ and the length $l_1 = 2\,l$. Link 2 has the mass $m_2 = m$ and the length $l_2 = l$. The spring has an unstretched length of L_0. Determine the spring stiffness k for an equilibrium at a given angle θ and a given force F.

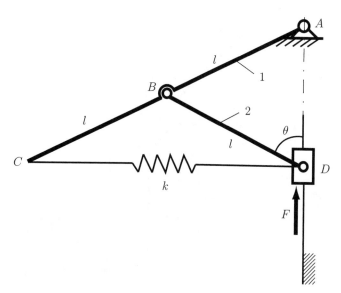

FIGURE II.7.20   *Mechanism for Problem II.7.7.*

II.7.8 The link BC shown in Figure II.7.21 has a mass m and is connected to two springs ($AB = BC = l$). Each spring has the stiffness k and the unstretched length of the two springs is L_0. Determine the spring stiffness k which will establish an equilibrium at a given angle θ. Use the following numerical application: $l = L_0 = 300$ mm, $m = 10$ kg, and $\theta = 60°$.

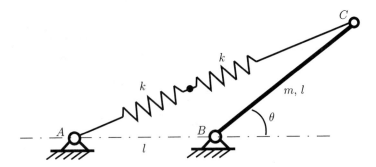

FIGURE II.7.21 *Problem II.7.8.*

II.7.9 The mechanism shown in Figure II.7.22 has the link BC with the mass m and the length l ($AB = AC = l/2$). The spring has the stiffness k and is unstretched when $\theta = 0$. Find the equilibrium value for the coordinate θ. Use the following numerical application: $l = 400$ mm, $m = 10$ kg, $F = 70$ N, and $k = 1.8$ kN/m.

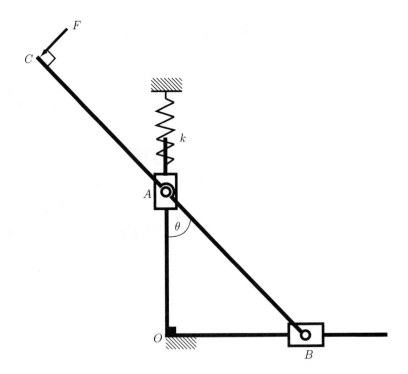

FIGURE II.7.22 *Mechanism for Problem II.7.9.*

II.7.10 The link of mass m and length l is connected to two identical horizontal springs, each of stiffness k, as shown in Figure II.7.23. The initial spring compression at $\theta = 0$ is d. For a stable equilibrium position at $\theta = 0$ find the minimum value of k.

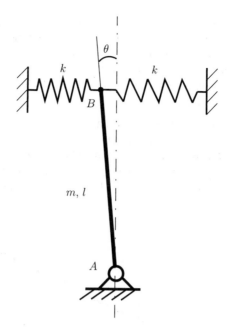

FIGURE II.7.23   *Problem II.7.10.*

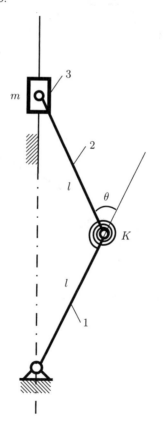

FIGURE II.7.24   *Mechanism for Problem II.7.11.*

II.7.11 The mechanism shown in Figure II.7.24 has two identical links, 1 and 2, each of length l and negligible mass compared with the mass m of the slider 3. The two light links have a torsion spring at their common joint. The moment developed by the torsion spring is $M = K\theta$, where θ is the relative angle between the links at the joint. Determine the minimum value of K that will ensure the stability of the mechanism for $\theta = 0$.

II.7.12 Figure II.7.25 shows a four-bar mechanism with $AD = l$. Each of the links has the mass m ($m_1 = m_2 = m_3 = m$) and the length l ($l_1 = l_2 = l_3 = l$). At B a vertical force F acts on the mechanism and the spring stiffness is k. The motion is in the vertical plane. Find the equilibrium angle θ. Use the following numerical application: $l = 15$ in., $m = 10$ lb, $F = 90$ lb, and $k = 15$ lb/in.

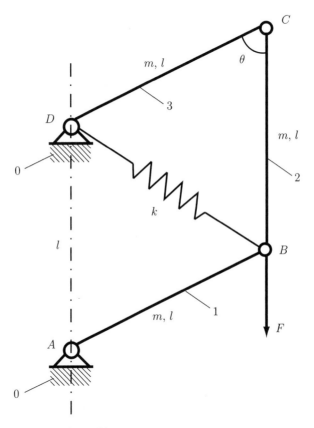

FIGURE II.7.25 *Mechanism for Problem II.7.12.*

II.8   Disk Friction and Flexible Belts

II.8.1   Disk Friction

The sliding surfaces are present in most machine components (bearings, gears, cams, etc.) and it is desirable to minimize the friction in order to reduce energy loss and wear. In contrast, clutches and brakes depend on friction in order to function. The function of a clutch is to permit smooth, gradual connection and disconnection of two elements having a common axis of rotation. A brake acts similarly except that one of the elements is fixed.

In pivot bearings, clutch plates, and disk brakes there is friction between circular surfaces under distributed normal pressure. Two flat circular disks are considered in Figure II.8.1. The figure shows a simple disk clutch with one driving and one driven surface. Driving friction between the two develops when they are forced together. The disks can be brought into contact under an axial force P. The maximum moment that this clutch can transmit is equal to the moment M required to slip one disk against the other. The elemental frictional force acting on an elemental area is

$$dF_f = \mu \, p \, dA,$$

where p is the normal pressure at any location between the plates, μ is the coefficient of friction, and $dA = r \, dr \, d\theta$ is the area of the element.

The moment of this elemental friction force about the shaft axis is

$$dM = \mu \, p \, r \, dA,$$

and the total moment is

$$M = \iint \mu \, p \, r \, dA,$$

where the integral is evaluated over the area of the disk.

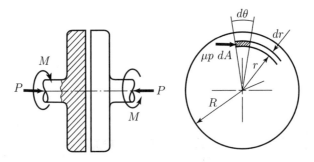

FIGURE II.8.1 *Simple disk clutch.*

The coefficient of friction, μ, is assumed to be constant. If the disk surfaces are new, flat, and well supported it is assumed that the pressure p is uniform over the entire surface so that

$$P = \pi R^2 p.$$

The total frictional moment becomes

$$M = \iint \mu \frac{\mu P}{\pi R^2} r \, dA = \frac{\mu P}{\pi R^2} \int_0^{2\pi} \int_0^R r^2 \, dr \, d\theta = \frac{2}{3} \mu P R. \qquad \text{(II.8.1)}$$

The total moment is equal to a friction force μP acting at a distance $2R/3$ from the shaft center. If the friction disks are rings, as shown in Figure II.8.2, the frictional moment is

$$M = \frac{\mu P}{\pi R^2} \int_0^{2\pi} \int_{R_i}^{R_o} r^2 \, dr \, d\theta = \frac{2}{3} \mu P \frac{R_o^3 - R_i^3}{R_o^2 - R_i^2}, \qquad \text{(II.8.2)}$$

where R_o and R_i are the inside and outside radii.

It is reasonable to assume that after the initial wearing-in period is over, the surfaces retain their new relative shape and further wear is therefore constant over the surface. This wear depends on both the pressure p and the circumferential distance traveled. The distance traveled is proportional to r. Therefore the following expression may be written:

$$r p = K,$$

where K is a constant that is determined from the equilibrium condition for the axial forces

$$P = \int p \, dA = K \int_0^{2\pi} \int_0^R dr \, d\theta = 2 \pi K R.$$

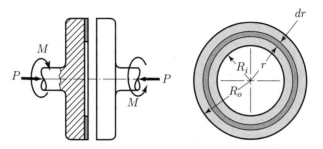

FIGURE II.8.2 *Disk clutch with ring friction disks.*

The constant K is

$$K = \frac{P}{2 \pi R}.$$

With $p r = P/(2 \pi R)$, the frictional moment is

$$M = \int \int \mu\, p\, r\, dA = \frac{\mu P}{2 \pi R} \int_0^{2\pi} \int_0^R r\, dr\, d\theta = \frac{1}{2} \mu P R. \qquad \text{(II.8.3)}$$

The frictional moment for worn-in plates is, therefore, only $(1/2)/(2/3) = 3/4$, as much as for new surfaces. If the friction disks are rings of inside radius R_i and outside radius R_o, the frictional moment for worn-in surfaces is

$$M = \frac{1}{2} \mu P (R_o + R_i). \qquad \text{(II.8.4)}$$

Actual clutches employ N friction interfaces transmitting torque in parallel. The number of friction interfaces N is an even number. For a clutch with N friction interfaces, Eq. (II.8.4) is modified to give

$$M = \frac{1}{2} \mu P (R_o + R_i) N. \qquad \text{(II.8.5)}$$

The ratio of inside to outside radius is a parameter in the design of clutches. The maximum moment for a given outside radius is obtained when [7]

$$R_i = R_0 \sqrt{\frac{1}{3}} = 0.58\, R_0, \qquad \text{(II.8.6)}$$

and the proportions commonly used range from $R_i = 0.45\, R_0$ to $R_i = 0.80\, R_0$.

Disk clutches can be designed to operate either "dry" or "wet" with oil. Most multiple-disk clutches, including those used in automotive automatic transmissions, operate wet.

II.8.2 Flexible Belts

In the design of belt drives and band brakes the impending slippage of flexible cables, belts, and ropes over sheaves and drums is important. Figure II.8.3(a) shows a drum subjected to the two-belt tensions T_1 and T_2, the moment M necessary to prevent rotation, and a bearing reaction R.

Figure II.8.3(b) shows the free-body diagram of an element of the belt of length $r\,d\theta$. The forces acting on the differential element are calculated using the equilibrium of the element. The tension increases from T at the angle θ to $T + dT$ at the angle $\theta + d\theta$. The normal force which acts on the differential element of area is a differential dN. The friction force, $\mu\,dN$, is impending motion and acts on the belt in a direction to oppose slipping.

The equation for the equilibrium of forces in the t-direction gives.

$$T \cos \frac{d\theta}{2} + \mu\,dN = (T + dT) \cos \frac{d\theta}{2} \quad \text{or} \quad \mu\,dN = dT, \qquad \text{(II.8.7)}$$

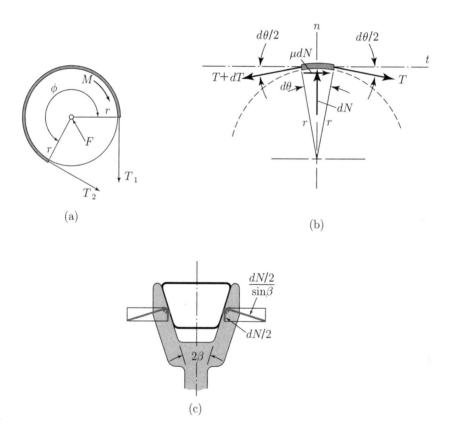

FIGURE II.8.3 *(a) Drum subjected to belt tensions; (b) free-body diagram of an element of the belt; (c) V-belt of angle β.*

where the cosine of the differential quantity is unity in the limit ($\cos d\theta/2 \approx 1$).

Equilibrium of forces in the n-direction gives

$$dN = (T + dT) \sin \frac{d\theta}{2} + T \sin \frac{d\theta}{2} \quad \text{or} \quad dN = Td\theta, \qquad \text{(II.8.8)}$$

where the sine of the differential angle is the angle in the limit ($\sin d\theta/2 \approx d\theta/2$) and the product of two differentials is neglected in the limit compared with the first-order differentials ($dT\, d\theta \approx 0$).

The two equilibrium relations Eqs. (II.8.7), (II.8.8) give

$$\frac{dT}{T} = \mu\, d\theta,$$

and integrating between corresponding limits T_1 and T_2 [with M in the direction shown in Fig. II.8.3(a) $\Longrightarrow T_2 > T_1$]:

$$\int_{T_1}^{T_2} \frac{dT}{T} = \int_0^\phi \mu\, d\theta \quad \text{or} \quad \ln \frac{T_2}{T_1} = \mu\, \phi,$$

where ϕ is the total angle of belt contact expressed in radians.

The tension T_2 is

$$T_2 = T_1\, e^{\mu\, \phi}. \qquad \text{(II.8.9)}$$

If a rope were wrapped around a drum n times, the total angle of belt contact is

$$\phi = 2\, \pi\, n.$$

Equation (II.8.9) also applies to belt drives where both the belt and the pulley are rotating at constant speed and describes the ratio of belt tensions for impending slippage (or slippage).

The centrifugal force acting on a flat belt creates a tension of [7]:

$$T_c = m'\, V^2 = m'\, \omega^2\, r^2, \qquad \text{(II.8.10)}$$

where m' is the mass per unit length of belt, V is the belt speed, and r is the pulley radius. Equation (II.8.9) becomes

$$\frac{T_2 - T_c}{T_1 - T_c} = e^{\mu\, \phi}. \qquad \text{(II.8.11)}$$

The centrifugal force tends to reduce the angles of wrap ϕ.

For a V-belt of angle β [see Fig. II.8.3(c)], Eq. (II.8.11) becomes

$$\frac{T_2 - T_c}{T_1 - T_c} = e^{\mu\, \phi/\sin \beta}. \qquad \text{(II.8.12)}$$

II.8.3 Examples

EXAMPLE II.8.1:

The automobile disk brake, shown in Figure II.8.4, consists of a flat-faced rotor and caliper which contains a disk pad on each side of the rotor. The inside radius is R_i and the outside radius is R_o. The forces behind the two pads are equal to P and μ is the coefficient of friction. The normal pressure p is uniform distributed over the pad. Show that the moment applied to the hub is independent of the angular span α of the pads.

Solution   The force acting on the pads is

$$P = pA = p \int_0^\alpha \int_{R_i}^{R_o} r\, dr\, d\theta = \frac{p}{2} \int_0^\alpha (R_o^2 - R_i^2)\, d\theta = \frac{p}{2} (R_o^2 - R_i^2)\alpha.$$

The moment applied to the hub is

$$M = 2 \int \mu p r\, dA = 2\mu p \int_0^\alpha \int_{R_i}^{R_o} r^2\, dr\, d\theta = \frac{2\mu p}{3}(R_o^3 - R_i^3)\alpha$$

$$= \frac{2\mu}{3} \frac{2P}{(R_o^2 - R_i^2)\alpha}(R_o^3 - R_i^3)\alpha = \frac{4\mu P}{3} \frac{R_o^3 - R_i^3}{R_o^2 - R_i^2}.$$

The expression of the moment M shows no dependence with the angular span α of the pads. The pressure variation with the angle θ would not change the moment M.

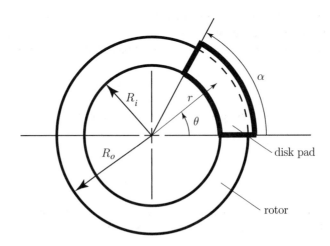

FIGURE II.8.4   *Disk brake for Example II.8.1.*

EXAMPLE II.8.2:

The basic disk clutch, shown in Figure II.8.2, has the outside disk diameter of 6 in. The kinetic coefficient of friction is 0.3 and the maximum disk allowable pressure is 100 psi. The disk clutch is designed to transmit a moment of 400 lb·in. Determine the appropriate value of the inside diameter and the clamping force.

Solution The maximum moment for a given outside radius is obtained from Eq. (II.8.6):

$$R_i = 0.58 R_0 = 0.58 (3) = 1.74 \text{ in.}$$

The greatest pressure occurs at the inside radius. The design of a clutch of inside radius R_i and allowable pressure p_{max} is based on

$$p\,r = K = p_{max}\,R_i. \qquad (\text{II}.8.13)$$

The total moment that can be developed over the entire interface is

$$M = \int \mu\,p\,r\,dA = \int_{R_i}^{R_o} \mu\,(p\,r)\,(2\pi\,r\,dr) = \int_{R_i}^{R_o} 2\pi\,\mu\,p_{max}\,R_i\,r\,dr$$

$$= \pi\,\mu\,p_{max}\,R_i\,(R_o^2 - R_i^2), \qquad (\text{II}.8.14)$$

or

$$M = \pi\,(0.3)\,(100)\,(1.74)\,(3^2 - 1.74^2) = 979.421 \text{ lb} \cdot \text{in.}$$

For $R_i = 1.74$ in. and $p_{max} = 100$ psi, the clutch is overdesigned based on the output moment by a factor of $979.421/400 = 2.448$.

Accepting the overdesign, the clamping force is calculated from Eq. (II.8.4) for $M = 400$ lb·in. as

$$P = \frac{2M}{\mu\,(R_i + R_o)} = \frac{2\,(400)}{0.3\,(1.74 + 3)} = 562.588 \text{ lb.}$$

EXAMPLE II.8.3:

Determine the force F on the handle 1 of the differential band brake [Fig. II.8.5(a)] that will prevent the wheel 2 from turning on its shaft. The external moment $M = 200$ N·m is applied to the shaft. The coefficient of friction between the band and the wheel of radius $r = 100$ mm is 0.45. The following dimensions are given: $l = 500$ mm, $h = 80$ mm, and $\theta = 30°$.

Continued

EXAMPLE II.8.3: *Cont'd*

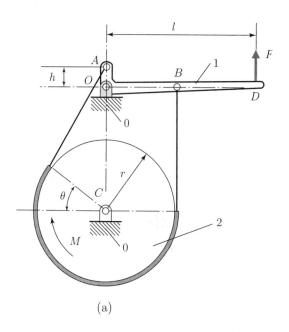

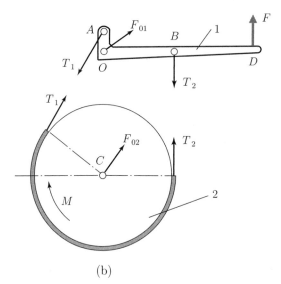

FIGURE II.8.5 *(a) Differential band brake for Example II.8.3; (b) free-body diagrams.*

Solution The free-body diagrams for the handle 1 and the wheel 2 are given in Figure II.8.5(b). For the band the tension T_2 is given by Eq. (II.8.9):

$$T_2 = T_1 e^{\mu \phi} = T_1 e^{7\pi/6} = 5.203 \, T_1, \qquad (\text{II.8.15})$$

where $\phi = \theta + \pi$.

EXAMPLE II.8.3: *Cont'd*

For the wheel 2 the sum of the moments with respect to its center C gives

$$\sum M_C^{(2)} = M - r(T_2 - T_1) = 200 - 0.1(T_2 - T_1) = 0. \qquad \text{(II.8.16)}$$

The tensions T_1 and T_2 are obtained from Eqs. (II.8.15) and (II.8.16):

$$T_1 = 475.791 \text{ N} \quad \text{and} \quad T_2 = 2475.79 \text{ N}.$$

For the link 1 the sum of the moments with respect to point O gives

$$\sum M_O^{(1)} = r T_2 - l F - h T_1 \sin\theta = 0,$$

and the force F is

$$F = \frac{r T_2 - h T_1 \sin\theta}{l} = \frac{0.1(2475.79) - 0.08(475.791)\sin 30°}{0.5} = 457.095 \text{ N}.$$

EXAMPLE II.8.4:

A 3000 rpm motor drives a machine through a V-belt with an angle $\beta = 18°$ and a unit weight of 1.75 N/m (Fig. II.8.6). The pulley on the motor shaft has a 0.1 m pitch radius and the angle of wrap is 170°. The maximum belt tension should be limited to 1000 N and the coefficient of friction is at least 0.3. Find the maximum power that can be transmitted by the smaller pulley of the V-belt drive.

Solution   The speed of the belt in m/s is

$$V = \frac{\pi d n}{60} = \frac{\pi (0.2)(3000)}{60} = 31.415 \text{ m/s}, \qquad \text{(II.8.17)}$$

where $d = 2r = 2(0.1) = 0.2$ m and $n = 3000$ rpm.

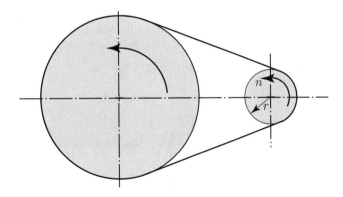

FIGURE II.8.6   *V-belt drive for Example II.8.4.*

Continued

EXAMPLE II.8.4: *Cont'd*

Equation (II.8.10) gives the tension created by the centrifugal force:

$$T_c = m' V^2 = \left(0.178\frac{\text{kg}}{\text{m}}\right)\left(31.415\frac{\text{m}}{\text{s}}\right)^2 = 176.063 \text{ N}, \qquad \text{(II.8.18)}$$

where m' is the mass per unit length of belt:

$$m' = \frac{1.75 \text{ N/m}}{9.81 \text{ m/s}^2} = 0.178 \text{ kg/m}. \qquad \text{(II.8.19)}$$

From Eq. (II.8.12), with $T_1 = T_{max} = 1000$ N, the tension T_2 is

$$T_2 = T_c + \frac{T_1 - T_c}{e^{\mu \, \phi/\sin \beta}} = 176.063 + \frac{1000 - 176.063}{e^{0.3\,(170)\left(\frac{\pi}{180}\right) / \sin\left[18\left(\frac{\pi}{180}\right)\right]}}$$

$$= 222.292 \text{ N}.$$

The moment on the pulley is

$$M = (T_1 - T_2)\, r = (1000 - 222.292)\,(0.1) = 77.770 \text{ N} \cdot \text{m}. \qquad \text{(II.8.20)}$$

The power transmitted by the pulley is

$$H = \frac{M \, n}{9549} = \frac{77.770\,(3000)}{9549} = 24.433 \text{ kW}. \qquad \text{(II.8.21)}$$

EXAMPLE II.8.5:

A 30 hp, 2000 rpm electric motor drives a machine through a multiple V-belt as shown in Figure II.8.7. The belts have an angle $\beta = 18°$ and a unit weight of 0.012 lb/in. The pulley on the motor shaft has a diameter of 6 in. and the angle of wrap is 165°. The maximum belt tension should be limited to 110 lb and the coefficient of friction is at least 0.2. Determine how many belts are required.

Solution   The speed of the belt in m/s is

$$V = \frac{\pi \, d \, n}{60} = \frac{\pi\,(6)\,(2000)}{60} = 628.319 \text{ in./s}.$$

Equation (II.8.10) gives the tension created by the centrifugal force:

$$T_c = m' V^2 = \left(0.000031\frac{\text{lb} \cdot \text{s}^2}{\text{in.}^2}\right)\left(628.319\frac{\text{in.}}{\text{s}}\right)^2 = 12.260 \text{ lb},$$

EXAMPLE II.8.5: *Cont'd*

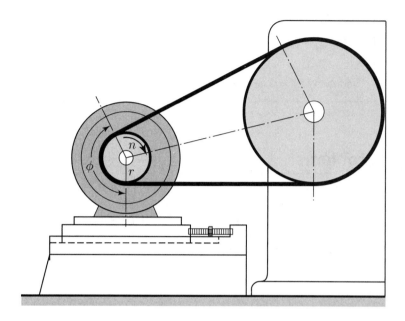

FIGURE II.8.7 *V-belt drive for Example II.8.5.*

where m' is the mass unit length of belt:

$$m' = \frac{0.012 \text{ lb/in.}}{(32.2 \text{ ft/s}^2)(12 \text{ in./ft})} = 0.000031 \text{ lb} \cdot \text{s}^2/\text{in}^2. \qquad \text{(II.8.22)}$$

From Eq. (II.8.12), with $T_1 = T_{max} = 110$ lb, the tension T_2 is

$$T_2 = T_c + \frac{T_1 - T_c}{e^{\mu \phi / \sin \beta}} = 12.260 + \frac{110 - 12.260}{e^{0.2 \,(165)\left(\frac{\pi}{180}\right) \Big/ \sin\left[18\left(\frac{\pi}{180}\right)\right]}}$$

$$= 27.417 \text{ lb}.$$

The moment on the pulley is

$$M = (T_1 - T_2)\, d/2 = (110 - 27.417)\,(6/2) = 247.748 \text{ lb} \cdot \text{in.}$$

The power per belt transmitted by the pulley is

$$H = \frac{M\,n}{5252} = \frac{247.748\,(2000)}{5252\,(12)} = 7.862 \text{ hp/belt}.$$

Continued

EXAMPLE II.8.5: *Cont'd*

The number of belts is

$$N = \frac{30}{7.862} = 3.815,$$

and four belts are needed.

II.8.4 Problems

II.8.1 The circular disk 1 is placed on top of disk 2 as shown in Figure II.8.8. The disk 2 is on a supporting surface 3. The diameters of 1 and 2 are 10 in. and 14 in., respectively. A compressive force of 100 lb acts on disk 1. The coefficient of friction between 1 and 2 is 0.30. Determine: (a) the couple that will cause 1 to slip on 2; (b) the minimum coefficient of friction between the disk 2 and the supporting surface 3 that will prevent 3 from rotating.

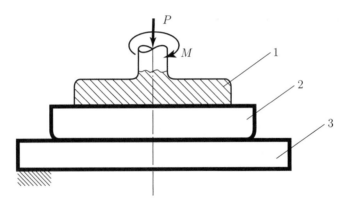

FIGURE II.8.8 *Friction disks for Problem II.8.1.*

II.8.2 A shaft and a hoisting drum are used to raise the 600 kg load at constant speed as shown in Figure II.8.9. The diameter of the shaft is 40 mm and the diameter of the drum is 300 mm. The drum and shaft together have a mass of 100 kg and the coefficient of friction for the bearing is 0.3. Find the torque that must be applied to the shaft to raise the load.

II.8.3 The disks shown in Figure II.8.10 can be brought into contact under an axial force P. The pressure p between the disks follows the relation $p = k/r$, where k is a constant. The coefficient of friction mu is constant over the entire surface. Derive the expression for the torque M required to turn the upper disk on the fixed lower in therm of P, μ, and the inside and outside radii R_o and R_i.

II.8.4 The cable reel in Figure II.8.11 has a mass of 300 kg and a diameter of 600 mm and is mounted on a shaft with the diameter $d = 2r = 100$ mm. The coefficient of

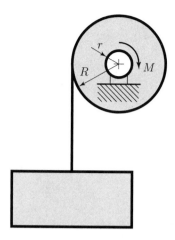

FIGURE II.8.9 *Hoisting drum for Problem II.8.2.*

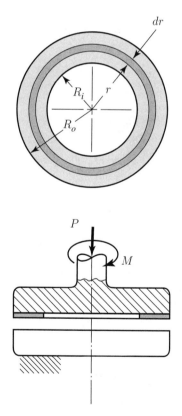

FIGURE II.8.10 *Friction disks for Problem II.8.3.*

friction between the shaft and its bearing is 0.20. Find the horizontal tension T required to turn the reel.

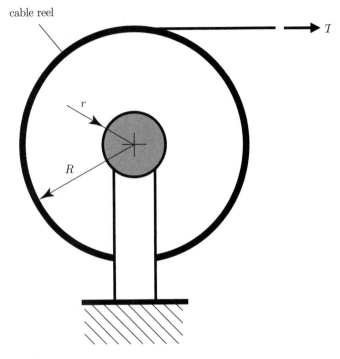

FIGURE II.8.11 *Cable reel for Problem II.8.4.*

II.8.5 For the V-belt in Figure II.8.3(c) derive the expression among the belt tension, the angle of contact β, and the coefficient of friction when slipping impends.

II.8.6 A cable supports a load of 200 kg and is subjected to a force $F = 600$ N which makes with the horizontal axis the angle θ, as shown in Figure II.8.12.

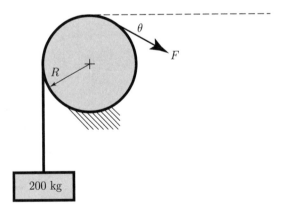

FIGURE II.8.12 *Cable support for Problem II.8.6.*

The coefficient of friction between the cable and the fixed drum is 0.2. Find the minimum value of θ before the load begins to slip.

II.8.7 A band brake is shown in Figure II.8.13. The band itself is usually made of steel, lined with a woven friction material for flexibility. The drum has a clockwise rotation. The width of the band is b, the coefficient of friction is μ, and the angle of band contact is ϕ. Find the brake torque and the corresponding actuating force F if the maximum lining pressure is p_{max}. Use the following numerical application: $b = 80$ mm, $r = 300$ mm, $h = 150$ mm, $l = 800$ mm, $\phi = 270°$, $p_{max} = 0.6$ MPa, and $\mu = 0.3$.

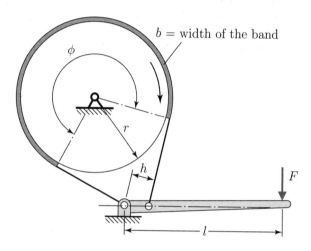

FIGURE II.8.13 *Band brake for Problem II.8.7.*

II.8.8 Figure II.8.14 shows a simple band brake operated by an applied force F of 250 N. The band is 30 mm wide and is lined with a woven material with a coefficient of friction of 0.4. The drum radius is $r = 550$ mm. Find the angle of wrap ϕ necessary

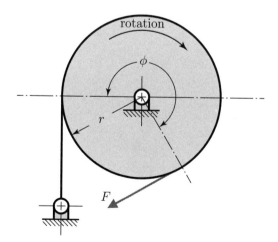

FIGURE II.8.14 *Simple band brake for Problem II.8.8.*

to obtain a brake torque of 900 N·m and determine the corresponding maximum lining pressure.

II.8.9 A 25 hp, 1800 rpm electric motor drives a machine through a multiple V-belt as shown in Figure II.8.6. The belts have an angle $\beta = 18°$ and a unit weight of 0.012 lb/in. The pulley on the motor shaft has a diameter of 3.7 in. and the angle of wrap is 165°. The maximum belt tension should be limited to 200 lb and the coefficient of friction is at least 0.3. Determine how many belts are required.

II.8.10 A 3500 rpm motor drives a machine through a V-belt with an angle $\beta = 18°$ and a unit weight of 2.2 N/m (see Fig. II.8.6). The pulley on the motor shaft has a 180 mm diameter and the angle of wrap is 160°. The maximum belt tension should be limited to 1300 N and the coefficient of friction is at least 0.33. Find the maximum power that can be transmitted by the smaller pulley of the V-belt drive.

References

[1]   M. Atanasiu, *Mechanics [Mechanica]*, EDP, Bucharest, 1973.

[2]   A. Bedford, and W. Fowler, *Dynamics*, Addison Wesley, Menlo Park, CA, 1999.

[3]   A. Bedford, and W. Fowler, *Statics*, Addison Wesley, Menlo Park, CA, 1999.

[4]   A. Ertas, J. C. Jones, *The Engineering Design Process*, John Wiley & Sons, New York, 1996.

[5]   A. S. Hall, A. R. Holowenko, and H. G. Laughlin, *Theory and Problems of Machine Design*, Schaum's Outline Series, McGraw-Hill, New York, 1961.

[6]   B. G. Hamrock, B. Jacobson, and S. R. Schmid, *Fundamentals of Machine Elements*, McGraw-Hill, New York, 1999.

[7]   R. C. Juvinall and K. M. Marshek, *Fundamentals of Machine Component Design*, 3rd ed., John Wiley & Sons, New York, 2000.

[8]   D. B. Marghitu, *Mechanical Engineer's Handbook*, Academic Press, San Diego, 2001.

[9]   D. B. Marghitu, M. J. Crocker, *Analytical Elements of Mechanisms*, Cambridge University Press, Cambridge, 2001.

[10] D. B. Marghitu and E. D. Stoenescu, *Kinematics and Dynamics of Machines and Machine Design*, class notes, available at www.eng.auburn.edu/users/marghitu/, 2004.

[11] C. R. Mischke, "Prediction of Stochastic Endurance Strength," *Transaction of ASME, Journal of Vibration, Acoustics, Stress, and Reliability in Design*, Vol. 109 (1), pp. 113–122, 1987.

[12] R. L. Mott, *Machine Elements in Mechanical Design*, Prentice Hall, Upper Saddle River, NJ, 1999.

[13] W. A. Nash, *Strength of Materials*, Schaum's Outline Series, McGraw-Hill, New York, 1972.

[14] R. L. Norton, *Machine Design*, Prentice-Hall, Upper Saddle River, NJ, 1996.

[15] R. L. Norton, *Design of Machinery*, McGraw-Hill, New York, 1999.

[16] W. C. Orthwein, *Machine Component Design*, West Publishing Company, St. Paul, 1990.

[17] I. Popescu, *Mechanisms*, University of Craiova Press, Craiova, Romania, 1990.

[18] C. A. Rubin, *The Student Edition of Working Model*, Addison–Wesley Publishing Company, Reading, MA, 1995.

[19] I. H. Shames, *Engineering Mechanics — Statics and Dynamics*, Prentice-Hall, Upper Saddle River, NJ, 1997.

[20] J. E. Shigley and C. R. Mischke, *Mechanical Engineering Design*, McGraw-Hill, New York, 1989.

[21] J. E. Shigley, C. R. Mischke, and R. G. Budynas, *Mechanical Engineering Design*, 7th ed., McGraw-Hill, New York, 2004.

[22] J. E. Shigley and J. J. Uicker, *Theory of Machines and Mechanisms*, McGraw-Hill, New York, 1995.

[23] A. C. Ugural, *Mechanical Design*, McGraw-Hill, New York, 2004.

[24] R. Voinea, D. Voiculescu, and V. Ceausu, *Mechanics [Mechanica]*, EDP, Bucharest, 1983.

[25] J. Wileman, M. Choudhury, and I. Green, "Computation of Member Stiffness in Bolted Connections," *Journal of Machine Design*, Vol. 193, pp. 432–437, 1991.

[26] S. Wolfram, *Mathematica*, Wolfram Media/Cambridge University Press, Cambridge, 1999.

[27] National Council of Examiners for Engineering and Surveying (NCEES), *Fundamentals of Engineering. Supplied-Reference Handbook*, Clemson, SC, 2001.

[28] * * * , *The Theory of Mechanisms and Machines [Teoria Mehanizmov i Masin]*, Vassaia Scola, Minsk, 1970.

[29] * * * , Working Model 2D, Users Manual, Knowledge Revolution, San Mateo, CA, 1996.

Index

clearance, 642

closed

   joint, 56

   kinematic chain, 57

clutch, 755

coefficient of friction, 39

coincident points, 182

column, 463

complex

   motion, 52

   chains, 59

component, 7

composite areas, 21

compression, 463

conjugate gear-tooth action, 639

connectivity table, 97

constant life fatigue diagram, 497

contact

   angle, 589

   ratio, 646

contour, 57

   equations, 181

Coriolis acceleration, 151, 181

couple, 25

coupler, 60

crank, 60

critical load, 463

cross product, 12

D

d'Alembert principle, 213

dedendum, 642

deflection, 454, 458

degenerate mechanism, 70

degrees of freedom, 51, 61

design charts, 619

diametral

   clearance, 586

   pitch, 643

direct dynamics, 213

direction

   cosines, 10

   of a vector, 3

differential, 652

driver, 61

driven link, 61

dot product, 11

dyad, 68

dynamic

   load, 660

   viscosity, 607

E

efficiency, 545

endurance limit, 491

epicyclic gear train, 647

equilibrium, 34

equivalence of systems, 26

Euler column formula, 465

F

family, 61

fatigue, 491

   failure, 491

   strength, 491

first moment, 16

fixed

   reference frame, 141

   vector, 4

fluctuating stresses, 494
follower, 61
force closed joint, 56
frame, 53
free-body diagram, 36, 215
free vector, 4
friction force, 38
full joint, 54
fundamental kinematic chain, 67

G

gear, 641
Gerber parabolic relation, 500
gradient factor, 493
ground, 53, 60
Goodman lines, 497

H

half-joint, 55
higher joint, 56
hydrodynamic, 583
hydrostatic, 583

I

idler, 646
inertia
    force, 213
    moment, 214
inertial reference frame, 151
independent
    parameters, 51
    contour, 78
inverse dynamics, 213

involute, 639
    pressure angle, 640
interference, 644

J

joint, 52
    class, 52
    constant, 550

K

kinematic
    chain, 57
    pair, 52

L

lead, 537
    angle, 539
Lewis equation, 658
life requirement, 596
line of action of a vector, 3
linear cumulative damage rule, 498
link, 52
load
    factor, 494
    intensity, 443
loop, 57
lower joint, 56

M

magnitude, 3
major diameter, 537
$Mathematica^{TM}$, 261

mass
  center, 18
  moment of inertia, 211
mechanism, 59
median life, 596
method of decomposition, 17, 19
Miner's rule, 498
minor diameter, 537
mixed kinematic chain, 58
mobility, 61
mobile reference frame, 141
modified Goodman diagram, 498
modifying factors, 492
module, 643
Mohr's circle, 436
moment, 21
multiple-threaded screw, 537
multigrade oil, 611

N

Newton's
  first principle, 27
  second law, 27
  third law, 27

O

open kinematic chain, 58
order of a joint, 56
orientation of a vector, 3

P

Petroff's equation, 612
pinion, 641

pitch, 537
  circle, 641
  diameter, 537, 584, 641
  point, 641
planet gear, 648
planetary gear train, 647
plane of symmetry, 17
point of application of a
        vector, 4
Poisson
  formulas, 143
  ratio, 441
position vector, 15
power screws, 541
preload force, 549
pressure angle, 642
primary reference frame, 141
principal
  directions, 437
  stresses, 437
proof
  strength, 554
  load, 554

R

radial equivalent force, 598
rated capacity, 596
rating life, 596
redundant support, 37
reliability factor, 597
resultant, 6, 7
Reynolds equation, 618

rigid body, 51
ring gear, 648
rocker, 60
rolling bearing, 583
rotation, 51

S

safety factor, 555
scalar, 3
    product, 11
    triple product, 13
self-locking, 545
sense of a vector, 3
separating force, 550
screw thread, 537
shear, 435
    strain, 441
shock force, 599
simple chains, 59
simple couple, 25
singularity functions, 444
slenderness ratio, 466
sliding
    bearing, 583
    vector, 4
slope, 109
S−N diagram, 491
spring
    constant, 550, 729
    index, 728
    ends, 730
square thread, 541
strain, 441

standard life, 596
stress, 435
    fatigue factor, 662
structural diagram, 97
sun gear, 648
surface
    factor, 493
    endurance, 662
Somerfeld number, 615
system groups, 67

T

tension, 435
tensile stress area, 552
thread angle, 538, 545
three-force member, 37
torque, 25
torsion, 452
translation, 51
transmissible vector, 4
two-force member, 37

U

unit vector, 6

V

vector, 3
    product, 12
    triple product, 14
velocity, 142
    ratio, 641

virtual
   displacement, 738
   work, 739
viscometer, 609

W

wear tooth load, 660

wrench, 31
Working Model, 419

Z

zero vector, 5